MARINE ECOLOGY

A Comprehensive, Integrated Treatise on Life in Oceans
and Coastal Waters

MARINE ECOLOGY

A Comprehensive, Integrated Treatise on Life in Oceans and Coastal Waters

Editor
OTTO KINNE
Biologische Anstalt Helgoland
Hamburg, Federal Republic of Germany

VOLUME V
Ocean Management
Part 1

A Wiley–Interscience Publication

1807   1982

175 YEARS OF PUBLISHING

1982
JOHN WILEY & SONS
Chichester · New York · Brisbane · Toronto · Singapore

Library of Congress Cataloging in Publication Data (Revised):

Kinne, Otto.
  Marine ecology.
  Vol. 5–   published: Chichester, New York,
Wiley.
  Includes bibliographies.
  CONTENTS: v. 1 Environmental factors 3 v.—
v. 2 Physiological mechanisms 2 v.—[etc.]—v. 5
Ocean management.
  1. Marine ecology—Collected works. I. Title.
QH541.5.S3K5   574.5′2636   79-121779

ISBN 0 471 27997 8 (v. 1)   AACR1

British Library Cataloguing in Publication Data:

Marine ecology.
  Vol. 5: Ocean management
  Part 1
  1. Marine ecology
  I. Kinne, Otto
  574.5′2636   QH541.4.S3   80-42065

  ISBN 0 471 27997 8

Typeset by Preface Ltd, Salisbury, Wiltshire
Printed in Great Britain at The Pitman Press, Bath, Avon.

FOREWORD

to

VOLUME V: OCEAN MANAGEMENT

'Ocean Management', the last volume of *Marine Ecology*, describes and evaluates all the essential information available on structures and functions of interorganismic coexistence; on organic resources of the seas; on pollution of marine habitats and on the protection of life in oceans and coastal waters. The volume consists of three parts:

Part 1

Zonations and Organismic Assemblages

Part 2

Ecosystems and Organic Resources

The culmination of *Marine Ecology*, Volume V has its roots in, and draws much of its basic substance from, the preceding volumes of the treatise:

Volume I ('Environmental Factors') is concerned with the most important environmental factors operating in oceans and coastal waters and their effects on microorganisms, plants and animals.

Volume II ('Physiological Mechanisms') reviews the information available on the mechanisms involved in the synthesis and transversion of organic material; in thermoregulation, ion- and osmoregulation; in evolution and population genetics; and in organismic orientation in space and time.

Volume III ('Cultivation') comprehensively assesses the art of maintaining, rearing, breeding and experimenting with marine organisms under environmental and nutritive conditions which are, to a considerable degree, controlled.

Volume IV ('Dynamics') summarizes and critically evaluates the knowledge available on the production, transformation and decomposition of organic matter in the marine environment, as well as on food webs and population dynamics.

Of necessity somewhat heterogeneous in concept and coverage, 'Ocean Management' introduces the readers to fields of applied marine ecology, i.e. to man's use of oceans and coastal waters for his own ends. In order to provide a solid basis for a sound assessment of the sea's man-supporting qualities, Parts 1 and 2 deal comprehensively with the basic multi-specific units encountered and with the resources which they constitute. After summarizing our current knowledge on the large variety of organismic groupings in the form of zonations and assemblages, and after considering the structures and functions of major marine ecosystems, the significance of these units and their components as resources utilizable for food or for raw materials are reviewed in depth. Part 3, finally, evaluates man's potentially destructive impact. It focuses on the different facets of pollution and critically assesses measures—currently applied and considered practicable in the future—for protecting life in oceans and coastal waters from detrimental human influences.

Selected as early as 1965, the title 'Ocean Management' still seems too ambitious, if not somewhat misleading—even though, as anticipated, a host of new data and many new and important insights into the machinery of ecological systems have been brought to light in the meantime. Environmental management requires the concerted, judicious, responsible application of science and technology for the protection and control of those properties of ecosystems, species, resources, or areas that are regarded as absolute requirements for the continued support of civilized human societies[*]. Our knowledge has remained insufficient for an objective, exact definition of such properties. Hence, maintenance of a high degree of organismic and environmental diversity, and maximum possible conservation of natural conditions is deemed essential to avoid or to reduce irreversible long-term damage.

The means and aims of ocean management receive detailed attention in Part 2. Our capacity for true management is still restricted to narrowly-defined areas and to specific organisms, i.e. to a few heavily exploited or otherwise especially endangered species. Marine ecosystem management remains problematic; it is in need of much more basic ecological knowledge than is at present available.

The recent trend of using English as the international scientific language—in itself of great significance for international communication and cooperation—has often been disadvantageous to scientists with insufficient command of that language. It has frequently diminished the representativeness and distorted the emphasis of the total information actually available. While organizing and editing *Marine Ecology*, I have therefore attempted to include scientists from countries which made important contributions to the field of marine ecology, but whose scientists largely use non-English languages. In this way I wanted to underline the international status and significance of marine ecological research and the need to draw from different sources in order to provide the best possible representation of the state-of-the-art. Of course, I had to pay for this: an enormous amount of time and effort had to be invested in translation and manuscript improvement.

While I am writing this last foreword of the treatise, *Marine Ecology* is nearing completion. The remaining parts of Volume V will be appearing shortly. It is my sincere wish to thank again all those who have supported me during the many years of planning, carrying out and finalizing this *magnum opus*. From 1965 to 1981, the work on *Marine Ecology* has taken up most of my evenings, weekends and holidays. Who can blame me for feeling relieved?

As was the case with previous volumes of this treatise, I have received much help and advice while working on Volume V. With profound gratitude I acknowledge the close and fruitful cooperation with all contributors; the support, patience and confidence of the publishers; and, last but not least, the technical assistance of Monica Blake, Alice Langley, Julia Maxim, Seetha Murthy, Sherry Stansbury and Helga Witt.

O.K.

[*]KINNE, O. 1980: *14th European Marine Biology Symposium 'Protection of Life in the Sea'*: Summary of symposium papers and conclusions. *Helgoländer Meeresunters.*, **33**, 732–761.

CONTENTS
OF
VOLUME V, PART 1

CONTRIBUTORS
TO
VOLUME V, PART 1

KINNE, O., *Biologische Anstalt Helgoland (Zentrale), Notkestrasse 31, 2000 Hamburg 52, Federal Republic of Germany.*

PERES, J. M., *Station Marine d'Endoume et Centre d'Oceanographie, Rue de la Batterie-des-Lions, Marseille (7e), France.*

OCEAN MANAGEMENT

Marine Ecology Vol. 5, Part 1
Edited by Otto Kinne
© 1982 John Wiley & Sons Ltd

1. INTRODUCTION TO PART 1—
ZONATIONS AND ORGANISMIC
ASSEMBLAGES

O. KINNE

(1) General Aspects

Ever since the first ecologists turned their attention to the seas, nature's seemingly boundless diversity has induced attempts to develop a conceptual framework for describing and understanding the patterns of organismic distribution and interorganismic organization in space and time. Over the years, marine ecologists have proposed a variety of classifications and systems—often similar, sometimes divergent, and occasionally contradictory. Dismissed by some as too speculative and subjective, research into the organization of organismic coexistence in multi-species systems has received new impetus, directive and significance from recent developments in modern ecosystem research, from man's increasing demand for marine resources (Part 2), as well as from man-made pollution and deformation of the marine environment and the urgent need to develop adequate measures for the protection of life in the seas (Part 3).

Although considerable information has become available, especially within the last two decades, the picture slowly emerging has still to be focused and often suffers from insufficiently backed claims and interpretations. There is great need for developing a precise, internationally standardized, analytical instrumentarium and for achieving basic conceptual agreement.

In an attempt to outline the state of our present knowledge, we have therefore chosen to illuminate the situation from two, partially overlapping perspectives: the description, analysis, and evaluation of: (i) zonations and assemblages, i.e. of groups of species (populations) living together in a defined place or environment; and (ii) ecosystems, i.e. entities of interacting and interdependent biotic and abiotic components, linked by coordinated patterns of energy flow and material cycling, as well as by homeostatic control mechanisms*. While the grouping of organisms into zonations and assemblages stresses the fact that they are found together in a certain locality, the ecosystem concept emphasizes the functional interaction and integration of the different types of organisms concerned. Pelagic and benthic organisms, for example, will always have to be considered as belonging to different assemblages, but they may functionally be so closely linked as to be viewed as members of one and the same ecosystem. The chapters devoted to zonations and assemblages emphasize detail and situational specificity; those concerned with ecosystems stress perspectives of synthesis and generalization.

*e.g. KINNE, O. (1977). Summary, conclusions and closing. In *Int. Helgoland Symp. 'Ecosystem Research'*, *Helgoländer wiss. Meeresunters.*, **30**, 709–27.

In contrast to Volumes I, II and III, Volume V continues to elaborate in many of its chapters the more holistic-dynamic approach introduced and developed in Volume IV. While the first three volumes largely emphasize environment–organism relations at the individual and species levels, i.e. the responses and activities of single organisms or populations, Volume IV and, especially, Volume V pay tribute to organisms acting in concert. They describe and explore the supra-individual and multi-species levels and stress the fact that new properties may result when subsystems organize into larger systems. Volume V documents the interdisciplinary nature and integrating capacity of modern ecological research, its paramount significance for man's long-term future, and the urgent need for managing man's growing impact on the seas.

Part 1 of Volume V of *Marine Ecology* is devoted to reviewing our present knowledge of zonations and organismic assemblages. In an attempt to restrict undesirable repetition and unnecessary diffuseness, I have invited one reviewer to author all chapters. A Frenchman, Professor J. M. PERES, has also filled, or at least reduced, a linguistic gap—the sometimes insufficient coverage in *Marine Ecology* of French papers. As are all previous tomes of this series, the present book is the result of numerous, intensive and detailed exchanges between author and editor. Without so much patience, determination and good will on the side of the author, 'Zonations and Organismic Assemblages' would never have been published.

Drawing from numerous, often widely scattered sources and employing to the fullest his own professional experience of many years, J. M. PERES has compiled the most comprehensive treatment available on the subject. The difficulties in proposing a classification of organismic assemblages in oceans and coastal waters are extraordinary. They are not only due to the almost irritating degree of diversity and variability, but also to lack of sufficient conceptional directivity and methodological congruence among the investigators involved.

In the following 'Comments on Chapters' it proved impossible to do justice to the tremendous amount of information presented and discussed. I have, therefore, firmly restricted myself to very general summaries, attempting to select only some of the most important aspects presented and to highlight, for the benefit of the reader, the basic conceptual framework of 'Zonations and Organismic Assemblages'.

(2) Comments on Chapters 2 to 9

Chapter 2: Zonations

The term 'zonation' refers to the occurrence of certain organisms in areas (zones) distinct in their own organismic composition from that of the adjacent territorium. Chapter 2 reviews all essential information at hand on vertical and horizontal zonations of marine organisms, both in the pelagial and the benthal. It emphasizes and critically evaluates the significance and major role of food availability in controlling organismic zonations.

Both in the pelagial and benthal, characteristics of, and changes in, assemblage composition provide the most useful ecological criteria for describing and analysing zonation patterns. While environmental factors, such as temperature, have been used by other investigators—and are known to act as important stimuli—changes in assemblage

composition tend to mirror the total resulting ecological impact and its biological consequences.

In the pelagial, the following major vertical zones are distinguished: epipelagic zone, mesopelagic zone, infrapelagic zone, bathypelagic zone and abyssopelagic zone.

In the benthal, all vertical zonations proposed thus far are considered more or less inadequate. Employing general features of organismic assemblages controlled by environmental factors acting in concert, a general scheme of vertical zones is proposed which is assumed to be applicable in all regions of the World Ocean. Among the environmental factors involved, humidity, light and hydrostatic pressure—often re-ferred to as 'climatic factors'—control the composition of benthos assemblages in largely the same way in the whole World Ocean.

The World Ocean as a whole can be divided vertically into two main ecological systems: the phytal, which can support photoautotrophs, and the aphytal which cannot. Both phytal and aphytal may comprise several vertical zones, subzones or horizons. Through its control over primary production, light controls food supply, i.e. the master factor governing all essential aspects of vertical and horizontal zonations, both in the pelagial and the benthal.

Horizontal zonation receives only brief attention. In the pelagial, horizontal zones are often narrowly related to the phenomenon of 'ecological succession'. On a small scale, pelagic zonations depend on the combined effects of four groups of factors: (i) life cycle, life span and reproductive potential; (ii) predator–prey relations and losses due to sinking; (iii) passive drifting of an assemblage with its supporting water mass; (iv) the amount of energy available at the lowest level of the trophic pyramid, i.e. the level of local primary production. On a large scale, they depend on the physical and chemical properties, as well as the movements, of the water masses concerned. In the benthal, horizontal zonation is largely controlled by the nature of the substrate (which, in turn, is a function of water movement and the water content of suspended particles) and depends decisively on trophic inputs from the pelagial.

Previously assumed to represent an important source of organic matter available for deep-sea organisms, the concept of 'organic rain' has now been abandoned. Present knowledge indicates complete recycling of the organic production in the upper 250-m water layer, except for large particles such as faeces of big animals and bodies of large-sized plankters; these sink deeper and may reach the sea bottom. A comparable pattern prevails in the remineralization of dissolved organic carbon. The most produc-tive sea areas tend to be those near the coast, i.e. areas peripheral to the main oceanic water bodies. The World Ocean itself appears to be rather poor in nutrients over some 90% of its area.

Chapter 3: General Features of Organismic Assemblages in Pelagial and Benthal

The first attempts to comprehend patterns of organismic distribution in space concen-trated on qualitative aspects, i.e. the listing of different kinds of plants and animals which inhabit a given type of environment (e.g. the sea bottom). The lists compiled were then compared with other lists recorded from related, but different environments, in an attempt to define those species narrowly related in their distribution to the main characteristics of the environment concerned. Later, quantitative aspects dominated the scene; in the benthal, PETERSEN counted the number of individuals occupying a defined bottom area and, emphasizing species which predominated in weight ('pilot species'), developed the community concept.

Subsequently, the community concept was increasingly confused with that of 'biocoenosis'—a term introduced by MÖBIUS and redefined by ALLEE and SCHMIDT as an association of living things inhabiting a uniform division of the biosphere, its members being dependent upon each other and thus forced into biological balance. The members of a biocoenosis usually breed in the area occupied by the biocoenosis concerned. This area—the 'biotope'—is characterizable in terms of relatively uniform, principal habitat features. Since the terms 'community' and 'biocoenosis' are well-defined only for the benthos, the more general term 'organismic assemblage' was adopted here for describing and analysing organismic groupings both in the pelagial and benthal.

The term 'organismic assemblage' refers to a group of species which live together in a defined environment—irrespective of the sampling method employed and of the method used for analysing the sample. In contrast, the related expression 'association' comprises, according to its redefinition by MARGALEF, a group of different species collected together by means of one and the same sampling method and the subsequent analysis of the data thus obtained.

Organismic assemblages are more difficult to study and to categorize in the pelagial than in the benthal. A tridimensional 'environment in motion', the pelagial lacks conspicuous, overt differences in environmental characteristics, and harbours mostly small, short-lived organisms which—due to vertical migrations, seasonal variations in climate and changing predator–prey relations—often undergo extensive fluctuations, both in terms of quantity and quality. Consequently, detailed investigations of pelagic assemblages require such a tremendous amount of data that the investigator must, at least at present, restrict his attention to a few, ecologically most important species ('indicator species').

In contrast, the much more stable benthal is, in essence, a bidimensional environment with its vertical extension rarely exceeding a few decimetres. It is characterized by more easily detectable differences in environmental features, especially with regard to the substratum. Many benthic plants and animals attain longer life spans and larger body sizes than their pelagic counterparts. Vertical migrations are less extensive and less pronounced; most macrobenthic plants and many animals are fixed or sedentary, with limited locomotory ranges. The percentage of immigrants and visitors is smaller in the benthal than in the pelagial, and seasonal variations are often less evident. All this leads to less prominent fluctuations in organismic distributions on and in the sea bottom than in the free water above.

Among the factors affecting the distribution and abundance of the macrobenthos, humidity, light and hydrostatic pressure ('climatic factors') are of particular significance. Another group of abiotic factors which act locally on the sea bottom and/or its inhabitants are referred to as 'edaphic factors', e.g. substrate characteristics, water movement, salinity change, accumulation of organic matter. These can modify or sometimes even overrule the effects of climatic factors. Biotic factors (e.g. competition; variations in predator–prey relations; migrations; changes in substrate characteristics due to organismic activities) can modify the effects of both climatic and edaphic factors, as well as directly affect the composition and dynamics of the assemblage concerned.

Chapter 3 further reviews and discusses methods of delimiting neighbouring groups of organisms (biocoenoses, communities, assemblages) and the continuum concept in the benthal. For the pelagial, delimitation is much more problematic. The concept of organismic assemblage must be applied here with caution.

Chapter 4: Structure and Dynamics of Assemblages in the Pelagial

The tridimensional pelagial is in continuous motion, and is largely inhabited by small-sized organisms. Its assemblages appear more immediately integrated than those of the benthal. Frequently, direct pathways exist from phytoplankters to carnivores of the highest trophic level. In addition, a much larger percentage of the primary production is directly consumed by animals here than in the benthal. Consequently, the energy transfer to commercially used, large-sized carnivorous fishes tends to be more efficient. This may explain the fact that landings of pelagic fishes are double those of demersal fishes.

The most complete size classification of organisms in the pelagial has been proposed by SIEBURTH and co-authors (1978)[*]. For true plankters, the size categories adopted by most authors are as follows: *nanoplankton*, 2·0–20 μm (including photoautotrophs and heterotrophs); *microplankton*, 20–200 μm; *mesoplankton*, 0·2–20 mm; *macroplankton*, 2–20 cm; *megaloplankton*, 20–200 cm. Organisms smaller than 2·0 μm are usually classified as *ultraplankton* (or '*ultrananoplankton*').

Horizontal and vertical water movements and horizontal and vertical migratory activities—both modified by diel and seasonal changes in environmental factor intensities—largely determine the distribution patterns of pelagic organisms in time and space, and affect their trophic interrelations and reproductive performance.

Especially in surface and subsurface waters, small-sized organisms predominate. They represent the first steps of the trophic pyramid; nano-, micro- and mesoplankters may constitute a larger total biomass than the nectonic, macro- and megaloplanktonic animals, particularly at higher latitudes. Small organisms have, in general, a shorter life span and hence turn-over time than large ones. Consequently, the productivity of nano-, micro- and mesoplankters is very high. Pronounced successions in time and space facilitate assemblage structuration. They accentuate differences and variability in intra- and interassemblage dynamics.

The major factors which have been shown to affect the structure and dynamics of organismic assemblages in the pelagial include light, nutrients, temperature, salinity, water movement, sinking and grazing. Chemical interactions between coexisting organisms remain to be investigated. Presumably, they play a role in coordinating interspecific relations and in promoting and directing system integration.

[*]Sieburth, J. Mc. N., Stemacek, V. and Lenz, J. (1978). *Pelagic Ecosystem Structure: Heterotrophic Compartments of the Plankton and their Relationship to Plankton Size Fractions, Limnol. Oceanogr.*, **23**, 1256–63.

Heterogeneity in spatial distribution, i.e. patchiness, is of considerable ecological significance. In the pelagial, patchiness primarily results from water movement, predation and the capacity of the water body concerned to support essential life processes of its inhabitants. In the open sea, phytoplankton patches tend to occur either in the form of strips a few metres wide but hundreds of metres or more long, or of ellipses with a mean diameter of usually between 10 and 50 km. Zooplankton patches are more difficult to study. Their occurrence is often inversely related to phytoplankton patches. Zooplankton patches are apparently an essential prerequisite for the adequate development of pelagic carnivores, including larval stages of fishes.

The delimitation and classification of organismic assemblages is less advanced in the pelagial than in the benthal. Our present knowledge is, to a large extent, based on information obtained by Soviet Russian and French scientists.

The development of a plankton assemblage may essentially proceed in three different ways: (i) exponential population growth causes depletion of nutrients and growth-promoting substances; the assemblage collapses; (ii) the assemblage maintains itself at a juvenile stage (continuous, moderate growth and exploitation), retaining a low degree of organization; (iii) the assemblage develops a more sophisticated organization and thus controls growth: increasing amounts of energy and matter are diverted from growth into system integration, the major losses resulting from transfer of energy and matter from one trophic level to the next and to energy channelled into migrations and locomotion in general.

Where two assemblages of different organizational states come into contact, the more mature system exploits the other. Such exploitation tends to increase the degree of organization of the more mature assemblage, providing its predators are sufficiently diversified and specialized and thus able to control the prey populations present. In contrast, exploitation by unspecialized predators may result in a decrease of assemblage organization, either because it favours species with higher growth rates or because it causes the exploitation intensity to exceed the threshold of system homeostasis (overgrazing, overfishing). Surface-water assemblages suffer continuous losses of energy and matter from the sinking of organic particles below the compensation depth; hence they may never attain the climax stage.

Where originally separate, different assemblages become organizationally connected, and the resulting new system tends to attain a higher level of integration. Zones of contact between neighbouring ecological systems deserve special attention from marine ecologists.

Chapter 5: Structure and Dynamics of Assemblages in the Benthal

In essence bidimensional, the non-moving benthos is more heterogeneous than the pelagic environment and hence offers support to a larger variety of different kinds of organisms. Apart from the micro-organisms, about 98% of all the species known to exist in oceans and coastal waters belong to the benthos. While we are still in the process of discussing and evaluating the significant features of benthos assemblages, certain basic types of structures and functions of benthos assemblages are becoming manifest, and a world-wide classification of benthos assemblages is proposed in Chapters 8 and 9.

Life in the benthos is characterized by severe spatial problems. Settling stages of plants and animals usually encounter considerable difficulties in finding suitable vacant spaces for attachment and development. This is especially so with solid substrates. On soft substrates, the surface is never totally covered by meio- or macrobenthic organisms. While it seems less difficult for settlers to find a vacant place on or in a soft substratum, the risk of being preyed upon and eaten tends to be even larger here than in the pelagial because most benthic organisms aggregate in the immediate vicinity of the water/sediment interface.

The heterogeneity of the benthic environment is largely due to physico-chemical differences (hardness, chemical properties, texture, granulometric characteristics, penetrability) between the non-living and living substrates encountered. Whilst the properties and distribution of solid substrata depend primarily on geological processes, those of soft substrata are largely a function of water movement and biological activities.

Differences in water-movement patterns further affect nutrient supply and sedimentation rate, as well as metabolic exchange processes, thus adding further environmental diversity.

In determining the composition of a benthic assemblage, mainly attached or sedentary species are taken into account. Swimming and floating animals are less directly and less intimately linked to the sea bottom. In general, the percentage of benthic animal species which move actively or are passively transported over large distances is lower than in the pelagial. The fact that long-lived and large-sized organisms are more abundant in the benthal than in the pelagial tends to reduce the extent of biological fluctuations.

Nevertheless, the diversity of ecological niches and the complexity of the trophic connex are much larger than in the pelagial. This and the much higher number of different types of species present makes detailed analyses of trophic relationships in the benthal considerably more difficult. In general, living benthic plants are of much less significance as immediate food providers than are pelagic plants. However, decomposing plant material appears to play a more important role in the benthal than has been hitherto assumed. There are numerous different mechanisms employed by benthic animals for obtaining food. The high diversity in feeding types (see also Volume II, Part 1) mirrors the large variety of food materials available.

Nutritionally, the micro- and meiobenthos constitute an almost autarchic system which does not make a significant amount of energy available for the macrobenthos, except to detritus-feeders. The fact that living macrophytes are hardly consumed by benthic animals results in an increase in the number of detritus feeders. A larger number of bottom-living animals are rarely preyed upon (protection due to hard exoskeleton, repellents, toxic substances or large size) than in pelagic assemblages. While benthic assemblages are, in general, more mature than pelagic assemblages, the total amount of energy available at the base of the trophic pyramid which finally reaches the top-predator level is approximately three times higher in pelagic than in benthic assemblages. This difference is reflected in fishery landings of pelagic and benthic species, respectively.

Patchiness in organismic distributions has been less thoroughly investigated in benthic than in pelagic assemblages. Ecological dynamics in benthic assemblages appear to be primarily affected by predator–prey relationships and by a number of aperiodic and periodic changes, of both short- and long-range character. In Chapter 5 these are considered in considerable detail.

Chapters 6 and 7: Major and Specific Pelagic Assemblages

The categorization of, and the differentiation between, organismic assemblages are not free of subjective elements. In addition, in the pelagial vertical migrations of many plankters and considerable methodological differences in plankton sampling complicate the collection of reliable information and its analysis.

While quantitative aspects have been included in the reviewer's considerations wherever necessary, the classification of pelagic assemblages proposed is mainly based on qualitative aspects, i.e. on diffferences in species composition. Owing to the different ecological conditions in the pelagial and the benthal, the classification of major and specific assemblages requires somewhat different approaches in these two principal habitats of the sea. The major pelagic assemblages are treated separately for different oceans: Southern, Pacific, Indian and Atlantic Oceans, and the Arctic Basin. Specific assemblages are distinguished for neuston, pleuston, floating *Sargassum* and for discoloured waters.

Chapters 8 and 9: Major and Specific Benthic Assemblages

The major benthic assemblages proposed are reviewed by beginning with the shallowest and ending with the deepest waters studied. The vertical zones distinguished are: supralittoral, midlittoral, infralittoral, phytal, circalittoral, bathyal, abyssal and hadal. Within each zone, the reviewer considers first hard- and then soft-substrate assemblages. Providing the essential substrate properties are the same, the assemblages of different biogeographic regions turn out to be similar in each vertical zone. Employing the concept of assemblage and iso-assemblage, this fact facilitates a world-wide classification of organismic groupings in the benthal.

Much of the information presented and evaluated is based on the benthal of the north-eastern Atlantic Ocean and the European Mediterranean Sea—the two most thoroughly investigated sea areas.

The specific benthic assemblages distinguished concern the microbenthos and meiobenthos of soft-bottom environments. Particular attention is paid to the psammic system, the sulphide system and the coastal subsoil environment.

Marine Ecology Vol. 5, Part 1
Edited by Otto Kinne

2. ZONATIONS

J. M. PERES

(1) Vertical Zonation

This chapter successively reviews the most general features involved in the vertical zonation of marine life in the pelagial and benthal—first from a qualitative point of view in order to delimit the different vertical zones and their boundaries, then from a quantitative point of view. Among the factors which govern the different aspects of vertical zonation, the quantity and quality of food resources receives special attention, as does the problem of specific diversity.

(a) Vertical Zonation in the Pelagial

Although many pelagic organisms, even the true plankters, may perform vertical migrations, we propose here a general scheme of vertical zonation, mainly with respect to the plankton. Such a scheme must first be considered from a qualitative point of view: quantitative aspects will be treated later.

Numerous schemes of zonation in the pelagial have been proposed. As an example we mention here the concept of BRUUN (1957); he named the upper water layers where photoautotrophic organisms can develop the epipelagic zone, and subdivided the water layers below according to their temperatures. Below the mean compensation depth he distinguished: (i) the mesopelagic zone with the 10 °C isotherm as lower limit; (ii) the bathypelagic zone between the 10 and 4 °C isotherms (BRUUN assumed that the latter would be a sharp boundary for many cold-loving species); (iii) the abyssopelagic zone represented by all water masses with temperatures below 4 °C, except for those in trenches deeper than 6000 to 6500 m; these are classified—in spite of the fact that their temperature is not very different from that of the abyssopelagic zone—as the hadopelagic zone. For several reasons it is problematic to accept such a scheme: (i) at different latitudes the 10 and 4 °C isotherms may occur at very different depths; (ii) phenomena such as convergences, divergences and upwellings disturb the vertical temperature gradient as well as the distribution of plankton assemblages.

It seems that a vertical zonation based on the most significant changes in assemblage composition is of more general value and, in addition, has the advantage of being somewhat parallel to the proposed vertical benthos zonation. Hence, for establishing such a scheme I have utilized and slightly modified information presented first by BIRSHTEIN and co-authors (1954) and later by VINOGRADOV (1968, 1970, 1977, and in Volume V) on plankton assemblages in the 9575-m deep Kuril-Kamchatka Trench (Fig. 2-1). In regard to plankton distribution, this is the most thoroughly investigated area in the whole World Ocean. Moreover, this area harbours fairly rich plankton

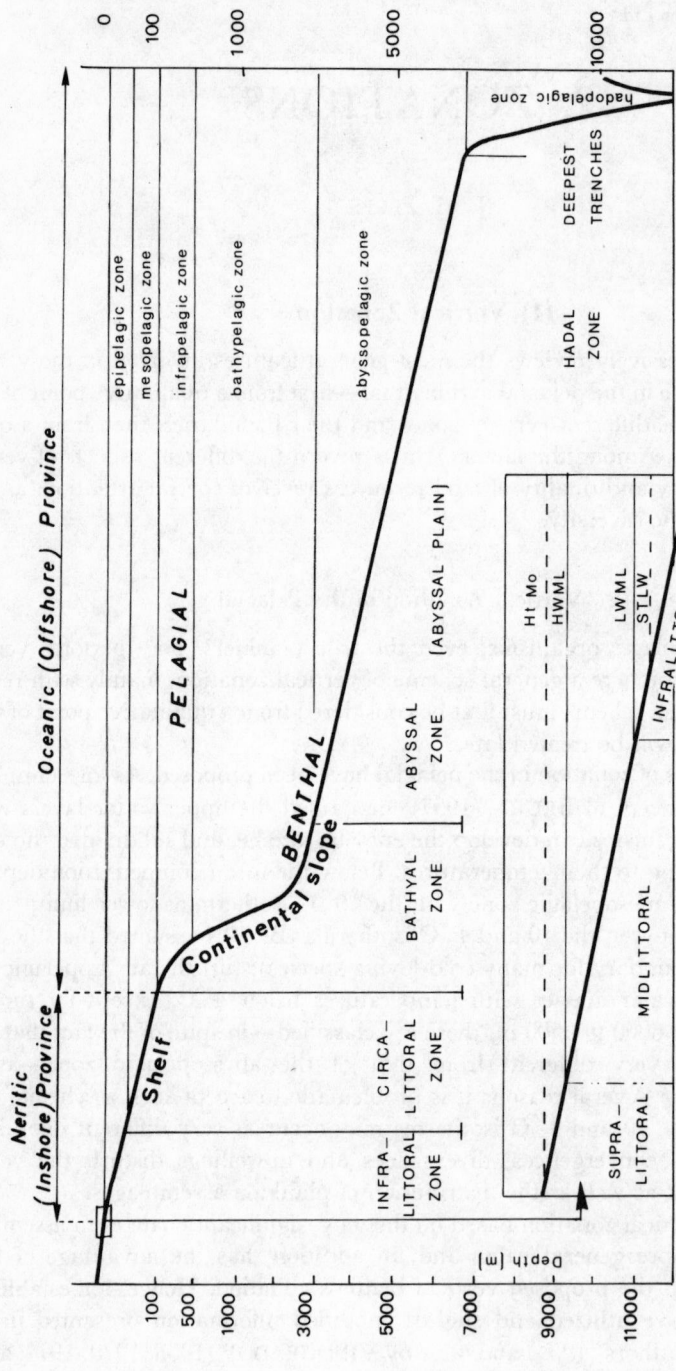

Fig. 2-1: Main vertical and horizontal divisions of the marine environment. HtMo: highest level of moistening by high water spring tide or waves and sprays; HWML: high water mean level; LWML: low water mean level; STLW: spring tide low water. (Original.)

populations and thus makes it easier to demonstrate changes in their composition related to water depth.

Five vertical zones may be distinguished in the Kuril-Kamchatka Trench:

(i) **The epipelagic zone** is inhabited by the epiplankton and exactly corresponds to the euphotic layer, i.e. its lower boundary marks the mean compensation depth for photoautotrophic phytoplankters (in the area considered here: mainly diatoms and some dinoflagellates). Since the compensation depth depends on light penetration, the epipelagic zone may be only 15 to 20 m deep in rather turbid waters, but up to 120 m in very transparent waters; its average depth is about 50 m.

(ii) **The mesopelagic zone** corresponds to the upper part of the oligophotic layer where, due to insufficient illumination, the amount of energy obtained by phytoplankters through photosynthesis is lower than necessary to meet their metabolic expenses. This means that photoautotrophs which have sunk down to this layer cannot develop but only survive until energy has been exhausted and they finally die or transform into a resting stage.

Such potential for temporary survival is important to phytoplankters since subsurface waters may be subject to mixing with the upper-lying epipelagic zone, for example, due to wind-generated upwelling or winter cooling of the euphotic layer. Hence, phytoplankters which have sunk to the mesopelagic zone may escape degeneration and death if dynamic processes transport them back in time to the euphotic layer.

The zonation proposed by BIRSHTEIN and co-authors (1954) takes into consideration such ecological perspectives by treating these two zones (surface and subsurface) together; they are characterized by the presence of photoautotrophic phytoplankters. Moreover, many zooplankton species perform vertical nychthemeral migrations between epiplankton and mesoplankton. Thus, they are often considered together. Except for local conditions (for example abundance peaks of cladocerans in areas with reduced salinities), or transient—meteorological or dynamic—situations, copepods almost always predominate in assemblages associated with euphausiids in the higher latitudes or with foraminiferans in tropical or equatorial areas.

The lower boundary of the mesopelagic zone lies between 200 and 300 m. Incidentally, during the day a depth of 250 to 300 m approximately corresponds to the upper boundary of the main deep scattering layer.

(iii) **The infrapelagic zone** is located between 200 to 300 m and 600 to 700 m; in the middle latitudes the average temperature at the lower depth limit is about 10 °C; this corresponds to the lowest boundary ascribed by BRUUN (1957) to his 'mesopelagic zone'. In the Kuril-Kamchatka Trench, BIRSHTEIN and co-authors (1954) consider this zone as transitional between, on the one hand, the surface and subsurface waters, with a large annual temperature range featuring very low winter values and, on the other hand, oceanic intermediate waters, less cold in winter and with a less pronounced annual temperature range. They noted within this zone a marked change in the fauna, chiefly with respect to narcomedusae, euphausiids and fishes (several species of myctophids, stomiatids, *Argyropelecus*, etc.). Among the fishes, many species may rise up at night to shallower depths. Usually not occurring in deeper waters, these species may be considered true inhabitants of the infrapelagic zone. Presumably, some of them rise up at night in search of food. Members of other species also migrate at night into the infrapelagic zone—probably for the same reason; among the latter, BIRSHTEIN and co-authors (1954) mention, for example, juveniles of the mysid *Gnathophausia gigas* and of the prawn *Hymenodora frontalis*, the amphipods *Scina borealis* and *S. incerta*, the copepods *Scaphocalanus magnus*, *Pseudochirella spinifera*, *Spinocalanus stellatus*, *Haloptilus pseudoxy-cephalus*. That the infrapelagic zone appears to be a temporary 'meeting place' where its own inhabitants encounter intruders from deeper layers is certainly its most interesting feature and possibly a widespread phenomenon: I had many opportunities to observe it during dives with a bathyscaphe.

(iv) **The bathypelagic zone** extends from 600 to 700 m down to 2000 to 2500 m and is inhabited by the bathyplankton. At middle latitudes, its lower boundary approximately corresponds to the 4 °C isotherm. Copepods predominate in this zone as in the mesopelagic and infrapelagic zones; however, they are mostly represented by different species, except for immigrants from these two zones. Many new species of euphausiids, stomiatids and myctophids occur—mainly below 100 m—often in populations of greater abundance than in the infrapelagic zone. The first representatives of typical deep-sea pelagic forms appear, such as sergestid shrimps, holothurioids of the family Pelagothuridae and fishes of the family Ceratiadae. For the Kuril-Kamchatka Trench, BIRSHTEIN and co-authors (1954) mention an assemblage of medusae, which I have also observed in the northern part of the Japan Trench: *Crossota rufobrunnea*, *Pentachogon haeckeli*, *Halicreas minimum*, *Atolla bairdi*, *Periphylla hyacinthina*, etc. They also collected some Siphonophora of the family Dinophyiidae, the chaetognath *Eukrohnia fowleri*, the mysid *Eucopia grimaldii*, many amphipod species (e.g. *Koroga megalops*, *Cyclocaris guilelmi*, *Eusirella multicalceola* and several species of the genera *Scina* and *Lanceola*), the prawns *Hymenodora frontalis* and *Gennadas borealis*, together with juveniles of *Hymenodora gracialis*, etc.

(v) **The abyssopelagic zone** extends from 2000 to 2500 m down to 6000 to 6500 m, i.e. the depth corresponding to the upper boundary of the trenches. The abyssoplankton is characterized by a decrease in both species number and general abundance. According to BIRSHTEIN and co-authors (1954), copepods no longer predominate; the most important groups are now chaetognaths, mysids and decapod crustaceans—all carnivores or scavengers. However, other peracarids—namely amphipods—were probably underestimated in the samplings; this was demonstrated later after employing more efficient collecting equipment. Still deeper, in the Kuril-Kamchatka Trench, the chaetognath *Eukrohnia fowleri*, which is still abundant in the upper levels of the abyssopelagic zone, almost disappears, and the medusa *Crossota rufobrunnea*, characteristic of the bathypelagic zone, is completely absent. On the other hand, new species may be considered characteristic of the abyssoplankton such as the mysids *Eucopia australis* and *Bentheuphausia amblyops* and several shrimps of the genus *Acanthephyra*. The copepod fauna exhibits a marked change in composition; several characteristic genera appear such as *Cephalophanes* (BIRSHTEIN and co-authors, 1954).

Formerly, Soviet scientists tentatively distinguished a fifth, hadopelagic zone comprising waters filling the trenches in depths exceeding 6000 to 6500 m. In comparison with the abyssoplankton, the hadoplankton turned out to be greatly impoverished both in number of species and individuals; large-sized carnivorous and scavenging groups such as decapod crustaceans and fishes seemed to be missing, whereas amphipods largely predominated and appeared to play a very active role; next in importance were ostracods and copepods. However, more recent data strongly suggest that the existence of a hadopelagic zone ranking equivalent to the abyssopelagic zone is questionable.

Near the sea floor, the plankters of the greatest oceanic depths are difficult to sample with nets; baited traps placed on the bottom may catch planktonic species attracted from the immediate upper layers, as well as nektobenthic species whose habitat is restricted to the 1- to 2-m thick bottom-water layer.

However, every time I dived with the French bathyscaphes 'FNRS 3' and 'Archimede' in abyssal and upper hadal depths, I noticed a striking increase in abundance and diversity of plankton organisms near the bottom, i.e. in the last 10 or 20 m and, especially, in the last 2 m above the sea floor. Recently, BELAYEV (1976) gave an interesting account of the pelagic and near-bottom fauna in maximal oceanic depths for which he identified 57 species. Crustaceans largely dominate this fauna; they are represented by 23 copepods, 26 amphipods, 5 ostracods and 3 mysids. BELAYEV also recorded one species each of chaetognaths and siphonophores (*Chuniphyes* sp.), as well as the trachylid medusa *Voragonema profundicola*. Several of the species mentioned were also collected from abyssal or even bathyal depths. However, a few are also known to inhabit the epipelagic and mesopelagic zones: the copepods *Calanus plumchrus* (sometimes recorded even at the surface and seemingly a bipolar species), *Metridia ochotensis* (recorded up to 50 m) and *Heterorhabdus compactus* (present in all oceans from 8500 m up to 100 m). The chaetognath *Eukrohnia fowleri* mentioned by BELAYEV, is also known to be eurybathic and psychrophilic. Of the species listed by BELAYEV, about 50% are hadal endemics with endemisms of generic rank accounting for around 10%.

(b) Vertical Zonation in the Benthal

General Aspects

After more than 30 years of sampling, mostly over all of the Mediterranean Sea, but also in the Atlantic and Indian Oceans, I am convinced that all vertical zonations of benthos assemblages thus far proposed are more or less inadequate. Two examples: (i) The 'sublittoral' of many authors evidently includes organismic assemblages of which the primary level and consequently the trophic network are very different. (ii) The 'intertidal' cannot be considered to represent an ecologically significant unit, because common guidelines must be adopted for shores with or without tides, and because the brevity of emersion and immersion periods near the low and high water of spring tide levels cannot significantly affect the organisms living at these levels.

Table 2-1 summarizes the most important systems of vertical zonation thus far proposed. The vertical zonations proposed by PERES (1957) and PERES and PICARD (1959) represent an attempt to establish a general framework which can be used in all regions of the World Ocean (PERES, 1976, p. 224 ff.). Instead of measurements of environmental factors the basic principle of the scheme is to use general features of the organismic assemblages which result from cumulative, interferent or antagonistic influences of these factors on the organisms and the assemblages they constitute. In several cases, discontinuity in the assemblages corresponds to a sharp change in local geomorphology.

Three factors—humidity, light and hydrostatic pressure—may be considered to con-

Table 2-1

Major systems of vertical zonation in the benthal. SST: largest range spring tides; MST: mean range spring tides; UF: upper fringe of the infralittoral zone (After HEDGPETH, 1957; slightly modified; reproduced by permission of the Geological Society of America)

Spray zone	Between tide-marks or seasonal levels (High water SST MST / Low water MST SST)	To limit of large algae or hermatypic corals	To edge of continental shelf	Continental slope	Abyssal plain	Trenches	Source
	'littoral'		'flachsee'	'tiefsee'	- - -	- - -	WALTHER (1893)
	littoral	littoral	coastal	deep-sea	- - -	- - -	PRUVOT (1897)
	littoral	sublittoral	elittoral	archibenthal	abyssal		GILSEN (1929, 1930)
	littoral	littoral	sublittoral	bathyal	abyssal	ultraabyssal	ZENKEVICH (1956)
supralittoral	eulittoral	sublittoral	sublittoral	archi-benthal	abyssal-benthal		ALLEE and co-authors (1943)
supralittoral	littoral	sublittoral	sublittoral	bathyal	abyssal	hadal	HEDGPETH (1957)
supralittoral	midlittoral	infralittoral UF	circalittoral	bathyal	abyssal	hadal	PERES (1957)

trol the composition of benthos assemblages in largely the same way in the whole World Ocean. These factors are called 'climatic' factors (p. 52).

With regard to the uppermost levels, the humidity gradient represents the major controlling factor (Volume I, Chapter 4). It includes such aspects as true humectation by spray and brief immersion due to waves, as well as longer periods of immersion due to tides or meteorological events (winds and atmospheric pressure fluctuations).

The factor light (Volume I, Chapter 2) is of much more general importance. It controls the primary production of photoautotrophic organisms, i.e. the synthesis of organic matter, except for the rather limited amount produced by chemoautotrophic protophytes (Volume II, Chapter 2). Thus, the World Ocean as a whole may be divided into two main systems: phytal (or 'littoral' in terms of vertical zonation in the benthal) where photoautotrophic organisms may exist, and aphytal where these cannot live. Moreover, based on the light gradient—decreasing irradiance with increasing distance from the water surface down to compensation depth—marine organisms, and consequently the assemblages they constitute, may be classified in conformity with their light requirements or tolerances, into photophilic and sciaphilic forms. Obviously, numerous intermediate levels of requirement and tolerance may exist between these two opposite groups. Of course, the distribution of photophilic and sciaphilic assemblages not only depends on water depth: topographic peculiarities of shallow water rocky biotopes, such as overhangs, crevices or caves, affect the average illumination; hence sciaphilic species or assemblages may occur as 'enclaves' amidst photophilic ones (Fig. 2-2).

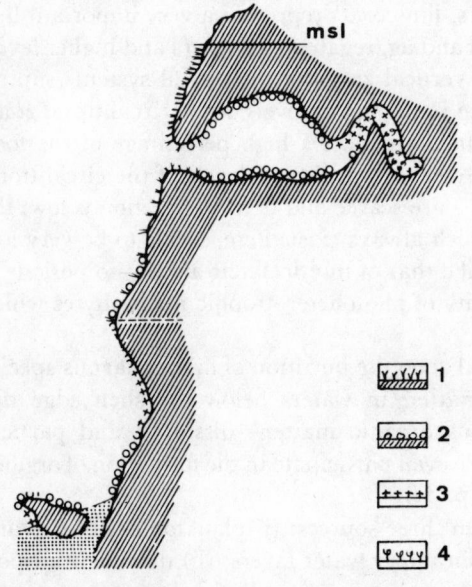

Fig. 2-2: Influence of illumination on the distribution of benthos assemblages on craggy rock substrates. msl: mean sea level; 1: photophilic assemblages with predominant seaweeds; 2: sciaphilic assemblages with scattered seaweeds; 3: assemblages of semi-obscure caves; 4: assemblages of quite dark caves. (Original.)

Regarding the third factor, hydrostatic pressure (Volume I, Chapter 8), our present knowledge is still rather limited. Referring to benthic species, MENZIES and GEORGE (1972, p. 36) notice that 'eurybathial benthic species are few in number and constitute only a small per cent of the total deep sea fauna (less than 1%). They are found most abundantly in regions of isothermal conditions from the surface to the sea bed.'

Due to the generally small number of specimens available, their taxonomy is poorly investigated, and hence they could possibly consist of two or more species limited to more restricted depth ranges. MENZIES and GEORGE do not agree with BRUUN (1957), who denies any significant influence of hydrostatic pressure on the deep-sea fauna.

Since in the phytal system the illumination exerts a direct influence only on epibionts, the effect of light seems questionable with regard to the vertical zonation of infaunal assemblages. However, the illumination indirectly influences the distribution of many infaunal detritivorous macrobenthos species in the phytal, because an important percentage of their food requirements is met by the microphytobenthos. Therefore, the quantity and spectral composition of the light reaching the sea bottom may be used for delimiting macrobenthos assemblages in the phytal system not only with regard to those occurring on the bottom surface but also to infaunal assemblages.

Since the deep-sea environment is quite dark (except for light of biological origin), below a depth of 300 to 900 m—depending on latitude, season and especially differences in suspended-matter content—it is quite obvious that illumination cannot influence the distribution of benthic animals which inhabit deeper water.

In fact, if we neglect direct light effects on organisms (Volume I, Chapter 2), illumination, due to its influence on primary production, may be considered as a single aspect of a much more general factor: food supply. In almost all assemblages, microphagous species (suspension feeders, detritus feeders, limivores) represent a very important link between primary or paraprimary (bacteria and aggregates) producers and higher levels of the trophic pyramid. In the first three vertical zones of the phytal system (supra-, medio- and infralittoral), and probably also in the upper levels of the circalittoral zone, the *in situ* primary and paraprimary productions meet a high percentage of the food requirements of microphagous species. In contrast, at deeper levels of the circalittoral zone, multicellular algae—if present at all—are scarce and their production is low; the productivity of the microphytobenthos, which always exists here, seems to be very low and discontinuous since it is restricted—like that of multicellular algae—to periods of sufficient light penetration. The productivity of photoheterotrophic microphytes which live a little deeper seems negligible.

Thus, in the lower half of the circalittoral zone the nutrition of microphagous species depends in part on exogenous organic matter; in waters below the shelf edge this dependence becomes total. The exogenous organic matter—dissolved and particulate—may be metabolized by bacteria; these even participate in the formation of organic aggregates formed from dissolved matter (p. 37).

The input of organic material stems from three sources: (i) plankton organisms and their detritus which sink to the bottom from upper water layers; (ii) detritus transport from land or shallow benthos; and (iii) organismic migration. In the case of the first two sources, the distribution of such exogenous food resources depends on water currents, mostly those near the sea floor. The currents also influence the nature of the substratum: strong currents prevent sedimentation and leave bare hard substrates or coarse sediments, whereas weak currents—or stagnant water—result in high sedimentation rates and thus lead to silty or muddy substrates (see also Volume I, Chapters 5, 7 and 10).

In more general terms, this means that below a depth which approximately corresponds to the middle of the circalittoral zone vertical range, the main features of the vertical zonation in the benthos depend on the ability of the local organisms to meet their food requirements through complementation of resources produced *in situ* by exogenous resources—a phenomenon which, in turn, largely depends on the pattern of water movement (Volume I, Chapter 5).

The third source of energy input into soft-bottom assemblages (probably restricted to infralittoral and circalittoral zones) is based on the circadian migration of small-sized benthic organisms, mostly peracarid crustaceans, which during the day are buried in the sediment and at night rise up into the epiplanktonic—or even hyponeustonic water layer—feeding upon plankton resources before returning to the sea floor. This phenomenon mainly concerns sea bottoms and depths ranging from a few metres to about 80 to 100 m; not a function of water movement, the amount of energy input—corresponding to that contained in the food collected from the plankton of upper water layers—depends on the population density of these migrating species and, in part, also on the lunar cycle. However, endogenous rhythms play a role in migration behaviour: the richer the surface plankton assemblage, the more important the energy input to the bottom.

Both main systems, phytal and aphytal, may be subdivided into several vertical zones. A vertical zone (in French *étage*) is the depth interval of the benthic domain where the ecological conditions related to the main environmental factors mentioned above (climatic factors) are homogeneous, or exhibit a gradient between two critical levels which corresponds to the boundaries of the zone. Each vertical zone may reveal in its organismic assemblages some diversity related to the effects of other factors which are of less general significance, such as substrate nature, temperature, chemical composition of the sea water, water movement, etc. However, except when the influence of the latter factors largely predominates over that of the three climatic factors, the boundary between two adjacent vertical zones always corresponds to a sharp change in assemblage composition.

Vertical zones may sometimes be divided into subzones and the latter, in turn, into levels of lesser vertical range which may be called 'horizons'.

Zonation in the Phytal System

In the phytal system four vertical zones may be distinguished. Their names always include the suffix 'littoral' (Fig. 2-1).

(i) **The supralittoral zone** is inhabited by organisms which are never (or very rarely) immersed, but require a relatively high degree of humidity. On shores where the tidal range is large, immersion only occurs at high water of spring tides. In areas with small tidal ranges, such as the Mediterranean Sea, immersions are very irregular since they are mainly related to heavy storms and, to a lesser extent, to changes in atmospheric pressure, as well as wind direction and strength. In exposed places, mainly on rocky coasts, the vertical range of the supralittoral zone is larger than in sheltered areas, due to the fact that permanent or almost permanent wave breaking results in sprays which keep a sufficient degree of humidity sometimes several metres above the mean sea level (atidal shores) or the mean high tide level (tidal shores).

(ii) **The midlittoral zone** comprises assemblages whose constituting species require or tolerate a slightly prolonged emergence but cannot tolerate permanent or almost permanent immersion. Hence the midlittoral zone includes several of the 'intertidal' assemblages, especially those which appear to be most characteristic because they occur at levels where emergence and immersion alternate most frequently.

According to STEPHENSON and STEPHENSON (1949), most authors agree that the intertidal zone corresponds to an ecological unit composed of three levels: supralittoral fringe, midlittoral zone and infralittoral fringe. While I agree with the STEPHENSONS' midlittoral zone, I think their supralittoral fringe, corresponding to the 'spray zone', really constitutes the supralittoral zone mentioned above. In other respects, due to the rather striking similarity between the assemblages which inhabit the STEPHENSONS' infralittoral fringe and those which occur a little deeper on and within very shallow—but never emerged—substrates, I suggest that this fringe be considered only as part of the infralittoral zone.

On shores with a large tidal range—i.e. more than 1·5 m at spring tides—the upper limit of the midlittoral zone corresponds to the level of the high water of small neap tides, and its lower limit to the low water of small neap tides. On shores with a small tidal range—e.g. most Mediterranean shores—the upper limit of the midlittoral zone corresponds to the highest level of wave immersion or when the sea surface is smooth, to a rising of its mean level; its lowest limit corresponds to the level of 'normal' emergence. However, some exceptional circumstances, for example, the combination of high atmospheric pressure and wind blowing towards the sea, may cause the highest levels of the infralittoral zone to emerge. Such a phenomenon is as exceptional and brief as the emergence of the above-mentioned infralittoral fringe during low water of spring tide level on tidal shores.

(iii) **The infralittoral zone** is characterized by an upper limit corresponding to the highest level inhabited by species which cannot endure a slightly prolonged emergence (see above); its lower limit corresponds to the maximal depth consistent with the survival of sea grasses and photophilic algae, i.e. with their compensation depth. Obviously, this depth depends on light penetration which, in turn, is closely related to sea-water turbidity. In unpolluted waters of the English Channel the infralittoral zone does not extend below 20 m, while in the Mediterranean it extends down to 35 m in the Western Basin and a little deeper still in the Eastern Basin. In some tropical areas with very clear water, the infralittoral zone might reach down to 60 to 70 m. However, the significance of local sea grass as an indicator of the lower boundary of the infralittoral zone requires some qualifications since *Halophila stipulacea* is the only phanerogam species capable of existing below true photophilic algae, possibly down to 100 m.

In the infralittoral fringe, the vertical range of which is a few decimetres on shores with small or no tides, but larger on shores with a more extensive tidal range, the general composition of the plant and animal assemblage is not very different from that occurring below the lowest emergence level. However, it must be pointed out that while some species never occur in this fringe, others are much more common due to competition removal.

Possible subdivisions of the infralittoral zone must wait until more detailed and accurate investigations have been carried out. RIEDL (1964), who carefully investigated rocky substrates on Adriatic Sea shores, suggested that wave-caused water movement might exhibit three different patterns (for details consult Volume I: RIEDL, 1971): (i) from the water surface down to a depth of about 3 m, i.e. in the breaker zone, water movements are multidirectional; (ii) from 3 m to about 11 m, water movements are oscillating, viz. bidirectional; (iii) from 11 m to 20–25 m (sometimes down to 35 m), the water flows in a direction which depends on that of the waves (in the absence of a permanent current). Layers which correspond to the three patterns could be separated by two critical levels: the first boundary between (i) and (ii) corresponds to a depth of about $2 \cdot 5 \, h$, where h is the average wave amplitude; the second boundary, between (ii) and (iii), corresponds to a depth value between $\lambda/2$ and λ, where λ is the swell wavelength. In my opinion, the first critical level may be recognized anywhere on soft substrates. Furthermore, I believe that the existence of the third layer might be illustrated by the orientation perpendicular to the unidirectional current of fan-shaped colonies of some invertebrate species (gorgonians, bryozoans) in the deepest levels of the infralittoral zone.

(iv) **The circalittoral zone** extends from the lowest boundary of the infralittoral zone down to the maximal depth consistent with the survival of photoautotrophic multicellular algae. It must be emphasized that such algae do not exist in all types of substrate in the circalittoral zone; for example, muddy bottoms are usually unvegetated, but it is generally easy to attribute such sea bottoms to the circalittoral zone, taking into account other bottoms at the same depth, either rocky or soft—but with scattered hard substrates—which are more or less vegetated. The circalittoral zone approximately corresponds to the lower shelf and the shelf edge.

While FERGUSON-WOOD (1966) reported living benthic multicellular algae from a depth of about 250 m in the Florida Strait, I believe that true photoautotrophic metaphytes cannot exist for long below 150 m, the maximum depth they reach, for example, in the Eastern Basin of the Mediterranean Sea. The multicellular algae observed by FERGUSON-WOOD and those I have sometimes observed from deep-sea vehicles or in dredge samples, must, in my opinion, be considered to be damaged or dying. As regards the benthic diatoms (see p. 528) observed at a depth of 350 m off Marseilles by PLANTE-CUNY (1969), one may assume photoassimilation of dissolved organic matter (acetates, lactates, succinates), the energy requirement of which is about one tenth that of photosynthesis.

The physiology of multicellular algae which inhabit the Arctic Basin shelf in the infralittoral zone beneath the sea ice—more or less covered with snow—is still a matter of debate. Obviously, the illumination there is very poor; it corresponds to conditions usually prevailing in the circalittoral zone: some measurements demonstrated that the energy these algae receive in August would be less than approximately 1/5 of that received by the ice surface in June. Since these algae do not seem to exhibit growth reduction in winter, WILCE (1967) assumed that they might survive and develop through heterotrophy.

Earlier, I discussed the possible existence of a **bathylittoral** zone intercalated between circalittoral and bathyal zones (PERES, 1967). The main feature of such a zone would be, according to ERCEGOVIC (1957), the absence of multicellular algae, the plant component in the assemblage being represented only by unicellular species. Such uni-cells have been shown to exist on soft bottoms, but their physiology and ecology are insufficiently documented. On hard bottoms, the existence of unicellular algae is likely, although it has not been demonstrated yet. As has already been pointed out, several physiological peculiarities of algae living in poorly illuminated environments appear to differ from those of true photoautotrophs.

Recent information on the offshore rocky-bottom assemblage (p. 459) suggests placing this in the circalittoral zone, not only because it comprises numerous species of the coralligenous assemblage (p. 453), but also because it contains some poorly developed Lithothamnieae and even lower invertebrates still harbouring a few zooxanthellae, the latter probably playing a negligible part in the food supply of the host.

In general, it seems that the sea floor between depths of 125 to 150 m and 250 to 300 m almost invariably reveals transitional formations between the well-characterized circa-littoral and bathyal organismic assemblages. On soft substrates in the Mediterranean Sea and on the southern Portugal coasts, I have often observed a striking minimum of both biomass and species richness. In contrast, hard substrates sometimes exhibit more abundant and diversified assemblages, because some mixing occurs there between circalittoral species (chiefly those of the shelf-edge rocky bottoms, e.g. *Dendrophyllia cornigera*, *Antipathes fragilis*, *Poecillastra compressa*) which extend downwards and bathyal species which extend upwards. Among the latter, in regions of western Europe, we mention the brachiopod *Gryphus vitraeus* (= *Terebratula vitrea*) which also occurs on pebbles and the echinoid *Cidaris cidaris* which is highly eurytopic and inhabits rocky substrates (where it grazes on sponges) as well as soft (generally rather coarse) substrates.

Nevertheless, several authors (see review by REX, 1976, pp. 982–3) suggest that some soft bottoms on the upper slope exhibit a comparatively high overall diversity related to some spatio-temporal patchiness. However, I believe such diversity is not a general phenomenon and is possibly restricted to situations where the food resources—even if available irregularly—are sufficiently important. The very low density and the general scarcity I have observed in the areas mentioned above might be ascribed to their oligotrophy.

In summary, it seems that a true bathylittoral zone with specific organismic assemb-lages does not exist. There appears to be only some transitional formation, a 'contact margin' (REYSS, 1970) between circalittoral and bathyal zones. This contact margin exactly corresponds to ZENKEVICH's (1956) 'transitional horizon'.

It must also be pointed out that in the Antarctic region a very important mixing of circalittoral and bathyal species occurs over a vertical range of several hundred metres. This is due to the fact that the very cold Antarctic surface water, flowing downwards along the slope to form the Antarctic bottom water, makes the hydrological conditions very homogeneous. Moreover, the Antarctic surface water also transports towards the depths of a bulk of suspended organic particles—originating from the most thriving infralittoral and upper circalittoral assemblages—which provides an abundant food supply to typical circalittoral species, mainly suspension feeders. The homogeneous hydrological conditions, due to the Antarctic surface water descending along the shelf and the slope, also promote an upward spreading of many bathyal species which compete and mix with circalittoral species. This results in a large transitional belt between the two vertical zones and thus in a sometimes almost indistinct boundary between them.

Zonation in the Aphytal System

The aphytal system may be divided into three vertical zones: bathyal, abyssal and hadal (the latter is also known as ultra-abyssal). These three zones correspond to the three main geomorphological entities of the sea floor: the continental slope, the abyssal plain and the trenches.

(i) **The bathyal zone** comprises assemblages occurring on the continental slope and its canyons below the transitional formations discussed above. At the lower part of the slope, the inclination often becomes more gentle and results in another transitional zone where some mixing of bathyal and abyssal assemblages may occur.

In the bathyal zone appear the first typical deep-sea groups, such as sponges of the class Hexactinellidae, holothurioids of the order Elasipoda, and fishes of the family Macrouridae. We can also notice many new genera and species of numerous other taxa such as decapod crustaceans, asteroids and echinoids.

The lower boundary of the bathyal zone is marked by striking changes in the fauna. For example, almost all the hexactinellids disappear (only four genera were recorded from the abyssal zone). With regard to the holothurioids of the order Elasipoda, the bathyal families Deimatidae and Laetmogonidae are replaced by the abyssal families Psychropotidae and Elpidiidae. The asteroid species of the archaic order Phanerozonia also change between 3000 and 3500 m. Moreover, at about the same depth almost all eurybathic shelf species which are not sufficiently eurythermal (i.e. in this case, psychrophilic or at least cold-tolerant) disappear.

Using these faunistic changes as a basis to discriminate the bathyal zone from the abyssal zone seems to be more realistic than using the 4 °C isotherm proposed by BRUUN (1957); the depth of this isotherm, in context with the geomorphological features which usually separate the continental slope from the abyssal plain, may vary considerably at different latitudes.

(ii) **The abyssal zone** comprises sea floors at depths between 3000 to 3500 m and 6500 to 7000 m. Since the horizontal distances in this zone are very large, it is usually called the abyssal 'plain'. Its lower boundary corresponds to the gradient increase which occurs at the edge of the trenches.

In general, besides the faunistic changes mentioned above in context with the bathyal zone, the abyssal assemblages exhibit a sharp decrease in both species number and biomass. However, a maximum in specific diversity may sometimes be observed in the lowest, gentle part of the slope (named abyssal rise). In this transitional belt some mixing may occur between bathyal and abyssal species. In contrast, the decreasing biomass gradient seems to remain unaffected by this mixing.

In general, it seems that many abyssal species may exist over the whole vertical range of the zone. According to NYBELIN (1957), however, the number of fish species recorded exhibits a marked decrease below 4500 m, as, apparently, does fish abundance. WOLFF (1977, p. 785) has commented on this by pointing out that, at abyssal depths, fishes and amphipods represent the overall dominating scavengers; however, unlike fishes, lysianassid amphipods (sometimes represented by 'giant' forms) remain common or possibly become commoner below 4500 m, even down to the hadal zone. VINOGRADOVA (1958) also noticed some rather important faunal changes in the Pacific Ocean at a depth of about 4500 m. In the present state of investigation, it is difficult to decide whether the abyssal zone must be divided into two subzones at this depth, or whether the sea bottoms situated between 3000 and 4500 m correspond in particular areas to an unusual extension of the transitional belt mentioned above between the bathyal zone and the true abyssal. The latter interpretation cannot be precluded since, with increasing depth, environmental gradients diminish while the depth range of any transitional belt between two vertical zones increases, except in cases of a sudden discontinuity in bottom geomorphology.

(iii) **The hadal zone**, or ultra-abyssal zone, comprises trenches deeper than 6000 to 7000 m. The walls of these trenches are generally rather steep, but their bottom is flat. In comparison with the abyssal fauna, the hadal fauna exhibits a higher degree of impoverishment with respect to both number of species and biomass. Much more striking is the fact that many animal groups well represented in the abyssal are absent in the hadal zone—mostly those at the top of the trophic pyramid (i.e. predators), such as decapod crustaceans, pycnogonids, cephalopods, asteroids and fishes. However, using baited traps it was recently demonstrated that scavengers may be very abundant in the hadal zone. For example, thousands of lysianassid amphipods attracted to fish bait have been collected at 9600 m in the Philippine Trench (WOLFF, 1977). These very active amphipods have apparently fully assumed the role of fishes as scavengers; other examples of such substitution are the deepest—but abyssal—areas of the Norwegian and Greenland Seas (p. 504).

Nevertheless, other groups—mainly microphagous species—also disappear at the lower limit of the abyssal zone, e.g. members of the Gorgonia, Pennatularia, cirriped crustaceans, ophiuroids and echinoids.

Barophilic bacteria which can grow only under pressures exceeding 600 to 700 bar might also be considered to be characteristic of the hadal zone.

WOLFF (1977) has emphasized the important role played in the deepest bottom assemblages by giant (i.e. of macrofaunal size) agglutinating foraminiferans hitherto neglected. These are remarkably abundant in some apparently highly oligotrophic, areas. The xenophyophores, usually several centimetres in diameter, may feature pseudopodia which can extend to 10–15 cm and hence be of considerable trophic significance. Members of the superfamily Komokiacea are distributed world-wide, with a relatively high abundance in hadal trenches and the abyssal oligotrophic north Pacific Ocean. In the latter, their volume exceeds that of metazoans and equals that of other foraminiferans (WOLFF, 1977, p. 784).

Conclusions

The vertical zonation scheme adopted in the preceding pages does not differ significantly from the schemes proposed by various other authors (Table 2-1).

The main differences are the following: (i) the supralittoral fringe of the 'intertidal zone' of other authors is separated as supralittoral zone; (ii) the infralittoral fringe of the 'intertidal zone' is joined to the shallow-water bottoms which never emerge and the true midlittoral zone restricted to levels where immersion and emergence alternate; (iii) the sublittoral of many authors is divided into two zones: the well-illuminated infralittoral zone inhabited by photophilic plants with an important benthic primary production and the less well-illuminated circalittoral zone where the benthic primary production by sciaphilic plants is less important and the animal biomass is generally larger than the plant biomass. As regards the aphytal system, the scheme adopted follows that proposed by Russian scientists based on their investigations in the Pacific, Indian and Atlantic Oceans. Incidentally, the infralittoral and circalittoral zones in the benthal approximately correspond to the epiplanktonic and mesoplanktonic zones in the pelagial.

Finally, it should be emphasized that the vertical zonation adopted for the benthal never refers to isobaths for separating vertical zones, but only to fundamental changes in the nature and composition of the benthal organismic assemblages concerned.

Although flora and fauna of polar, temperate and tropical seas are very different—at least on the shelf and the upper parts of the continental slope—we can recognize within the assemblages they constitute some general structural features which indicate parallelisms between assemblages from different biogeographic regions. The concept of 'isocommunities' proposed by THORSON (1975) was the seed of this vertical zonation scheme.

As will be seen later (p. 373ff) almost all organismic assemblages conceived by the reviewer can easily be accommodated in this scheme, irrespective of the biogeographic region inhabited. This fact suggests that the scheme can be of world-wide value.

(c) Vertical Changes in Biomass or Abundance

Plankton

If we assume the plankton biomass represents a value of 1 in the epipelagic and mesopelagic zones—considered together because of the mixing of assemblages due to vertical circadian migrations—the biomass would be 1/3, 1/25, 1/50 and 1/500 in the infra-, bathy-, abysso and hadopelagic zones respectively (Fig. 2-3). The corresponding decrease in abundance (i.e. number of individuals per unit volume) with increasing water depth is probably even more marked because the body size of planktonic organisms generally increases as a function of depth.

These general relationships are rough approximations, but they hold irrespective of vertical circadian variations, which are uncommon for species normally inhabiting waters below the lower level of the infrapelagic zone.

Changes in the almost exponential overall decrease in biomass with increasing depth are related mainly to three phenomena:

(i) At middle—and sometimes also rather low latitudes—convergences originating at high latitudes may cause sinking of cold-water surface assemblages. Obviously (for example, in the northern Pacific Ocean, see p. 237), not all 'expatriated' surface zooplankters can adapt to the new environmental conditions met in the infrapelagic zone or deeper. Therefore their population strength progressively decreases. Presumably, this is not primarily a consequence of temperature—which decreases with increasing depth to about 4 °C—but rather of inadequate nutrition: the food resources tend to become quantitatively or qualitatively insufficient for some of the members of the original surface assemblage.

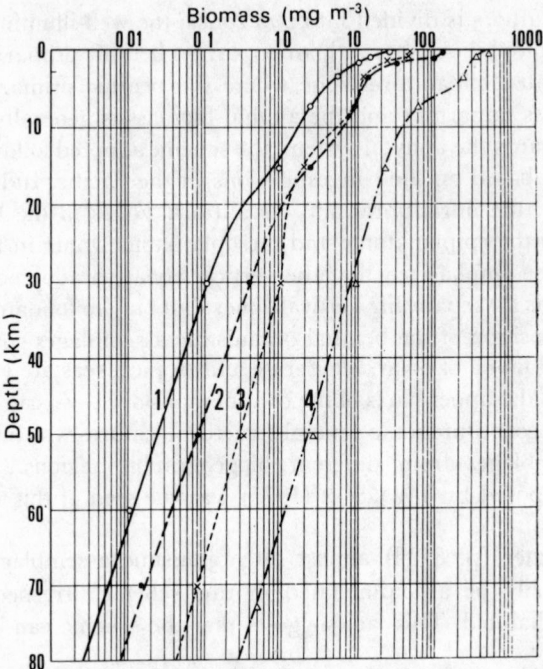

Fig. 2-3: Vertical distribution of zooplankton biomass
in the deep layers of the Pacific Ocean. 1: Marianas
Trench; 2: Bougainville Trench; 3: Kermadec Trench;
4: Kuril-Kamchatka Trench (average values from six
stations). (Based on information provided by
VINOGRADOV, 1968.)

Other sinking populations adapt to life in the deep sea and even continue to multiply
as long as the food resources support them; this results in higher abundance and biomass
values in the infrapelagic zone—and sometimes (in somewhat lower latitudes) also in
the bathypelagic zone—than in upperlying waters. A prerequisite for such spreading of
surface and subsurface zooplankters towards lower latitudes and greater depths is
presumably the ability of these species—especially the filter feeders among them—to
modify their feeding habits with regard to the nature, and possibly also the size, of
the particles consumed; apparently, they become more and more omnivorous.

The ability of epi- and mesoplanktonic species to adapt to the deep-sea pelagial
reaches a maximum in those cold-loving species which exhibit the phenomenon called
'equatorial submergence'. Such species inhabiting surface and/or subsurface layers at
high latitudes of both hemispheres are found at deeper levels in the middle and low
latitudes (for example: the chaetognath *Eukrohnia hamata*, p. 238). It seems that the very
rare cases of equatorial submergence involve only relatively large-sized predators. Zooplankters on the whole tend to become larger with increasing depth; they must also be
very active swimmers and prey seekers because plankton abundance decreases with

depth. The latter fact probably explains why such species apparently do not occur below about 4000 m. Probably such predators can breed even in the inhabited deeper layers.

With regard to sinking species which do not exhibit equatorial submergence, it seems that their populations cannot be considered to consist only of 'expatriated' individuals. Presumably, they can breed as long—and as deep—as food resources (which, in general, decrease with increasing depth and decreasing latitude) suffice.

(ii) At high latitudes, where the primary production—and consequently the abundance of epiplanktonic and mesoplanktonic populations—are restricted to spring and summer, many species descend during winter due to cooling and food shortage in the surface and subsurface layers. (Fig. 2-4). This phenomenon which results in a winter

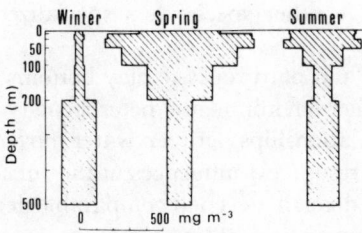

Fig. 2-4: Vertical distribution of zooplankton biomass (mg m^{-3}) in the north-western part of the Pacific Ocean and the southern part of the Bering Sea at different seasons (average values from 67 stations). (Based on information provided by VINOGRADOV, 1968.)

increase of biomass and abundance in the infrapelagic zone and often in the upper levels of the bathypelagic zone also is well documented in the boreal and boreo-arctic areas of the north-eastern Atlantic Ocean (p. 269) as well as in the Antarctic region (p. 194). Such overwintering seems to involve mostly filter feeders; since they feed intensively— mainly on phytoplankton—during the warm-season peak in the surface layers, they accumulate a bulk of nutritive reserves; thus, the food-resource problem is not critical for them. However, we cannot exclude the possibility that they continue to feed, at a very reduced rate, on non-living particles and aggregates.

(iii) A third phenomenon—of more local significance than the two preceding ones —which can modify the vertical quantitative distribution of zooplankters in the excessive decrease in the oxygen content of intermediate waters (occurring in enclosed basins such as fjords but also in the open ocean) at depths of minimal horizontal motion outside areas with strong coastal upwelling. According to LONGHURST (1976, p. 131):

'below about 0·2 ml/l oxygen or in the presence of H$_2$S there may be exclusion of plankton . . . though exclusion is seldom complete and at oxygen levels of 0·2–0·5 ml/l, diel migrants and seasonal "hibernators" may establish their day or longer-term residence depths.'

For example, in the Arabian Sea (p. 261) layers from 125 to 200 m to 1000 to 1200 m, with oxygen contents sometimes lower than 0·15 ml l^{-1}, exhibit the lowest biomass values (1·7 to 3·4 mg m^{-3}) between depths of 200 to 500 m, whereas below 1000 to 1500 m, biomasses of 5 to 10 mg m^{-3} may be observed.

Benthos

Changes in biomass or abundance of benthic organisms in relation to water depth are much more variable in different places than in plankters. It seems that there are two main reasons for this:

(i) In general, the number of species is much greater in the benthos than in the plankton (possibly 98% of all multicellular marine organisms belong to the benthos). Moreover, interspecific competition must be assumed to be more severe in the benthal—an almost bidimensional environment—than in the pelagial. Furthermore, except for the more ubiquitous species, most benthic plants and animals tend to occupy specific places—in regard to the quality of the substrate, surrounding water and depth—a fact which enables them to outcompete other species less specialized for prevailing environmental conditions.

(ii) Substrate diversity, ranging from compact and hard rocks to clay bottoms with particles smaller than 1 μm, results in a much higher environmental heterogeneity than in the pelagial. The direct and indirect (mainly relationships between water movement and substrate characteristics; importance of food resources) influences of the substrate on the organisms often cause a striking organismic diversity, even at comparable depths. Both species composition and biomass values within an assemblage may vary considerably. Thus, tables usually presented in general books listing average biomass values over large depth ranges, such as 0 to 200 m, 200 to 3000 m or below 3000 m, can contribute but little to attempts at accurate analysis of biomass changes as a function of water depth. Moreover, important additional data, obtained during the last twenty years, may modify the interpretations made thus far.

On supralittoral hard bottoms, the biomass is always low, usually reaching from a few grams up to 100 to 150 g m^{-2} (w.w.) and sometimes attaining values of over 400 g m^{-2} on soft substrates.

In the midlittoral zone, we can generally observe, on hard substrates, a progressive increase in biomass from upper to lower levels. On temperate shores (mainly tidal) the algal standing crop (often several kg m^{-2}) is more important than that of animals (usually less than 100 to 200 g), except for bottoms primarily occupied by barnacles or sessile pelecypods such as mussels or oysters. In midlittoral mussel beds, the standing crop may often range from several to tens of kg m^{-2}. Mussel shells and other calcareous material of the associated flora and fauna generally represent about 70 to 75% of the total standing crop. On soft substrates the biomass usually ranges from a few grams up to a few hundred g m^{-2}.

In the infralittoral zone on hard, well-illuminated substrates, irrespective of those occupied by hermatypic reef-building corals, the total biomass is generally higher than that of the midlittoral zone—mostly on soft bottoms—and algae often still predominate on hard bottoms. If the substrate nature does not change with increasing depth, a progressive decrease tends to occur in plant biomass and an increase in animal standing crop. Since the latter change is generally less pronounced than the former, the total standing crop tends to decrease. Table 2-2 exemplifies this general trend, summarizing data obtained by ALEEM (1956) on the assemblage of the giant kelp *Macrocystis pyrifera* at the coast of California (USA).

Table 2-2

Changes in total biomass and that of phyto- and zoobenthos in the *Macrocystis pyrifera* assemblage on California coasts (After ALEEM, 1956; reproduced by permission of the American Association for the Advancement of Science)

Depth (m)	Biomass (g m^{-2})		
	Phytobenthos	Zoobenthos	Total
1·5	4667	125	4792
7	2490	325	2819
15	1972	392	2364
22	606	317·2	983·2

Where mytilid assemblages occur in the infralittoral zone, their biomass is usually of the same order of magnitude (often 10 to 20 kg m^{-2}) as that of midlittoral beds, even at lower levels of the zone. As regards coral reefs, which represent a complex of assemblages, the average biomass seems to be about several tens of kg m^{-2}, 80 to 90% of which is calcareous material.

On unvegetated infralittoral soft bottoms, the biomass values largely depend on substrate nature, but the biomass always seems to be higher than in soft bottoms of the midlittoral zone and, in some assemblages of temperate areas, may be as high as a few kg m^{-2}.

The best-known vegetated soft bottoms are sea-grass beds. On the north-western shores of the Pacific Ocean (p. 446), *Zostera* beds exhibit plant and animal standing crops ranging from 3500 to 7300 g m^{-2} and from 500 to more than 3000 g m^{-2} respectively. Such enrichment of the animal component is mainly due to an increase in infauna elements which may be 2·5 times more abundant in *Zostera* beds than in similar but unvegetated neighbouring sediments. In general, the total biomass of soft-bottom vegetation consisting of large algae seems to be much lower than that of sea-grass beds.

On circalittoral hard bottoms, the plant standing crop seems to be almost always lower (excluding, where pertinent, the calcareous matter of Corallinaceae) than the animal standing crop. The total biomass sharply decreases since the percentage of calcareous material is also high in the animal component, especially in typical coralligenous assemblages (p. 453). Here the biomass usually tends to range between 500 and 1000 g m^{-2}, including at least 50% calcareous matter. The highest biomass values seem to occur in rather cold (subarctic and subantarctic) areas. In the Antarctic region, where the most recent investigations suggest that the biomass does not change much from this level down to the upper slope, biomasses as high as 3000 g m^{-2} have been recorded (p. 459) between depths of 100 to 500 m; however, these consisted mainly of siliceous sponges whose organic-matter content comprises a fairly low percentage of the total biomass.

On circalittoral soft bottoms, the standing crop—consisting exclusively or almost exclusively of animals—usually decreases to 250 to 500 g m^{-2} in cold temperate and subarctic areas, but may be about 10 times lower in true Arctic areas and most of the intertropical regions. However, where large-sized species such as ophiuroids, echinoids or gastropods are abundant, the biomass may be as high as 1000 or even 2000 g m^{-2}.

In the bathyal zone, the biomass of solid-substrate assemblages is unknown. On soft substrates (with the exception of particular areas such as the Californian coast, where biomasses as high as 200 g m^{-2} have been observed), the values seem to range between 2·5 and 17 g m^{-2}; large variations occur as a function of local conditions, mainly influenced by the fertility of upperwater and shelf assemblages. In the Arctic Basin at a depth of 1000 to 2500 m, the biomass of the bathyal mud is very low (0·04 g m^{-2}).

The standing crop of abyssal-mud assemblages depends much more on nutrient input from the pelagial or from the neighbouring slope and shelf. Thus, in the western half of the northern Pacific Ocean, at depths of 3000 to 6000 m, biomass values in the north-western, south-western, north-eastern and south-eastern quadrants are 0·2 to 1, 0·1 to 0·2, 0·5 to 0·8 and 0·01 g m^{-2} respectively. Nearer to the Alaskan coast, biomasses at depths between 4200 and 4700 m may rise to 2·2 g m^{-2}. This fact indicates that the nearer the land, the higher the standing crop. In contrast, a comparison between the northern and southern quadrants mentioned above indicates the influence of the productivity in the upperlying pelagial.

For the hadal zone few data are available. In the Pacific Ocean biomass values range from 0·007 to 0·022 g m^{-2}. An exceptionally high biomass (0·6 g m^{-2}) was recorded from the Romanche Trench in the Atlantic Ocean.

It must be remembered that the biomasses mentioned for the abyssal and hadal zones do not generally take into account the possible presence in an individual sample of a large-sized invertebrate such as a holothurioid.

In spite of the fact that the biomass values listed in Table 2-3 are approximate, the following conclusions may be drawn: (i) The values recorded in supra- and midlittoral soft substrates are similar and depend more on the inhibition of the substrate—related to granulometric characteristics—than on its depth level. (ii) Maximum biomass values clearly correspond to the infralittoral zone. (iii) Benthos biomass seems to decrease from the permanently immersed level down to the greatest water depths, but not exponentially as is generally assumed for most plankton assemblages; instead (except for local, particular circumstances), two marked discontinuities are revealed.

Table 2-3

Approximate mean biomass values (g m^{-2}) in different vertical zones of the benthal. ve and uv: vegetated and unvegetated soft bottoms, respectively; v: biomass mostly consisting of plants except for coral reefs and mussel beds; c: mostly calcareous material (Compiled from various sources)

Zone	Solid substrates	Soft substrates
Supralittoral	5–150	5–400
Midlittoral	100–5000 (v)	5–400 (uv)
Infralittoral	10,000–20,000 (v)	500–7000 (ve)
		100–3000 (uv)
Circalittoral	500–1000 (c)	200–500
Bathyal	?	2–17
Abyssal	?	0·01–2
Hadal	?	0·007–0·02

Therefore, the following oversimplified scheme might be proposed. The first discontinuity separates the infralittoral zone from the circalittoral zone. At this level, the mean total biomass suddenly decreases from values of kg m^{-2} to less than a few hundred g m^{-2} (less than several tens of g m^{-2} in the poorest bottoms); this discontinuity corresponds to the sudden decrease in metaphyte—possibly also microphytobenthos—primary production. The second discontinuity occurs near the shelf edge where the biomass suddenly decreases from usually several hundred g to less than 20 g m^{-2} (often only about 2 g m^{-2}). Of the two discontinuities, the first one involves a decrease of about 10 times, and the second of about 20 times. In general, beyond the shelf edge the decrease in biomass with depth in the bathyal and abyssal zones seems to be progressive. Below the abyssal, the data available do not suggest any more discontinuities in the biomass distribution between the abyssal and the hadal zone. However, the data at hand are too scarce to be really significant. If real, the absence of an abrupt biomass decrease between bathyal and abyssal, on the one hand, and between abyssal and hadal, on the other, might be due to a decrease in environmental diversity with increasing depth; this also pertains to the amount of food present. Furthermore, the more gentle slope (abyssal rise) corresponding to the transitional belt between continental slope and abyssal plain may also contribute to the apparently smooth biomass decrease.

The first discontinuity, between infralittoral and circalittoral zones, is probably due to a marked decrease in metaphyte abundance. To some extent, the two zones appear to parallel the epipelagic and mesopelagic zones; however, the two pelagial zones cannot really be separated from an ecological point of view, because of numerous diel zooplankton migrants and because plankton previously sunk to the oligotrophic layer may be lifted back into the euphotic layer by dynamic processes. Since the productivity can change abruptly neither in the underlying waters nor in the microphytobenthos at the shelf edge, the pronounced discontinuity there is much more difficult to explain. One possible explanation could be an increase in current speed perpendicular to the isobaths resulting in a decrease of organic-matter input on both sides of the shelf edge, thus causing local food shortage. This hypothesis gains support from the relative scarcity of the fauna both before the edge in the offshore detritic assemblage (p. 479) and beyond it in the transitional belt between circalittoral zone and upper bathyal levels (p. 21).

In the preceding pages we have considered biomass as a function of depth. Unfortunately, in the past the abundance (number of individuals per unit space) has often been confused with biomass, and as emphasized by WOLFF (1977), this has led to the generally accepted opinion that with increasing water depth and distance from land the benthic fauna consists of fewer individuals which belong to fewer species. However, application of new equipment—such as anchor dredge and epibenthic sled—together with careful washing of the samples obtained through fine-meshed sieves has greatly increased the actual numbers of individuals caught, thereby increasing the number of species recorded. In addition, taxonomists have recognized that the samples contained numerous closely related species (WOLFF, 1977). While the diversity problem will be discussed below, we must mention here that three sled samples from the northwestern Atlantic Ocean collected from 1400, 2900 and 4700 m contained 25200, 12100 and 3700 individuals belonging to 365, 310 and 196 species respectively. The hitherto most efficient abyssal sample was taken with a large trawl in the eastern Pacific Ocean, 3600 m deep, and contained 2100 individuals belonging to 132 species (WOLFF, 1977, p. 781).

In order to compare population size as well as diversity on and in deep-sea and shallow-water soft substrates, SANDERS (1968, 1969) carried out, in different coastal areas, epibenthic sled samplings which were processed with identical techniques. For estimating numerical abundance, he counted polychaetes and bivalves which represented about 80% of the total fauna. In a tropical shallow-water environment (Bay of Bengal), SANDERS found about 2800 individuals and 80 species and in a boreal shallow-water environment (Buzzards Bay, north-eastern United States coast), about 2500 individuals and 25 species. Compared with the data obtained from a continental slope area (2800 individuals, approximately 80 species), these values document that some bathyal soft bottoms may reveal a numerical organismic abundance of the same order of magnitude as does the shelf. The sharp decrease in biomass near the shelf edge (Table 2–4) may be ascribed to the fact that in the bathyal benthos, the percentage in the whole fauna of small-sized species (mainly peracarids, polychaetes and pelecypods) is higher in comparison with that on the shelf. Thus, increasing depth would affect abundance slightly, biomass considerably.

Table 2-4

Decrease in biomass with depth in the pelagial and benthal expressed as a fraction of the maximum value (=1), which usually occurs in the shallowest water layer or near the sea bottom. Neuston and pleuston systems in the pelagial and supra- and midlittoral zones in the benthal have been omitted (Original)

Depth (m)	Pelagial zones	Relative biomass		Benthal zones
		pelagial	benthal	
0 50	epipelagic		1	infralittoral
100 150 200	mesopelagic	1	≃1/10	circalittoral
250 500	infrapelagic	≃1/3		
1000 1500 2000	bathypelagic	≃1/25	≃1/250	bathyal
3000 4000 5000 6000	abyssopelagic	≃1/50	≃1/5000	abyssal
7000 8000 9000 10,000 11,000	hadopelagic	≃1/500	≃1/50,000	hadal

The general significance of this hypothesis was recently debated by LAVIE and KRIPKE (1977). By means of trawling and deep-sea photographs, they investigated the megafauna (decapods, echinoderms, fishes) from 497 to 2780 m in the western Atlantic Ocean, south of New England. These authors do not agree with the usually adopted

concept that megafaunal biomass is small in comparison with meiofaunal and macrofaunal biomass. They state that fish biomass, which represents approximately 40% of
the whole megafaunal biomass in the depth range sampled, varies between 0·63 and
4·66 g m^{-2}. Taking into account other data with respect to the macrofauna at the same
depths in the same area, LAVIE and KRIPKE found the mega- and macrofaunal biomasses to be essentially of the same order of magnitude.

> 'The assumption that the biomass of the larger animals is relatively insignificant
> seems to be unfounded. Unless turnover rates in the macrofauna are relatively
> extremely high—and all the evidence suggests that this is not the case—the mega-
> faunal biomass cannot be supported by the macrofauna.' (LAVIE and KRIPKE,
> 1977, p. 141.)

Observations with a baited deep-sea camera, as well as studies of the feeding habits of
Coryphaenoides arcuatus, a macrourid on the continental rise which accounts for as much as
80% of the fish biomass there, have suggested to the authors mentioned that large
pelagic animals and fast-falling bodies (e.g. of fishes, whales, squids and decapods)
constitute the main food resource of the deep-sea megafauna. Since such food items are
scarce and occur randomly and since the metabolic requirements of a large animal are
less per unit weight than those of a small one, selection for large, efficient food seekers
appears to prevail in the deep-sea megafauna. This assumption of LAVIE and KRIPKE is
supported by the 'dramatic increase' in the relationship between mean individual fish
weight and water depth at the beginning of the rise near 2000 m. Finally, LAVIE and
KRIPKE refute the usual assumption that small forms dominate food-limited associations. They suggest that both forms of life are possible in the deep sea, particularly where
the faunal components are relatively independent of one another. Large body size
appears to be selected for where high motility prevails, while small size predominates
where the ability to live in a fine-grained sediment and to feed on a thin organic layer is
important.
    In summary, irrespective of the specific diversity problem and thus limiting the
discussion to biomass and population size, we conclude that in both the pelagial and
benthal the amount of food available represents a very important, if not the most
important, factor controlling these two parameters as well as the species composition of
the assemblages studied.

(2) Horizontal Zonation

    A detailed treatment of horizontal zonation of both organisms and the assemblages
they constitute must consider biogeographical aspects. Since these are beyond the scope
of the present chapter and receive attention in other parts of Volume V, we shall restrict
our discussion here to patterns of horizontal zonation within a biogeographic region.
Relationships between sea-water temperature and specific richness which largely control
horizontal zonation are dealt with.
    In the pelagial, we must distinguish between plankton organisms, the assemblages
formed by them, and the peculiarities of individual nektonic species participating in a
pelagial system. With regard to the plankters concerned, horizontal zonation depends on

the combined effects of four groups of factors: (i) the life cycle, life span and multiplication rate of the species; (ii) the predator–prey relationships within each assemblage and
losses due to sinking; (iii) the passive drifting of an assemblage with the water mass it
inhabits due to water movement (current speed and direction); (iv) the energy available
at the lowest level of the trophic pyramid, i.e. local primary production. In other words,
horizontal zonation of plankton organisms (and assemblages) is narrowly related to
'ecological succession' (p. 106). Factors (ii), (iii) and (iv) also influence the composition
and abundance of larval stages and juveniles in the nektonic fraction of a pelagial
assemblage; for adults one has to take migratory behaviour into account.

The combined effects of these factor groups may be exemplified by referring to the
progressive space shifting of different pelagial species and assemblages as they are
transported farther and farther from an area with maximum primary productivity, for
example, either northwards or southwards (meridian shifting) of an equatorial
divergence (Fig. 2–5) or if the intensity of the upheaval decreases along the same
latitude (latitudinal shifting), e.g. westwards in the equatorial divergence of the Pacific
Ocean (Fig. 2–6).

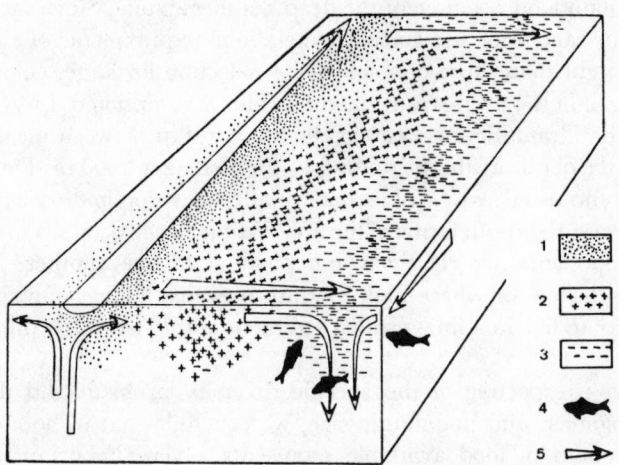

Fig. 2-5: Block diagram of the horizontal zonation of different
pelagial assemblages. Successively predominant species
pertain to increasingly higher levels of the trophic pyramid,
in proportion to increasing distance from the equatorial
divergence. 1, 2, 3, 4: areas of phytoplankton, mesozooplankton, macroplankton, fish maximum, respectively; 5: direction
of currents. (Based on information provided by VINOGRADOV
and VORONINA, 1963.)

In the benthal, the horizontal zonation of macrobenthos organisms and the assemblages they constitute is mainly controlled by the nature of the substrate, which depends
on both the water content in suspended particles and water movement. In assessing the
role of primary production for affecting horizontal zonation in the benthal we must bear
in mind that: (i) the benthal surface area which facilitates photosynthesis represents

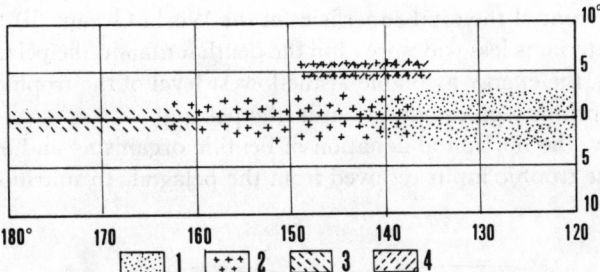

Fig. 2-6: Scheme of the horizontal distribution of different
plankton assemblages in the equatorial divergence of
the Pacific Ocean. 1, 2, 3 and 4: assemblages in which
phytoplankters, mesozooplankters, migrant and surface
macroplankters, respectively, predominate. (Based on
information provided by VINOGRADOV and VORONINA,
(1964.)

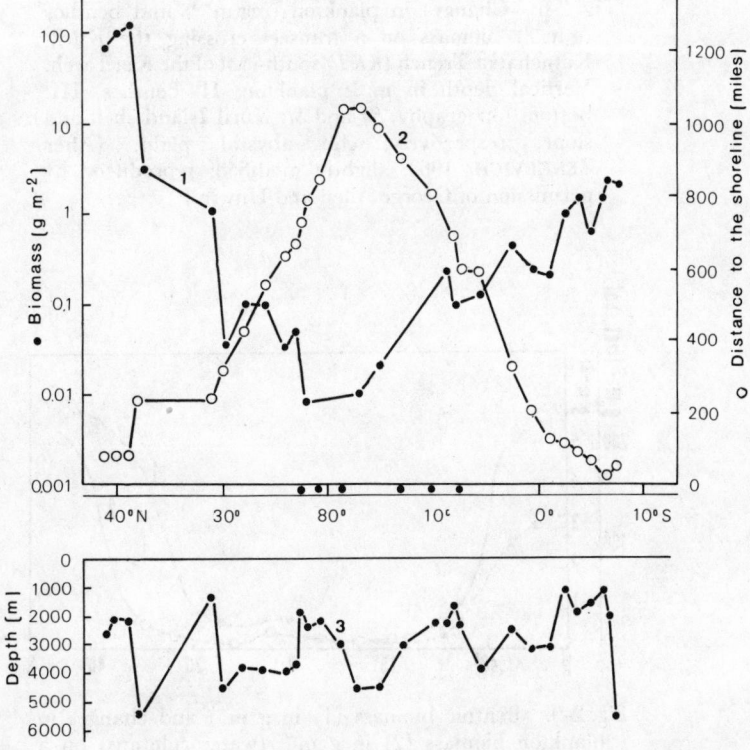

Fig. 2-7: Changes in benthos biomass (g m^{-2}) in relation to distance-to-
shoreline and water depth on a transect from Hokkaido to New Guinea.
(Based on information provided by BELAYEV, 1960.)

only a few per cent of the total superficies of the World Ocean; (ii) the consumption of fresh plant material is less widespread in the benthal than in the pelagial. Consequently, in the benthal, the energy available at the lowest level of the trophic pyramid depends less on *in situ* primary production than in the pelagial.

In summary, the horizontal zonation of benthic organisms and assemblages largely depends on the trophic input received from the pelagial. In nutritionally poor benthic

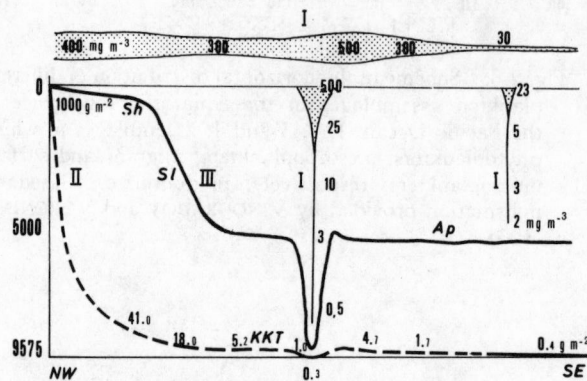

Fig. 2-8: Changes in plankton (mg m⁻³) and benthos (g m⁻²) biomass on a transect crossing the Kuril-Kamchatka Trench (*KKT*) south-east of the Kuril arch. Vertical depth in m. I: plankton; II: benthos; III: bottom topography. *Sh* and *Sl*: Kuril Island shelf and slope, respectively; *Ap*: abyssal plain. (After ZENKEVICH, 1963; slightly modified; reproduced by permission of George Allen and Unwin.)

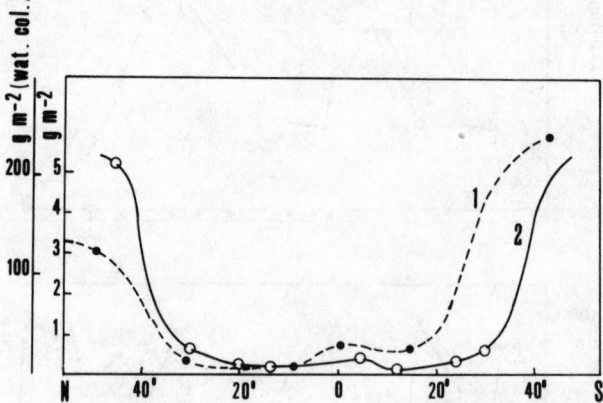

Fig. 2-9: Benthic biomass (1) in g m⁻² and changes in plankton biomass (2) in g m⁻² (water columns) on a meridian transect in the Pacific Ocean. (After VINOGRADOV, 1977; based on information provided by VINOGRADOV, 1960 and FILATOVA, 1969.)

areas, nutrients from more productive benthos assemblages may also be of importance. In this case, the intensity and patterns of water movement, especially currents and tides, tend to attain decisive significance.

Due to both organic-matter input from continents and higher productivity of plankton and benthos assemblages of the shelf, the distance to shoreline exerts more controlling influence than water depth on the amount of organic matter received by the sea bottom and hence the horizontal distribution of benthic organisms and assemblages (Figs 2–7 and 2–8); see also the example provided for benthic assemblages of the North Pacific Ocean abyssal plain (p. 492). The almost complete decomposition of organic matter sinking down the water column above great depths dampens the effect of high productivity in the surface waters of the equatorial divergence. Hence the resulting increase in bottom-fauna biomass is fairly limited (Fig. 2–9).

(3) Food: the Major Factor Controlling Vertical and Horizontal Zonations in Pelagial and Benthal

(a) General Aspects

The information presented in the preceding pages documents that vertical changes in composition biomass and abundance of assemblages, both in the pelagial and benthal, are always closely related to the amount of food available. However, not only the quantity of food, but also its quality is important. We must further consider the feeding habits of the species concerned, i.e. the degree of their specialization for utilizing a given food source or their versatility for making use of a wide array of different foods.

A comparison of vertical changes in biomass of the pelagial and benthal (Table 2–4) reveals that the depth-dependent biomass decrease is much more marked in the benthal than in the pelagial. Actually, the benthic biomass data listed only represent the macrobenthos; it is a well-documented fact that the biomass ratio macrobenthos/meiobenthos decreases with increasing depth at least in the mean abyssal depths down to values of about 1/1. Nevertheless, even if we take into account the meiofaunal biomass in the abyssal and hadal zones (the meiofauna in the latter is poorly investigated), the abyssal and hadal benthos biomasses would be approximately one hundred times lower than in the homologous vertical zones of the pelagial.

Such discrepancy in the depth-dependent biomass values of pelagial and benthal— which may become even more accentuated when sufficient productivity values become available—arises from several convergent causes, the most obvious of which are discussed below.

(b) Distribution and Fate of Primary Production

Since primary production in the marine environment constitutes the major part of the organic matter available at the base of the trophic system, we must first point out that benthic and planktonic primary production exhibit very different characteristics. The infralittoral zone is that section of the benthal which may actually be compared with the euphotic layer (epipelagic zone); its lower boundary rarely exceeds 40 m. Since most coastal waters are more turbid due to their richer plankton and the proximity of the shoreline, the productive macrobenthic flora—which cannot develop on all infralittoral bottoms—occupies presumably less than 1 to 2% of the total superficies of the World Ocean. If we add to this the microphytobenthos production, the above estimate would probably only have to be slightly modified. In contrast, the euphotic layer extends over the whole World Ocean. Its thickness varies between 10 to 30 m in eutrophic areas with production rates over 0.5 g C m^{-2} d^{-1}, and 50 to 80 m in very clear waters of oligo-trophic areas with production rates below 0.5, often less than 0.1 g C m^{-2} d^{-1} (for example central parts of the oceanic gyres). In the latter case the chlorophyll a maximum occurs near the base of the euphotic layer (LORENZEN, 1976).

True grazers, except those which feed on microphytobenthos, are not common in the benthos macrofauna. Hence, a very important part, possibly as much as 90%, of primary production from the benthic macrophytes becomes available to secondary consumers only in the form of detritus. Moreover, it is assumed that the detritus itself cannot be digested and that only attached microorganisms—mainly bacteria and fungi—provide energy for consumers.

We may expect some parallelism between epi- plus mesopelagic zones in the pelagial, on the one hand, and infra- plus circalittoral zones in the benthal, on the other. In each of these pairs, the first-mentioned zone contains more primary producers, the second more consumers. In reality, however, such parallelism is fallacious. In the pelagial, exchange prevails between the two zones: algal cells sink or are transported upwards, e.g. by turbulence or vertical mixing; the nocturnal grazing of diel zooplankton migrants temporarily increases phytoplankton production; phytoplankters such as some dino-flagellates conduct diel migrations. In contrast, in the benthal, the two zones are not superimposed but adjacent; exchange between them only arises from passive transport by water currents (whatever their origin) or from migrations, mainly seasonal, involving fairly motile species.

Let us now summarize the information available on the fate of primary production.

(i) Regarding **benthic macrophytes**, except for a small fraction directly consumed by fresh-plant grazers and an insufficiently documented fraction represented by release of dissolved organic compounds, almost the whole production enters the system as detritus, which is decomposed by bacteria and fungi. Detritus plus attached living microorganisms constitutes an important food source for microphagous animals.

The amount of detritus transported beyond the shelf edge depends on local circum-stances, such as the amount of macrobenthic primary production, the width of the shelf, the geomorphology of the continental slope (gradient, canyons), the occurrence of descending water currents or episodic turbidity currents. Obviously, high macrophytic production, narrow shelf, steep continental slope, as well as strong and frequent down-ward currents tend to increase the chances for decaying material from benthic macro-phytes to reach greater depths.

(ii) The fate of **phytoplankton production** depends on the balance achieved bet-ween phytoplankton and zooplankton populations and, especially, on their respective growth rates. Some phytoplankters cannot be consumed by local zooplankters, for example, because their cells are too small and hence may be consumed by ciliates as intermediate link in the food web, or they are too large and hence sink when dying. Most of the phytoplankton is consumed by herbivorous zooplankters. The efficient utilization of primary production by these consumers depends on environmental factors (Volume

I) and the balance between phytoplankters and zooplankters within the assemblage. CUSHING (1971) has demonstrated that transfer coefficients between primary and secondary producers decrease from about 18% to 4% with increased primary production from about 0·1 to 2 g C m^{-2} d^{-1}. The higher the primary production, the higher the fraction that remains unconsumed, dies or sinks, or is consumed but incompletely digested and/or assimilated. Referred to as 'overgrazing', this phenomenon does not involve a total loss since phytoplankton cells damaged by copepods tend to initiate fast nutrient recycling, and faeces from overgrazers contain more undigested organic material and, therefore, have a higher nutritional value for animals inhabiting the lower water layers.

On the shelf—especially in the circalittoral zone—sinking of dead phyto- or zooplankters, faeces, exuviae and other debris can represent an important part of the food supply available to microphagous bottom dwellers. Offshore, where the water-column height increases, organic-material input from the euphotic layer rapidly diminishes with increasing depth. According to LORENZEN (1967) chlorophyll a concentrations at 1·3 to 3 times euphotic-zone depths are exceedingly low, and the vertical distribution of particulate organic carbon suggests that surface-produced particulates do not reach depths exceeding a few hundred metres.

However, due to their higher sinking rate, larger particles such as faeces from herbivores may reach greater depths. This leads us to the different pathways which the organic matter produced may travel in offshore areas—i.e. beyond the shelf edge—and towards oceanic depths. We shall further consider energy from sources other than the euphotic layer which might enter the base of the trophic pyramid.

(c) Food Resources in Offshore Areas and Deep Waters

Organic Material Sinking from the Euphotic Layer

In the past, passive sinking from the euphotic layer—the 'organic rain'—was considered an important source of organic matter for the deep sea. It was assumed that the amount of sinking dead cells would be directly related to the production rate in the euphotic layer, inasmuch as the efficiency of grazing does not increase in proportion to the size of the phytoplankton population. This concept has now been completely abandoned: sinking, dying or dead cells, as well as small debris and particles, are quickly mineralized by bacterial activities, mostly at the level of the upper and lower thermocline, since sinking speed is strongly reduced there due to the abrupt change in sea-water density. We now assume a complete recycling of the organic production in the upper 250 m or so of the water column, i.e. within the epi- and mesoplanktonic zones. Data from various areas of the World Ocean suggest that the content in particulate organic carbon of the sea water does not exhibit significant changes below a few hundred metres, and the same seems to be true as regards organic aggregates. Obviously, particles larger than the usual size of phytoplankton cells—such as faeces from big animals, bodies of large-sized dead zooplankters or vertebrates—can sink deeper than smaller-sized particles and may even reach the sea bottom depending on both the ratio of their weight/volume and local water depth.

The sea-water content of dissolved organic carbon (DOC) seems to follow the same pattern as that of the particulate organic carbon. BARBER (1968) demonstrated that an almost homogeneous DOC value of about 0·60 mg l^{-1} occurs between depths from 400 to 6000 m, but he also recorded a slight increase in the deepest layers in places of higher primary production in the euphotic layer, e. g. in the area of the Peru Current. The latter observation suggests that this dissolved organic matter is rather inert from a biochemical point of view. In fact, BARBER observed that deep waters, which he concentrated 5 to 7 times via a technique that allows the bacterial flora to survive, did not exhibit any decrease in the DOC content after two months, whereas surface waters processed in the same way exhibited a 50% DOC decrease within one month. Therefore, the organic compounds dissolved in deep waters seem to be highly resistant to microorganism activities, at least as regards bacteria, the abundance of which exponentially decreases, approximately from the lower thermocline downwards. It is possible, however, that this assumption cannot be applied to the microflora inhabiting the superficial film of the deep-sea muddy bottoms. For further details consult Volume IV: WANGERSKY (1978) and SOROKIN (1978).

'Ladder of Migrations' Hypothesis

Although the preceding section indicates that sinking cannot cause a substantial increase in the amount of deep-sea particulate organic matter, VINOGRADOV (1977) (Fig. 2–9) has suggested that the biomass of the abyssal macrofauna is significantly higher in the Pacific Ocean just below the equatorial current system; this might be related to important divergences occurring there. Thus, we may assume that another pathway exists for the downward transport of organic material originating in the euphotic layer.

The 'ladder of migrations' hypothesis, proposed by VINOGRADOV (1953, 1955), might be one of these pathways. Recently discussed by LONGHURST (1976), the hypothesis states that significant amounts of organic matter are actively transported downwards by coprophagy and predation due to overlapping patterns of vertical migration extending to at least 4000 to 5000 m below the water surface. Among the arguments advanced by VINOGRADOV, we quote the following: (i) The organic matter contained in dead bodies of medium-sized sinking plankton organisms should be rapidly reduced with increasing depth; for example, a pteropod shell contains less than 20% of the initial organic matter at a depth of about 500 m and 0% at 1500 m. (ii) Some abyssal organisms have mandibles which seem to be adapted to microphagous diets, e.g. members of the copepod genus *Gaetanus*, many peracarids (e.g. the mysids *Boreomysis* and *Hyperamblyops*, as well as isopods of the genus *Eurycope*), and euphausiids of the genus *Bentheuphausia*; in the north-western Pacific Ocean, CHINDONOVA (1959) observed that the gut of these abyssal crustaceans contained radiolarians; this fact led her to hypothesize that they can perform short-term migrations from 4000 to 5000 m up to the infra- or mesopelagic zone. (iii) The stomach of some mysids and *Bentheuphausia*, sampled at depths of 5000 to 7000 m, contained slightly broken diatoms with well preserved chloroplasts.

However, according to LONGHURST (1976), VINOGRADOV's hypothesis of the 'ladder of migrations' might be criticized for the following reasons.

(i) BOLTOVSKOY and LENA (1970) demonstrated that decomposition rates of organic matter contained in dead planktonic foraminiferans are much slower than assumed by VINOGRADOV in the case of pteropods; it may require several weeks or even months. Moreover, the sinking speed of foraminiferans is much lower than expected. According to BOLTOVSKOY and LENA, at depths between 500 and 1500 m, foraminiferan tests—which obviously sink more slowly than larger-sized pteropods and thus are exposed to bacterial activities for longer in the subsurface layers—generally contain 30 to 40% of their initial organic matter. With regard to these protozoans, sampling of empty tests cannot be taken as evidence for total decomposition: they are often a consequence of reproduction.

To some extent, an intermediate position between that held by VINOGRADOV, on the one hand, and BOLTOVSKOY and LENA, on the other, is expressed by HARDING (1974, p. 151) who believes that

'evidence presented on the decomposition of copepods in surface waters . . . suggests that little of this material reaches the deep sea with its original organic content. . . this material may serve as a substrate for bacteria and organics which in turn might supplement the diet of deep-sea, filter-feeding copepods.'

HARDING (1974, p. 152) further suggests that ingestion of organic particles cannot completely support a deep-sea filter-feeding copepod population. He considers protists to be the most important food source for copepods. These heterotrophs, in turn, appear to be supported by dissolved and particulate organics, bacteria and—possibly—detrital remains from organisms present in deep waters.

(ii) Adaptation to microphagous diets cannot be considered *per se* as evidence for the existence of a 'ladder of migrations'; it may suggest a 'ladder of coprophagy', since zooplankters exist at any depth and produce faeces which obviously sink to a greater depth. Thus it may be assumed that zooplankters in a given depth layer feed—at least in part—on faeces from animals inhabiting the layers above, their own faeces, in turn, being consumed by species inhabiting the depths below. It must be pointed out, however, that due to the depth-related exponential decrease in plankton abundance and biomass, the amount of faeces available decreases proportionately.

(iii) In the north-western Pacific Ocean, VINOGRADOV (1968) noticed that the abundance of aulacanthid radiolarians in the deeper (5000 to 6000 m) abyssopelagic zone must be ascribed to their sinking from the surface layers aggregated into clumps.

(iv) In spite of the broken diatoms recorded in the gut of some deep-sea plankters by VINOGRADOV (1968), it seems that the greenish mass sometimes found in the stomach of several crustaceans more often represents olive-green cells, which are ubiquitous at great depths (LONGHURST, 1976). We shall discuss the nature of the olive-green cells in the next section.

Finally, as emphasized by LONGHURST (1976, p.121), upward vertical diel migrations from 'far below the lighted zone' have not yet been ascertained. According to WATERMAN and BERRY (1971), samples from opening–closing hauls centred at 600, 1000, 1400 and 1700 m revealed that 25 of the 207 species collected exhibit a nocturnal rise.

'Species of *Eukrohnia*, *Eucopia*, *Acanthephyra* and *Gennadas* rose from 1400 to 1000 m at dusk and two species of chaetognaths descended from 1700 m at dawn. This

appears to be the greatest depth for which we have good evidence of diel migration.'
(LONGHURST, 1976, p. 121.)

The food input to the deep layers of the pelagial which, at high latitudes, may be
ascribed to zooplankters performing seasonal vertical migrations—such as those
involved in the overwintering phenomenon (p. 88)—seems to result in food exchanges in
almost the same depth range as that presently assumed for diel migrants, i.e. from
surface or subsurface layers down to a little below 1000 m.

Olive-green Cells

Olive-green cells were first observed and described by HENTSCHEL (1936) from the
South Atlantic Ocean during the 1925–27 'Meteor' expedition and since then have been
intensively investigated mainly by FOURNIER (1966, 1970, 1971, 1973). In samples from
the aphotic layer of the North Atlantic Ocean, electron microscopy enabled FOURNIER
(1970) to distinguish two main groups in these pigmented microorganisms: the first
group is composed of cells of extremely low organization, not more complex than that of
procaryotes; the second group, present at lower concentrations than the first one, is
clearly eucaryotic and contains chloroplasts.

The organisms were recorded from the Pacific Ocean by FOURNIER (1966) and
HAMILTON and co-authors (1968) from 300 down to 3000 m. FOURNIER (1971) investigated the distribution of the olive-green cells in the Atlantic Ocean between latitudes of
10 ° and 40 ° N and in the Western Basin of the Mediterranean Sea. He found them
throughout most of the 5000 m water column sampled.

Olive-green cells are spherical or ellipsoidal. Their size ranges from 1 to 15 μm in
diameter. Olive-green cells are almost totally absent from the euphotic layer, apparently
because of intensive grazing—an assumption which must, however, be considered with
some reservations. Beginning at about 50 m, maximum abundance is attained in the
300- to 500-m layer; they occur here mainly attached to organic debris which is particularly frequent at these depths. Attachment and/or clumping probably facilitates their
downward transport from surface layers—their presumed place of origin. Below 300 to
500 m—a layer often corresponding to a strong pycnocline (whose presence probably
explains the abundance peak mentioned above)—the abundance of olive-green cells
progressively decreases.

'Average estimates of the carbon represented by these cells yielded a range of 0·59
to 0·05 μgC/liter, which accounted for about 10% of the total particulate organic
carbon present at 400 and 4800 m, respectively.' (FOURNIER, 1971, p. 952.)

Olive-green cells are evidently heterotrophic and may draw the organic nutrients consumed from the debris they are attached to. Utilization of dissolved organics cannot be
excluded. The large number of stations (44) sampled by FOURNIER documents regional
abundance differences (which, however, do not affect the vertical abundance gradient).
The differences appear to be related to phytoplankton activities in the euphotic layer.

The role of olive-green cells in the trophic food web of the deep-sea plankton remains
uncertain. FOURNIER (1971) suggests that they are consumed by foraminiferans,

radiolarians and ciliates and these, in turn, by copepods. More recently, FOURNIER (1973, p. 38) found many olive-green cells in pelagic tunicates (*Pyrosoma* and salps) collected between the surface and 750 m of the Subtropical Pacific Ocean in an area south-west of the island of Oahu, as well as in the gut of a pedunculate ascidian of the genus *Culeolus* from the Indian Ocean. Employing staining techniques he demonstrated that these cells contain considerable amounts of carbohydrates and may thus represent a source of particulate organic carbon for deep-sea organisms. Considering the low biomass they represent, the real significance of olive-green cells as food source requires estimation of turnover rates. If such turnover exists, it is probably rather low. There are indications that some olive-green cells are able to divide (possibly those of the first type mentioned above). However, at present, it seems more likely that most of them, especially those of the second type, represent resting and/or encysted stages of unidentified phytoplankters.

Other Heterotrophic Deep-sea Microorganisms (Excluding Bacteria)

In the Mediterranean Sea and the Indian Ocean, BERNARD (1953, 1959, 1960, 1964, 1966) observed—much below the euphotic layer and sometimes as deep as 2000 m—microorganisms which he described as various stages of coccolithophorids. This interpretation was contested by FOURNIER (1968) who, through measurements of particulate organic carbon at different depths, obtained values corresponding to the abundance values of *Coccolithus* mentioned by BERNARD. Now, it is a well-documented fact that phytoplankton represents on average only 20% of the total particulate carbon. Moreover, regarding the abundance variation of *Coccolithus*, BERNARD worked with a value of 10 to 20 times, whereas it is generally assumed that the total particulate organic carbon value fluctuates seasonally by a factor of only 4. On the basis of these considerations, FOURNIER (1968) suggests that the objects observed by BERNARD were probably debris of dead cells and inorganic particles.

In the Atlantic Ocean, on a transect from Narragansett to Gibraltar and also on the north-eastern coast of South America, FOURNIER (1966) recorded colourless flagellates of 3 to 5 μm with two flagellae, possibly belonging to the Chrysophyceae. Their abundance ranged from 14200 to 220000 cells l^{-1} with a maximum of about 1000 m (average: 107000 cells l^{-1}). The abundance decreased upwardly to 15000 cells l^{-1} (at 50 m) and downwardly to 39000 cells l^{-1} (5000 m). Furthermore, the abundance also decreased with decreasing distance from the continent. FOURNIER suggests that these flagellates are heterotrophic and that their small size facilitates the utilization of dissolved organics.

Trophic Input from Land

The trophic input into the marine environment from land almost exclusively originates from river run-off and land drainage (see also Volume IV: WANGERSKY, 1978). Organic input from the atmosphere mainly benefits the neuston and should be regarded as negligible.

The input from terrestrial fresh water consists of mineral nutrients and organic material, either dissolved or particulate (and, sometimes, larger debris). Dissolved substances tend to be confined to superficial layers due to the lower density of fresh water. In the area of mixing between fresh water and sea water, particulate material, except for the smallest size (clays), tends to flocculate with the organics adsorbed. Thus most of the latter material tends to enrich shallow-water bottoms. The enrichment of the pelagial in dissolved nutrients—sometimes excessive (eutrophication) in inshore areas—tends to spread farther from the shoreline. It supports coastal plankton assemblages which, in turn, indirectly, i.e. through sinking of live or dead material, support benthos assemblages on the lower shelf or even the slope. Only the lightest clay particles are transported farther away from the shoreline over tens or sometimes hundreds of miles. They still seem to keep part of the organics which are adsorbed over some 50 to 70 nautical miles; these decrease with time and transport distance.

These considerations indicate that the fertilization of offshore and deep-sea waters depends on the run-off or land drainage. Furthermore, the narrower the shelf, which tends to trap fertilizing substances, the more will the offshore pelagial and the near deep-sea benthal benefit from terrestrial fertilizing input. We also cannot neglect large floating objects such as tree trunks and branches which may be transported far from the shoreline, progressively become water-saturated and finally sink. For instance, diving to a depth of 7300 m, 100 nautical miles off the northern shore of Puerto Rico, I have observed large quantities of leaves, branches and wood debris—together with newspapers and other terrestrial garbage.

The extent and shape of the marine area which is influenced by river run-off also depends on the prevailing water-current regime in the vicinity of the river mouth. For example, river mouths not 'swept' by strong currents—such as those of the Mississippi, Nile (before the building of the Aswan Dam) and the Pô—markedly fertilize neighbouring marine areas. In contrast, Amazon run-off, rapidly diluted by the Guianas Current, does not seem to establish increased fertility near river-mouth areas.

In summary, the richest and most productive marine areas which export nutrients for the benefit of offshore areas, deep waters and sea floors are (except for the equatorial current systems, the fertility of which arises from divergence processes): (i) shelf areas or coastal areas near the shelf where upwelling occurs; (ii) coastal areas receiving fertilizing input from land, such input mainly benefiting the shelf and its immediate vicinity. The biomass represented by olive-green cells, a possibly significant primary resource in the deep pelagial, is related to primary production in the euphotic layer. Heterotrophic bacteria also establish the densest populations in shelf and coastal areas; their abundance mainly depends on the available amount of degradable organic matter. Therefore, the World Ocean appears to be rather poor in nutrients over some 90% of its surface area; nutritionally rich or very rich waters amount to only some 10% and are almost exclusively restricted to coastal, i.e. 'peripheral', areas.

This situation emphasizes the paramount importance of peripheral areas for ecological processes in the World Ocean. Such emphasis is supported by a comparison of fishery landings from peripheral and central ocean areas. In fisheries statistics, the boundary between shelf and offshore areas—i.e. the 200-m isobath—does not exactly correspond to the ecologically more significant boundary suggested in the preceding pages. In spite of this discrepancy, it is striking to note that landings from areas inside and outside the 200-m isobath represent some 87% and 13% of the total fishery harvest respectively.

Literature Cited (Chapter 2)

ALLEE, W. C., EMERSON, A. E., PARK, O., PARK, T. and SCHMIDT, K. P. (1949). *Principles of Animal Ecology*, Saunders Co., Philadelphia and London.

ALEEM, A. A. (1956). Quantitative underwater study of benthic communities inhabiting kelp beds off California. *Science, N.Y.*, **123** (3188), 183.

BARBER, R. T. (1968). Dissolved organic carbon from deep waters resists microbial oxidation. *Nature, Lond.*, **220**, (5164), 274–5.

BELAYEV, G. M. (1960). Some regularities of the quantitative distribution of the bottom fauna in the western Pacific (Russ.). *Trudy Inst. Okeanol.*, **41**, 98–105.

BELAYEV, G. M. (1976). The pelagic and near-bottom fauna of the maximal ocean depths. (Russ.) *Trudy Inst. Okeanol. Akad. Nauk. SSSR*, **99**, 178–96.

BERNARD, F. (1953). Rôle des flagellés calcaires dans la fertilité et la sédimentation en mer profonde. *Deep Sea Res.*, **1**, 34–46.

BERNARD, F. (1959). Fertilité élémentaire en Méditerranée de 0 à 1.000 m, comparée avec l'Océan Indien et l'Atlantique tropical africain. *First Int. oceanogr. Congr.* (Abstr. papers), **1**, 832.

BERNARD, F. (1960). Plancton unicellulaire récolté dans l'Océan Indien par le "Charcot" (1950) et le "Norsel" (1955–56). *Bull. Inst. océanogr. Monaco*, **1166**, 1–59.

BERNARD, F. (1964). Le nannoplancton en zone aphotique des mers chaudes. Pelagos. *Bull. Inst. oceanogr. Alger.*, **2**, 5–32.

BERNARD, F. (1966) Abondance du nannoplancton dans les couches aphotiques des Océans—conséquences probables pour la productivité profonde des mers chaudes. *Second. Int. oceanogr. Congr.* (Abstr. papers), **2**, 38–9.

BIRSHTEIN, J. A., VINOGRADOV, M. E., and CHINDONOVA, YU. G. (1954) Vertical zonation of plankton in the Kuril-Kamchatka Trench. (Russ.) *Dokl. Akad. Nauk SSSR*, **95**, 389–92.

BOLTOVSKOY, E. and LENA, H. (1970). On the decomposition of the protoplasm and the sinking velocity of the planktonic foraminifers. *Int. Rev. ges. Hydrobiol.*, **55**, 797–804.

BRUUN, A. F. (1957). Deep sea and abyssal depths. In J. W. Hedgpeth (Ed.), *Treatise on Marine Ecology and Paleocology*. Geological Society of America. (*Mem. geol. Soc. Am.*, **67**, 641–72.)

CHINDONOVA, YU. G. (1959). Nutrition of certain groups of deepwater macroplankton in the north-west Pacific Ocean. *Trudy Inst. Okeanol.*, **30**, 166–89.

CUSHING, D. H. (1971). Upwelling and the production of fish. *Adv. Mar. Biol.*, **9**, 255–334.

ERGEGOVIC, A. (1957). Principes et essai d'un classement des étages benthiques. *Recl. Trav. Stn mar. Endoume* (*Bull.* 13), **22**, 17–21.

FERGUSON-WOOD, E. J. (1966). Plants of the deep oceans. *Zeitschr. f. allgem. Mikrobiologie*, **6**, 177–9.

FILATOVA, Z. A. (1969). Quantitative distribution of deep-sea benthic fauna (Russ.). In L. A. Zenkevich (Ed.), *Biologiya Tikhogo Okeano*. Izdat. Nauk, Moskva. pp. 202–216.

FOURNIER, R. O. (1966). North Atlantic deep-sea fertility. *Science, N.Y.*, **153** (3741), 1250–2.

FOURNIER, R. O. (1968). Observations of particulate organic carbon in the Mediterranean Sea and their relevance to the deep-living coccolithophorid *Cyclococolithus fragilis*. *Limnol. Oceanogr.*, **13**, 693–7.

FOURNIER, R. O. (1970). Studies on pigmented microorganisms from aphotic marine environments. I. *Limnol. Oceanogr.*, **15**, 675–82.

FOURNIER, R. O. (1971). Studies on pigmented microorganisms from aphotic marine environments. II. North Atlantic distribution. *Limnol. Oceanogr.*, **16**, 952–61.

FOURNIER, R. O. (1973). Studies on pigmented microorganisms from aphotic environments. III. Evidence of apparent utilization by benthic and pelagic Tunicata. *Limnol. Oceanogr.*, **18**, 38–43.

GISLEN, T. (1929). Epibioses of the Gullmar Fjord. I. Kristinebergs Zool. Sta. 1877–1927, utg. *K. Svenska Vetenskapakad.*, **3**, 1–123.

GISLEN, T. (1930). Epibioses of the Gullmar Fjord. II. Kristinebergs Zool. Sta. 1877–1923, utg. *K. Svenska Vetenskapakad.*, **4**, 1–380.

HAMILTON, R. D., HOLM-HANSEN, O. and STRICKLAND, J. D. H. (1968). Notes on the occurrence of living microscopic organisms in deep water. *Deep Sea Res.*, **15**, 651–6.

HARDING, G. C. (1974). The food of deep-sea copepods. *J. mar. biol. Ass. U.K.*, **54**, 141–55.

HEDGPETH, J. W. (1957). Classification of marine environments. In J. W. Hedgpeth (Ed.), *Treatise on Marine Ecology and Paleoecology*. The Geological Society of America. (*Mem. geol. Soc. Am.*, **67**, 17–28.)

HENTSCHEL, E. (1936). Allgemeine biologie des Süd-Atlantischen Ozeans. *Wiss. Ergebn. dt. atlant. Exped. "Meteor"*, **11**, 1–343.

LAVIE, P. and KRIPKE, D. F. (1977). Megafauna in the deep sea. *Nature, Lond.*, **269** (5624), 141–3.

LONGHURST, A. R. (1976). Vertical migration. In D. H. Cushing and J. J. Walsh (Eds), *The Ecology of the Sea*. Blackwell, Oxford. pp. 116–37.

LORENZEN, C. J. (1967). Vertical distribution of chlorophyll and phaeopigments Baja California. *Deep Sea Res.*, **14**, 735–6.

LORENZEN, C. J. (1976). Primary Production in the sea. In D. H. Cushing and J. J. Walsh (Eds), *The Ecology of the Seas*. Blackwell, Oxford. pp. 173–85.

MENZIES, R. J. and GEORGE, R. V. (1972). Eurybathial benthic organisms. In R. W. Brauer (Ed.), *Barobiology and the Experimental Biology of the Deep Sea*. University of North Carolina, pp. 33–6.

NYBELIN, O. (1957). Deep sea bottom fishes. *Rep. Swedish Deep-Sea Exped. 1947–1948*, **2–3**, 247–345.

PERES, J. M. (1957). Le problème de l'étagement des formations benthiques. *Recl. Trav. Stn mar. Endoume (Bull.* 12), **21**, 4–21.

PERES, J. M. (1967). The Mediterranean benthos. *Oceanogr. mar. Biol. A. Rev.*, **5**, 449–533.

PERES, J. M. (1976). *Précis d'Océanographie Biologique*, Press. Univ. France, Paris.

PERES, J. M. and PICARD, J. (1959). On the vertical distribution of benthic communities. *First Int. oceanogr. Congr.* (Abstr. Papers), **1**, 322.

PLANTE-CUNY, M. R. (1969). Recherches sur la distribution qualitative et quantitative des diatomées benthiques de certains fonds meubles du Golfe de Marseille. *Recl. Trav. Stn mar. Endoum (Bull.* 45), **61**, 87–197.

PRUVOT, G. (1897). Essai sur les fonds et la faune de la Manche Occidentale comparés à ceux du Golfe du Lion. *Arch. Zool. Exp.*, **3**, 559–87.

REX, M. A. (1976). Biological accommodation in the Deep-Sea Benthos. Comparative evidence on the importance of predation and productivity. *Deep Sea Res.*, **23**, 975–87.

REYSS, D. (1970). *Bionomie Benthique de Deux Canyons sous-marins de la Mer Catalane: le Rech du Cap et le Rech Lacase-Duthiers*, Thèse Doct., Faculté des Sciences, Paris.

RIEDL, R. (1964). Die Erscheinungen der Wasserbewegung und ihre Wirkung auf Sedentarier im mediterranen Felslitoral. *Helgoländer wiss. Meeresunters.*, **15**, 294–352.

SANDERS, H. L. (1968). Marine benthic diversity: a comparative study. *Am. Nat.*, **102**, 243–82.

SANDERS, H. L. (1969). Benthic marine diversity and the stability–time hypothesis. *Brookhaven Symp. Biol.*, **22**, 71–81.

SOROKIN, Yu. I. (1978). Decomposition of organic matter and nutrient regeneration. In O. Kinne (Ed.), *Marine Ecology*, Vol. IV, Dynamics. Wiley, Chichester, pp. 501–616.

STEPHENSON, T. A. and STEPHENSON, A. (1949). The universal features of zonation between tide-marks on rocky coasts. *J. Ecol.*, **37**, 289–305.

THORSON, G. (1957). Bottom communities (sublittoral or shallow shelf). In J. W. Hedgpeth (Ed.), *Treatise on Marine Ecology and Paleoecology*. The Geological Society of America. (*Mem. geol. Soc. Am.*, **67**, 461–534.)

VINOGRADOV, M. E. (1953). The role of vertical migration of the zooplankton in the feeding of deep sea animals. *Priroda, Moskva*, **6**, 95–6.

VINOGRADOV, M. E. (1955). Vertical migrations of zooplankton and their importance for the nutrition of abyssal pelagic fauna. *Trudȳ Inst. Okeanol.*, **13**, 71–6.

VINOGRADOV, M. E. (1960). Quantitative distribution of the deep-sea plankton in the western and central parts of the Pacific Ocean (Russ.). *Trudȳ Inst. Okeanol.*, **41**, 55–84.

VINOGRADOV, M. E. (1968). *Vertical Distribution of Oceanic Zooplankton*, Izdat. Nauka, Moskva.

VINOGRADOV, M. E. (1970). Vertical distribution of zooplankton in the Kuril-Kamchatka region of the Pacific Ocean (after data from the 39th cruise of the vessel Vitjaz). (Russ.) *Trudȳ Inst. Okeanol.*, **86**, 99–116.

VINOGRADOV, M. E. (1977). *Biology of the Ocean*, Vols 1, 2 (Russ.), Izdat. Nauka. Moskva.

VINOGRADOV, M. E. and VORONINA, N. M. (1963). Plankton distribution on the Equatorial Current's waters of the Pacific Ocean (Russ.). *Trudȳ Inst. Okeanol.*, **71**, 22–54.

VINOGRADOV, M. E. and VORONINA, N. M. (1964). Plankton distribution on the Equatorial Current's waters of the Pacific Ocean (Russ.). *Trudȳ Inst. Okeanol.*, **65**, 58–76.

VINOGRADOVA, N. G. (1958). Vertical distribution of the deep-sea bottom fauna of the Ocean. *Trudȳ Inst. Okeanol.*, **27**, 87–122.

WALTHER, J. (1893). Einleitung in die Geologie als historische Wissenschaft, I. Bionomie des Meeres, II. Die Lebensweise der Meeresthiere, G. Fischer, Jena.

WANGERSKY, P. J. (1978). Production of dissolved organic matter. In O. Kinne (Ed.), *Marine Ecology*, Vol. IV, Dynamics. Wiley, Chichester. pp. 115–220.

WATERMAN, T. H. and BERRY, D. A. (1971). Evidence for diurnal vertical plankton migration below the photic. *NSF Grant Report*, 1971.

WILCE, R. T. (1967). Heterotrophy in Arctic sublittoral seaweeds: a hypothesis. *Bot. Mar.*, **10**, 185–97.

WOLFF, T. (1977). Diversity and faunal composition of the deep-sea benthos. *Nature, Lond.*, **267** (5614), 780–5.

ZENKEVICH, L. A. (1956). *The Seas of USSR—Their Flora and Fauna*. Gosud. Uchebno-pedagog. Izdat, Moskva.

ZENKEVICH, L. A. (1963). *Biology of the Seas of URSS* (Russ.), Izdat. Akad. Nauk SSSR.

REFERENCES

Marine Ecology Vol. 5, Part 1
Edited by Otto Kinne
© 1982 John Wiley & Sons Ltd

3. GENERAL FEATURES OF ORGANISMIC ASSEMBLAGES IN PELAGIAL AND BENTHAL

J. M. PERES

Organismic assemblages in the pelagial and the benthal exhibit a number of characteristically different features. The pelagial harbours a vast moving medium which at first sight appears rather homogeneous. Most of the inhabitants are small-sized and many have a short life cycle. At a given place, the changes in the components of flora and fauna may be considerable. These facts have caused ecologists to concentrate on investigating the relationships between physical and chemical factors characteristic of a given body of water and its living inhabitants.

The benthal is characterized by a conspicuous diversity of environmental features especially in regard to the nature of the substratum (Volume I, Chapter 7) and the different patterns of water movement (Volume I, Chapter 5). Most inhabitants, of both plant and animal origin, attain larger body sizes than their pelagic counterparts. At a given place, the changes observed in organismic assemblages tend to be smaller than in the pelagial. These facts have led ecologists to focus their attention on the structural and spatial variability of bottom assemblages, and on attempts to detect and formulate rules and laws which govern macrobenthos distribution as a function of substrate characteristics.

Investigations in the pelagial require the application of a considerable variety of sampling techniques covering the entire water column. The concept selected for analysing essential features of pelagic assemblages must take a number of points into account. (i) Since the pelagial is an 'environment in motion', transport displacement, dilution and dispersion may significantly and rapidly affect and change the organismic assemblage studied. (ii) The pelagial is a strongly tridimensional environment. (iii) Organisms capable of appreciable vertical migrations are common in both the plankton and nekton. (iv) The short life cycles of most plankton organisms often cause intensive short-term fluctuations in assemblage composition. (v) Seasonal changes in the abundance of primary producers are often large—except in tropical oligotrophic areas—a fact which tends to induce subsequent changes in successive trophic levels of the assemblage. (vi) Rather pronounced changes in phytoplankton populations may further occur over short periods of time as a function of intensive grazing by abundant herbivores, especially if grazing peaks coincide with reduced algal growth rates.

Complete detailed investigations of pelagic assemblages require a tremendous amount of data and hence are, at present, impractical. The marine ecologist devoted to comprehending pelagial assemblages must restrict his attention to the study of species which appear to be functionally the most important. Of particular ecological significance are 'indicator' species, i.e. species which respond sensitively and reliably to a given change in their environment.

Investigations in the benthal go back to such early workers as MARION (1883a,b) who described the nature of different bottoms and the species composition of their flora and fauna on the whole shelf of the Gulf of Marseilles (France) and extended his comparative analysis to include both the nature of the bottom and the faunal composition down to a water depth of 2000 m. Like all his contemporaries, MARION was interested only in qualitative aspects, i.e. the list of species found on or in a given type of bottom. He did not attempt to pin-point the factors actually affecting the distribution of macrobenthic species.

Investigations concentrating on qualitative aspects are based on lists of plants and animals which inhabit a given type of sea bottom. These are then compared with other lists recorded from bottoms with different characteristics (e.g. depth, texture, water movement). This comparative method aims at singling out those species which appear to be narrowly related to the main features of the particular bottom environment inhabited.

More than three decades later, another method for analysing bottom assemblages was proposed by PETERSEN (1911, 1913). He concentrated on quantitative rather than qualitative data, i.e. he focussed on evaluating the number of individuals present per bottom-surface area and on the weight of the living matter present. Since PETERSEN investigated soft bottoms only with bottom-samplers often penetrating the sediment to varying degrees, the number of individuals (abundance) and their weight (biomass) recorded should be related to units of volume rather than to those of surface. Placing paramount importance on those species which predominate in weight, PETERSEN developed his community concept.

Unfortunately, later, marine ecologists increasingly confused the concept of community with that of biocoenosis and often did not carry out their investigations as carefully as PETERSEN had done. This had led to considerable argumentation and misconception. Hence, before proceeding, we must briefly refer here to the concept of biocoenosis.

Coined by MÖBIUS (1877), the term 'biocoenosis' has been defined by ALLEE and SCHMIDT (1951, p. 167) as an

'association of living things inhabiting a uniform division of the biosphere . . . The members of a biocoenosis are dependent upon each other and are thus forced into biological balance, which is self-regulating and fluctuates about a mean.'

Biocoenosis members usually breed at the locality occupied by the biocoenosis. This locality is termed 'biotope'. The biotope represents a topographic unit characterized by relative uniformity of principal habitat features.

Some authors have interpreted the concepts of community and biocoenosis as being opposed to each other and mutually exclusive. In reality they are complementary. This will be shown below. In spite of the fact that the terms community and biocoenosis are well defined with regard to the benthos and thus can be used for analysing the distribution of benthic organisms, it has been preferred to adopt here a more general and objective term for both the benthal and the pelagial, namely that of **organismic assemblage** (in French: *peuplement*). This term refers to all those species which live together in an homogeneous environment. The term 'assemblage' does not imply anything in regard to the sampling method employed nor to the method (qualitative or quantitative) used for studying the sample collected. A related term—more often used by botanists than by zoologists—is 'association'. According to MARGALEF's (1977, 1978) redefinition, association refers to the populations of several species collected together by means of a single sampling method and to the subsequent processing and interpretation of the data obtained.

In the benthal, the relationships between organisms and sea bottom are usually very close and intimate. The specificity of such correlations provides for a firm frame of reference. Typically, classifications and zonations in the benthal are based on the nature of the substratum which may display a wide diversity of living conditions—much wider in fact than is the case in the free water above the sea bottom.

A number of other factors allow benthos assemblages to be classified more readily than pelagos assemblages: (i) The vertical extension of the benthos assemblage is much smaller, even where it includes both epifauna and infauna its vertical extension usually does not exceed a few decimetres. (ii) The percentage of immigrants and visitors—due to both diurnal and seasonal dynamics—is much lower. (iii) The longer mean life span of macrobenthic organisms causes less pronounced short-term fluctuations in the total number of species present and in their abundance. (iv) The motility of macrobenthic animals is less pronounced; many are fixed or sedentary; even numerous motile crustaceans, cephalopods and fishes which preferably or exclusively feed on macrobenthic prey can usually be readily caught, counted and classified as to whether they are characteristic members of the biocoenosis, transient components—often occurring only seasonally—or rare guests. (v) Changes in benthos assemblages—which usually concern species abundance rather than changes in species composition—tend to be less pronounced than in pelagos assemblages. Furthermore, they often reveal regular patterns and hence can be more readily detected without obscuring the understanding of basic structural and functional assemblage dynamics. (vi) Seasonal variations, where they occur, tend to be less extensive. This is due to the fact that changes in primary production are less effective because herbivorous animals which depend on fresh plant materials are rather scarce in the benthos; in contrast, detrivorous animals which feed on decaying plant matter accumulating at the bottom-water surface, as well as on its epiphytic or associated microflora, are very numerous. In other words, the food resources for primary consumers are both more diverse—either in the fresh condition or at various stages of decomposition—and subject to less quantitative temporal fluctuations.

(1) Organismic Assemblages in the Benthal

Since organismic assemblages can be more easily classified in the benthal than in the pelagial (see preceding section) we shall begin with the former. The benthal comprises considerable environmental diversity and numerous microhabitats. This fact makes it necessary first to pay some attention to terminological aspects and to the factors affecting the distribution and abundance of the macrobenthos.

(a) Terminology

The substratum (or substrate) inhabited by benthic assemblages can have quite different properties (Volume I, Chapter 7). It may be solid (e.g. rocks, concrete, bricks, iron or wood), consist of living or dead organisms, or be soft, i.e. be composed of sediments such as sand or mud. For details consult Volume I: GERLACH (1972). In sheltered shallow waters, substrates which appear soft in terms of their texture and structure must be considered solid or hard in regard to their ecological significance. While composed of fragments, they have effects on organisms similar to a continuous rock surface providing the fragments are never moved (i.e. not shifted or turned by water movement), because they are located in a very sheltered area. In such a case, an organismic assemblage living on the substratum may be quite similar to another one inhabiting a continuous rocky surface, except for those species which dwell in the interstices between the fragments.

Organisms living on a substratum are called epibionts. While epibiotic plants (epiphytes) are always fixed, epibiotic animals (epizoans) may include sessile, sedentary and motile forms. Many epiphytes and epizoans thrive on, others are entirely restricted to, living substrates. While epiphytes often occur on a large variety of plant substrates, i.e. are euryplastic in terms of substrate selection, many epizoans inhabit rather specific substrates, i.e. are stenoplastic in regard to substrate selection. For examples consult p. 156.

Macrobenthic animals which live inside a substrate are named 'endobionts' (infauna). Taking into account the nature of the substratum, we call organisms associated with sand 'psammon' and, accordingly, may differentiate between epipsammon (organism living on the sand) and endopsammon (organisms living within the sand). Parallel terms for rock-living forms are 'epilithon' and 'endolithon'.

It must be specifically pointed out that all terms described and defined above concern only plants and animals of the macrobenthos, that is, species with body sizes larger than 1 mm (in some classifications either 0·5 or 2 mm). Depending on their body-size range smaller organisms are referred to as 'micro'- or 'meiobenthos'. Both micro- and meiobenthos make up the 'mesobenthos'. On rocky substrates the equivalent term is 'mesolithon'.

The term 'stratum' refers to the specific properties of a layer contained within the total vertical extension of an assemblage of benthic organisms. In many assemblages an upper and a lower stratum can be distinguished, sometimes also a middle one. Where the upper stratum modifies the ecological conditions at the level of the lower stratum, the latter may be inhabited by an organismic assemblage different from that which inhabits the upper stratum. Such a situation prevails if the upper stratum consists of large metaphytes which greatly diminish the degree of illumination received by the assemblage occurring on the sea bottom itself. For example, on Mexican and Californian shores, at a depth of 10 m under the canopy of the giant kelp *Macrocystis pyrifera*, where the degree of illumination is only 0·1% of that at the surface, DAWSON and co-authors (1960) reported an assemblage of sciaphilic algae (species of *Leptocladia, Cystoseira, Prionites, Bossiella, Acrosorium*), whereas outside the kelp bed, on bottom areas which receive an illumination equivalent to 5% of that at the surface, photophilic algae predominate. A comparable phenomenon occurs in the densest and deepest *Posidonia* beds (p. 440) of the Mediterranean Sea: the upper stratum, i.e. the sea-grass leaves, harbours an assemblage which belongs to the well-illuminated infralittoral zone (p. 18), whereas the erected stems lying under the leaf canopy reveal an assemblage which belongs to the coralligenous group (p. 441) of the circalittoral zone. In contrast, an upper stratum made of slender forms, such as the gorgonians, does not modify the illumination on the lower stratum and hence maintains the unity of the whole assemblage. Sometimes the canopy of metaphytes, if it is sufficiently dense, reduces the water movement (Volume I, Chapter 5) just above the bottom. This affects the lower-stratum assemblage where species dominate which prefer a sheltered microenvironment.

To some extent such influence of a dense canopy of large-sized metaphytes may be considered as a specific aspect of the more general concept of 'enclave', which refers to the presence, due to microclimatic factors, of one organismic assemblage within another one. In shallow waters, for example, submarine caves, overhangs or crevices are inhabited by sciaphilic organismic assemblages, whereas the neighbouring well-illuminated rocky surface areas exhibit photophilic assemblages. Another example is that of the rockpools of the midlittoral zone on tidal shores which keep up the water at low tide and are inhabited by an assemblage which has some similarity to that existing in the infralittoral zone.

In some biocoenoses (*sensu* MÖBIUS, 1877) the predominance of a given ecological factor may cause luxuriant development of one or a few species (which may or may not be characteristic of the biocoenosis concerned) but fail to cause any substantial change in the specific—i.e. qualitative—composition of the biocoenosis. The result of such a change is called a facies. In the intertidal, the moisture gradient often induces the formation of facies which here are named belts, due to their upper and lower boundaries being parallel to the shore line (e.g. Fig. 8-3, p. 380). However, where the species which predominates in such a belt establishes a very dense population, one may sometimes observe the disappearance of other species which usually participate in the assemblage. Some shelf assemblages, particularly in areas in which the annual temperature range is large (e.g. 10–15 °C), may exhibit seasonal facies, especially in shallow waters (p. 380).

(b) Factors Acting upon the Distribution and Abundance of the Macrobenthos

The presence of a species in a given assemblage and its population abundance are the result of complex interactions between numerous factors, both abiotic and biotic. Three major abiotic factors affect the depth distribution of the macrobenthos and, consequently, the pertinent assemblage composition: humidity—i.e. the degree of moistening in bottoms which may emerge; light—which is required or tolerated to different degrees by the species concerned and, at the same time, provides the energy for primary production in the photoautotrophic phytobenthos; in the absence of light, exogenous organic material from the phytal substitutes as energy source; and hydrostatic pressure.

These three main abiotic factors—humidity, light and pressure—are often called 'climatic'. In addition to these three factors, a variety of other factors, some abiotic, others biotic, control or affect macrobenthos assemblages.

'Edaphic' factors (from the Greek εδαφοσ which means soil) compromise a group of abiotic factors which act locally on the soil and/or its inhabitants. They can modify or even dominate the effects of climatic factors. If an edaphic factor becomes predominant, this may lead to the substitution of species and, consequently, result in an assemblage which is better adjusted to the local environment than an assemblage distribution-ally controlled by climatic factors. Edaphic factors are, for example: (i) substratum characteristics such as hardness of rock, softness of sediments, grain size or changes in sedimentation rate; (ii) excessive intensity of water movement, due to waves or currents, as well as excessive water stagnation; (iii) irregular freshwater inflow or excessive evap-oration resulting in a salinity increase; (iv) excessive accumulation of organic matter due to natural phenomena or man-made pollution.

The much larger number of benthal assemblages—as compared with the pela-gial—is, at least in part, a consequence of the practically infinite number of possible combinations between abiotic factors either climatic or edaphic.

If climatic factors predominate over edaphic ones, the assemblage concerned remains restricted to one vertical zone of the benthal. In contrast, if one or several edaphic factors predominate over climatic ones over a large depth range, the assemblage concerned may extend over several vertical zones. A moderate influence of one or several edaphic factors upon a climatic-controlled assemblage results in the appearance of facies, but if the influence becomes more and more manifest the original assemblage tends to 'degrade' and, finally, a new assemblage controlled by the edaphic factor(s) prevails.

Edaphic factors may also be responsible for breaks or regressions which may occur in a climatic succession which, therefore, cannot reach its ultimate stage of climax.

Finally, some edaphic factors can modify the intensity of a climatic factor. For instance, an increase in water turbidity which diminishes light penetration tends to induce a rise of the lower limit of the infralittoral and circalittoral zones (pp. 18 and 19). Differences in the degree of wave exposure can affect both the upper and lower boundaries of the midlittoral zone and its vertical range. Obviously, such modifications in the extension of the vertical zones result in parallel changes in the extension of their assemblages.

Biotic factors are inherent to the assemblage itself. They can modify both the effects of some abiotic factors as well as composition and dynamics of the assemblage. Examples of biotic factors are: (i) Biologically induced modifications in substrate characteristics. Organisms which elaborate calcareous matter—such as calcareous rhodophytes, some anthozoans and bryozoans—may induce a consolidation of an originally soft, but coarse bottom, finally leading to biogenic rock ('concretionment'). In such cases, the original soft-bottom assemblage will be replaced by a hard-bottom assemblage. An inverse process may be brought about by increasing populations of boring organisms. A dense canopy of metaphytes progressively diminishes the degree of substrate illumination and sometimes also modifies other characteristics in the local environment, e.g. decreases bottom-water movement or increases the settlement of suspended organic matter. Such modifications may affect the lower-stratum assemblage and change its composition. A further example is the release of large amounts of faecal pellets by suspension-feeding invertebrates, which tends to cause a progressive substitution of detritus-feeders for sestonophages. (ii) Competition for space, e.g. related to both the different settling periods of gametes or larvae and the different growth rates of competing species illus-trated by the second- and third-degree epibioses phenomenon and, on vacant substrate space, by the successive appearance of different assemblages initiated by 'pioneer' species (Fig. 5-11). (iii) Modifications in the relative abundance of predator and prey populations (p. 171); these tend to influence community composition but in general not that of the biocoenosis. (iv) Migrations induced by requirements for particular breeding environments or by the search for food.

(c) Delimitation of Macrobenthos Biocoenoses Based on the Qualitative Method

An ecologist who undertakes investigations with the purpose of delimiting neighbouring macrobenthos biocoenoses must first of all collect samples from all substrate types present at different water depths. The quality of the information obtained increases with the number of strategically placed stations. The samples must provide knowledge, as complete as possible, on the whole flora and fauna present. While motile species may be used to define a biocoenosis, it is more advantageous to select sedentary or fixed species.

A comparative analysis of the lists of species recorded at each station, together with an analysis of data on environmental parameters, allows classification of the species found in three principal groups: (i) 'Characteristic' species, which preferably* occur in a particular biotope, irrespective of the abundance of their populations. (ii) 'Accompanying' species which occur in several biotopes and are often also referred to as being 'eurytopic'. A less adequate term is 'ubiquitous' because no species is truly ubiquitous, i.e. inhabits the total range from the shallowest waters to the greatest depths and from polar seas to equatorial shores. (iii) 'Casual' species which are characteristic of a biocoenosis other than that in the investigated biotope. Populations of casual species are always scarce and/or sporadic, and often sterile because they are outside their normal biotope.

The methods employed for sampling macrobenthos biocoenoses vary as a function of habitat and organisms. Since practically all sampling methods have their shortcomings, it is useful to employ as many different methods as possible and to use underwater photography as a complementary tool.

In biotopes which can be directly studied either because they emerge during low tide or because they are accessible to scuba diving or underwater vehicles, methods can be used for epiflora and epifauna studies which are similar to those used by terrestrial phytosociologists. The latter make use of two scales, abundance–dominance and sociability, marked from 1 to 5. On the first scale, a species which is present but very scarce is noted by the sign (+) (Fig. 3-1). As additional information, the investigator can also indicate whether a species inhabits the upper (US) or the lower stratum (LS) or whether it lives epizoic or epiphytic. In colonial invertebrates each colony is considered one individual; however if a single colony covers a significant part of the substrate surface, the sign (+) is added to assessments of abundance–dominance which correspond to the minimum degree in sociability. The superficies adopted for making up the statement must be inhabited by an homogeneous assemblage; in general, a superficies of one square metre is sufficient, but sometimes it may be better to use superficies up to 8 to 10 m^2.

Statements like that in Table 3-1 assist in determining species characteristic of the biocoenosis concerned. For the same locality statements, made at successive time intervals, allow recognition of changes in the assemblage characteristics, for example, due to

---

*Species which are strictly limited to only one biotope scarcely exist.

season or pollution. A statement may be complemented in the course of further sampling, this yielding information on the degree of sexual maturity, biomass, etc.

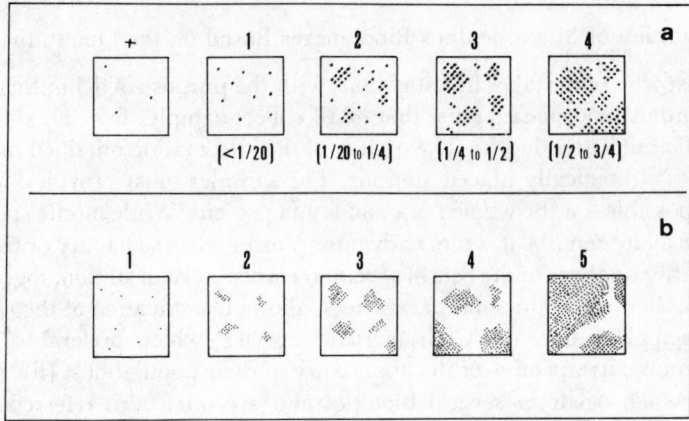

Fig. 3-1: Marking of abundance–dominance (a) and sociability (b) according to the phytosociological method. (a) +: very scarce; 5: species covering more than 3/4 of the substrate superficies. (b) 1: scattered individuals; 5: dense population. (After PÉRÈS and PICARD, 1964; reproduced by permission of Station Marine d'Endoume et Centre d'Océanographie.)

Table 3-1

Example of phytosociological statement regarding algal populations in the upper fringe of the infralittoral zone on the French Mediterranean coast. Superficies: 1 m^2; depth: 0.4 m; percentage of algal covering: 90%; substrate slope: 5 (p. 100). On each line the first and second numbers indicate abundance, dominance or sociability. (After PÉRÈS and PICARD, 1964, p. 24, reproduced by permission of Station Marine d'Endoume et Centre d'Océanographie)

3.4	*Halopteris scoparia* (US)
2.3	*Jania rubens* (US–LS)
1.1	*Cladostephus verticillatus* (US)
1.1	*Cystoseira abrotanifolia*
+	*Padina pavonica*

In most cases, a comparison of all species present at stations regarded to belong to a certain biotope (on the basis of similar habitat conditions) makes it rather easy to establish a list of characteristic as well as accompanying species for a given biocoenosis. Nevertheless, the delimitations of two adjacent biocoenoses remain somewhat subjective. In order to increase the objectivity of such delimitation, several mathematical methods were proposed. Phytosociologists often use the SORENSEN similarity coefficient (QS):

$$QS = \frac{2c}{a+b} \times 100 \qquad (3.1)$$

where a and b are species numbers of the two biocoenoses A and B, and c is the number of species common to both biocoenoses. GAMULIN-BRIDA (1960) proposed using a similar coefficient:

$$QS_1 = \frac{2c_1}{a_1 + b_1} \times 100 \qquad (3.2)$$

which takes into account not the species numbers but an approximate estimation of their respective abundances (a_1, b_1, c_1) which are marked arbitrarily: 1 = 1 to 5 individuals; 2 = 6 to 25; 3 = 26 to 100; 4 = 101 to 500; 5 = more than 500. The greater the difference between QS and QS_1, the more dissimilar are the two biocoenoses. FAGER (1963) has rightly criticized the equation, pointing out that a QS of 0·5 may be obtained from 10 individuals of a given species as well as from several hundreds; futhermore, if $a + b$ and c remain of the same value, QS remains constant even if the respective abundances of a and b vary. FAGER suggested a new similarity coefficient based upon the geometric mean in the proportions of simultaneous occurrence, corrected by taking into account the number of occurrence. The resulting equation is

$$QS = \frac{c}{\sqrt{(ab)} - \frac{1}{2}\sqrt{b}} \qquad (3)$$

where a and b are the numbers of occurrence cases of species A and B; c is the number of cases of simultaneous occurrence; A and B must be selected in such a way that $a \leqslant b$.

Another method has been proposed by PICARD (1965), who investigated the deep shelf and upper slope soft bottoms in the Marseilles area (France). Firstly, the area to be investigated is quickly surveyed in order to obtain approximate delimitations of different biotopes which are assumed to harbour homogeneous biocoenoses. Secondly, in each biotope, the investigator collects (with a 'heart-urchin dredge' which is rather similar to an anchor dredge) as many samples as is necessary for evaluating the 'minimum volume' of the sediment, i.e. the volume which must be sampled for recording at least 95% of the macrobenthos species present. The minimum volume is mostly 50 dm^3 but may be as high as 500 dm^3 in biocoenoses harbouring a scattered fauna. Thirdly, ten samples are collected, each corresponding to the minimum volume; the whole collection is then sieved in order to separate and count all live individuals of all macrobenthos species present. In this way we obtain the abundance of each species as well as the ratio between the abundance of a given species and the total number of individuals of all the species, i.e. the dominance value, which is expressed as a percentage.

This method also allows the determination of the degree of affinity of a mixed assemblage for two or three neighbouring, well-defined biocoenoses. As an example, let us suppose that a mixed assemblage includes species which belong to three different biocoenoses A, B, C; N_A, N_B, N_C are then the average dominances of all characteristic species existing in the biocoenoses A, B and C, and n_A, n_B and n_C are the average dominances of the characteristic species of A, B, C collected in the mixed assemblage. However, the number of characteristic species may be very different in the biocoenoses mixed in the bottom area investigated. Hence, correction coefficients must be calculated for each biocoenotic stock. For biocoenoses A, B and C these are:

for A
$$C_A = \frac{N_A \times 100}{N_A + N_B + N_C}$$

for B
$$C_B = \frac{N_B \times 100}{N_A + N_B + N_C}$$

for C
$$C_C = \frac{N_C \times 100}{N_A + N_B + N_C}$$

The degree of affinity of the mixed assemblages for the biocoenoses A, B and C is expressed by $n_A \times (C_B + C_C)$, $n_B \times (C_A + C_C)$ or $n_C \times (C_A + C_B)$.

The method described above is useful for studying transitional assemblages, subject to change either as a function of time (e.g. due to variations in environmental factors) or space. In regard to space, it is a well-known fact that two neighbouring biocoenoses are not usually separated by a sharp boundary but by a transition zone, the width of which depends on the gradient of prevailing environmental factors (p. 20).

(d) Delimitation of Communities Based on the Quantitative Method

General Aspects

The investigations which led PETERSEN (1911, 1913) to the community concept were prompted by an applied problem: the feasibility of increasing the yield of plaice by means of transplantation of young individuals from the North Sea into shallow areas of the Danish coast. Plaice live on soft bottoms and feed on small-sized microphagous invertebrates. PETERSEN (1913) assumed that most of the energy available at the bottom of the trophic pyramid was represented by decaying material originating from *Zostera marina* beds and that most prey of plaice feed upon this organic matter either directly or with an intermediate link (bacteria, micro- and meiobenthic species, etc.). The role of *Zostera* beds as predominant primary producer for the benthic tropic system is currently considered questionable (p. 445). Nevertheless, PETERSEN succeeded in his attempt.

The major aspect of the community concept is that one has to select as 'pilot' species in the community not the characteristic species (i.e. those which do not exist in another community) but those which predominate in weight and number of individuals. According to PETERSEN, pilot species of a community must be: (i) sufficiently abundant to be collected by means of bottom samplers; (ii) permanently present, i.e. not of seasonal occurrence; (iii) regularly distributed on the bottom, i.e. not in patches.

THORSON (1957) proposed the addition of three further qualifications: (i) to select, within the group of pilot species, animals of various feeding habits (e.g. filter feeders, detritus feeders, deposit feeders) but to exclude predators; (ii) to relate the number of pilot species selected to peculiarities of their life cycle: the shorter their average life-span, the greater must be the number of pilot species considered; (iii) to select individuals of average body sizes; excessive body size can disproportionately increase the biomass percentage where the species in question is represented only by one or two individuals. Very small-sized animals which are difficult to weigh exactly can more easily be counted and then be weighed as a whole.

The above-mentioned points (i) and (iii) were exemplified by SANDERS (1958) in a study on a silt-clay community at Buzzards Bay, near Woods Hole, Massachusetts (USA). Expressed as percentage of the total number of individuals, the predominant species (which represented as a whole 90% of the total number) were the infauna components *Nucula proxima*, 59%; *Nephthys incisa*, 17%; *Ninoe nigripes*, 4·3% (all deposit feeders); and the epifauna components *Cylichna orzya* (deposit feeder), 4·1%; *Callocardia morrhuana*, (suspension feeder) 3·4%; *Hutchinsoniella macracantha* (deposit feeder), 2%. In contrast, the 13 specimens representing larger-sized species (individual weight over 0·1 g), amounted to more than 55% of the total biomass. Hence it appears that for a meaningful comparison of biomasses from different communities, large-sized animals should not be selected as pilot species.

The biomass—also referred to as standing crop—is usually expressed as weight unit m^{-2}. THORSON (1957) recommended using a depth-dependent reference system: 0·1 m^2 surface area from 0 to 200 m, 0·2 m^2 from 200 to 2000 m, and 1 m^2 for depths in excess of 2000 m.

Processing of Samples

Typically, all individuals of all species in a given sample are counted and the representative of each species weighed. The rough or wet weight (w.w.) thus obtained corresponds to the weight of living organisms cleared of excess water, i.e. except that within the living tissues. However, rough weight is a poor measure of organic-matter content. The tissue-water content of plants and animals may vary considerably, and the mineral parts (shell, skeleton, spicules, etc.) cannot be taken into account in the evaluation of useful biomass, i.e. the amount of organic matter utilizable by consumers. Hence, the wet material—fresh or conserved in alcohol or formalin solution—must first be decalcified and then carefully dried, first at 60 °C, then at 100 to 110 °C until the weight remains constant. In this way we obtain the dry weight (d.w.), a valuable expression of the organism's organic-matter content.

While the dry weight of decalcified organisms is the most significant parameter of their biomass (standing stock), its determination requires so much time and care that dry weight values are seldom used. Since wet weight is much easier to calculate, wet weight/dry weight conversion tables have been worked out. Such tables are of some value only for a single species and even in this case some irregularities may result in relation to age classes, stages of sexual cycle, or moulting conditions in crustaceans. Tables proposed for taxa above the species level are of very limited, if any, value. Moreover, we cannot disregard potential errors introduced in both w.w. and d.w. due to the weight of gut contents. This may be very important in limivorous species which engulf sediment; in the polychaete *Ophelia bicornis*, for example, the gut content may be up to 24% of the wet—or alcohol—weight. In microphagous, non-limivorous invertebrates, errors introduced into biomass estimations by the gut-content weight are generally higher in detritus feeders than in suspension feeders.

It seems that PETERSEN (1913), whose attention was at first focussed on applied aspects, did not consider his communities to be true ecological units. However, later authors who more or less exactly followed his method, increasingly interpreted his concept in terms of ecological units. We must further emphasize here that, in spite of the principles introduced by PETERSEN and later THORSON (1957), the selection of pilot species of a community remains somewhat subjective. BROTSKAJA and ZENK-EVICH (1939) were aware of this when they studied benthos communities in the Barents Sea and suggested substituting the PETERSEN communities by new units which they called 'complexes'. The selection of pilot species in such a complex is exactly defined: each species has a density index expressed by the number of individuals × total weight; if the product obtained is too high, its square root is used. Then the density index of each species is compared with the density index of the whole community. Species which have a higher density index—between 45% and 98% in the paper mentioned—are considered pilot species. While this method is more objective than that used by PETERSEN, it is rather intricate and hence was later abandoned. More recently some attempts have been made to use mathematical models for establishing lists of pilot species.

Obviously, the community concept can also be applied to hard-bottom assemblages. However, sampling difficulties—except where divers are employed—have largely restricted the quantitative method to the exploration of soft bottoms. With regard to the latter, analyses of the sediment itself (granulometry, $CaCO_3$ content, organic matter, etc.) are necessary to determine the relationships between community and substrate as well as those between substrate and organisms, on the one hand, and the near-bottom water, on the other. For ecological research, the local micro- and meiobenthos (p. 523) must also be studied.

PETERSEN's concept of bottom communities characterized by species which predominate in terms of numbers and weight has been questioned by several authors who argued that quantitative data on very few of the species present have been analysed subjectively while other potentially important species have been neglected. However, STEPHENSON and co-authors (1972, pp. 405, 406), who analysed PETERSEN's original data by employing various computer techniques, confirmed the existence of 'PETERSEN-type' communities in the majority of the cases examined. Nevertheless, they emphasize that the disparities between the classifications obtained by them 'imply that a compromise involving numbers and weights is not possible'; instead, communities based on numbers must be distinguished from those based on weight. STEPHENSON and co-authors further point out that quantitative analyses are necessary, and that it remains to be shown whether the extra labour involved in obtaining weight data is worthwhile.

(e) Comparison of Qualitative and Quantitative Methods for Delimiting Benthos Assemblages

The two preceding sections outlined the differences between qualitative and quantitative methods employed for investigating assemblages. We shall now compare the usefulness of these methods in relation to specific research aims.

The qualitative method employs—for bottom areas which cannot be sampled directly—towed equipment such as dredges and trawls (for a description see, for example, HOLME and MCINTYRE, 1971). The equipment used differs in sampling efficiency. The results obtained are usually a function of several factors including towing speed, gear weight, mesh size, population density, substrate texture and water depth. Thus the sample obtained does not necessarily reflect the true abundance or the respective dominance of the different species present. Moreover, the quantity of organisms dredged in relation to the surface supposed to have been sampled is much less than in collections carried out by quantitative gears (e.g. Table 3-2). The gear accumulates the organisms along the haul and cannot provide information about their previous distribution on the bottom (random distribution, patchiness, etc.). The method provides a basis for a fairly

Table 3-2

Collection efficiency of bottom samplers. Comparison of average numbers of macrobenthic animals collected from 0·1 m² surface area. Sandy bottom; 6-m depth; Gulf of Marseilles, France (After REYS and co-authors, 1966; reproduced by permission of the authors)

Device	Dredge	Orange-peel bottom sampler	Suction sampler
Volume of sediment collected (dm³)	1	4·8	20
Crustaceans	0·4	1·9	6
Molluscs	6·2	7·3	13·2
Polychaetes Errantia	0·7	46·1	187·2
Polychaetes Sedentaria	0·4	8·1	7·4
Phoronids	0	39·7	352·8

complete list of species which inhabit the bottom, including the most motile representatives, provided that different equipment (dredges and trawls) were used. If the distance covered by a haul is too wide, overlapped sampling may occur, i.e. different neighbouring assemblages may be sampled. The qualitative method usually allows the mapping out of the general distribution of biocoenoses; the accuracy of the map obtained depends largely on the number of samplings: the narrower the network of stations, the more accurate the chart.

The quantitative method employs equipment which concentrates on a small defined, small surface area: the bottom-sampler opening. The number of organisms per sample is small, and statistically significant information on the community can be obtained only with at least 20 to 70 collections. Such intensive sampling reduces the range of potential errors due to irregularities in species distributions. It cannot, however, prevent an information gap with regard to motile species which may escape the sampler and to organisms whose size exceeds that of the sampler. Among the quantitative samplers we can distinguish mechanical devices (generally equipped with jaws of different shapes) and suction samplers (e.g. REYS and co-authors, 1966; HOLME and MCINTYRE, 1971). Evidently, the volume of sediment collected by mechanical samplers —and consequently the number of organisms caught—depends on penetration depth, which is a function of the weight of the device and the consistency of the sediment. Such restrictions do not prevail in suction samplers, but to date these have been employed only by divers. Both accuracy and efficiency are much higher in suction samplers than in the mechanical bottom samplers used thus far (Table 3-2).

Suction samplers have two further advantages: (i) simultaneous sampling and selection of depth; (ii) low average variability between samples, i.e. only about 20% (100–150% with mechanical samplers).

With the possible exception of deep-sea macrobenthos assemblages (mainly in the abyssal and hadal zones), a single set of samplings collected at a given time usually cannot provide reliable information. The dynamic processes which control organismic occurrence—e.g. seasons, p. 161—necessitate repeated samplings over a 15-month period, at least in all four seasons, where possible every month. Short-interval sampling is especially necessary in communities harbouring species with life-spans of one year or less.

The concepts of biocoenosis and community open up different avenues for analysing characteristics of assemblages. The biocoenosis concept seeks to comprehend the laws and rules which control the spatial distribution of organisms in relation to abiotic and biotic factors and thus to delimit the assemblages concerned, together with obtaining as extensive a list as possible of all participating species. The community concept is more restricted and pertains mainly to applied ecological perspectives. It usually focusses on the amount of food available for species belonging to the first carnivorous level of the trophic pyramid, with special attention to demersal fishes. Consequently, these two avenues for analysing the composition of assemblages are not opposed to each other, but complementary.

In the opinion of the reviewer, it would be better to retain the meanings of biocoenosis and community outlined on p. 48, since both are essential for studying macrobenthos assemblages. For a more detailed definition of 'organismic assemblage', it is suggested that the definition of community offered by MILLS (1969, p. 1415) be summarized:

'a group of organisms occurring in a particular environment, presumably interacting with each other and the environment, and separable by means of ecological survey from other groups',

keeping in mind that numerical methods can be used to define communities (*sensu* PETERSEN) as homogeneous groups (FIELDS, 1971).

(f) The Concept of Continuum in the Benthal

The presence of definable assemblages in the benthal has been questioned by several authors (CURTIS and McINTOSH, 1951; CURTIS, 1959; MAYCOCK and CURTIS, 1961) who investigated changes in the distribution of different species in relation to environmental factors. From the fact that the distribution curves obtained for different species do not coincide, these authors infer that the assemblage is continuously changing on a space scale and, therefore, that one cannot distinguish consistent units of ecological value. In other words, they consider the distribution of organisms as constituting a complex 'continuum' of their respective populations in relation to gradients of environmental factors rather than discontinuous units. Within a space continuum some areas may be recognized in which several species are more abundant than elsewhere; such an area is named a 'nodum' (see BOUDOURESQUE, 1970a, b).

The concept of continuum was developed mainly by plant ecologists, for example in regard to the relative abundance of different tree species along an environmental gradient on a mountain slope. However, in the sea, such gradients can, in the reviewer's opinion, only prevail on deep-sea bottoms, especially in the abyssal zone. It is possible that in such areas, the wide grid of sampling, which is unavoidable in deep-sea investigations due to the time required for each haul, prevents a potential continuum from being recognized. For example, the four macrobenthos assemblages distinguished in the abyssal zone of the northern Pacific Ocean (p. 21) could in reality represent four isolated points in the spatial continuum of a single assemblage, possibly related to progressive changes in food resources.

Environmental gradients occur especially on shelf bottoms for example, with respect to illumination and wave movement as a function of increasing water depth. However, on many macrobenthos organisms these factors exert effects which are not linearly related to the intensity of the stimulus. For example, a given macrobenthic plant cannot live beyond its characteristic critical depth. Another example is the distribution of attached species depending on wave movement (multidirectional, bidirectional or unidirectional; see p. 19 and Volume I, Chapter 5). Geomorphological features and the nature of the sea floor—e.g. rocky, sandy or muddy, often closely related to prevailing water currents and the amount of terrigenous input—together with the role played by currents in the transport of particulate food for microphagous macrobenthos can cause a striking heterogeneity of the benthal. Such heterogeneity, in turn, may induce discrete boundaries between neighbouring assemblages. Even in parts of the bathyal zone in which the geomorphology of the slope displays some marked irregularities we may observe discrete boundaries in the distribution of macrobenthic animals.

Discrete boundaries between macrobenthos assemblages are rather widespread in the upper levels of the shelf. Discrete boundaries prevail, for example, on hard bottoms between shallow-water assemblages in sheltered and exposed areas, or between assemblages in well-illuminated surfaces and those in the crevices or caves of the rocky substrate. In general, discrete boundaries also occur between assemblages inhabiting, at the same depth, a bare muddy sand bottom and a sea-grass bed respectively.

In other areas of the shelf, a wider transition zone prevails between two adjacent assemblages, e.g. between an assemblage in fine sand and one in a muddy substrate along a considerable gradient of increasing mud content in the sediment. Another example of a wide transition zone is the area between the deepest levels of a sea-grass bed, in which the plants become more and more scattered due to decreasing illumination and the beginning of the assemblage inhabiting the deeper and soft, unvegetated sea floor.

The transition zone of tension between two adjacent assemblages is known as 'ecotone'. In an ecotone the outposts of each assemblage

'are maintaining themselves in environments that are increasingly unfavourable. The tension may result chiefly from a struggle with physical conditions or from a direct competition between certain species.' (CLARKE, 1954, p. 411).

In the benthal, ecotones may be much wider than in the cases mentioned above—e.g. on the Antarctic slope with a rather homogeneous macrobenthos assemblage ranging from 50 to 100 m down to a depth of 500 to 700 m (pp. 20 and 464). Since in the latter example some progressive changes in the qualitative and quantitative composition of the assemblage can be observed, this slope environment might be considered to some extent as a continuum.

The qualitative and quantitative composition of the fauna in an ecotone—irrespective of the predominant influence the progressive change in the substrate's nature—seems to depend mainly on three factors: (i) the percentage of species, in either of the two adjacent assemblages, which are sufficiently versatile (i.e. sufficiently eurytopic) for participating in the outposts; (ii) the percentage of species with a mobility which is sufficient for enabling transitory ecotone invasion, e.g. in terms of food acquisition; (iii) the amount and nature of the food available within the ecotone.

The low abundance and specific richness the reviewer has noticed (p. 20) on soft bottoms in several areas of the Mediterranean Sea and south Portugal coasts in the ecotone between the deeper circalittoral macrobenthos assemblage and the upper bathyal one are probably due to a combination of influences from the factors mentioned in (i), (ii) and (iii). In contrast, in most inshore areas of the Mediterranean Sea, which provide more abundant and diverse food, we have observed an ecotone between the rocky substrates covered with algae on the one hand and the sea-grass (*Posidonia*) beds on the other where scattered boulders or flagstones—both vegetated—alternate with soft-bottom patches with scattered sea-grass heads. In this ecotone motile species (e.g. shrimps, prawns, crabs, fishes) of both adjacent assemblages largely intermix.

According to CLARKE (1954, p. 412), who chiefly considered the continental environment, physical factors in the ecotone area are different (usually intermediate) from those in either of the bordering assemblages. As a consequence, various plants and animals which do not occur or are rare in the bordering assemblages may become abundant in the ecotone. In the benthal two parameters appear to be of paramount importance for controlling the distribution of macrobenthos species and hence affect their classification into 'characteristic', 'accompanying' or 'casual' species (p. 53): energy resources (light and nutrient salts for plants, food for animals) and the nature of the substrate. In fact, I could not find any example in the marine benthal where members of a species occurring in an ecotone were not present in the bordering assemblages. Possible exceptions might be some sponges—e.g. the hexactinellids *Hyalonema thomsoni* and *H. lusitanicum*—which seem to exist, on the Atlantic Ocean slope off France and northern Spain, only in two large ecotones: just below the most developed populations of deep-sea corals (p. 485) and at the lowest levels of the slope where the slope inclination becomes more gentle (abyssal rise). In both these ecotones, due to a reduced bottom current, the sedimentation rate of the smallest-sized and hence the lightest organic particles increases, thus producing favourable conditions for sponge nutritions; in the former location most particles available to the sponges originate from the rich 'white-coral' assemblage.

With regard to the abundance displayed in the ecotone by species from bordering assemblages, it seems that most of them qualify as 'accompanying' species which are more eurytopic than species characteristic of the assemblage. Of the conditions which support a given species, apparently the most significant ones are vacant space on hard substrates, and a qualitatively and quantitatively adequate food supply on soft substrates. In both cases, additional support is obtained from reduction or absence of competition with species characteristic of the bordering assemblages, the latter often being unable to adjust to the environmental conditions of the ecotone. Nevertheless, even in hard-substrate ecotones, the development of abundant populations requires the availability of sufficient energy, as well as of sufficient and regular input either of planktonic larval stages—where such stages occur in the life cycle—or of invading adults, where the forms are sufficiently motile; the latter is also true for soft-substrate ecotones.

KREBS (1978) discusses the significance of the concepts of continuum proposed by the 'individualist school', and of assemblages, i.e. integrated units with discrete boundaries, proposed by the 'organismic school', which apparently mainly took into account the associations of continental plants. He concludes (p. 406) that 'the information available leans more toward the individualistic interpretation' and points out that the classification of assemblages is for the convenience of man, but not a description of the fundamental structure of nature. This, of course, is true for many if not most biological classification concepts. Nevertheless, such concepts are necessary because they are prerequisites for comprehending a nature whose complexity by far exceeds man's analytical and synthetical potentials. Krebs' interpretation appears to be valid in regard to the plant associations he investigated, but less so as far as the distribution of marine macrobenthos is concerned.

In my opinion, the organismic concept—besides being more convenient for many practical purposes—seems to be more suitable for studying the benthal for the following reasons: (i) Pronounced gradients are less widespread and less extensive—except at abyssal depths—than those in environments investigated by terrestrial botanists. (ii) Ecologists of the individualistic school base their system on fewer dominant species—usually the larger and more conspicuous ones such as trees—while marine benthologists use both plants and animals as 'pilot' species of the assemblage; within the assemblage, they further discriminate, for example, by means of the 'reproductive index' (BOUDOURESQUE, 1970a, b), 'casual', 'characteristic' and 'accompanying' species. (iii) Presence and distribution of macrobenthic species are markedly influenced by the nature of their substrate; in addition, the marine substrates are more diversified over small spatial scales than terrestrial soils. (iv) Despite such diversity, marine substrates often exhibit common features on a worldwide basis: each type of substrate, (e.g. rock, boulders, pebbles, sand, mud, clay) displays physical features which are very similar all over the World Ocean and—as THORSON (1957, p. 504) has recognized—exhibit 'isocommunities', i.e. communities in which different species belonging to the same genus—or occupying the same ecological niche—replace each other as characterizing species.

In accordance with my suggestion that the term 'organismic assemblage' be substituted for the more restrictive term 'community' (*sensu* PETERSEN), I suggest that the term 'isoassemblage' be adopted as a substitute for THORSON's term 'isocommunity'.

(2) Assemblages in the Pelagial

In the pelagial, both terminology and delimitation of units of ecological significance are much more problematic than in the benthal. In general, it seems that three main units of ecological significance can be distinguished: association (see definition on p. 49), assemblage and ecosystem.

In the pelagial, it is necessary to employ many different sampling devices and methods for collecting the different types of organisms present. Their size and motility are so diverse and the organisms are often so irregularly distributed over a water mass that it is almost impossible to speak of a 'biocoenosis', especially if one takes into account the patchiness of plankton and the schooling of many nektonic species. In addition, sampling such large areas introduces uncertainties as to whether the investigator is still dealing with one and the same biocoenosis.

It is generally impossible to obtain reliable quantified data in the pelagos; hence we can hardly speak of 'community' in the meaning used for benthos assemblages. For example, measurements of chlorophyll a, amylases or proteases made in order to assess biomasses of algae, herbivores or carnivores are too general and approximate for a detailed analysis. Furthermore, they correspond only to a transient stage—a 'snapshot'—of the assemblage present. Only the use of as many different sampling methods as are necessary for obtaining sufficient knowledge on both flora and fauna in a given volume of the water column considered a sufficiently homogeneous environment, followed by accurate sample processing and data interpretation, might allow the ecologist to apply the concept of 'assemblage'.

Certainly, the ecosystem concept would be the best to use in the marine pelagial (Volume V: VINOGRADOV, in press), since the environment exerts a more direct and intimate influence on pelagic organisms than on their benthic counterparts. For example, the generally smaller size and thinner teguments of phytoplankters and zooplankters augment the immediate impact of the environment and tend to increase material exchanges with the surrounding sea water. In addition, most planktonic species (holoplankton and many nektonic forms) complete their whole life cycle in the free water. Nevertheless, I have chosen to retain the assemblage concept for the pelagial also, for conceptual reasons of symmetry with the benthal and because in many investigations data on physical and chemical environment parameters are not available or insufficient for analysing ecosystem dynamics.

However, the concept of organismic assemblage must be applied with caution in the pelagial. There are two major reasons for this.

(i) As MARGALEF (1977, p. 860; see also Volume IV) has pointed out, the spatial scale of samplings must be carefully considered since the distribution of different plankton species is frequently inhomogenous. In an apparently homogeneous water volume, not only may different assemblages exist but also various types of organismic distribution. The most frequent structural pattern possibly involves small zones of reduced diversity in which a few species are represented by large populations (or even predominate), their population density decreasing exponentially toward peripheral zones; here the diversity is higher or the predominant species less numerous.

(ii) Being a tridimensional environment, the pelagial features transitional zones which may be different from those most often observed in the benthal, due to the heterogeneous distribution which characterizes the plankton in general. MARGALEF (1977, pp. 856, 857) has discussed this problem imagining watertight walls placed vertically or horizontally within the pelagial. If the horizontal wall separates an upper, well-illuminated layer from a deeper, less well-illuminated one, the upper layer will tend to accumulate deposited organic material on the horizontal imaginary wall, the mineralization of which will restitute within the layer itself a new material cycle over a smaller vertical range. In the deeper layer, provided the dividing wall is transparent and the illumination still sufficient for some photoautotrophic organisms, the extent of the material cycle will be reduced, and decaying organic matter will accumulate on the bottom. According to MARGALEF, such separation is markedly asymmetrical, since it isolates two subsystems which normally represent parts of a larger system, capable of self-regulation, which has now been suppressed by the wall. Let us hypothesize the presence of a vertical wall in the water volume. At first sight this would not seem to induce any pronounced changes, except that the wall reduces or inhibits the horizontal transport of populations—and of material in general—in the same direction. This would result in the separation of two (or several in the case of more than one wall) compartments of the ecosystem that are similar to each other and relatively independent. According to MARGALEF, such separation is symmetrical in type. However, if the vertical wall is located within a spatial succession, for example between an area—say a divergence—in which the plankton production is high, and a peripheral area toward which this production is 'exported', we would again find a separation of an asymmetrical type. In effect, without the wall the extra production of the fertile area, i.e. the production not consumed in the area itself, will benefit consumers in peripheric areas, conversely nutrients from recycling progressively return from the periphery to the place of production. Where the hypothetical watertight wall extends so deep that it hinders such return, the isolated ecosystem would rapidly deteriorate because the maintenance of its dynamic balance depends on both its capacity to export the extra production and its chances of receiving new nutrients.

MARGALEF (1977) further proposed discriminating as true 'ecotones' the transition zones, the limits of which are generally rather sharp, from the 'ecoclines' which represent more gradual transition zones with less discontinuous limits.

Literature Cited (Chapter 3)

ALLEE, W. C. and SCHMIDT, K. P. (1951). *Ecological Animal Geography* (2nd edn), Wiley, London.

BOUDOURESQUE, C. F. (1970a). *Recherches de Bionomie Analytique, Structurale et Expérimentale sur les Peuplements Benthiques Sciaphiles de Méditerranée Occidentale (fraction algale)*, Thèse Doct., Université de Aix-Marseille.

BOUDOURESQUE, C. F. (1970b). Recherches sur les concepts de biocoenose et de continuum au niveau de peuplements benthiques sciaphiles. *Vie Milieu*, 21 (1B), 103–36.

BROTSKAJA, V. A. and ZENKEVICH, L. A. (1939). Quantitative evaluation of the bottom fauna of the Barents Sea. *Trudȳ VNIRO*, 4, 5–98.

CLARKE, G. L. (1954). *Elements of Ecology*, Wiley, London.

CURTIS, J. J. (1959). *The Vegetation of Wisconsin*, University of Wisconsin Press, Madison.

CURTIS, J. J. and MCINTOSH, R. P. (1951). An upland forest continuum in the prairies forest border region of Wisconsin. *Ecology*, 32, 472–96.

DAWSON, E. Y., NEUSHUL, M. and WILDMAN, R. D. (1960). Seaweeds associated with kelp beds along Southern California and Northwestern Mexico. *Pacif. Nat.*, 1 (14), 81.

FAGER, E. W. (1963). Communities of organisms. In M. N. Hill (Ed.), *The Sea*, Vol. II, Wiley, New York. pp. 415–37.

FIELD, J. G. (1971). A numerical analysis of changes in the soft-bottom fauna along a transect across False Bay, South Africa. *J. exp. mar. Biol. Ecol.*, 7, 215–53.

GAMULIN-BRIDA, H. (1960). Primjena Sørensenove metode pri istraživanju bentoskih populacija. *Biol. Glasn.*, 13, 21–41.

GERLACH, S. A. (1972). Substratum. General introduction. In O. Kinne (Ed.), *Marine Ecology*, Vol. I, *Environmental Factors*, Part 3, Wiley, London. pp. 1245–50.

HOLME, N. A. and McINTYRE, A. D. (1971). *Methods for the Study of Marine Benthos* (*IBP Handbook* 16). Blackwell, Oxford.

KREBS, C. J. (1978). *Ecology—The Experimental Analysis of Distribution and Abundance* (2nd edn), Harper & Row, New York.

MARGALEF, R. (1977). *Ecología*, Omega, Barcelona.

MARGALEF, R. (1978). General concepts of population dynamics and food links. In O. Kinne (Ed.), *Marine Ecology*, Vol. IV, *Dynamics*. Wiley, Chichester. pp. 617–704.

MARION, A. F. (1883a). Esquisse d'une topographie zoologique du Golfe de Marseille. *Ann. Mus. Hist. nat. Marseille*, **I** (1), 1–108.

MARION, A. F. (1883b). Considérations sur les faunes profondes de la Méditerranée. *Ann. Mus. Hist. nat. Marseille*, **I** (2). 1–50.

MAYCOCK, P. F. and CURTIS, J. J. (1961). The phytosociology of boreal Conifer-Hardwood forests of the Great Lakes region. *Ecol. Monogr.*, **30**, 1–35.

MILLS, E. L. (1969). The community concept in marine zoology, with comments on continua and instability in some marine communities. A review. *J. Fish. Res. Bd Can.*, **26**, 1415–28.

MÖBIUS, K. (1877). *Die Auster und die Austernwirtschaft*, Wiegandt, Hempel and Parry, Berlin.

PERES, J. M. and PICARD, J. (1964). Nouveau manuel de bionomie benthique de la Mer Méditerranée. *Recl Trav. Stn mar. Endoume*, **47**, 1–137.

PETERSEN, C. G. J. (1911). Valuation of the sea. I. Animal life of the sea-bottom, its food and quantity. *Rep. Dan. biol. Stn*, **20**, 1–81.

PETERSEN, C. G. J. (1913). Valuation of the sea. II. The animal communities of the sea bottom and their importance for marine zoogeography. *Rep. Dan. biol. Stn*, **21**, 1–44.

PICARD, J. (1965). Recherches qualitatives sur les biocoenoses marines des substrats meubles dragables de la région marseillaise. *Recl Trav. Stn mar. Endoume*, **52**, 1–244.

REYS, J. P., TRUE, M. A. and TRUE-SCHLENZ, R. (1966). Un nouvel appareil de prélèvement quantitatif des substrats meubles. *Int. oceanogr. Congr.*, **2**, 350.

SANDERS, H. L. (1958). Benthic studies in Buzzards Bay. I. Animal sediment relationships. *Limnol. Oceanogr.*, **3**, 245–58.

STEPHENSON, W., WILLIAMS, W. T. and COOK, S. D. (1972). Computer analyses of Petersen's original data on bottom communities. *Ecol. Monogr.*, **4**, 387–415.

THORSON, G. (1957). Bottom communities (sublittoral or shallow shelf). In J. W. Hedgpeth (Ed.), *Treatise on Marine Ecology and Paleoecology*, The Geological Society of America. (*Mem. geol. Soc. Am.*, **67**, 461–534.

VINOGRADOV, M. E. (in press). Ecosystem of the open ocean. In O. Kinne (Ed.), *Marine Ecology*, Vol. V, Ocean Management, Part 1. Wiley, Chichester.

Marine Ecology Vol. 5, Part 1
Edited by Otto Kinne
© 1982 John Wiley & Sons Ltd

4. STRUCTURE AND DYNAMICS OF ASSEMBLAGES IN THE PELAGIAL

J. M. PERES

(1) General Aspects

The pelagial features several specific characteristics: it is tridimensional, in continuous motion, largely dominated by small-sized organisms and contains more immediately integrated assemblages than does the benthal.

As a tridimensional environment, the pelagial reveals appreciable stratification in the qualitative and quantitative composition of its assemblages. These are often related to diel and seasonal migrations of its organismic components.

In continuous motion, large scale horizontal water currents can significantly modify the spatial distribution and abundance of pelagic organisms. The combination of an undercurrent and a surface current (e.g. the Cromwell current under the southern equatorial current in the Pacific Ocean) may alternately transport zooplankters which perform diel migrations in opposite directions by day and night. Convergence processes can also control the distribution of pelagic assemblages. For example, the sinking of the Antarctic surface water, which becomes the Antarctic intermediate water along the Antarctic convergence, transports the surface-water plankton assemblage downwards; this affects the composition of the assemblage as it drifts northwards. Furthermore, vertical ascending currents (divergence, upwelling, doming) may cause rising of plankters. Still more important, vertical water movement and mixing that occur during winter in temperate regions can induce enrichment in nutrient salts of the euphotic layer.

In the surface and subsurface layers (epi- and mesopelagic zones) small-sized organisms largely predominate, at least from a functional point of view, since they represent the first two steps of the trophic pyramid. Nano-, micro- and mesoplankton may even constitute a greater biomass than that represented by macro- and megaloplanktonic and nektonic animals especially at high latitudes. Since small-sized organisms have a shorter turnover time than large-sized ones, the potential productivity of nano-, micro- and mesoplankters is very high. Moreover, their short life-spans may result in a pronounced time succession of populations of different species and consequently affect the assemblages these populations constitute. In turn, the combination of this time succession of assemblages with their transport due to currents—mainly in surface waters—may result in a spatial disjunction of assemblages which chronologically follow one another (Figs 2-5, 2-6 and 6-26).

Trophically, pelagic assemblages often appear to be more immediately integrated than their benthic counterparts. Direct pathways often lead from the phytoplankton to carnivores of the highest trophic level. In the pelagial the part of primary production which is consumed alive by herbivorous species is much more important than in the benthal, even when we take into account in the latter the consumption of micro-phytobenthos by macrobenthic animals (often wrongly referred to as 'detrivores'). The energy transfer to macrobenthos assemblages from the micro- and meiobenthos psammic system is probably very low (p. 565), the latter being considered—at least in shallow depths—as an almost autarchic system. Hence, many macrobenthos micro-phages, namely the suspension feeders, mainly utilize plankton organisms. 'Blind alleys' in the trophic food web—i.e. pathways leading to decomposition and mineraliza-tion of dead organisms, before their energy reached the potential terminal food link—are also less numerous in the pelagial than in the benthal. Therefore, energy transfer to the large-sized carnivorous fishes is more immediate in the pelagial than in the benthal; this explains the fact that landings of pelagic fish are double those of demersal fish.

(2) Phytoplankton

(a) Time and Space Distributions of Phytoplankton

As is well documented, life-span tends to increase with increasing body size. Further-more, the smaller the size—and thus the higher the surface/volume or surface/mass ratio—the better the buoyancy. All photoautotrophic phytoplankters, which must remain within the euphotic layer as much as possible, are small-sized and have a short life-span. As a consequence, they often exhibit successions in which populations of different species succeed each other, depending on fluctuations in different factors. These will be summarized below.

In temperate regions, for example, the spring bloom begins with an abundance peak of diatoms, followed by a stage in which dinoflagellates increase in abundance, and finally by a phase of flagellate (often mainly coccolithophorid) predominance. The number of annual successions depends either on the degree of seasonality or on meteorological sequences.

Time and space changes in the organization of phytoplankton assemblages have been investigated by MARGALEF (1962), in the areas off Castellón and the Ebre Delta (Spain). MARGALEF distinguished three successive stages: (i) Predominance of opportunistic diatom species, such as *Bacillaria paradoxa, Chaetoceros affinis, C. compressus, Leptocylindrus danicus, Skeletonema costatum, Thalassionema nitzschioides* and of species belonging to the naked flagellate genus *Exuviella*. (ii) Increase in diatom diversity associated with several species of dinoflagellates, such as *Ceratium furca, C. tripos, C. fusus, Peridinium brochi* and *P. trochoideum*. At the end of the second stage, species of diatom genera as yet uncommon appear (e.g. *Bacteriastrum, Rhizosolenia*). (iii) At the third stage, some diatoms such as *Hemiaulus hauckii* and various species of the genus *Chaetoceros*, still exist, but dinoflagel-lates (represented by species of the genera *Ceratium, Dinophysis, Gonyaulax* and *Ornithocer-cus*) predominate. At the end of this last stage, if the succession does not become dis-turbed, other dinoflagellates appear which belong to the genera *Ceratocorys, Oxytoxum, Pyrophacus, Podolampas*, etc. (to a certain extent these may be considered as representing a fourth stage).

According to MARGALEF (1962), the most complete succession in the area he investi-gated occurs between May and September, i.e. during the period of best weather.

However, there are at least two other successions beginning in February and October; these are restricted to the first and second stages, since progressive changes in the phytoplankton populations may become fully achieved only if the waters remain sufficiently stable. Whenever some turbulence occurs—as is often the case in late winter or autumn—water movement causes dispersion, and consequently loss of phytoplankters. The latter must be compensated for by increase in population growth, promoted primarily by an increase in mineral nutrients originating from mixing processes. Chlorophyll *a* content and production per unit biomass attain maxima at the first stage, and minima at the end of the succession. In contrast, the specific diversity index

$$d = \frac{S - 1}{\log_e N}$$

in which N and S are the numbers of individuals and species, respectively, progressively increases along with the succession, as well as the pigment diversity index (D_{430}/D_{665})—the ratio of yellow to green pigments expressed as optical density at 430 and 530 nm. The D_{430}/D_{665} ratio increases with assemblage age in proportion to an increase in substances originating from chlorophyll degradation. The latter stem from increasing numbers of plant cells which die and subsequently decay or are consumed by grazers.

In the euphotic layer of the World Ocean, a spatial distribution related to the amount of phytoplankton nutrients available corresponds to this succession. Diatoms always predominate in eutrophic regions such as estuaries, areas of divergence, upwelling or doming, and also at high latitudes. In oligotrophic areas, dinoflagellates predominate in number of species and sometimes in biomass. In ultra-oligotrophic areas, such as the central parts of large oceanic gyres, coccolithophorids are the most abundant, followed by dinoflagellates.

(b) Major Factors Affecting Phytoplankton Distribution and Abundance

Light

The mean illumination level—and thus the amount of radiant energy available for photosynthesis—depends on latitude and season (Volume I, Chapter 2). In relation to water depth, illumination depends on turbidity which, in turn, mainly depends on the content of suspended particles including the phytoplankton itself (Volume IV: FINENKO, 1978). It is generally admitted that the lower limit of the euphotic layer corresponds to the 1% level of incident illumination. In this context, some additional statements must be made: (i) Not all phytoplankters require or tolerate the same illumination levels: some are photophilic, others—mainly dinoflagellates—more or less sciaphilic. (ii) Since excessive illumination is detrimental to most phytoplankters and results in photo-inhibition, maximum phytoplankton abundance usually occurs at a depth corresponding to about one third of the incident radiant energy. (iii) Motile species, such as dinoflagellates and naked flagellates, can adjust the depth inhabited to their specific light requirements.

Point (ii) deserves further attention in context with the concept of maximum production P_{max}. Where the illumination is weak, production is proportional to radiant energy;

further increase in illumination leads to light saturation corresponding to P_{max} (often also called P_{sat}); beyond the threshold of light saturation (Volume I: HELLEBUST, 1970), plant production decreases with increasing illumination. The ratio of the production P for a given level of light energy to P_{max} is the 'relative photosynthesis'. Appropriate curves (Fig. 4-1) based on data from various authors who experimented with members of the three main phytoplankton groups (green flagellates, diatoms and dinoflagellates) make it clear that: (i) light saturation occurs at a very low energy level in green flagellates, but at a maximum level in dinoflagellates, while diatoms occupy an intermediate position; (ii) the diatom curve reveals a marked plateau; this means that these algae tolerate variations of about twice the light energy received without exhibiting any significant change in their P_{max} value.

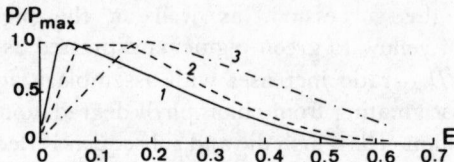

Fig. 4-1: Relative photosynthesis in the most common phytoplankton groups. 1: green flagellates; 2: diatoms; 3: dinoflagellates; E: light energy in cal g^{-1} cm^{-2} min^{-1}. (Original.)

Nutrients

The influence of nutrient contents on phytoplankton abundance and distribution has been briefly referred to above. Either nitrogen or phosphorus must be considered as the most important limiting factor; this uncertainty probably corresponds to different conditions of mineral nutrient input, renewal and recycling in different regions of the World Ocean. For example, in some tropical regions, Cyanophyceae (*Trichodesmium*) capable of utilizing molecular nitrogen may enrich the water with various combined forms of this element and, therefore, phosphorus will act as limiting factor; in contrast, coastal waters which receive domestic sewage contain large amounts of phosphorus and, consequently, nitrogen may act as limiting factor. In fact, while the amount of nutrient salts is taken into account, the nitrogen/phosphorus (N/P) atomic ratio is about 15 to 16, i.e. the same as that in living phytoplankters; N/P ratios between 10 and 20 appear to be required for adequate development of phytoplankton assemblages where diatoms usually predominate. Such conditions occur in all areas in which nutrients in the euphotic layer stem from mineralization of decaying plankton material, for example, either where dynamic processes (divergence, upwelling, doming) occur or in temperate seas after the winter mixing. This makes available in spring the nutrients originated from mineralization of organic material produced by plankton populations of the preceding year in the surface layers.

In areas in which most mineral nutrients originate from continental environments, the N/P ratio often differs considerably from the optimum value. In general, land drainage or unpolluted river input usually tend to augment the N/P ratio; the consequences of such change for phytoplankters are still insufficiently investigated, especially with regard to values over 25 to 30. Shelf areas receiving water from polluted rivers or domestic sewage from large coastal cities usually exhibit very low N/P ratios (sometimes below 1); in such eutrophic—or better 'dystrophic'—areas, phytoplankton assemblages are quantitatively rich (large biomass) but the specific diversity decreases. Many phytoplankters characteristic of dystrophic areas utilize organic substances instead of carbon dioxide as carbon source; this leads to a selection of species maximally adapted to polluted environments which are usually characterized by a tremendous multiplication rate.

In different regions of the World Ocean, differences in phytoplankton abundance and turnover depend largely on the availability of nutrient salts and only to a small extent on the ratio nitrogen/phosphorus compounds. Silica, though not a true nutrient, may also be a limiting factor in diatom populations. Variations in the combination between nutrients and illumination can cause striking differences in phytoplankton abundance between oceanic regions in which both these factors reach limiting values. For example, the highest latitudes always feature plenty of nutrient salts, but photo-autotrophic phytoplankters develop only during a brief period of maximum illumination, i.e. from late spring to summer. In intertropical offshore surface waters, on the other hand, radiant energy always abounds but strong, low-depth thermoclines—except in divergent areas—prevent rising of nutrients from subsurface or intermediate waters into the euphotic layer; consequently, the local phytoplankton is characterized by low, almost constant biomass values. In areas with highly transparent waters, the thermocline may be located above the compensation depth. This supports high phytoplankton abundance just below the thermocline; motile organisms such as dinoflagellates are particularly able to make the best of such a situation.

Nutrient input into the euphotic layer from subsurface or intermediate waters is not generally continuous; it often underlies periodic variations—pluriannual or of shorter intervals. For example, the intensity of coastal upwellings and doming depends on wind force which, in turn, depends on large-scale meteorological phenomena. River floods, which induce nutrient enrichment of coastal areas, generally exhibit a fairly regular occurrence related to rainy seasons; their influence on phytoplankton depends on the magnitude and frequency of the rainfall. In temperate regions the winter cooling of surface waters leads to vertical water mixing. Providing radiant energy and some other essential parameters—such as temperature (see p. 69 and Volume I: GESSNER, 1970, p. 394)—remain adequate, input of nutrient salts into the euphotic layer causes a bloom.

In general, the kinetics of nitrogen uptake are used for determining the rate of photosynthetic carbon assimilation applying Dugdale's expression

$$\frac{V_{max} N}{K_s + N}$$

where V_{max} is the maximum rate of nitrogen uptake in the absence of nutrient limitation, N is the concentration of inorganic N, and K_s is the half-saturation coefficient, i.e. the nutrient concentration at which uptake is half the maximum rate (STEELE and FROST, 1977; see also Volume IV: FINENKO, 1978, p. 23).

Other Dissolved Substances

In addition to nutrient salts, other dissolved substances (Volume I, Chapter 10; Volume IV, Chapter 4) may control phytoplankton multiplication. Some of them are heavy metals, for example, iron: it is well documented that delimitations between neritic and oceanic phytoplankton assemblages are strongly correlated with the iron content of sea water, since neritic species require more iron than oceanic ones.

Organic substances involved in biochemical interactions, such as chelators, vitamins (cyanocobalamine, thiamine, biotine, etc.) and, more generally, growth-promoting or growth-inhibiting substances, are also very important (Volume I: COLLIER, 1970, p. 72; Volume III: KINNE, 1977, p. 616). Consequently, successions of phytoplankters may be related not only to mineral nutrients, but also to prevailing levels of growth-promoting or growth-inhibiting substances. For example, during the French 'Antiprod' cruise in the Antarctic divergence in 1977, phytoplankton was rather poor and primary production very low, even though nutrient salts, light and temperature—as was supported by experiments—could not be considered as limiting factors. This suggests that growth-promoting substances may have been missing or growth-inhibiting substances may have been present.

Temperature

Temperature (Volume I, Chapter 3) exerts a paramount effect on metabolic rate; in particular it influences rates of photosynthetic assimilation and respiration (Volume I: GESSNER, 1970, p. 394). A delay sometimes observed at the beginning of a plankton bloom, in spite of abundant nutrients and adequate radiant energy, has been attributed by some authors to low temperature. Since carbon assimilation rate increases more rapidly than respiration rate in the lower part of the species-specific temperature range, we may suppose that a temperature which is rather low favours assimilation more than respiration, and that the reverse is true for temperatures which are too high. In fact, in temperate or warm seas low temperature in surface waters usually originates from a dynamic process that results in a rather marked instability in the water column, the latter being probably more responsible than the temperature *per se* for the delay in the phytoplankton outburst.

Salinity

In oceanic waters, salinity does not seem to be an important factor in phytoplankton distribution and abundance. For details consult Volume I, Chapter 4.

Water Movement

In phytoplankton, water movement (Volume I, Chapter 5) tends to affect population density as well as assemblage composition. The changes induced depend on the original state of the plankton assemblage (pp. 68, 87, 219) at the onset of transport as well as transport duration and distance. Disjunction in space of the different assemblages involved in the time succession (pp. 68, 87; Fig. 6-26) results from the increasing average periods of time which the populations of the constituting species require to develop, proportional to the system's age.

On a large space scale, long-range horizontal drifting—e.g. from the starting point of the Kuroshio current off the Philippines (latitude 10–15° N) to a latitude of about 50° N—involves large changes in several environmental parameters, most significantly temperature. Progressive temperature decrease results in gradual disappearance of thermophilic species and simultaneous abundance increase of eurythermal species—also of psychrophilic and/or opportunistic neritic species entering the Kuroshio water mass if the latter contacts or mixes with coastal waters or the Oyashio water mass.

In temperate regions vertically descending water movement originating from autumn or early winter gales causes a dilution of nutrients in the euphotic layer due to subsurface water influence. Episodic but much more marked and influencing deeper waters is the gliding of large volumes of surface water, where the isopycnes are unequally disturbed by turbulence in adjacent areas. For example, in March 1970, on a transect through the Gulf of Lion from the French coast (Fig. 4-2, Stn 1) to the south (Fig. 4-2, Stn 5,6), the turbulence was gentle inshore, i.e. in the southern and northern parts of the transect, but strong offshore (Fig. 4-2, Stn 2); hence, it generated an unbalanced density distribution ($\sigma_t > 29$ offshore; $\simeq 28 \cdot 7$ inshore). The return to balanced conditions, through gliding of heavier water along the isopycnes, explains the occurrence at 200 to 300 m of a large lens of water with a chlorophyll a content exceeding $0 \cdot 2$ mg m^{-3} (NIVAL, 1976).

Effects on phytoplankton abundance and distribution due to distortions in intermediate waters, such as divergence, upwelling and doming, are based on a combination of nutrient input into the euphotic layer and instability generated by the upheaval itself. An example is the doming of the Ligurian Sea almost in the middle of a transect from Nice to Calvi (Corsica) repeatedly investigated by French oceanographers. Winter conditions lead to the formation of Mediterranean deep water thus indirectly affecting the spring bloom: while the nutrient salts introduced in some areas (Fig. 4.3) should support abundant phytoplankton and high primary productivity, turbulence partly counteracted such beneficial effect and phytoplankton cells could not multiply before attainment of stability in time or space. As exemplified in Fig. 4-4, phytoplankton development in winter is restricted to areas which are marginal relative to the focus of doming, and consequently becomes sufficiently stabilized. In April, progressive warming of the surface layers increasingly stabilizes the water masses; this process starts in peripheral areas and proceeds towards the doming focus; it is accompanied by increasing phytoplankton abundance.

In the euphotic layer, phytoplankton abundance from August, 1962, to August, 1964, especially in 1963, corresponds to the phytoplankton growth in marginal—i.e. coastal—areas early in the year (Fig. 4-5). Progressive stabilization of the water mass gradually causes the zone of maximum phytoplankton abundance to move increasingly offshore to its summer location in the doming area itself. According to NIVAL (1976) phytoplankton growth would be maximal (up to 2 mg m^{-3} chlorophyll a) for a stability, expressed by $\Delta\sigma_t$(0 to 100 m), of 100 to 150 $\times$ 10^{-3}, such a chlorophyll content being approximately about ten times higher than that usually encountered in summer in Mediterranean surface waters.

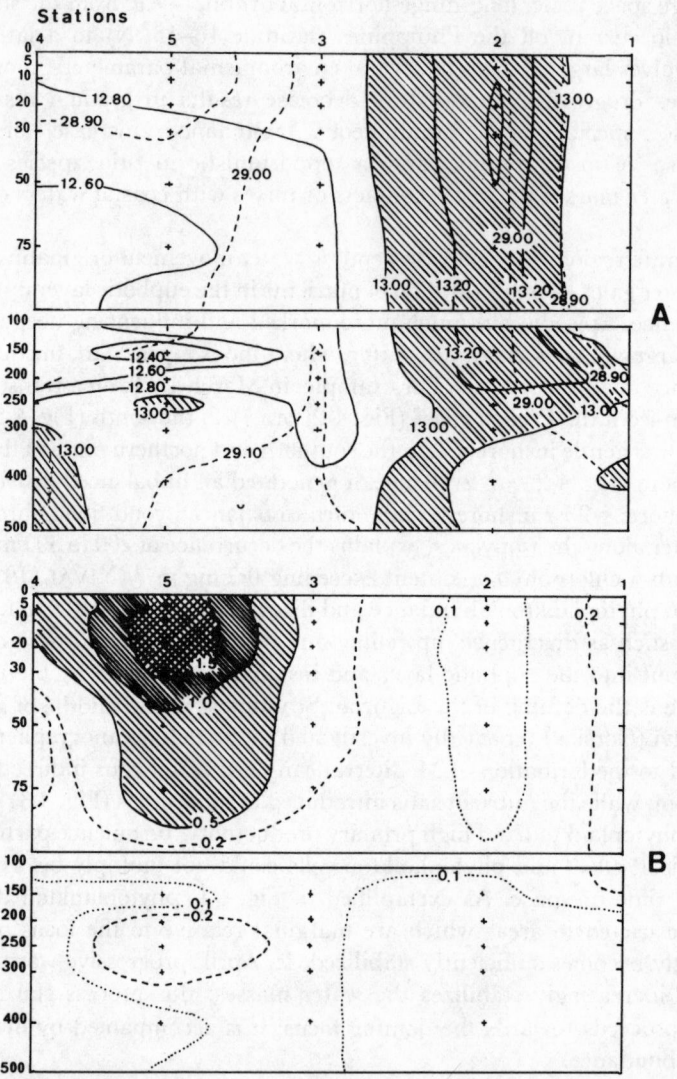

Fig. 4-2:   Transect in the Gulf of Lion, Mediterranean Sea, in March 1970. A: temperature (°C) and density (6). B: chlorophyll *a* (μg l^{-1}). For comments see text. (After NIVAL, 1976; reproduced by permission of the author.)

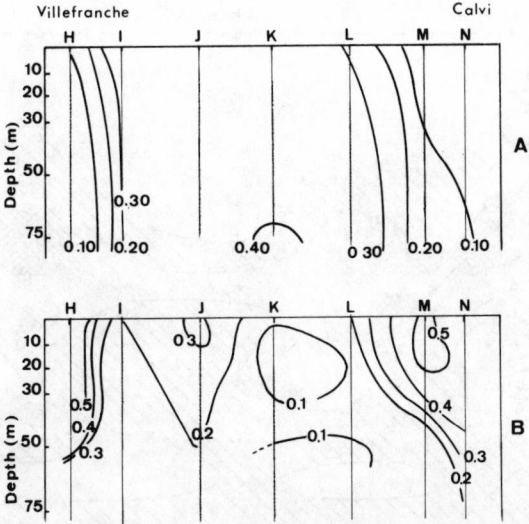

Fig. 4-3:   Repartition of PO_4–P (μg at l^{-1}) and chlorophyll a (mg m^{-3}) on the transect Villefranche (near Nice) to Calvi (Corsica) in March 1963. Offshore surface waters, although rich in nutrients, do not exhibit a high chlorophyll a content; coastal waters, while poorer in nutrients, are more stable and contain much more chlorophyll a. (After NIVAL, 1976; reproduced by permission of the author.)

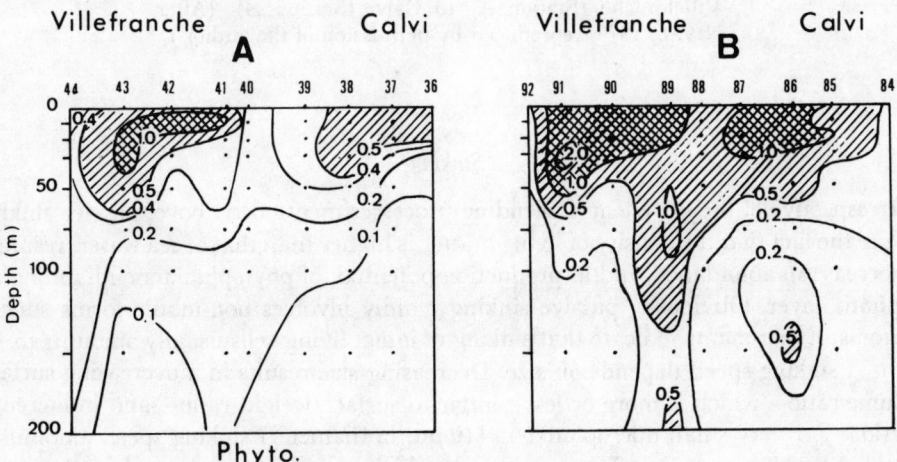

Fig. 4-4:   Phytoplankton biomass (expressed in chlorophyll a mg m^{-3}) in the 0 to 200-m layer, on a transect from Villefranche to Calvi. A: March 13, 1959; B: April 15, 1969. (After NIVAL, 1976; reproduced by permission of the author.)

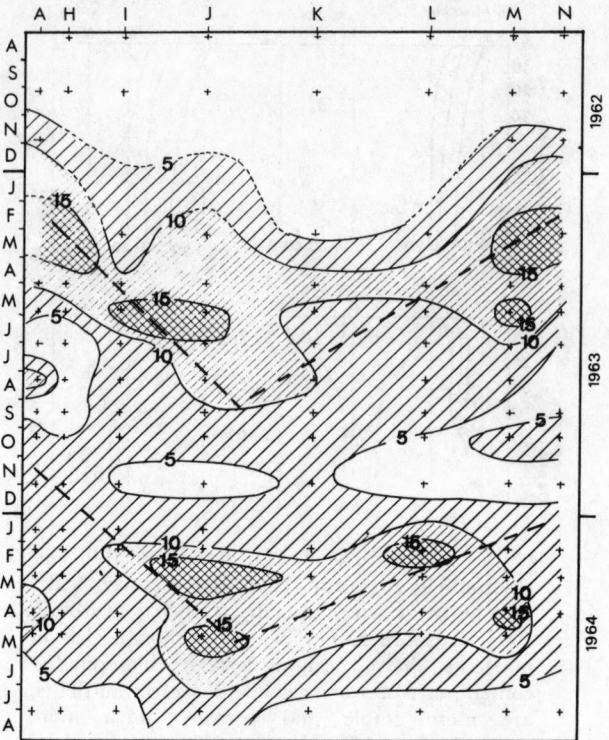

Fig. 4-5: Phytoplankton repartition (mg chlorophyll
a m^{-2}) in the euphotic (0–75-m layer) on a transect from
Villefranche (Station A) to Calvi (Station N). (After
NIVAL, 1976; reproduced by permission of the author.)

Sinking

Irrespective of the turbulent descending processes mentioned above, passive sinking,
due to the fact that the density of living matter is higher than that of sea water, results in
a decrease in abundance, and in production potential, of phytoplankters inhabiting the
euphotic layer. Obviously, passive sinking mainly involves non-motile forms such as
diatoms. Experiments indicate that sinking of intact living cells usually amounts to 1 to
10 m^{-1}; sinking speed depends on size. Decreasing size results in a decreasing surface/
volume ratio—which is more or less similar to surface/weight ratio—and in increased
friction. For very small microplankters (10 μm in diameter) sinking speed amounts to
1 m d^{-1}. Sinking can also affect growth rate which may, in turn, depend on the rate of
nutrient uptake. According to STEELE and FROST (1977), nutrient depletion in the
water near the cell may, in nutrient-poor waters, decrease the rate of nutrient uptake;
this phenomenon may be compensated for by the cell sinking through the water.

The maximum water depth attained by sinking living cells depends on hydrologic
features: a strong thermocline occurring at moderate depths (40–80 m)—often observed

in intertropical regions—can stop the sinking, and the cells can survive if the thermocline lies above the compensation depth. Such a 'conservative mechanism' of a strong thermocline is essential in oligotrophic areas. In such areas, matter and energy losses in the homogeneous surface layer due to sinking can be compensated for only through fast assimilation of nutrients from recycling processes occurring within the layer itself.

Where no thermocline (or only a very weak one) occurs, sinking cells quickly die below the compensation depth. This is proved by the fact that 'chlorophyll concentrations at 1·3 to 3 times euphotic zone depths are exceedingly low' (LORENZEN, 1976, p. 180). In such a case, not only does the energy represented by sinking phytoplankton fail to benefit consumers below a few hundred metres, but also the surface layers become poorer and poorer in nutrient salts due to deficiency in the recycling process. Therefore, phytoplankton populations can be maintained there only through nutrient input from outside, thus allowing 'new' production to substitute for the production from recycling.

Grazing

Taking into account both phytoplankton populations and the impact upon consumers, herbivore grazing significantly affects phytoplankton abundance and distribution and consequently requires special treatment.

(3) Phytoplankton Populations and Zooplankton Grazing

(a) General Aspects

The term 'grazing' refers to the intake of particulate food by zooplankters. The most important pertinent feeding types are briefly summarized below (see also Volume II: PANDIAN, 1975, and Volume IV: CONOVER, 1978, p. 246 ff): (i) Filtration of a water current without the assistance of a sticky secretion, e.g. in many copepods and euphausiids, especially those feeding on small particles. In these crustaceans, different appendages generate the water current and act as filtering mechanism. A similar feeding type is known from appendicularians, the water current being generated by tail-beating within the gelatinous lodge inhabited. (ii) Filtration of a water current generated by ciliary activities which direct the particles either directly to the mouth—e.g. in young larval stages of most molluscs and echinoderms—or to a pharyngeal chamber where they become agglutinated due to a sticky, mucous secretion before being filtered through the branchial screen, e.g. in all the Thaliacea. (iii) Raptorial feeding: the grazer crushes the large-sized cells with its mandibles and subsequently ingests the fragments, e.g. in several copepod genera. Some copepods are able both to filter, but less perfectly than species of group (i), and to catch individual cells, their feeding strategy depending on the size of the food particles present.

The preceding classification of grazers only takes into account metazoans. Phagotrophic protozoans, such as tintinnids and some colourless dinoflagellates, probably contribute to control of phytoplankton abundance. Usually 20 times more abundant than these two groups, ciliates play a still more important role. However, on the following pages we concentrate on filtering forms—usually considered as mesozooplankton—and particularly on copepods which are by far the best investigated group as far as grazing is concerned. Phagotrophic species are discussed separately (p. 90).

In addition to phytoplankton, particulate food in the pelagial comprises aggregates composed of bacteria living on bubbles and utilizing dissolved organic matter as well as organic debris surrounded by decomposing bacteria (Volume IV: CONOVER, 1978, p. 241; SOROKIN, 1978). Bacterioplankton consists almost exclusively of such attached bacteria which may represent important food for the grazers. In the Azov Sea, for example, the copepods *Acartia clausi, A. latisetosa, Centropages kroyeri* and larval stages of *Balanus* feed almost exclusively on phytoplankton (*Cyclotella caspia, Exuviella cordata* and *Prorocentrum micans*) in summer, whereas in spring and summer when these phytoplankters become less abundant, their diet mostly consists of aggregates (KUDELINA and ZHULAREVLA, 1963). True grazers—i.e. filter feeders—obtain their energy (in addition to grazing on phytoplankton and aggregates) also from small-sized consumers such as ciliates and even very small larval stages (e.g. nauplii) as long as these are larger than the mesh size of the filtering device.

(b) Grazing Effects on Phytoplankton Populations

Most investigations on zooplankton grazing deal with crustaceans, especially copepods (Volume IV: CONOVER, 1978, p. 246 ff.) and euphausiids. Both these groups are filter feeders. The volume of the surrounding water cleared from suspended particles is known as 'filtered volume' (V). It corresponds to the product of water-current speed generated and the surface area of the filtering device, as well as to the energy spent for food collection (NIVAL, 1976). Since experiments testify that the food available is often not completely consumed, the concept of 'apparent filtered volume' (V_a) has been introduced. V_a is always lower than V. The ration (R) is the weight of food consumed; to some extent it depends on food density. Since the ration cannot increase indefinitely in proportion to increasing food biomass (B), there is a maximum ration (R_{max}), i.e. the highest ration an animal can collect and/or digest within a given unit of time.

In most experiments, cell densities were used which were much higher than those usually occurring in the natural environment. In spite of the rather pronounced variability in the grazer's response to experimental prey density changes, we may assume that changes in filtration rate, in relation to cell density, attain a maximum around values which represent optimum prey exploitation. In suboptimal densities, filtration rate sharply decreases with decreasing cell density, while in higher densities the decrease in the filtration rate is much slower and exhibits a narrower range of variability (Fig. 4-6). These facts suggest that copepods are, to some extent, able to adjust their food-collecting strategy to changes in prey density. The adjustments appear to aim at optimizing the energy budget.

The energy budget may correspond to optimal expenses, i.e. $E_i = E_c + E_m + E_u$, where E_i is ingested energy, E_c energy spent for collecting particles, E_m energy required for maintenance of the general structures of the organism, E_u energy invested in reproduction or accumulation of nutritive reserves. The energy budget may also just balance the minimum expenses: $E_i = E_c + E_m$. The amount of food caught in the filtered volume (V) per unit time supplies energy E_i, which is the product of the assimilable energy within a particle (ε) and concentration (N) of these particles. Then, food density (B) is $B = \varepsilon N$ and $E_i = BV$.

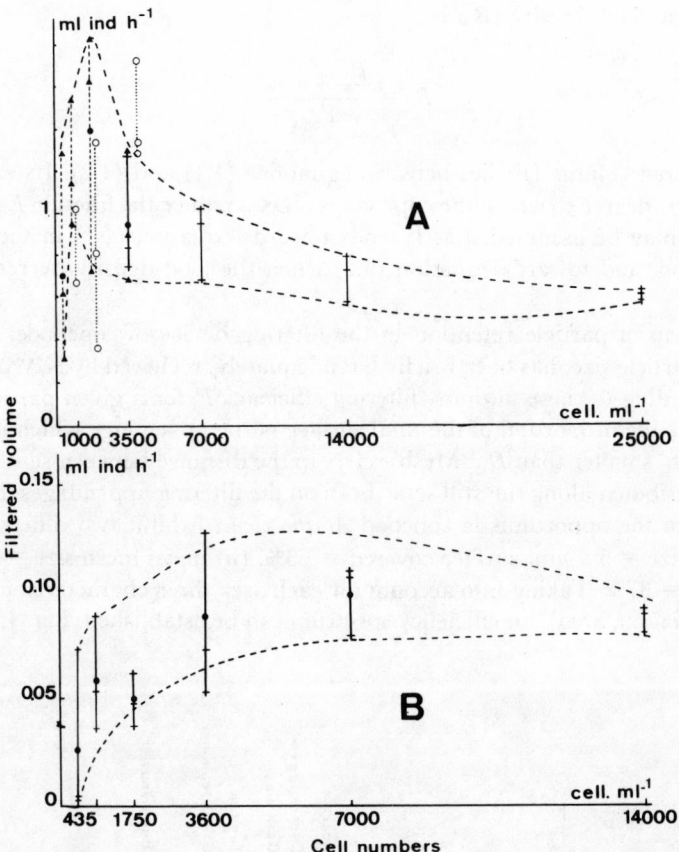

Fig. 4-6: Changes in volume filtered by one individual of *Temora stylifera* (A) and *Acartia clausi* (B) in relation to food abundance (diatom *Skeletonema costatum*). There is a hyperbolic relation above 3500 cells ml^{-1} for *T. stylifera*. (After NIVAL, 1976; reproduced by permission of the author.)

The energy a copepod spends for filtration increases with increasing filtered volume. Assuming a laminar flow through the appendages' setae of the filtering device, $E_c = KV$; where the energy budget is balanced, $BV = KV + E_m + E_u$. Therefore, the value of

$$V = \frac{E_m + E_u}{B - K} \qquad (4.1)$$

agrees with the hyperbolic curve in Fig. 2-6A. If the copepod restricts its energy expenses to the minimum consistent with survival, the filtered volume is

$$V = \frac{E_m}{B - K} \qquad (4.2)$$

and the critical food density (B_c) is

$$B_c = \frac{E_m + V_{max}}{V_{max}} \tag{4.3}$$

The real filtered volume (V) lies between equations (4.1) and (4.2). Its exact location depends on the degree to which the copepod is able to reduce the fraction E_u of its energy expenses. It may be assumed that V tends towards equation (4.1) in the presence of abundant food, and towards equation (4.2) when the food density decreases (NIVAL, 1976).

The problem of particle retention in the filtering device of copepods, mainly as a function of particle size, has been briefly but adequately, reviewed by NIVAL and NIVAL (1976). According to these authors, filtering efficiency F_i for a given particle size D_i is determined as the proportion of the total surface of the appendage which is covered by areas of 'mesh' smaller than D_i. 'Mesh' refers to the distance between setules which are regularly distributed along the stiff setae born on the filtering appendages. For example, appendages of the opportunistic copepod *Acartia clausi* exhibit two different areas: (i) mean mesh size = $5\cdot7$ μm; surface covered = 63%; (ii) mean mesh size = $9\cdot2$ μm; surface covered = 37%. Taking into account for each area three characteristics (mesh size, standard deviation, area), an efficiency spectrum can be established. Fig. 4.7 shows that

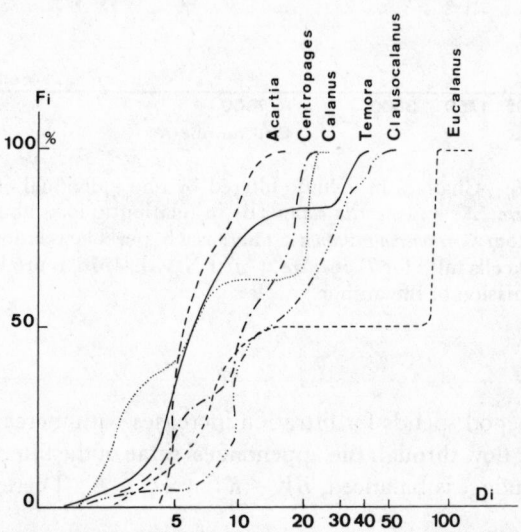

Fig. 4-7: Filtration efficiency (Fi) in copepods. Efficiency curves for maxillae and maxillipeds. Variability not taken into account. Di: average diameter of food particles. (After NIVAL and NIVAL, 1973; reproduced by permission of the American Society of Limnology and Oceanography.)

A. clausi filters most efficiently particles between 5 and 10 μm; *Clausocalanus* is more efficient than *Centropages* for particles smaller than 8 μm whereas the reverse is true for particles between 8 and 40 μm. Therefore, the abundance of the two grazers may be related to the abundance of particles in either of these size classes.

The amount of food collected by a grazer (the ration) depends on the efficiency of its filtering appendages and their total surface. According to NIVAL and NIVAL (1973), for *Acartia clausi* and *Eucalanus elongatus*, the filtering surface of the two pairs of appendages—expressed as $cm^2 . 10^{-4} - mx2 + mxp$—falls between 14·32 and 206·94, respectively. In the natural environment, the ration is further related to the size spectrum of the particles present. For example, in waters with an identical particulate-matter biomass, *Euphausia pacifica* will collect more food if particle mean size is 32 μm than if it is only 8 μm.

In many cases, phytoplankton outbursts coincide with the breeding period of a grazer; hence larval stages may coexist with adult individuals. NIVAL and NIVAL (1976) estimated filtering efficiency across a spectrum of particle sizes for adult and copepodite stages of *Acartia clausi*. From measurements of the mesh size between setules of the maxillae, they computed that the average size of the smallest particle retainable increases with age—ranging from 3 μm in Copepodites I to 7 μm in the adults. The particle size for which the filtering efficiency is theoretically 100% ranges from 5 μm in Copepodites I to 12 μm in adults. This corresponds to the general slope of the spectra illustrated in Fig. 4-8. The computation was borne out by a grazing experiment on a

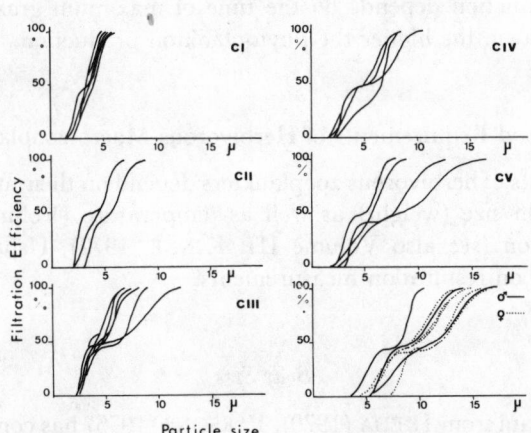

Fig. 4-8: *Acartia clausi* spectra of filtration efficiency (computed from the measurements of setae) for each developmental stage. Each line represents the spectrum of one individual. Curves interrupted when reaching 100% efficiency. (After NIVAL and NIVAL, 1976; reproduced by permission of the American Society of Limnology and Oceanography.)

natural population of phytoplankton, mainly *Skeletonema costatum* (see also POULET, 1973).

In summary, the experiments conducted suggest the following:

(i) Phytoplankton diminution due to grazing probably cannot occur during the exponential phase of population growth, but only during a period of reduction in growth, for example, because of nutrient shortage.

(ii) Under certain conditions, grazing that reduces the growth of potentially dominant species may become a major factor in increasing diversity. For example, intensive grazing on a young stage of phytoplankton succession—mainly involving small-sized species—will tend to reduce the number of the latter, whereas populations of large-sized species, weakly exploited, will tend to increase. Such changes in the size spectrum subsequently result in increasing population strength of other grazing species with efficiency spectra better fitted to larger-sized food.

(ii) Optimum phytoplankton utilization is accomplished by coexisting zooplankters which graze on particles of different sizes, resulting in decreased interspecific competition. This in turn may explain both the time-scale succession and space-scale disjunction of fine and coarse filter feeders.

(iv) Discontinuous nocturnal grazing may increase phytoplankton production relative to a population continuously grazed day and night, due to better utilization of the growth potential of phytoplankters (less primary production lost to phytoplankton respiration; unimpeded plant growth during daylight) (for details consult discussion by LONGHURST, 1976, pp. 134–5). According to PETIPA and MAKAROVA (1969), discontinuous grazing by diel migrant herbivores results in a sinusoidal curve of phytoplankton abundance, whereas continuous grazing leads to parabolic decrease. The intensity of phytoplankton production depends on the time of maximum grazing: the closer this maximum to the dawn, the higher the phytoplankton production.

(c) Food Requirements of Herbivorous Macro-zooplankters

Food requirements of herbivorous zooplankters depend on their metabolic rate which, in turn, depends on size (weight) as well as temperature (Volume I: KINNE, 1970, p. 443) and nutrition (see also Volume III: KINNE, 1977). Usually, metabolic rate evaluation is based on respiration measurements.

Body Size

Tabulating the data from IKEDA (1970), PARSONS (1976) has compiled Table 4-1 for boreal zooplankton species, assuming a growth rate of 7% d^{-1} and a growth efficiency of 80%. The range of animal weight listed includes the difference between a small nauplius and the adult of a large-sized copepod, as well as representatives of species of different adult sizes. It shows that the smaller the grazer, the greater its food requirements. PARSONS (p. 92) concludes:

'The disadvantage of being small in terms of growth efficiency may be in part offset by a greater reproductive ability among small animals ... the amount of energy devoted to reproductive tissue (generative growth) is much greater per unit body weight among small species compared with larger ones.'

Table 4-1

Effect of body size on metabolic rate, growth and food intake of boreal zooplankters, assuming an assimilation efficiency of 80% (Based on data from IKEDA, 1970; reproduced by permission of Hokkaido University)

Animal wet weight (mg)	Metabolism	Growth (assumed)	Total	Food intake	Growth efficiency (K_1 %)
		(% of body weight d^{-1})			
0·005	27	7	34	42	17
0·05	13	7	20	25	28
0·5	6·4	7	13·3	16·8	42
5·0	2·8	7	9·8	12·3	57

Temperature and Nutrition

Within the thermal range tolerated, metabolic rate tends to increase (Volume I: KINNE, 1970, p. 322). Zooplankton respiration rate per unit body dry weight is higher in areas with abundant phytoplankton than in nutritionally poor areas. In the presence of large amounts of food, grazers are more active, while food shortage induces minimization of energy expenses (NIVAL, 1976). The concentration of particulate organic carbon required by herbivores of different sizes for supporting their metabolic requirements are listed in Table 4-2, assuming that filtration rate is directly related to the grazer's body size.

Table 4-2

Concentration of particulate organic carbon required to support the growth and metabolic requirements of different-sized zooplankton (Based on data from IKEDA, 1970; reproduced by permission of Hokkaido University)

Herbivore wet weight (mg)	Food intake (% of body weight d^{-1})	Animal C (μg)	Food intake (μg C ind.$^{-1}$ d^{-1})	Filtration rate (ml d^{-1})	Food conc. required (μg C l^{-1})
0·005	42	0·385	0·168	0·2	790
0·05	24	3·75	0·94	2·0	420
0·5	16·8	37·5	6·3	20·0	315
5·0	12·3	375	42·3	200·0	215

The food concentrations required by herbivores of different sizes represent rough approximations, possibly because the increase in filtration rate with increasing size is lower than admitted by PARSONS (1976). It depends on both the size of the filtering surface area which seems to increase more slowly than the body size and the speed of the water current flowing through the screen. PARSONS (1976, p. 93) also emphasizes that the calculated concentrations listed in the last column of Table 4-2 'appear to be very high when compared with the levels of particulate organic carbon reported in most oceanic areas', but they are of the same order of magnitude as the concentrations used in experiments. Presumably, sampling by large bottles—usually practised for evaluating chlorophyll content, i.e. phytoplankton abundance—provides underestimated data because phytoplankton distribution is very heterogeneous. Moreover, one may assume that grazers are capable of detecting microdistributions of algal cells within a volume corresponding to that scanned by the largest of their appendages, i.e. the antennulae (CUSHING, 1959).

Nature of Food Resources

The evaluation of biomass and production of phytoplankton in terms of carbon units induced parallel expressions of biomass and production in herbivores. However, a grazer neither eats nor assimilates carbon, but only different kinds of—usually living—organic matter. In different groups of phytoplankters, marked differences exist in the biochemical composition of their living matter; for example, diatoms contain much more carbohydrate than dinoflagellates which, in turn, contain a higher percentage of protein. Moreover, biochemical composition may be modified in the asymptotic phase of population growth compared to that during the exponential phase.

The recently introduced method for comparing biomasses of herbivores and carnivores by measurements of major digestive enzymes (amylases and proteases) represents an important advance for investigating zooplankton assemblages. Further progress requires discrimination of successive populations and assemblages.

An example of such a succession in copepods with different feeding habits has been described by MARGALEF (1967) for the western region of the Mediterranean Sea. MARGALEF distinguishes three groups of copepods: (i) species which are mainly herbivorous, e.g. those of the genera *Oithona*, *Metridia*, *Acartia* and *Calanus*; (ii) species which must be considered omnivorous since the vegetal fraction in their diet decreases with age, e.g. those of the genera *Centropages*, *Temora*, *Gaidius* and some *Euchaeta*; (iii) raptorial species (food: either large-sized phytoplankters or other zooplankters), sometimes strictly carnivorous, e.g. of the genera *Candacia*, *Tortanus*, *Oncaea* and some *Euchaeta*.

Stage I of the phytoplankton succession (p. 68), which includes small-sized flagellates and diatoms (*Skeletonema costatum*, *Rhizosolenia delicatula* as well as several species of the genera *Thalassiosira* and *Chaetoceros*) is mainly grazed on by small-sized animals of group (i), for example, *Oithona* sp. stage II in which *Chaetoceros* species largely predominate, associated with dinoflagellates (*Prorocentrum*, *Peridinium*, *Ceratium*), is mainly exploited by a mixture of copepods belonging to groups (i) and (ii), such as species of *Centropages*, *Calanus* and *Paracalanus*, which also feed upon ciliates, the latter being fairly abundant at this stage. Copepods of group (iii), mixed with some members of group (ii) (e.g. *Temora*), exhibit a life cycle which allows their adults to feed on Stages III and IV of MARGALEF's phytoplankton succession: at Stage III, on the largest diatom species of the genera *Coscinodiscus*, *Rhizosolenia* and *Bacteriastrum* as well as on dinoflagellates; at Stage IV, on large-sized dinoflagellates (p. 68) which predominate together with coccolithophorids and sometimes Cyanophyceae (diatoms became scarce: *Hemiaulus*, some species of *Rhizosolenia*). At Stages III and IV it seems that the utilization of algae becomes strongly reduced; the zooplankters feed mainly on ciliates, as well as on nauplii and other larval stages.

(4) Changes in Relative Abundance of Phytoplankton and Zooplankton

Phytoplankton and zooplankton populations may be directly or inversely proportional to one another. In the former case, the zooplankton maximum coincides with that of the phytoplankton (Figs 4-9, 4-10(f)), whereas in the latter it is more or less delayed in comparison with the phytoplankton peak. Occurrence of the zooplankton maximum before the phytoplankton bloom is quite exceptional (Fig. 4-10(g)). Irrespective of the circumstances in which zooplankton scarcity during a bloom may be ascribed to the release of repellent substances by phytoplankters, inverse relationships can only originate either from grazing or from differences in the growth rate of plants and animals (Fig. 4-10).

Grazing influence on phytoplankton populations has been discussed above and some additional aspects will be analysed below, but differences in the growth rate of phytoplankton and zooplankton certainly control their fluctuations in abundance much more than grazing. According to NIVAL (1976), growth rate of adult copepods is low ($r = 0.00278\ h^{-1}$) even at maximal ration, whereas phytoplankton growth rate is high ($K = 0.01\ h^{-1}$). Under certain circumstances, phytoplankton generations can follow one another in a daily rhythm. In herbivorous copepods, generation time is at least four weeks, sometimes longer. In carnivorous zooplankters, both development to the adult stage and life-span are still longer than in herbivores. Moreover, the time lag required to reveal a demonstrable response to a change in environmental conditions amounts to hours in phytoplankton, days in herbivores and weeks in carnivores.

Obviously, the causative basis and starting point of the dynamics of a pelagial assemblage are changes in phytoplankton abundance and composition in response to time or space variations of the parameters analysed above.

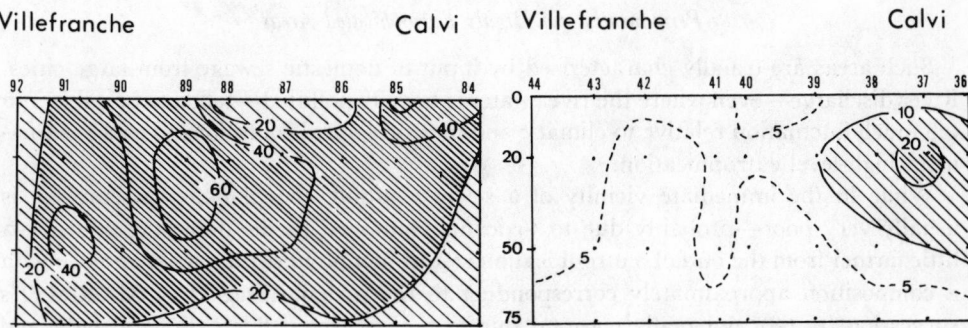

Fig. 4-9: Coincidence of phytoplankton (left; see Fig. 4-4) and zooplankton (right) maxima (mg m^{-3} d.w.) on the transect from Villefranche (Station A) to Calvi (Station N). Zooplankton endures the same constraints from turbulence as phytoplankton. (After NIVAL, 1976; reproduced by permission of the author.)

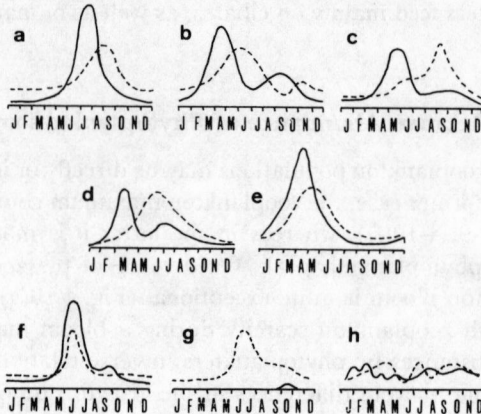

Fig. 4-10: Seasonal changes in relative biomass
of phytoplankton (solid line) and zooplankton
(broken line) in various regions of the world
ocean. a: Arctic Seas north of Siberia;
b: neritic areas of the Norwegian Seas;
c: Baltic Sea; d: Japan Sea; e: tropical neritic
area where a rainy season results in only one
annual bloom; f: neritic areas of the Bering
Sea; g: oceanic areas of the Bering Sea;
h: offshore tropical areas. (After HEINRICH,
1962; modified; reproduced by permission of
the author.)

(a) Effects of Environmental Factors on Phytoplankton–Zooplankton Relationships

Permanently and Highly Eutrophicated Areas

Such areas are usually characterized by input of domestic sewage from large cities.
River discharge—even where the river water is heavily polluted—tends to be subject to
too much fluctuation relative to climatic and meteorological conditions to cause permanent, high-level eutrophication.

While in the immediate vicinity of a sewer's outfall, the plankton assemblage is
usually very poor—probably due to toxicants contained in the domestic effluent—a
little farther from the outfall, eutrophication results in a phytoplankton assemblage with
a composition approximately corresponding to the juvenile Stage I of MARGALEF's
succession (p. 68); not really constant due to both seasonal fluctuations in light and
temperature and dispersion by currents and casual gales, species abundance remains
very high, species diversity low. Turbidity generated by suspended sewage material and
high phytoplankton abundance decreases compensation depth. Herbivorous zooplankters also reveal a low degree of diversity; their abundance is augmented but less so than

that of phytoplankton. Energy transfer from the first to the second level of the trophic pyramid is low, leading to underconsumption of phytoplankters and, in turn, to increased sinking of dying or dead cells. The latter sometimes causes a lowering, if not depletion, of the oxygen content in the subsurface layers and on the bottom. The third trophic level, i.e. carnivores feeding on herbivores, is usually missing or poorly represented. In these permanently and highly eutrophicated areas, the phytoplankton–zooplankton relationships are usually direct.

Oligotrophic Areas

In oligotrophic areas, e.g. in subtropical latitudes, the upper homogeneous and warm water layer is sharply separated from the subsurface layers by a permanent thermocline which inhibits upward diffusion of nutrient salts from infrathermoclinal depths. Thus, mineral nutrients available to phytoplankters stem from recycling processes which are intensified by high temperature. Qualitatively, the local phytoplankton is highly diversified (dinoflagellates and coccolithophorids predominate as in Stages III–IV of MARGALEF's succession); quantitatively it is very poor. There are only slight changes in abundance and biomass on an annual scale (Fig. 4-10(h)). In the homogeneous layer there are usually two maxima of phytoplankton abundance. The first maximum, at depths of 5 to 10 m, is due to optimum radiant-energy levels—assimilation in the uppermost layers being diminished due to photo-inhibition—enhanced by high temperatures (p. 72). The second maximum lies in the vicinity of the thermocline: either just above it, resulting from nutrient-salt enrichment caused by weak, occasional diffusion processes through the thermocline; or just below it, where the homogeneous layer is sufficiently transparent to allow the euphotic layer to extend beyond the thermocline. In oligotrophic areas the zooplankton is likewise quantitatively very poor, but it is highly diversified; while herbivores are somewhat scarce, carnivores are abundant. The resulting assemblage appears to be well balanced and highly stable; phytoplankton–zooplankton relationships are direct. A comparison between eutrophic and ultraoligotrophic areas regarding the energy transfer from the first to the second level of the trophic pyramid supports CUSHING's assertion that transfer coefficients between primary and secondary producers diminish from about 18% to 4% with increased primary production from about $0 \cdot 1$ to 2 g C $m^{-2} d^{-1}$ (quoted by PARSONS, 1976, p. 95).

Limited, Generally Seasonal, Nutrient Enrichment

Such enrichment may arise, for example, from: (i) river floods; (ii) vertical mixing; (iii) coastal upwelling.

(i) River floods due to pronounced rainy seasons initially cause conditions similar to those induced by domestic sewage, the starting point being a phytoplankton development which corresponds to the first stage of MARGALEF's succession. However, input of nutrient salts from the river more or less quickly diminishes as the flood decreases. Simultaneously, marine phytoplankters contribute to the overall decrease in nutrients. Both phenomena combine to induce, in the vicinity of the river's mouth, a temporal succession of phytoplankton and zooplankton assemblages like that analysed by MARGALEF (1962, 1967; see also pp. 68 and 84). This temporal succession results in spatial disjunction of successive stages as the water mass initially fertilized near the river's mouth drifts away due to coastal currents. In such a situation phytoplankton–zooplankton abundances are generally inversed.

Three additional points require attention: (a) The components of plankton assemblages become less and less abundant while drifting away from the river mouth due to progressive mixing of enriched waters with nutrient-poorer marine waters. (b) Shallowness and turbulence near the river mouth support the re-occurrence of resting stages of plankters temporarily deposited in sediments thus increasing species diversity. (c) An upper layer of fresh water may support freshwater plankters—especially diatoms—often in high abundance. These die and disintegrate progressively as the salinity increases. The decaying cells release chlorophyll which transforms into phaeopigments. Therefore, the content in phaeopigments of the waters in a dilution area cannot be considered, without qualification, to represent an integrated value of grazing and photo-oxidation on a time scale of a week or less as suggested by LORENZEN (1976, p. 181). Near the mouth of the Rhône, for example, most phaeopigments originate from decaying freshwater phytoplankters (SAUTRIOT, personal communication). In general, as far as zooplankters are concerned, the upper reduced-salinity layer is inhabited by euryhaline species, with the true marine surface plankters occupying somewhat deeper (e.g. 0·5 m) levels.

(ii) Vertical mixing due to winter cooling of surface waters is a most important phenomenon in temperate regions. It results in spring phytoplankton outbursts and subsequent zooplankton increase. To some extent, the autumn bloom—arising partly from recycling processes acting upon summer plankton and partly from turbulence generated by autumn gales—exhibits a pattern similar to, but less marked than the spring bloom (Fig. 4-10(b)).

In the spring bloom, which is always more pronounced, we may distinguish two different situations. (a) In shelf areas, winter mixing often involves the whole water column and therefore can return to the surface—which becomes warmer and more illuminated in spring—nutrients not only from subsurface layers but also from mineralization of organic detritus on the bottom. It further brings to the surface resting stages of phytoplankters and zooplankters deposited on the bottom since the end of the preceding year's plankton succession. Hence, in shelf areas the time succession from Stages I to IV and their spatial distribution by subsequent drifting are very similar to those near the mouth of a river (see above). Due to the differences in growth rate between phytoplankton and zooplankton, the abundance increase of the latter lags behind. Such temporal sequence also prevails in the autumn bloom. (b) In offshore areas the consequences of vertical mixing on phytoplankton outbursts are the same as in shelf areas. However, where water depth exceeds a few hundred metres, it is unlikely that resting stages are lifted to the surface layer once they had sunk below the lower boundary of the mixed layer. Nevertheless, especially in cold-temperate areas like the Norwegian Sea and the Antarctic region, many zooplankters have adjusted to overwintering. After intensive feeding and accumulation of nutritional reserves, most of the copepods, for example, migrate downward (copepodite Stage V), sometimes below 1000 to 1500 m. After 6 to 8 months they migrate upwards again in the following spring (adult stage) for feeding on a new phytoplankton bloom and for breeding. In this situation, the phytoplankton–zooplankton relation seems to be related mainly to the time required for the upward migration of herbivores relative to the time of phytoplankton outburst: in the Barents Sea the zooplankton peak is delayed for about one month relative to maximal phytoplankton bloom; in some areas of the Antarctic region, maximum copepod abundance from 0 to 10 m occurs early in spring, thus coinciding with the phytoplankton bloom. In oceanic areas of the Bering Sea, maximum zooplankton abundance precedes the phytoplankton peak, possibly due to significantly unbalanced ecological conditions.

Oversummering has been observed in the common copepod *Calanoides carinatus* by PETIT and COURTIES (1976) on the coasts of the Congo. Not a tropical but a warm-temperate species, *C. carinatus* was the only survivor of the warm season (copepodite stage V). Stage V comes up from 800 m to the surface layers of the shelf and the near offshore areas as soon as the cooling of the surface waters begins. Possibly, six generations follow one another from June to October during the upwelling season, but the formation of the copepodite stage V begins at the end of July and progressively intensifies with increasing temperatures.

Finally, brief meteorological events such as strong storms can also induce vertical mixing and thus small blooms, the subsequent evolution of which depends on both the importance of the mixing and the subsequent return to hydrological stability.

(iii) Upheaval of intermediate waters into the euphotic layer may be caused by doming, upwelling or divergence. The terms divergence and upwelling are often confused and considered to be synonymous, although the former should preferably be used for describing open-ocean processes and the latter for coastal, but not nearshore, wind-affected processes (i.e. 50–150 nautical miles from the shore). A divergence is a permanent but often pulsating process, whereas upwelling may be both seasonal or permanent—in the latter case always pulsating.

A permanent upheaval like that occurring, for example, in the equatorial divergence of the Pacific Ocean, generates a belt with a width of 5° to 10° in latitude, in which the organismic assemblage remains at, or near, the youngest (I) stage of MARGALEF's succession, but in a dynamic equilibrium oscillating in accordance with the pulsations of the upheaval (for details, see Chapter 6, p. 217ff). Since the latitudinal current exhibits two meridian components—opposite in direction, on both sides of the divergence line—one successively encounters while moving, increasingly northward or southward from the divergence, a more and more advanced stage of MARGALEF's succession (Figs 2-5 and 6-26) until a convergence line is reached which causes the sinking of the surface assemblage. Coastal upwellings exhibit the same space shifting of successive assemblages, but with an asymmetrical pattern related to the Coriolis acceleration which in both hemispheres tends to force upwelled waters further and further away from the shoreline; between upwelling and shoreline are inshore waters.

The three examples suggest that the balance achieved between the first two levels of the trophic pyramid clearly depends on the time lag of grazer appearance in an assemblage already enriched with phytoplankton. Oligotrophic systems which permanently include grazers seem to be balanced best. In systems with more or less periodic eutrophication, the less delayed the increase in grazers, the better the balance: a grazer population established from resting stages results in less efficient phytoplankton utilization than that of immigrants—moving upward after overwintering providing the immigrants reach the surface layer before the bloom declines. However, in excessively eutrophicated assemblages—often referred to as 'dystrophic' state—lack of balance does not come from a delay in herbivore appearance (these usually exist permanently in the system), but is due to insufficient grazer abundance relative to the phytoplankton mass present.

(b) Overgrazing

During pronounced phytoplankton blooms it has sometimes been observed that filtering grazers—e.g. copepods or euphausiids—can collect and destroy many more phytoplankters than they are able to ingest and assimilate. This 'overgrazing', also called 'superfluous feeding', has been analysed by BEKLEMISHEV (1966), and later contested by CONOVER (1966; see also Volume IV: CONOVER, 1978).

It is generally assumed that the food requirements of copepods approximately correspond to 10 to 14% of their own weight; overgrazing would correspond to a consumption of about 40%. It is further agreed that the number of faecal pellets a copepod can produce cannot exceed 2 to 3 h^{-1}. Based on experiments with *Calanus hyperboreus* offered diatoms, CONOVER (1966) demonstrated that the assimilation rate neither depends on food concentration nor on the food quantity ingested. Therefore, although overgrazing has been observed, the grazer cannot benefit from it.

The consequences of overgrazing for plankton assemblages are still being discussed. Obviously, organic matter from damaged cells is more quickly mineralized than that of undamaged, dead cells. This would tend to accelerate the recycling of nutrient salts. Dissolved organic substances released from damaged cells can be metabolized by bacteria epiphytic on aggregates. The latter, enriched with living matter, can subsequently be utilized as food by microphages.

Another, possibly positive, consequence of overgrazing would be the part played by faecal pellets as food for pelagic and benthic organisms. Coprophagy is a well-documented phenomenon in the benthal because faecal pellets always contain undigested organic material which can be converted by bacteria into living matter and thereafter be eaten by 'detrivorous' benthos invertebrates. It has also been assumed that faecal pellets of plankton organisms, the sinking speed of which is greater than that of smaller organic particles, represents a food source for macrozooplankters inhabiting deeper layers and/or benthos invertebrates providing the water column is not too high. Plankton consumers may be coarse filter feeders and raptorial species or fine filter feeders, in the latter case after the pellets have been divided into smaller fractions by formerly active species. However, as has already been mentioned, it is generally admitted that the organic particles originating from the euphotic layer disappear completely below a depth of about 250 m. Therefore, one may assume that faecal pellets of surface zooplankters can play some part in the food supply of other zooplankters only in subsurface layers (mesopelagic zone).

(c) Phagotrophic Herbivores

The existence of phagotrophic microzooplankters, which feed on small-sized phytoplankton and probably also on other organic particles has been mentioned on p. 77 (see also Volume II: PANDIAN, 1975, p. 74). In the preceding pages, it has been demonstrated that any species at the adult stage is characterized by an efficiency spectrum as regards the filtering of particles of different sizes. However, even though the minimum and maximum prey sizes that a species is able to collect may cover a large range, there is always a smaller size range within which the copepod can filter more efficiently. POULET (1973), for example, observed that *Pseudocalanus minutus* is capable of feeding on particles ranging from 4 to 100 μm, but prefers sizes between 25 and 57 μm. Since *P. minutus* is a small-sized copepod, one may assume that the smallest particle size preferred by copepods in general would be about 20 μm.

According to RILEY (1957), nanoplankton usually represents 70% to 80% of the total phytoplankton (expressed as chlorophyll *a*) in oceanic areas. Both in oceanic waters below 25 m and in neritic waters (even near the surface), this high percentage tends to decrease to about 40% of the total biomass present (MALONE, 1971). Since most copepods are imperfectly adapted for feeding on small cells, we may assume that the nanoplankton would be 'undergrazed' unless utilized by other consumers. Investigating the plankton in the Bay of Villefranche (Mediterranean coast of France), RASSOUL-ZADEGAN (1975) reports that oligotrich ciliates, tintinnids and nauplii, represent 70·7%, 5·5% and 23·7%, respectively, of the total microzooplankton volume (80 mm^3 m^{-3}) in the surface layer (0 to 20 m). Other data from many parts of the World Ocean suggest that ciliates predominate often, if not always, in the microzooplankton. A mathematical analysis of fluctuations in different groups of nanoplankton and microplankton in the Bay of Villefranche led RASSOULZADEGAN (1979) to conclude that large-sized ciliates mainly feed on nanoflagellates (5–15 μm), the small-sized ones on bacteria (ultraplankton). In the Gulf of Marseilles, TRAVERS (1971a,b) noticed that tintinnids may reach abundances of more than 1000 ind. l^{-1}. During the cold season, maximum abundance most often coincides with that of diatoms (direct relationship), both groups exhibiting two peaks in late winter and early spring. However, during the warm season, diatom abundance decreases and the relationship becomes indirect; the quick decrease of small phytoplankton blooms occurring in this period suggests that it may be ascribed to grazing by these phagotrophs.

Foraminiferans and radiolarians are other microzooplankters which feed on nanophytoplankton. Obviously, the consumption of nanoplankters by phagotrophic microzooplankters, which, in turn, may be consumed by filtering macrozooplankters—e.g. copepods, euphausiids, thaliaceans (and non-filtering forms such as fish larvae and chaetognaths)—must affect the trophic chain.

Regarding the first two links of the trophic chain, RYTHER (1969) distinguishes two main patterns: (i) in coastal areas, which are usually more or less eutrophic, phytoplankton is mainly consumed by meso- and macrozooplankton; (ii) in oceanic oligotrophic areas an additional link occurs, mainly represented by ciliates feeding upon nanophytoplankters while consumed by meso- and macrozooplankton. The difference between these two patterns does not appear to be as marked as assumed by RYTHER: it is well documented that, at the beginning of an enrichment process, small-sized cells with growth rate higher than those of larger cells usually abound and therefore can induce an increase in the ciliate population. This has been observed by RASSOULZADEGAN (1975) in areas of the Bay of Villefranche (Mediterranean coast of France) which are rendered eutrophic by domestic sewage.

In general, a proper distinction between true herbivores and species able to change their diet from plant to animal food appears more difficult than usually assumed. This has been demonstrated by BOUCHER and SAMAIN (1975) in the north-west African upwelling system of Mauritania. The authors investigated the feeding behaviour of zooplankton in the natural environment (on a 30-nautical-mile transect from nearshore areas with maximum upwelling to offshore), using measurements of specific enzymatic activity (amylase and proteases) as nutritional index. Since the authors found no correlation between amylase and protease activities, the diet of the populations studied may be represented by the ratio of these two enzymatic activities.

BOUCHER and SAMAIN (1975) also compared amylase specific activities with the 'grazing index'* proposed by LORENZEN (1967) who assumed that the most important source of phaeopigments should be ascribed to zooplankton grazing on phytoplankters. LORENZEN's hypothesis is supported both by many observations on the relative distribution of phytoplankton and zooplankton and by the positive correlation between grazing index and zooplankton abundance. However, while LORENZEN's index is applicable in the Mauritania upwelling region, where no river input occurs, it has been demonstrated by BLANC and LEVEAU (1973) that death and subsequent decay of freshwater phytoplankters transported to the sea is also an important source of phaeopigments. This hinders the use of the Lorenzen equation as grazing index.

In the samples studied by BOUCHER and SAMAIN (1975), copepods, which represent 80 to 95% of the total number of individuals present, belong to the genera *Calanus* (*sensu lato*), *Clausocalanus*, *Euchaeta*, *Temora*, *Centropages*, *Pleuromamma*, *Candacia*, *Acartia*, *Oithona*, *Oncaea*, *Corycaeus* and *Euterpina*. The highest biomass values usually correspond to a large predominance of a single species, e.g. *Calanoides (Calanus) carinatus*, *Temora stylifera* and *Acartia clausi*; often the cladoceran *Podon intermedius* is also abundant. In all species mentioned above, both amylase and protease activities have been observed. Hence these species are considered to be potential mixed-diet feeders. Further samplings were carried out in the same water mass marked by a drifting buoy, with two different plankton nets. In Figs 4-11 and 4-12, the specific activities of amylase and proteases have been plotted against time for three main genera of microphages. Neither amylase nor protease activity changes significantly in members of the two genera *Calanus* and *Temora* which appear to be true mixed-diet feeders. However, the curve in Figure 4-12 suggests that *Podon intermedius* is omnivorous with a marked preference for animal food where available.

Comparing data obtained at a nearshore (Stn 72) and an offshore (Stn 68), BOUCHER and SAMAIN (1975) found a peak of amylase specific activity clearly marked during the night at the latter (Fig. 4-13). A comparison between the specific composition of the zooplankton assemblages at the two stations, together with a more detailed study on amylase and protease activities of *Temora* and *Calanus*, suggests that these two genera which predominate in the mesozooplankton at station 72—where the temperature is rather low and the instability of the water mass quite marked—are responsible for most of the grazing pressure exerted in nearshore areas. At station 68, where the water mass is warmer and better stabilized, *Temora* and *Calanus* populations are scarce but *Acartia clausi*, being here preferentially herbivorous, represents 35% to 40% of the mesozooplankton.

---

* $\log\left(\dfrac{\text{chlorophyll } a}{\text{chlorophyll } a + \text{phaeophytin}}\right)$

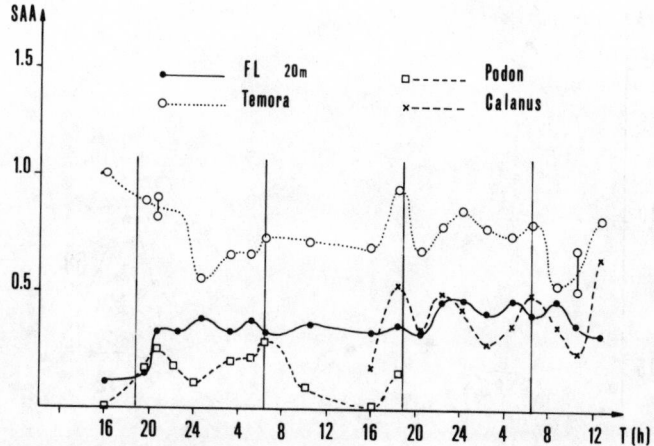

Fig. 4-11: Changes in specific activity of amylase (SAA) for members of three copepod genera samples with a W.P.2 net, mesh: 200 μm; FLH 20 m, changes in specific activity of amylase in total plankton samples with F.A.O. net (mesh: 500 μm); T: time (hours) of samplings along the movement of the water mass labelled with a drifting buoy; vertical lines correspond to dusk and dawn, respectively. (After BOUCHER and SAMAIN, 1975; reproduced by permission of the authors.)

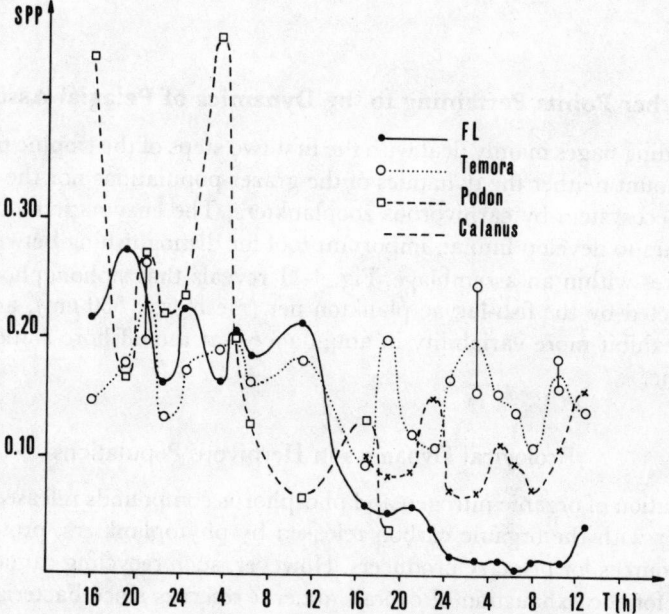

Fig. 4-12: Changes in specific activity of proteases (APP). Same conditions as in Fig. 4-11. (After BOUCHER and SAMAIN, 1975; reproduced by permission of the authors.)

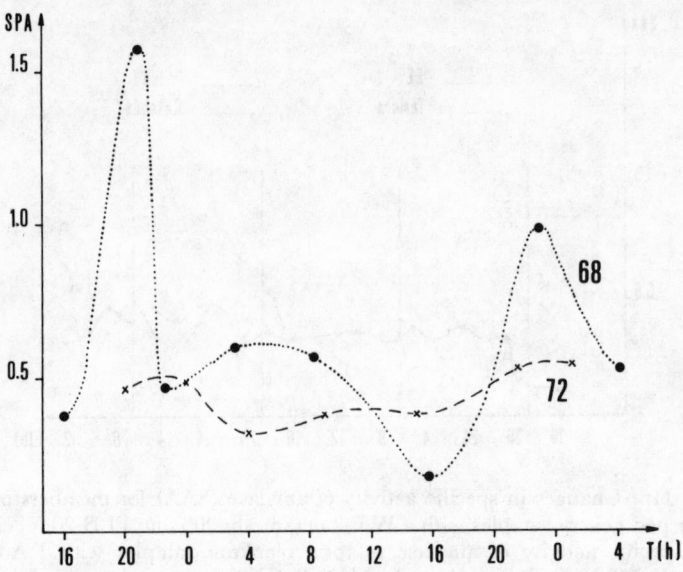

Fig. 4-13: Changes in specific activity of amylase (SPA) in the total sample. 68: offshore station; 72: nearshore station. (After BOUCHER and SAMAIN, 1975; reproduced by permission of the authors.)

(5) Further Points Pertaining to the Dynamics of Pelagial Assemblages

The preceding pages mainly deal with the first two steps of the trophic pyramid. They take into account neither the dynamics of the grazer populations nor the role played in the pelagial ecosystem by carnivorous zooplankters. The enzymatic method just mentioned appears to develop into an important tool for distinguishing between herbivores and carnivores within an assemblage. Fig. 4-11 reveals that siphonophores, which are mainly collected by the fish-larvae plankton net (mesh size: 500 μm), are more abundant—and exhibit more variability in abundance—at the offshore station than at the nearshore one.

Ecological Dynamics in Herbivore Populations

Mineralization of organic nitrogen and phosphorus compounds released by zooplankters, together with the organic carbon released by phytoplankters, provide additional nutritive resources for primary producers. However, such recycling cannot significantly compensate for the exhaustion of overall mineral reserves since bacteria which utilize carbohydrates as a carbon source also require nitrogen and therefore compete with photoautotrophs. Moreover, the nature of the algal food available changes progressively: diatom cells increase in size, a fact which may impede their utilization by some grazers. The relative abundance of dinoflagellates increases and thus changes the carbohydrate/

protein ratio, since dinoflagellates contain more protein and less carbohydrate than diatoms.

Where food quantity and/or quality become limiting factors, the grazers' reproductive rate decreases. The grazers' population may become further reduced because their first larval stages which require food may not find a sufficient number of small-sized algal cells. However, the larvae may partly compensate for this by switching to bacteria or small-sized bacterial aggregates—an alternative fostered by an increase in bacterial abundance supported by organic substances released from phytoplankters and zooplankters. In these cases of nutrient enrichment, restricted to a rather short period of time, the stability of the assemblage tends to increase.

Where nutrient enrichment is permanent—as in open ocean divergences—or prevails for a long time (e.g. several months)—as in most coastal upwellings—the assemblage concerned tends to remain at a youthful stage of development and thus exhibits a low degree of stability. More mature and more stable stages occur farther from the upwelling zone, i.e. in the presence of decreasing mineral nutrients. Simultaneously, the youngest succession stages tend to become more and more dispersed and unused mineral nutrients more and more diluted. Both these processes result in the substitution of young by older and more mature succession stages. A combination of succession in time with drifting and dispersion processes induces a more marked and sometimes almost permanent spatial distribution of the different stages considered.

In general terms, it seems that an assemblage mainly composed of phytoplankters and herbivores will tend to be unstable if the grazers have a wide efficiency spectrum and a high filtering rate. Such is the case in the Western Basin of the Mediterranean Sea during winter, when copepods of the genera *Calanus* and *Eucalanus* predominate (NIVAL, 1976). To some extent, also pteropods, which collect their food particles on a mucous sheet and, therefore, may utilize a wider range of particles than copepods (STEELE and FROST, 1977) should be able to increase the instability of an assemblage.

Thaliacea species, which often exhibit very dense populations—due to the quick increase in blastozoid number via budding—seem to be particularly able to increase the instability of the assemblage since they intensively and indiscriminately filter particles of a very large size spectrum (NIVAL, 1976). In contrast, during summer, the predominant copepod species belong to the genera *Acartia, Microsetella* and *Euterpina*. These have a narrow food-particle-size spectrum and low filtering rates resulting in a more stable assemblage. Especially where certain parts of the particle-size spectrum are insufficiently controlled by grazers, the assemblage can accommodate types of herbivores capable of exploiting the vacant niches (NIVAL, 1976). To some extent, phagotrophic microzooplankters occupying the level between primary producers and filtering macrozooplankters exemplify the tendency of an ecosystem toward increased stability. This will be discussed below with respect to carnivores which may control herbivore populations. However, where carnivores are absent in an assemblage, STEELE and FROST (1977, p. 529) suggested that

> 'the failure of a particular herbivore population is related to absence of the right size of food at a particular stage of its development, rather than a generally low rate of energy intake over the whole life cycle. Such deprivation occurs at the extremes of its size range either as lack of food for the adults, preventing reproduction, or as lack of food for the first feeding stage after development from eggs to nauplii.'

Interference by Carnivores

The preceding pages have concentrated on algal cells and grazers. However, ecosystems devoid of carnivores practically never exist in nature. Food links in the pelagial have been analysed by WYATT (1976) who has also contributed ideas to the following considerations.

As has already been indicated, a clear distinction between herbivores and carnivores is difficult and can, at best, be approximate: many zooplankters are mixed-diet feeders. As adults, most small-sized zooplankters—both filter feeders whose filter-mesh size retains plant as well as animal prey, and non-filterers collecting prey with mucous sheets or filaments—must be considered mixed feeders. For example, copepods with their wide-range filter mechanism can collect algal cells, ciliates and nauplii; young herring (7–12 mm in length) not only feed on dinoflagellates (species of *Peridinium*, *Ceratium*, *Prorocentrum*) but also on tintinnids, cypris larvae of cirripeds, larval stages of molluscs. The general significance of mixed feeding becomes even more evident if we consider the feeder's whole life cycle. Few, if any, animals spend their whole life cycle within a single trophic level, even if such a level is broadly defined (WYATT, 1976). HARDY (1924) found that young herring (12–42 mm) feed mainly on representatives of *Pseudocalanus* and, incidentally, *Temora*; young herring (42–130 mm) also feed on other copepods (*Acartia* and *Centropages*) and mysids; herring over 150 mm in length take in a diet in which copepods (*Temora* and *Calanus*) are largely replaced by large-sized prey such as pteropods, peracarids (*Nyctiphanes*, hyperiids *Apherusa*), *Oikopleura*, *Sagitta* and young *Ammodytes*. At the same time the feeding behaviour changes progressively: hunting of large prey increases while the percentage of small-sized food particles retained by the branchiospines decreases.

In general, it seems that changes in diet composition and feeding behaviour can be related to three main factors: energy requirements, relative size of predator and prey, and prey availability.

(i) Energy requirements increase, in spite of the decrease of metabolic rate with age and size: a fully grown individual requires more food than a juvenile of the same species. Adult herring, for example, can no longer live on small-sized prey; their filtering device (gill-rakers, also called branchictenia or branchiospines) is insufficient for satisfying their food requirements. Hence the herring must switch to raptorial feeding which secures more energy from bigger prey, thus improving the collecting-cost versus energy-gain relation. According to CUSHING (1964), the daily ration for an adult herring amounts to approximately 3% of its body weight; this cannot be obtained by filtering, except possibly unless the herring feeds within a plankton patch particularly rich in *Calanus finmarchicus*. Incidentally, from this we can also infer that herring do not perform superfluous feeding (p. 90).

(ii) Predator–prey size relations tend to follow the rule: the larger the predator, the larger its prey (exception: e.g. whales which feed on zooplankton). Within a given species this relation is exemplified by the herring: its prey size increases as the herring grows. A parallel example is the increasing prey size in growing *Sagitta elegans* (RAKUSA-SUSZCZEWSKI, 1969): the smallest individuals mainly feed on small-sized copepods (e.g. *Oithona* spp., *Paracalanus parvus*, *Acartia clausi*); between approximately 10 and 16 to 18 mm, *Temora longicornis* and *Pseudocalanus elongatus* predominate in the diet; above 16 to 18 mm, copepodites IV and V and adults of *Calanus finmarchicus* (size over 2 mm) represent the most important part of the diet. Mysticete whales are striking examples of large-size animals apparently predetermined to act as 'top predators', but actually feeding at trophic levels much below those exploited by related genera. The adaptive strategy of Mysticeti appears to have followed two different paths leading to two different feeding habits (NEMOTO, 1970). In the Southern Ocean, the largest whales, such as *Balaenoptera musculus* (blue whale), *B. physalus* (fin whale), are 'swallowers', almost exclusively feeding on krill *Euphausia superba*. The humpback whale *Megaptera nodosa* also belongs to the swallowing type but mainly consumes copepods and small fishes. Swallowing whales only occur in areas with very abundant zooplankton, i.e. mainly in the Antarctic region. In contrast, whales of the 'skimming type' such as *Balaena mysticetus* and *Eubalaena glacialis* swim and engulf plankters of the surface layers and can exploit rather scattered pelagial plankton assemblages. WYATT (1976) lists further examples: the basking shark *Cetorhinus maximus*, the whale shark *Rhyncodon typicus*, sunfishes of the genus *Mola* and rays of the genus *Manta*.

(iii) Prey availability often affects the feeding strategy of the predator. For example, the buccal appendages of 'omnivorous copepods' have a filtering device with a generally large mesh size. But these coarse filter feeders also have mandibles with strong teeth and can switch to raptorial feeding if the size spectrum of the prey available markedly exceeds that of their filter mesh. Another example of an adaptive feeding strategy is that of plaice (*Pleuronectes platessa*) larvae which, up to yolk-sac resorption, feed on phyto-plankton, but later turn almost exclusively to *Oikopleura* until metamorphosis, which occurs two months later. Where *Oikopleura* is absent, the larvae prefer unicellular algae to other zooplankters.

Other significant changes in diet and feeding behaviour have been noticed in *Euphausia pacifica* both in the North Pacific Ocean and Japan Sea. According to PONOMAREVA (1955, 1963), phytoplankters as well as zooplankters can be found in their thoracic basket; however, phytoplankters do not play an important role except in spring. LASKER's (1966) experiments support these observations. He demonstrated that *E. pacifica* feeds on algae and small nauplii, preferring the latter. Therefore, this species is a typical mixed-diet feeder. In areas with few, small phytoplankters and a scarcity of small-sized animal prey, *E. pacifica* switches to raptorial feeding. Since euphausiids are less abundant in subarctic areas than in the Antarctic region, *Balaenoptera musculus*— which feeds almost exclusively on *E. superba* in the latter region—also consumes euphausiids (mainly *Thysanoessa inermis* and *Meganyctiphanes norvegica*) in subarctic latitudes supplementing its diet with pteropods (*Clio* and *Limacina*). *B. physalus*, mainly a krill-eater in the Antarctic region, feeds in the North Pacific Ocean on copepods (chiefly *Calanus cristatus* and *C. plumchrus*), schools of young fishes (*Clupea, Theragra, Mallotus*) and even small squid.

Ecological Consequences of Predation Pressure

Presumably, predation pressure in pelagial assemblages involves at least three main effects: the buffering effect exerted by mixed-diet feeders on populations belonging to the next two successive levels of the trophic pyramid; the protective effect regarding small-sized prey due to progressive increase in the size of prey required by growing predators; and the effect of various adaptive mechanisms on the number of trophic levels between primary producers and top carnivores.

(i) Buffering effect: A predator which consumes algae and microzooplankters, such as protozoans and nauplii, tends to reduce its prey populations and thus to enhance the development of small-sized algae usually consumed by microzooplankters providing that sufficient nutrient salts are available. The predator further tends to reduce copepod populations whose nauplii it consumes.

(ii) The protective effect has been clearly demonstrated by WYATT (1976) in North Sea herring. As herring growth proceeds, smaller prey organisms are relieved of predation pressure while larger ones become subject to it. WYATT suggests that interspecific competition may be very transient or even infrequent in natural communities. Furthermore, successive spawning is an important strategy for reducing interspecific competition, especially during the larval and post-larval stages, with their high and often specific nutritional requirements.

(iii) Effect of adaptive mechanisms: In general, food webs are restricted to about five or six trophic links. Moving further and further away from the initial link—i.e. the primary producer—the body size of the species involved increases while metabolic efficiency decreases. As WYATT (1976, p. 356) puts it:

'A large part of the energy young fish gain from their food is used for growth. As they become bigger, this proportion declines, until eventually it approaches zero. All the intake is then used in maintenance and for reproduction and bodily growth practically ceases.'

The same is true for other species: in general, the smaller a species is in size, the shorter its life-span, and the larger the amount of energy obtained from food which is utilized for growth. Therefore, the shorter the food chain, the greater the part of energy provided by the primary producer which reaches the terminal link.

Probably the most efficient food chain in the World Ocean is that of the blue whale and fin whale, the largest living vertebrates on earth: phytoplankton → krill → whale. Such a direct food chain is facilitated by special adaptations of these swallowing-type whales for locating and exploiting krill concentrations (p. 197), as well as by the fact that in the Antarctic region, interspecific competition diminishes due to the reduced specific richness and diversity in these very cold waters. The biomass and productivity of krill possibly exceed those of any other herbivorous species in the World Ocean, while the consumption by important competitors (birds of the order *Impennae* and various fishes) is probably relatively insignificant compared with the large amounts consumed by the big whales—at least in the past, about 50 years ago, when the whale stocks were still almost unaffected by man. Areas in which the secondary level of the trophic pyramid is less productive are inhabited by smaller-sized whales. These either feed simultaneously on herbivores and carnivores, e.g. *Balaenoptera borealis*, or occupy a position at the top of the pyramid phytoplankters → ciliates → filter-feeding crustaceans → whales. Obviously, such food chains are less conducive to efficient energy transfer from primary producers up to top predators.

In the preceding examples, the adaptive mechanism for shortening the food chain is of a functional nature, but a mechanism of conjunctural nature—i.e. related to casual changes in the composition of the plankton assemblage—may also exist, i.e. in plaice larvae which eat algae rather than other zooplankton where *Oikopleura* is absent (WYATT, 1976).

As an adaptive mechanism for shortening the food chain, we may also consider to some extent feeding habits of larvae in some clupeiform fishes where the adult mainly feeds on phytoplankton. As is well documented, the youngest larval stages of most fishes, at least in part, consume algae. Such preference is commonly taken into account in fish-farming experiments. With increasing age, animal prey constitutes a growing part of the diet (Volume III: KINNE, 1977; see also KINNE and ROSENTHAL, 1977). An example of the reverse situation is the Peruvian anchovy. For many years it was assumed that Peruvian anchovy—the landings of which are the most important of all exploited marine species—would exclusively consume phytoplankton, except when upwelling processes were suppressed by the El Niño Current. However, it is now well established that juveniles (up to 40 mm in length) feed exclusively upon zooplankton, mainly eggs and young stages of copepods, while adults mainly eat phytoplankton. Presumably anchovy shoals move actively into areas with pronounced upwelling and hence abundant phytoplankton. If the phytoplankton abundance decreases—often in May—adults return to zooplankton feeding and sometimes to raptorial feeding even consuming euphausiids of 15 to 20 mm in length.

Pilchard *Sardinops ocellata* and anchovy *Engraulis capensis* from south-west African coastal waters are both non-selective filter feeders. Here, juveniles are zooplanktophagous, consuming mainly calanoid copepods, whereas adults are phytoplanktophagous, feeding chiefly on diatoms (KING and MACLEOD, 1976). Food switching in the pilchard occurs at approximately 100 mm standard length, in the anchovy at about 80 mm. The authors attribute the switching in both species to a smaller porosity of the filter mechanism due to the overlap of gill-rakers of the second, third and fourth arches and the swallowing action. The higher proportion of phytoplankters found in the diet of the adult pilchard, in comparison with that of the adult anchovy, was attributed to differences in number and structure and adult gill-rakers.

(6) Food Web and Food Chain

Food-web and food-chain dynamics have been discussed in several papers. For details consult especially WYATT (1976) and Volume IV: MARGALEF (1978). Many accounts ignore: (i) that the potential food links do not usually occur simultaneously; (ii) seasonal variations; (iii) changes with predator size; and (iv) the possibility of switching to unusual diets during times of acute food shortage. Of course, the total number of food links is restricted by the different types of potential food organisms available at any given time and place. Hence, the dietary variability for a given predator may be less than the usual type of predator–prey diagram (such as that in Fig. 4-14) would indicate. According to MARGALEF (1978), a food web can be dissected into a number of links which make up a number of knots. These knots can be represented by a species or a group of organisms. Basic factors in the evolution of a food web are structural and functional properties of the organisms involved—adaptation, competition, imprinting, learning and opportunism (KINNE, 1977, pp. 722–3). A food chain, i.e. a series of linear food

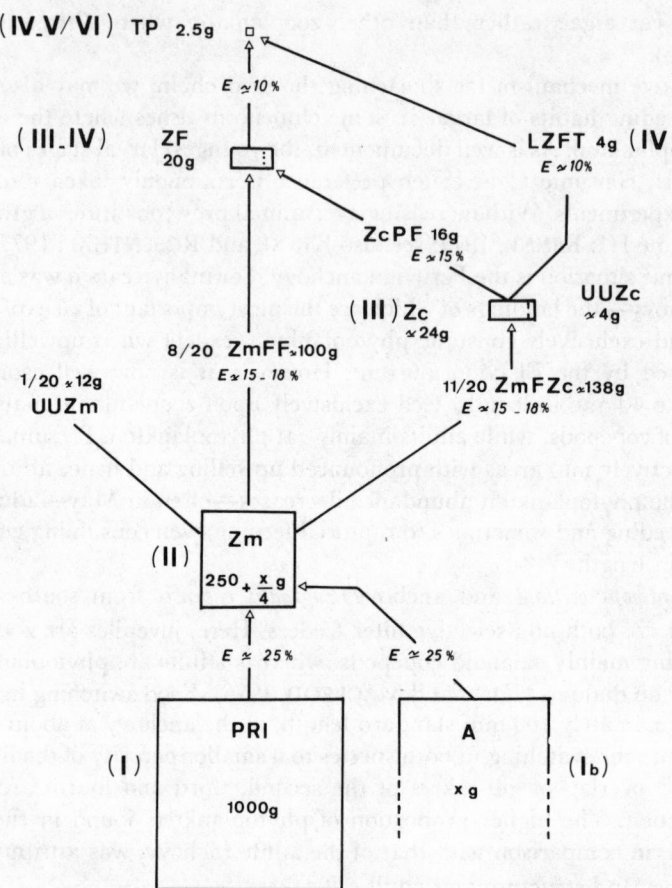

Fig. 4-14: Tentative scheme of production in the trophic food pyramid (g m^{-2} y^{-1}) of the pelagial. Fishes feeding on phytoplankton disregarded. I. I[b], II, III, IV, V and VI: successive steps (largest-sized top predators feed on smaller-sized) of trophic pyramid. A: aggregates, detritus and micro-organisms; E: assumed percentage of energy transfer; LZFT: large-sized zooplankters fed by animals other than zooplanktonophagous fishes; PRI: phytoplankton really ingested; TP: top predators; UUZc: unedible or unconsumed Zc; UUZm: unedible or unconsumed Zm; Zc: production of carnivorous zooplankton (2nd-level consumers); ZcPF: Zc fed by planktonophagous fishes; ZF: zooplanktonophagous fishes; Zm: microphagous mesozooplankton; ZmFF: Zm fed by planktonophagous fishes; ZmFZc: Zm fed by carnivorous zooplankteres. (Original.)

links—neither convergent nor divergent—is considered an extreme type of a simplified or degenerated food web (MARGALEF, 1978).

Basically, in a food web, populations at a given trophic level tend to regulate those at the level immediately below and, in turn, are controlled by the population at the trophic level immediately above. This is a principle of organismic coexistence, and a prerequisite for control and balance in ecosystems (see also Volume IV and Volume V, Part 2).

(7) Patchiness

(a) General Aspects

Homogeneous distribution of organisms in plankton assemblages prevails, at best, over geographical ranges extending over hundreds or thousands of kilometres. Over smaller ranges the distribution of both microscopic plants and small zooplankters seems to display much greater variability than the surrounding environment (STEELE, 1976). Heterogeneity in spatial distribution has become known as patchiness, a phenomenon of considerable ecological significance. Detection and analysis of patchiness depend on sampling strategy and sampling equipment. Patchiness mainly depends on water movement, since plankters are passive drifters, but the living conditions met within a given body of water are also of importance.

(b) Phytoplankton Patches

Phytoplankton patches tend to occur in two main categories: (i) strips a few metres wide but hundreds of metres or more in length; and (ii) roughly elliptical aggregations with a mean diameter usually between 10 and 50 km.

Strips, especially the very long narrow ones in the open ocean, are related to the 'Langmuir circulation' (Vol. I, p. 1528). They originate from a strictly mechanical process in an environment the main characteristics of which are homogeneous. Other strips

'may occur at boundaries between waters of different salinity such as is found when river or estuarine water flows in the sea. This circulation can serve to maintain the sharpness of the boundary and also to concentrate the particulate material along the edge.' (STEELE, 1976, pp. 105–6)

Very similar are elongated patches in the open ocean, for example, along a convergence line. In the last two cases patchiness is associated with physical discontinuity.

The roughly elliptical phytoplankton aggregations may originate from different causes. For example, in front of a river mouth or the sewer of a big city, alternating input peaks due to short-term flow pulsations can result in 'lenses' of freshened water. If the salinity of a lens is sufficiently different from that of the surrounding water, mixing processes can be very slow. Hence, such a lens can drift to distant neritic areas without marked changes in its physical and chemical characteristics. Such long-lived lenses, enriched with nutrient salts, can support a bloom which, if the growth rate is sufficiently high, results in a phytoplankton patch. In temperate regions, where a spring thermocline can trigger a phytoplankton bloom by initially trapping a few plant cells in the euphotic layer, the thermocline may not always form at the same depth or time; thus we may expect spatial variation in phytoplankton concentration during the outburst period (EVANS and co-authors, 1977). It seems that this second type of patchiness occurs only in areas which are sufficiently eutrophic.

Assuming that a phytoplankton patch is sharply separated from the subsurface layer by a strong pycnocline and that endemic grazers do not exist, the fate of the patch will depend on the balance between two factors: (i) the rate of horizontal diffusion; and (ii) phytoplankton growth rate. Demonstrated by several authors and reviewed by STEELE (1976), apparent horizontal dispersion is, in fact, a combination of vertical mixing and horizontal movements. STEELE stresses that the coefficients used in relation to biological processes are simplifications of very complicated three-dimensional events. Considering from a theoretical point of view the fate of a phytoplankton patch, KIERSTEAD and SLOBODKIN (1953) conclude that small patches will more or less rapidly disappear due to mixing effects, whereas in large patches the phytoplankton growth will more than counteract the effects of dispersion and, thus, abundance will increase. The critical diameter l_c of the patch, above which this increase would occur, has been calculated as

$$l_c = 4.8\sqrt{(k/a)} \tag{4.4}$$

where k is the diffusion coefficient and a is the phytoplankton growth rate.

Taking into account some other examples of pronounced phytoplankton blooms, mainly in areas of coastal upwelling, STEELE (1976) concludes that such blooms can develop only if grazing is zero or very low. This could be the case where the phytoplankters produced cannot be utilized by local herbivores, because of size or chemical factors inhibiting or even repelling the grazers. The latter tend to occur during red tides (p. 366), in which an increasingly crowded dinoflagellate population eliminates all zooplankters—possibly due to the release of toxic substances—and finally becomes unispecific. In the North Sea, herring avoid areas of diatom outbursts and in the Indian Ocean, mackerel (genus *Rastrelliger*) spawn near *Noctiluca* patches. This dinoflagellate produces substances which repel predators that usually feed on the mackerel eggs and larvae. Investigating the plankton distribution in whaling areas near South Georgia, HARDY (1956) noticed that: (i) the distribution of zooplankters which do not perform diel migrations is unrelated to phytoplankton patches; (ii) zooplankters which keep away from phytoplankton patches display a classic diel migration pattern, i.e. rising at dusk and sinking at, or before, dawn; (iii) zooplankters with maximum abundance in areas of dense phytoplankton populations are also diel migrants but ascend earlier in the afternoon and remain up for a longer time.

Experiments carried out in order to determine time-to-death in starving copepods and effects of discontinuous food availability on egg production demonstrate that different species are adapted to different scales of patchiness. *Acartia tonsa*, *A. clausi* and *Centropages typicus*, for example, depend on continuous food availability and respond sensitively to small-scale patchiness. In contrast, *Pseudocalanus minutus* and *Calanus finmarchicus* are more independent of such small-scale variability in their food resources. To some extent, ecological success of copepods may be related to the duration of plant-food abundance in the euphotic layer providing access to the food source is secured either by passive drifting or active vertical migration (DUFF, 1977).

(c) Zooplankton Patches

Microdistribution in zooplankters is difficult to study. Usually collected by nets hauled through the water for distances ranging from tens of metres to kilometres, replicate tows indicate the same general statistical feature of overdispersion. Evidence for zooplankton patches is mostly restricted to larger scales (STEELE, 1976). In general, phytoplankton and zooplankton abundance are inversely related. Where herbivorous zooplankters exist we can use phytoplankton net growth as the value of a (Equation (4.4)). STEELE argues that since the difference between production and loss due to grazing is not negligible, a will be smaller and hence l_c larger than the value deduced from phytoplankton only; if $a = 0$, no patchiness is possible.

However, EVANS and co-authors (1977), who constructed a mathematical model in which vertical processes such as turbulent mixing and zooplankton migration can affect the distribution of plankton, conclude that vertical current shear can combine with vertical migration to produce zooplankton patches.

In his valuable review, STEELE (1976) emphasizes the ecological significance of patches as areas of abundant food for many pelagic fishes and larval stages. They may facilitate food consumption at an increased rate. In the North Sea, shoals of young herring (up to standard length 15 cm), which are exclusive filter feeders, apparently seek zooplankton patches (mainly those particularly rich in *Calanus finmarchicus*). This predation itself may be another source of plankton patchiness. The presence of zooplankton patches may be even more important for fish larvae. These require relatively high concentrations of small zooplankters and need these above-average concentrations for periods of one or two weeks. They can find such conditions predominantly in large patches: the larger the patch the slower the water exchange through its circumference. Therefore, patchiness must be assumed to be a regular feature in the pelagial.

(8) Delimitation and Classification of Assemblages in the Pelagial

(a) General Aspects

In the preceding sections we have outlined and discussed the difficulties for delimiting and classifying organismic units of ecological significance in the pelagial. In view of these difficulties, it is not surprising that planktologists have rarely attempted to propose a classification of plankton assemblages.

The most stimulating contribution to this problem is probably BEKLEMISHEV's (1966). He suggested that, in the open ocean, turbulence gyres containing organismic units of ecological significance display three different scales of magnitude: (i) a few tens of metres, corresponding to the smallest type of patchiness; (ii) a few tens of kilometres, represented by areas in which a bloom can start and develop, for example, due to locally increased vertical mixing followed by stabilization; (iii) several thousand kilometres, corresponding to large oceanic gyres, such as the Kuroshio and the North Pacific Drift, the Gulf Stream and the North Atlantic Drift. According to BEKLEMISHEV, units of ecological significance of the first two types can exist only as long as environmental conditions remain sufficiently constant. Organismic units of the third type should be considered as communities, and would be parallel to my 'organismic assemblages' (p. 4). BEKLEMISHEV considers as 'primary communities' those of type (iii) extending from one side of an ocean to the other. He complements his scheme with two further categories: (iv) a 'secondary community', which may occur in the contact zone between two 'primary communities', and whose composition depends on the transport of organisms from adjacent primary communities; (v) in the transitional areas between 'inshore communities' and primary communities, 'distant neritic communities' may exist which combine features of offshore secondary communities and true inshore communities.

BEKLEMISHEV's (1966) two community types (iv) and (v) correspond to the composite populations—ecotones and ecoclines—inhabiting the transition zone, the main features of which have been summarized on p. 65 and are discussed below in more detail. In the reviewer's opinion, neither types (i) and (ii) nor the linear patches—which he has omitted—can be considered as true assemblages, i.e. as units of ecological significance. They all correspond to patchiness (p. 101) in the wider sense. Since patches can be recognized only by using some specific sampling devices and/or methods, patches in general might be better considered as 'associations' *sensu* MARGALEF (1977; see also Volume IV: MARGALEF, 1978).

The reviewer also cannot accept BEKLEMISHEV's (1966) definition of primary communities as those extending from one side of an ocean to the other. It is well documented that the plankton assemblage of the North Atlantic Drift on the meridian 45° W—and even that of the Gulf Stream to the south of Newfoundland—is quite different from that harboured by the Gulf Stream at its starting point in the Florida Strait. Apparently, changes in the composition of plankton assemblages throughout a large oceanic gyre originate from several but narrowly related and simultaneously effective factors. The period of time required by a water body and its assemblage to be transported through the whole gyre and back to its starting point is very long, because drifting is extremely slow relative to the distance to be covered. Considering the brief life-span of most plankters, a succession of generations must be involved. A regular succession of generations can occur only where the environmental conditions remain acceptable to assemblage members; e.g. the thermal requirements for breeding must be met. However, the latitudinal temperature range along a large oceanic gyre usually covers at least 25 to 30 °C and consequently must induce considerable changes in surface temperature. Thus, 'expatriated' species tend to disappear progressively as the ambient temperature exceeds their lower, or upper, thermal limits. Other changes throughout a large oceanic gyre concern nutritional aspects. Even if, in inshore areas, the water's physical and chemical characteristics remain largely unchanged, the peripheral part of an oceanic gyre which comes into contact with a nutritionally enriched area—for example a distant neritic area or the margin of a coastal upwelling—receives some additional energy. Two examples of energy exchange between adjacent assemblages in a contact zone are discussed later (p. 109). Here it must suffice to mention that, in a contact zone, the assemblage of a large oceanic gyre, usually fairly mature and well organized, gains energy at the expense of a less mature assemblage. Hence, productivity of the former should increase in proportion to the amount of energy received from the latter. However, the energy quickly disperses in the very large water volume of an oceanic gyre and is distributed and stored in the biomass of higher trophic levels. Ultimately, the support received by the assemblage is rather small and generally restricted to parts close to the contact zone.

It then seems that we consider the plankton assemblage inhabiting the water mass of an oceanic gyre to be homogeneous, at least from a qualitative point of view, only if essential physical and chemical parameters remain sufficiently constant over the time and space scales involved. Obviously, such constancy depends on both the size of the gyre and the range of latitude encompassed. For example, one can consider the Kuroshio and its derivative which turns to the right before reaching Hawaii, as well as the gyre constituted by a portion of the Gulf Stream between the Florida Strait and about 35° N, on the one hand, and the western peripheral areas of the Sargasso Sea, on the other, to be inhabited by a homogeneous pelagial assemblage. Such assemblages can also be recognized in water masses which are well characterized from a hydrological point of view but do not flow in a roughly circular or elliptical path, such as the north equatorial current. However, while relative constancy of environmental conditions is a necessary prerequisite for a water mass to be inhabited by a homogeneous assemblage, it is not the only one. It is of paramount importance that the energy resources at the primary level of the system do not fall below a critical value in certain parts of the water mass considered. For example, in the north equatorial current of the Pacific Ocean, the most fertile areas are located off the American coast and the gradual decrease in fertility towards the west results in the progressive disappearance of many species—obviously those with the highest food requirements, irrespective of the level they occupy in the trophic pyramid. The same situation can be observed in the Sargasso Sea when compared with the assemblage typical of the Gulf Stream off Florida. These changes in pelagial assemblages—essentially due to a progressive decrease in primary and para-primary production as the water mass drifts away from the area of maximum fertility—also result in a progressive decrease in the total biomass of the assemblage.

Returning to BEKLEMISHEV's (1966) classification, we shall attempt to evaluate his concept of secondary community (p. 103). It seems that his secondary community narrowly corresponds to the contact zone. In a contact zone mixing occurs between different assemblages. In the open ocean, the extension of a contact zone depends on the dynamics of the water masses which come into contact. Examples of contact zones are: the area between antarctic and subantarctic plankton assemblages in the vicinity of the Antarctic convergence (p. 187); the area in the north-western = (boreo-arctic) Pacific Ocean between subarctic water and north-western central water. In such very long contact zones, which correspond to MARGALEF's (1977) ecoclines, it has often been observed that some species from an adjacent ecosystem may display greater abundance than in the ecosystem they originated from. Moreover, it seems that in both ecoclines mentioned, some endemic species occur: zooplankters in the north-eastern Pacific Ocean transition zone and phytoplankters in the Antarctic convergence. On the other hand, not a single endemic species has been recorded from the transitional area between Kuroshio and Oyashio water masses in the north-western Pacific Ocean. To some extent, the occurrence of the chaetognath *Sagitta elegans* in the 'mixed waters' around the British Isles is also an example of an endemic species characteristic of the transition zone. Endemic species apparently never exist in transition zones of the benthal.

Finally, BEKLEMISHEV (1966) distinguished inshore communities and distant neritic communities (in the contact zone between a primary community and an inshore community). Distant neritic communities will be treated below in the context of a general discussion on ecological and energetic aspects in contact zones. As regards the inshore community, it seems desirable to retain the term 'inshore' to denote pelagial assemblages in shallow waters. I propose that these organismic units be known as 'inshore pelagial assemblages'.

Inshore pelagial assemblages are characterized by (usually seasonal) enrichment in nutrient salts which recur over intervals of less than a year. Examples include vertical winter mixing on the shelf and freshwater input from estuaries. Every cycle (sometimes also pulsating) in the enrichment process results in a phytoplankton outburst which usually initiates MARGALEF's (1967) ecological succession (p. 68) whose stages may easily be recognized on both time and space scales. Time succession, as well as space disjunction arising from drifting and dispersion of the once-enriched water volume, might possibly be considered as an example of continuum (certainly a better example than those proposed by several authors in the benthal). The accuracy of MARGALEF's (1967) analysis suggests considering each stage of the succession as a biocoenosis. However, in the pelagial, too much depends on the sampling strategy applied on the one hand, and the patchiness encountered on the other. Hence, I suggest employing the term 'inshore pelagial assemblage' for the whole succession generated by a coastal enrichment process, and distinguishing within such a unit, subassemblages based on characteristic and/or predominant species at each dynamic stage.

(b) Contact Zones between Pelagial Assemblages

General Aspects

A recent synthesis by FRONTIER (1977; 1978) of phenomena in the contact zone between two pelagial assemblages allows reconsideration of several as yet incompletely solved problems. Most of the following pages are based on Frontier's article.

In the pelagial two assemblages may come into contact which are quite different from an ecological point of view, although sometimes their respective physical environment differs only slightly in such factors as temperature, salinity or nutrient salts. Mixing of water masses with sufficiently different densities (σ_t) is generally rather slow compared with their respective movements. Therefore, the extension of a contact zone may largely depend on the pattern of observation and sampling. In reality, the contact zone is not bi- but tridimensional. Turbulent mixing processes result in heterogeneous zone structure (for example, water lenses of different hydrological characteristics) rather than in a regular gradient of physical and chemical parameters. This heterogeneous structure—a 'volume mosaic'—represents an additional argument against BEKLEMISHEV's (1966) concept of secondary community (p. 104).

It has long been recognized that a thermal surface front which separates a cool, nutritionally rich water mass from a warm, nutritionally poor one initiates an increase in biomass and productivity. Based on his investigations off California, GRIFFITHS (1965) suggested that such enrichment may result from: (i) mechanical concentration of plankters in a convergence area resulting in a local eutrophication of surface waters generated by an increase in the recycling processes of decaying organic material; or (ii) lifting of nutrients from deeper layers in divergence or upwelling areas, which augments primary production. Taking into account the concept of ecological succession developed by MARGALEF (1967, 1968) and ODUM (1971), FRONTIER (1978) suggests a more comprehensive interpretation of the phenomena occurring in the contact between two ecosystems differing in their degree of maturity.

Abrupt and pronounced changes in environmental conditions which surpass the system's resistance to deformation result in its partial or total destruction. 'Vacant' niches become available for recolonization. Where significant portions of the system have been destroyed, a 'pioneer' assemblage composed of a few species (low diversity index) with high growth and multiplication rates and short life-cycles can establish itself. Such an assemblage can be considered as a juvenile stage of an ecological succession *sensu* MARGALEF (1967). With time, more additional species join the 'pioneer' assemblage or replace other species so that the assemblage increases in complexity in terms of trophic interactions. Such progressive 'maturity' of the assemblage may end in the climax stage which corresponds to a dynamic equilibrium with a high diversity index. The main changes occurring, according to FRONTIER (1977), as the ecological succession develops from the juvenile stage to the climax, the last stage of an ecological succession.

I would like to make two additional remarks at this juncture. First, it seems that there is an exception to the general rule that biomass increases up to the climax stage: where a once-enriched water volume—for example, an isolated lens—maintains its individuality during drifting, biomass can attain maximum values very close to the point of initial enrichment but its minimum stays at the most mature stage. Second, increasing specific diversity and complexity of the food web results in a more mature assemblage, consisting of producers, consumers (herbivores and carnivores of two or three trophic levels), as well as various decomposers, assuring a more complete recycling of all organic matter. Ultimately, the ecosystem becomes practically 'saturated', i.e. all its ecological niches become occupied. In contrast, a juvenile ecosystem is unsaturated, i.e. it requires significant amounts of matter and energy from outside and often loses matter to adjacent ecosystems. In such a system the biotic diversity is lower than its maximum potential diversity (MARGALEF, 1977, p. 876).

In the marine environment, oligotrophy and maturity are directly related. FRONTIER (1978) assumes this relation to be based on dilution and dispersion processes by turbulent mixing which proceed as the water mass drifts further away from the place where it received the original input initiating the ecological succession. It appears, then, that juvenile and mature pelagial assemblages differ considerably from one another in the degree of their organization ('structuration'). Young assemblages, which tend to colonize suitable environments as quickly as possible, exhibit a low degree of organization and a rapid increase in biomass. In the process of maturation, the assemblage becomes increasingly diversified with regard to specific composition, food-web and biochemical interactions, and comprises more specialized ecological niches; in such a system the degree of organization is high and the biomass low. The increase in the degree of organization is directed towards an increase in the system homeostasis, i.e. augmented mutual control of the populations constituting the system, which tends towards a state of dynamic stability mentioned earlier.

Exploitation Among Adjacent Assemblages

In general, the term 'ecosystem exploitation' signifies an export of biomass irrespective of the recipient. Examples of such export are: dead or living organisms sinking from a surface ecosystem to deeper layers; drifting away of part of the biomass due to a current or grazing on phytoplankters in a contact zone by herbivores from an adjacent ecosystem or assemblages. The influence of exploitation on an ecological succession depends on its intensity. Slight exploitation, such as moderate grazing by herbivores on phytoplankton, may temporarily retard or stop the succession and keep the ecosystem at a juvenile stage. Excessive exploitation of a juvenile coastal ecosystem, such as that arising from a strong wind-generated current in an offshore direction, causes 'dilution' of the plankton assemblages thus retrograding it to its initial stage. In coastal areas, a new succession will start only after the wind ceases and a subsequent and sufficiently long period of hydrological stability prevails.

According to FRONTIER (1978), the development of a juvenile planktonic assemblage, sufficiently provided with nutrients, may proceed in three different ways:

(i) Exponential growth leads to the exhaustion of nutrients and/or growth-promoting substances (such as vitamins or chelators), and results in abrupt biomass disappearance as well as negative consequences, such as oxygen depletion, hydrogen sulphide formation, which are characteristic of dystrophic situations.

(ii) The assemblage maintains itself at a juvenile stage as long as excess energy and matter are exported due to exploitation (e.g. permanent but moderate grazing on phytoplankton populations). In this case the system retains a low degree of organization; it cannot degenerate, since grazing and recycling maintain an almost constant level in both phytoplankton abundance and nutrient salts. However, such a system can neither develop nor augment its organization since it has no surplus in energy required for creating and maintaining new structures.

(iii) An ecological organization develops which can control growth. At first sight, this seems to imply improvement since the more organized the system's biomass, the more probable its maintenance under somewhat fluctuating environmental conditions. However, the system must 'pay' for increased security: an increasing portion of the matter and energy fixed is diverted from growth processes as the organization becomes increasingly sophisticated. The most important losses arise from transfer of matter and energy from one trophic level to the next, and from energy expenditure due to migrations and locomotion in general.

It is well established that if two assemblages with different degrees of maturity come into contact, the more mature system becomes enriched at the expense of the less mature one. In other words the former assemblage exploits the latter. Since a more mature system is more diversified and more organized, it contains more diversified consumers belonging not only to the first level (herbivores) but also to higher levels of the trophic pyramid. The carnivorous species involved are also more motile and hence more able to move into the contact zone for feeding on organisms belonging to the less mature assemblage and to distribute the energy drawn from the contact zone to large volumes of their more mature ecosystem. Therefore, the less organized assemblage tends to retain the same degree of maturity as long as its nutrient resources are sufficient and its exploitation is not excessive; the more mature system, however, gains energy and consequently grows (increase in carnivore biomass, appearance of new predators, etc.) faster than before.

As MARGALEF (1968) has emphasized, the flow of energy and matter is always directed from the less mature towards the more mature ecosystem or system part.

According to FRONTIER (1978), exploitation tends to increase the degree of organization of the more mature ecosystem, providing that the predatory species are diversified and specialized and consequently able to control the abundance of the different prey populations. In contrast, exploitation by unspecialized predators might result in a decrease in the degree of assemblage organization, either because it favours species with higher growth rates, or causes the exploitation to exceed the threshold of system homeostasis (e.g. overgrazing, overfishing). FRONTIER also emphasizes that the role played by motile species in the contact zone is of paramount importance not only for the transport and subsequent distribution of biomass, but also for increasing the primary production within the more mature ecosystem thanks to recycling of their organic exudates and detritus.

As noted by FRONTIER (1978), investigations on ecological successions in the marine environment are more difficult than those in terrestrial environments. In the sea mixing and drifting processes frequently bring into contact different systems—or system parts—representing different degrees of organization and consequently affecting the dynamics of ecological succession. Moreover, in the pelagial, surface-layer assemblages suffer continuous loss of energy and matter, due to sinking of plant cells below the compensation depth, and, more generally, of organic particles. Consequently, marine surface assemblages must compensate for heavy loss (due to grazers plus sinking). Obviously, losses due to sinking are reduced in the most oligotrophic areas with a permanent and well-marked shallow thermocline, for example in the subtropical regions and anticyclonic gyres of the open ocean. Hence, FRONTIER assumes that pelagial surface assemblages may never reach the climax stage.

Transfer of Biomass and Energy Through a Contact Zone

Let us now attempt to analyse the changes induced by transfer of biomass and energy in the dynamics of two adjacent organismic assemblages. FRONTIER (1978) selected and discussed three examples: estuarine assemblages, upper thermocline assemblages and front assemblages.

Estuarine assemblages (ecosystems). In an estuary, the fertilizing effect due to river run-off during, or following, the rainy season initiates a young-stage ecological succession. In typical estuaries, providing that the run-off is not excessive, the freshwater inflow leads to a brackish surface layer which drifts away from the shore. To some extent, this layer drags with it a part of the more saline waters below. Therefore, the total water volume transported is much greater than the freshwater input. This flow pattern is compensated for by a deep counter-current of sea water flowing towards the shore.

Investigating an estuarine ecosystem of this type on the north-western coast of Madagascar (rainy season from November to April), FRONTIER (1978) observed that in November and December the first important run-off, together with high temperature, resulted in excessive eutrophication because the two-layer circulation had not yet become established. Hence, some of the dystrophic events mentioned above (p. 107, point (i)) could be observed. In January, when the two-layer circulation begins to work, the system gains some dynamic stability because part of its biomass is forced offshore in the surface drift. Thanks to this 'physical exploitation' (*sensu* MARGALEF, 1977), the upstream system—i.e. that which occupies the area near the river mouth—reaches the juvenile stage characterized by abundant phytoplankton (mostly diatoms), small-sized grazer populations, a low diversity index, and scarcity of carnivores.

In contrast, that part of the ecosystem which drifts away from the shore with the surface water, which progressively loses its stratification and increases its salinity, features an increasing degree of organization. The ecosystem increases its diversity index and develops a more sophisticated food web. The later stage of this development mainly occurs on outer shelf areas of the bay, where the system includes zooplankters performing vertical (diel or ontogenic) migrations. These migrants may be brought back to the upstream area by the deep counter-current and then exploit the youngest stage of the assemblage. Obviously, the benthos assemblages inhabiting the outer-shelf bottoms also benefit from the sinking material or migrating species of the plankton assemblage, but also provide larval stages. In April, the run-off ceases, but by then the ecosystem has already developed such a large biomass and stable structure that it remains almost unchanged over several months, until September when pronounced hydrological changes prevail, causing an abrupt decrease in biomass.

Upper thermocline assemblages. In intertropical areas the shallow-water thermocline often occurs within the euphotic layer. The same has been observed in the Mediterranean Sea in a doming area (JACQUES, 1974). In such a situation, the thermocline represents a contact zone between the homogeneous (nutrient-poor) upper layer and the underlying waters with high nutrient contents. The general scheme of the dynamics there seems to be the following. On the one hand, the upward diffusion of nutrient salts from subsurface and intermediate waters is somewhat inhibited at the level of the pycnocline; providing the illumination is sufficient this leads to locally increased primary productivity. On the other hand, the sinking rate of phytoplankters and detritus becomes retarded at the pycnocline; this leads to an increased abundance of heterotrophic bacteria which mineralize the decaying organic material and thus provide nutrients for a recycling primary production. Therefore, a juvenile stage of the ecological succession may appear in the vicinity of the thermocline. In spite of the oligotrophy which generally characterizes the homogeneous surface layer, its zooplankters sometimes represent a biomass larger than that expected on the basis of local primary production. This fact may be attributed to the zooplankters exploiting the net production of the thermoclinal layer and thereafter distributing their own secondary production within the whole suprathermoclinal layer. As usual, exploitation of the less mature assemblage by the more mature one tends to keep the former at a juvenile stage, but allows the latter to reach a higher degree of organization.

As emphasized by FRONTIER (1978), the vertical—mainly daily—migrations of zooplankters constitute an important aspect in the organization of the surface system. These migrations increase the flux of nutrient salts from the infrathermoclinal layer upwards to the surface layer. Consequently, a large biomass may be maintained in the surface layer, in spite of the permanent loss of matter due to sinking. In other words, part of the energy provided by primary production is diverted from its normal path, namely temporary storage in biomass, and utilized by moving zooplankters. Their motility may be considered an improvement of the structuration of the assemblage. The young subassemblage quickly provides for new primary productivity, while the more mature one utilizes the product immediately for maintaining a high biomass in an oligotrophic environment.

Front assemblages. Thermal fronts in the surface water originate either from a convergence between two different surface water masses or from the rising of a thermocline. They often exhibit high values of animal biomass especially of second- or third-level carnivores. FRONTIER (1978) studied the equatorial front on the West African coast off Cape Lopez (Gabon) which—between June and October—separates the warm (>24 °C) and less saline (<35%) 'Guinean waters' from colder and more saline waters originating from the Lomonosov undercurrent. This thermal front received much attention because it supports large populations of yellowfin tuna. Adult yellowfin inhabit the oligotrophic Guinean waters in which they conduct large migrations; spawning and larval development occur in the same water mass. Presumably, larvae and fry live in the suprathermoclinal layer the zooplankters of which exploit the rich infrathermoclinical layer. At the time of the appearance of the front in the surface waters, the yellowfin aggregate in the vicinity of the contact zone between oligotroph warm waters and cold waters which contain about three times as much phyto- and zooplankton. The tuna aggregation is clearly related to the location of the large pelagic biomass they feed on; the latter depends on the contact zone characterized by increased local transfer of energy and matter.

Similar aggregations along thermal fronts which separate two different water masses have also been observed in large-sized cetaceans.

According to FRONTIER (1978), several assemblages or parts thereof—each with its own organization, degree of maturity and dynamics—can become connected; such a connection results in a new system with an integration level higher than that of each of its constituting subsystems. To some extent, such views might support BEKLEMISHEV's concept of 'secondary communities' providing we speak of 'secondary assemblages' rather than of 'secondary communities'. In fact, it is not yet clear whether populations inhabiting a contact zone of a thermal front—which, in the example chosen by FRONTIER, is a transitory hydrological structure—really represent a relatively independent assemblage or only a transient mixing stage of two adjacent, well-characterized pelagial assemblages.

(c) Long-term Changes in Pelagial Assemblages

The increasing danger of critical global marine pollution (Volume V, Part 4, see also KINNE, 1980) has stimulated concerted research on natural large-scale fluctuations in marine populations. Long-term (years, decades) monitoring programmes are being carried out in various areas. They involve abiotic factors (e.g. temperature, salinity, solar radiation) and biotic ones (mainly plankton composition and abundance, primary production, etc.) as well as indirect information based on fish-landing statistics. The data produced by these programmes indicate that climatic changes significantly affect marine populations. According to LONGHURST and co-authors (1972), climatic effects may manifest themselves in biological trends which extend over many years as well as in drastic changes of population dynamics (population explosions or collapses). While long-term fluctuations can also occur in the benthal, they are more conspicuous in the pelagial, due to the shorter life-span of most plankters. The monitoring of changes in zooplankton biomass at an ocean weather station in the north-eastern Pacific Ocean between 1957 and 1968 revealed that the biomass was approximately three times larger between 1958 and 1960 and from 1966 to 1968 than it was from 1962 to 1964, while no long-term changes occurred in the total zooplankton crop in the north-temperate mid-Pacific Ocean. LONGHURST and co-authors (1972) attributed these variations in biomass to changes in surface salinities, which may be considered as indicators of vertical water mixing which depends, in turn, on shiftings in the Aleutian atmospheric low-pressure cell.

Effects of long-term climatic changes on plankton populations have been investigated most thoroughly in the North Atlantic Ocean and the North Sea. In these regions, monthly plankton samplings have been carried out since 1948, using the continuous plankton recorders towed from merchant ships and ocean weather ships. According to LONGHURST and co-authors (1972), the data reveal two major patterns: a roughly linear downward trend with peaks in the early fifties and late sixties, and a period of low abundance between (Fig. 4-14). Representing an important percentage of the total zooplankton in the north-western areas of the North Sea, the copepod *Pseudocalanus elongatus*—in addition to a seasonal cycle (low abundance in winter, high in summer)—exhibits a series of well-marked abundance peaks and troughs which a spec-

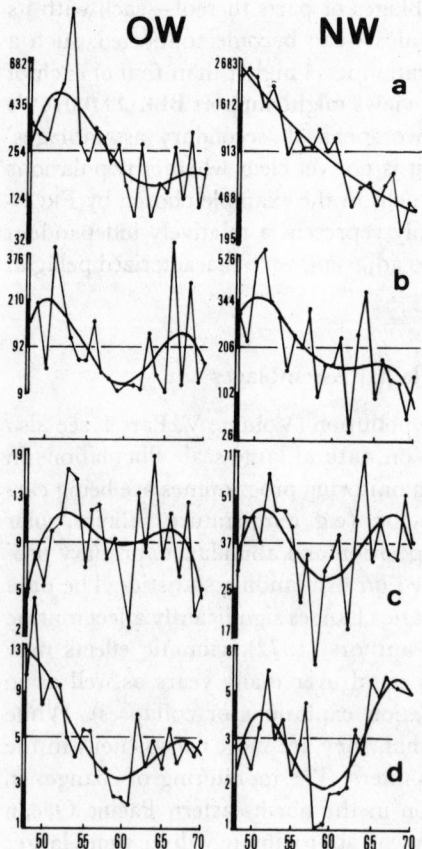

Fig. 4-15: Annual abundance fluctuations in 4 zooplankters for 2 areas. Dominant patterns of recent year-to-year changes. Selected from data for about 50 species. Heavy curve: trend derived by statistical analysis. OW: Offshore waters, Atlantic Ocean 55° to 59° N, 9° to 19° W; NW: neritic waters, eastern North Sea. (a) *Pseudocalanus*; (b) *Siratella*; (c) *Calanus*; (d) euphausiids. (After LONGHURST and co-authors, 1972; reproduced by permission of *New Scientist*.)

tral analysis showed to correspond to a rhythm with a wavelength of 3 to 3·5 years (Fig. 4-15).

Other studies revealed similar rhythms in various areas of the North Atlantic Ocean. They suggest the existence of another rhythm of about 8 to 10 years, a coincidence between the 3- to 3·5-year rhythms of changes in plankton populations in the North Sea on the one hand and changes in atmospheric circulation on the other, associated with a high-pressure cell over Greenland from 1956 to 1970, as well as with sea-surface temperature anomalies—especially in winter—in the north-eastern areas of the Atlantic Ocean. Although the continuous plankton recorder survey is not really adjusted for conducting phytoplankton sampling—the 250-μm mesh size is too large—it indicates during the same period in both the north-eastern Atlantic and the North Sea a trend towards increased algal abundance since the forties. This trend has tentatively been ascribed to a decrease in copepod numbers and the resulting relaxation of grazing pressure. However, such an assumption seems questionable. Recent, more accurate investigations on North Sea plankton populations suggest that abundance changes in different groups of phytoplankters have been well marked over the last decades. Apparently, diatoms exhibited a pronounced decrease between 1966 and 1972 in the whole North Sea, as well as in the English Channel and the Celtic Sea. This decrease was matched by an increase in small-sized species, such as microflagellates, uncounted and unregistered by the continuous plankton recorder. This situation possibly explains that in the North Sea the

'patterns of annual variation in the herbivorous zooplankton were not related to those in the phytoplankton. In the Southern North Sea, the decrease in copepod numbers between 1960 and 1966 did not correspond with a change in the diatoms registered by the Continuous Plankton Recorder: the decline in this latter group did not start before the mid-sixties.' (GIESKES and KRAAY, 1977, pp. 361–2)

It has also been assumed that the decrease in zooplankton biomass in the North Sea and the north-eastern Atlantic Ocean might be ascribable to a delay in the advent of the phytoplankton spring bloom which progressively shifted—with some small-range irregularities—from early March in 1948 to late April in 1972. This led to a shortening of zooplankton growing season by about one month (PARSONS and co-authors, 1977).

Before beginning the continuous plankton recorder survey (1948), the Marine Biological Association laboratory had been carrying out regular investigations on plankton and fish populations in the western English Channel since the twenties. These showed that, from about 1920/1924 to the early sixties, zooplankton abundance decreased and the assemblage of the mixed waters characterized by *Sagitta elegans* almost disappeared. This led to a marked decrease in herring landings, whereas pilchard fishery increased. Since about 1965 the trend seems to have reversed: the primary production continuously increased between 1965 and 1970 from 140 to 212 g C m^{-2} y^{-1}: diatoms (especially *Biddulphia sinensis* and *Coscinodiscus concinnus*) became much more abundant in spring and early summer and the total zooplankton had greatly increased by 1970, particularly *Calanus*, paralleled by a reduction in spawning intensity of pilchard. RUSSELL and co-authors (1971, p. 469) noticed that the western Channel area investigated corresponds.

'to a well marked boundary between boreal forms and warm-water (lusitanian) species. Thus, the years of paucity of plankton and abundance of pilchard may correspond to a retreat northward of the boreal fauna during the recent ameliora- tion of climate in the northern hemisphere. That boreal forms are now returning during a period of worsening climate is borne out by the evidence for recent increase in numbers of adult fish such as cod and ling, while haddock have appeared for the first time.'

Without going into more detail, we can conclude that long-term (i.e. extending over several decades) changes in plankton abundance and composition in the North Atlantic Ocean and its adjacent seas appear to be related to climatic changes in the Arctic. These seem to affect the current regime, especially the northward extension of Atlantic waters in the North Sea through the English Channel—enhanced or reduced by Arctic water warming or cooling respectively. Consequently, in pelagial assemblages of the north- eastern Atlantic Ocean and adjacent seas, we may tentatively expect long-term changes with a wavelength of about 30 to 35 years. Within such long waves, shorter waves of about 10 to 11 years (possibly adjusted on the sunspot cycle) and 3 to 3·5 years can be distinguished. All these biological changes seem to be related to climatic changes. Some of the changes in pelagial assemblages also affect the benthos, especially species with life-cycles which include a planktonic stage. It may further be suspected that biological changes of still longer wavelengths, i.e. about one century, might occur.

Although not really an example of long-term change, the 'El Niño cycle' deserves mention in this context. As is well known, north-westerly winds in the South Pacific Ocean induce a strong upwelling along the Chilean and Peruvian coasts. Any marked slackening of the normal intensity of these trade winds results in a southward extension of equatorial waters; such extension is called 'El Niño' since it normally occurs around Christmas. Whatever the origin of this phenomenon—possibly a decrease in the slope of the sea surface across the equatorial Pacific Ocean—its ecological consequences are drastic: (i) pronounced decrease in phytoplankton abundance, mainly as regards diatom populations; (ii) occurrence of shoals of the anchovy Engraulis ringens in waters deeper (e.g. 100 m) than during inter-Niño periods; (iii) a tendency of anchovy shoals to migrate southwards; (iv) decrease in recruitment of anchovy due to failure in sexual maturation and food shortage in juveniles (true larval stages seem to be more carni- vorous).

The last El Niño in 1972/1973 caused the annual anchovy landings to fall from about 12 to 2 million metric tons. Populations of seabirds, especially those feeding mainly on anchovy, decline drastically during El Niño periods. Temporary instabilities in atmos- phere–ocean systems which induce El Niño are fairly common and have probably occurred about 10 to 12 times during the present century. Pelagial ecosystems and bird populations have particularly suffered when there were two peaks of intensity (as in 1957/1958 and 1972/1973); the single peak in 1965 induced less damage.

Literature Cited (Chapter 4)

BEKLEMISHEV, K. V. (1966). Large-scale pattern of distribution of oceanic plankton communities. *Second Int. oceanogr. Congr. (Abstr. papers)*, **2**, 28.

BLANC, F. and LEVEAU, M. (1973). *Plancton et Eutrophie. Aire d'Epandage Rhodanienne et Golfe de Fos (Traitement Mathématique des Données)*, Thèse Doct., Université Aix-Marseille (II).

BOUCHER, J. and SAMAIN, J. F. (1975). Etude de la nutrition du zooplancton en zone d'upwelling par la mesure des activités enzymatiques digestives. In H. Barnes (Ed.), *Proc. 9th European Marine Biological Symposium*. Aberdeen University Press. pp. 329–41.

COLLIER, A. W. (1970). Oceans and coastal waters as life-supporting environments. In O. Kinne (Ed.), *Marine Ecology*, Vol. I, Environmental Factors, Part 1. Wiley, London. pp. 1–93.

CONOVER, R. J. (1966). Factors affecting the assimilation of organic matter by zooplankton and the question of superfluous feeding. *Limnol. Oceanogr.*, **11**, 346–54.

CONOVER, R. J. (1978). Transformation of organic matter. In O. Kinne (Ed.), *Marine Ecology*, Vol. IV, Dynamics. Wiley, London. pp. 221–449.

CUSHING, D. H. (1959). On the nature of production in the sea. *Fishery Invest., Lond.*, **22**, 1–40.

CUSHING, D. H. (1964). The work of grazing in the sea. In D. J. Crisp (Ed.), *Grazing in Terrestrial and Marine Environments*. Blackwell, Oxford. pp. 207–25.

DUGG, M. (1977). Some effects of patchy environments on copepods. *Limnol. Oceanogr.*, **22**, 99–107.

EVANS, G. T., STEELE, J. H. and KULLENBERG, G. E. B. (1977). A preliminary model of shear diffusion and plankton populations. *Scot. Fish. Res.*, **9**, 1–20.

FINENKO, Z. Z. (1978). Production in plant populations. In O. Kinne (Ed.), *Marine Ecology*, Vol. IV, Dynamics. Wiley, London. pp. 13–88.

FRONTIER, S. (1977). Réflexions pour une théorie des écosystèmes. *Bull. Ecol.*, **8**, 445–64.

FRONTIER, S. (1978). Interface entre deux écosystèmes: examples dans le domaine pélagique. *Ann. Inst., océangr., Monaco*, **54**, 95–106.

GESSNER, F. (1970). Temperature: plants. In O. Kinne (Ed.), *Marine Ecology*, Vol. I, Environmental Factors, Part 1. Wiley, London. pp. 363–506.

GIESKES, W. W. G. and KRAAY, G. W. (1977). Continuous plankton records: changes in the plankton of the North Sea and its eutrophic southern bight from 1948 to 1975. *Neth. J. Sea Res.*, **11**, 334–64.

GRIFFITHS, R. C. (1965). A study of oceanic fronts off Cape San Lucas, Lower California. *Spec. scient. Rep. U.S. Fish. Wildl. Serv.*, **449.**

HARDY, A. C. (1924). The herring in relation to its animate environment. Part I: the food and feeding habits of the herring with special reference to the east coast of England. *Fishery Invest., Lond.*, **7**, 1–53.

HARDY, A. C. (1956). *The Open Sea, its Natural History: The World of Plankton*, Collins, London.

HEINRICH, A. K. (1962). The life histories of plankton animals and seasonal cycles of plankton communities in the oceans. *J. Cons. perm. int. Explor. Mer*, **27**, 15–24.

HELLEBUST, J. A. (1970). Light: plants. In O. Kinne (Ed.), *Marine Ecology*, Vol. I, Environmental Factors, Part 1. Wiley, London. pp. 125–58.

IKEDA, T. (1970). Relationships between respiration rate and body size in marine plankton animals as a function of the temperature of the habitat. *Bull. Fac. Fish. Hokkaido Univ.*, **21**, 91–112.

JACQUES, G. (1974). La thermocline dans l'écologie du phytoplancton. *Oceanis*, **1**, 51–76.

KIERSTEAD, H. and SLOBODKIN, L. B. (1953). The size of water masses containing plankton blooms. *J. mar. Res.*, **12**, 141–7.

KING, D. P. F. and MACLEOD, P. R. (1976). Comparison of the food and the filtering mechanism of pilchard *Sardinops ocellata* and anchovy *Engraulis capensis* off South West Africa, 1971–72. *Investl Rep. Div. Fish. Un. S. Afr.*, **111**, 1–29.

KINNE, O. (1970). Temperature: animals. Invertebrates. In O. Kinne (Ed.), *Marine Ecology*, Vol. I, Environmental Factors, Part 1. Wiley, London. pp. 407–514.

KINNE, O. (1977). Cultivation of animals. In O. Kinne (Ed.), *Marine Ecology*, Vol. III, Cultivation, Part 2. Wiley, Chichester. pp. 579–1293.

KINNE, O. (1980). Closing address. *14th European Marine Biology Symp. Protection of Life in the Sea. Helgoländer Meeresunters*, **33**, 732–61.

KINNE, O. and ROSENTHAL, H. (1977). Commercial cultivation (aquaculture). In O. Kinne (Ed.), *Marine Ecology*, Vol. III, *Cultivation*, Part 3. Wiley, Chichester. pp. 1321–98.

KUDELINA, E. N. and ZHULAREVLA, S. K. (1963). Food of copepods and balanus larvae in the Azov Sea. (Russ.) *Trudy Azosk. nauchno-issledovat. Inst. Rybn. Khozyaistva*, **6**, 71–82.

LASKER, R. (1966). Feeding, growth and carbon utilization of a euphausiid crustacean. *J. Fish. Res. Bd Can.*, **23**, 1291–317.

LONGHURST, A. R. (1976). Vertical migration. In D. H. Cushing and J. J. Walsh (Eds), *The Ecology of the Seas*. Blackwell, Oxford. pp. 716–37.

LONGHURST, A. R., COLEBROOK, M., GULLAND, J., LE BRASSEUR, R. J., LORENZEN, C., and SMITH, P. (1972). The instability of ocean populations. *New Scient.*, **1**, 4.

LORENZEN, C. J. (1967). Vertical distribution of chlorophyll and phaeopigments: Baja, California. *Deep Sea Res.*, **14**, 735–46.

LORENZEN, C. J. (1976). Primary production in the sea. In D. H. Cushing and J. J. Walsh (Eds), *The Ecology of the Seas*. Blackwell, Oxford. pp. 173–85.

MALONE, T. C. (1971). The relative importance of nanoplankton and net plankton as primary producers in the California current system. *Mar. Biol.*, **10**, 285–9.

MARGALEF, R. (1962). Organisation spatiale et temporelle des populations de phytoplancton dans un secteur du littoral méditerranéen espagnol. *Pubbl. Staz. zool. Napoli*, **32**, 336–48.

MARGALEF, R. (1967). Some concepts relative to the organization of plankton. *Oceanogr. mar. Biol. A. Rev.*, **5**, 257–89.

MARGALEF, R. (1968). *Perspectives in Ecological Theory*. University of Chicago Press.

MARGALEF, R. (1977). *Ecología*. Ediciones Omega, Barcelona.

MARGALEF, R. (1978). General concepts of population dynamics and food links. In O. Kinne (Ed.), *Marine Ecology*, Vol. IV, Dynamics. Wiley, Chichester, pp. 617–704.

NEMOTO, T. (1970). Feeding pattern of baleen whales in the ocean. In J. H. Steele (Ed.), *Marine Food Chains*. Oliver and Boyd, Edinburgh. pp. 241–52.

NIVAL, P. (1976). *Relations Phytoplancton–Zooplancton; Essai de Modélisation*, Thèse Doct., Université Pierre et Marie Curie, Paris.

NIVAL, P. and NIVAL, S. (1973). Efficacité de filtration des copépodes planctoniques. *Ann. Inst. océanogr., Monaco*, **49**, 135–44.

NIVAL, P. and NIVAL, S. (1976). Particle retention efficiencies of an herbivorous copepod, *Acartia clausi* (adult and copepodite stages): Effects on grazing. *Limnol. Oceanogr.*, **21**, 24–38.

ODUM, E. P. (1971). *Fundamentals of Ecology*, W. B. Saunders, Philadelphia.

PANDIAN, T. J. (1975). Mechanisms of heterotrophy. In O. Kinne (Ed.), *Marine Ecology*, Vol. II, Physiological Mechanisms, Part 1. Wiley, London. pp. 61–249.

PARSONS, T. R. (1976). The structure of life in the sea. In D. H. Cushing and J. J. Walsh (Eds), *The Ecology of the Seas*. Blackwell, Oxford. pp. 81–97.

PARSONS, T. R., TAKAHASHI, M. and HARGRAVE, B. (1977). *Biological Oceanographic Processes*, 2nd ed., Pergamon Press, Oxford.

PETIPA, T. S. and MAKAROVA, N. P. (1969). Dependence of phytoplankton production on rhythm and rate of elimination. *Mar. Biol.*, **3**, 191–5.

PETIT, D. and COURTIES, C. (1976). *Calanoides carinatus* (Copépode pélagique) sur le plateau continental Congolais. I: Aperçu sur la répartition bathymétrique, géographique et biométrique des stades; générations durant la saison froide 1974. *Cah. ORSTOM, Océanogr.*, **14**, 177–99.

PONOMAREVA, L. A. (1955). Food and repartition of the Japan Sea Euphausiids (Russ.). *Zool. Zh.*, **34**, 85–97.

PONOMAREVA, L. A. (1963). Euphausiids of the northern half of the Pacific Ocean; extension and ecology of the most abundant species. (Russ.) *Izdat. Akad. Nauk.*, 142 pp.

POULET, S. A. (1973). Grazing of *Pseudocalanus minutus* on naturally occurring particulate matter. *Limnol. Oceanogr.*, **18**, 564–73.

RAKUSA-SUSZCZEWSKI, S. (1969). The food and feeding habits of Chaetognatha in the seas around the British Isles. *Polskie Arch. Hydrobiol.*, **16**, 213–32.

RASSOULZADEGAN, F. (1975). *Ecologie et Relations Trophiques du Microzooplancton dans un Ecosystème Néritique*, Thése Doct., Université Pierre et Marie Curie, Paris.

RASSOULZADEGAN, F. (1979). Cycles annuels de la distribution de différentes catégories de particules du seston et essai d'identification des principales poussées phytoplanctoniques dans les eaux néritiques de Villefranche-sur-Mer. *J. exp. mar. Biol. Ecol.*, **38**, 41–56.

RILEY,, G. A. (1957). Phytoplankton of the North-Central Sargasso Sea, 1950–1952. *Limnol. Oceanogr.*, **2**, 252–70.

RUSSELL, F. S., SOUTHWARD, A. J., BOALCH, G. T. and BUTLER, E. I. (1971). Changes in biological conditions off Plymouth during the last half century. *Nature, Lond.*, **234**, 468–70.

RYTHER, J. H. (1969). Photosynthesis and fish production in the sea. The production of organic matter and its conversion to higher forms of life vary throughout the world ocean. *Science, N.Y.*, **166**, 72–6.

SOROKIN, Yu. I. (1978). Decomposition of organic matter and nutrient regeneration. In O. Kinne (Ed.), *Marine Ecology*, Vol. IV, Dynamics. Wiley, Chichester. pp. 507–676.

STEELE, J. H. (1976). The role of predation in ecosystem models. *Mar. Biol.*, **35**, 9–11.

STEELE, J. H. and FROST, B. W. (1977). The structure of plankton communities. *Phil. Trans. R. Soc.*, **280**, 485–534.

TRAVERS, M. (1971a). *Le Microplancton du Golfe de Marseille: Études Quantitative, Structurale et Synécologique; Variations Spatio-, Temporelles*, Thése Doct., Université Aix-Marseille (II).

TRAVERS, M. (1971b). Diversité du microplancton du Golfe de Marseille. *Mar. Biol.*, **8**, 308–43.

WYATT, T. (1976). Food chains in the sea. In D. H. Cushing and J. J. Walsh (Eds), *The Ecology of the Seas*. Blackwell, Oxford. pp. 341–58.

Marine Ecology Vol. 5, Part 1
Edited by Otto Kinne

5. STRUCTURE AND DYNAMICS OF ASSEMBLAGES IN THE BENTHAL

J. M. PERES

(1) General Aspects

The most characteristic features of assemblages in the benthal have been summarized on p. 49ff. In the following sections we shall detail these characteristics, as has already been done for pelagial assemblages. For organizational reasons, we deal first with macrobenthos assemblages, and then with soft-bottom, microbenthos and meiobenthos assemblages (p. 523).

(2) Macrobenthos Assemblages

(a) Surface Area and Volume of Macrobenthos Assemblages

In the marine environment, a macrobenthos assemblage occupies a surface area which is generally very large in comparison with the volume occupied. In other words, the vertical extension of macrobenthos assemblages is usually very small; in fact, they may be considered to be almost bidimensional. However, many macrobenthos assemblages on both hard and soft substrates exhibit some vertical stratification in two main layers, represented by the infauna on the one hand and the epiflora and epifauna on the other. The latter, in turn, may be further stratified: some organisms lie flat on the substrate—often by most of their body surface (lower stratum)—whereas the bodies of others are more or less erect and thus constitute a higher stratum. Motile forms can participate in either of these strata.

The quasi-bidimensional structure of macrobenthos assemblages tends to cause severe spatial problems. Seeds and larvae of assemblage members usually encounter considerable difficulties in finding a vacant place for settling and subsequent development. In contrast, in the pelagial—even if we consider organisms restricted to only one vertical zone (p. 47ff)—it is practically always possible to find living space. The thickness of any zone always amounts to several tens of metres, often much more. However, living space alone is not enough; there must also be sufficient nutritional support. The algal cell or the newly hatched larva may later starve to death, but such danger also exists in benthos assemblages.

On soft substrates, the surface area is never totally covered by organisms. Hence, finding a vacant place may not seem much more difficult than in the pelagial. However, the risk of a settling stage being eaten when it reaches the bottom, or when it hatches there, is greater than in the pelagial because most members of the benthos assemblage concentrate in the immediate vicinity of the water/sediment interface. The danger of starvation also tends to prevail longer than in the pelagial.

Finding a vacant place is much more difficult on hard substrates than on soft ones as the substrate is usually completely occupied. However, there are exceptions: (i) in coastal areas, extreme wave intensities, sometimes combined with abrasion by suspended sand, tend to remove or injure attached organisms; (ii) in abyssal and hadal zones, fauna scarcity often prevails on rocky substrates; (iii) in totally dark, shallow waters, e.g. remote parts of submarine caves, darkness effects a marked decrease in sessile benthic invertebrate abundance—a phenomenon still to be explained.

Three major circumstances contribute to the space problems on hard substrates: (i) The presence of epibionts* of the second and third degree. (ii) The rapid covering of any vacant hard substrate by epiflora and epifauna. 'Fouling' of man-made substrates (ships' hulls, pipes) is an example of quick colonization of hard substrates. (iii) The succession of populations and transient assemblages on vacant substrates. These are discussed on p. 157.

Once an organism has settled on a vacant space, subsequent development tends to occur in two ways, depending on morphological peculiarities and physiological (mainly nutritional) requirements: (i) horizontal occupation of the lower stratum of the assemblage with a tendency to cover some of the neighbouring organisms, resulting in an epibiosis of the second degree; (ii) vertical extension, combining the utilization of a very restricted surface area for attachment with access to upperlying waters. Erect ascidians, for example, extend their inhalent and exhalent siphons considerably beyond the level of the substrate surface, while many sponges, hydroids, gorgonians, etc. increase their body surfaces, thus augmenting their chances for exchange (oxygen, carbon dioxide) and food intake.

(b) The Benthal as a Heterogeneous Environment

As has already been emphasized, both environment and organisms of benthal assemblages display a much larger range of diversity than in pelagial waters. For example, in any area or depth, hard and soft substrates entertain quite different assemblages (see also Volume I, Chapter 7).

Solid substrates differ mainly in regard to the degree of their hardness. To some extent, this is linked to their chemical nature and their texture. Hardness influences both abundance and composition of the assemblage. Texture refers to the degree of homogeneity and the distribution of different constituents within a given substratum. For example, a ship's hull, volcanic glass, granite or sandstone have very different textures, resulting in differences in the microstructure of the substrate/water interface. The interface can be very even, as in volcanic glass, or it can reveal numerous small holes and crannies which facilitate settlement of seeds and larvae.

Soft substrates differ in their granulometric characteristics both in mean grain size (pebbles to the finest clays) and degree of grain-size mixing. Sometimes the chemical

*For a discussion of the ecological terminology, see KREBS (1978).

nature of the particles also influences organismic distribution; for example, organic-matter content, which often controls the composition of an assemblage, depends, in turn, on silt and/or clay content. Some species have been shown to be more abundant in sands with increased amounts of calcareous particles. Granulometric characteristics (Volume I: GERLACH, 1972) also influence the composition of the infauna. They largely determine burrowing resistance and filter-feeding conditions; the filtering device may become clogged if the silt and/or clay content of the sediment is too high. In soft bottoms, granulometric characteristics largely depend on water movement (Volume I, Chapter 5); the latter is also of paramount importance in regard to input and renewal of suspended organic particles which often represent the most important—sometimes even the only—food resource for many filter-feeding invertebrates.

(c) The Benthal as a Non-moving Environment

The strong connection between bottom characteristics and local assemblage characteristics underlines the fact that the benthal is a non-moving environment. Any significant prolonged change in the composition of a macrobenthos assemblage indicates marked changes in the sea bottom, particularly on soft substrate. However, where an assemblage occupies a very large area, its qualitative homogeneity (biocoenosis aspect) may reveal space-scale changes in the abundance of different species; such quantitative changes (community aspect) usually originate from a lack of synchronism in predator–prey relationships in the different parts of the area concerned (p. 180).

In order to determine the composition of an assemblage as accurately as possible, mainly fixed or sedentary species are taken into account. Swimming animals—belonging to the nektobenthos—are less intimately linked to the bottom. They may be included as characteristic and/or predominant species, at least as transient hosts, especially if they exploit the assemblages in terms of food resources or shelter during at least part of their life-cycle. Food exploitation prevails, for example, in fishes and shrimps of seagrass beds and outer reef slopes. Many small-sized peracarids seek shelter; they bury themselves in shallow-water soft bottoms during the day, rising in the pelagial at night for feeding. Similarly, some fishes such as the hake of the north-eastern Atlantic Ocean (*Merluccius merluccius*) and some large-sized penaeid prawns of the continental slope are mainly nektobenthic—or even truly benthic—by day and more nektonic at night.

Although the benthal is a non-moving environment, the waters overlying it usually move. The influence of water movement on both the characteristics of the soft bottoms themselves and those of the assemblages existing on and within them has received detailed attention in Volume I: RIEDL (1971a,b), SCHWENKE (1971). On hard bottoms water-movement effects seem to be less marked than on soft bottoms. However, water movement plays a significant role in the transport and renewal of mineral nutrients for plants and of organic particles for filter feeders. In addition, waves and water currents prevent or reduce the deposition of mineral particles on horizontal or gently inclined hard substrates and thus keep them 'clean' and suitable for settling of seeds and larvae. In the upper levels, a sufficiently frequent or permanent, non-excessive exposure to waves hinders stratification of the water, resulting in local rising and hence abundance of stenotopic species, i.e. those which cannot tolerate severe changes in surface-water characteristics, e.g. excessive warming or cooling due to thermal transfer from air, or excessive decrease in salinity due to heavy rain.

A sufficiently high upper stratum, for example, that formed by large, ribbon-shaped blades of some sea-grasses or by kelp thalli can strongly reduce the intensity of bottom currents as well as that of wave-induced turbulence (Volume I: SCHWENKE, 1971). Beneath the plant canopy, a more sheltered environment prevails.

(d) Benthal Migrants and Motile Species in General

The percentage of benthic species which move actively over considerable distances or can be passively transported is very low. Passive transport has been demonstrated for some red algae of the genus *Peyssonnelia* (p. 468) and seems also to exist in some compound ascidians which—pulled out from their attachment site—can survive for a long time.

Diel migrants moving between benthal and pelagial have been mentioned above. Some burrowing species also exhibit a diel rhythm but remain benthic. For example, the Norwegian lobster *Nephrops norvegicus*, which inhabits soft bottoms of the lower shelf and upper slope, remains buried during the day but moves over the bottom at night, i.e. *N. norvegicus* moves between two different strata of the same assemblage. Further examples of diel migrations between two adjacent strata of the same assemblage are provided by many sedentary animals of *Posidonia oceanica* beds (p. 439). Heterogeneous sea bottoms may feature on limited surface areas a number of very different microenvironments. On coral-reef flats, for example, algal tufts, holes and crevices of dead corals, sandy patches, living coral heads and bushes, etc. are narrowly intermixed. Here diel rhythms in assemblage composition, linked with a particular microenvironment, are very common. However, except for predators intruding at night from the outer-reef slope, the overall composition of reef-flat assemblages remains fairly constant. To some extent, the example referred to above parallels circadian changes in plankton assemblages of epipelagic and mesopelagic areas (p. 11). Migrations due to a combination of ontogenetic and trophic factors exhibited by some nektobenthic species are not sufficiently numerous to induce substantial changes in the composition of benthic assemblages. For example, in benthos assemblages of brackish-water lagoons, visits of juvenile and immature penaeid prawns, grey mullets or pompanos (genus *Trachinotus*) do not significantly affect the assemblage.

(e) Life-Span of Benthos Organisms

On average, the life-span of macrobenthic organisms is much longer than that of plankton organisms. This difference is mainly related to the fact that the mean body size of the former is larger than that of plankters. The life-span of nektobenthos animals, especially fishes, is of the same order of magnitude as that of true nektonic species and the respective body sizes of both groups are quite similar.

While the life-span of micro- and meiobenthos appears to be of the same order of magnitude—or a little longer—than that of plankton organisms, the life-span of macro-benthic organisms ranges from a few weeks for nematodes, small-sized polychaetes and peracarid crustaceans, to several years or even a few decades for large-sized invertebrates and fishes.

In addition to the relationship between life-span and body size, we must take into account two other factors which influence the life-span: temperature and food availability. The lower the mean annual temperature, the longer the life-span tends to be, and the later the onset of sexual maturity. This general rule is exemplified by numerous invertebrates in high-latitude seas and deep-sea bottoms, as well as by laboratory studies (Volume I: KINNE, 1970). Food availability appears especially to affect small-sized invertebrates usually considered as detritus feeders. For example, on deep-sea bottoms the amount of organic matter which can easily be metabolized—either by these invertebrates themselves or by bacteria subsequently utilized by them—generally seems to be small. It has been assumed that some of these small-sized invertebrates—e.g. pelecypods—are permanently undernourished and thus suffer from a very low growth rate and delayed (possibly up to several decades) sexual maturity. Their life-span is possibly twice as long as the time required for attaining sexual maturity.

In summary, the longer life-span of most macrobenthic organisms compared with that of plankters results in a greater stability of assemblage composition. Assemblage stability also depends on the range of fluctuations in environmental parameters (p. 157ff); it may be commensurate with values intermediate between the mean life-span of macrobenthic member species and the mean period required to reach sexual maturity.

(3) Producers and Consumers in the Benthal

(a) General Aspects

Before we can attempt to analyse the ecological peculiarities of plants and animals in the benthal, as has been done for the pelagial (p. 67ff), we must emphasize that trophic links in the benthal are much more difficult to unravel because benthic organisms represent about 98% of all the species known to exist in oceans and coastal waters. Compared with the pelagial, such tremendous specific richness, together with the heterogeneity of the substrate, results in a much higher diversity of ecological niches and, hence, in a much more complex trophic connex.

(b) Plants in the Benthal

As we began our analysis in the pelagial with the phytoplankton, we shall now first consider the benthic plants, especially those which are photoautrophic, i.e. true primary producers. Hence, we shall take into account neither the fungi, which seem to play an insignificant role in benthic assemblages—except as decomposers—nor the lichens whose marine representatives are restricted to the supralittoral zone and the upper midlittoral subzone.

Unicellular algae on and in soft bottoms will be considered separately in the section devoted to micro- and meiobenthos (p. 523ff). There is much less information available regarding the ecological role of unicellular algae on hard substrates. Attached diatoms often predominate on permanently immersed substrates. Frequently they also cover living substrates, mainly metaphytes but sometimes also external parts of animals, e.g. shells of molluscs, tunics of ascidians.

Cyanophytes, either epiphytic or endophytic (the latter restricted to calcareous substrates such as limestone, shells, dead corals), are particularly abundant in the midlittoral zone and, to a lesser extent, in the upper levels of the infralittoral zone. Their abundance seems to decrease quickly with increasing water depth, and they probably largely disappear a little below the upper levels of the continental slope. However, on 1000- to 2400-m-deep muddy bottoms of the Mediterranean Sea, PSHENIN (1965) observed the blue-green alga *Coelosphaerium benticum*. This alga cannot grow on mineral nutrients alone—although it contains some chlorophyll in illuminated cultures—but grows well on glucose and can fix dissolved nitrogen from ambient sea water.

Unicellular chlorophytes mix with cyanophytes in the supralittoral and midlittoral zones and tend to substitute progressively for them on deeper rocky sea bottoms of the shelf. They probably disappear in the vicinity of the shelf edge.

Although unicellular algae probably account for a significant part of the primary production in the benthal, especially on soft substrates, I shall mainly deal here with macrophytes (also known as metaphytes).

Sea-grasses are represented in the marine environment by about 45 species, and distributed mostly in warm-temperate and intertropical areas. They are photophilic and thus restricted to the infralittoral zone (p. 438). Except for one species of the genus *Phyllospadix*, they inhabit soft bottoms. Unlike most algae, they require true soil from which they draw at least part of their mineral nutrients. However, they can also assimilate nutrient salts dissolved in the ambient water. Epiphytic algae (mainly diatoms), which often largely cover their blades, participate in matter exchange between the sea-grass and the sea water. Depending on the morphology of their blades and the density of their lawns, sea-grasses offer both extra superficies for sessile and sedentary small-sized crawling invertebrates and—beneath their canopy—shelter for many benthic species, sedentary or motile. Compared with similar but unvegetated sea bottoms, sea-grass beds display an increased number of ecological niches. Moreover, sea-grass facilitates sediment fixation (Volume I: DEN HARTOG, 1972, p. 1284).

The requirements of different sea-grass species for soil quality (mineral nutrients) varies. Since the decaying organic matter from the plant itself and its associated flora and fauna tends to increase the content of the soil in humic substances, over a long period of time a succession of species may occur in a given place. 'Pioneer' species, more tolerant to poor soil, become progressively replaced by species which require richer soil. Dense beds of some species (e.g. of the genera *Thalassia*, *Posidonia* and *Zostera*) with wide and ribbon-shaped blades can increase the settling of suspended particles and thus form terraces raised above surrounding unvegetated bottoms.

There are many more multicellular algae in the marine environment than there are sea-grasses. Except for a few species, e.g. the chlorophyte *Caulerpa prolifera*, which seem to be able to draw some nutrients from a soft substrate, all multicellular marine algae are attached to solid substrates or, less often, to living organisms. They obtain all their mineral nutrients from the ambient water. Some kelp species (e.g. of the genus *Macrocystis*) develop on small solid substrates (such as gravel or dead shells) scattered on soft sea bottoms; the adults remain fixed by an anchor-like ball of compacted mud enclosed by their ramified haptera.

Differences in life-span allow distinction of perennial, annual and seasonal species (Volume I: GESSNER, 1970, p. 395). Seasonal species survive in the form of resting stages. Some perennial algae can persist several years without any marked change in their external habitus whereas the thallus of others partly degenerates during stress conditions. Sometimes the whole thallus is lost with only a basal disc or some rhizoids remaining. In general, the life-span of most perennial algae seems to be shorter than that of invertebrates or fishes participating in the same assemblage. Phaeophytes live longest. Some species of the genus *Cystoseira* live 10 years, *Ascophyllum nodosum* 12 to 19 years, and *Nereocystis luetkana* and *Pterygocystis californica* up to 24 years. The three main groups of multicellular algae (chlorophytes, phaeophytes, rhodophytes) are represented in every part of the phytal system of the World Ocean, but phaeophytes are more common on temperate and cold-temperate shores, especially the large-sized species, whereas the number of rhodophyte species increases with the increasing mean temperature. The ratio number of rhodophyte species to the number of phaeophyte species is about 1·5 in the Arctic, 2 in the North Sea, 3 in the Mediterranean Sea, and 4·6 in the Caribbean Sea. For further details consult Volume I: GESSNER (1970).

On the basis of differential light requirements, different species of algae can be classified into photophilic forms inhabiting the infralittoral zone and sciaphilic forms living in the circalittoral zone. Light effects of marine plants have received attention in Volume I by HELLEBUST (1970). The pigments which determine the characteristic colour of brown and red algae can absorb solar radiation of wavelengths which cannot be efficiently absorbed by chlorophyll. Hence, it has long been assumed that major differences in the vertical distribution of chlorophytes, phaeophytes and rhodophytes are related to differential light-absorption characteristics. This, however, is not corroborated by field data. There is no clear relation between the depth inhabited and algal pigmentation. Table 5-1 testifies that the number of chlorophyte species affected by increasing water depth is less than that of brown and red algae. Apparently both light intensity and quality affect the vertical zonation of benthic unicellular algae (Volume I: HELLEBUST, 1970, pp. 149–50).

Table 5-1

Number of chlorophyte, phaeophyte and rhodophyte species found at two different water depths. American coast of the Tropical Atlantic Ocean (After TAYLOR, 1960; modified; reproduced by permission of University of Michigan Press)

Depth (m)	Chlorophytes	Phaeophytes	Rhodophytes	Total
30	56	15	60	131
90	12	2	9	23

The sensitivity of multicellular algae to temperature (annual values and extremes) depends on their biogeographical distribution. Within a given area, the degree of eurythermy tends to decrease with increasing water depth. The same is true with regard to the degree of euryhalinity (Volume I: GESSNER, 1970; HELLEBUST, 1970). On cold shores, the ability of an alga to withstand freezing is directly related to its degree of euryhalinity. In temperate regions with their large annual temperature ranges, the upper levels of the benthal reveal seasonal successions of short-lived species. On the northern coast of the western Mediterranean Basin, for example, the rocky-substrate winter flora, which must sometimes endure temperatures below 10 °C in the 0- to 1-m range, mainly comprises boreal species, whereas in summer subtropical species thrive at the same places where temperatures as high as 26 °C now prevail.

The nutritional requirements of multicellular marine algae are still poorly investigated as are the relationships between the effects of nutrient salts, temperature and light. In the immediate vicinity of the shoreline, the amounts of mineral nitrogen and phosphorus seem always to exceed the requirements of shallow-water algae. Phosphate content usually fluctuates considerably (SEOANE-CAMBA, 1965). According to CONOVER (1958), the influence of light, temperature and nutrient salts on the period of maximum growth of estuarine seaweeds varies largely according to the species concerned. Some species, such as *Cladophora gracilis* f. *tenuis, Enteromorpha linza, E. plumosa, Stylophora rhizoides* and *Ulva lactuca* var. *latissima*, grow maximally under strong illumination (more than 500 cal g cm^{-2} d^{-1}) at moderate temperatures (18 to 20 °C). *Agardhiella tenera, Enteromorpha compressa* and *Gracilaria verrucosa*—as well as the phanerogams *Ruppia maritima* and *Zostera marina*—display faster growth at lower illumination levels (less than 400 cal g cm^{-2} d^{-1}) and temperatures above 24 °C. CONOVER also demonstrated that a decrease in production rate may be related to a decrease in inorganic phosphorus content (less than 0·4 μg at. l^{-1}) of the sea water in spring (April–May) and of nitrates (less than 0·2 μg at. l^{-1}) in autumn (October) in many estuarine algae, such as *Enteromorpha clathrata, E. intestinalis, Ectocarpus siliculosus, Gracilaria verrucosa, Polysiphonia novae-angliae, Punctaria plantaginea* and *Scytosiphon lomentarius*. The growth of multicellular algae may further be influenced by the fluctuations of growth-promoting substances such as vitamin B$_{12}$ (Volume I: COLLIER, 1970, p. 72; Volume III: BONOTTO, 1976, pp. 473–7). Apparently, growth-promoting substances are released mainly by bacteria and blue-green algae but rarely by multicellular seaweeds. In common with some phytoplankters, several seaweeds of eutrophicated areas—e.g. species of *Ulva* and *Enteromorpha*—are able to utilize low-molecular-weight organic substances as a source of carbon and nitrogen instead of carbon dioxide and inorganic nitrogen compounds.

Water movement (Volume I, Chapter 5) reduces or eliminates water stratification and thus tends to minimize temperature and salinity gradients. Furthermore, it replenishes nutrients and disperses excretory products. However, mechanical effects of excessive wave movements can cause damage, e.g. detachment and/or tissue injuries.

(c) Accessibility of Primary Benthos Production to Consumers at the
Second Trophic Level

True Herbivores

Although pelagial filter feeders can utilize non-living particles, evidence has been provided in the preceding section that they largely graze on living phytoplankters wherever these are sufficiently abundant and of convenient size. In contrast, true herbivores are relatively infrequent in the benthal. Of the two types of plants available unicellular algae are more often consumed alive than multicellular algae.

On solid substrates, both non-living and living unicellular algae are intensively grazed upon by many prosobranch gastropods (limpets, littorinids, *Gibbula* and many small-sized species of *Bittium, Rissoa, Rissoina*, etc.) and most of the polyplacophores. These grazers also feed on the youngest stages of multicellular algae. On British coasts, SOUTHWARD (1964) demonstrated that cyclic abundance variations of *Patella vulgata* and brown algae of the family Fucaceae are inversely related; on average, one *P. vulgata* grazes $1·5 \text{ cm}^2 \text{ h}^{-1}$. On the coast of Oregon (USA), a total volume of littorinids of $0·2 \text{ cm}^3$ dm^{-2} (i.e. about three individuals) suffices to keep the substrate quite bare (CAS-TENHOLZ, 1961). In summer diatom density in the supralittoral zone seems to be controlled mainly by environmental factors, in the upper midlittoral by grazing. In winter, the influence of grazers seems to be lower since the activity of the gastropods decreases. Grazing by large-sized gastropods, such as limpets, can cause rock erosion if the latter is sufficiently soft: on the limestone cliffs of Dover (English Channel), erosion rate due to limpets has been estimated to be $1·5 \text{ mm yr}^{-1}$.

On soft substrates, unicellular algae (mainly diatoms), both attached on sediment particles and free living, are consumed by various invertebrates usually considered as 'detritus feeders', but really consuming both living algae and true detritus, i.e. non-living organic particles covered by epiphytic bacteria. In any case, 'detritus feeders' cannot avoid ingesting mineral particles together with organic ones. Whereas suspension feeders exploit moving water layers above the bottom and hence can often obtain sufficient food, most of the so-called 'detritus feeders' (which are at least in part herbivorous) must move about in order to collect their food. On soft bottoms, it seems that the only organisms we can consider strictly herbivorous are the 'sand-lickers'—e.g. some tanaids—which graze the microphytic (unicellular algae and bacteria) cover of sand grains. However, their role in the assemblage usually seems to be insignificant.

Multicellular algae are less frequently grazed upon by benthic herbivores than unicellular algae. However, *Littorina obtusata*, a very common species in algal belts of rocky shores of the midlittoral zone in the English Channel, feeds on almost all the brown algae of the family Fucaceae, with a striking preference for *Fucus vesiculosus*. On soft mixed bottoms, some large-sized intertropical prosobranchs—such as species of the genera *Strombus* and *Lambis*—exlusively feed on small pieces of multicellular algae (e.g. *Cladophora*, *Hypnea* and *Polysiphonia*) fixed on small solid particles scattered over the bottom or on sea-grass leaves. However, *Strombus* can also feed on microphytobenthos via ingestion of gravel and coarse sand particles which are thrown out again after the digestive enzymes have removed the epiphytic algae; in contrast, the meiofauna ingested together with the gravel and sand does not seem to be digested. Among opisthobranch gastropods, species feeding on large seaweeds are much more common. Some of them belong to the Tectibranchiata, e.g. *Aplysia* which feeds mainly on red algae, and the biggest species (weighing up to 6 kg) *Tethys californicus*, which feeds on *Codium*. Many nudibranchs (e.g. species of *Hermaea*, *Elysia*, *Acteonia* and *Limapontia*) are grazers, especially utilizing green algae, e.g. *Codium*, *Ulva*, *Enteromorpha*, *Cladophora* and *Chaetomorpha* (MILLER, 1961; SWENNEN, 1961).

Apart from gastropods, only a few benthic animals are herbivorous. Like gastropods, regular echinoids are to some extent able to control algal populations on solid substrates. This has been demonstrated, for example, in the Plymouth area (English Channel) for *Echinus esculentus* which grazes on *Hildenbrandtia prototypus*, cyanophytes and small-sized chlorophytes at the rate of $2·9 \text{ cm}^2 \text{ h}^{-1}$. It has recently been demonstrated that these echinoids, like other echinoderms (crinoids, asteroids and ophiuroids), can also utilize dissolved organics and small organic particles, both living (e.g. diatoms) and non-living. These are digested by enzymatic exudates from the epithelium, especially in the grooves of the spines (Volume II: PANDIAN, 1975, p. 98). In the Caribbean Sea and Gulf of Mexico, *Lytechinus variegatus* seems to feed exclusively on the blades of the sea-grass *Thalassia testudinum* ('turtle-grass') which is obviously also the main, if not the exclusive, diet of the marine green turtle *Chelonia mydas* and the sirenian *Trichecus manatensis*. *T. testudinum* is also consumed by a fish (genus *Sparisoma*). In the Mediterranean Sea another sparid fish (*Sarpa salpa*) feeds on brown algae of the genus *Cystoseira* which occupy the upper levels of the infralittoral zone on rocky substrates. *S. salpa* probably ingests the phytal fauna (p. 143) together with the algae; it is not yet certain whether the fish is able to digest the vegetal material ingested. Fishes of the family Siganidae (rabbit-fishes) must be considered omnivorous rather than strictly herbivorous (Volume III: KINNE, 1977, p. 1007). Herbivorous nutrition is also very rare in decapod crustaceans: at the Kerguelen Islands (South Seas), the diet of the spiny lobster *Jasus lalandei* mainly consists of multicellular algae; in the Florida mangrove swamps, the crab *Cardisoma guanhumi* mainly feeds on leaves and fruit of the mangrove trees. For further details on marine animal nutrition consult Volume II: PANDIAN (1975) and Volume III: KINNE (1977).

In summary, in benthic assemblages direct utilization of living vegetal matter is rare. Herbivorous nutrition is much more frequent in the pelagial than in the benthal. Since photosynthesis is the primary source of organic-matter production—except for organics synthetized by chemoautotrophic bacteria (Volume II: SCHLEGEL, 1975)—a critical analysis of the pathways of energy and matter between pelagic and benthic assemblages is of paramount ecological significance. The natural starting point of all such pathways is the organic matter, both dissolved and particulate, synthetized by photoautotrophs. In a different context, these aspects have received attention in Volume IV: CONOVER (1978), FINENKO (1978), GREZE (1978), SOROKIN (1978) and WANGERSKY (1978).

Dissolved Organics as Nutritive Resource for Macrobenthos Invertebrates

The nutritional significance of dissolved organic matter (DOM) has received detailed attention in Volume II: PANDIAN (1975, p. 72) and Volume IV: CONOVER (1978) and WANGERSKY (1978). Numerous macrobenthic invertebrates have been shown to take up and metabolically utilize DOM. However, the importance of this phenomenon in terms of nutritional support under conditions *in situ* remains a matter of debate. Presumably, the nutritional utilization of DOM is insignificant when compared with the total energy budget of most benthic invertebrates. However, it must be remembered that pogonophores which have no gut probably fulfil all their energy requirements through DOM absorption and skin digestion; NØRREVANG (1965) observed that the tentacles of *Siboglinum ekmani* bear pinnulae which support microvilli embedded in a mucous sheet, possibly for entrapping very small particles (up to macromolecules?).

Macrobenthic invertebrates clearly benefit indirectly from the DOM locally available: via transformation of dissolved substances by bacteria. Drifting or surface-attached bacterial aggregates are an important food source of benthic microphages.

Nutritional Significance of Detritus

An accurate evaluation of the energy transfer from the first to the second step of the trophic pyramid remains difficult. The best way for such transfer seems to be the utilization of detritus, especially particulate matter from metaphytes at different stages of decomposition. All organic detritus particles are covered with microorganisms (especially bacteria, fungi, colourless flagellates and ciliates) (Volume IV: CONOVER, 1978, p. 301), THAYER and co-authors (1977, p. 120) report for the macrodetrital fraction of *Zostera marina* leaves a progressive decrease in the C/N ratio relative to dead leaves and thus an improvement of nutritional quality for potential consumers. It appears that the consumer benefits mainly, if not exclusively, from the living epiphytic microorganisms not from the dead organic matter. Since the efficiency of energy transfer from non-living organic matter is estimated to be about 30 to 40%, this means that about 60 to 70% of the energy stored in the vegetal biomass is lost for consumers at the second step of the trophic pyramid which cannot utilize fresh vegetal material.

In summary, for most 'detritus feeders', the non-living organic parts of the detritus serve as substrate for aggregating their real food: microorganisms. An exact differentiation between organisms which utilize the non-living components of the detritus particles and those which utilize the attached microorganisms is difficult, if not impossible. A mechanism for discrimination and selected uptake of these two components does not seem to exist.

Obviously, the superficial film of the sediment represents both the layer richest in organic detritus and that most accessible to animals. CONOVER (Volume IV, 1978, p. 295) combined as 'interface-feeders' consumers which crawl or swim over soft bottoms, hard bottoms and attached structures. Because the latter two probably feed exclusively on living material, I prefer to consider them as 'grazers' (either utilizing plants, p. 326, or animals). On soft-bottom interfaces the food resources of the so-called detritus feeders are a combination of particulate materials both organic—living and non-living—and inorganic, i.e. they differ considerably with regard to the nature of their energy content, collecting mechanisms required, and the technique of transfer to the consumer.

Regarding the feeding behaviour of detritus feeders, five principal groups may be distinguished. These differ with respect to the mechanism of food collection and the sediment layer actually fed upon:

(i) Food collection by mechanical means only. Examples are amphipods of the genus *Corophium* which scrape the surface film round their burrows with their antennae; the decapod *Galathea dispersa* and some crabs of the genus *Uca*—possibly also some pagurids—scrape the sediment, usually with their third maxillipeds, the material gathered in this way being later sorted by various other buccal appendages.

(ii) Food collection by ciliary currents together with particle aggregation due to a sticky secretion. This feeding behaviour is rather common in echinoderms. Some ophiuroids, mainly of the family Amphiuridae (which can also entrap suspended particles from the bottom water), have podia terminating in a small club with mucus cells and are thus capable of picking up film particles. Irregular echinoids display a variety of feeding behaviour usually involving podia and/or spines. According to CHESHER (1963) in *Moira atropos*, a burrowing species of the Florida coasts, the front ambulacrum is probably the most active part of the test which participates in food collection. However, dorsal podia can also extend up to the sediment surface and collect particles from the surface film; the spatulate spines of the front ambulacrum transport the particles—previously wrapped in mucus—down to the peristome, forming a mucous string which is later broken up by buccal podia before being ingested. According to McGINITIE and McGINITIE (1949) and GOODBODY (1960) who studied *Dendraster excentricus* and *Mellita sexiesperforata*, the role of the podia in the feeding of these irregular echinoids (family Scutellidae) is insignificant: the epithelium of the dorsal parts of the test bears 'golf-club shaped' spines which elaborate mucus agglomerating the particles; these are then transmitted to the ramified ambulacral areas and finally to the mouth; the podia appear to have only a tactile function in this family, with the possible exception of *Mellita sexiesperforata* which can catch very small (1 μm) particles.

To some extent, the most primitive pelecypods of the order Protobrachiata can also be considered to belong to this group. For example, species of the genera *Nucula* and *Yoldia* extend over the sediment surface their labial palps, the ciliary groove of which transports film particles to the mouth; here the particles are sorted, inedible material being rejected through the exhalent siphon. However, according to STASEK (1965) this feeding behaviour is combined with pumping actions generated by the ctenidial cilia; these result in the formation of small balls which are transmitted by ciliary currents either directly to the mouth (the smaller ones) or to the labial palps (the larger ones).

(iii) Food collection by picking up. To some extent, the part played by podia in the feeding of some ophiuroids and echinoids can be considered as a 'pick-up' technique (examples: holothurioids of the orders Aspidochirota and Elasipoda). Differences in the GW/BW ratio* recorded by SOKOLOVA (1959, 1972) suggest that the ability of buccal podia to select edible particles from the sediment largely differs according to the species involved; species of SOKOLOVA's group (i & ii) are rather selective, whereas those of his group (iii) (p. 132) must be considered limivorous; they are whole-sediment engulfers rather than detritus feeders.

(iv) Food collection by pumping and subsequent filtering. This feeding mechanism is typical of detritus-feeding pelecypods, such as most species of the families Tellinidae, Garidae, Scrobiculariidae, Pandoridae, etc. . In fact, the principal food-collecting mechanism is almost the same in both detritus- and suspension-feeding pelecypods (Volume II, Part 1, 1975: PANDIAN, p. 75); it will be summarized later in the section dealing with suspension-feeding invertebrates in general. However, the siphons of detritus-feeding pelecypods are more extensible than those of sestonophagous species, especially the inhalent one. Some prosobranch gastropods inhabiting soft bottoms, such as species of the genus *Aporrhais*, also seem to be filtering species feeding on the surface film; the same is possibly true for some species of the genus *Turritella*, although some of these have been described as suspension feeders.

In other detritus feeders water currents are generated not by ciliary currents but by the animal moving within its burrow. Figs 5-1 and 5-2 illustrate two examples of this behaviour: an echiurid of the genus *Urechis* and the polychaete *Arenicola marina*. The feeding behaviour of *Urechis* has been described by McGINITIE and McGINITIE (1949,

---

*GW/BW ratio stands for weight-of-the-gut content/weight of the animal ratio (expressed in 1/10,000, i.e. $^0/_{000}$).

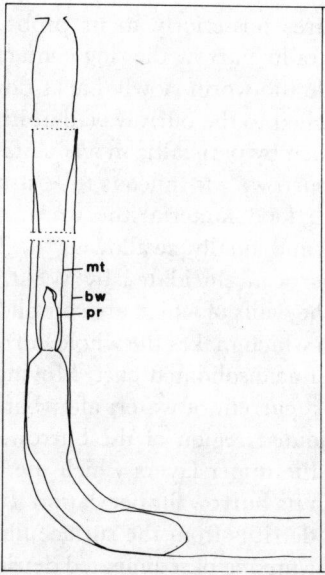

Fig. 5-1: *Urechis caupo*. Feeding behaviour during pumping phase. bw: burrow wall; mt: mucus tube; pr: proboscis. (After McGinitie, and McGinitie 1949, simplified; reproduced by permission of McGraw-Hill Publications.)

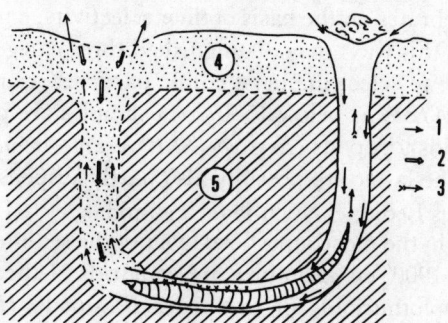

Fig. 5-2: *Arenicola marina*. Feeding behaviour. 1: water current; 2: input of particles originating from upper sand layer and near-bottom water; 3: rejection of faeces; 4: upper-sand oxidized layer; 5: lower-sand reduced layer. (After Wells, 1966; modified; reproduced by permission of Netherlands Institute for Sea Research.)

p. 186–7). This worm features posteriorly to its proboscis a ring of mucus glands; moving toward one entrance of its burrow the ring contacts the walls of the burrow and begins to secrete mucus while the worm slowly backs down its burrow thus forming a transparent mucus tube attached to the burrow wall at its upper end and to the body of the worm at its lower end. Then by peristaltic movements along its body, *Urechis* pumps a water current through its burrow, 'the mucous tube straining all the particles'. When the tube becomes filled with food material the worm contracts, thus detaching the mucous tube from its body and finally swallowing it. The feeding behaviour of the lugworm *Arenicola marina* has been elucidated by WELLS (1945, 1949); the L-shaped burrow of this polychaete—the walls of which are consolidated by mucus—is completed by an unconsolidated portion which makes the whole burrow more or less U-shaped; the mouth of the worm is in the unconsolidated part. Moving into the L-shaped part of its burrow the worm generates a current of water; after bathing the gill-tufts, the current flows through the unconsolidated region of the burrow. This results in a mixture of superficial film and sand of the upper layers which the worm ingests; at intervals the lugworm moves backwards in its burrow for defecation. It is not certain that *Urechis* and *Arenicola* feed exclusively on detritus from the surface film; in both species one cannot exclude the possibility of some mixing of sedimented detritus with particles suspended in the bottom water.

(v) Food collection by sediment swallowing. Bulk-swallowing of sediment (limivorous species) is widespread, mainly in deep-sea invertebrates, but also occurs in inhabitants of shallower-water bottoms, e.g. some polychaetes, holothurioids and irregular echinoids. Many deep-sea ophiuroids and asteroids are also assumed to be mud eaters, but from the bathyscaphe I have observed that some species are often narrowly associated with sponges and this would suggest that they could be filter-feeding in the current generated by the former.

TURPAEVA (1953) and SOKOLOVA (1959) investigating the invertebrate bottom fauna, mainly in the depths of the northern Pacific Ocean, attempted to classify sediment-feeding invertebrates on the basis of their selectivity, employing measurements of the GW/BW ratio. In this way, they distinguished three categories: (i) Species which are, to some extent, able to select the most edible particles occurring in the superficial sediment layer; their GW/BW is below $1550^0/_{000}$; examples are polychaetes (*Amphicteis*, *Brada* and some terebellids), isopods (*Eurycope*, *Storthyngura*, *Antarcturus*, etc.), ophiuroids of the genus *Amphiura* (the latter are possibly also partly suspension feeders) and many pelecypods of the families Ledidae, Nuculidae, Tellinidae, etc. (ii) Species whose ability to select edible particles in the most superficial layer of the sediment is less marked; their GW/BW ratio is about $1900^0/_{000}$; examples are echiurids of the genera *Thalassema* and *Tatjanellia* and many holothurioids of the order Elasipoda (*Elpidia*, *Scotoplanes* and *Psychropotes*). (iii) Species unable to perform any selection within the sediment; their GW/BW ratio averages $3200^0/_{000}$; examples are many polychaetes (*Travisia*, some species of the families Maldanidae, Capitellidae and Opheliidae), some sipunculids (genera *Aspidosiphon* and *Phascolosoma*), some asteroids (*Ctenodiscus*, *Eremicaster*, *Thoracaster*, *Vitjazaster*) and echinoids (*Brisaster* and various species of the families Spatangidae and Pourtalesiidae) as well as many holothurioids: *Trachostoma*, *Molpadodema*, *Chiridota*, *Pseudostichopus*, *Parastichopus* and *Paelopatides*. With regard to the holothurioids, GW/BW ratios as high as $6335^0/_{000}$ have been recorded by TURPAEVA; species of group (iii) can be considered limivorous.

A special kind of detritus available and consumed by detritus feeders is represented by faecal pellets from both planktonic and benthic organisms; their significance as a food resource has been discussed extensively by PANDIAN (Volume II, Part I, 1975, pp. 202ff.). Aggregates formed in the pelagial and thereafter deposited on the bottom constitute another food source for detritus feeders.

In summary, we may assume that the energy available to benthic consumers at the first, i.e. vegetal, level of the benthic food pyramid is smaller than in the pelagial. Furthermore, the energy produced by benthic primary producers is less access-ible—including the microphages. Such deficit in benthal energy supply must be com-pensated for by energy income from the pelagial: sinking particles, dead or alive, not only support detritus feeders on the sea bottom but can also be 'intercepted' on their way down by a variety of suspension feeders.

Suspension Feeding

Suspension feeding has been reviewed in Volume II by PANDIAN (1975, p. 75) and in Volume IV by CONOVER (1978, e.g. p. 277). Quantitatively, suspension feeders (or sestonophages) require either a sufficiently high density of digestible suspended particles or a sufficient supply through water movement. Examples are: the South European mussel *Mytilus galloprovincialis* which attains high abundancies only in protected and rather eutrophic areas or on more oligotrophic but sufficiently exposed open coasts; the 'goose-crab' *Emerita analoga* which is restricted to the breaker zone of warm-temperate Brazilian beaches; intertropical species of the genus *Donax* which permanently migrate upward and downward, thus remaining in the breaker zone of the beach.

On deep-sea bottoms sestonophages feed on bacteria, aggregates, and faecal pellets from deep-sea plankters, all these particles being by far less abundant than in shelf areas. Hence, suspension feeders mainly occur in places with high-speed bottom cur-rents. Examples are crinoids of the genus *Leptometra* which establish dense populations on soft bottoms near the shelf edge and in the upper levels of the slope (p. 479) and sestonophages on offshore rocky bottoms (p. 463), e.g. sponges, ahermatypic scleracti-nians, alcyonarians, gorgonians, antipatharians. Diving with the bathyscaphe on the lower levels of the continental slope of Madeira Island, I have observed large ses-tonophages such as anthozoans and stalked crinoids on the rocky steps which emerge from the sediment. Investigating bottom assemblages of the Pacific Ocean down to the greatest depths, SOKOLOVA (1959) found that suspension feeders are better represented, and sometimes predominant where the bottom profile is convex, possibly due to local increase in the speed of bottom currents. During bathyscaphe observations on the abyssal plain west of the Azores archipelago, I have noticed sestonophages such as glass-sponges and anthozoans, almost exclusively inhabiting small steps (about 25 cm high) in the muddy bottoms.

In principle, suspension feeders can be classified grossly into 2 major categories: (i) species which exploit existing water movement, i.e. which are unable to generate cur-rents on their own; (ii) species which actively produce water movement utilized for feeding, respiration and excretion.

Among the species unable to generate water currents are certain gastropods of the family Vermetidae. For example, *Vermetus gigas* elaborates a large, sticky mucous sheet which entraps floating particles; once sufficiently loaded with food, the sheet is ingested. Some ophiuroids of the genus *Ophiothrix* collect their food in a similar way employing mucus filaments between the spines of their arms. Holothurioids of the order Dendrochirota possess ramified buccal podia which constitute a large unciliated filtering basket; the prey is agglomerated in mucus secreted by pharynx and oesophagus. Ophiuroids of the family Gorgonocephalidae seem to be unable to agglomerate small prey; the Red Sea species *Astroboa nuda* seizes prey by the hanger-shaped ends of the finest ramifications of its arms (TSURNAMAL and MARDER, 1966).

The most widespread mechanism of suspension feeding combines current generation and filtration. Macrobenthos invertebrates solved this problem in two different ways.

(i) Some of them feature an extensive filtering, basket-like, structure of rather low efficiency combined with rather weak ciliary currents, e.g. sabellid and serpulid polychaetes, bryozoans and other lophophorates, hydroids which, in addition, possess nematocysts. In general, in this group the water to be filtered enters the basket at its top and flows out at its base between the proximal parts of the tentacles which form the basket.

(ii) Other suspension feeders generate stronger inhalent currents due to a more developed ciliated surface and a narrower inhalent opening; after passing over the gills which retain particulate food (as well as dissolved oxygen), the water is forced out through an exhalent opening together with undigestable particles; examples are pelecypods, including detritus-feeding forms (Volume II, PANDIAN, 1975, p. 80 and Volume IV, CONOVER, 1978, p. 280) but excluding members of the order Septibranchiata, some species of the orders Protobranchiata and Ascidians. The latter retain particles by a combination of filtration and entrapping in a particularly heavy mucus string; the same is true for the suspension-feeding gastropod *Crepidula fornicata* (Volume II, PANDIAN, 1975, p. 87). Some gastropods of the family Vermetidae collect their food by cilia-generated currents and subsequent filtering—either combined with mucus entrapping, e.g. genus *Petaloconchus*, or without, e.g. members of the genera *Dendropoma* and *Aletes*. According to GOSSELCK and co-authors (1978), *Branchiostoma senegalense* is a non-selective suspension feeder; its size-spectrum of filtering efficiency ranges from 1 to 300 μm. In the second group of suspension feeders, sponges appear to be particularly efficient in collecting the smallest particles (up to 0·1 μm) due to the combination of forced-water circulation in their body cavities and a high filtering capability of the choanocytes.

Among species credited to be suspension feeders, non-parasitic cirripeds (both pedunculate and thoracic), which collect suspended particles by rhythmic beating of their setose cirri (Volume II: PANDIAN, 1975, p. 90), seem to occupy a particular position; their feeding behaviour 'is perhaps more nearly raptorial than filter feeding' (Volume IV: CONOVER, 1978, p. 291).

In shallow-water areas, macrobenthos suspension feeders mainly feed on living phytoplankters and aggregates which cover a wide size scale. Apparently, these are usually sufficiently abundant for meeting the energy requirements, even in species whose filtering efficiency spectrum is rather narrow. Hence, they did not need to develop adaptive mechanisms, such as those which allow some plankton microphages to substitute raptorial feeding for coarse filter feeding when the size of available prey increases too much. However, on deep-sea bottoms where suspended particulate matter is often exceedingly scarce, typical suspension feeders can display an adaptive behaviour which allows them to substitute the food resources in the sediment surface film for those in the bottom water. For example, the pennatularian *Umbellula* can bend its stalk and collect food from the water layer immediately above the bottom ('boundary layer') and possibly from the sediment surface film. According to PASTERNAK (1964), the antipatharian *Bathypates lyra* is also able to exploit the boundary layer by extending the main axis of its colony supported by small branches immediately above the bottom. From my own observations during bathyscaph dives, I conclude that some stalked crinoids exhibit similar behaviour. A further example are the pelecypods of the order Septibranchiata, which also changed their feeding behaviour from microphagy to macrophagy. Another noteworthy adaptive mechanism is that of ascidians of the family Octacnemidae—closely related to the Cionidae—which can utilize fine nutritive particles and small crustaceans, thanks to their ability to direct the oral basket formed by the inhalent siphon towards the water stream, e.g. in members of the genus *Dicopia*, whereas in the genus *Octacnemus*, small prey—mainly crustaceans—are transported into the branchial sac by eight membranaceous and muscular lobes of the oral siphon (MONNIOT and MONNIOT, 1975). Much more amazing is the adaption of tunicates of the new class Sorberacea whose branchial sac disappeared when the inhalent siphon developed into a prehensile apparatus; their diet includes foraminiferans, nematods, polychaetes, crustaceans as well as ophiuroids (MONNIOT and MONNIOT, 1975).

Conclusions

In essence, animals exclusively or predominantly belonging to the second level of the trophic web can be classified into 4 categories: grazers, detritus feeders, limivores and suspension feeders. A disadvantage of such classification is that it does not take into account possible combinations of the feeding types distinguished. Hence, the predominant feeding type should be used as the primary criterion.

However, even this may be difficult on intertidal soft bottoms. The pelecypod *Macoma balthica*, for example, feeds on detritus at low tide but is a suspension feeder at high tide (BRADFIELD and NEWELL, 1961). Also *Cochlodesma praetenue*, another pelecypod, is able to feed both on detritus and suspended particles. The prosobranch gastropod *Nassarius obsoletus*, very common on Atlantic coasts of North America, is mainly a film eater but can also feed on dead animals, decaying algae and related materials.

The distribution of the major trophic groups of bottom invertebrates (sestonophages, selective deposit feeders, non-selective deposit feeders, limivores) depends on the availability of their specific food resources which, in turn, depends on both the bottom profile and the degree of eutrophy (for details consult Figs 5-3, 5-4 and 5-5).

(d) Carnivores

Classifying macrobenthos carnivores is also difficult. Firstly, some microphages, especially amidst the coarse filter feeders, ingest both vegetal and animal prey, if both are simultaneously available, as do many omnivorous zooplankters; examples are many colonial cnidarians (hydroids, alcyonarians, pennatularians, gorgonians, antipatharians) whereas scleractinians and crinoids of the family Gorgonocephalidae seem

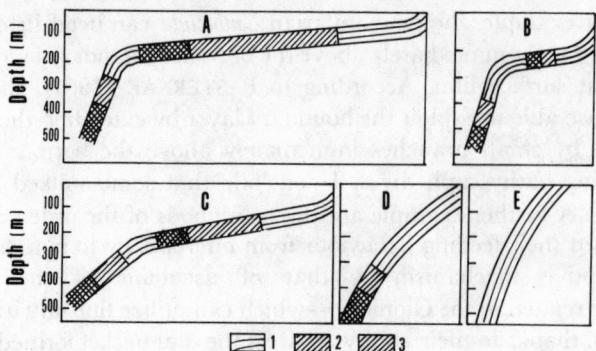

Fig. 5-3: Distribution of predominant trophic groups on shelves of different widths and profiles (A–E). 1: sestonophages; 2: selective deposit feeders; 3: non-selective detritophages and limivores. (Based on information provided by VINOGRADOV, 1977.)

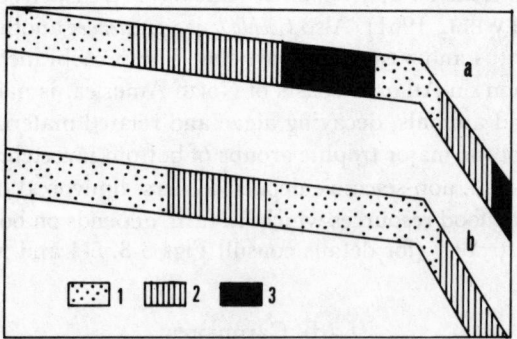

Fig. 5-4: Schematic zonation of predominant trophic groups on a flat shelf in eutrophic (a) and oligotrophic (b) areas. 1: sestonophages; 2: selective deposit feeders; 3: non-selective detritophages. (Based on information provided by VINOGRADOV, 1977.)

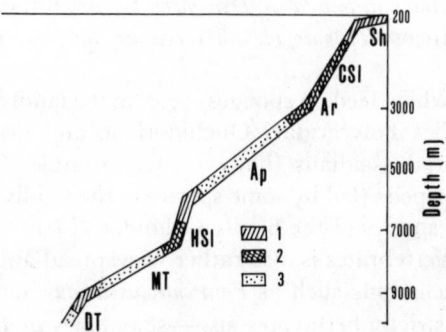

Fig. 5-5: Schematic zonation of predominant trophic groups related to profile of the ocean bottom. Sh: shelf (see Figs 5-3 and 5-4 for detail); CSl: continental slope; Ar: abyssal rise; Ap: abyssal plain; HSl: slope into trenches: MT: mean depth trenches; DT: deepest trenches. 1: sestonophages; 2: mixing up of sestonophages, detritophages and carnivores; 3: non-selective detritophages and limivores. (Based on information provided by SOKOLOVA, 1959.)

to be strictly carnivorous. However, many cnidarians harbour zooxanthellae which meet at least a part of their energy requirements (p. 410). Secondly, certain macrobenthos species feed on micro- and meiozoobenthos; although possibly not very numerous—at least on shelf bottoms—these cannot be completely disregarded. Thirdly, some species are both microphagous and macrophagous, for example, some actinians which can feed on living or dead large-sized prey, as well as on small-sized prey such as *Metridium senile*; the ciliated epithelium transports the particles up to the top of the tentacles which bend to transfer the prey into the mouth. Fourthly, many macrophages indiscriminately feed on living prey and dead animals. Nevertheless, I suggest classifying benthic macrophagous carnivores into four types: grazers feeding on living colonial invertebrates; exclusive predators; combined predators and scavengers, scavengers.

Grazers Feeding on Colonial Invertebrates

Gastropods of the suborder Nudibranchiata are without doubt a taxonomic unit which by far includes the greatest number of grazers on colonial invertebrates, especially hydroids. According to SWENNEN (1961), who investigated their feeding habits on Netherland coasts, some of them feed without preference on various hydroids, e.g. *Dendronotus frondosus, Doto coronata, Facelina coronata*, whereas others specialize more on prey of only one genus (*Coryphella verrucosa* on *Eudendrium* spp., *Tergipes despectus* on

Laomedea spp.,) and sometimes on only one species, for example, *Precuthona peachi* on *Hydractinia echinata* and *Trinchesia foliata* on *Dynamena pumila*. Other Cnidaria can also be preyed upon by nudibranchs: *Tritonia plebeia* feeds on alcyonarians, and most of the aeolidids on actinians.

There are also grazers which feed on sponges—e.g. in the family Glossoridae—and on bryozoans—in the families Polyceridae, Onchidoridae and Okeniidae. More rarely grazed upon are compound ascidians (however, for example, *Gonoidoris castanea* only feeds on *Botryllus*) and cirripeds (fed by some species of the family Onchidorididae); fish eggs are preyed upon by species of the family Calmidae (MILLER, 1961).

Grazing on colonial invertebrates is also rather widespread amidst echinoderms. For example, some regular echinoids such as *Paracentrotus lividus* and *Psammechinus miliaris* which are credited to be strictly herbivores also—sometimes preferentially—feed upon sessile invertebrates, e.g. hydroids, bryozoans and compound ascidians, on an algal thallus or sea-grass blade. Amazingly, this diet seems to be more suitable to them, resulting either in better growth or in a greater gonadal index (REGIS, 1978). This behaviour is more common in asteroids which exhibit a marked preference for grazing on sponges. Feeding exclusively on sponges are the tropical species *Oreaster reticulatus* and *Pteraster tessellatus*. The small-sized *Asterina gibbosa*, very common on the western European coasts, can feed on sponges but prefers compound ascidians. The 'crown-of-thorns' *Acanthaster* only feeds on scleractinians, greatly damaging many coral reefs in the last fifteen years mainly in the southern tropical regions of the Pacific Ocean.

Exclusive Predators

Exclusive or real predators are so numerous that it is almost impossible to give anything more than a few selected examples from different phyla or classes. For example, among polychaetes predators exist in the families Glyceridae, Eunicidae, Nephthydidae and Nereidae; these feed on small-sized invertebrates. Jaws in the pharynx cannot be considered as evidence for predatory behaviour : all nereids have jaws but some of them are microphages, whereas among nephthydids, the proboscis of which does not bear jaws, some species are detritus feeders (e.g. *Nepthys incisa*) and other carnivores, such as *N. cirrosa* and *N. hombergi* (CLARK, 1962). The amphinomid *Hermodice caruculata* only feeds on scleractinians which it sucks with its extroversed and largely dilated buccal cavity. All the nemertines are predators and feed mainly on polychaetes captured with the proboscis. The turbellarian *Pseudostylochus ostreophagus* bores the shell of oysters of the genus *Crassostrea* by purely chemical means feeding on the soft parts of the body.

In gastropods, macrophagous behaviour is very exceptional in the opisthobranchs. However, the cephalaspid *Navanax inermis* mainly feeds on other opisthobranchs which it detects chemotactically and then engulfs. Apparently, fishes might also be preyed upon by *N. inermis* (PAINE, 1963). Predation by prosobranch gastropods, which are very numerous and display a variety of feeding behaviour, has been extensively reviewed by PANDIAN (Volume II, 1975, p. 92) and CONOVER (Volume IV, 1978, p. 305).

Almost all benthic cephalopods seem to be predators, mainly feeding on decapod crustaceans and fishes which they actually hunt. However, PILSON and TAYLOR (1961) noticed that, in the northern Pacific Ocean, some species of the genus *Octopus* can bore gastropod and pelecypod shells with their radula; the authors assume that the cephalopod injects through this hole—much smaller than from boring snails—a venomous substance secreted by the posterior salivary glands which kills the prey.

Predation by large-sized asteroids was previously, if only briefly, mentioned by PANDIAN (Volume II, 1975, p. 100) and CONOVER (Volume IV, 1978, p. 314). However, it deserves some additional comments. Macrophagous asteroids display two main feeding mechanisms:

(i) In many families the sea star extroverts its stomach over the prey and performs extraoral digestion, as mentioned above for some grazing species. However, since most of the prey organisms consist of pelecypods strongly protected by a shell, the sea star must first open that shell. This '*Asterias* type' of feeding begins with an 'attaching phase', i.e. the sea star covers the pelecypod and firmly fastens its tube feet on the shell; in the second, 'stretching phase', the predator opens its arms causing an elongation of the podia; finally, in the 'pulling phase', the tube feet contract, thereby opening the prey's shell. The force involved in this mechanism can be as strong as 4 to 5 kg and can be sustained over several hours. Sea stars of the '*Asterias* type' may also start feeding on their prey by inserting one or more stomach lobes through any opening into digestible tissues (FEDER and MØLLER-CHRISTENSEN, 1966, p. 103). This mechanism can combine with the pulling: a gap of 0·1 mm is sufficient for *Asterias forbesi* to introduce its stomach into the shell of *Mytilus edulis*. It must also be pointed out that, in spite of the fact that pelecypods are most commonly preyed upon by sea stars of the '*Asterias* type', some other animals may also be consumed, such as gastropods and, especially, thoracic cirripeds. Finally, even in sea stars where feeding is usually extraoral, some small prey—e.g. young or small-sized pelecypods and barnacles ripped off the substrate—can be found in the stomach.

(ii) A feeding mechanism, almost generally adopted by all species of the family Astropectinidae, consists of engulfing whole prey animals. These predators actively move, mainly at night, on soft substrates with their conical, sucker-lacking tube feet in search of burrowed prey; once it has detected its prey, the sea star begins to dig, using the long and strong spines in its arms and finally engulfs the prey. Most of its prey are molluscs (pelecypods, gastropods, scaphopods) and some small echinoids. The digging-for-food by these sea stars is so efficient that investigating their stomach content often yields a better knowledge of the ambient infauna assemblage than dredging. The prey is slowly digested in the stomach, and undigestible parts are later rejected through the mouth because the stomach is blind (no anus). Some pelecypods, especially those with very low metabolic rates, may withstand digestion in the stomach with their shell firmly closed and thus may be thrown out still alive. Without going into too much detail, it is also interesting to note that among the rather specialized sea stars, some feed mainly—sometimes exclusively—on other echinoderms. For example, some, not all, species of the genus *Luidia* feed only on echinoids (e.g. *L. clathrata*); others, such as *L. ciliaris* (eastern Atlantic Ocean and Mediterranean Sea) feed on members of various classes of the phylum: echinoids, ophiuroids, asteroids and holothurioids. The feeding behaviour of some species of the genera *Solaster* and *Crossaster*, whose diets are restricted to other asteroids, is particularly amazing because they induce the prey to autotomize an arm which they afterwards consume.

Combined Predators and Scavengers

Some species of the genus *Conus* may be classified in this group. For example, *Conus californicus*—which is able, as are other conids, to feed on various living gastropods (e.g. species of *Nassarius, Polinices*) and polychaetes—seems in the natural environment to feed mainly on dead molluscs, polychaetes and fishes. Many decapod crustaceans—particularly large-sized Reptantia such as lobsters, spiny lobsters—are also predators and scavengers and the same is true for some demersal fishes.

As regards echinoderms, it has been assumed that some ophiuroids, particularly on deep-sea bottoms meet their food requirements either by scavenging or by predation on very small animals (meiobenthos?); their blind stomach prevents them from swallowing mud—a condition which is widespread among deep-sea invertebrates and, especially, other echinoderms. It may also be mentioned here that some regular echinoids, usually credited to be rather strict grazers, can feed on dead animals. For example, although largely provided with algae, *Paracentrotus lividus* seems to prefer broken mussels or crabs and even dead fish (REGIS, 1978). According to FUJI (1967), the gut of the Japanese *Strongylocentrotus intermedius*, the main food of which consists of brown algae (chiefly *Laminaria japonica*), often contains small-sized amphipods, isopods and molluscs; possibly these echinoids can only utilize dead animals, at least as far as the swimming amphipods and isopods are concerned. Sometimes, predatory and scavenging behaviour may alternate according to the season: for example, the antarctic sea star *Odontaster validus* is a predator in summer, but a scavenger in winter (p. 407).

Scavengers

It seems that prosobranch gastropods comprise more species, such as the genera *Buccinum, Neptunea*, credited to be more strictly necrophagous than any other taxonomic group. Most species of the family Nassidae cover, often almost completely, large-sized dead animals and suck these out with their long proboscis. Large-sized and actively swimming peracarids which inhabit deep-sea oligotrophic bottoms seem to be exclusively scavengers (p. 22).

(e) Species Collecting Food by means of Several Mechanisms

Finally, some species collect non-dissolved food by employing different mechanisms, each usually providing food which may vary with regard to origin, nature and size.

In addition to the pelecypod *Macoma balthica*, I mention here (p. 135) the following examples. Several polychaetes, e.g. *Owenia fusiformis*, can (i) collect small suspended particles thanks to the ciliary epithelium on their tentacle crown; and (ii) sort the sand with their lips when burrowing, ingesting particles smaller than 0·2 mm but rejecting larger ones or using these for tube building (DALES, 1957). The enteropneust *Balanoglossus gigas* usually feeds by means of water currents generated by the ciliary epithelium of its proboscis, the mucus-trapped particles then being sorted on the collar with the small-sized ones selected for ingestion; *B. gigas* can also swallow sediment by sorting it with its proboscis and collar. Possibly, a third mechanism exists: screening of the respiratory water current and transfer of the particles retained directly to the oesophagus. According to FEDER and MØLLER-CHRISTENSEN (1966, p. 96), the asteroid *Patiria miniata*—which is rather common on US Pacific coasts—'is mainly an omnivorous scavenger of both plant and animal materials that are enveloped by its protusible and remarkably large cardiac stomach', and it is also able to use 'its stomach as a ciliary (flagellary) mucous organ' by secreting, on its oral side raised from the substrate, a mucous sheet for water filtration. This makes it a suspension feeder. Aquarium observations have further shown *P. miniata* to be able to feed on sessile diatoms by licking them with its stomach.

According to MAGNUS (1964), the Red Sea ophiuroid *Ophiocoma scolopendrina*—which mainly inhabits, in very dense populations up to 50 individuals m^{-2}, the upper levels of the infralittoral zone either in crevices of the rocky bottoms or between rhizomes of sea-grass—is a very versatile feeder. It is omnivorous and can collect its food in three different ways: (i) its tube feet collect and agglomerate in a mucous secretion microalgae and detritus covering the substrate and transmit the resulting balls to the branchial groove which, in turn, transports them to the mouth; (ii) the spread tube feet form two comb-like structures on each arm, thus filtering the suspended particles; (iii) where the intertidal substrate dries up at low tide, the oral face is turned up when the tide returns and the tube feet catch particles—mainly dead unicellular algae—which float up to the air–sea interface.

Ophiocomina nigra, very common on all coasts of western Europe, is both microphagous (see (i) below) and macrophagous (points (iii), (iv), (v)). It exhibits five different feeding behaviours: (i) detritus feeding on the sediment surface film; (ii) suspension feeding by employing mucous filaments spread between the terminal spines of its upwardly bent arms, the collected particles being agglomerated by ciliary currents and transported by the podia to the mouth; (iii) grazing on sessile algae and dead animals; (iv) capturing by curling the arms for obtaining rather large prey such as motile invertebrates or debris; (v) capturing by tube feet of smaller prey (FONTAINE, 1965).

(4) Relationships Among Consumers in the Benthal

(a) Comparison between Food Web and Food Chains in Pelagial and Benthal

This section is devoted to species whose entire life-cycle is completed in the benthal and to the benthic stages of species whose life-cycle begin with planktonic larval stages. In order to appreciate the specific trophic structure of the benthal, it seems useful to refer to an example of a linear food chain in the pelagial, as it occurs in surface and subsurface layers of rather oligotrophic areas (Fig. 5-6).

Fig. 5-6 requires four specifications: (i) filter-feeding fishes and whales, which are restricted to sufficiently eutrophic areas are not considered; (ii) intercalated additional links can exist (e.g. 1b and 3b); (iii) the links 2, 4 and 5 are considered in a global sense, irrespective of additional links which may appear within the system concerned; for example, a carnivorous copepod or a fish larva can be consumed by a chaetognath, which, in turn, will be consumed by a fish at link 3b or 4.

Such a linear food chain in an oligotrophic region of the pelagial is characterized by three factors: (i) Average body size progessively (i.e. without hiatus) increases from link

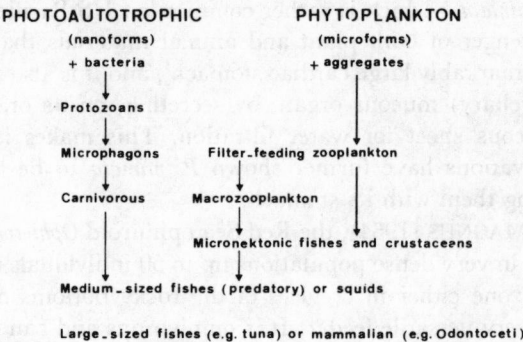

Fig. 5-6: Scheme of linear food chains in the pelagial.
(Original.)

1 to link 6, combined with decreasing abundance, decreasing turnover rate and increasing life-span. The range of the size scale seems to be larger in the pelagial than in the benthal, i.e. from a small coccolithoporid (3–5 μm in diameter) to a big bluefin (3·5 m in length) or to very large sharks which are all nektonic or nektobenthic. Moreover, from the trophic point of view, the whole pelagos in a given area constitutes a single ecosystem, from the smallest organisms to the largest ones, no matter what the number of links in the linear food chain: five or six as in Fig. 5-6, or only three as in the food chain of eutrophic Antarctic regions: diatoms → euphausiids → blue whale. (ii) Unedible organisms are very rare in the pelagial; this means that through the various pathways described above—sometimes rather indirect—part of the energy stored in a living pelagic organism practically always becomes transferred to a subsequent trophic link; in other words, in the pelagial trophic food web blind alleys seem to be rather exceptional. (iii) Since dead pelagic organisms and detritus sink more or less rapidly to deeper layers and scatter within a progressively increasing water volume, thanks to turbulence and currents, detritus eaters and scavengers play a rather insignificant role in the pelagic foodweb.

The trophic structure of the benthal is characterized by a reverse situation. This can be summarized in three points: (i) While at first sight the range of the benthal size scale seems to fall into the same order of magnitude as that in the pelagial, such similarity is misleading from the functional point of view, as will be emphasized below. The micro- and meiobenthos constitute an almost autarchic ecosystem which does not provide a significant amount of energy to the macrobenthos, except for the part contributed by the microphytobenthos in the diet of the so-called 'detritus-feeding invertebrates'. Moreover, taking into account only the range of the size scale, it is obvious that the largest species which can be considered truly benthic, such as the flatfish genus *Hippoglossus* and the giant crab genus *Macrochirus*, are very few and not as large as the biggest pelagic animals. (ii) The practical non-utilization of fresh, i.e. living, plants by consumers at the second link of the benthal trophic chain as opposed to the situation in the pelagial, has been discussed above (p. 129). This results in an increased percentage either in species usually called 'detritus feeders' (although in the phytal zone they feed on both detritus and microphytobenthos), or species definitely restricted to detritus and its epiphytic cover. (iii) Inedible or rarely-preyed-upon invertebrates are much more numerous in the benthal than in the pelagial. Many benthal invertebrates are protected by a hard exoskeleton (sometimes with spines, tubercules, etc.), by elaborate repellents, and sometimes by toxic substances. Although some carnivorous macrobenthos animals —which feed for example on sponges, anthozoans, ascidians and echinoderms—are known, many species of these phyla are rarely preyed upon and thus represent blind alleys of energy transfer. The size of many benthos invertebrates also protects them from predation by most demersal fishes.

(b) Significance of Small-sized Organisms as Food Source for Macrobenthos Animals

The hypothesis that the amount of energy transferred from the micro- and meiobenthos to the macrobenthos is very small requires verification. While this hypothesis may be roughly acceptable with regard to soft-bottoms, it remains questionable for deep-sea muddy bottoms (p. 567) and for shallow-water hard bottoms. As a matter of fact, both micro- and meiobenthos of hard bottoms, even in the midlittoral and infralittoral zones, are poorly investigated. However, many macrobenthos invertebrates, such as polyplacophores, prosobranch gastropods and echinoids, feed on microphytobenthos of hard (rocky) or of living substrates and certainly contribute more to a direct energy transfer from the first to the second trophic link than grazing on metaphytes. Moreover, the epiphytic felt formed by this microphytobenthos, either on rock or on a living substrate (e.g. a sea-grass leaf, p. 441), is inhabited by micro- and meiofauna elements, mainly nematods, harpacticoids, small polychaetes, small gastropods, etc. .

Consideration of the potential pathways of energy and material from the micro- and meiobenthos to the macrobenthos must include, for hard substrates, the very particular assemblage represented by the 'phytal water', i.e. the water within the algal tufts. Obviously multicellular algal species can support epiphytic diatoms or other unicellular algae and also sessile or sedentary invertebrates, except for algae whose thallus exudes slimy or repellent substances which tend to inhibit epiphyte settling; clearly the morphology and consistency of the thallus or phanerogam leaf is also of paramount importance as regards the composition and abundance of the epiphytic cover. The so-called 'phytal fauna' also includes the non-sessile animals, i.e. those which are swimming, walking and crawling within the algal tufts.

According to WIESER (1951), characteristics of the surrounding water between algal tufts influences composition and abundance of the phytal fauna, but less than the morphological characteristics of the host alga itself. Considering both the degree of ramification of the thallus and the characteristics of its surface—e.g. its slimy nature—WIESER has proposed to use an 'adsorption index' he inferred from the difference in weight between the algal tuft without and with the film of water which is retained by the thallus just removed from water (expressed in percentage of the weight of the latter). The more ramified the thallus, the higher the adsorption index (AI). Examples of AI quoted by WIESER are: *Cladophora suhriana*, 88%; *Jania rubens*, 72%; *Gelidium corneum*, 65%; *Cystoseira mediterranea*, 68%; *C. abrotanifolia*, 24%; *Fucus serratus*, 19%; *Pelvetia canaliculata*, 19%. On an average, the phytal fauna abundance seems to be approximately related to the AI value; for example, *C. abrotanifolia* (IA, 24%), *C. mediterranea* (IA, 68%) and *Gelidium corneum* (IA, 65%) exhibit 21, 68 and 134 individuals per gram of thallus, respectively. *C. spicata*, which inhabits rather exposed areas, exhibits a rich phytal fauna, whereas *C. abrotanifolia* whose biotope is very similar but whose thallus less ramified, has a much poorer associated fauna.

In fact, apart from morphological thallus peculiarities, several other parameters control phytal-fauna abundance, particularly the degree of exposure. For example, in the northern Adriatic Sea ZAVODNIK (1965) found significant differences in biomass of the phytal fauna in algal species whose biotope differs as regards the degree of wave exposure (Table 5-2). Also the taxonomic composition of the fauna changes as a function of exposure: amphipods are more abundant in exposed places whereas copepods and nematodes—and sometimes polychaetes—predominate on algal species inhabiting more sheltered places. However, even if some non-crustose multicellular algae usually small-sized and forming thick and short tufts—resist wave stress in very exposed localities, the phytal fauna tends to decrease in both abundance and specific richness.

Table 5-2

Increase in biomass and abundance of phytal water fauna with the degree of exposure (After ZAVODNIK, 1965; reproduced by permission of the author)

Algal species	Algal biomass (g m^{-2}; w.w.)	Animal biomass (g m^{-2}; w.w.)	Number of individuals
Fucus virsoides	4750	22	107,000
Cystoseira abrotanifolia	4333	30	121,000
C. barbata	7500	94	491,000
C. spicata	9375	290	1,020,000
Halopteris scoparia	3088	608	2,265,000

Another important parameter controlling the abundance of the phytal fauna is the content in organic matter—both living and non-living—of the surrounding waters. The most abundant phytal fauna recorded by ZAVODNIK (1965) occurs in *Halopteris scoparia* tufts, with very strong sedimentation of organic and inorganic particles (Table 5-2). In the same way, MAKKAVEEVA (1964), who compared the composition of the phytal shore fauna of *Cystoseira* assemblages in the Adriatic, Aegean and Black Seas, found it to be very similar to the same predominant species: young *Platynereis dumerili*, *Leptochelia dubia* and *Rissoa splendida*. However, both in number of individuals and biomass, the phytal fauna is much more abundant in the Black Sea, where the surface-water plankton is richer than in the other two seas.

On tidal shores, water movement and organic-matter content combine their positive effects. HAGERMANN (1966), for example, recorded up to 4,600,000 individuals m^{-2} of algal vegetation in the *Fucus serratus* belt of Øresund shores (Denmark).

Finally, the phytal fauna exhibits important seasonal changes in both abundance and composition. This has been demonstrated by ZAVODNIK (1967b) on Adriatic Sea shores and by HAGERMANN (1966) in the fauna associated with *Fucus serratus*. According-ing to HAGERMANN, species feeding on epiphytic diatoms, such as some nematodes, ostracods, and harpacticoids and small-sized gastropods, always represent a rather important percentage (38–68%) of the whole local assemblage. This suggests that detailed assessment of seasonal fluctuations in abundance and biomass of the phytal fauna would require better knowledge of seasonal fluctuations in the microphytobenthos epiphytic on multicellular algae.

(c) Some Aspects of Energy Input from Pelagial to Benthal

The accessibility of energy from benthos primary producers to consumers in the second link of the food chain has been reviewed above (p. 127ff) from a rather descriptive and analytical point of view. The next step is to discuss this problem again in a more global way.

As pointed out below (p. 531), on soft bottoms of the infralittoral zone, the amount of organic carbon synthetized by the microphytobenthos often falls in the same order of magnitude as in the upperlying water. Its accessibility to consumers is very similar to that in the pelagial. However, irrespective of the fact that part of this primary production is utilized by the meiozoobenthos, there are two factors which suggest that utilization of soft-bottom unicellular algae by the macrobenthos is less efficient than grazing in the pelagial: (i) Macrobenthos detritus feeders are, on average, much larger than zooplankton grazers; the amount of energy they obtain from microphytobenthos and detritus which is required for maintaining their biomass is much higher than in the latter; the longer life-span of macrobenthos consumers only increases this difference. (ii) The nutritive particles ingested by detritus feeders are always mixed with inorganic particles which must be thrown out again, resulting in additional energy losses.

On hard bottoms the amount of organic carbon synthesized by unicellular algae is still poorly investigated and the number of grazers is relatively low. However, the role of the phytal fauna in energy transfer from the first to the second link of the food chain appears to be rather insignificant in terms of the total energy requirements of consumers at this second link of the macrobenthos subsystem.

As regards the macrophytes, both on rocky shores and on mixed and soft bottoms, it is fairly certain that they synthesize a much larger amount of organic carbon—mainly in the lower midlittoral subzone and the infralittoral zone—than the microphytobenthos and phytoplankton. However, as has been explained above (p. 128), this material is insufficiently accessible to consumers.

Many macrobenthos animals at the second trophic level cannot really thrive unless they receive some energy from the upperlying waters—i.e. from the pelagial—most of this energy being represented by sinking organic particles both living and non-living. Some of the sinking particles sediment on the bottom and thus benefit both the meiozoobenthos and the detritus-feeding macrobenthos; other particles are intercepted by suspension feeders. Since most macrobenthic suspension feeders are afterwards consumed by macrobenthos carnivores, these sestonophagous species exploit the pelagial to the ultimate benefit of the benthal. Another pathway macrobenthic animals use to exploit the pelagial is indirect development including planktotrophic larvae (sensu THORSON, 1950); larval stages feed on plankton algae or animals and grow in the pelagial before settling to the bottom where they bring the energy gained during their planktonic life. The same is true as regards benthic species—e.g. many peracarid crustaceans which stay in their burrows by day but migrate at night to the pelagial where they feed on plankton organisms.

In fact, except for deep-sea assemblages, the benthal can also provide some energy to the pelagial, e.g. through floating eggs released by bottom invertebrates or fishes, which may be utilized by consumers in the plankton. However, this energy input from the benthal to the benefit of the pelagial is very low in comparison with that flowing the opposite way.

MASSE (1971) observed a striking example of the inability of a macrobenthos assemblage to develop unless it received sufficient energy from the pelagial. In some areas of the Gulf of Marseilles (France), larval stages in the assemblage inhabiting well-sorted fine sands (p. 435) in the infralittoral zone can be transported by currents to areas where the sediment characteristics are quite suitable to adults, but where the upperlying waters are exceedingly oligotrophic. Once metamorphosed, the young benthic stages begin to grow; they compete more and more for food as they grow till finally the total amount of food becomes insufficient and thus stagnates their growth (e.g. the pelecypod *Venus gallina*). The assemblage can survive only due to a periodic input of larval stages from less oligotrophic areas.

To some extent, one can assume that coral reefs, which represent a complex of assemblages (the biomass and species richness of which are higher than in any other ecosystem of the benthal) could not have thrived so well, unless they could develop two adaptive mechanisms to offset the insufficient energy amount they obtain from the tropical pelagial, of which the extreme oligotrophy is well documented. The most efficient mechanism is certainly symbiosis between hermatypic scleractinians and zooxanthellae (p. 412). However, the massive release of mucus by many coral-reef invertebrates (particularly scleractinians and hydrocorallians) results in a bulk of nutritive particles for filter-feeding organisms in waters flowing across the reef. These particles may be considered as compensating for the scarcity of true plankton.

As regards the exploitation of particles sinking from the pelagial to the benthal, intraspecific competition for food between different size classes may occur in macrobenthic suspension feeders, chiefly pelecypods. For example, on sandy bottoms densely populated with full-grown individuals of *Cerastoderma edule*, very young individuals more or less starve because the larger ones catch most of the sinking particles since their inhalent currents are much stronger. This intraspecific competition can be related to the fact that in suspension feeders such as pelecypods and ascidians the mesh size of the filter is rather similar in young and fully-grown individuals. Moreover, especially in ascidians, the part played in particle retention by mucous filaments—which do not exist in filtering crustaceans—tends to restrict the proper role of mesh size itself.

(d) Influence of Predator–Prey Relations on the Structure of Benthos Assemblages

In the pelagial the more mature the system, the higher the percentage of predatory species it includes (p. 108). As the eutrophy of the euphotic-layer system decreases progressively because the enrichment process is a transient one, or because the system is drifting further and further away from the area of fertilization, abundance and species number of predators increase. Therefore, there is a tendency toward temporal or spatial disjunction of the different steps in the ecological succession (Figs 2-5, 2-6, 6-26). Only systems which are permanently oligotrophic, such as those of the suprathermoclinar layer in subtropical areas of the open ocean, exhibit a rather high and steady percentage of carnivorous species.

In comparison with the pelagial, the macrobenthos seems to exhibit several peculiarities as regards the influences exerted by predators on its structure and dynamics. Tentatively, these peculiarities are: (i) In comparison with the total number of animal species, the macrobenthos subsystem generally includes a higher percentage of predator species than the pelagial, that is, in accordance with the higher degree of maturity of the former. (ii) Since on average, the life-span of macrobenthos animals is longer than that of zooplankters—including carnivores—predators at the adult stage occur in the macrobenthos subsystem more permanently than in the pelagial. (iii) The structure of a macrobenthos assemblage and the population dynamics of the different species which may be preyed upon within the assemblage are influenced by a complex combination of several factors: the recruitment of populations of the various prey organisms; the recruitment of predator populations at all levels—including the highest ones—of the food pyramid; the peculiarities in the feeding behaviour of different predators which either enable them to meet their energy requirements on various kinds of food or, on the contrary, lead them to select a well-defined type of prey (specialized predators, p. 172).

Experiments carried out by BRUNSWIG and co-authors (1976) on population dynamics of a soft-bottom macrobenthos assemblage (25-month study of the fauna in 1 m^{-2} metal boxes filled with sterilized sediment) exemplify the influence of predators on their prey populations. The first stage of repopulation is characterized by a fauna predominated by the cumacean *Diastylis rathkei* and epibenthic polychaetes (e.g. *Harmothoe sarsi*) and, a little later, by sedentary polychaetes, such as various spionids and *Pectinaria koreni*; at the second stage (12–18 months after the experiment was initiated),the abundance of polychaetes strongly decreased, possibly due to an increase in the nemertean population; at the third stage, the authors observed a dominance of pelecypods (chiefly *Abra alba*) together with errant polychaetes (*Nephthys* sp.). This was followed by a fourth stage, characterized by an increase in the predator populations, among them some polynoid polychaetes, *Carcinus maenas*, and—in terms of biomass—mainly *Asterias rubens* which tends to exhaust the population of pelecypods (especially that of *Abra alba*). According to BRUNSWIG and co-authors, predators are of eminent importance in succession control, in spite of the fact that abundance data—especially for *Abra alba*, *Pectinaria koreni* and *Diastylis rathkei*—suggest that differences in recruitment from year to year also influence succession.

On hard artificial substrates (pilings) in the subtidal at Beaufort, North Carolina (USA), experimental manipulations of densities of the echinoid *Arbacia punctulata* by KARLSON (1978) exemplify the effects on the assemblage composition of a combination between prey recruitment rate and the predatorial feeding behaviour. *A. punctulata* which is a grazer on algae in other localities at Beaufort, appears as a generalized carnivore with some preference for certain sponges (e.g. *Microciona*), rarely feeding on the colonial hydroid *Hydractinia echinata*, but never on the sponge *Xestospongia halichondroides* (KARLSON). In the absence of *A. punctulata* the composition of the assemblage on the pilings is highly variable, with *H. echinata* being very rare. In contrast, on piling assemblages which include *A. punctulata*—especially those which had been built up more than ten years ago—*H. echinata* can occupy up to 30% of the substrate although its recruitment rate is much lower than that of most other species constituting the assemblage (*Balanus* spp., species of *Ostrea, Bugula, Schizoporella, Styela plicata*, etc.). According to KARLSON (p. 237), 'the importance of vegetative growth as a mechanism for utilizing substratum under favourable conditions is clearly indicated by this long accumulation of *Hydractinia*;' these conditions provide through *Arbacia* grazing unoccupied substrates favouring both vegetative growth and recruitment of the hydroid.

Predator–prey relations and the influence they exert on the population dynamics of the prey species are mentioned below with an eye on unspecialized, as well as specialized predators. However, as regards the latter, one has to take into account the particular behaviour of some species such as many carnivorous gastropods (*Thais, Acanthina*) which only feed either on cirripeds or on sessile pelecypods.

MURDOCH (1969) demonstrated that in assemblages in which these two kinds of prey attain similar abundances, some individuals of *Acanthina* exhibit a transient preference for one or the other prey (Table 5-3).

Table 5-3

Acanthina spirata: preferential predation as a function of adaptation to one or the other prey (After MURDOCH, 1969; reproduced by permission of Duke University Press)

Acanthina spirata	Number of *Balanus glandula* preyed	Number of *Mytilus edulis* preyed
Non-adapted to one or the other prey (control)	73	98
Adapted (accustomed) to feeding on *Balanus*	130	18
Adapted (accustomed) to feeding on *Mytilus*	12	168

(e) Food Web and Food Chain in the Benthal

When we now attempt to synthetize the preceding considerations regarding food webs and food chains in the benthal, some peculiarities become apparent. These may be summarized as follows: (i) Microenvironments—or 'habitats' as redefined by WHITTAKER and co-authors (1973)—are much more diversified in the benthal than in the pelagial: while in the latter all organisms come into contact only with sea water, in the benthal they also come into contact with the bottom. This may vary considerably, for example in the degree of hardness or granulometric characteristics, and this affects organismic distributions. (ii) The specific richness of flora and fauna is much higher in the benthal than in the pelagial. Thus, in the benthal, both food and feeding behaviour are much more diversified.

Both points (i) and (ii) combine to increase the number of ecological niches in the benthal; the term 'niche' is well used here as redefined by WHITTAKER and co-authors

(1973); it refers to the role of an organism within a community, i.e. its place in the biotic environment, as well as its relations to food and enemies.* Therefore, trophic relations tend to be more numerous and more intricate in the benthal than in the pelagial.. (iii) Taking into account the two types of 'negative interactions' (KREBS, 1978) between species, namely competition for the same food source and predation, it seems that the parts they respectively play in the structure of the ecosystem would be different in the benthal and the pelagial.

Considering global efficiencies in regard to that percentage of the energy provided at the base of the system—i.e. the amount of energy made available by primary production—which reaches the terminal link of the longest food chain in the pelagial, two different situations may occur. In the first one, primary productivity is low and so is the content in non-living organic particles—which, to some extent, may substitute for phytoplankton in the diet of some 'herbivorous' zooplankters. This results in food shortage for animals at the first consumer level; hence the abundance of first-level consumers is low, but their specific diversity tends to be high, the interspecific competition between them being buffered by an increase in the abundance and/or diversity of predatory species. However, the energy (food) available at the second level ('herbivores') and the energy loss in the transit from the second to the third (first-order carnivores) and from the third to the fourth levels (second-order carnivores) of the trophic pyramid, usually leads to the absence of the fifth and sixth trophic levels within these ultraoligotrophic pelagos systems.

In pelagos systems with their high primary production, the plankters at the first consumer level usually cannot utilize all the vegetal food, except in regions where pelagos assemblages include dense populations of sufficiently large-sized herbivores, e.g. krill and some species of anchovy, consisting of sinking, dying phytoplankters, faecal pellets etc.; such 'organic rain' largely benefits the macrobenthos, providing the depth does not exceed 200–250 m. In these eutrophic pelagos systems, carnivorous species are neither abundant nor diversified and the food chain is short. Top predators—i.e. those at the fifth and sixth steps of the trophic pyramid—in general do not exist. However, they do occur farther away from the most eutrophic area or—if in the area itself—after cessation of eutrophic-layer enrichment, because in both cases the carnivores of the third and fourth steps of the trophic pyramid they feed upon become sufficiently abundant.

In the benthal at least on the upper part of the shelf (down about 50–60 m), the total amount of energy available at the base of the trophic pyramid seems almost always

*The concept of epibiontic life requires some qualification. While the terms 'epiflora' and 'epifauna' correspond to plants and animals whose body extends completely or partly above the bottom/water interface, the term 'epibiont' refers to sessile species of both plants (epiphytes, epiphytic) and animals (epizoans, epizoic) attached not to the hard substrate itself but to another (usually also sessile) organism. This relationship is called 'epibiosis'. An epibiosis of the second degree corresponds to such a relation between only two 'superimposed' organisms, for example: a brown alga (e.g. *Fucus* sp.) whose thallus supports small hydroids, bryozoans or small-sized serpulid polychaetes (e.g. *Spirorbis* sp.); a sea-grass blade which supports small multicellular algae (e.g. *Polysiphonia* sp.), hydroids, bryozoans, small compound ascidians (e.g. didemnids or botryllids); a sponge partly covered with the zoantharian *Parazoanthus axinellae*; a simple ascidian whose tunic supports bryozoans, serpulids, small compound ascidians, or even small mussels (e.g. *Modiolaria* sp.) partly embedded within the tunic. An epibiosis of the third degree corresponds to the superimposition of three different organisms; such a combination is often observed in cases in which the organism attached to the non-living substrate is a large seaweed. Much more rarely, three animals combine: I have once observed a pelecypod of the genus *Arca* supporting a compound ascidian (*Didemnum*) which, in turn, supported a kamptozoan of the genus *Loxosoma*.

sufficient for meeting the requirements of the macrobenthos because: (i) production from algal and sea-grass populations on shallow-water hard and soft bottoms is high; (ii) production from microphytobenthos, especially on soft bottoms, is sufficient; (iii) particulate organic matter, living or non-living, sinks down from upperlying waters. As mentioned above, the latter energy source may benefit bottom invertebrate assemblages of the shelf and sometimes also assemblages somewhat below (200–250 m depth) in areas where plankton production is high, for example, in the region of the California current upwelling. Although many macrobenthos consumers possess mechanisms for feeding detritus, it is obvious that the energy content of detrital material is lower than that of fresh plants from which the detritus originated.

However, there is practically little, if any, competition between animals representing these two ethological types. To some extent, due to the fact that first-level benthos consumers derive their food from different sources (suspended or deposited materials), such lack of competition compensates for the energy loss ascribable to the limited consumption of fresh plant material characteristic of macrobenthic animals of the first consumer level. Competition between suspension feeders and detritus feeders is also reduced because the larval stages of both groups tend to settle on that type of soft bottom where the food sources available will be most suitable to adults. Suspension feeders prefer rather coarse bottoms (sand or gravel) with efficient water renewal; detritus feeders prefer muddy bottoms with high sedimentation rates. On deep-sea bottoms all the energy available at the base of the macrobenthos subsystem originates either from the upperlying waters or from the organic material suspended in descending currents, sometimes along the neighbouring continental slope, sometimes by advection from higher latitudes related to convergences. The nature of the food resources available there to species of the first consumer level remains poorly documented; their utilization by macrobenthos species possibly involves prior reworking by species of the micro- and meiobenthos.

We turn now to the part played by carnivores in the benthal. Compared with the pelagial, macrobenthos systems exhibit the following peculiarities:

(i) For any consumer level of the trophic pyramid, macrobenthos species which have developed protective mechanisms to restrict predation pressure are much more numerous than in the pelagial. Some of these adaptations are morphological in nature (hard skeleton, spiny body), others behavioural in nature (e.g. ability to elaborate, and sometimes to exude, repellent or toxic substances and mainly the ability to live in burrows or holes and crevices, both in soft and hard substrates). In contrast, in the pelagial, except for cases of mimicry and defence by nematocysts in the Cnidaria, most species can only escape from some predators—the 'hunters'—through downward migration by day. However, the latter adaptation also exists in some burrowing benthos animals which rise from shelf bottoms to the epiplanktonic layer at night and for the cryptic fauna of coral reefs (p. 427). Thus, it might be assumed that predation pressure is lower in macrobenthos assemblages than in pelagial assemblages.

(ii) The above assumption requires specifications. As a counter measure to adaptations developed by endangered species, macrobenthos carnivores have developed numerous adaptive mechanisms—both morphological and behavioural—which may be related to the much higher number of taxa and species presented. Such diversification enables the pool of 'carnivores' in the benthal to fully exploit all types of preys encountered. However, in a macrobenthos assemblage, predation pressure largely depends on the general richness of the fauna. For example, in cold seas with a qualitatively poor fauna, first-level consumers are more abundant than carnivores. In contrast, in tropical and subtropical regions, greater faunal richness results in a higher number of carnivores, many of them specializing on a well-defined prey type.

(iii) Taking into account all available prey and all carnivores in the macrobenthos, it seems that, in general, the resources consumed by invertebrates compared to those consumed by fishes are much higher in the benthal than in the pelagial. This may be ascribed to two factors: (a) In general, especially in temperate and intertropical regions, carnivorous invertebrates of about fish size which participate in a given assemblage tend to comprise a higher percentage of the total number of individuals in the macrobenthos subsystem than in the pelagial. (b) Macrobenthos invertebrates have higher metabolic rates and, thus, higher food requirements than fishes. For example, on soft shelf bottoms of the north-east Atlantic Ocean, invertebrates at the second consumer level (first-order carnivores) typically utilize three to four times more of the energy available at the first consumer level than do fishes.

(iv) Let us now consider the influence exerted on the structure and dynamics of macrobenthos assemblages by the tendency of some predators to restrict themselves to prey organisms which are well defined in terms of taxonomy and, sometimes, also size. In general, specialized carnivores tend to be less common in the pelagial than in the benthal. Microphagous carnivores—e.g. medium-sized North Sea herring—employing a filtering device indiscriminately ingest all organisms retained by their filter screen and may select a particular type of food only in relation to their ability to seek patches inhabited by their favourite prey and to stay within such patches as long as density remains sufficient or as long as they need to satisfy their appetite.

Small and medium-sized raptorial zooplankters, such as carnivorous copepods and euphausiids, usually seem to select prey according to size. The prey usually selected is smaller than, or at the most equal to, their own size. The same is true for big nektonic fishes; in their diet the percentage of fish compared to that of smaller prey such as peracarid crustaceans and shrimps increases with their own body size as well as with increasing age of the biggest individuals (e.g. in tunas or bill-fish). Only some chaetognaths can sometimes feed on prey which exceed their normal body size (e.g. small-sized or young fishes).

In contrast, in the benthal it seems to be easier to distinguish specialized and unspecialized predators. In a macrobenthos assemblage with low specific richness (both prey organisms and predators), highly specialized predators can induce pronounced changes. In such cases there are often a few species at the first consumer level which largely predominate in abundance and/or biomass of the assemblage. If the predator can almost exhaust the population of its favourite prey—e.g. naticid gastropods feeding on some pelecypods (p. 172)—then the global composition of the assemblage may become deeply modified. In macrobenthos assemblages the specific richness is high; the populations of different species—prey organisms or predators—are always less abundant. Therefore, exploitation of the prey population preferred by the carnivore is less easy than in the preceding case, particularly if the predator is not highly motile. Where the specialized predator would be able to almost exhaust the population of its favourite prey, this would not markedly change the global composition of the assemblage due to its specific richness and the low abundance of the populations of most species, compared to the total number of individuals of all species concerned. However, this is not true in cases of 'demographic explosion' of a highly specialized predator such as the asteroid *Acanthaster planci* which during the last ten years has devastated the populations of scleractinians in many coral reefs of Pacific Ocean shores.

As regards non-specialized or (better) non-exclusive predators, it is difficult to evaluate their potential influence of macrobenthos assemblages with a high specific richness, such as those constituting the coral reef biocoenotic complex, because they represent a minority there compared to the total number of predatory species, most of them more or less specialized. However, their influence seems to be rather insignificant. In macrobenthos assemblages with an average or low specific richness, it seems that two main groups must be distinguished among the non-specialized predators: microphages which are partly 'detritus feeding' and macrophages which indiscriminately feed on living or dead organisms. Species of the first group, e.g. many ophiuroids (p. 141) as well as several irregular echinoids are sometimes able to feed on 'detritus'—i.e. living or non-living organic matter in the superficial sediment film—and sometimes also on post-larval stages of various bottom invertebrates just settling on the bottom. The predation exerted by this group, chiefly on post-larvae of some pelecypods, may substantially reduce the recruitment of their population. These non-specialized predators can survive long periods of poor food supply between periods of massive sinking of small-sized living prey and may induce more marked changes in the composition of a macrobenthos assemblage than specialized predators. For the second group which combines predation and scavenging, several examples have been mentioned above. In general, these species which occupy a higher level in the trophic pyramid and thus constitute populations of lesser abundance than species of the first group tend to influence changes in their prey populations to a lesser degree.

Finally, a comparison between trophic relationships in the pelagial and benthal, discussed from several points of view in the preceding pages, suggests that the macrobenthos subsystem exhibits several striking peculiarities:

(i) The amount of energy transfer from the micro- and meiobenthos subsystem to the macrobenthos subsystem seems to be rather significant, possibly except for abyssal and hadal assemblages.

(ii) There is a low-level utilization of fresh plant material by animals of the first consumer level; this is, in part, compensated for by the adaptation of many microphages feeding either on detritus or on sinking seston.

(iii) The macrozoobenthos as a whole developed an amazing diversity in its feeding ecology; this may be ascribed partly to the higher specific richness of the benthos in comparison with the pelagos (about 98% as against 2% for the whole marine fauna) and partly to the higher diversity of ecological niches in the benthal. Providing a given taxon exhibits a sufficient degree of organization and is still represented in the present marine environment by a significant number of species, it is clear that the older the taxon—on a geological time scale—the more numerous the opportunities its species had to diversify their feeding behaviour. This statement may be exemplified by a comparison between the two phyla Porifera and Echinodermata which appear simultaneously as early as the lower Cambrian. Present Porifera species with a low degree of organization utilize only suspended particles, with the possible exception of species inhabited by unicellular algae (symbionts). However, all extant classes of the echinoderms were already present as early as the Middle Ordovician, and these now exhibit a diversification in feeding which seems to be higher than in any other phylum.

(iv) Apart from interspecific diversity in feeding ethology, many macrobenthos animals may meet their energy requirements by employing several (from two to five, p. 140) different mechanisms, each corresponding to the exploitation of a particular compartment of the food resources available within the biotope inhabited. Sometimes the compartments exploited belong to the same level of the trophic pyramid, in other cases to different levels. In other words, omnivores are more common in the benthal, particularly those usually occupying the first and second consumer steps of the trophic pyramid—e.g. regular echinoids, some ophiuroids and asteroids. This versatility also pertains to taxons almost exclusively represented by carnivores; for example, among the decapod crustaceans, many crabs feed indifferently on polychaetes, pelecypods, carnivorous gastropods, echinoids, etc., which occupy the first and second consumer level(s) of the trophic pyramid. To some extent, species which are both predatory and scavenging may be associated with these opportunist predators.

(v) In the benthos, there are many short food chains ending in an inedible—or rarely consumed—organism. While in the pelagial, this type of food chain almost exclusively ends in jelly-like animals (Cnidaria, Ctenaria, Thaliacea); in the benthal it often ends in large-sized invertebrates with protective adaptations—either morphological (e.g. shelled molluscs, echinoids, asteroids, cirripeds), or behavioural (infaunal species). The common occurrence of such 'blind alleys' in the benthic food web results in storage of energy and matter in the biomass of the consumers when occupying the terminal link of these short food chains, for example: detritus → detritus-feeding pelecypod → asteroid or crab; phytoplankton → suspension-feeding pelecypod → boring gastropod → asteroid or crab. Dead bodies, unless intercepted by scavengers, progressively decompose and thus provide energy for bacteria—which may be fed in turn by filter feeders—and also release dissolved organics which may be assimilated by various invertebrates. The terminal stage of decomposition is mineralization which provides mineral nutrient to photoautotrophic organisms and thus starts a new cycle. Whereas in the pelagial, populations at any consumer level regulate those at the level immediately below, one might suspect that, in the benthal, populations of unpreyed carnivores at the end of short food chains could not be so regulated. However, this is not true. In my opinion, in these food chains the lack of a subsequent link occupied by predators might well be counterbalanced by reciprocal feeding (mutual control) between carnivores of blind alleys. Examples are an asteroid feeding on another species of the same class, or a crab feeding on an echinoid or a boring gastropod. Higher motility, strong claws and hard exoskeleton appears to assist decapod crustaceans—particularly Brachyura—in controlling populations of more sedentary and less well-armed predators.

(vi) Looking back at the food web and food chain concepts discussed for the pelagial, one may admit that simple and linear food chains may also be recognized in the benthal during the whole life history of any species, provided space and time scales are sufficiently restricted. The life-cycle of many macrobenthos animals begins with a planktonic larval stage. For example, trophic aspects of the life-cycle of sole or plaice may be summarized as follows: carnivorous planktonic larval stage feeding on ciliates, rotifers and small-sized crustaceans utilizing photoplankton → metamorphosis into a young benthic juvenile which feeds on such small-sized soft prey as harpacticoids, tanaids, or small polychaetes. Most of these prey organisms probably belong to the first consumer level, as do those utilized by older fish (mainly polychaetes and pelecypods). Prey size increasing with fish age, it seems that even adult individuals mainly feed on first-level consumers, even though their gut may sometimes contain some carnivorous polychaetes. Other fishes, such as the sea bass *Dicentrarchus labrax*, exhibit more complex trophic relations: while the diet of planktonic larvae is the same as in sole and plaice, after metamorphosis—at least on Mediterranean shores—benthic juveniles mainly feed on small-sized amphipods and isopods, but progressively turn to larger crustaceans such as *Crangon*, *Macropipus* and *Carcinus*, i.e. carnivores of the second or even third consumer level, although amphipods always represent an important percentage of their diet. Finally, individuals over 300 to 400 mm standard length turn more and more to fishes, either demersal (*Gobius*) or nektonic: sprat, sardine, anchovy. Fully grown sea bass exploit both benthic and pelagic food resources. The sea bass itself is preyed upon by bluefin tuna when the latter migrates near the shoreline. The life history of the North Atlantic Ocean hake *Merluccius merluccius* is, from a trophic point of view, rather similar to that of the sea bass. Here, fully grown individuals are nektobenthic by day, but nektonic at night, mainly feeding on macroplankton and nekton organisms.

Although the schemes (Figs 4-14 and 5-7) provided for the trophic pyramids in pelagial and benthal are largely hypothetic, it cannot be denied that the percentage of the total energy available at the base of the system which reaches to the top-predator level in the pelagial is approximately three times higher than in the benthal. This discrepancy approximately corresponds to the quantitative difference between landings of pelagic and demersal species respectively. I do not believe that this discrepancy in the benthal can be ascribed to poor utilization of fresh plant material by first-level consumers, since: (i) detritus feeders are well adapted to utilize optimally all kinds of plant detritus at any stage of decomposition; and (ii) suspension feeders exploit, for the benefit of the benthal, particulate materials sinking from the pelagial—either surplus of primary production or true detritus (dying cells or animals, faecal pellets, etc.).

On the other hand, it is a well-documented fact that, in general, benthic assemblages are more mature than pelagical assemblages. Further, it is well known that the more mature an ecosystem, the more diversified and the more abundant the predators become in comparison with the total number of consumers present. The higher specific richness of benthic assemblages and the increased number of ecological niches, particularly with regard to carnivores, result in the development of amazingly diversified adaptations which allow the organism concerned to exploit fully all animal food resources. However, the greater body size and the longer life-span characteristic of macrobenthic animals in comparison with zooplankton and micronekton species, result in greater energy expenses required for maintenance of the biomass. The occurrence of many blind alleys in the food web of the benthal enhances this trend even more: energy is either dissipated or diverted from the most direct paths of transport to the highest step of the trophic pyramid; in the pelagial the greatest waste of energy occurs in passing from the base to the first step of the trophic pyramid (first-order consumers), but in the benthal it occurs in passing from first to second and from second to third consumer levels.

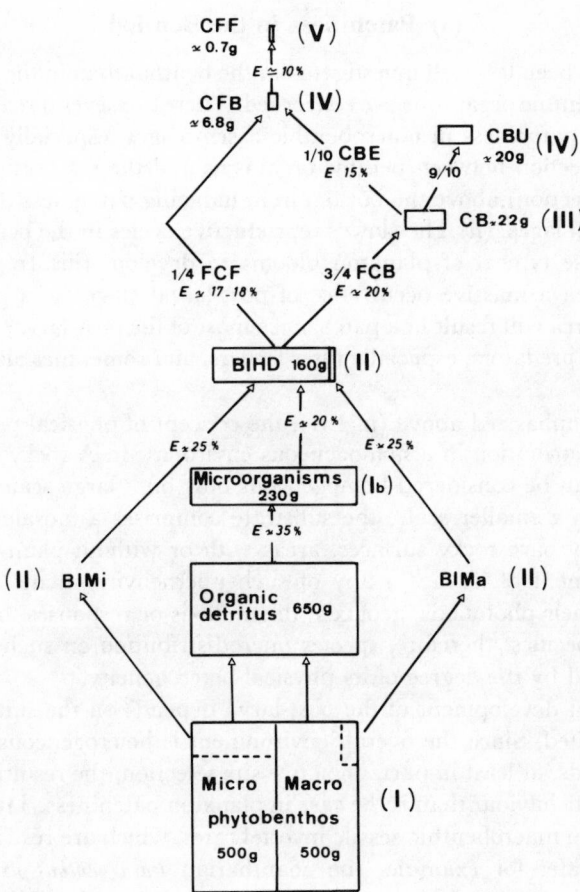

Fig. 5-7: Tentative scheme of the trophic pyramid (g m^{-2} yr^{-1}) in the benthal. E: assumed percentage of energy transfer. I, II, III, IV, V: successive steps in the pyramid; Ib; paraprimary production by microorganisms feeding on detritus; BIMi: herbivores feeding on unicellular algae; BIMa: herbivores feeding on fresh metaphytes; BIHD: estimated production of herbivorous and detritivorous bottom invertebrates; FCF: primary consumers (II) utilized by fish; FCS; primary consumers utilized by carnivorous bottom invertebrates; CB: production of carnivorous invertebrates edible and unedible by fishes, respectively; CFB: production of fishes feeding on bottom invertebrates at Steps II and III; CFF: production of fishes feeding on fishes at Step IV. (Original.)

(5) Patchiness in the Benthal

Patchiness has been less well investigated in the benthal than in the pelagial—at least as far as macrobenthic organisms are concerned. There are several reasons for a reduced manifestation of patchiness in macrobenthic assemblages, especially the following: (i) The strong connection between benthic organisms and the sea bottom prevents water movement (convection) above the bottom from inducing patchiness the same way as in planktonic assemblages. (ii) The slower reproductive cycles in the benthal do not allow patches like those typical of plankton blooms to develop. (iii) In the benthal, it is unlikely that even a massive occurrence of post-larval stages of a given species in a limited bottom area will result in a patch since most of the post-larvae will be consumed by macrobenthic predators, especially filter feeders, and sometimes also by adults of the same species.

As has been emphasized above (p. 101), the concept of physical patchiness involves heterogeneous distribution in a homogeneous environment. A rocky bottom, although well definable, can be considered homogeneous only on a large scale, e.g. over tens of square miles. On a smaller scale, the substrate comprises a mosaic of microenvironments: practically bare rocky surfaces, areas with or without plant growth, crevices, etc. . The settlement of larvae in any of such microenvironments depends on their behaviour, e.g. their phototaxis, geotaxis, thigmotaxis or responses to chemical stimuli from adult conspecifics; therefore, species microdistribution on such rocky bottoms is directly controlled by the degree of its physical heterogeneity.

The subsequent development of the post-larva depends on the suitability of the environment inhabited. Since the overall environment is heterogeneous and since larval settlement depends, at least in part, on active site selection, the resulting distribution is affected more by behaviour than is the case in plankton patchiness. True patchiness also does not prevail in macrobenthic sessile invertebrates, which are restricted to a particular living substrate: for example, the zoantharian *Parazoanthus axinellae* to certain sponges; some polychaetes of the genus *Spirorbis* to certain algal species (e.g. *S. borealis* to *Fucus serratus* and *S.* cf. *tridentatus* to *Corallina* sp.); some bryozoans (*Alcyonidium hirsutum, A. polyoum, Flustrella hispida*) largely to *Fucus serratus*. Crowding, i.e. behaviourally determined aggregation, also cannot be considered as true patchiness. Aggregations of scavengers—such as gastropods of the family Nassidae or amphipod crustaceans—on a dead prey or aggregations of gastropods of the family Cerithiidae during low tide on tropical beaches or mud flats (increase of desiccation resistance) are transient phenomena generally concerning a single or a few species. Crowding is due to a behavioural response and ceases with the effectiveness of the inducing stimulus (Volume II: SCHÖNE, 1975).

On the other hand, patchiness on soft substrates is well documented (and may render questionable biomass evaluations with quantitative bottom samplers). For example, on the compact blue mud in Buzzards Bay (Atlantic coast of the USA), GRAY (1974) noticed large variations between samples, especially in large-sized invertebrates such as *Nephthys incisa, Pandora trinileata, Eupleura caudata, Cytherea convexa* and *Mactra lateralis*. The same phenomenon occurred in nearby *Zostera marina* beds at North Falmouth (USA) with regard to the distribution of *Ophioderma brevispina*. However, in both these examples, the environment is apparently quite homogeneous.

In the English Channel *Ophiothrix fragilis* often occurs in dense patches. WARNER (1969) assumes that such aggregation allows *O. fragilis* to achieve positional stability in the water current from which it filters suspended food particles. In the Mediterranean Sea, *O. quinquemaculata*, a common member of coastal detritic bottom assemblages, exhibits a patchy distribution. Sometimes *O. quinquemaculata* seem to aggregate on the bottom itself, but more frequently they cluster on a small, isolated hard substrate area, either non-living (e.g. small boulder, large shell) or living (e.g. large-sized ascidians of the genera *Microcosmus*, *Polycarpa*, *Styela* or *Phallusia*) and, more rarely, on sponges. Such aggregation may increase both the positional stability of the brittle-stars on the bottom and facilitate food collection since the water current is stronger in the 5- to 10-cm layer above the bottom than in the immediate vicinity of the interface. Since *O. quinquemaculata* aggregations may occur on non-living substrates, the water currents generated by the ascidian do not seem significantly to improve food acquisition. On the other hand, in the North Sea the octocorallian *Alcyonium digitatum* occurs more often in the immediate vicinity of large-sized sponges, possibly benefiting from the sponge's feeding currents.

Sometimes changes in sediment characteristics generated by one given macrobenthic organism can cause patchiness in another. For example, the holothurioid *Molpadia oolitica*, which inhabits muddy clay deeper than 22 cm, collects only fine particles and accumulates its faecal pellets at the sediment surface thus generating a granulometric gradient (RHOADS and YOUNG, 1971). The faecal pellet hillocks become colonized and consolidated by tube-building polychaetes of the genus *Euchone* and these therefore exhibit a patchy distribution.

(6) Dynamics of Benthic Assemblages

(a) General Aspects

The qualitative and quantitative compositions of benthic assemblages (p. 49ff) usually reveal a snap-shot situation. Investigations carried out earlier or later can lead to quite different results. However, qualitative aspects often do not change significantly where the organismic assemblage has already reached its 'climax' stage, providing the environmental conditions remain sufficiently constant between the two samplings.

Serial observations on a benthic assemblage extending over a period of time may yield valuable information not only on successive stages (biocoenotic succession), but also on cyclic changes relative to the normal fluctuations in the intensity patterns of abiotic and biotic factors. The investigation of such changes provides an important tool for determining production rates, especially in invertebrate populations (p. 56). The study of the dynamics of an assemblage requires that this assemblage is sufficiently homogeneous over the whole area inhabited and that it change synchronously. The time intervals between samplings must be chosen as a function of fluctuations in abiotic factors and the life-span and life-cycle of the most abundant species. Long-range (sometimes up to many thousand years) changes in assemblages due to long-range natural changes in environmental conditions, or sudden changes following an abrupt change in abiotic factors, provide models for studying man-made alterations of marine biota.

The dynamics of benthic assemblages may be classified on a time-scale basis. Considering only macrobenthic species, two generalizations seem in order: (i) most species have an average life-span of at least one year, sometimes a little shorter, but often longer, especially in colder environments (high latitudes, deep sea); (ii) shallow-water benthic assemblages exhibit the most pronounced changes, some seasonal, others extended over several years. Hence, one year may be a convenient time-scale basis. This allows a distinction between short-range (one year or less) and long-range (more than one year) dynamics. In each group we must distinguish periodic and aperiodic processes. The examples briefly referred to below will be discussed later (see p. 157) from a different point of view.

(b) Short-range Dynamics

Aperiodic Changes

Heavy storms may exert at least four types of effects.

(i) Casting up to the shoreline pelecypods normally buried in shallow soft bottoms; examples are: *Cerastoderma edule* on the coast of the Wadden Sea and *Lentidium mediterraneum* on the French Mediterranean coast (PICARD, 1965). EGGLESTON and HICKMAN (1972) observed mass stranding of pelecypods at Te Waewae Bay, Southland (New Zealand) caused by a combination of heavy inshore winds and low air temperatures coupled with an increased flow of cold fresh water across the beach; the total number of animals stranded was estimated at more than 20 million, 90% of which were large individuals of *Mactra discors* from lower shore areas.

(ii) Pronounced numerical fluctuations in populations of small peracarids (*Eurydice, Bathyporeia, Cumopsis*) which inhabit highly exposed intertidal or very shallow sandy bottoms when the sea is smooth, but migrate into deeper waters during heavy storms (LAGARDERE, 1966; MASSE, 1971).

(iii) Waves pulling off sessile organisms offering sufficiently large volumes and/or superficies; examples are clusters of mussels (p. 383) as well as large seaweeds such as Fucaceae, *Sargassum* and Laminariaceae. Where two closely related species coexist and exhibit different capacities for resisting wave effects, each of them may predominate at different places depending on the degree of wave exposure or the irregular occurrence of heavy storms. For example, on Californian coasts, the two mussels *Mytilus californianus* and *M. edulis* occupy the same ecological niche; *M. edulis* tends to cover (and sometimes finally to eliminate) *M. californianus* in sheltered areas or during calm weather periods. However, its byssus is weaker than that of *M. californianus*; hence, a heavy storm may clear the *M. edulis* cover and thus support growth of *M. californianus* (HARGER, 1970).

(iv) Hurricanes and typhoons can greatly injure coral reef assemblages of the outer reef flat (p. 412). Strong waves may rip off boulders from the reef front and throw them shoreward. Moreover, suspended mineral particles, both lifted from neighbouring soft bottoms by wave action and contributed by land drainage due to heavy rain, slow down coral growth (increased energy expenses due to ciliary cleaning). Sometimes the degree of sedimentation may be high enough to cover and kill the corals and prevent planulae from settling until the hard substrate of dead corals becomes cleaned again by wave action.

Long periods of calm weather tend to correspond to a certain season and cannot be considered on the basis of their periodic recurrence. However, on the French Mediterranean coast, long periods of high barometric pressure together with seaward winds, causing a lowering of the mean sea level, may occur during any season. Hence, the upper levels of the photophilic algal (*Jania rubens*) vegetation (p. 395) may emerge for several days and become greatly damaged. In such a particular meteorological situation, PERES and PICARD (1964) have also demonstrated the disjunction of macrobenthic animals characteristic of the midlittoral assemblage on sandy beaches (p. 387).

In tropical regions 'red tides' may cause marked changes in benthos assemblages. In zones located just below the discoloured waters, benthic assemblages are often totally wiped out due to oxygen depletion and subsequent occurrence of hydrogen sulphide. In peripheral areas the toxics released by dinoflagellates can lead to selective mortality and thus distort assemblage composition (p. 356).

In some small bays of the Mediterranean Sea, storms may induce **sporadic eddies** which concentrate large amounts of fine particles (silt and mud) over coarse sandy bottoms; the particles sediment when the sea becomes smooth again. Thalli of red algae of the family Squamariaceae float on the resulting layer of fluid mud; such dynamic changes are usually restricted to a limited area.

Heavy rainfall usually does not greatly affect permanently immersed assemblages. The salinity decrease is mainly restricted to a very thin superficial water layer; moreover, water currents and waves quickly achieve mixing. Even intertidal assemblages do not seem to be damaged by heavy rainfall. In soft substrates exposed to rain during low tide, the salinity of interstitial waters is reduced only in the uppermost few centimetres, and most of the stenohaline organisms can dig deeper and thus avoid the diluted layer. Most large algae on temporarily exposed, hard substrates are not damaged by rainfall because they are highly euryhaline (HOFMAN, 1959; Volume I: GESSNER, 1970). Intertidal animals are usually also fairly euryhaline and sometimes protected against rain during low tide by a cover of large algae. However, occasional rain damage does occur in benthic assemblages. In very sheltered harbour areas, for example, heavy rain may induce a sudden decrease in salinity (5–10%) which can kill many animals of the hard substrate assemblage such as *Ciona intestinalis*, *Bugula neritina*, etc. (p. 430). Salinity decrease due to heavy rainfall during hurricanes and typhoons induces hermatypica corals to release their symbiotic zooxanthellae; the resulting 'whitening' causes a decrease in growth and building activities until the stock of zooxanthellae is reconstituted through division of the remaining cells. This requires several weeks, sometimes months (p. 410).

Since average seasonal temperature fluctuations are periodic, they are not considered here. Brief periods of exceptionally high or low air temperatures exert very limited, if any, influence on sea water and its inhabitants. In contrast, prolonged influences of **exceptional temperatures** (Volume I: KINNE, 1970, p. 407ff.) can affect populations of some species negatively. For instance, during a very hot period near the Tortugas Islands, MOORE (1958) observed catastrophic mortalities of both sedentary (*Fissurella* and *Diadema*) and motile animals (*Octopus* and fishes). On European Atlantic coasts, the very cold winter of 1962/63 greatly injured shallow-water benthic assemblages. According to CRISP (1964) and CRISP and co-authors (1964), 'Lusitanian' elements were more injured than 'Celtic' species, and these were more than boreo-arctic species. ZIEGEL-MEIER (1964) found maximum mortalities in pelecypods (except for *Nucula nitida* and *Macoma balthica*), echinoderms (mainly *Amphiura filiformis*) and *Branchiostoma lanceolatum*; polychaetes and crustaceans seemed to be more tolerant, but among the fishes, a striking mortality was observed in *Solea solea*.

Prolonged exceptionally low temperatures can lead to long-term changes which may be considered a secondary outcome of the above-mentioned seasonal effects. On the French coast of the English Channel, populations of *Carcinus maenas*, *Mytilus edulis* and *Psammechinus miliaris* which were almost completely destroyed during the severe winter of 1962/63 did not exhibit any appreciable decrease during the following summer (BERNARD and co-authors, 1963). Populations of other species were more abundant in the 1963 summer than before, e.g. calcareous sponges, Phyllodocidae, *Polydora* species and *Balanus balanoides*. It is difficult to explain this phenomenon. No evidence could be produced for increased breeding activities. In some cases, competition between two species occupying the same ecological niche was proposed as an explanation. For example, high mortality among *Ensis ensis* and *Pholas dactylus* could have promoted the development of *Solen marginatus* and *Barnea candida* respectively. It has been suggested that the abundance peak of *Balanus balanoides* could have been caused by the exceptionally great phytoplankton outburst in spring 1963, related to the increased vertical mixing generated by the cold winter, which resulted in abundant diatom food for cypris larvae. In turn, it seemed that the abundance of young barnacles resulted in a striking increase in populations of the predatory gastropod *Lamellidoris bilamellata*.

Aperiodic short-range changes in benthic assemblages ascribed to **biological events** mainly arise from trophic relationships; examples are scavengers crowding on dead animals. Temporarily aggregating scavengers are various species of shrimps, prawns and crabs, the small pagurid *Diogenes pugilator*, and gastropods of the family Nassidae, e.g. *Nassa reticulata*, *Nassarius pygmaeus* (MASSE, 1971) and *Cyclonassa neritea*. In the latter species, which mainly feeds on broken or dead *Macoma tenuis*, PICARD (1965) noticed maximum population abundance in autumn, concentrating on beaches that had been frequented by numerous bathers during preceding weeks. Regular holiday activities thus make originally aperiodic crowding effects periodic.

Diogenes pugilator populations also exhibit aperiodic fluctuations which are related not to predator–prey relationships but to the availability of empty gastropod shells. On exposed beaches, the number of shells available is a function of meteorological events which either favour shells cast up to the shallowest depths or bury the shells.

Periodic Changes

Circadian rhythms. In *Posidonia oceanica* beds in (p. 443) the Mediterranean Sea, LEDOYER (1962) observed that numerous animals, in particular isopods and amphipods but also gastropods and occasionally small echinoids, migrate at night to the upper part of the leaves. During the day, all these animals remain on the sea-grass stems close to the sea bottom. Experiments by LEDOYER demonstrate that the nightly ascent is triggered mainly by an increase in dissolved CO_2 concentration due to interrupted photosynthesis. In some upper-level algal communities on the coast of Yugoslavia, ZAVODNIK (1976) recorded circadian migrations of the whole 'phytal fauna' (p. 144), i.e. small-sized polychaetes, copepods, amphipods, decapods and gastropods, related to variations of O_2 and CO_2 contents in the sea water caused by fluctuations in gas-exchange activities of the algae.

Both examples elucidate temporary stratification dynamics rather than overall composition changes of the assemblage concerned. In contrast, in several soft-bottom communities of the shelf, a significant number of peracarids (mainly amphipods and tanaids) turn planktonic or even hyponeustonic at night. In such a case, sediment sampling by day or at night will result in different impressions of assemblage composition.

Tide-dependent influences on short-term fluctuations in benthic assemblages are largely restricted to the intertidal zone. In general, the most motile species leave their high-water biotope with the ebb tide and return with the flood. Such behaviour is very common for natantian decapods and fishes which inhabit soft bottoms (without metaphytes and sea-grass beds which emerge at low tide) and coral-reef flats. On rocky substrates with heavy algal vegetation, which are common on temperate shores, in addition to the animals mentioned above, numerous others (amphipods, for example) leave the biotope at ebb tide, whereas others, such as nematodes, molluscs and echinoderms, remain under the leaf canopy. In the northern Adriatic Sea *Fucus virsoides* community, ZAVODNIK (1967a) noticed interesting differences depending on whether low-tide samplings were made at night or by day. At night, nematodes, copepods and mites are much more abundant. Such tide-dependent migrations are less pronounced during cold periods.

In my opinion, we cannot consider as true tide-dependent assemblage dynamics regular up and down migrations such as are exhibited by the goose-crab *Emerita analoga* (tropical American shores), crustaceans of the genera *Gastrosaccus* and *Hippa* on the coast of Ghana, or those of many pelecypods of the genus *Donax* on tropical beaches, which by permanently moving up and down assure optimum feeding conditions in the surf zone. Some of these migrating species belong to midlittoral assemblages (p. 387ff), others to upper clean-sand infralittoral assemblages (p. 432). However, such migrating behaviour seems to be absent in other species of each of these assemblages.

Seasonal fluctuations in benthos assemblages are mostly due to temperature changes. These are large in temperate seas and small in intertropical or polar areas. This classical, simple relationship is, however, only a rough approximation of reality.

In polar seas, freezing is usually combined with a salinity increase. High-latitude assemblages have been investigated almost exclusively in summer. Hence, our knowledge of seasonal fluctuations is incomplete. Except for the upper levels, where the substrate may be covered (and sometimes abraded) by sea ice, fluctuations in assemblages are very slight. Apparently assemblage members are well adapted to their environmental conditions. Some metaphytes may have heterotrophic capabilities: even during the dark winter period, large algae may become fertile below 3-m thick ice. The freezing temperature of the blood (fishes) or haemolymph (invertebrates) decreases in winter (Volume I: KINNE, 1970). The percentage of scavengers is higher than in any other area of the World Ocean; in some species necrophagy is the only means of nutrition in winter, for example, in the asteroid *Odontaster validus* (ARNAUD, 1970, 1974).

However, on Antarctic shores heavy blooms of benthic diatoms and small filamentous algae may occur in summer on shallow-water rocky substrates.

In intertropical regions the annual temperature range, in contrast to common assertions, may be pronounced, predominantly due to cessation or reversal of seasonal winds. This brings different water masses into contact with a given assemblage. Such a process is often combined with significant salinity fluctuations when landwards blowing winds cause heavy rains which, returning from the land, bring large amounts of warmed-up fresh water (and suspended material) to the coastal areas. A striking example of seasonal dynamics of benthic assemblages related to changes in hydrological climate was observed by LE LOEUFF and INTES (1969) on the Ivory Coast shelf. From August to October the shallowest (0–30 m) shelf areas are strongly cooled due to upwelling, which substitutes subtropical temperate and saline waters for the suprathermoclinal or 'Guinean' waters, which are warmer but of lower salinity (p. 281). At 30 m depth, the annual temperature range may be as large as 9 to 10 °C. In contrast, below 60 m, the hydrological climate only changes slightly on an annual scale. On soft bottoms beyond the upper wave-stirred levels, LE LOEUFF and INTES distinguished two assemblages. The deeper assemblage between 30 and 60 m is more diversified but it includes only species which cannot tolerate warm waters. The members of the shallow-bottom (5–30 m) assemblage are not affected by fluctuations in the hydrological climate; however, the gastropod *Aplysia fasciata* occurs only during the warm season when it feeds on thriving seaweeds, whereas the shrimps *Palaemon hastatus* and *Hypolysmata hastatoides* occur only in the cold season when they leave the coastal lagoons for breeding in more saline waters.

Such examples demonstrate how difficult it is to discriminate between temperature and salinity effects in terms of seasonal fluctuations. Deatails of temperature–salinity interaction have been dealt with in Volume I especially by KINNE (1970, 1971) and ALDERDICE (1970), as well as by ALDERDICE and FORRESTER (1971).

Possibly only one case of macrobenthic aggregation can really be referred to as patchiness similar to that prevailing in the plankton: members of the red alga genus *Peyssonnelia* floating on a fluid mud layer, the thallus of the alga supporting a particular assemblage of macrobenthic invertebrates (p. 468). Such facies, which may follow gales as the sea calms, can persist for several months.

More generally, it seems that occasional periods of excessive muddy and/or clayey sedimentation on a restricted bottom area—for example due to lenses of highly turbid diluted waters originating from a spasmodic river flood—can induce the formation of transient patches on sandy substrates. The fine-particulate matter allows members of a muddy-bottom assemblage to settle. The duration of such patchiness depends on meteorological events which may or may not clean the bottom from the fine-particulate matter.

The ability of larval stages to discriminate sediment characteristics suitable for adults is well documented. The stimulus which induces larval settling and subsequent metamorphosis seems to be chemical in nature and is probably related to organic exudates from microorganisms epiphytic on sediment particles (Volume II: CREUTZBERG, 1975). In the polychaete *Scolelepis fuliginosa*, GRAY (1971) demonstrated that chemical and physical treatment made the sediment unattractive for potential settlers. GRAY's observations are of particular interest because the patches of *S. fuliginosa* he studied were localized and their density (280 ind. m^{-2}) was extremely low. In fact, *S.*

fuliginosa and *Capitella capitata* are practically the only inhabitants of the 'polluted zone' (BELLAN, 1967a,b, 1976) in which both species often attain a density of several thousand individuals per square metre. However, sometimes, especially in summer, the environmental conditions (particularly as regards the oxygen content of the bottom water) exceed the tolerance range of these species, resulting in massive kills of whole populations. Later, the populations of both the species tend to recover again on the whole bottom area formerly occupied. The recovery process is supported by undamaged, peripheral areas or by scattered small patches to which some individuals escape, as soon as the environmental conditions return to normal. It may be assumed that GRAY (1971) observed the beginning of this recolonization process of a polluted area.

Patchiness seems to be rather common in macrobenthic assemblages of the abyssal zone but possibly only concerns large-sized non- or poorly-motile invertebrates, the only ones appearing on deep-sea photographs and directly observable from deep-sea vehicles. The distribution of some suspension feeders on small bottom elevations (which result in an increase in the speed of bottom-water currents) have been mentioned on p. 133. However, patchiness also occurs on quite flat bottoms in which 'oases' of bottom life may be observed, separated from each other by large, apparently non-populated areas.

Finally, I would like to mention a phenomenon which cannot really be considered as true patchiness; it has been observed in dense populations of suspension-feeding pelecypods. For example, on sandy bottoms inhabited by particularly dense populations of fully grown *Cerastoderma edule*, intensive feeding of adults generating strong inhalent currents deprives the youngest stages of food, so that the latter finally starve and die. This results in populations of predominantly only one or two age classes. These become more dispersed, thus initiating the start of a new cycle.

If we exclude the effects of seasonal migrations, the dynamics of benthic assemblages—especially with respect to quantitative parameters—are mainly affected by factors which are more or less temperature dependent: (i) recruitment and growth; (ii) mortality (irrespective of cause, including predation). Fluctuations in the numbers of individuals or colonies originate from the balance between rates of recruitment and mortality. The most trivial case on temperate coasts concerns species which reveal a pronounced increase in reproduction and abundance in spring or early summer, resulting in a summer or autumn standing crop increase due to individual growth. Of the many papers on this subject we mention here that by BEUKEMA (1973), who for over six years frequently sampled three intertidal stations on a tidal flat area in the westernmost part of the Wadden Sea (Netherlands) and found that biomass fluctuated with a regular annual pattern. Maximum values were recorded at each station from July to September, minimum values from December to March. The steep increases during spring were mostly due to fast growth of animals already present in winter. Spat fall contributed only a minor percentage to the annual biomass increases. Declines in autumn were attributable both to decreases in numbers and in individual weight. The latter dominated in the large and deep-living individuals of *Mya arenaria* and *Arenicola marina* which comprised about half the total biomass of the benthos (BEUKEMA, p. 106).

The same situation prevails on subtidal bottoms. This is exemplified by studies of BUCHANAN and co-authors (1978), who investigated the soft-bottom macrofauna from 20 m down to 80 m off the coast of Northumberland (England). The seasonal changes in abundance and biomass (Fig. 5-8) appeared to be independent of the assemblage composition.

The most important seasonal fluctuations concern species with a life-span of about one year or less and with several successive generations during the most suitable season. Each chain of successive generations originates either from individuals produced in the last spawning of the preceding year, which endured the seasonal environmental pessimum, or from resting stages such as winter buds (sponges, ascidians), colony portions (sponges) or basal thallus parts (many multicellular algae).

Such seasonal fluctuations in population abundance depend directly on temperature which largely determines: (i) the sequence of breeding activities (maturation, gamete release, fertilization, development duration); and (ii) the formation of resting stages, where these exist. Temperature may also modify predation pressure. The availability of food can also influence abundance: supranormal recruitment in a given year can result in food shortage, leading to a decrease in growth and reproductive capacity of adults, which, in turn, diminishes in the following year's recruitment.

Seasonal temperature changes can also modify micro-environment characteristics. On rocky shores of the northern Adriatic Sea, ZAVODNIK (1967b) noticed only small variations in the abundance of small-sized phytal faunal elements (p. 144) on and between large seaweeds. The one exception was *Cystoseira* species whose thallus is greatly modified in winter. The seasonal disappearance of forms with brief life-cycles may lead to the predominance of other groups (e.g. nematodes). The spring and early summer bloom of small-sized epiphytic algae (cyanophytes, diatoms, etc.) results in a striking development of the phytal micro- and meiofauna.

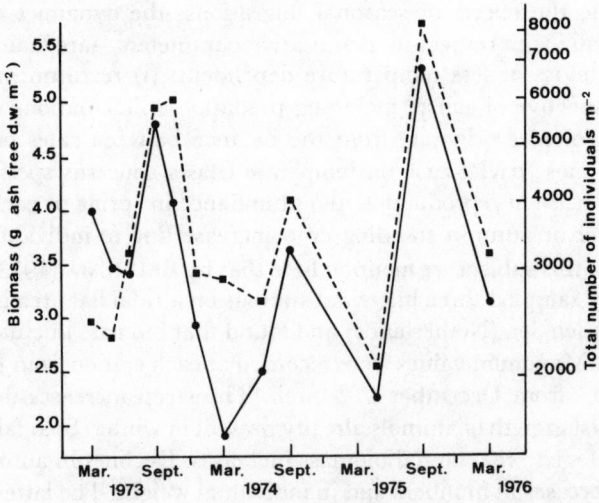

Fig. 5-8: Seasonal variation in biomass (circles) and total number of individuals (squares) during the period 1973 to 1976 on the Northumberland coast; depth 20 m; percentage of silt less than 20. (After BUCHANAN and coauthors, 1978; reproduced by permission of Cambridge University Press.)

In addition to temperature, other factors undergo seasonal changes which may induce fluctuations in benthic assemblages. In the Mediterranean Sea, for example, sand and gravel shelf bottoms inhabited by the 'Amphioxus assemblage' (p. 450) are characterized in summer by both increased illumination and decreased wave-induced, bottom-water movement. In this season, the following algae thrive: the phaeophytes *Arthrocladia villosa* and *Sporochnus pedunculatus* and the rhodophytes *Brongniartella byssoides* and *Polysiphonia* sp. (FELDMANN, 1937). At the same time, the abundance of animals such as *Psammobia costulata* and *Euthalenessa dendrolepis* which are adapted to water movement decreases, whereas *Sphaerechinus granularis*, mainly represented by young individuals, becomes much more common.

Seasonal fluctuations of wave activities, in combination with those of illumination (see above), exert a larger influence at the upper levels, both through mechanical and desiccation effects (especially in the supra- and midlittoral zones). For example, on the rocky shores of São Paulo State (Brazil), stronger and more frequent wave contact in winter supports a temporary settlement in the upper midlittoral subzone of young *Mytilus*, which die in early spring. On exposed locations of rocky shores of the northern Mediterranean Sea, the algae *Bangia fuscopurpurea* and *Porphyra leucosticta* can only exist in winter, in the upper and lower midlittoral subzones, respectively (p. 378).

In some cases, influence of seasonal variations in the tidal cycle has been documented. On the coast of Ghana, for example, maximum low water of each tidal cycle occurs by day in winter but at night in summer. Hence, the stress of emergence is less severe in summer, and some sessile species, such as the alga *Hypnea musciformis*, may be found at higher levels in summer than in winter.

Finally, modifications in assemblage dynamics may result from seasonal changes in food availability. For instance, at shallow depths in the Mediterranean Sea, PICARD (1965) observed a winter peak in the populations of some amphipods of the genus *Atylus*. This peak corresponds to the greater amount of decaying plant material transported by currents from the sea-grass beds of *Posidonia oceanica* which loses its leaves in autumn. The amphipods feed upon this decaying organic matter.

(c) Long-range Dynamics

Long-range dynamics in benthic assemblages, like the short-range fluctuations reviewed above, may be aperiodic if caused by casual circumstances. They may also be more or less periodic; however, the periodicity is often irregular and its causes are insufficiently investigated.

Aperiodic Changes

Natural disasters which damage benthic assemblages may reveal valuable information on their long-term dynamics. For instance, coral reefs in which scleractinians are killed regenerate very slowly if at all. On Queensland shores, coral reefs destroyed in 1918 by an exceptional salinity decrease exhibited only negligible evidence of reconstitution in 1953. Five years after similar severe salinity damage, Hawaiian reefs with a 60 to 70% surface cover of hermatypic corals were composed largely of zoantharians and young oysters and only 1% corals (JOHANNES, 1972). It seems that when all scleractinians are killed, settling chlorophytes (among them boring species) quickly occupy the substrate, thus hindering the fixation of planulae from neighbouring undamaged reefs. Possibly herbivorous fishes which graze on algae also destroy the young coral colonies. Moreover, calcareous suspended particles originating from the destruction of dead reef organisms also hinder reef regeneration. In any case, regeneration probably requires several decades, especially when large areas of the reef have been destroyed.

Aperiodic changes caused by biotic factors can, for example, be due to population outbursts of predators. Thus the asteroid *Acanthaster planci* ('crown-of-thorns') can cause considerable sudden damage to coral reefs. The conditions which support outbreaks of the predator's population are unknown. Among the possible explanations proposed, we mention here: (i) long-range but slight changes in environmental conditions which favour reproduction and/or larval survival; (ii) decrease in the numbers of large gastropods which feed on *A. planci*; the gastropods may be overfished by the mother-of-pearl industry, or their larvae killed by pollutants. Once the scleractinians have been destroyed by *Acanthaster*, vacant areas in the reefs are quickly invaded by other sessile species (algae, mainly chlorophytes; zoantharians, etc.).

Diseases may also cause aperiodic changes in the dynamics of benthic assemblages (e.g. KINNE, 1980). Further examples of aperiodic changes, the study of which contributes to better knowledge of the long-term dynamics of benthic assemblages, can be found in DOTY (1967), who studied communities on the Hawaiian Islands on lava flows of different ages. According to DOTY, a lava assemblage requires at least five to ten years to reach its climax. Further changes may occur after the climax: 100-year old lava assemblages differ from those of prehistoric age.

Progressive, irreversible evolution of the assemblages results from long-term changes in climatic conditions. The following examples concern the Mediterranean Sea.

The cooling of coastal waters prevailing since the last warm period (Tyrrhenian II), especially in northern areas subjected to strong northerly winds, has progressively altered the *Posidonia oceanica* beds, possibly because flowering as well as fruit and seed development requires high temperatures, especially in late spring and early summer. In the Gulf of Genoa, comparisons between present assemblages and Quaternary thanatocoenoses reveal a general trend towards increased siltation and mud formation on shelf and slope substrates. The beginning of increased land drainage is perhaps related to modifications in the continental vegetation induced by climatic changes in the Quaternary period and, later, to the wood clearing reported since historic times. Pine trees, which increasingly replaced oaks, are less able to retain the finest soil particles. A comparable siltation process which probably reached its climax many centuries ago, when almost all the forests had already disappeared, may be observed in Iraklion Bay, Crete.

On a smaller time scale, I also mention here KLIMOVA's (1976) observation on progressive changes in Peter the Great Bay (Sea of Japan) between 1925 and 1970: at depths of 50 to 90 m increased bottom siltation resulted in the substitution of an *Echiurus echiurus* biocoenosis for a *Venus fluctuosa* biocoenosis; the *Solariella* biocoenosis was replaced by a *Macoma calcarea–Cucumaria calcigera* assemblage.

Harnessing of rivers, like that carried out over the last 20 years in the Rhône valley, may have similar consequences. Silt and clay were formerly transported mainly by floods from the estuary to the open sea and deposited far from the shoreline; now the more regular run-off due to harnessing allows greater amounts of the finest particles to deposit on shallow shelfs. Therefore, siltation of coarse detritic shelf bottoms in the Gulf of Lions has increased over the last 25 years. Such evolution may require several decades; it is as inevitable as that in the Genoa and Iraklion Gulfs.

Period Changes

Climatic or cosmic events. Long-term climatic changes, which may extend over years, decades or more than a century, probably arise from periodic changes in solar activity. Progressive warming of the Arctic Ocean since at least 1880 to 1890 has been documented, and the melting of glaciers in the Alps suggests that this warming involves the whole of Europe. Thus, many algae which inhabit temperate shores extend their distributional boundaries farther to the north. Large stocks of cod are at present exploited on the west coast of Greenland, where this species was not observed a few decades ago. However, some signs suggest that the warming process in the Arctic Ocean has tended to reverse in the last 10 to 15 years.

A shorter period of cyclic changes corresponding to solar activity has a duration of about 11 years. Related to this cycle, fluctuations in laminarian abundance have been observed along the coast of Scotland. At the upper levels of rocky coasts around Australia, GUILER (1960) observed pronounced changes in local assemblages during the same 11-year period. At Dodges Ferry the succession was: *Galeolaria caespitosa* (polychaete) belt, *Mytilus planulatus* beds, *Pyura stolonifera* (ascidian) cover; at Colls Bay, GUILER observed alternating predominance of the alga *Hormosira banksii* and of *Mytilus planulatus* beds.

These cyclic changes occur mainly in the upper assemblages on rocky substrates, where settling and growth is supported by prevailing climatic fluctuations. Predator–prey relationships probably also play some indirect role.

On subtidal soft bottoms off the coast of Northumberland, significant modifications may occur not only in biomass and abundance (Fig. 5-8) but also in assemblage composition (Fig. 5-9) (BUCHANAN and co-authors, 1978). BUCHANAN and co-authors have related the changes to climatic variations. The data from monitoring station M_1 sampled regularly from 1972 to 1976 (Table 5-4) and from station P (Table 5-5) led BUCHANAN and co-authors to suggest that there is evidence of a turnover in species ranking, particularly obvious in the top-ranking dominants.

Within a group of 18 species, three subgroups can be recognized; these are defined by the nature of their population changes over the five-year period (Table 5-6): (i) 'gaining' species exhibit net gains over the period and significant gains from year to year in at least four of the five years; (ii) 'losing' species show significant losses in at least four years; (iii) 'neutral' species reveal no significant change over the period. Population trends of the subgroups are shown in Fig. 5-9, which illustrates the progressive changeover of abundance between gaining and losing groups (BUCHANAN and co-authors, 1978, p. 204).

BUCHANAN and co-authors (1978) hypothesize that the change in the composition of the assemblages might be ascribed to temperature anomalies during the winter (December to May). The main response appears to have been the replacement of some

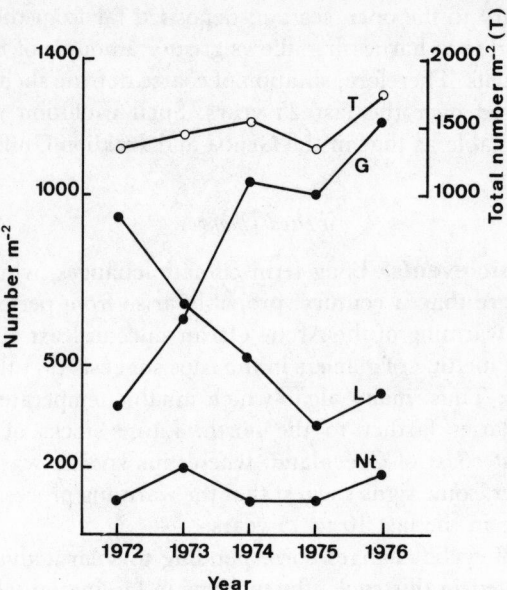

Fig. 5-9: Contribution of the top-ranked 17 species to the total number of individuals present in March samples, divided into gaining species (G), losing species (L) and neutral species (Nt), with the resulting total (T); same station as in Fig. 5-8. (After BUCHANAN and co-authors, 1978; reproduced by permission of Cambridge University Press.)

of the larger-sized species, e.g. *Ammotrypane aulogaster* and *Chaetozone setosa*, by species of smaller body size, e.g. *Paraonis gracilis* and *Prionospio malmgreni*. This change was accompanied by a general rise in the total number of individuals present. No species was completely eliminated from the fauna list. The authors assume that bottom communities evolved an adaptive strategy to cope with environmental instability.

'This strategy involves a broad spectrum of species. From within this spectrum, different suites of species enjoy competitive advantage and thrive quantitatively at different points of the temporal range of temperature variation. These changes are reversible and under the control of density-dependent recruitment and mortality factors.' (BUCHANAN and co-authors, 1978, pp. 207–8)

Table 5-4

Changes in species ranking: coast of Northumberland; station M_1; depth: 50 m. List of numerical sequence of the five top-ranking species for each year (After BUCHANAN and co-authors, 1978; reproduced by permission of Cambridge University Press)

Species	Year				
	1972	1973	1974	1975	1976
Spiophanes bombyx	1	—	—	—	—
Chaetozone setosa	2	2	5	—	—
Amphiura filiformis	3	5	—	5	—
Prionospio malmgreni	4	1	2	1	1
Nephthys hombergi	5	4	—	—	—
Ampelisca tenuicornis	—	3	—	—	—
Phoronis muelleri	—	—	1	—	—
Paraonis gracilis	—	—	3	2	2
Magelona minuta	—	—	4	3	—
Tharyx multibranchis	—	—	—	4	—
Thyasira flexuosa	—	—	—	—	3
Owenia fusiformis	—	—	—	—	4
Myriochele oculata	—	—	—	—	5

Table 5-5

Changes in species ranking: coast of Northumberland; station P; depth: 80 m. List of numerical sequence of the five top-ranking species for each year (After BUCHANAN and co-authors, 1978; reproduced by permission of Cambridge University Press)

Species	Year					
	1971	1972	1973	1974	1975	1976
Heteromastus filiformis	1	1	2	2	2	2
Ammotrypane aulogaster	2	—	—	—	—	—
Chaetozone setosa	3	3	5	—	4	—
Prionospio cirrifera	4	4	3	5	3	—
Paraonis gracilis	5	2	1	1	1	1
Prionospio malmgreni	—	5	4	—	—	3
Oligochaetes	—	—	—	3	5	4
Pseudeurythoe hemuli	—	—	—	4	—	—
Abra nitida	—	—	—	—	—	5

Table 5-6

Population changes in three subgroups of a total of 18 species. Gaining, losing and neutral species at station M_1 in the period 1972 to 1976. Gains and losses are expressed in number of individuals per square metre (After BUCHANAN and co-authors, 1978; reproduced by permission of Cambridge University Press)

Gaining species		Losing species		Neutral species
Paraonis gracilis	238	Sphiophanes bombyx	219	Ampharete finmarchia
Prionospio malmgreni	144	Chaetozone setosa	129	Glycera rouxi
Myriochele oculata	129	Ampelisca tenuicornis	112	
Thyasira flexuosa	115	Amphiura filiformis	50	Goniada maculata
Owenia fusiformis	104	Nephthys pombergi	31	
Magelona minuta	100	Scoloplos armiger	26	
Tharyx multibranchis	27	Rhodine gracilior	24	
		Ammotrypane aulogaster	17	

Changes over a much longer period have been observed in soft-bottom assemblages in the North Sea by URSIN (1952). Their dynamics are more difficult to explain. URSIN compared his own samples with those of DAVIS (1923, 1925) who used identical methods and gear. The differences are summarized in Table 5-7. According to URSIN,

Table 5-7

Long-term differences between species abundance in soft-bottom assemblages. Based on investigations by DAVIS (1923, 1925) and URSIN (1952) (After URSIN, 1952; reproduced by permission of Macmillan Journals Ltd)

Animals	Abundance (ind. m^{-2})	
	1922 (DAVIS)	1951 (URSIN)
Spisula subtruncata	472	5
Mactra corallina	11	1
Other pelecypods (mostly Tellina fabula)	4	43
Polychaetes	4	70
Echinoderms	4	28
Varia sp.	8	65

the striking decrease of the warm-temperate *Spisula subtruncata* and *Mactra corallina* might have been caused by the very cold winters of 1940/41 and 1947. Such an assumption can hardly be accepted. Since larvae of both species have a fairly long planktonic stage, larvae from unaffected nearby populations could easily restock the depopulated sea bottoms. More likely, the great amount of faecal pellets released by the dense population of *S. subtruncata* caused an increase in the finer fraction of the sediment, thus providing a convenient habitat for detritus feeders such as *Tellina fabula* and some polychaetes. The simultaneous increase in echinoderms suggests the influence of predator–prey relationships (p. 435).

Long-range fluctuations due to **biological events** are related to: (i) intraspecific and interspecific relationships; (ii) species-specific characteristics such as fecundity, growth, tolerance to environmental stress or starvation.

On the following pages, interactions between individuals of one and the same or of different species are dealt with in the following subsections: (i) predator–prey relationships; (ii) competition for space; (iii) modifications in substrate nature due to plant or animal activities; (iv) competition for food; and (v) autecological factors.

(i) Predator–prey relationships. On rocky substrates, for example in mussel beds of the lower midlittoral (p. 384) of European coasts, long-term (several years) fluctuations occur. These are very intricate because they involve the two predators *Nucella lapillus* and *Asterias rubens* and the two prey species *Mytilus edulis* and *Balanus balanoides*. A sound interpretation of these fluctuations requires further search taking into account the concept of size-limited predation recently developed by PAINE (1976). Through long-term (12 years) observations *in situ* and transplantation experiments concerning predator–prey relationships between *Pisaster ochraceus* and *Mytilus californianus*, PAINE demonstrated that a size threshold exists which protects mussels from being eaten by the asteroid. In other words, the mussels attain a 'refuge' through size increase so that prey and predator can coexist in close proximity on condition that the surviving prey attains a large size, thus making a reproductive contribution which compensates for loss in total abundance.

In the Mediterranean Sea, mussel and barnacle populations of the infralittoral zone exhibit less marked fluctuations than on European tidal shores, but *Nucella* is absent, and *Asterias rubens* is replaced by *Marthasterias glacialis*. The latter asteroid is not abundant at the upper levels which the mussels preferably inhabit; moreover, this predator is less specialized than *A. rubens* which obviously can feed on shelled molluscs, as well as on living echinoids and all kinds of dead animals (TORTONESE, 1965). Other predators are not abundant there. Hence, the long-term stability of the mussel assemblage seems to be higher here because of balanced gain (settlement and growth of new individuals) and loss (disappearance of fully grown individuals, e.g. through natural death or removal by storm waves). In the vicinity of Naples, PARENZAN (1940) described a several-year succession of assemblages dominated by echinoderms or *Phallusia*. Since predators feeding upon simple ascidians are rare, another group not taken into account by PARENZAN may have attained maximum development synchronous with echinoderm population minima. In the area of Marseilles (France), shallow-water, sandy bottoms exhibit alternate periods of pelecypod and echinoderm (e.g. *Echinocardium mediterraneum*) predominance (for another example consult p. 435).

According to ZIEGELMEIER (1974), who from 1949 until 1960 studied standing crop and abundance in German Bight soft bottoms, the relationships between dynamic fluctuations of the main groups are not obvious. Only three generalizations seem permissible: (i) fluctuations in the abundance of echinoderms (*Echinocardium* and ophiuroids) follow those of pelecypods but are less marked; (ii) with regard to polychaetes (mainly *Spiophanes bombyx* and *Magelona papillicornis*) and pelecypods (*Nucula nitida* and *Tellina (Angulus) fabula*), the long-term fluctuations are inverse but seasonal fluctuations are direct; (iii) populations of the predator *Natica (Lunatia) nitida* increase with the increasing pelecypod abundance. Pursuing his investigations until 1973, ZIEGELMEIER (1974) observed that the decrease in the pelecypod population, caused (at least in part) by predation pressure, resulted in complete disappearance of *N. nitida* in 1970, and was followed by recovery as soon as the pelecypod population increased again in 1972 (Fig. 5-10). These observations support the general opinion that highly specialized predators, such as naticid gastropods, completely disappear when their prey density is reduced below a critical level.

The populations of such specialized predators are probably much less abundant than those of their prey. Versatile predators such as *Amphiura* species, which feed on such prey as post-larvae or very young pelecypods but can switch to detritus and microscopic organic particles, represent a permanent threat to their prey, especially since such species which occupy a lower level of the trophic pyramid entertain populations much more abundant than those of more specialized predators. However, many predators (numerous ophiuroids, drilling prosobranchs, nudibranchs which graze on hydroids, etc.) stop feeding during their breeding period thus allowing the larvae of their prey species to settle and to begin to develop, providing their breeding season is sufficiently adjusted to such transient predator passivity (THORSON, 1960).

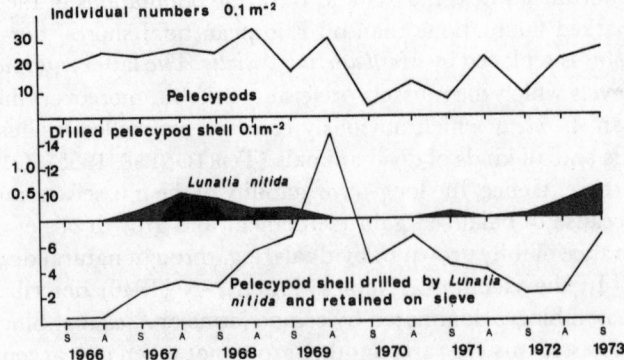

Fig. 5-10: Abundance comparison between pelecypods, boring naticid *Lunatia nitida* and pelecypod shells drilled by *L. nitida*; spring 1966 to spring 1973; S: spring; A: autumn. (After ZIEGELMEIER, 1974; reproduced by permission of Biologische Anstalt Helgoland.)

Another remarkable example of predation-pressure effects on the dynamics of a benthic assemblage concerns the grazing by young plaice on siphons of the pelecypod *Tellina tenuis* (O group). Such grazing shortens the inhalent siphon and hinders food uptake until the siphon is regenerated. Moreover, the regeneration process consumes energy which otherwise would be available for growth and breeding. Thus heavy grazing may cause a *Tellina* population to attain a point of no return. Below a certain threshold in the density of the *Tellina* population, the plaice lose the drive of siphon predation, a fact which facilitates recovery of *T. tenuis*. Usually, three years are sufficient for such recovery, and then a new cycle may start (TREVALLION and co-authors, 1968).

Predator–prey relationships may determine, or interfere with, the delimitation of assemblages and their rate of production. On Scottish coasts CONNELL (1961) demonstrated that the gastropod *Nucella lapillus* can drill only a limited number of *Balanus balanoides* every day, thus favouring the selection of the biggest individuals. Now the older and bigger a barnacle, the slower its growth rate and the wider the surface area occupied. Thus, since overcrowding tends to hinder standing crop increase, in a sense, predator action enhances the overall production of the barnacle population. In the midlittoral zone (p. 382) the upper limit of the vertical distribution of *B. balanoides* is related to the effects of environmental factors, whereas the lower limit depends on biotic factors: competition for space and food, as well as predator–prey relationships. Factor fluctuations are more effective near the upper and lower boundaries of the population, while recruitment and mortality are less variable at intermediate levels.

Predator–prey relationships may interfere with other factors. For example, in the Gulf of Finland SEGERSTRÅLE (1960) observed over a period of several years a very interesting inverse relation between the respective abundance of *Macoma balthica* and *Pontoporeia affinis*: where bottom currents are sufficiently strong to prevent mud sedimentation young *Macoma* are often abundant, while on deeper and muddier bottoms *Pontoporeia* is abundant and young *Macoma* very sparse. SEGERSTRÅLE assumes that post-larval and young *M. balthica* whose growth rate is low are easily caught by *P. affinis*. Furthermore, *P. affinis* is planktonic at night and must burrow a new hole every dawn when it returns to the bottom. Such activity may bury young *M. balthica* which then die in the poorly oxygenated sediment.

(ii) Competition for space. Competition for space, especially for the surface available, is a very important biotic factor on marine hard substrates. In the Mediterranean Sea and many other regions, succession of assemblages on a vacant substrate has been observed many times both on natural and man-made rocky substrates since this was first reported by HUVE (1953a,b; see also Fig. 5-11). Another example of assemblage succession was observed by POORE (1968) for $3\frac{1}{2}$ years during wharf-pile studies at Lettelton (New Zealand). In the midlittoral zone, the piles were initially colonized by the barnacle *Elminius modestus* which, after $2\frac{1}{2}$ years, was replaced by the mussel *Modiolus neozelanicus*, the latter attaining an unstable climax liable to storm injury. In the infralittoral zone the amphipod *Corophium acherusicum* and the bryozoan *Bugula* sp. predominated during the initial settlement; this was followed five months later by a colonization of solitary ascidians. The latter became dominant after $1\frac{1}{2}$ years and provided new niches for superficial foulers, crevice dwellers and motile species.

The succession of assemblages varies according to region, depth and wave exposure. Nevertheless, some generalization seems possible: (a) The succession may exhibit slight

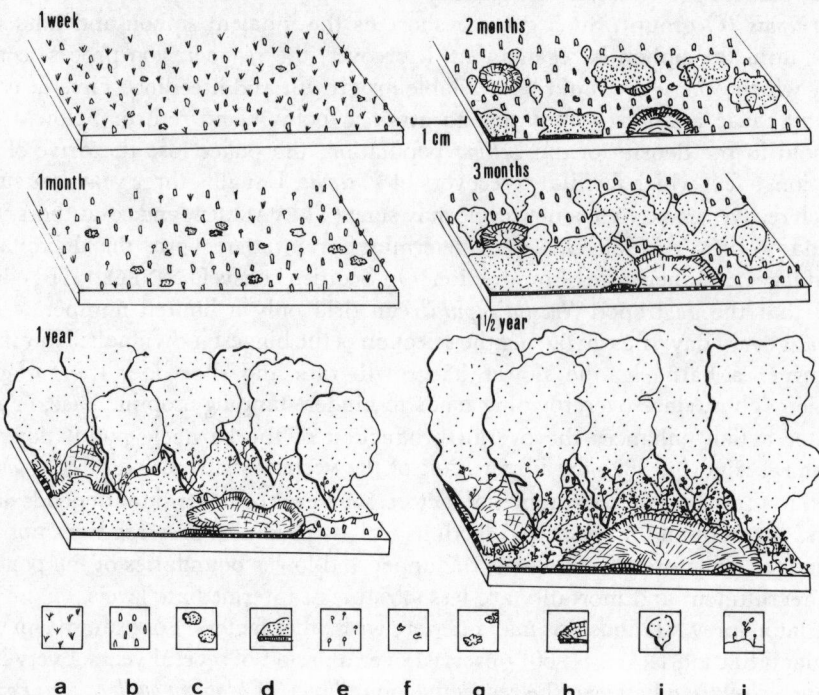

Fig. 5-11: Stages of colonisation of an artificial substrate by sessile flora and fauna in the infralittoral zone (depth 0·6 m) of the French Mediterranean Sea coast. a: colonial diatoms; b: small-sized multicellular algae; c and d: encrusting calcareous rhodophytes, settled early and later, respectively; e and j: hydroids settled early and later, respectively; F: polychaetes (mainly of the *genus Spirorbis*); g and h: bryozoans, settled early and later, respectively; i: *Ulva*, sp. (After HUVE, 1953a; reproduced by permission of *Rec. Trav. Stn. mar. Endoume*.)

changes depending on the starting time of observations or of the experiments, and to the breeding periods of the species involved in the succession. (b) When an advanced stage succession exhibits an upper stratum sufficiently dense to induce a significant decrease in illumination on the substrate itself, the original photophilic assemblage may be affected and its composition modified. On the other hand, it must be emphasized that the development of an upper stratum results in an increase in standing crop and production. Encrusting assemblages are often more or less torpid, but the development of an upper stratum corresponds to a three-dimensional assemblage. (c) Wave exposure is a very important factor in assemblage dynamics: on exposed hard substrates, wave-caused damage may sometimes arrest a succession or induce regression to earlier stages. Where sufficient suspended material is allowed to settle, the development of tubicolous animals (polychaetes, amphipods) may result in complex assemblages, often with a mixture of hard and soft substrate components. (d) Data from succession experiments on artificial substrates must be critically analysed; unpublished investigations by the reviewer demonstrated that successions at one and the same location can be quite different on panels with surface areas of 100 or 625 cm^2.

On soft substrates, competition for space is not as intense as on hard substrates. However, the infauna of assemblages in the infralittoral zone may become impoverished when sea-grass species develop there, for example, in communities of the inner reef flat and on muddy sands in sheltered areas of the Mediterranean Sea where the entanglement of *Cymodocea nodosa* rhizomes in the sediment becomes very heavy.

(iii) Modification in substrate nature due to plant or animal activities. In benthic assemblages, sediment turnover by burrowing animals or modification of grain size by faecal-pellet production may result in more or less cyclic changes. At Barnstable Harbor (Massachusetts, USA), the predominant species of an assemblage was the small gastropod *Nassarius obsoletus* during the summer of 1959, but in the next year amphipods of the genus *Ampelisca* (MILLS, 1968) predominated. Possibly sediment turnover by *Nassarius* induced a decrease in infauna elements and thus enhanced the settling of *Ampelisca* species. Later, feeding and tube-building of the amphipods caused an increase in the finest sediment fraction; the resulting faeces-enriched fluid sediment surface can easily be resuspended at intervals by tidal currents or waves thus pushing the amphipod population away. In such a case, the amphipod assemblage seems unable to reach a climax stage, but this situation cannot be generalized without further qualification, and presumably other assemblages with a different dominating species of the genus *Ampelisca* may be more stable.

In *Spisula subtruncata* populations the oldest individuals—through faecal and pseudo-faecal pellet production—increase the amount of the finest sediment particles thus rendering the sediment unsuitable for conspecific larval settlement. Presumably, patches of *S. subtruncata* which are often observed in the North Sea and comprise only individuals of the same size (and age class) result from inhibition of post-larvae settlement due to a modification in sediment characteristics (p. 170). After all the individuals have died, faecal pellets may be resuspended and progressively carried away by bottom currents until finally a new *S. subtruncata* population appears.

In the preceding examples the changes observed mainly correspond to differences in the abundance of species considered to belong to a single assemblage. In other cases modifications in substrate nature may result in a succession of assemblages which must be considered as distinct ecological units.

Sometimes effects on assemblage composition induced by changes in substrate characteristics due to plant or animal activities may be modified by other factors such as predator–prey relationships and/or casual events such as gales. This was demonstrated by EAGLE (1975) in the Horse Channel (Great Britain). On shallow-water muddy sand, EAGLE observed an irregular alternation between assemblages, the most important of which were dominated by *Pectinaria koreni*, *Abra alba* (sometimes together with spionid polychaetes) and *Lanice conchylega* (sometimes together with *Eulalia sanguinea* and *Phyllodoce mucosa*). As is well known, non-selective deposit feeders such as *A. alba* and *P. koreni* rework the sediment in a layer several centimetres thick and the resulting effect on sediment stability that such deposit feeders at densities of 1000 ind. m^{-2} or more will exert can easily be appreciated. Feeding activities of young plaice, sole and dab are presumably also significant in such sediment reworking. EAGLE believes that while tidal currents can disturb the sediment surface in shallow areas, they are not directly responsible for the erratic changes in the fauna of the deeper muddy sand. Substrate erosion can be caused by superimposition of severe wave-induced, bottom-water turbulence and tidal forces. EAGLE further suggests that the frequent dense populations of *Pectinaria* or *Abra* can be washed out during gales because the sediment had previously been loosened by animal feeding activities. Where species of *Pectinaria* and *Abra* dominated, their feeding activities prevented the settlement of heterospecific spat; this resulted in an assemblage with a low specific diversity. In contrast, where a surface-deposit and filter-feeding species such as *Lanice conchylega* dominates—although perhaps ingesting spat—it does not prevent settlement of juveniles, and the species number of the assemblage increases. In spite of the fact that the somewhat more diverse *Lanice* assemblage 'appeared to be potentially more stable', the worms can be removed—probably because they cannot extend their anchoring tubes as deep as they do in beach sands (EAGLE, p. 875) and because they are eaten by fishes which suck the whole worm from the tube, leaving it empty in the substrate.

(iv) *Competition for food.* It is often difficult to demonstrate assemblage dynamics exerted by interspecific food competition. Investigating the intertidal sand flats at Barnstable Harbor (Massachusetts, USA), SANDERS and co-authors (1962) observed that large populations (up to 40,000 ind. m^{-2}) of the small pelecypod *Gemma gemma* exclude the large clam *Mya arenaria*, probably because *G. gemma* consumes most of the available food. The data from VATOVA (1949) on soft-bottom communities in the northern Adriatic Sea suggest not only inverse fluctuations due to predator–prey (echinoderm–pelecypod) relationships (p. 180), but also inverse variations between two detritus-feeder groups: pelecypods and gastropods of the genera *Turritella* and *Aporrhais*.

On hard substrates, CONNELL (1970) found a striking example of seasonal change in food competition. At San Juan Island (Washington, USA), three species of *Thais* feed on the *Balanus glandula* which inhabit midlittoral rocky substrates. Barnacle larvae may settle on the entire midlittoral zone, but fully grown individuals exist only at the highest levels except for very sheltered places, where both *B. glandula* and *Thais* are very scarce, and on the most exposed rocks, where *Thais* cannot live. In summer intense competition prevails between the three *Thais* species which all feed on young barnacles in the lower midlittoral. In autumn only *T. emarginata*, which can better withstand emergence than its competitors, migrates upwards and feeds on older barnacles, but in the following spring it returns to the lower midlittoral and resumes competition with the other two species. The behaviour of *T. emarginata* involves a decrease in the competition for a common food stock with the other two species. In spite of the consumption of part of its barnacle population in autumn and winter by *T. emarginata*, the upper midlittoral constitutes a recruiting zone which, through the larvae released, will repopulate the lower subzone the following spring.

(v) *Autecological factors.* Obviously autecological factors may play some part in the dynamics of benthic assemblages, especially those related to the reproductive cycle. Fundamental investigations by THORSON (1946) revealed that most life-cycle phases (maturation, gamete release, fertilization, early developmental stages, larval life, settlement on the substrate) do not involve important losses, except for those suffering from the risks encountered during the planktonic larval stage, mainly due to predator activities. THORSON noticed that species with direct development, or with short-period planktonic larvae, exhibit only very small annual abundance fluctuations. In contrast, species with larvae which spend several weeks or months in the pelagial tend to undergo significant fluctuations in recruitment, depending mainly on the suitability of the prevailing conditions for the larvae during their planktonic phase (Fig. 5-12).

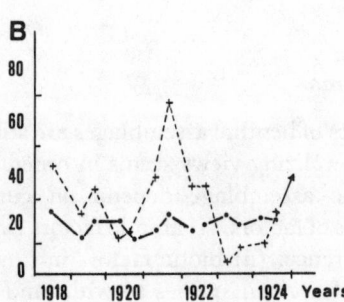

Fig. 5-12: Annual variation in biomass (B, in g m^{-2}) of bottom invertebrates recruited partially by larvae with a long planktotrophic pelagic life (*Abra alba*: Full line) or with non-pelagic development (*Macoma calcarea*: broken line). (After THORSON, 1946, based on data from BLEGVAD, 1925; reproduced by permission of Danish Institute for Fishery and Marine Research.)

THORSON's (1946) hypothesis mainly concerns species which breed every year, with a life-span a little longer than one year, or extending over several years. However, there are exceptions to such annual breeding. For example, North Sea echinoderms exhibit two different patterns: some, such as *Amphiura filiformis* and *Brissopsis lyrifera* which grow quickly and have a brief life-span, breed only once and then die; hence recruitment occurs every year; others, such as *Amphiura chiajei*, *Echinocardium cordatum* and *Cucumaria elongata* which grow more slowly and have a longer life-span feature more than one breeding period; recruitment is sporadic and sometimes two subsequent spawnings are separated by several years.

Even in short-lived (annual or a few years) species which breed every year, a marked random decrease in recruitment may cause pronounced changes within the assemblage. BUCHANAN and co-authors (1974), for example, studied terrigenous mud assemblages on the Northumberland shelf over a period of four years; they found that the number of species and the total estimated production at the secondary trophic level remained essentially stable. However, the number of individuals more than doubled within the test period. At the beginning of the investigation in 1971, *Ammotrypane aulogaster* and *Abra nitida* contributed to the total production estimates (1738 mg m^{-2} yr^{-1}) 20·5% and 6%, respectively (477 mg m^{-2} for both). Later they exhibited an abrupt decrease both in abundance and biomass. In 1974 both species contributed only 43 mg m^{-2} yr^{-1}, i.e. 2·3% of the total production.

A second group of species among the production dominants rapidly increased in number, biomass and production. This group included the polychaetes *Heteromastus cirriformis*, *Prionospio cirrifera*, *Paraonis gracilis* and *Glycera rouxi* which contributed 999 mg

m^{-2} to the total production (1853 mg m^{-2} yr^{-1}) in 1974. Therefore, in 1974 the increased numbers of four polychaete species have together contributed 27·5% of total production thus apparently compensating almost exactly for the loss encountered by the elimination of *Abra* and *Ammotrypane* (BUCHANAN and co-authors, 1974, p. 793). A third group of active producers revealed no response, remaining stable in both numbers and biomass during the investigation; these included *Calocaris macandreae*, *Lumbrinereis fragilis*, *Spiophanes kroyeri* and *Chaetozone setosa*. Since *Ammotrypane* was the community's top producer in 1971, BUCHANAN and co-authors believe that the variability in its population density has played a crucial role in the dynamics of the community as a whole.

Conclusions

Generalizations on the structure and dynamics of benthic assemblages are still premature. However, a tentative outline of some generalizing views seems in order.

The structure and dynamics of a benthic assemblage depend on cumulative, antagonistic and interfering effects of two groups of factors: (i) abiotic factors extrinsic to the assemblage, of aperiodic or periodic occurrence; (ii) biotic factors intrinsic to the assemblage, some autecological—e.g. ability of a given species to withstand environmental stress, intraspecific competition for food—and others synecological (interspecific relations).

The influence of **abiotic factors** depends on: (i) the nature, intensity and aperiodic or periodic occurrence of the disturbance; and (ii) the ratio between the whole area occupied by the assemblage and the part of this area damaged. Aperiodic changes due to random meteorological conditions can be exemplified by a coral reef. Damaged from heavy rain it can repair itself within a few months, but the combined efforts of a strong typhoon, heavy waves, rain and turbidity may deform the assemblage for several decades.

Until about a decade ago, long-term ($>$ 1 yr) periodic fluctuations in benthic assemblages received little, if any, attention; probably because our knowledge regarding the causative factors was insufficient. Now, the impacts of cosmic and climatic fluctuations on periodic changes in pelagial assemblages (p. 111) are recognized—if not fully explained—and it is obvious that they have repercussions on the benthal, for example, with regard to differences in annual recruitment of different species, which in turn result in changes in interspecific relations, such as predation and competition for space or food.

Changes in the substrate nature related to long-term changes in silt and clay input involve decades or even centuries, as do those originating from organismic activities, e.g. the ecological succession resulting in *Posidonia* bed climax.

The impact of short-term and middle-term periodic changes in environmental factors which depends largely on autecological characteristics of the species will be dealt with below. Regarding point (ii), it is obvious that if the damage concerns only a very limited area of the assemblage, reconstitution will be faster.

Biotic factors are so numerous and their synergistic effects so complex that a synthesis of their specific influence on macrobenthic assemblage structure and dynamics is difficult. In ecological terms, the factors involved may be subdivided into four groups: (i) autoecological factors; (ii) factors related to the size of the area occupied by the assemblage and to currents in upper water layers; (iii) synecological factors related to intra-specific and interspecific relationships: competition for space or food and predation.

Among the autecological factors are life-span, capability of forming dense populations (asexual reproduction, fecundity and rhythm of recruitment) and/or high biomasses (growth rate), gregariousness and behavioural characteristics. With regard to the latter, short-term periodic changes of less than 24 h (tidal and circadian rhythms) may have particular effects. Amphipods and shrimps which leave the algal canopy of the rocky intertidal at low tide do not significantly change the composition and dynamics of the assemblage. The same is true for bottom-dwelling amphipods or tanaids which rise to the hyponeuston layer at night. Such phenomena involve a particular behaviour pattern of a limited number of motile species and can be considered as 'microdynamics'. However, regular migrations between pelagial and benthal involve energy transfer from the pelagos, with its often higher productivity, to the benthos.

Middle-term changes, i.e. seasonal fluctuations, may involve two patterns: (i) Organisms responding directly and quickly to changes in an environmental factor (e.g. temperature, salinity, time of low tide, wave exposure). To this group belong migrators, but also forms periodically detached by winter storms, especially at the upper levels of rocky substrates. (ii) Organisms responding indirectly and more slowly. To this group belong organisms revealing recruitment peaks, which play an important role in assemblage structure and dynamics. Successful recruitment in species having larvae with a very long planktonic phase depends much more on the conditions prevailing in the pelagial than on the number of eggs produced.

Hence, long-term fluctuations may interfere with seasonal fluctuations in assemblages related to changes in predator–prey relationships. For example, if a predator stops feeding because the temperature is too high (or low), or because it begins to breed, the population of its prey tends to increase. The main characteristic of seasonal fluctuations in an assemblage is their regular occurrence.

Factors related to area size and currents in upper water layers. When environmental conditions are uniform over large bottom areas inhabited by a single assemblage, it seems that dynamic events do not occur simultaneously over the whole area. Thus, differences in the abundances of predators and prey observed at the same time in different parts of the area may seem to be a consequence of lack of fluctuation synchronism. Such lack of synchronism is probably necessary for the reconstitution of a prey population locally depleted by a predator. In his *Turritella profunda zoocoenosis* of soft bottoms of the northern Adriatic Sea, VATOVA (1949) recorded the following percentages—western subarea: pelecypods, 6·8; echinoderms, 75·1; southern subarea: pelecypods, 91·6; echinoderms, 1·3. These values indicate that major predators and prey exhibit almost opposite abundance relationships at the same time in the two subareas. Possibly the western subarea was sampled after the echinoderms had almost exhausted the pelecypod population, while the southern subarea was investigated during an abundance minimum of the echinoderms which permitted thriving development of the pelecypod population. Subsequent samplings over many years in both subareas may reveal progressive changes in the pelecypod : echinoderm ratio, up to a quantitative relation inverse to that recorded by VATOVA.

The capability of an assemblage to regenerate, after disturbance or destruction, in part of the area previously inhabited obviously depends on its breeding potential and settling opportunities of nearby recruiting stocks. However, it also depends on the ability of species of other assemblages to occupy the vacant space. (For example, substitution of chlorophyte and zoantharian assemblages for hermatypic corals, p. 165.) If the area inhabited by the injured assemblage is small, organisms transgressing from surrounding assemblages may hinder the settling of the species characteristic of the damaged assemblage in different ways, e.g. by faster occupation or by monopolization of the food available.

It seems that areas occupied by different assemblages tend to be restricted when edaphic factors (such as grain size, bottom currents, organic matter content of the sediment) reach an excessive intensity. Then interpenetration of different assemblage occurs which may result in a mosaic-like structure (CABIOCH, 1961; GLEMAREC, 1964, 1965; PEARSON, 1970), especially on shallow or semi-enclosed soft bottoms.

If an assemblage has been impoverished or largely destroyed, its regenerations will proceed much faster if water currents provide large amounts of seeds and larvae from undamaged areas. In general, regular, tidal currents (unless too intense) support existing assemblages. Observations by MASSE (1971) on both sides of the Marseilles area (France) emphasize the impact of the combined influence of wind-induced currents and shelf width on the dynamics of some soft-bottom assemblages. To the east of Marseilles, the shelf is very narrow and the slope fairly steep; northerly winds induce small upwellings. Thus, intermediate waters are substituted for coastal waters pushed offshore with their larvae (mainly photopositive ones), and the recruitment of shallow water assemblages becomes impoverished. To the west of Marseilles, temporary upwellings are less effective because the shelf is wider and the general westward drift (cyclonic gyre of the Mediterranean Sea's western basin) can transport larvae from eastern areas to similar bottoms located farther to the west.

Obviously, the influence of **synecological factors** partly depends on autecological characteristics of the species participating in the assemblage. For example, species capable of asexual reproduction colonize a solid substrate faster than those which make use only of sexuality. Dense populations of filter-feeding species whose pumping rate and screening efficiency are high will result in starvation of either their own offspring or other species with less efficient filtering devices and mechanisms. The role of predator–prey relationships in structure and dynamics has been discussed on p. 148ff.

From the examples provided it appears that the balance of an assemblage depends on the degree of harmony between fluctuations of abiotic and biotic factors. The possibility of achieving such a harmony depends on the variability of environmental factors and the organismic ability for making the necessary adjustments.

Literature Cited (Chapter 5)

ALDERDICE, D. F. (1970). Factor combinations: responses of marine poikilotherms to environmental factors acting in concert. In O. Kinne (Ed.), *Marine Ecology*, Vol. I, Environmental Factors, Part 3. Wiley, London. pp. 1659–722.

ALDERDICE, D. F. and FORRESTER, C. R. (1971). Effects of salinity and temperature on embryonic development of petrale sole (*Eopsella jordani*). *J. Fish. Res. Bd Can.*, **28**, 727–44.

ARNAUD, P. M. (1970). Frequency and ecological significance of necrophagy among the benthic species of Antarctic coastal waters. In M. W. Holdgate (Ed.), *Antarctic Ecology*, Vol. I. Academic Press, London. pp. 256–66.

ARNAUD, P. M. (1974). Contribution à la bionomie marine benthique des régions antarctiques et subantarctiques. *Téthys*, **6**, 465–656.

BELLAN, G. (1967a). Pollution et peuplements benthiques sur substrat meuble dans la région de Marseille. I. Le secteur de Cortiou. *Revue int. Océanogr. méd.*, **6**, 53–87.

BELLAN, G. (1967b). Pollution et peuplements benthiques sur substrat meuble dans la région de Marseille. Deuxième partie. L'ensemble portuaire marseillais. *Revue int. Océanogr. méd.*, **8**, 51–95.

BELLAN, G. (1976). La pollution par les tensio-actifs. In J. M. Peres (Ed.), *La Pollution des Eaux Marines*. Gauthier-Villars, Paris. pp. 31–50.

BERNARD, F., CAMON, E. and CLERET, J. J. (1963). L'hiver 1962–63 en Basse-Normandie. Effets immédiats et différés sur la faune intertidale. *Bull. Soc. linn. Normandie*, **4**, 97–110.

BEUKEMA, J. J. (1973). Migration and secondary spatfall of *Macoma balthica* (L.) in the western part of the Wadden sea. *Neth. J. Zool.*, **23**, 356–7.

BLEGVAD, H. (1925). Continued studies on the quantity of fish-food in the sea bottom. *Rep. Dan. Biol. Sta.*, **31**, 27–56.

BONOTTO, S. (1976). Cultivation of plants: multicellular plants. In O. Kinne (Ed.), *Marine Ecology*, Vol. III, Cultivation, Part 1. Wiley, Chichester. pp. 467–529.

BRADFIELD, A. E. and NEWELL, G. E. (1961) The behaviour of *Macoma balthica* (L.). *J. mar. biol. Ass. U.K.*, **41**, 81–7.

BRUNSWIG, D., ARNTZ, W. E. and RUMOHR, H. (1976). A tentative field experiment on population dynamics of macrobenthos in the western Baltic. *Kieler Meeresforsch.*, 49–59.

BUCHANAN, J. B., KINGSTON, P. F. and SHEADER, M. (1974). Long-term population trends of the benthic macrofauna in the offshore mud of the Northumberland coast. *J. mar. biol. Ass. U.K.*, **54**, 785–95.

BUCHANAN, J. B., SHEADER, M. and KINGSTON, P. F. (1978). Sources of variability in the benthic macrofauna off the South Northumberland Coast, 1971–1976. *J. mar. biol. Ass. U.K.*, **58**, 191–209.

CABIOCH, L. (1961). Étude de la répartition des peuplements benthiques au large de Roscoff. *Cah. Biol. mar.*, **2**, 1–40.

CASTENHOLZ, R. W. (1961). The effect of grazing on marine littoral diatoms on the western coast of Norway. *Sarsia*, **29**, 237–56.

CHESHER, R. H. (1963). The morphology and function of the frontal ambulacrum of *Moira atropos* (Echinoidea: Spatangoida). *Bull. mar. Sci. Gulf Caribb.*, **13**, 549–73.

CLARK, R. B. (1962). Observations on the food of *Nephthys*. *Limnol. Oceanogr.*, **7**, 380–5.

COLLIER, A. W. (1970). Oceans and coastal waters as life-supporting environments. In O. Kinne (Ed.), *Marine Ecology*, Vol. I, Environmental Factors, Part 1. Wiley, London. pp. 1–93.

CONNELL, J. H. (1961). The influence of interspecific competition and other factors on the distribution of the barnacle *Chthamalus stellatus*. *Ecology*, **42**, 710–23.

CONNELL, J. H. (1970). A predator–prey system in the marine intertidal region. I. *Balanus glandula* and several predatory species of *Thais*. *Ecol. Monogr.*, **40**, 49–78.

CONOVER, J. T. (1958). Seasonal growth of benthic marine plants as related to an estuarine environment. *Diss. Abstr.*, **19**, 1180.

CONOVER, R. J. (1978). Transformation of organic matter. In O. Kinne (Ed.), *Marine Ecology*, Vol. IV, Dynamics. Wiley, Chichester. pp. 221–449.

CREUTZBERG, F. (1975). Invertebrates. In O. Kinne (Ed.), *Marine Ecology*, Vol. II, Physiological Mechanisms, Part 2. Wiley, London. pp. 709–916.

CRISP, D. J. (1964). The effects of the winter 1962–63 on the British marine fauna. *Helgoländer wiss. Meeresunters.*, **10**, 311–27.

CRISP, D. J., MOYSE, J. and NELSON-SMITH, A. (1964). General conclusions. In D. J. Crisp (Ed.), The Effects of the Severe Winter of 1962–64 on Marine Life in Britain. *J. Anim. Ecol.*, **33**, 165–210.

DALES, R. P. (1957). Some quantitative aspects of feeding in sabellid and serpulid fan worms. *J. mar. biol. Ass. U.K.*, **36**, 309–16.

DAVIS, F. M. (1923). Quantitative studies on the fauna of the sea bottom. I. Dogger Bank. *Fishery Invest., Lond.*, **6**, 7–54.

DAVIS, F. M. (1925). Quantitative studies on the fauna of the sea bottom. II. South of the North Sea. *Fishery Invest., Lond.*, **8**, 1–50.

DOTY, M. S. (1967). Pioneer intertidal population and the related general vertical distribution of marine algae in Hawaii. *Blumea*, **15**, 95–105.

EAGLE, R. A. (1975). Natural fluctuations in a soft bottom benthic community. *J. mar. biol. Ass. U.K.*, **55**, 865–78.

EGGLESTON, D. and HICKMAN, R. W. (1972). Mass stranding of molluscs at Te Waewae Bay, Southland, New Zealand. *N.Z. Jl mar. Freshwat. Res.*, **6**, 379–82.

FEDER, H. and MØLLER-CHRISTENSEN, A. (1966). Aspects of asteroid biology. In R. A. Boolotian (Ed.), *Physiology of Echinodermata*. Wiley, London. pp. 81–127.

FELDMANN, J. (1937). Recherche sur la végétation marine de la Méditerranée: la côte des Albères. *Revue algol.*, **1937**, 1–350.

FINENKO, Z. Z. (1978). Production in plant populations. In O. Kinne (Ed.), *Marine Ecology*, Vol. IV, Dynamics. Wiley, Chichester. pp. 13–88.

FONTAINE, A. R. (1965). The feeding mechanisms of *Ophiocomina nigra*. *J. mar. biol. Ass. U.K.*, **45**, 373–85.

FUJI, A. (1967). Ecological studies on the growth and food consumption of Japanese common littoral sea urchin *Strongylocentrotus intermedius* (A. Agassiz). *Mem. Fac. Fish. Hokkaido Univ.*, **15**, 83–160.

GERLACH, S. A. (1972). Substratum: general introduction. In O. Kinne (Ed.), *Marine Ecology*, Vol. I, Environmental Factors, Part 3. Wiley, London. pp. 1245–50.

GESSNER, F. (1970). Temperature: plants. In O. Kinne (Ed.), *Marine Ecology*, Vol. I, Environmental Factors, Part 1. Wiley, London. pp. 363–406.

GLEMAREC, M. (1964). Bionomie benthique de la partie orientale du Golfe du Morbihan. *Cah. Biol. mar.*, **5**, 33–96.

GLEMAREC, M. (1965). La faune benthique dans la partie méridionale du Massif Armoricain, Etude préliminaire. *Cah. Biol. mar.*, **6**, 51–66.

GOODBODY, I. (1960). The feeding mechanism in the sand dollar *Mellita sexiesperforata* (Leske). *Biol. Bull. mar. biol. Lab., Woods Hole*, **119**, 80–6.

GOSSELCK, F., KELL, V., and SPITTLER, P. (1978). On the feeding of *Branchiostoma senegalense* (Acrania: Branchiostomidae). *Mar. Biol.*, **46**, 175–9.

GRAY, J. S. (1971). Factors controlling population localizations in polychaete worms. *Vie Milieu*, **11** (*Suppl.* 22), 707–22.

GRAY, J. S. (1974). Animal–sediment relationships. *Oceanogr. mar. Biol. A. Rev.*, **12**, 223–61.

GREZE, V. N. (1978). Production in animal populations. In O. Kinne (Ed.), *Marine Ecology*, Vol. IV, Dynamics. Wiley, Chichester. pp. 89–114.

GUILER, E. R. (1960). The intertidal zone forming species on rocky shores of the East Australian coast. *J. Ecol.*, **48**, 1–28.

HAGERMANN, L. (1966). The macro- and microfauna associated with *Fucus serratus* L., with some ecological remarks. *Ophelia*, **3**, 1–43.

HARGER, J. R. (1970). The effect of species composition on the survival of mixed populations of the sea mussels *Mytilus californianus* and *Mytilus edulis*. *Veliger*, **13**, 147–52.

HARTOG, C., DEN (1972). Plants: multicellular plants. In O. Kinne (Ed.), *Marine Ecology*, Vol. I, Environmental Factors, Part 3, Wiley, London. pp. 1277–89.

HELLEBUST, J. A. (1970). Light: plants. In O. Kinne (Ed.), *Marine Ecology*, Vol. I, Environmental Factors, Part 1. Wiley, London. pp. 125–58.

HOFMAN, C. (1959). Études écologiques et physiologiques de quelques algues de la Mer Baltique. *Colloq. int. Cent. natn Rech. scient.*, **1957**, 205–18.

HUVE, P. (1953a). Compte rendu préliminaire d'une expérience de peuplement de surfaces immergées. *Recl. Trav. Stn mar. Endoume*, **8**, 173–92.

HUVE, P. (1953b). Étude expérimentale du peuplement de surfaces rocheuses immergées en Méditerranée occidentale. *C. r. hebd. Séanc. Acad. Sci., Paris*, **236**, 419–22.

JOHANNES, R. E. (1972). Coral reefs and pollution. In *Marine Pollution and Sea Life*: F.A.O. *Tech. Conf. on Marine Pollution and its Effects on Living Resources and Fishing*, 1970. Fishing News (Books) Ltd, London. pp. 364–75.

KARLSON, R. (1978). Predation and space utilization patterns in a marine epifaunal community. *J. exp. mar. Biol. Ecol.*, **31**, 225–39.

KINNE, O. (1970). Temperature: invertebrates. In O. Kinne (Ed.), *Marine Ecology*, Vol. I, Environmental Factors, Part 1. Wiley, London. pp. 407–514.

KINNE, O. (1971). Salinity: invertebrates. In O. Kinne (Ed.), *Marine Ecology*, Vol. I, Environmental Factors, Part 2. Wiley, London. pp. 821–995.

KINNE, O. (1977). Cultivation of animals: research cultivation. In O. Kinne (Ed.), *Marine Ecology*, Vol. III, Cultivation, Part 2. Wiley, Chichester. pp. 579–1293.

KINNE, O. (Ed.) (1980). *Diseases of Marine Animals*, Vol. I, General Aspects, Protozoa to Gastropoda. Wiley, Chichester.

KLIMOVA, V. L. (1976). Changes in the distribution of the trophic zones of benthos in Peter the Great Gulf between the 30' and 70' years (Russ.). *Okeanologiya*, **16**, 343–5.

KREBS, C. J. (1978). *Ecology—The Experimental Analysis of Distribution and Abundance* (2nd ed.), Harper and Row, New York.

LAGARDERE, J. P. (1966). Recherches sur la biologie et l'écologie de la macrofauna des substrats meubles de la côte des Landes et de la côte basque. *Bull. Cent. Étud. Rech. scient.*, Biarritz, **6**, 143–209.

LEDOYER, M. (1962). Étude de la faune vagile des herbiers superficiels de Zostéracées et de quelques biotopes d'algues littorales. *Recl. Trav. Stn mar. Endoume*, **25**, 117–235.

LE LOEUFF, P. and INTES, A. (1969). Premières observations sur la faune benthique du plateau continental de Côte d'Ivoire. *Cah. ORSTOM, Océanogr.*, **7** (4), 61–6.

MCGINITIE, G. E. and McGINITIE, N. (1949). *Natural History of Marine Animals*, McGraw-Hill, New York.

MAGNUS, D. B. (1964). Gezeitenströmungen und Nahrungsfiltration bei Ophiuren und Crinoiden. *Helgoländer wiss. Meeresunters.*, **10**, 104–17.

MAKKAVEEVA, E. G. (1964). Turf biocoenoses of the Adriatic Sea (Russ.). *Trudy sevastopol. biol. Sta.*, **17**, 39–47.

MARGALEF, R. (1978). General concepts of population dynamics and food links. In O. Kinne (Ed.), *Marine Ecology*, Vol. IV, Dynamics. Wiley, Chichester. pp. 617–704.

MASSE, H. (1971). *Contribution à l'Étude Quantitative et Dynamique de la Macrofaune des Peuplements des Sables Fins Infralittoraux des Côtes de Provence*, Thèse Doct., Université Aix-Marseille.

MILLER, M. C. (1961). Distribution and food of the nudibranchiate Mollusca of the south of the Isle of Man. *J. Anim. Ecol.*, **30**, 95–116.

MILLS, E. L. (1968). The community concept in marine zoology with comments on continua and instability in some marine communities. A review. *J. Fish. Res. Bd Can.*, **26**, 1415–28.

MONNIOT, C. and MONNIOT, F. (1975). Feeding behaviour of abyssal tunicates. In H. Barnes (Ed.), *Proceedings of 9th European Marine Biological Symposium, 1975*. Aberdeen University Press. pp. 357–62.

MOORE, H. B. (1958). *Marine Ecology*, Wiley, New York.

MURDOCH, W. W. (1969). Switching in general predators: experiments on predator specificity and stability of prey populations. *Ecol. Monogr.*, **39**, 335–54.

NØRREVANG, A. (1965). Structure and function of the tentacle and pinnules of *Siboglinum ekmani* Jägersten (Pogonophora), with special reference to feeding problems. *Sarsia*, **21**, 37–47.

PAINE, R. T. (1963). Food recognition and predation on opisthobranchs by *Navanax inermis*. *Ecology*, **46**, 603–19.

PAINE, R. T. (1976). Size-limited predation: an observational and experimental approach with the *Mytilus–Pisaster* interaction. *Ecology*, **57**, 858–73.

PANDIAN, T. J. (1975). Mechanisms of heterotrophy. In O. Kinne (Ed.), *Marine Ecology*, Vol. II, Physiological Mechanisms, Part 1. Wiley, London. pp. 61–249.

PARENZAN, P. (1940). Biocenologia bentonica dei fondi marini a Fango (Golfo di Napoli). *Boll. Idrobiol. Cacc. Pesca Afr. orient. ital.*, **1**, 117–42.

PASTERNAK, F. A. (1964). Deep-Sea pennatularians (Octocorallia) and antipatharians (Hexacorallia) collected during the R/S Vitjaz in the Indian Ocean and the resemblance between the faunas of Pennaturalians in the Indian and Pacific Oceans. *Trudy Inst. Okeanol.*, **69**, 183–215.

PEARSON, T. H. (1970). The benthic ecology of loch Linnhe and loch Eil, a sea loch system on the west coast of Scotland. I. The physical environment and distribution of the macrobenthic fauna. *J. exp. mar. Biol. Ecol.*, **5**, 1–34.

PERES, J. M. and PICARD, J. (1964). Nouveau manuel de bionomie benthique de la Mer Méditerranée, *Recl. Trav. Stn mar. Endoume*, **31**, 1–137.

PICARD, J. (1965). Recherches qualitatives sur les biocoenoses marines des substrats meubles dragables de la région marseillaise. *Recl Trav. Stn mar. Endoume*, **36**, 1–160.

PILSON, M. and TAYLOR, P. B. (1961). Hole drilling by *Octopus*. *Science, N.Y.*, **134**, 1366–8.

POORE, C. B. (1968). Succession of a wharf-pile fauna at Lytellton, New Zealand. *N.Z. Jl mar. Freshwat. Res.*, **2**, 577–90.

PSHENIN, L. N. (1965). *Coelosphaerium benticum* (n. sp.) and *Rhynchomonas metabolita* from deep-sea muds of the Mediterranean in nitrogen-fixing mixed cultures (Russ.). *Trudy sevastopol. biol. Sta.*, **15**, 8–27.

REGIS, M. B. (1978). Croissance de deux Echinoides du Golfe de Marseille (*Paracentrotus lividus* [Lmk] et *Arbacia lixula* L.). *Aspects Écologiques de la Microstructure du Squelette et de l'Évolution des Indices Physiologiques*, Thèse Doct., Université Aix-Marseille.

RHOADS, D. C. and YOUNG, D. K. (1971). Animal-sediment relations in Cape Cod Bay, Massachusetts. II. Reworking by *Molpadia oolitica* (Holothurioidea). *Mar. Biol.*, **11**, 255–61.

RIEDL, R. (1971a). Water movement: general introduction. In O. Kinne (Ed.), *Marine Ecology*, Vol. I, Environmental Factors, Part 2. Wiley, London. pp. 1085–9.

RIEDL, R. (1971b). Water movement: animals. In O. Kinne (Ed.), *Marine Ecology*, Vol. I, Environmental Factors, Part 2. Wiley, London. pp. 1123–56.

SANDERS, H. L., GOUDSMIT, E. M., MILLS, E. L. and HAMPSON, G. E. (1962). A study of the intertidal fauna at Barnstable Harbor, Massachusetts. *Limnol. Oceanogr.*, **7**, 63–79.

SCHÖNE, H. (1975). Orientation in space: animals. General introduction. In O. Kinne (Ed.), *Marine Ecology*, Vol. II, Physiological Mechanisms, Part 2. Wiley, London. pp. 449–553.

SCHLEGEL, H. G. (1975). Mechanisms of chemo-autotrophy. In O. Kinne (Ed.), *Marine Ecology*, Vol. II, Physiological Mechanisms, Part 1. Wiley, London. pp. 9–60.

SCHWENKE, H. (1971). Water movement: plants. In O. Kinne (Ed.), *Marine Ecology*, Vol. I, Environmental Factors, Part 2. Wiley, London. pp. 1091–121.

SEGERSTRÅLE, S. G. (1960). Investigations on Baltic populations of the bivalve *Macoma baltica* (L.). I. Introduction. Studies on recruitement and its relation to depth in Finnish coastal waters during the period 1922–59. Age and growth. *Commentat. biol.*, **23** (2), 1–72; (9), 1–19.

SEOANE-CAMBA, J. (1965). Estudios sobre las algas bentonicas en la costa sur de la Peninsula Ibérica (litoral de Cadiz). *Investigación pesq.*, **29**, 3–216.

SOKOLOVA, M. N. (1959). On the distribution of deep water bottom animals in relation to their feeding habits and the character of sedimentation. *Deep Sea Res.*, **6**, 1–4.

SOROKIN, Yu. I. (1978). Decomposition of organic matter and nutrient regeneration. In O. Kinne (Ed.), *Marine Ecology*, Vol. IV, Dynamics. Wiley, Chichester. pp. 501–616.

SOUTHWARD, A. J. (1964). Limpet grazing and the control of vegetation on rocky shores. In D. J. Crisp (Ed.), *Grazing in Terrestrial and Marine Environments*. Blackwell, Oxford. pp. 263–73.

STASEK, C. R. (1965). Feeding and particle-sorting in *Yoldia ensifera* (Bivalvia: Protobranchiata) with notes on other nuculanids. *Malacologia*, **2**, 349–66.

SWENNEN, C. (1961). Data on distribution, reproduction and ecology of the nudibranchiate molluscs occurring in the Netherlands. *Neth. J. Sea Res.*, **1**, 191–240.

TAYLOR, W. R. (1960). *Marine Algae of the Eastern Tropical and Subtropical Coasts of the Americas*, University of Michigan Press, Ann Arbor.

THAYER, G. W., ENGEL, D. W. and LaCROIX, M. W. (1977). Seasonal distribution and changes in the nutritive quality of living, dead and detrital fractions of *Zostera marina* L. *J. exp. mar. Biol. Ecol.*, **30**, 109–27.

THORSON, G. (1946). Reproduction and larval development of Danish marine bottom invertebrates, with special reference to the planktonic larvae in the sound (Oresund). *Meddr Kommn Danm. Fisk.—og Havunders.* (ser. Plankton), **4**, 1–523.

THORSON, G. (1950). Reproductive and larval ecology of marine bottom invertebrates. *Biol. Rev.*, **25**, 1–45.

THORSON, G. (1960). Parallel level-bottom communities, their temperature adaptations, and their balance between predators and food animals. In Buzzati-Traverso (Ed.), *Perspectives in Marine Biology*. University of California Press, Berkeley. pp. 67–86.

TORTONESE, E. (1965). Echinodermata. In *Fauna d'Italia*, Vol. VI. Ediz. Calderini, Bologna. pp. xiii–422.

TREVALLION, A., STEELE, J. H. and EDWARDS, R. R. C. (1968). Dynamics of a benthic bivalve. In *Symposium on Marine Food Chains, 23–26 July*, Contrib. 13. University of Aarhus, Denmark.

TSURNAMAL, M. and MARDEC, J. (1966). Observations on the basket star *Astroboa nuda* (Lyman) on coral reefs at Elat (Gulf of Akaba). *Israel J. Zool.*, **15**, 9–17.

TURPAEVA, E. F. (1953). Food and trophic groups of marine bottom invertebrates. (Russ.) *Trudy Inst. Okeanol.*, **7**, 259–99.

URSIN, E. (1952). Change in the composition of the bottom fauna of the Dogger Bank area. *Nature, Lond.*, **1952**, 170–324.

VATOVA, A. (1949). La fauna benthonica dell'Alto e medio Adriatico. *Nova Thalassia*, **1**, 1–110.

VINOGRADOV, M. E. (Ed.) (1977). *Biology of the Ocean*, Part 2, Biological productivity of the Ocean. Izdat. Nauka, Moscow.

WANGERSKY, P. J. (1978). Production of dissolved organic matter. In O. Kinne (Ed.), *Marine Ecology*, Vol. IV, Dynamics. Wiley, Chichester. pp. 115–220.

WARNER, G. F. (1969). Brittle-star beds in Torbay, Devon. *Underwater Association Report*, **4**, 81–85. (T. G. W. Industrial and Research Promotions Ltd, Carshalton, Surrey).

WELLS, G. P. (1945). The mode of life of *Arenicola marina* (L.). *J. mar. biol. Ass. U.K.*, **26**, 170–207.

WELLS, G. P. (1949). The behaviour of *Arenicola marina* in sand, and the role of spontaneous activity cycles. *J. mar. biol. Ass. U.K.*, **28**, 465–78.

WELLS, G. P. (1966). The lugworm (*Arenicola*). A study in adaptation. *Neth. J. Sea Res.*, **3**, 294–313.

WHITTAKER, R. H., LEVIN, S. A. and ROOT, R. B. (1973). Niche habitat and ecotope. *Am. Nat.*, **107**, 321–38.

WIESER, W. (1951). Über die quantitative Bestimmung der Algen—Mikrofauna felsiger Meeresküsten. *Oikos*, **3** (1), 124–31.

ZAVODNIK, D. (1965). Quelques résultats des recherches actuelles sur les peuplements phytaux dans l'Adriatique du Nord. *Rapp. P.-v. Réun. Commn int. Explor. scient. Mer Méditerr.*, **18**, 101–6.

ZAVODNIK, D. (1967a). The community of *Fucus Virsoides* (Don.) Ag. on a rocky shore near Rovinj (northern Adriatic). *Thalassia jugosl.*, **3**, 105–13.

ZAVODNIK, D. (1976b). Dynamics of the littoral phytal on the west coast of Istria. *Diss. Acad. sci. art. Slov.*, cl. IV, **10**, 5–71.

ZIEGELMEIER, E. (1964). Einwirkungen des kalten Winters 1962–63 auf das Makrobenthos im Ostteil der Deutschen Bucht. *Helgoländer wiss. Meeresunters.*, **10**, 276–82.

ZIEGELMEIER, E. (1974). Untersuchungen über die Bodenfauna der Deutschen Bucht. In *Jahresbericht 1974*, Biologische Anstalt Helgoland, Hamburg.

Marine Ecology Vol. 5, Part 1
Edited by Otto Kinne
© 1982 John Wiley & Sons Ltd

6. MAJOR PELAGIC ASSEMBLAGES

J. M. PERES

(1) General Aspects

As already emphasized, differentiation between organismic assemblages, qualitative and quantitative, cannot follow the same scheme in both pelagial and benthal. For the pelagial, I have attempted to differentiate assemblages on the basis of the water masses inhabited. Consequently, descriptions of pelagic assemblages follow the generally accepted concept of the distribution of water masses (Volume I: COLLIER, 1970, p. 1 ff.). In other words, I have adopted a geographical framework. In the pelagial, information on vertical distributions of different assemblages is sometimes marked or complicated by: (i) vertical migrations of many planktonic organisms; (ii) methodological aspects of plankton sampling, with regard to vertical hauls which sometimes result in a mixture of organisms inhabiting wide ranges, e.g. over 0 to 200 m or even 0 to 500 m.

Quantitative data on biomass and production have been expressed by different authors in different ways: e.g. wet weight (w.w), dry weight (d.w.), or carbon weight. Unless specifically mentioned, all the biomass data referred to in this chapter correspond to wet weight. In contrast, benthic assemblages will be classified according to the general scheme of vertical zonation proposed on p. 9. In each zone, assemblages are differentiated according to the nature of the substrate.

I have mostly preferred to differentiate assemblages on the basis of their specific compositions, i.e. qualitatively (biocoenosis concept). However, quantitative aspects will also be considered wherever it may reasonably be assumed that, in a given place, biocoenosis and community correspond to an 'assemblage' as defined on p. 64.

(2) Southern Ocean

(a) General Aspects

The Southern Ocean (Volume I: COLLIER, 1970, p. 41 ff.) is not well delimited from a geographical point of view; its conventional boundaries are based upon the distribution of its water masses and the characteristics of their movements. It is usually admitted that the northern boundary of the Southern Ocean is the subtropical convergence which corresponds to the contact zone between the south central waters of the three oceans, and subantarctic surface waters. South of this convergence the whole Southern Ocean may be divided from a biogeographical point of view into two zones, subantarctic and antarctic, separated by the Antarctic convergence (South Polar front). Since westerlies predominate south of Latitude 35° to 38° S, with maximum velocities between 50° and

60° S, while easterlies predominate in the immediate vicinity of the Antarctic continent, the Antarctic divergence occurs between the eastward and westward drifts generated by the westerly and easterly winds, respectively. The divergence separates southern and northern Antarctic subzones, each with slightly different biological characteristics.

(b) Phytoplankton

Composition and Distribution

According to HASLE (1969) the phytoplankton of the Southern Ocean (Pacific sector) is more diversified than previously assumed. She identified approximately 100 diatom, 70 dinoflagellate, 5 coccolithophorid and 5 silicoflagellate taxa; unclassified monads and naked flagellates are numerous. The latter, together with diatoms and dinoflagellates, are regular elements of the standing stock within the area, where each of the diatom genera *Chaetoceros*, *Coscinodiscus*, *Fragilaria*, *Nitzschia* and *Rhizosolenia* are represented by 10 to 15 species and the dinoflagellate genus *Peridinium* by about 20 species. *P. applanatum* is the most abundant thecate dinoflagellate. Among the athecate species, only *Exuviella baltica* and some gymnodinians are rather abundant. The silicoflagellate *Dictyocha speculum* seems to appear in larger concentrations in the Southern Ocean than in any other area of the World Ocean. Although HASLE's investigations were restricted to the Pacific sector, her results are of general value for the whole Southern Ocean; she noticed that this flora consists of both cosmopolitan species and those endemic to the Southern Ocean.

In most parts of the area investigated, diatoms are more numerous than the other groups; the largest populations occur in the Antarctic zone. Diatom abundance, expressed as log average concentration for the 0- to 25-m layer, and Secchi disc readings are negatively correlated. This means that diatoms are mainly responsible for the turbidity. Dinoflagellates are everywhere subordinate to the other groups, tending to be particularly scarce in the southern Antarctic subzone. Coccolithophorids, which are mainly restricted to the subantarctic zone, are sometimes the most abundant group within this zone, whereas unclassified flagellates may predominate at some stations near Graham Land. The Antarctic convergence is a very important biogeographical boundary.

North of the Antarctic convergence, the subantarctic surface water is inhabited by the subantarctic phytoplankton assemblage, the indicator species of which are *Rhizosolenia delicatula*, *Amphidinium amphidinoides*, *Oxytoxum variabile* and *Phalacroma pulchellum* (WALSH, 1971). HASLE (1969) reports the following dominant species: *Chaetoceros neglectus*, *Nitzschia barkleyi* and *Coccolithus huxleyi*. In addition to the characteristic species, one may also observe in the subantarctic zone some allochthonous forms, apparently introduced from the north—in spite of the boundary represented by the Antarctic convergence.

South of the Antarctic convergence, the most characteristic species of the Antarctic zone assemblage are *Chaetoceros bulbosum*, *C. dichaeta*, *Dactyliosolen antarctica* and *Eucampia zoodiacus* (WALSH, 1971). Taking into account the dominant species, HASLE (1969) distinguishes two subzones: (i) in the northern subzone, the two species of the genus *Chaetoceros* mentioned above are abundant, together with *Fragilariopsis nana* and *Nitzschia closterioides*; (ii) in the southern subzone *Fragilariopsis curta*, *F. cylindrus* and *Nitzschia subcurvata* predominate; many species appear in or on the sea ice and some may produce resting spores (endospores).

HENDEY (1937) suggested that there might be also a special flora associated with the Antarctic convergence, very abundant in genera and species, particularly in diatoms. Obviously, the flora contains species belonging to both phytoplankton assemblages separated by the convergence. However, it also includes endemic species, such as *Corethron criophilum 'inerme'*, *Coscinodiscus excentricus*, *Rhizosolenia rhombus* and *R. simplex*. The existence of this 'Antarctic convergence flora' has been confirmed in the Indian sector of the Southern Ocean by EL-SAYED (1970) who collected the same characteristic species and also noticed that diatoms are especially abundant.

At the Antarctic convergence, the antarctic phytoplankton assemblage sinks with the Antarctic surface water it inhabits and progressively disappears, except where the convergence moves (e.g. by change in strength of westerly winds); this results in vertical mixing which brings back the phytoplankters into the euphotic layer.

The vertical distribution of phytoplankters has been insufficiently investigated to date. According to HASLE (1969), maximum concentrations of diatoms occur most frequently at a depth of about 25 m in the subantarctic zone and in the layer 10 to 25 m in the Antarctic zone; most diminutive 'monads and flagellates' are fairly close to the surface; diminution in dinoflagellate populations seems to occur a little deeper than that of other groups. In general, the upper layers exhibit a low specific diversity; maxima in species number occur deeper than those in biomass (WALSH, 1969, 1971). Local hydrographic conditions also play an important part both regarding abundance and vertical distribution of phytoplankton: the more marked the salinity and density gradients, the more abundant the diatoms (sometimes more than 1.5×10^6 cells l^{-1}), monads and flagellates. In stratified waters, phytoplankton distribution is stratified and its distribution is more homogeneous. Phytoplankters are also sometimes more abundant at lower levels in areas with pronounced vertical circulation.

WALSH (1971) attempted to analyse the relative importance of habitat variables for predicting distribution patterns of phytoplankton. Phosphate and nitrate—which appear to be present in excess of metabolic requirements—provide very little information on changes in standing crop, species number and individual species distribution. WALSH suggests that some biochemical parameters—e.g. vitamin B_{12}, glycolic acid or enzyme activity—might provide a better basis for interpretation and prediction.

During the 1977 French cruise 'ANTIPROD', it was observed that in the Antarctic surface water, phytoplankton exhibited growth rates in terms of cell numbers that were inversely correlated with the cell's content in nitrogen and carbon. In other words, cells divide without increase in biomass. This might be ascribed to molybdenum deficiency required for nitrate reductase activity. Molybdenum is an enzyme catalysing the reduction of nitrates which are very abundant in Antarctic surface water. However, this inhibition of primary production might also be ascribed either to the low content in iron, or, more probably, to the very low content in vitamin B_{12} (approximately four times less than in Antarctic neritic waters).

Abundance and Production

As pointed out by HASLE (1969), the average phytoplankton standing stock per square metre is larger in the Antarctic zone than in the subantarctic zone, mainly due to the striking abundance of diatoms as well as 'monads and flagellates' in the former. In contrast, dinoflagellates are somewhat more abundant in the subantarctic zone. On an average, during the productive season, the abundance is higher than 10^4 cells l^{-1} in most areas of the Antarctic zone, mainly in the vicinity of the Antarctic divergence; it is between 10^2 and 10^4 cells l^{-1} in many areas closer to the Antarctic continent. In the subantarctic zone, the abundance usually decreases to between 10^2 and 10^4 cells l^{-1}, except for a belt of abundance higher than 10^4 cells l^{-1}, approximately between 70° W and 50° E. According to HASLE (1969, p. 159), seasonal changes in abundance are more marked in diatoms: their maximum density occurs in the subantarctic zone in mid-December, in the middle areas of the Antarctic zone in mid-January and still further south at the end of January and the first half of February. Seasonal changes in dino-flagellate abundance parallel those of diatoms, except for the second half of the period when maximum densities are found more to the north. In mid-February the dinoflagel-late population declines, while the diatom population is still at its height. With regard to 'monads and flagellates', maximum densities occur farther to the south in the Antarctic zone and the populations apparently increase in mid-January.

Our present knowledge on primary productivity is still quite limited. However, chiefly according to several papers by EL-SAYED (1970, 1971) and EL-SAYED and co-authors (1964), the following facts may be considered as rather well established: (i) Primary production and phytoplankton standing crop are higher in the Antarctic zone than in the subantarctic zone. (ii) At the Antarctic convergence a striking drop occurs in both chlorophyll a content and ^{14}C uptake (Fig. 6-1). (iii) In general, primary productivity in the Atlantic sector is higher than in the Pacific sector; the Indian sector is somewhat intermediate (EL-SAYED, 1970). (iv) Striking differences have been noticed between the

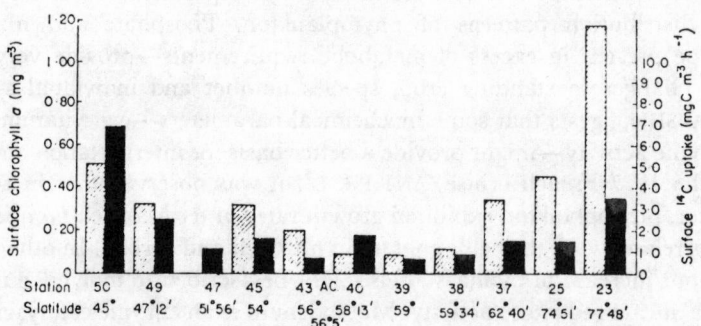

Fig. 6-1: Distribution of chlorophyll a in mg m^{-3} (hatched) and ^{14}C uptake in mg C m^{-3} h^{-1} (black) in surface waters on a transect between Tasmania and Ross Sea (Eltanin Cruise 27). (After EL-SAYED, 1970; reproduced by permission of Academic Press, London.)

productivity of oceanic and neritic areas in both Antarctic and subantarctic zones. For example, in the Atlantic sector, values of chlorophyll a and ^{14}C uptake in some neritic areas are five times higher than those in oceanic areas. Among the most productive regions are: the Bransfield Strait; the area north of the South Shetlands; the area between the South Orkneys and South Georgia. In contrast, primary production seems to be particularly low in the southern and eastern areas of the Weddell Sea. (v) The shape of the annual primary production curve is still poorly known. For a long time it was assumed that a sharp increase in primary production occurs early in November; its decrease in autumn seems to be more progressive and a rather important photosynthetic activity might be observed in February. However, in the USSR sector of the Antarctic, KLIACHTORIN (1964) observed the rise of primary production early in January, with a peak (200–300 mg C m^{-3} d^{-1}) between January 20 and 25 and a sharp fall down to about 10 mg C m^{-3} d^{-1} at the end of that month. In fact, the seasonal primary production curve is probably quite variable, depending on local conditions and yearly climatic differences. (vi) Among the factors which control primary production, light appears to be the most important, mainly below 50 m; this suggests that no light limitation occurs above the 10% light level.

'In general, the nutrient distribution shows higher concentration south of the convergence than north of it. However, it is interesting to note that although the concentration of nutrient elements north of the convergence is lower than in the Antarctic waters proper, even the lowest levels of concentrations are higher, in general, than are the winter maxima of temperate regions' (EL-SAYED, 1970, p. 133).

(vii) Antarctic waters are rich only in coastal and inshore regions, not in oceanic regions (EL-SAYED, 1970); WALSH (1969) suggests that, in spite of the very strong peak which corresponds to the summer in the Southern Ocean, the overall production of this ocean and that of the temperate or subtropical areas would be of the same order of magnitude (Fig. 6-2).

HORNE and co-authors (1969), who measured primary production in an inshore area in the South Orkney Islands using ^{14}C techniques *in situ* during two consecutive seasons, observed the peak of phytoplankton activity in early February; the authors estimated primary productivity as 250 mg C m^{-2} h^{-1} and over the season as about 130 g C m^{-2}.

The significance of nanoplankton for primary production of the Ross Sea was investigated during the 1972 austral summer in four different places (FAY, 1973): Subantarctic region, Antarctic convergence, Antarctic region and Ross ice shelf. Standing crop values integrated for the 0- to 200-m layer were 25·62, 35·08, 51·41 and 96·37 mg chlorophyll a m^{-2}, respectively; nanoplankton accounted for 76·2%, 88·8%, 61·1% and 64·9% of the standing crop in the respective areas. Total phytoplankton production in the euphotic layer for the Subantarctic region, Antarctic region and Ross ice shelf averaged 5·72, 6·18 and 9·97 mg C m^{-2} h^{-1}, while nanoplankton production accounted for 90·2%, 67·0% and 54·0% of the total in the three regions. FAY attributes the increased standing crop and primary production south of the Antarctic convergence to an increase in the net plankton fraction.

Phytoplankton blooms seem to be rather rare in the Antarctic zone. EL-SAYED (1971) observed a very striking bloom in the Weddell Sea covering an area of about 15,000 km^2 featuring yellow-brown patches interspersed with narrow, long bands of rather clear

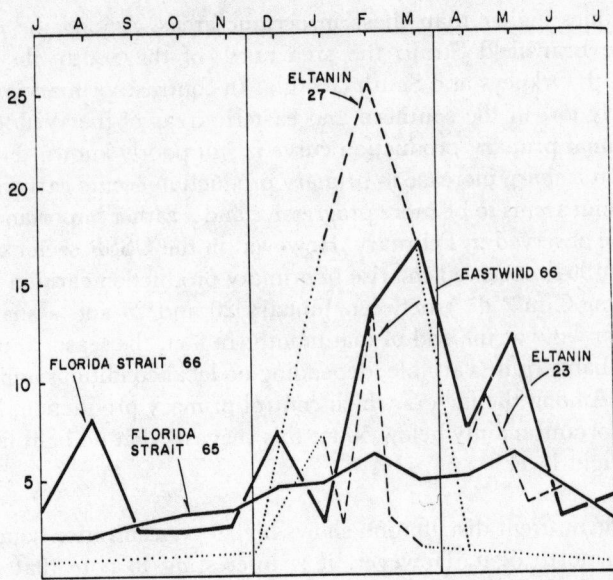

Fig. 6-2: Seasonal and year to year changes in standing crop
(number of chlorophyll fluorescent cells × 10^{12} m^{-2}) in
Florida Strait (solid lines) and Southern Ocean (broken
lines: observed; dotted lines: postulated). (After WALSH,
1969; reproduced by permisssion of American Society of
Limnology and Oceanography.)

water. The most abundant organism responsible for the bloom was the diatom *Thalassiosira tumida*. However, other diatoms also existed within the bloom, together with dinoflagellates and tintinnids. In spite of the exceptionally high chlorophyll *a* content (190 mg m^{-3}), primary production was only $1 \cdot 5$ g C m^{-2} d^{-1} because the euphotic layer was very thin (10 m). Phytoplankton was also surprisingly abundant on the underside of sea ice and pancake ice. Other blooms have been recorded by several authors; they were all attributable to diatoms: often to *Thalassiothrix antarctica* but sometimes also to *Chaetoceros sociale* (up to 25 × 10^6 cells l^{-1}), *Chaetoceros criophilum*, *Corethron criophilum*, *Biddulphia weissfloggi* and others.

Data on global primary production of the Southern Ocean remain somewhat unreliable, mainly in neritic areas. The contribution of unicellular algae, which live either upon and inside the ice or in the water columns beneath the ice remains to be exactly determined. BUNT (1964) believes that the production of the ice flora proper corresponds to rather low values of illumination and salinity. According to FUKUSHIMA and MEGURO (1966) primary production in neritic areas of the Antarctic zone appears to be significantly affected by phytoplankters living inside the ice where these are released upon the ice melting. FUKUSHIMA and MEGURO found up to 78 ml of diatoms in a 10-1 lump of ice sampled near the bottom, i.e. 100 times more than in the most superficial ice and about 560 times more than in the water itself. They also noticed that the predominant species of the three assemblages are different.

According to his observations, especially on Antarctic shores, ROUND (1971) suggested in a recent review that two different ice floras ('cryophilic associations') exist. The first, attached to the undersurface of the ice, mainly consists of benthic diatoms. The second, essentially of planktonic nature, includes not only diatoms but also dinoflagellates and *Phaeocystis*; it occurs in snow compacted on top of ice and in interstitial unfrozen water between ice crystals. Both these very abundant floras have a very high productivity (0.5–10 metric ton C d^{-1}) in the 2.6 million km^2 of brown ice around the Antarctic continent. The large populations often formed by epicryotic algae might be related to the very high stability of the habitat; negative factors such as wave action or grazing are non-existent or negligible; hence large populations can develop in spite of low temperatures and light intensities. How can unicellular planktonic algae survive during the 'polar night'? This interesting question is still to be solved. Some experiments have demonstrated their ability to survive 12 to 24 weeks in total darkness; such ability seems to be enhanced by low temperatures; during the winter resting period, the structure of the cell exhibits some changes towards a 'spore-like' stage which begins to develop again as soon as it is exposed to light.

(c) Bacterioplankton

In the Southern Ocean bacterioplankton is generally assumed to be rather poor, at least in the euphotic layer. SIEBURTH (1959, 1960b) assigned this scarcity to the acrylic acid released by *Phaeocystis poucheti* (Haptophyceae). The presence of this antibacterial substance (up to 7 μg^{-2} during a bloom) has been confirmed by GUILLARD and HELLEBUST (1971) who pointed out that this alga also releases several polysaccharides. SIEBURTH (1960a), who observed that the gut of seabirds was almost bacteria-free, assumed that the acrylic acid may be transmitted throughout the food chain: *Phaeocystis* → *Euphausia* → sea birds. The scarcity of bacterioplankton in surface waters of the Southern Ocean requires further investigation. According to MORITA and co-authors (1971), there is a significant amount of microbial activity in Antarctic waters. The bacteria they studied are obligate psychrophiles and exhibit a very high ability to utilize organic substrates. Significant mineralization occurs in the ice-laden offshore water. The highest activity was observed near the Antarctic convergence and in the vicinity of the iceshelf, whereas the microbial activity north of the Antarctic convergence was almost immeasurably small.

According to SEKI and KENNEDY (1969), bacteria and other heterotrophic microorganisms in the Strait of Georgia may represent an important part of the zooplankter's food supply during the winter. High microbial activity was recorded in the euphotic layer, especially on particulate organic matter, less than 25 μm in diameter; the standing crop was estimated to be between 40 and 3000 mg C m^{-2} and the zooplankton may be able to concentrate the bacteria by a factor of about 10^5. It seems that only the highest levels of microbial production are sufficient for providing more than a subsistence level for copepods.

(d) Zooplankton

Composition and Distribution

In general, the Antarctic convergence appears as a rather sharp boundary between two different zooplankton assemblages in the subantarctic and Antarctic zones. This seems to result mainly from the characteristics of the water circulation controlled by meteorological conditions, because most of the true antarctic species have a far wider range of tolerance of temperature and salinity than should be expected according to the intensities usually encountered in the Antarctic surface water. However, chaetognaths appear to occupy a somewhat special position from the biogeographical point of view. Most inhabit both Antarctic and subantarctic zones, the main biogeographical boundary for chaetognaths corresponding to the Subtropical convergence.

The Southern Ocean zooplankton contains a very high percentage of endemic species also exhibiting a striking uniformity in all its sectors, probably due to the circulation conditions arising from the westerly and easterly wind regimes. Meroplankton is almost absent, because the development of most benthic species is direct or involves brood care by the female. Similar to the situation described for the phytoplankton (p. 188) within the Southern Ocean zooplankton, two main assemblages can be distinguished which are usually referred to as 'communities'.

The subantarctic surface water is inhabited by the *Calanus simillimus* community; its most characteristic species are *C. simillimus*, *Clausocalanus laticeps*, *Cl. arcuicornis*, usually associated with *Eucalanus longiceps*, *Euaetideus australis*, *Conchoecia* sp., *Limacina balea*.

The Antarctic surface water is inhabited by the *Calanoides acutus* community; apart from *C. acutus*, the most characteristic species are: *Calanus propinquus*, *Euphausia superba* (krill), *Eukrohnia hamata*, *Limacina* sp., *Vanadis antarctica*; associates are *Pareuchaeta antarctica*, *Haloptilus ocellatus*, *Tomopteris carpenteri* and *Diphyes antarctica*. In spite of patchiness of some species and although the abundance is liable to vary in different areas the general composition of the community remains quite uniform (TAKEDA-AKIYAMA and ARIA, 1968). Some species such as the copepod *Metridia gerlachei* may be encountered in both the subantarctic and antarctic communities. It seems that a near-shore neritic community also exists but it has been as yet poorly investigated; its characteristic species appear to be *Antarctomysis maxima* and *Euphausia crystallorophorus*.

VORONINA (1963) demonstrated that in the Antarctic sector of the Indian Ocean, between the meridians 20° E and 98° E, the position of the boundary between the two oceanic communities mentioned above depends on the position of the South Polar front and on the characteristics of its vertical circulation. Under conditions of prolonged good weather, the boundary between the *Calanoides acutus* and *Calanus simillimus* communities is sharp and follows the line of convergence. If westerly winds increase in strength, or if the zone of maximal wind strength moves, changes in circulation result, as well as in a translation of the polar front. The translation induces a mixing of the two communities in the area between the old and new positions of the convergence. When the westerly wind is weak, some species of the subantarctic community submerse into the circumpolar Antarctic water (subantarctic intermediate water) flowing southwards in the intermediate layers up to the Antarctic divergence. Subsequently, the ascending waters at the divergence carry them into the Antarctic surface water. The respective distribution of antarctic and subantarctic zooplankton species thus appears to represent a valuable indicator of the location of the South Polar front and of changes in its position and width related to hydrometeorological fluctuations.

It must be mentioned here again that some species of the antarctic assemblage may sink at the Antarctic convergence and thereafter drift northwards with the Antarctic intermediate water. According to YAMANAKA (1976) carnivores are very abundant in summer (December) in the Antarctic intermediate water; this may be related to the abundant populations of filter feeders occurring at this time in the Antarctic surface water, the sinking of which at the convergence provides an abundant food supply for carnivorous zooplankters.

Macroplankton distribution, except for krill (p. 200) is still poorly known. BAR-KHATOV and co-authors (1973) collected samples in spring and winter 1968 and 1969 with a 10-ft Isaacs–Kidd midwater trawl in both subantarctic and Antarctic regions over the 0- to 100-m layer. They conclude: (i) in the Subantarctic region, the areas off south-eastern New Zealand are richer than the areas off south-western Chile; (ii) macroplankton is much more scarce in the Antarctic region than in the Subantarctic one; (iii) salps and cnidarians always predominate; next in abundance are euphausiids; (iv)

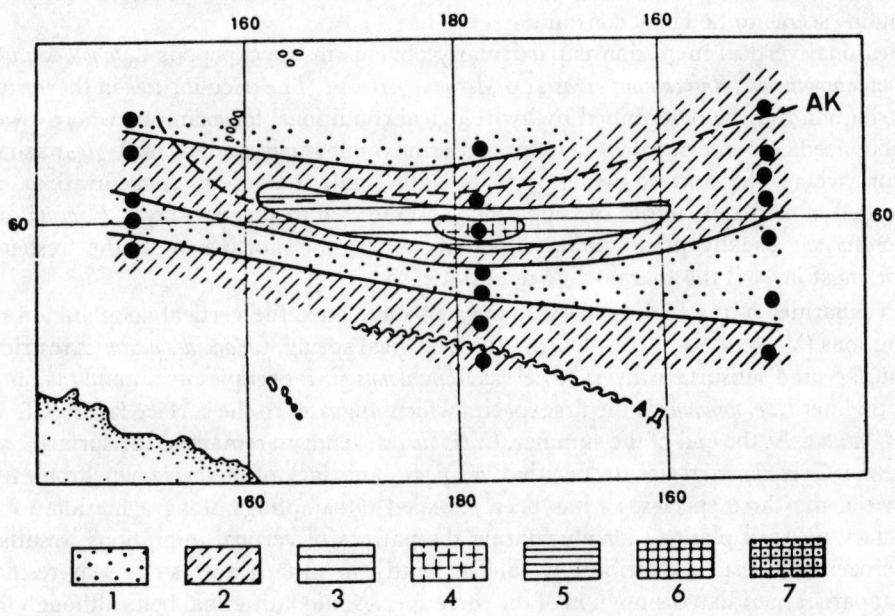

Fig. 6-3: Distribution of macroplankton biomass (g 1000 m^{-3}) in the Pacific sector of the Southern Ocean (November–December 1968). 1: <1; 2: 1–5; 3: 5–10; 4: 10–20; 5: 20–50; 6: 50–500; 7: >500; AK: Antarctic convergence. (Based on information provided by BARKHATOV and co-authors, 1973; reproduced by permission of American Geophysical Union, Washington, D.C.)

salps are usually more abundant in divergence zones, and siphonophores in convergence zones; (v) an increase in the total macroplankton biomass may occur in some areas close to the Antarctic convergence (Fig. 6-3).

Microzooplankters, e.g. protozoans, were also almost disregarded. BALECH (1973) briefly noticed that ciliates of the Hypotricha group are rather abundant in neritic areas of the Bellinghausen Sea. In the Antarctic sector of the Indian Ocean, the genus *Strombilidium* (Oligotricha) is common and tintinnids are represented by three endemic genera: *Haackmaniella*, *Protocymatocilis* and *Cymatocyles* (HADA, 1970).

Vertical Distribution and Migration of Mesozooplankton

Zooplankton vertical distribution and migration comprise specific phenomena. Although vertical migrations were observed long ago (FOXTON, 1964), they are still incompletely understood. In general, zooplankters inhabit deeper layers in winter, while they are more or less restricted to the phytoplankton-rich surface layers in summer despite the higher illumination occurring in these layers. In the Pacific sector of the Antarctic from November to January, the highest percentages (48–71%) of zooplankton biomass occur in the top 250 m of the 0- to 100-m water column, whereas in winter (May–October) the percentages for the 0- to 250-m layer decline to between 4 and 33% (HOPKINS, 1971). From March to October, 27 to 38% of the biomass of the 0- to 2000-m water column is found in its lower half. Despite the seasonal changes in biomass recorded at different depths in the 0- to 2000-m water column, the total biomass it contains seems to be fairly constant.

Seasonal vertical migration is particularly obvious in the copepods *Calanoides acutus*, *Calanus propinquus*, *Rhincalanus gigas* and *Metridia gerlachei*. The peculiarities in the vertical distribution may be determined by hydrological conditions; for example, where a well-pronounced cold-intermediate layer prevails, maximum aggregations of *R. gigas* usually occur; where the surface waters are homothermal, maximum concentrations are recorded immediately under the uppermost—somewhat freshened—layer. *C. acutus* and *C. propinquus* usually predominate either above the thermocline or in the freshened uppermost layer if the thermocline does not exist.

Peculiarities of the biological cycle may also influence the vertical zooplankton distributions (VORONINA, 1972). During the biological spring, *Calanoides acutus* is restricted to surface and subsurface layers, whereas *Rhincalanus gigas* occupies its annual maximum depth; later *C. propinquus* is the first species which migrates to the surface layer, followed by *C. acutus*. At the end of the summer, *C. propinquus* tends to remain in the surface layer, whereas *C. acutus* migrates deeper than *R. gigas*. This lack of life-cycle synchronization between the three species, as has been revealed in samplings along a meridian from January to April planned for elucidating the pattern of vertical migrations, results in heterogeneous vertical distributions. Subsequently, in surface waters this may result in an apparent meridian disjunction of the three species into latitudinal belts although they really participate in the same assemblage.

In addition to temperature and salinity, several other factors can affect seasonal vertical migrations. Among these, fluctuations in phytoplankton abundance are the most important. The upward migration of herbivorous species during spring and summer is probably related to increased phytoplankton abundance. The reason for the downward migration in autumn and winter is still unknown. It has been assumed that the specific density of the herbivorous copepods increases due to intensive grazing in the surface layers during spring and summer, but winter cooling of the surface waters probably also plays a role, since it has been shown that copepods involved in this phenomenon are very sensitive to sharp temperature boundaries: thermal gradients exceeding 0.07 °C m^{-1} may impede their seasonal migrations.

A general scheme of seasonal migration in the Atlantic sector of the Southern Ocean has been presented by VORONINA (1975) for the six most abundant copepod species, all filter feeders: *Calanoides acutus*, *Rhincalanus gigas*, *Calanus propinquus*, *C. simillimus*, *C. tonsus*, *Eucalanus acus*. These species may be called interzonal (on a seasonal scale) and the distribution of the overwintering stock depends on the circulation of the deep water. Fig. 6-4 illustrates the changes in population structure of these species as a function of the season.

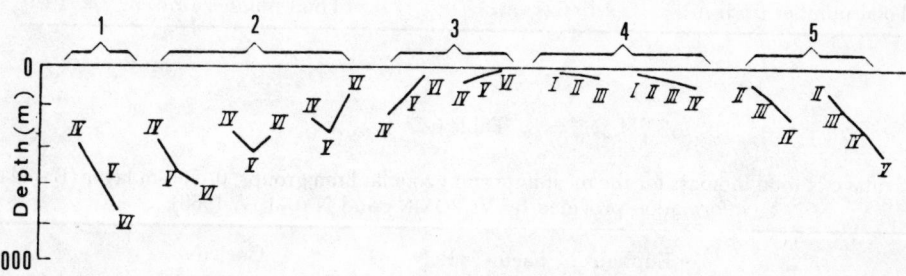

Fig. 6-4: Seasonal changes in the vertical structure of interzonal copepod populations. I to VI: average position of each copepodite stage. 1: winter; 2: beginning of spring migration; 3: full spring migration; 4: development of new generation; 5: autumn migration. (Based on information provided by VORONINA, 1975.)

According to YAMANAKA (1976), carnivorous copepods are more abundant in deep waters in winter and spring, probably because they feed there on overwintering filter feeders. Notwithstanding that our present knowledge of seasonal vertical migration is rather insufficient, we may conclude that it results in an isolation of the antarctic zooplankton community which is alternatively submitted to northward drifting in Antarctic surface water and southward drifting in subantarctic intermediate water (CURRIE, 1964).

Abundance and Production

Among the mesozooplankters, copepods always predominate, both in terms of total biomass percentage and number of individuals (VORONINA and NAUMOV, 1968; Table 6-2). Next in rank are chaetognaths and euphausiids. More detailed data on numbers of individuals for the most abundant species in surface waters are listed in Table 6-1.

Interesting evaluations and conclusions on biomass data for the Southern Ocean have been presented by VORONINA and NAUMOV (1968) (Table 6-2 and Fig. 6-5). Average mesoplankton biomass values, expressed in mg m^{-3}, are very similar in the three sectors

Table 6-1

Abundance (ind. m^{-2}) of the most common Pacific sector species (After TAKEDA-AKIYAMA and ARIA, 1968; reproduced by permission of Tokyo University of Fisheries)

Subantarctic zone Lat. S 50° 26′–51° 28′ January 5–7, 1965		Antarctic zone Lat. S 65° 13′–68° 26′ January 11–30, 1965	
Calanus simillimus	43·8	*Calanoides acutus*	58·8
Conchoecia spp.	21·8	*Calanus propinquus*	12·5
Clausocalanus arcuicornis	9·6	*Metridia gerlachei*	10·2
Rhincalanus gigas	5·1	Small salpids	6·3
Metridia gerlachei	4·5	*Eukrohnia hamata*	2·9
Parathemisto gaudichaudi	2·7	*Rhincalanus gigas*	2·7
Others	12·5	Others	6·6
Total number (ind. m^{-3})	19,627	Total number (ind. m^{-3})	3970

Table 6-2

Percentage of total biomass for the most important zooplankton groups; 0–500-m layer (Based on information provided by VORONINA and NAUMOV, 1968)

Zone	Copepods	Euphau-siids	Chaeto-gnaths	Poly-chaetes	Molluscs	Coelen-terates	Tunicates	Others
Antarctic	72·8	7·6	9·8	1·3	0·9	3·6	0·9	3·1
Subantarctic	61·5	7·2	13·2	1·5	2·4	2·1	4·9	7·2

of the Southern Ocean during the period of multiplication: Indian Ocean, 72·5; Pacific Ocean 0- to 100-m layer, 74·6; and Atlantic Ocean, 73·9 respectively, in the 0- to 500-m layer.

According to VORONINA (1966a,b) high zooplankton biomasses (more than 100 mg m^{-3}) may correspond to four main situations: (i) early spring, in a thin upper layer, related to copepod breeding; (ii) summer in the 0- to 200-m layer, with a prevalence of copepodite stages III–IV of dominant species; biomass maxima for different copepod species do not coincide in time and space—due to the different life-cycles; this results in an unsynchronized prevalence of these stages in the populations of different species; (iii) mechanical concentration of plankters into zones of convergence or divergence; in this case, composition and vertical distribution of the plankton bulk may depend on the season; (iv) neritic areas near islands. In addition, VORONINA and NAUMOV (1968) mention two further situations leading to high biomasses, observable only in extremely southern areas: (v) accumulation of copepods wintering in the area itself, prior to multiplication; (vi) more rapid development of the plankton as a result of early liberation from the ice.

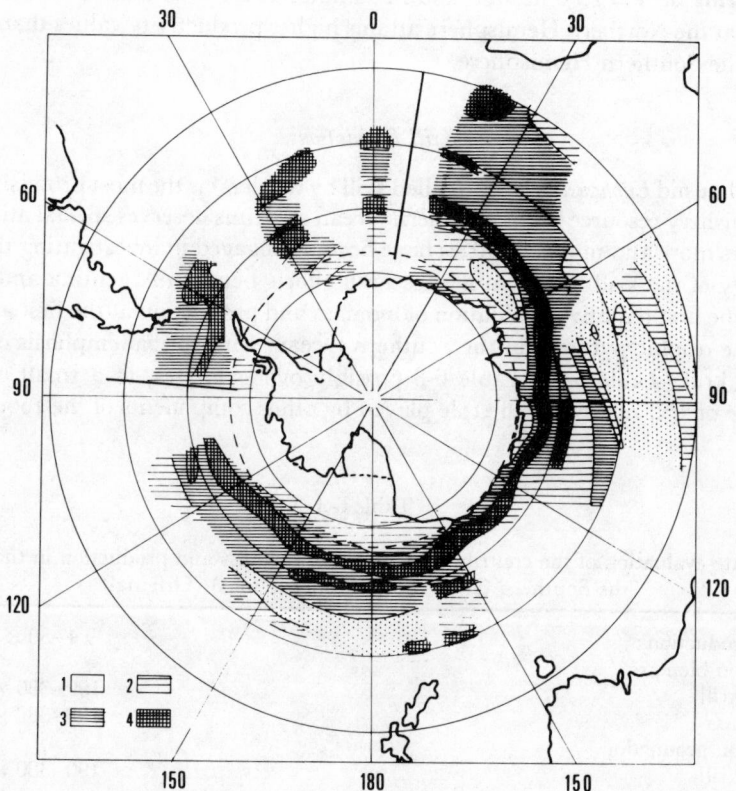

Fig. 6-5: Zooplankton biomass (mg m^{-3}) in the Southern Ocean, 0- to
100-m layer. 1: < 10; 2: 10–49·9; 3: 50–99·9; 4: >100. (Based on information provided by VINOGRADOV and NAUMOV, 1961.)

In areas of maximum water upheaval, the Antarctic divergence is characterized by
patches with a low plankton biomass, and also by surface spawning accumulation of
Euphausia superba. Between the Antarctic divergence and the Antarctic convergence, the
summer mesozooplankton biomass averages 200 to 300 mg m^{-3}; the highest values (500
to 800 mg m^{-3}) were observed north of the Antarctic divergence and south of the
Antarctic convergence (VINOGRADOV and NAUMOV, 1961).

That the plankton standing crop in surface waters (0 to 100 m) varies in different
locations as a function of the biological season at sampling has been confirmed by
VORONINA and ZADORINA (1974). During the early spring state of plankton development in the Antarctic zone, the average standing crop in the 0- to 100-m layer is
0·12 ± 0·17 ml m^{-3}; in the belt with maximum phytoplankton abundances it is
0·76 ± 0·11 ml m^{-3}. In the subantarctic zone during the zooplankton maximum, the
average standing crop is 0·56 ± 20 ml m^{-3}, i.e. larger than in the Antarctic zone. The

total standing crop in the 0- to 1000-m water column changes during the year, averaging in early spring $33 \cdot 4 \pm 25 \cdot 6$ ml m^{-3} and in summer $134 \cdot 1 \pm 55 \cdot 2$ ml m^{-3}. These values indicate that the Northern Hemisphere attains higher productivity values than the polar waters of the Southern Hemisphere.

Krill Populations

The euphausiid *Euphausia superba* (called krill by whalers) is the most promising—if not the only—fishery resource in the Southern Ocean and thus deserves special attention. In recent years more and more countries have become engaged in investigating the biology and ecology of the krill. Cooperative research efforts between Argentina and the USA produced the first tentative evaluation of biomass and production at the first and second levels of the trophic pyramid in the Southern Ocean, with special emphasis on the role played by krill (Table 6-3). Table 6-3 possibly overestimates, to a small extent, the importance of krill relative to the role played by other components of the zooplankton.

Table 6-3

Approximate evaluation of the contribution of krill to biomass and production in the pelagial of the Southern Ocean (data in metric tons) (Original)

Primary production	4–30×10^9 t yr^{-1}
Zooplankton biomass	
(except krill)	190–380×10^6 t
Krill biomass	190–380×10^6 t
Zooplankton production	
(except krill)	190–380×10^6 t yr^{-1}
Krill production	
(computed from biomass evaluation)	95–190×10^6 t yr^{-1}
Krill production	
(computed from maximum assumed primary production)	130–1000×10^6 t yr^{-1}

Krill is strictly antarctic. It occurs mainly in the immediate vicinity of the Antarctic divergence, to a lesser extent further to the north and in the westward drift generated by easterly winds blowing close to the Antarctic sector of the Southern Ocean between 60° W and 60° E. Elsewhere, krill population seems to be rather scattered. The densest krill shoals (up to 15 kg m^{-3}) occur in summer (January–April) in areas with strong surface-water downwellings associated with neighbouring upheavals. The sinking movement may be related either to dynamical processes arising from horizontal circulation or to descending currents near a shoreline (islands, Antarctic continent) or banks.

The thermal tolerance range of *Euphausia superba* is approximately $-1 \cdot 5$ to $+4 \cdot 0$ °C, and its life-span probably three to four years (maximum body length, 45 to 54 mm). Breeding probably occurs on the shelf during the third summer after hatching, but eggs appear to hatch more offshore in oceanic areas, possibly as deep as 2000 m. Larval development—nauplius, metanauplius and Calyptopsis I—takes place during the ascent; the youngest stage collected in surface waters is the Calyptopsis II. Periodicity in the occurrence of patches and variations in the size and shape have been observed; these may be related to light intensity and phytoplankton abundance. In general, krill is more abundant in surface waters during the day than at night. The stock of adult krill declines during the Antarctic summer, due to post-spawning mortality (SHEVTSOV and MAKAROV, 1969).

Euphausia superba plays a very important role in the trophic network of the Southern Ocean. The food chain leading, for example, to the largest whales (Mysticetes) is very direct (phytoplankton → krill → whale). Due to the short bloom characteristic of all polar waters, *E. superba* must resort to other particulate food when phytoplankton becomes scarce. Hence, krill represents a major pathway of energy transfer to the third link of the food chain. Nevertheless, even if we do not take into account the consumption by krill of particulate food other than living phytoplankters, the estimated krill production is in agreement with evaluations of primary production (Table 6-3); CHEKUNOVA and RYKOVA (1974) have demonstrated that *E. superba*, during its life of up to three years, consumes 60 g of phytoplankton (wet weight).

Euphausiids have the highest content in vitamin A of all invertebrates studied thus far. However, animals which feed on krill do not retain all the vitamin A they ingest; some vitamin A is released with the faeces and may thus be used by plankters (mainly copepods) or benthic filter feeders.

As is well known, catches of whales, which represented about 10% of the total fishery landings in the 1930s, currently amount to less than 0·4% due to overfishing of whales. The krill consumed by whale populations 40 to 50 years ago was estimated as 150×10^6 t yr^{-1}, that is, during the 3- to 4-month summer season; this amount is in the same order of magnitude as the values listed in Table 6-3. Therefore, although krill is utilized as food by other animals such as 'white-blood' fishes (Nototheniidae) and seabirds, the assumption appears reasonable that large amounts of krill could be caught without causing any damage to other populations in the Southern Ocean. It has been estimated that krill landings of about 70×10^6 t yr^{-1} would represent a reasonable value for sustainable yield. However, a commercially successful krill fishery requires adequate solution of problems such as location of krill patches, optimization of fishing techniques and proper product processing and marketing.

Deep-layer Plankton

Our present knowledge of deep-layer zooplankton in the Southern Ocean is still insufficient. The information collected in summer and fall (VINOGRADOV and VORONINA, 1974) may be summarized as follows.

In the subantarctic zone, below the biomass maximum in the 50- to 100-m layer, a strong decrease prevails to about 10 mg m^{-3} at 500 m, less than 5 mg m^{-3} at 500 to 1000 m, and less than 1 mg m^{-3} below 1500 m. At 4000 m the biomass is only about 0·3 mg m^{-3} (Fig. 6-6). A comparison of these data with those obtained in the northern parts of the Indian Ocean demonstrate that below 200 m the biomasses in both regions are of the same order of magnitude. In contrast, in the Kuril-Kamchatka Trench, the biomass at any depth below 200 m is about one order of magnitude higher. Fig. 6-7 documents changes in biomass of the three most important groups in the subantarctic zone as a function of water depth.

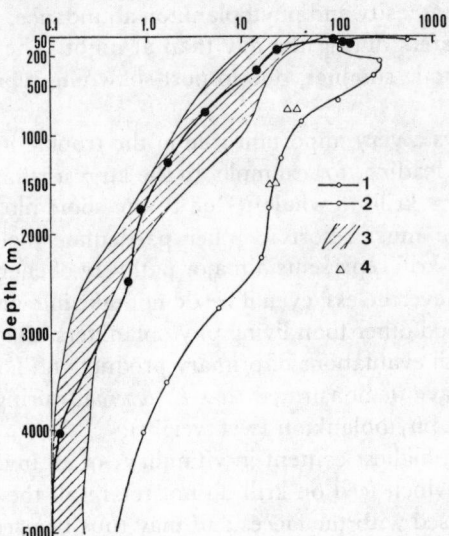

Fig. 6-6: Comparison of changes in biomass (mg m^{-3}) vs depth. 1: Station 894, 'Akademik Kurchatov', 4–5 Dec., 1971, 56°46′ S–24°53′ W. 2: Kuril-Kamchatka Trench. 3: Northern region of Indian Ocean. 4: Additional data for the layers 500 to 1000 and 1000 to 2000 m; 'Ob', 63°18′ S–135°13′ E, 22 March 1956; 63°50′ S–165°25′ E, 24 March 1958; 58°19′ S–160°04′ W, 7 April 1958. (Based on information provided by VINOGRADOV, 1970 and VINOGRADOV and VORONINA, 1964.)

In the Antarctic zone, data have been collected in April by VINOGRADOV and VORONINA (1974) at three stations. At 500 to 1000 m and 1000 to 2000 m, the biomass values recorded are 30 to 37 mg m^{-3} and 20 to 22 mg m^{-3}, respectively. These values are in the same order of magnitude as those obtained in the Kuril-Kamchatka Trench. In the Antarctic bottom water—which flows northwards, mainly from the Weddell Sea (but possibly also from the Ross Sea) and generates the deep and bottom waters of almost the whole World Ocean—endemic plankters seem to be rather rare.

Micronekton and nekton of deep waters are also insufficiently investigated. According to PARIN and co-authors (1974), the deep-water pelagic fish fauna of the subantarctic zone exhibits a high percentage of endemisms and is very different from that inhabiting the deep-water layers of the Antarctic zone. As regards the pelagic cephalopods, eurybathic deep-water species tend to occupy the whole water column, since there are no species characteristic of the surface layers. Therefore, the cephalopod fauna of inter-

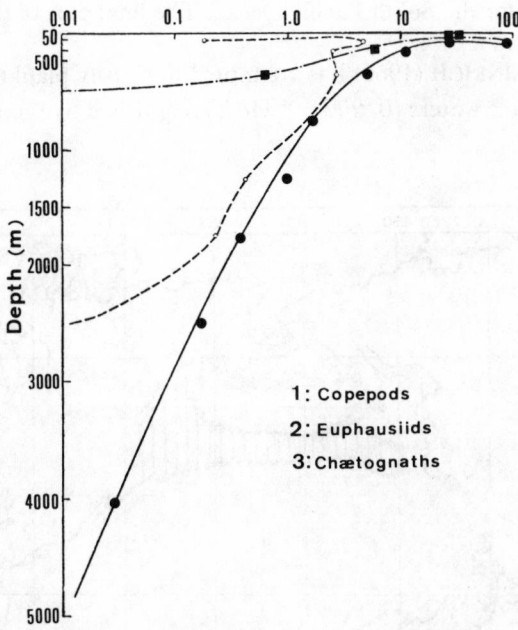

Fig. 6-7: Changes in biomass (mg m^{-3}) vs depth at 56° S and 24° W. (Based on information provided by VINOGRADOV and VORONINA, 1964.)

mediate waters is greatly impoverished (NESIS, 1974). In other respects, decapod crustaceans seem to be very rare in the meso- and infrapelagic zones.

In the past it was assumed that deep scattering layers (DSL) do not exist south of the Antarctic convergence where the waters are of very low temperature. However, NUÑEZ and NOVARINI (1976) found in the Antarctic convergence region three-layered regimes of DSL during daytime: (i) a surface layer extending from 35 to 110 m; (ii) a layer ranging from about 80 to 90 m; and (iii) two further layers, one from 0 to 110 m, the other, 40-m thick at 145 m. All layers were well defined and revealed vertical migration at dusk and dawn. Presumably, these DSL correspond to plankton inhabiting the Antarctic surface water, but NUÑEZ and NOVARINI could not find any scattering layer below 400 m, i.e. the approximate depth range (300–400 m) characterized by widely distributed DSL in the whole World Ocean.

(3) Pacific Ocean

Owing to the immense surface area (181 million km^2) of the Pacific Ocean and the inequality of the data available, it is difficult to assess the distribution of planktonic organismic assemblages within. While septentrional and central regions are rather well investigated in the Pacific Ocean, southern areas are still rather poorly documented. In this section we shall mainly consider surface and subsurface waters, first for the North Pacific Ocean, then for the South Pacific Ocean. The final part of the section then deals with deep-sea plankton.

Thus far, only HEINRICH (1962) has attempted to classify planktonic communities of surface and subsurface waters (0–200 m). He distinguished five communities (Fig. 6-8):

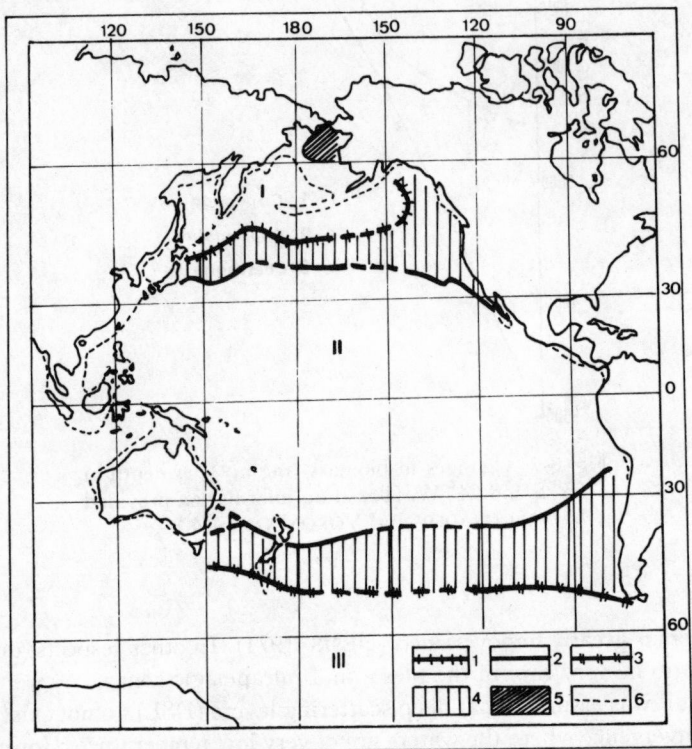

Fig. 6-8: Biogeographical regions of pelagic organismic assemblages in the Pacific Ocean. I: boreo-arctic; II: subtropical and intertropical; III: 'Antarctic' (pertains to the Southern Ocean). 1: southern limit of boreo-arctic region; 2: northern and southern limits of subtropical and intertropical region; 3: northern limit of 'Antarctic' region (Southern Ocean); 4: mixing zones of assemblages; 5: arctic assemblage of the northern Bering Sea; 6: shelf-edge. (Based on information provided by HEINRICH, 1962.)

northern community of the Bering Sea; offshore boreo-arctic community; neritic boreo-arctic community; offshore subtropical and tropical community; neritic subtropical and tropical community. He further distinguished an 'antarctic' community which really pertains to the Southern Ocean. According to HEINRICH, several plant and animal taxa are represented very differently in the three main regions (boreo-arctic, tropical and antarctic) separated by transitional areas in each hemisphere between the polar front and the Subtropical convergence (Table 6-4).

Table 6-4

Number of species for the most important taxa in surface and subsurface waters of the Pacific Ocean (Based on information provided by HEINRICH, 1962)

Organisms	Boreo-arctic region	Tropical region	Southern Ocean*
Diatoms (0–200 m)	69	131	86
Dinoflagellates (0–200 m)	29	226	7
Calanoid copepods (0–50 m)	23	190	8
Euphausiids (0–200 m)	7	53	7
Hyperiid amphipods (0–500 m)	24	75	16
Salpidae (0–200 m)	0	21	3
Appendicularians (0–200 m)	10	32	20
Chaetognaths (0–200 m)	7	26	7

*This part of the geographic Pacific Ocean does not correspond to the strictly oceanographic delimitation between Pacific Ocean and Southern Ocean.

As will be shown below, the tropical region (about 65–70° latitude) includes several subregions.

(a) Surface and Subsurface Waters of the North Pacific Ocean

Organismic assemblages of the North Pacific are by far the best known in this ocean. They are sufficiently different on western and eastern sides to require separate treatment.

North-western Pacific Ocean
Assemblages in the northern province of the Bering Sea

The plankton of the northern province of the Bering Sea has well-marked arctic characteristics. In summer phyto- and zooplankton biomasses are high, but seasonal fluctuations are marked and the community is unbalanced; it is weakly stratified and very poor in number of species. Predominant among the species present are the diatoms *Rhizosolenia haebetata* f. *hiemalis*, *R. alata* f. *curvirostris*, *Ceratium arcticum* as well as the copepods *Calanus glacialis*, *Metridia pacifica* and *Centropages abdominalis* in some inshore areas. These copepod species are monocyclic and grow up at the beginning of the phytoplankton bloom.

According to TANIGUCHI and co-authors (1976), the phytoplankton assemblage of the surface layer in the eastern Bering Sea exhibits marked changes in the early warming season: in May due to the sea ice melting, *Thalassiosira hyalina* and *T. nordenskiöldii* predominate together with *Fragilaria* spp. and *Navicula* spp.; in June *Thalassiosira* populations wither and sink into the bottom layer, but part of the *Fragilaria* and *Navicula* populations still remain in the subsurface layers. The thermocline seems to play an important role in the development of phytoplankton populations. Blooms in the different areas successively appear further and further from the innermost inshore areas as the thermocline extends towards the central parts of the Bering Basin.

In the central and western Bering Sea, MOTODA and TAKASHI (1974) estimated the average primary production as 200 to 300 mg C m^{-2} d^{-1} and observed that 80% of the copepod biomass in the 0- to 150-m water column occurs in the upper 80 m, the whole biomass being 200 to 500 mg m^{-3}. However, the zooplankton assemblage existing there in early to mid-summer cannot be considered as arctic, since about 50% of the zooplankton biomass in the 0- to 80-m layer consists of the two boreo-arctic copepods *Calanus cristatus* and *C. plumchrus*.

According to McRoy and GOERING (1974) the Bering Strait is one of the most productive in the northern part of the Bering Sea. McRoy and GOERING recorded there an integrated ^{14}C fixed value of 4·1 g C m^{-2} d^{-1} in June. In winter the average productivity in the ice-covered surface water is 1·20 mg C m^{-2} d^{-1}. The part played there in the productivity by sea-ice algae seems to be rather similar to that mentioned above for Antarctic coasts. On the undersurface of the ice there exists a community of algae and other microorganisms which is well developed in late winter and early spring; its productivity ranges from 2·2 to 4·8 mg C m^{-2} d^{-1} for standing stocks corresponding to 0·34 to 2·97 mg chlorophyll *a* m^{-2}. The authors suggest that the annual cycle of primary production begins with the development of the algal felt on the undersurface of the ice, followed by a bloom (89 mg C m^{-2}) at the ice front and finally by the spring bloom in open waters. Therefore, the annual increase in primary production in the Bering Sea begins in the northern and central regions which are ice-covered for six to seven months rather than in the southern regions. MOTODA and TAKASHI (1974) estimated the average primary production as 200 to 300 mg C m^{-2} d^{-1} in the Bering Sea.

Boreo-arctic assemblages

Before describing plankton assemblages of the boreo-arctic region, it must be remembered that the northern subtropical convergence separates the Oyashio waters (surface temperature 3 °C and 14–15 °C in winter and summer, respectively) from the more saline Kuroshio waters (surface temperatures 18–20 °C and up to 26 °C in winter and summer, respectively). Boreo-arctic assemblages inhabit the former, and tropical assemblages the latter. At the convergence, the subarctic Oyashio sinking water becomes the Arctic intermediate water. A transitional area occurs between these at about 2° latitude, extending beyond 42° N in summer and moving below 40° N in winter (Fig. 6-9). In the transition zone, boreo-arctic and tropical organismic assemblages mix; most of the Kuroshio species occur in subsurface layers. The predominant phytoplankter of the transition zone is the cosmopolitan diatom *Nitzschia seriata*, associated with less psychrophilic species of the offshore boreo-arctic community: *Thalassiosira subtilis*, *Chaetoceros affinis*, *Ceratium macroceros* and *C. tripos*.

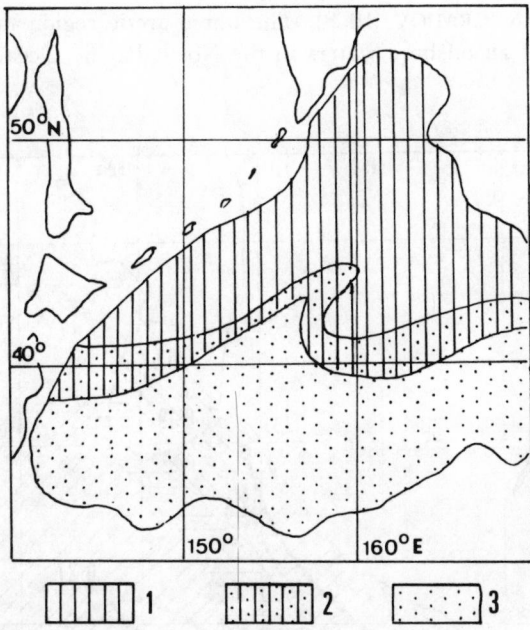

Fig. 6-9: Disposition—in July–August 1954—of the transition zone (2) between offshore boreo-arctic assemblage (1) and subtropical assemblage (3). (Based on information provided by BOGOROV and VINOGRADOV, 1955 and BOGOROV, 1958.)

Offshore boreo-arctic assemblage

The offshore boreo-arctic assemblage occurs in the oceanic province of the Bering Sea and the most northern parts of the Pacific Ocean itself. According to SMIRNOVA (1956) its phytoplankton contains about three times more diatoms than dinoflagellates, both in species number and biomass; most species there are characteristic of the cold-temperate Oyashio waters (5–10 °C), e.g. *Chaetoceros convolutus*, *C. atlanticus*, *C. concavicornis*, *Thalassiosira nordenskiöldii*, the latter generally accounts for up to 60 to 90% of the whole plant biomass. Other species which are more tolerant to higher temperatures predominate in the transition zone.

Three herbivorous copepod species contribute 80% of the total zooplankton biomass: *Calanus cristatus*, *C. plumchrus* and *Eucalanus bungii*. These species perform seasonal vertical migrations. The other important and rather characteristic species are: the copepods *Oithona similis* and *Metridia pacifica*, *Euphausia pacifica*, the pteropod *Limacina helicina*, and the chaetognath *Sagitta elegans*. Most of these boreo-arctic zooplankters of the northwestern Pacific Ocean display only a moderate intensity of vertical circadian migration: relatively large amounts of plankton remain in surface layers throughout the 24-h period even in autumn (VINOGRADOV, 1968). The boreo-arctic region includes the area of maximum fertility of all offshore waters in the North Pacific Ocean (Fig. 6-10). This

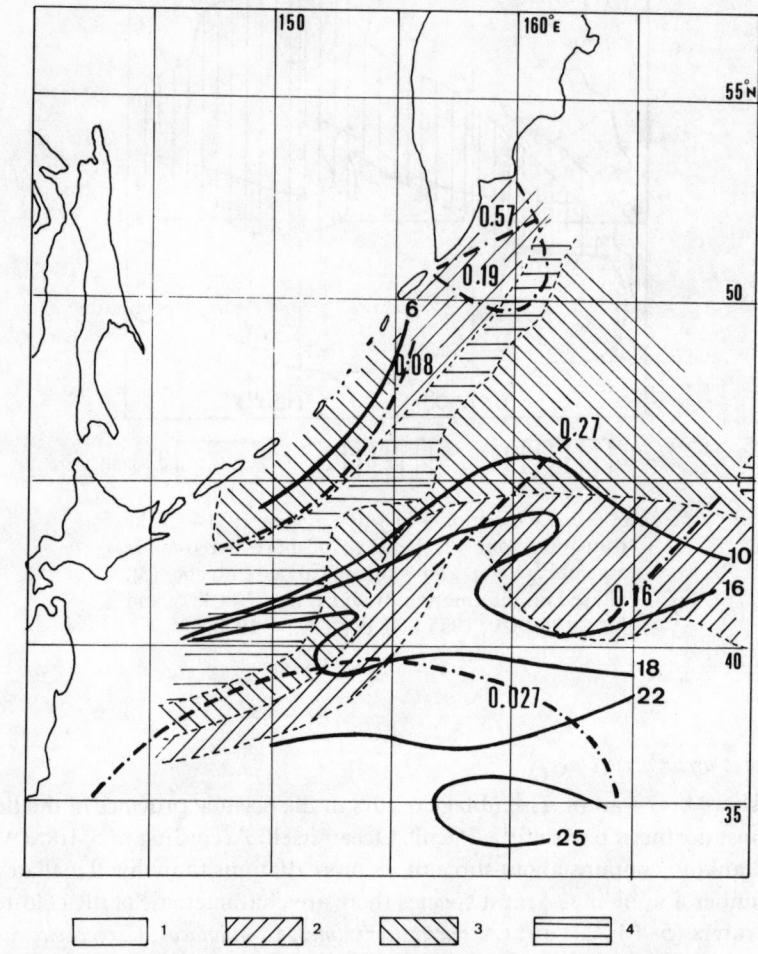

Fig. 6-10: Isotherms (solid lines) and average primary production (broken lines) (mg C l^{-1} d^{-1}) in the north-western Pacific Ocean from August–October 1954. Dotted lines delineate areas of different phytoplankton biomass (mg m^{-3}). 1: < 100; 2: 100–500; 3: 250–500; 4: >500. (Based on information provided by BOGOROV and BEKLEMISHEV, 1955; BOGOROV, 1958.)

area extends along the Kuril-Kamchatka arch about 50 miles from the shelf. The plant biomass, always in excess of 500 mg m^{-3}, may reach 2 to 3 g m^{-3}. Seasonal fluctuations are rather large. During the bloom period (May–June) primary production is about 250 mg C m^{-3} d^{-1}; biomass and primary production decrease towards both the shelf and the open ocean. Primary production was estimated by MOTODA and co-authors (1974) to be between 200 and 1000 mg C m^{-2} d^{-1}.

Areas of maximum zooplankton biomass (more than 500 mg m^{-3}) (Fig. 6-11) almost coincide with those of maximum primary productivity, but the zooplankton peak occurs

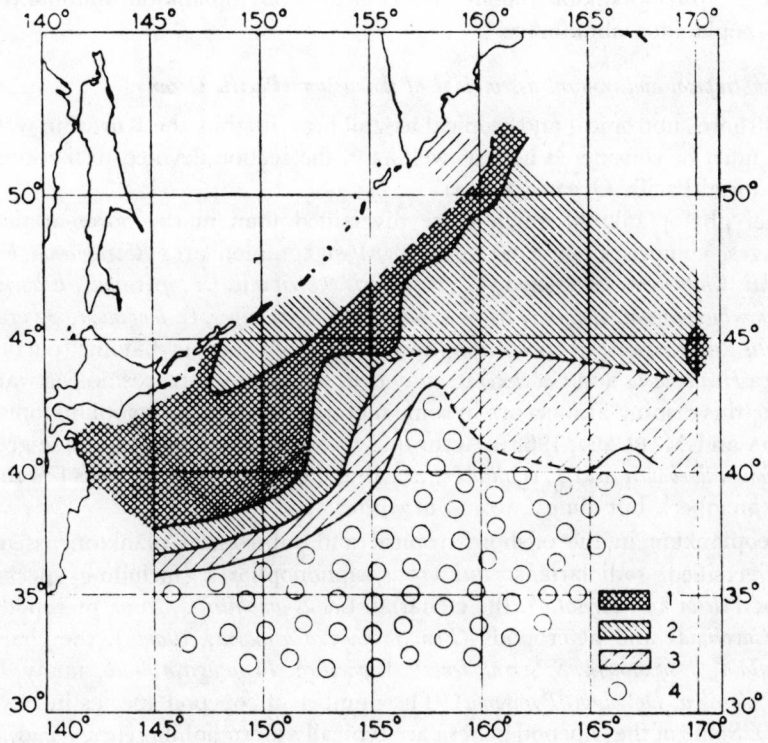

Fig. 6-11: Distribution of zooplankton biomass in the 0- to 100-m layer of the north-western Pacific Ocean, August–October 1954. 1: >500 mg m^{-3}; 2: 250–500 mg m^{-3}; 3: 100–250 mg m^{-3}; 4: <100 mg m^{-3}. (Based on information provided by BOGOROV and VINOGRADOV, 1960.)

a little later (end of June–July; BOGOROV and BEKLEMISHEV, 1955; BOGOROV, 1958). Most species successively participate in this peak according to their thermal requirements for maximum population growth. Seasonal fluctuations in zooplankton abundance are very large; at the end of the summer (September–October) the biomass quickly diminishes (summer: 100 mg m^{-3}; annual average: 25 mg m^{-3}). In general, this plankton ecosystem appears to be rather well balanced. Moreover, some of the predominant 'herbivorous' species are versatile as regards food supply. For example, *Calanus plumchrus* develops in winter when primary production is minimal, utilizing microzooplankton as food (LE BRASSEUR and KENNEDY, 1972).

Neritic boreo-arctic assemblage

In comparison with the offshore assemblage, the neritic boreo-arctic assemblage has a higher phytoplankton but a lower zooplankton biomass. The zooplankton peak, which is based on a small number of monocyclic or bicyclic copepod species, is almost synchronous with the phytoplankton bloom; it seems that phytoplankton abundance directly controls zooplankton abundance.

Offshore subtropical and tropical assemblage of the western Pacific Ocean

The offshore subtropical and tropical assemblage inhabits the Kuroshio water mass. Hence it must be considered here as well as in the section devoted to the intertropical and equatorial Pacific Ocean (p. 214).

The net phytoplankton, much more diversified than in the boreo-arctic offshore assemblages, comprises some 60 species. Most common are: *Rhizosolenia bergoni*, *R. stolterforthii*, *Chaetoceros lorenzianus*, *C. messonensis*, *C. eibenii*, *C. fortissimus*, *Ethmodiscus rex*, *Climacodium biconcavum*, *Ceratium tripos* f. *subsalsum*, *C. ramipes*, *C. longinum*, several species of *Pyrocystis* and *Ceratocorys*, coccolithophorids, etc. Some neritic diatoms (particularly in the genera *Hyalochaeta* and *Chaetoceros*) may drift out with the Kuroshio eastwards as far as 155° E; this drifting also occurs in some meroplanktonic and holoplanktonic animals (MOTODA and MARUMO, 1963). According to MARUMO (1975) the blue-green algae *Trichodesmium thiebautii* and *T. erythraeum* are abundant (10^2–10^3 filaments l^{-1}) in summer (June–September), but almost absent in winter.

The zooplankton in the offshore tropical and subtropical plankton assemblage is highly diversified: radiolarians, numerous siphonophores (including species of the pleustonic *Velella* and *Physalia*), the ctenarian *Cestus amphitrites*, some pteropods (*Euclio*, *Peraclis*, *Cavolinia*) and heteropods (*Pterotrachaea*, *Carinaria*, *Atlanta*), the chaetognaths *Sagitta enflata*, *S. hexaptera*, *S. serratodentata* f. *pacifica*, *Pterosagitta draco*, many Thaliacea (*Salpa*, *Cyclosalpa*, *Doliolum*, *Pyrosoma*). The number of copepod species increases up to about 200. Some of the copepod genera are typically thermophilic: *Acrocalanus*, *Undinula*, *Copilia*, *Sapphirina*, etc.; in general, these genera are represented by small-sized species. About 80% of the total biomass of the calanoid copepods is made up of some 15 species; none of the species surpasses 15% of the total biomass. All the copepods are polycyclic: in general, the mean number of generations per year increases southwards, up to about nine to ten in the warmest areas. There are also about ten species of euphausiids which belong to the genera *Euphausia*, *Nematoscelis* or *Stylocheiron*. The most complete species lists of this community originate from samplings west of the date-line.

IKEDA and MOTODA (1973) attempted an estimation of the production–consumption relationships of phytoplankton–herbivorous zooplankton–carnivorous zooplankton in the Kuroshio and its adjacent region, in the 0- to 150-m water layer, from the equator to 45° N and between the meridians 120° E and 150° E, i.e. in overlapping subtropical, tropical and equatorial areas; seasonal changes in the zooplankton composition were indistinct. In the area investigated the ratio of herbivorous zooplankton grazing to phytoplankton production ranges from 18 to 72% and the ecological efficiency from primary to secondary production from 5 to 22%. Herbivorous and carnivorous zooplankton daily productions are estimated at 9 to 57 mg C m^{-2} and 4 to 23 mg C m^{-2}, respectively. For all the calculated data the lesser range between minimum and maximum values has been recorded from the Kuroshio area located in the south-west of Japan; the lowest levels of herbivore (9 mg C m^{-2} d^{-1}) and carnivore (4 mg C m^{-1} d^{-1}) production have been recorded from the area located to the south of 20° N.

Over its tremendous spatial extension, the offshore subtropical and tropical plankton assemblage exhibits some heterogeneity. Taking into account only the subtropical region, one can differentiate two subassemblages within (BOGOROV, 1960): (i) In the northern subzone the average temperature is about 19 °C, the PO_4–P content high and the mean content of dissolved O_2 about 5.4 ml^{-1}. The phytoplankton is only slightly impoverished in comparison with that of the offshore boreo-arctic assemblage, and diatoms (62% of the total biomass) are more abundant than dinoflagellates. Although the spring primary production is only about $1/10$ of that in the offshore boreo-arctic assemblage because seasonal fluctuations are less marked and plant biomasses of 100 to 250 mg m^{-3} are not uncommon; the primary production in Kuroshio waters is estimated to be about 100 to 200 mg C m^{-2} d^{-1} (MOTODA and co-authors, 1974). The zooplankton may be rather rich just after phytoplankton peaks (160 mg m^{-3}), but only 25 mg m^{-3} in winter. (ii) In the southern subzone the average surface temperature is higher (21–25 °C), and the PO_4–P and O_2 contents lower (4.6 ml l^{-1}). The phytoplankton is more impoverished there and dinoflagellates (64%) more numerous than diatoms; the zooplankton is much poorer (20–30 mg m^{-3}).

In spite of the phytoplankton scarcity in the central and southern parts of the North Pacific gyre, VENRICK (1974) has observed some localized phytoplankton blooms there, accompanied and often dominated by an increase in the abundance of *Rhizosolenia* species harbouring the endophytic cyanophyte *Richelia intracellularis*, which may fix molecular nitrogen.

Plankton Assemblages of the North-eastern Pacific Ocean

General aspects

The distribution of plankton assemblages in the north-eastern Pacific Ocean, mainly between 140° W and 180°, has been reviewed by BEKLEMISHEV (1961). In essence, the assemblages are the same as in the north-western Pacific Ocean; however, some species are replaced by variants or local varieties. Offshore, the northern limit of the subtropical plankton lies at about 35° to 40° N; the southern limit of the boreo-arctic, between 50° and 45° N.

The transition zone between regions occupied by the pure boreo-arctic and subtropical assemblages exhibits some characteristics which are very different from those observed in the north-western transition zone. (i) The north-eastern transition zone is fan-shaped and stretches towards the east from the Gulf of Alaska to southern California. (ii) Instead of a simple mixture of species of the two assemblages, its planktonic assemblage contains some endemic species: the copepod *Eucalanus bungii californicus*; the euphausiids *Euphausia gibboides*, *Nematoscelis difficilis* and the bi-subtropical *Thysanoessa gregaria*; the fish *Myctophum californiense*. (iii) Some species from either of the two assemblages exhibit maximum abundance in the transition zone: the siphonophore *Weelia cylindrica* and the Salpidae. However, the pontellid copepods, which are very common in the hyponeuston of the subtropical region, can hardly tolerate the mixed waters of the transition zone; even the dominant species *Pontella tenuiremis* is very rare.

As in the north-west boreo-arctic offshore assemblage, vertical migration is of very moderate intensity. At ocean station 'P' (50° N; 145° W) on the south-eastern edge of the Alaskan gyre in July, vertical migration is restricted to a very small fraction of the species, i.e. primarily strong migrators (MARLOWE and MILLER, 1975). Upward migration of these species stops in the thermocline between 20 and 100 m; biomass and species number attain minimum values between 75 and 200 m.

East of 150° to 140° W, the mixed waters of the transition zone spread out. Northwards, they overlie boreo-arctic waters. Sometimes, subtropical forms such as radiolarians. *Velella* or salps may be found as far as 55° N. Some boreo-arctic species may be repelled by these invading mixed waters either to the west (the amphipod *Parathemisto japonica*) or towards deeper layers (e.g. *Calanus pacificus*). Southwards, the mixed waters and the boreo-arctic waters sink below the subtropical (central) water mass; along the American coast, the boreo-arctic species *Calanus cristatus* and *C. plumchrus* reach their southern limit at 45° N, *Metridia pacifica* and *Cololabis saira* at 40° N, but *Calanus pacificus* has been collected in sinking waters as far as 23° S. Other species, like *Euphausia pacifica*, do not show such submergence.

The Californian current system and its upwelling recently gave rise to several interesting investigations. According to MALONE (1971a) nanoplankters account for 60 to 99% of the observed primary productivity and standing crop, both inshore and offshore. Seasonal and geographic variations in the nanoplankton fraction are very stable. Therefore, the variations in phytoplankton productivity and standing crop are mainly due to net plankton. Higher values of the net plankton fraction are closely related to the occurrence of coastal upwelling; net plankton productivity and standing crop exceed those of nanoplankton only during the strongest upwelling period. According to MALONE (1971b) the nanoplankton fraction is significantly more abundant inshore than offshore, and naked dinoflagellates (*Chilomonas marina*, *Eutreptia*, etc.), together with *Coccolithus huxleyi*, predominate. The stability of nanoplankton abundance probably arises from drifting and sinking of surface water coupled with upwelling, and from a rather invariable grazing by short-lived microzooplankters. The importance of protozoans, mainly ciliates, has been emphasized by BEERS and STEWART (1970). In the 'pigment layer', protozoans account for 96 to 97% of the total microzooplankton numbers per cubic metre, but only for 24 to 37% of the microzooplankton biomass.

Net zooplankton may also be very abundant during the upwelling season. Off the Oregon coast, PETERSON (1976), recording in the upper 5- to 20-m layer of the near-shore zone, noted 20 to 80 individuals l^{-1} (d.w. biomass: 500–2000 mg m^{-3}); 90% of the biomass consisted of four copepod species: *Calanus marshallae*, *Pseudocalanus* sp., *Acartia clausi* and *A. longiremis*. According to PETERSON and MILLER (1977) zooplankton species composition exhibits large seasonal changes. Summer north winds produce a flow to the south and induce upheaval; dominant species are then those mentioned above, together with *Oithona similis*; all are primarily coastal forms with northern affinities and are present during the whole year. Winter south-west winds induce a northward and shore-ward flow that results in the appearance of two species with southern affinities, *Paracalanus parvus* and *Ctenocalanus vanus*, while the winter species tend to vanish.

Another interesting peculiarity is the abundance of the small-sized anomuran *Pleurocondes planipes* ('red crab') which accounts for more than 80% of the micronekton fraction and for up to 90% of the total plankton–nekton dry weight (LONGHURST, 1967a). LONGHURST assumes that *P. planipes*—usually consuming microzooplankton—mainly feeds there on phytoplankton employing a filter mechanism similar to that of the benthic *Porcellana*. He estimates grazing intensity to be 5·5 times that of other herbivorous species (copepods, euphausiids, mysids). According to WALSH and co-authors (1974) the red crab is a facultative herbivore which may participate in two different food chains: (i) During pre-upwelling, dinoflagellate (mainly *Gonyaulax polyedra*) → herbivorous copepod (*Acartia* dominant) → carnivorous red crab → dolphin (*Lagenorhynchus*) food chain. (ii) During the upwelling season (summer), additional nutrients in the euphotic layer lead to the substitution of diatoms for dinoflagellates, and the herbivorous copepod *Calanus helgolandicus* becomes the dominant species. WALSH and co-authors assume that as the larger diatoms become established, *P. planipes* reverts more to the herbivorous habits observed later in the season.

The two main offshore communities (boreo-arctic and subtropical) of the north-eastern Pacific Ocean are rather similar to those of the north-western Pacific. In the boreo-arctic offshore phytoplankton, diatoms predominate and primary production may surpass 100 mg C m^{-3} d^{-1} (250–650 mg C m^{-2} d^{-1}) in most of the Gulf of Alaska and off the Canadian coast; in the open ocean, primary production is 150 to 250 mg C m^{-2} d^{-1} (KOBLENTZ-MISHKE, 1961; Volume IV: FINENKO, 1978; Fig. 2-15, p. 57). Predominant groups in the zooplankton are copepods, amphipods, euphausiids and chaetognaths. Bacterioplankton is extremely scarce, mainly in the Gulf of Alaska, but bacterioneuston is one to four orders of magnitude more abundant and increases offshore (TSYBAN and TEPLINSKAYA, 1972; TSYBAN, 1973). In the subtropical (central) water masses, dinoflagellates predominate and primary production rarely exceeds 1·2 mg C m^{-3} d^{-1} (less than 100 mg C m^{-2}). As in the north-western Pacific Ocean, the zooplankton is highly diversified.

Macroplankton and micronekton

As usual, the distributional ranges of oceanic micronektonic fishes in the north-eastern Pacific Ocean are narrowly linked with more or less closed oceanic-current circulations. For example, the subarctic cyclonic gyre (Alaska current and Oyashio current) largely corresponds to the distributional area of several species, e.g. of the fish families Bathylagidae and Myctophidae, which usually have an expatriation area in the Californian current system. In the eutrophic upwelled waters off the Oregon coast, the fish biomass in the 0- to 1000-m water column is 3·6 mg m^{-3} (PEARCY and LAURS, 1966); in the area influenced by the Kuroshio waters north of the subtropical convergence, this biomass may rise up to 5·0 to 6·5 mg m^{-3}.

The abundance of mesopelagic and infrapelagic decapods seems to be correlated with production in the surface layers. In the deep Sagami Bay, Japan, their biomass is as high as 3·2 mg m^{-3}. According to AIZAWA (1974), the decapod biomass would be on the meridian 150° W: Oyashio waters, 0·44 mg m^{-3}; Kuroshio waters between 30° and 40° N, 1·24 mg m^{-3}; Kuroshio waters between 25° and 30° N, 0·74 mg m^{-3}; beginning of the Kuroshio at 20° to 25° N, 0·26 mg m^{-3}; north equatorial current between 10° and

$20°$ N, 0.19 mg m^{-3}, respectively. In the more productive Equatorial waters, the decapod biomass increases up to 0.66 mg m^{-3}.

(b) Nekton of the North Pacific Ocean

In the north-western Pacific Ocean distributional space–time dynamics of plankton assemblages are of paramount importance for the fisheries. Among the species of the offshore boreo-arctic community, *Euphausia pacifica*, which forms large shoals in spring off the northern coasts of Japan, represents an essential part of the food supply for some whales, several clupeiforms (*Sardinops*, *Engraulis*) mainly between 35 and $38°$ N, and also for scombriform fishes. The distribution of shoals of the saury (*Cololabis saira*) seems to reveal a close correlation with zooplankton abundance. The North Pacific herring (*Clupea pallasi*) occurs only north of 41 to $42°$ N; an area of winter aggregation exists in the eastern-central region of the Bering Sea at depths of 100 to 200 m; in spring and summer the herring migrates towards the shore for intensive feeding.

As is well documented, in the Northern Hemisphere the largest squid aggregations involving distant-neritic species of Ommastrephidae mainly occur in the vicinity of the North Polar front, especially in the western regions of the oceans. In the Pacific Ocean *Todarodes pacificus* is very abundant in frontal zones between the waters of the Oyashio and Kuroshio currents and their branches in the Okhotsk Sea and Japan Sea.

In the north-eastern Pacific Ocean the sardine (*Sardinops caerulea*) and anchovy (*Engraulis mordax*) constitute the major terminal link of the plankton food chain in the California current. Over the last decades a sharp decrease in the sardine stock occurred, while the anchovy stock increased. This phenomenon is partly due to sardine overfishing but also possibly to competition: while both species feed on the same planktonic assemblage, anchovy females are twice as prolific as those of sardine. Moreover, the minimum breeding temperature is as low as 11 $°$C in anchovy, but above 13 $°$C in sardine. Between 1949 and 1957, the California current showed extreme low surface temperatures due to an unusually strong and regular upwelling. This probably promoted anchovy populations.

(c) Intertropical Pacific Ocean

Offshore Tropical Assemblage

Over approximately 60 to 70 degrees latitude, the intertropical offshore plankton complex is uniform in its specific composition. Except for a small number of species which are more equatorial, the basic composition is the same from $30°$ N to $40°$ S. However, water-mass dynamics may modify numerical ratios between the different components of the plankton ecosystem, as well as its biomass and production, at different levels of the food pyramid.

SEMINA (1960, 1962) recorded about 210 phytoplankton species: 133 dinoflagellates, 72 diatoms and some cyanophytes and coccolithophorids; however, these figures are certainly too low. Among the dinoflagellates, the genus *Ceratium* is the most diversified (36 species); next in rank is *Oxytoxum* (18 spp.). The diatom genera which contain the largest numbers of species are *Chaetoceros* (13) and *Rhizosolenia* (10). Ten species of dinoflagellates and 16 species of diatoms were present in more than 50% and 20% of the samplings, respectively. Six species occur in almost all locations sampled: *Ceratium trichoceros*, *C. massiliense*, *Pyrocystis pseudonoctiluca*, *Ceratocorys horrida*, *Ethmodisus rex*, *Oscillatoria* (*Trichodesmium*) *thiebautii*. Very few common intertropical species extend their geographical distributions beyond the intertropical region: *Ceratium fusus*, *Rhizosolenia alata*, *Thalassionema nitzschioides* and *Halosphaera viridis*. Only two species seem to be restricted to the equatorial surface water: *Ceratium deflexum* and *Ornithocercus stenii*.

The vertical distribution of phytoplankton species conforms to the general scheme of tropical offshore areas, i.e. most of the diatoms occur in the surface layer (0–25 m), whereas dinoflagellates are often found in the 0- to 50-m layer (some even preferably between 50 to 100 m). However, there are some exceptions. For example, the diatoms *Planktoniella sol* and *Gossleriella tropica* which require a large supply of mineral nutrients (at least 8 mg m^{-3} PO_4–P), inhabit superficial water layers only in divergence areas; in convergence areas both species occur only between 50 and 100 m (SEMINA, 1960). According to ZERNOVA (1964), two other diatoms, *Coscinodiscus janischii* and *Thalassiosira subtilis*, exhibit a similar distribution.

The distribution of zooplankters is also uniform but less so than that of phytoplankters. This difference may be ascribed to five main reasons.

(i) The vertical migration significantly affects the distribution patterns of many zooplankters. VINOGRADOV and VORONINA (1963, 1964), who studied the distribution of 20 species (16 copepods and 4 pteropods) in the 0- to 500-m layer between 17° N and 16° S, proposed a general scheme which probably also holds for other groups (VINOGRADOV, 1968). VINOGRADOV and VORONINA distinguished three groups: (a) species which usually live in epiplanktonic and mesopelagic zones (0 to 200 m) and exhibit maximum abundance there: *Undinula darwini*, *Calocalanus pavo*, *Scolecithrix danae*, *Euchaeta marina*, *Sagitta ferox*, *S. bedoti*, *Creseis virgula*, etc.; (b) species which inhabit surface waters (about 25 to 100 m) at night, but descent to the infraplanktonic zone (p. 11) by day: *Neocalanus gracilis*, *Nannocalanus minor*, *Limacina inflata*, *Sagitta enflata*, *S. hexaptera*, *S. serratodentata*; (c) 'interzonal' species whose population maximum occurs in the infraplanktonic zone either permanently or during the day with no, or only weak, rising activities at night: *Rhincalanus cornutus*, *Haloptilus longicornis*, *Scolecithricella dentata*, *Sagitta lyra*. A fourth group may be added: hyponeustonic pontellid copepods, fish eggs and fish larvae. Irrespective of the group, all the true planktonic species exhibit a greater vertical migration range in the central water masses, but a smaller one in divergence areas (Fig. 6-12). Of course, the various types of vertical migration may lead representatives of some of these species to enter water currents of opposite directions by day and at night, thus modifying the composition of the assemblage.

(ii) Life-cycle length may modify distribution patterns. Long-lived species may drift northwards or southwards due to the meridian components of the equatorial currents. Hence their distributional area tends to be displaced in comparison with that of short-lived species (as illustrated in Fig. 6-33 for the Indian Ocean). Even in the latter, meridian components may cause a disjunction of their maximum abundance on both sides of the equatorial divergence: e.g. *Undinula darwini*, *Neocalanus gracilis*, *Eucalanus attenuatus*, *E. gracilis*, *Scolecithrix danae*, *Rhincalanus cornutus*, *Euchaeta marina*, *Pontellina plumata* (VINOGRADOV and VORONINA, 1963).

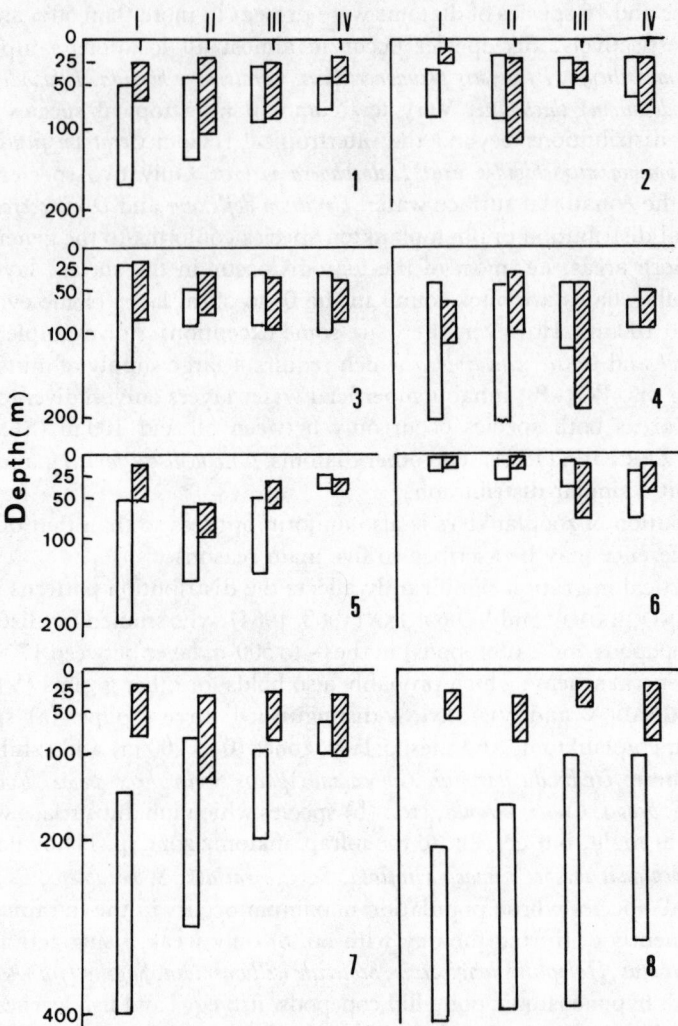

Fig. 6-12: Depths of maximal abundance of eight plankton species in
different water masses of the Pacific Ocean (white: day; hatched:
night). 1: *Undinula darwini*; 2: *Calocalanus pavo*; 3: *Euchaeta marina*; 4:
Scolecithrix danae; 5: *Pontellina plumata*; 6: *Creseis virgula*; 7: *Neocalanus
gracilis*; 8: *Limacina inflata*. I: central waters; II: waters of equatorial
counter-current and north equatorial current; III: waters of
south equatorial current, with the exception of equatorial
divergence area; IV: waters of tropical divergence between north
equatorial current and equatorial counter-current. (Based on
information provided by VINOGRADOV, 1968.)

(iii) Maximum primary productivity on the eastern side of the ocean may restrict some species to that side, e.g. five of the 14 pontellid copepods (VORONINA, 1964a), the copepod *Candacia pachydactyla* and the pteropod *Creseis virgula*.

(iv) Like the two dinoflagellates mentioned above, some zooplankton species seem to be restricted to equatorial waters, e.g. the euphausiids *Euphausia diomedae* and *Nematoscelis gracilis* and the chaetognath *Sagitta robusta*. As the meridian extension of the equatorial water narrows westwards, the distribution of these species follows the same pattern.

(v) The differences between tropical waters and equatorial waters may also affect breeding activities. It seems that breeding attains maximum values in equatorial waters in the copepods *Eucalanus attenuatus* and *Rhincalanus nasutus*, but north of $10°$ N and south of $10°$ S in *Haloptilus longicornis*, *Lucicutia flavicornis*, *Pleuromamma gracilis* and *Euaetideus giesbrechtii*.

Quantitative Heterogeneity in Subzones of the Intertropical Pacific Ocean

Apart from the southern subzone (p. 210) of the north-western central water mass, which overlaps the North Pacific and intertropical Pacific regions, the divergences and convergences related to the equatorial current system make the whole belt between $15°$ N and $10°$ to $15°$ S very heterogeneous.

According to BOGOROV (1960) three other subzones may be distinguished with very different fertilities: (i) The north equatorial current subzone, with a mean surface temperature of $27·6$ °C; it has the lowest PO_4–P content (surface, $2·3$; 100 m, $3·0$ mg m^{-3}), the lowest average number of phytoplankton cells (3100 l^{-1}), the highest percentage of dinoflagellates and the lowest zooplankton biomass (26 mg m^{-3} dry weight); its primary productivity seems to be rather low (about 100 mg C m^{-2} d^{-1}). (ii) The divergence subzone between the north equatorial current and the counter-current has a higher average surface temperature ($29·1$ °C); its PO_4–P content is the highest in the whole intertropical zone (surface, $9·0$; 100 m, 14.0 mg m^{-3}). In its phytoplankton, which is much more abundant (average: $19,900$ cells l^{-1}), diatoms and dinoflagellates attain almost equal abundances and the mean biomass of zooplankton increases (46 mg m^{-3}). Primary production may be estimated at 200 mg C m^{-2} d^{-1} in the western half of this zone. (iii) The divergence subzone of the south equatorial current with a mean surface temperature of 28.4 °C; although its PO_4–P content is rather high, (surface, $8·0$; 100 m, $10·0$ mg m^{-3}), its phytoplankton biomass is lower than in the second subzone (4300 cells l^{-1}); dinoflagellates predominate (63%). Its zooplankton biomass is lower than in the first subzone (12 mg m^{-3}). BEERS and STEWART (1971) here observed a striking abundance (47 mg^{-3} m^{-3}) of microzooplankton (mostly ciliates), suggesting that a significant part of the primary production passes through this supplementary link between algae and 'herbivorous' animals.

The marked increase in zooplankton volume near the equator (Fig. 6-13) emphasizes the paramount role of the south equatorial current divergence regarding zooplankton abundance in surface and subsurface layers. The intensity of the upheaval decreases westward as does primary production (Fig. 6-14). This general scheme of BOGOROV (1960) is based on a meridian section along $172°$ W; the quantitative data require confirmation. Firstly, the average value of primary production—for instance in subzone (iii)—which is only 100 to 150 mg C m^{-2} d^{-1} at $172°$ W rises to 150–250 at $120°$ W, and

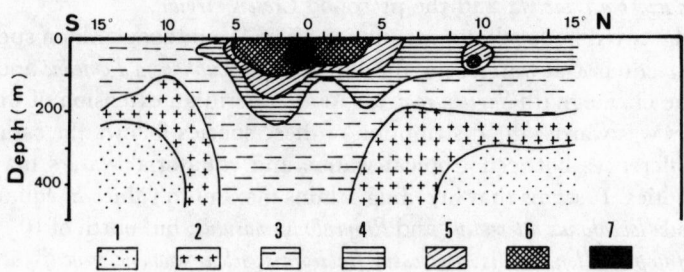

Fig. 6-13: Vertical distribution of zooplankton (ml. 1000 m^{-3}) on a
meridian (156° W) transect in the equatorial Pacific Ocean. 1: <5;
2: 5–10; 3: 10–25; 4: 25–30; 5: 50–100; 6: 100–250; 7: >250.
(Based on information provided by VINOGRADOV, 1968.)

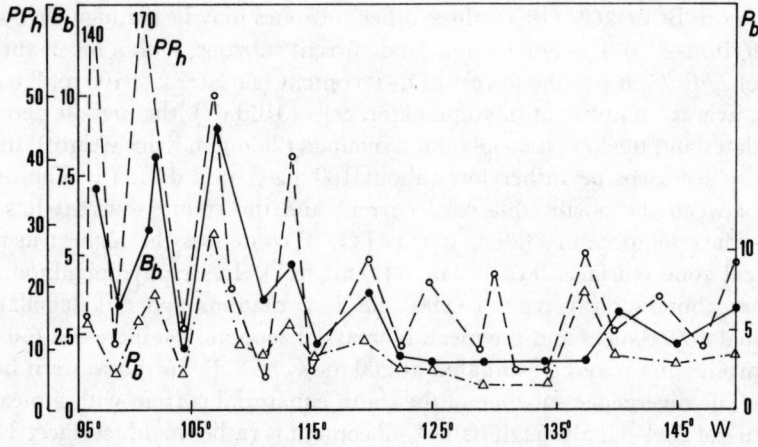

Fig. 6-14: Changes in phytoplankton primary production (PP_h), biomass
(B_b) and daily bacterioplankton production (P_b) in surface waters along an
equatorial transect of the Pacific Ocean. (Based on information provided
by SOROKIN and co-authors, 1975.)

the zooplankton biomass probably follows the same pattern. Secondly, the zooplankton
biomass values estimated by BOGOROV are largely approximations. According to
PONOMAREVA's (1960) data, obtained in the 0- to 200-m layer between New Guinea at
42° N along the meridian 142° E, biomass varies considerably due not only to vertical

migration—but also to season. In spring she found very high biomasses $(100-300 \text{ mg m}^{-3})$ in subzone (ii) and very low (less than 50 mg m^{-3}) in subzone (i). These differences disappear during summer (July–August).

For subtropical, tropical and equatorial offshore areas, PETIPA and co-authors (1975b) documented that in tropical divergence areas of the Pacific Ocean the nutritional requirements of copepods are fully satisfied mainly at the expense of bacteria, algae and protozoans; maximal copepod consumption greatly exceeds energy expenditure for respiration. The daily rations of copepods in divergence areas are very high (68–322% of the body weight) irrespective of whether the copepod is a predator or not. Owing to the abundant food supply in the surface layer, the copepods feed throughout the entire day and night periods without displaying consistent rhythm; however, at maximum food concentrations (10 g m^{-3}) the digestion time is shorter and assimilation drops to very low values. The increase in total food concentration upheaval areas favours rapid increase in body mass, the stored energy being $2 \cdot 5$ times higher (74% of body weight) than energy expenditure for respiration (31%). A reverse relationship may exist in oligotrophic areas. In fact, PETIPA and co-authors (1975) took into account only fine filter-feeder copepods. VINOGRADOV and co-authors (1969, 1972, 1973) investigated the plankton assemblage in the immediate vicinity of the equator in the 0- to 150-m layer at four different longitudes from east to west; here, upheaval intensity progressively decreases and the upper boundary of the thermocline sinks from 10–20 m to about 200 m; they took into account the whole ultra-, nano-, micro- and mesoplankton. The biomass values determined by them for the different assemblage elements separated decreases differentially from east to west (Table 6-5).

As the waters become more oligotrophic, the abundance of phytoplankton and bacteria—which together represent the food resource for 'herbivorous species (PETIPA and co-authors, 1974, 1975a,b)—decreases, but bacterial biomass changes less sharply than that of phytoplankters, especially as regards the larger-sized species. Also striking are the following facts: (i) the rather high percentage of protozoans—mainly infusorians —in the biomass, suggesting that they play an important role in the assemblage from a functional point of view; (ii) the high percentage of predators—45 to 59% of the zoo-plankton biomass, protozoans excluded—at all stations investigated. The authors attribute the latter to the 'intensive displacement of the young community of the upwelling region with the deeper waters of the Cromwell current that bear a more mature community with a larger quantity of predatory plankton' (VINOGRADOV and co-authors, 1969, p. 69). Based on their functional analyses VINOGRADOV and co-authors draw the following conclusions regarding assemblage development from highly productive to less productive regions:

(i) The decrease in food resources—phytoplankton and bacteria—available at the primary level results in a decrease in zooplankton production.

(ii) The degree of food-requirement satisfaction diminishes while the intensity of trophic relations, the balance between different trophic levels and the energy transmission through the system increase.

(iii) Due to the high concentration of predators the degree of cannibalism within the zooplankton is fairly large: 50 to 80% of the zooplankton production is consumed by the zooplankters themselves.

(iv) Increased cannibalism—mainly among protozoans—results in negative production values and phytoplankton: bacteria (P:B) ratio for all zooplankters. The zooplankton consumes itself to such an extent that its pressure on the lowest trophic levels, phytoplankton and bacteria, decreases. This, in turn, increases their production and P:B ratio (Table 6-5 and Fig. 6-16).

(v) Since part of the production of non-predatory mesoplankters is consumed at higher trophic levels (macroplankters and fish), one of the characteristics of the developing assemblage becomes very apparent: biomass increases and energy accumulates (net production quite positive) during early stages (station I) but energy is expended (net production negative) at the more mature stage (station III). Such energy accumulation and subsequent expenditure implies integrated succession in the planktonic assemblage at the equator.

Later, VINOGRADOV (1976), VINOGRADOV and SEMENOVA (1977) and VINOGRADOV and SHUSHKINA (1978) synthetized the ideas formerly developed. Admitting that upwelled waters originate from the depths of the upper part of the subsurface (Cromwell current) core, they emphasized that the lower trophic levels—i.e. bacterioplankton and phytoplankton—exhibit maximum development at the site of most intensive upheaval (97° W; station I; Table 6-5). As the plankton assemblage drifts westwards with the north equatorial or south equatorial current, the degree of its maturity increases. Since the Cromwell current originates from northern and southern tropical gyres inhabited by a rather mature assemblage which is transported eastwards by the current, assemblage elements mix up in the area of most intensive upheaval with members of the assemblage formed in the upwelled water. As the system becomes more mature, the percentage of carnivores in the total zooplankton increases as does the consumption of zooplankton by micronekton; consequently, the decrease in predation pressure at the lower trophic levels results in a secondary increase in their biomass and production (154° 55′ W; station IV; Table 6-5, Figs 6-15, 6-16).

The investigations summarized above (involving only four locations between 97° W and 154° 55′ E, but many taxa), corroborate results by GUEREDRAT and co-authors (1972) who investigated 34 different locations in the equatorial current system between 90° W and 162° E, but restricted their analysis to the diversity index in three zooplankton groups: copepods, euphausiids and micronektonic fishes. Sailing westwards from 90° W, the authors noticed that up to 105° W, the temperature in the surface waters is lower than 20 °C, the thermocline rather shaded, the upheaval strong and the content in nutrient salts rather high; therefore, net phytoplankton is abundant and chlorophyll a values are high. In these eastern areas herbivorous species always represent 90 to 95% of the total number of copepods; in this taxon, only one species, Eucalanus subtenuis, accounts for approximately 45% of all the specimens collected at the stations I to V. Carnivorous copepods represent only 5 to 10% of the total. The diversity index for the three groups studied is low.

Sailing west of 110° W, the authors found that this situation progressively changes, and farther than 145 to 150° W, the surface temperature surpasses 25 °C, a well-marked thermocline appears and the thickness of the homothermal surface layer increases; the upheaval is weak and sometimes even a convergence substitutes for it; nutrient salt contents, as well as net phytoplankton abundance are low and chlorophyll a, also less abundant, is distributed towards greater depths. Up to 165 to 170° E the global diversity

Table 6-5

Equatorial plankton assemblages. Biomass (kcal m^{-2}) of different groups of organisms in the water layer 0 to 150 m (Based on information provided by VINOGRADOV, 1977)

Organisms	Longitude W			
	Stn I 97°	Stn II 122° 30'	Stn III 139° 30'	Stn IV 154° 55'
Phytoplankton				
<15 μm	2·2	0·6	0·6	0·6
>15 μm	16·6	1·4	1·2	1·3
Total	18·8	2·0	1·8	1·9
Bacteria				
dispersed + aggregated	16·5	2·8	5·2	2·9
Protozoans				
flagellates	1·4	0·7	0·7	0·4
infusorians	0·2	1·3	0·4	1·7
radiolarians and foraminiferans	0·03	0·2	0·1	0·04
Total	1·6	2·2	1·2	2·1
Chiefly non-predatory animals				
fine filter feeders	0·02	0·04	0·06	0·08
small copepods	1·7	0·7	1·2	0·8
large calanoids	2·6	0·7	0·9	0·9
Total	4·3	1·4	2·1	1·8
Chiefly carnivorous animals				
cyclopoids	1·9	1·1	0·7	0·3
predatory calanoids	0·7	0·3	0·4	1·1
other predators	1·2	0·3	0·6	0·6
Total	3·8	1·7	1·7	2·0
Total zooplankton with protozoans	9·6	5·3	5·0	5·9
Total plankton	45	10	12	11
Percentage of protozoans in zooplankton	16	39	23	35

index is high as is that of all zooplankton groups studied. With regard to copepods, carnivorous species (mainly of the genera *Pleuromamma* and *Euchaeta*) usually account for 60 to 85% of the total biomass of the group. At most western stations, GUEREDRAT and co-authors (1972) recorded a small decrease in the diversity index; they attribute this to the temporal instability in local hydrological conditions, which may be related to a seasonal inversion in the vertical circulation originated from changes in the wind regime: south-easterly trades during the boreal summer and westerly monsoon during the boreal winter. Since their samplings succeeded each other on a time scale in the same direction as the flow of the equatorial current, GUEREDRAT and co-authors suggest that the changes in the zooplankton assemblage observed correspond to a spatio-temporal evolution of the system towards increased maturity and progressive organization.

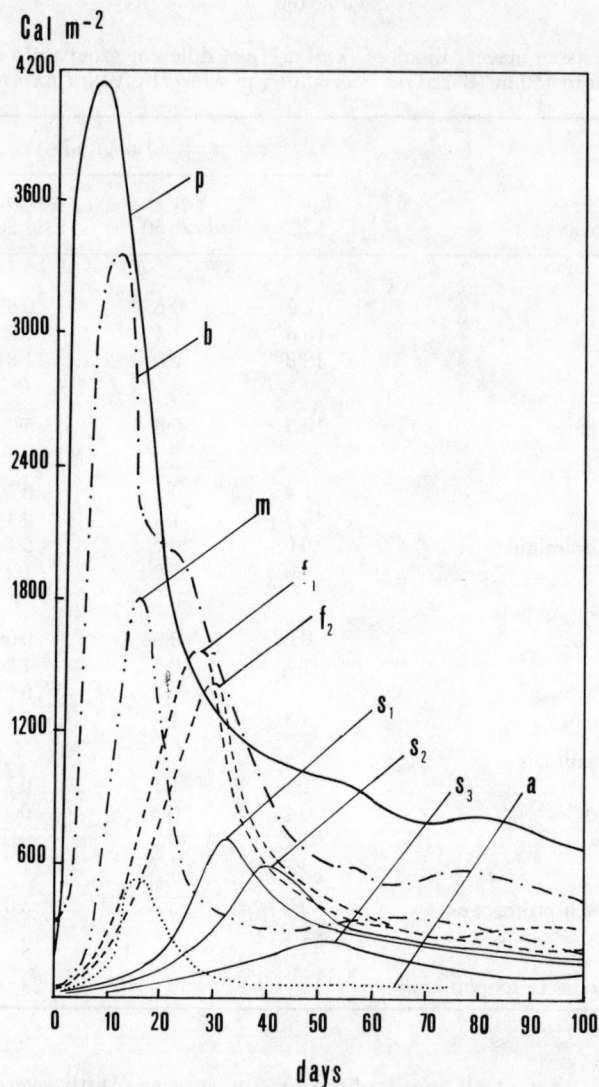

Fig. 6-15: Time changes in biomass (cal m^{-2}) of the most
significant components of the tropical pelagial ecosystem
in 0–150 m. p: phytoplankton; b: bacterio-plankton; a:
protozoans; m: microzooplankton; f_1 and f_2: small- and
large-sized phytophages; S_1: cyclopoid copepods; S_2:
predatory calanoids; S_3: chaetognaths. (Based on information provided by VINOGRADOV, 1977.)

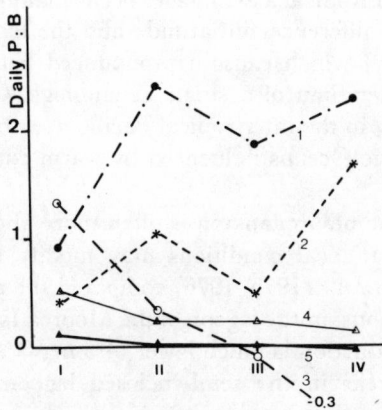

Fig. 6-16: Changes in specific production (P:B) of different assemblage components along an equatorial transect in the Pacific Ocean.
1: phytoplankton; 2: bacteria; 3:
protozoans; 4: phytophages and
euryphages; 5: predators. I: (1454)
01° 02′ S, 97° 03′ W; II: (1456)
00° 00′, 122° 05′ W; III: (1458)
00° 01′ S, 139° 41′ W; IV: (1461)
00° 00′, 154° 47′ W. (Based on
information provided by VINO-
GRADOV and co-authors, 1976.)

Sampling the upwelling region in the vicinity of 97° W, with 140-l bottles, VINO-
GRADOV and SHUSHKINA (1977, p. 390) analysed the vertical distribution of various
components of the assemblage inhabiting the upwelling area of the equatorial Pacific
Ocean. They 'frequently noticed the extremely uneven vertical distribution of zooplankton when the layers of high concentration alternated with poor, almost lifeless layers.'
Maximum aggregations of small, weakly migrating herbivorous copepods feeding
directly on the lowest trophic levels were generally confined to maximum concentrations
of the latter. Dominance of groups with similar feeding habits—e.g. appendicularians,
nauplii, small herbivorous copepods—may exhibit a distinct vertical divergence,
whereas layers of abundance of various carnivores usually coincide with each other and
with those of high concentrations of their potential prey organisms.

Neritic Tropical Assemblages

Due to the lower-temperature surface waters, related to the California current upwelling, neritic subtropical and tropical assemblages occur mainly on the western side of the Pacific Ocean. In view of differences in latitude and the local peculiarities (currents, estuaries, reef-lagoons, etc.) which cause a pronounced heterogeneity, it is better to speak of assemblages rather than of a single assemblage. Of course, neritic tropical assemblages occur not only in the intertropical Pacific itself but also in other regions of the North and South Pacific Oceans influenced by warm currents originated from the equatorial zone.

In these assemblages, the phytoplankton is often more abundant than in the neighbouring offshore areas, but local conditions may modify this general scheme. For instance, SOURNIA and RICARD (1975, 1976) compared the phytoplankton production in two polynesian atoll lagoons: in the lagoon of the Moorea Island Atoll, located behind a barrier reef, primary production is much lower ($4\cdot 3$ mg C m^{-3} d^{-1}) than in the outer slope waters ($13\cdot 9$), whereas in the semi-enclosed lagoon of Takapoto Atoll it is $9\cdot 7$ mg C m^{-3} d^{-1}. Some species may be endemic in a lagoon, e.g. at Rangiroa Atoll (MICHEL and co-authors, 1971) the algae *Nitzschia* cf. *seriata*, *Rhizosolenia calcaravis* and *Exuviella vaginula* do not exist in oceanic waters neighbouring the atoll. In the zooplankton, small-sized forms predominate as in offshore tropical assemblages, but the specific diversity decreases: in the Bikini Lagoon, for example, *Undinula vulgaris* alone accounts for 50% of the copepod biomass. In the Indonesian Sea three species, *Undinula vulgaris*, *Euchaeta marina* and *Labidocera acuta*, make up 60 to 80% of the total zooplankton. For copepods, the mean number of generations per year is 10. In general, seasonal fluctuations are more marked than offshore; the zooplankton peak immediately follows the phytoplankton bloom, and the plankton ecosystem often seems rather poorly balanced.

The plankton abundance in atoll lagoons largely depends on energy input from organic detritus and aggregates washed out by strong waves from peripheral coral reef assemblages during gales. According to MICHEL and co-authors (1971), part of this organic material is mineralized by bacteria; this results in nutrient enrichment which presumably can induce a bloom like that observed at Rangiroa Atoll. Another part of this exogenous organic material—especially aggregates—is fed by zooplankton. JOHANNES (1967) demonstrated that the guts of copepods from the Eniwetok Lagoon do not contain more than 1% recognizable plant material, and MICHEL and co-authors attributed the high zooplankton biomass they observed in Rangiroa Lagoon to these non-algal food resources, averaging 75 mg m^{-3}, i.e. five times higher than outside the lagoon, with a peak up to $2\cdot 75$ mg m^{-3}.

Another example of particular conditions in the neritic province of the intertropical Pacific Ocean has been presented by WICKSTEAD (1958) for the Java Sea, which is swept clean twice a year by alternating monsoons and occupied successively by Yellow Sea waters and Banda Sea waters. The planktonic assemblage includes a permanent component, as well as other components which either attain abundance only at certain seasons or are present in low numbers. For example, the copepods *Eucalanus* (?) *subcrassus*, *Euchaeta concinna* and *Labidocera* sp. were found throughout the year, whereas *Undinula vulgaris* and *Centropages* sp. were common only during the second quarter of the year (north-east monsoon), the two genera *Copilia* and *Sapphirina* in October (end of the summer monsoon). Total zooplankton peaks occur in February and March (north-east monsoon) and October (south-west monsoon).

Bacterioplankton in Tropical and Equatorial Regions of the Pacific Ocean

In the last ten years, SOROKIN (1970, 1971, 1973a,b) and SOROKIN and co-authors (1975) have paid special attention to the bacterioplankton in Pacific Ocean warm waters and its ecological role in both offshore and inshore plankton ecosystems (see also Volume IV: SOROKIN, 1978). In shallow waters of the western Pacific Ocean (coral reef lagoon) the average photosynthetic production is estimated at 10 to 20 mg C m^{-3} d^{-1}, i.e. equal to microbial production. According to SOROKIN (1973a), this means that microbial decomposition in waters over a reef would be 30 times lower than consumption by benthic assemblages of the reef itself.

SOROKIN (1971) suggested that two different situations may occur in equatorial and tropical regions of Pacific Ocean offshore waters. (i) In equatorial surface water, bacterial production is almost the same as primary production (560–600 mg C m^{-2} d^{-1}). (ii) In tropical surface water, bacterial production (B_p) is a little higher, but photosynthetic production (P_p) about half as much; thus the B_p/P_p ratio exceeds 2·5. In the immediate vicinity of the equator, SOROKIN found two maxima of bacterial biomass and production: at the upper boundary of the thermocline (70–80 m) in the euphotic layer and at a depth of 300 to 600 m in the main thermocline at the peak of the minimum oxygen layer (Fig. 6-17). SOROKIN postulates that a zooplankton biomass of 1·2 g C m^{-2} cannot be supported solely by primary production of about 0·2 to 0·3 g C m^{-2} d^{-1}, but that the sum of algal plus bacterial production would suffice (bacteria can form aggregates up to 5 µm in diameter; Fig. 6-18). In fact, it was observed that the cladoceran *Penilia* and most calanid copepods feed upon bacterial aggregates. Thus, it seems that microzooplankters feed almost entirely on bacteria and that bacteria account for about 30% of the diet in phytophagous filter-feeding crustaceans (Fig. 6-19).

Moreover, SOROKIN (1973b) also assumed that the dissolved organic carbon (DOC) which is oxidized in tropical surface water only partly originates from primary production *in situ*; another part would be provided by advection processes from productive zones at higher latitudes. This hypothesis has been challenged by BANSE (1974) who pointed out that important DOC gradients with increasing depth have never been observed and that bacterioplankton production is at least one order of magnitude lower than photosynthetic production, except in highly productive areas (divergence, blooms, etc.). SOROKIN's view concerning energy inflow into tropical pelagic systems in the form of dissolved organic matter produced in other areas of the ocean may be criticized because his measurements of bacterioplankton production with dark bottles kept at the *in situ* temperature for one to four days cannot be considered adequate. There are three main arguments against the assumed input of organic matter attributed to intermediate or deep layers by SOROKIN: (i) most bacteria found in ocean waters are in a dormant state; (ii) the main bulk of dissolved organic matter consists of rather inert material; (iii) solid surfaces, such as glass beads and walls of the glass bottles used, quickly promote bacterial activities by rendering dilute nutrients more available. In my opinion, bacteria

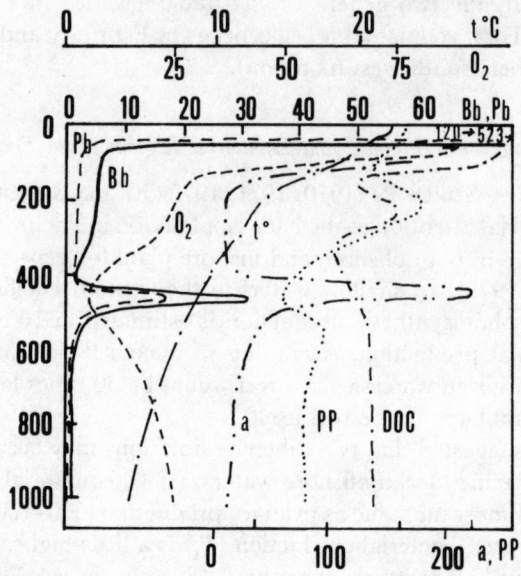

Fig. 6-17: Changes in bacterioplankton biomass (*Bb*, mg m^{-3}) and production (*Pb*, mg m^{-3} d^{-1}) with depth in the equatorial Pacific Ocean (latitude 01° 02′ S; longitude 97° 03′ W). O$_2$, oxygen in percentage of saturation. $t°$: temperature; *a*: percentage of bacterial activity; PP: bacteria potential production (mg m^{-3} d^{-1}); DOC: dissolved organic carbon (mg C l^{-1}). (Based on information provided by SOROKIN and co-authors, 1975.)

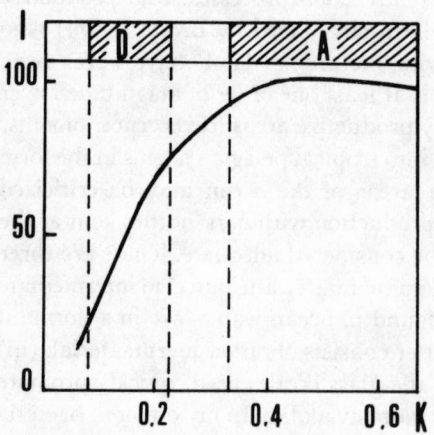

Fig. 6-18: Relationship between feeding intensity of a mixed population of calanoids and density of bacterioplankton populations in areas of strong (A) and weak (D) equatorial divergence at longitudes 97° W and 155° W, respectively. K: bacterial biomass (mg m^{-3}); I: assimilation of bacterioplankton in percentage of the maximum. (Based on information provided by SOROKIN and co-authors, 1975.)

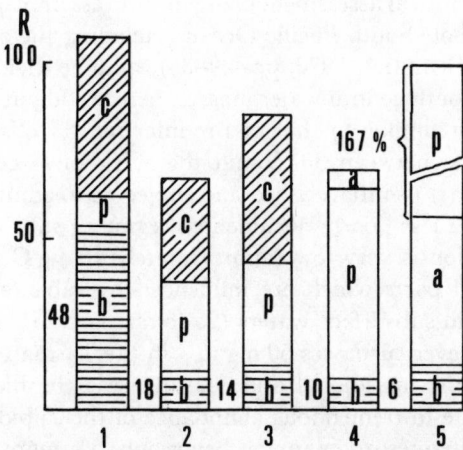

Fig. 6-19: Relative proportion of bacteria (b),
algae (p), ciliates (c) and small-sized crustaceans in the ration (R, in % of body
weight d^{-1}) of copepod species abundant
in the equatorial Pacific Ocean. 1: fine-filter
feeders, e.g. *Clausocalanus*; 2–3: mean-filter
feeders, e.g. *Paracalanus* and *Undinula*; 4:
coarse-filter feeders, e.g. *Eucalanus*; 5: mixed
feeders, e.g. *Acartia*. (Based on information
provided by SOROKIN and co-authors, 1975.)

in the open ocean are of minor importance as food for herbivorous zooplankters in
intermediate and deep waters. However, in surface and subsurface water layers, in
which the content of utilizable metabolites released by plankters is much higher, the role
of bacteria in the trophic pelagial network cannot be disregarded.

(d) South Pacific Ocean

General Aspects

Considering once more the general scheme of subdivision of the Pacific Ocean into
different regions, we can distinguish a South Pacific tropical and subtropical region
extending southwards as far as the subtropical convergence (35–45° S); in the South
Pacific central water mass, the surface temperature averages 21·7 °C, and the PO_4–P
content 4·9 mg m^{-3} (surface) and 7·1 mg m^{-3} (300 m). Phytoplankton abundance
amounts to about 3600 cells l^{-1}, as in the North Pacific tropical and subtropical region,
with a percentage of dinoflagellates decreasing southwards; the average zooplankton
biomass is smaller than in the symmetrical North Pacific region: 36 mg m^{-3}.

Unfortunately the data from this tremendous area of the Pacific Ocean are so scattered and vague that a critical assessment of organismic assemblages is quite impossible.

It seems that the whole South Pacific Ocean, excluding the areas influenced by the Peru current (Vol. I: COLLIER, 1970; pp. 29–35), may be divided into two subzones: (i) the subzone of the south central water masses, between about 15° S and 30° S where the assemblage is very similar to that of the intertropical offshore waters described above; (ii) the subzone between 30° S and the subtropical convergence, where the assemblage of subzone (i) is impoverished and subjected to admixture of other species.

The Tasman Sea is a less poorly documented region in subzone (ii); in its northern part, primary production is very low (approximately 10 mg C m^{-2} h^{-1}), whereas the central and meridional parts which are influenced by subantarctic waters are more productive, especially in subsurface waters (25 to 50 mg C m^{-2} h^{-1}; JITTS, 1965). The zooplankton biomass never surpasses 50 mg m^{-3} in the Tasman Sea and often remains below 25 mg m^{-3}. In the meridional and coastal areas the biomass may rise up to 200 mg m^{-3}, mainly due to tremendous abundance of the salpid *Thalia democratica*, the other components of the mesozooplankton being only 5% more abundant than in offshore areas. However, according to VOLKOV and co-authors (1972), macrozooplankton standing crop may be rather high (20–1000 mg m^{-3}) in the Tasman Sea, especially in the vicinity of New Zealand's western shores and still higher (50–3000 mg m^{-3}) in the areas south-east of New Zealand.

On temperate shores of the South Pacific Ocean, a seasonal succession of phytoplankton occurs, very similar to that well documented in the neritic areas of the Northern Hemisphere. In September, the spring bloom begins with small chain-forming diatoms (*Asterionella*, *Thalassiosira*, *Skeletonema*, *Nitzschia*, *Chaetoceros*); later, larger centric diatoms (*Ditylum*, *Leptocylindrus*, *Eucampia*, *Rhizosolenia*), together with the pennate genus *Melosira*, substitute for the preceding forms. In early summer (October) microflagellates, coccolithophorids and dinoflagellates appear and constitute the bulk of phytoplankton throughout the whole hot season. After two small peaks of cryptomonads (March) and silicoflagellates (April) a second diatom bloom occurs in autumn (May) dominated by species of *Leptocylindrus*, *Thalassiothrix* and *Rhizosolenia*. JEFFREY and CARPENTER (1974), who reported this succession, assume that it does not correspond to fluctuations in the nutrient salts content of neritic waters but arises from biochemical (vitamin) interrelationships between the algae.

Off Peru and northern Chile, the Peru current induces an upwelling that makes this region one of the most fertile in the World Ocean. Diatoms predominate in the phytoplankton; in the coastal part of the current, species of *Thalassiosira* and *Coscinodiscus* are more common in the north and *Corethron* in the south; the oceanic genera *Rhizosolenia* and *Planktoniella* are more abundant in offshore areas of the current, whereas species of *Chaetoceros* are generally abundant everywhere (GUNTHER, 1936). Diatom abundance in the Peru coastal current from about 23 to 4° S amounts to 5000 to 10,000 cells m^{-3}, whereas dinoflagellates generally account for only 100 to 300 cells m^{-3}, except for some areas north of 9° S. A decrease in cell number is often observed in most nearshore areas, except for the region around Trujillo. In other respects, phytoplankton abundance decreases gradually westwards—towards oceanic areas—further and further away from places with strong upheaval (SEMINA, 1971). Vertically, the maximum abundance of phytoplankton occurs above the pycnocline or in the pycnocline itself (RATKOVA, 1974).

The average size of phytoplankters is generally rather low, i.e. less than 50 μm, in the most fertile waters of the Peru coastal current, but increases north of 9° S and also in the less fertile waters of the Peru coastal current. On average, the size of phytoplankton cells also increases below the pycnocline.

It is generally admitted that primary production rates vary according to areas between 100 and 200 g C m^{-2} y^{-1}. Primary production seems to be higher in the Peru coastal current than in offshore waters and higher in spring than in winter (GUILLEN and IZAGUIRRE DE RONDAN, 1968). While the latter authors never found a daily primary production higher than 2·42 g C m^{-2} d^{-1}, VEDERNIKOV and STARODUBSTEV (1971) observed up to 6·08 g C m^{-2} d^{-1} from August to December 1968. RYTHER and co-authors (1970) recorded an average primary production of 10 g C m^{-2} d^{-1} over a five-day period, with a maximum of plant biomass on the third day; after the fifth day the biomass returned to its intitial value, possibly due to grazing. RYTHER and co-authors also assumed that the whole primary production cycle takes place in the 0- to 50-m layer; they could not observe any indication of 'organic rain' followed by accumulation and decay either in the underlying waters or on the bottom. Possibly the discrepancy between these results and those mentioned above can be ascribed to temporal and spatial differences in upwelling intensity.

Obviously, when the El Niño current develops, its southward flow tends to introduce warm (27 °C) and less saline (33·5% S) waters with relatively low nutrient contents, into the normally nutrient-rich waters supplied by upwelling. This irregularly occurring phenomenon results in a strong (5–10 times) decrease in primary production and, in turn, in the anchovy stock.

The zooplankton composition is also heterogenous. According to FAGETTI and FISCHER (1964) there is a predominance of copepods (73·4%) off the Chilean coast, followed by euphausiids (5·2%). The chaetognath *Sagitta bierii* seems to be characteristic of this assemblage in the Peru current waters, whereas *S. enflata* predominates in more oceanic (south-central) waters (FAGETTI, 1968, 1972). In upwelling areas of mesopelagic offshore waters, the chaetognaths *S. decipiens* and *Eukrohnis hamata* ascend into the epipelagic zone. Of course, in the Peru current hyponeustonic pontellid copepods are very rare: only two species, *Pontellopsis regalis* and *Labidocera acutifrons*, inhabit the most superficial layer, but they disappear near the coast. In the whole euphotic layer the dominant species is *Calanus australis*; it substitutes for pontellids in the nearshore hyponeuston. *Centropages brachiatus* is another rather common species in the euphotic layer (HEINRICH, 1971).

HEINRICH (1973) describes the horizontal distribution of copepods in the Peru current region between the equator and 30° S and eastwards up to 90° W. HEINRICH distinguishes three main copepod assemblages: (i) The 'coastal' region assemblage which inhabits a 150-mile wide zone from the coastline between latitudes 5 and 22° S. Here, the strongest upwelling pulses occur and, consequently, diatoms predominate; seston values usually fall between 0·5 and 1 ml m^{-3} but may often be higher. Three typical 'distant neritic' species—*Calanus australis*, *Eucalanus inermis* and *Centropages brachiatus*—exhibit a massive development; in the 0- to 200-m layer, average numbers of adult individuals reach 4970, 925 and 1760 m^{-3}, respectively. (ii) The north-western assemblage in an area mainly occupied by waters originating from equatorial regions. Here, the amount of seston rarely exceeds 1 ml m^{-3} in the 0- to 100-m layer. The predominant species of copepods and the average number of individuals per square metre in the 0- to 200-m layer are: *Acartia danae* (3050), *Undinula darwini* (2260), *Eucalanus subtenuis* (1270) and *Mecynocera clausi* (780). (iii) The southern region assemblage with the phytoplankton dominated by dinoflagellates and a seston volume less than 0.5 ml m^{-3}. All widely distributed copepod species with tropical affinities are less abundant here than in northern areas. The predominant species and their respective abundance of adult individuals per square metre in the 0- to 200-m layer are: *Nannocalanus minor* (1250), *Acartia negligens* (680), *A. danae* (500), *Mecynocera clausi* (230), *Calocalanus pavo* (100).

As regards the macrozooplankton, the Peru current region within the 200-mile zone exhibits the maximum standing crop (500 mg m^{-3}) between 13 and 17° S; in contrast only 20 mg m^{-3} are recorded in the subtropical convergence region (VOLKOV and co-authors, 1972).

The zooplankton distribution on a 130-mile transect from the Peruvian coast off-shorewards, at Latitudes 7 to 8° S, has been investigated by MIKHEYEV (1977) who analysed samplings from six stations: three above the shelf, one above the shelf edge and one above greater depths. MIKHEYEV suggests we distinguish between two zooplankton assemblages: (i) In the shelf waters (stations 1465 to 1467) with surface temperatures between 17 and 20 °C, *Acartia tonsa*, lichomolgids and larval stages of benthic polychaetes predominate, but with an important admixture of *Centropages brachiatus* and *Oithona similis* at station 1467. (ii) From the shelf edge towards offshore (stations 1468 to 1470), with surface temperature exceeding 22 °C, the copepods *Calanus australis* *Eucalanus inermis*, *Clausocalanus jobei* and *Corycaeus dubius* predominate. Abundance and species number of *Oncaea* also increase in this offshore region. In the small copepod assemblage (i) the proportion of meroplankton is large, and phyto- and euryphagous copepods predominate. The offshore assemblage (ii) features approximately three times as many copepod species as (i), with the dominants of the coastal assemblage being virtually absent. Large-sized organisms are found along with small-sized ones, and biting and swallowing predators appear. Biomass and specific diversity (Shannon's index) are positively correlated in the coastal (upwelled water) assemblage but negatively in the offshore assemblage. MIKHEYEV assumes that the striking maximum in zooplankton biomass at station 1468 might be ascribed to the exploitation (p. 108) of the younger assemblage (i) by the more mature one (ii).

Investigations on micronektonic fishes in the infrapelagic and bathypelagic zones along a 34° S transect by CRADDOCK and MEAD (1970) lead to a very similar conclusion and suggest that the meridian 80° W is an important boundary. From this longitude towards the east, some mixing takes place between warm waters (possibly equatorial) and subantarctic surface water. West of this line, South Pacific central water predominates and a permanent thermocline exists. Myctophids are the most abundant representatives of micronektonic fishes in both areas. The longitude 80° W clearly appears as a major biogeographical boundary: preponderance of myctophids is less noticeable west of 80° W and the same is true for gonostomatids. East of 80° W, mycto-phids increase in abundance and decrease in diversity with a marked percentage of cold-loving (even subantarctic) species. West of this boundary, the general abundance is lower, but the species number is higher, except for some endemics; most species are panoceanic or equatorial.

Macroplankton and Micronekton in the Intertropical and South Pacific Ocean

Many micronektonic species, oceanic or distant neritic, in the South Pacific Ocean are restricted to one water mass as in the north-east Pacific Ocean. This is exemplified by the differences between myctophid faunas on both sides of 80° W, mentioned above. Another example is the distant-neritic ichthyofauna of the eastern equatorial Pacific Ocean, represented by many endemic species of the families Myctophidae, Melamphaeidae, etc., some of them being expatriated in the Peru and California undercurrents. However, according to PARIN (1968, 1971, 1972, 1975), the distributional range of most micronektonic fishes encompasses more than one water mass. In subtropical, tropical and equatorial areas of the Pacific Ocean, the total fish biomass has been estimated by BOGOROV (1960) as follows: subtropical and tropical northern zones, $2 \cdot 0$ mg m^{-3}; north equatorial current, $0 \cdot 7$; divergence zone in the equatorial current, $1 \cdot 4$; subtropical and tropical southern zones, $1 \cdot 2$. BOGOROV's data certainly cannot be considered to be representative of the whole micronekton because he did not take into account crustaceans and squids. Moreover, even his data for fish biomass are probably too low.

Thus, VORONINA (1964b) who studied the distribution of macroplankton and micronekton between 15° N and 15° S on the meridians 160° E and 140° W, found in the 0- to 25-m layer an average biomass of 17.6 mg m^{-3}, the most important components being fishes (*Vinciguerria, Lampanyctus, Gonostoma*), decapods (*Oplophorus, Acanthephyra, Parapandalus, Systellaspis*) and euphausiids (*Thysanopoda*). Of course, some changes in the distribution of these animals may occur due to vertical circadian migrations.

According to PARIN and NESIS (in VINOGRADOV, 1977), at night the macroplankton and micronekton biomass (excluding jelly-like organisms) in the 0- to 600-m or 0- to 1000-m water column may be estimated to amount to 15 to 35% of the mesoplankton biomass. Maximum values ($14 \cdot 9–16 \cdot 8$ mg m^{-3}) have been recorded from the highly productive waters of the eastern equatorial regions of the Pacific Ocean. In less productive regions, e.g. central and western equatorial waters and the Sulu Sea, biomass values range between $4 \cdot 0$ and $8 \cdot 2$ mg m^{-3}, and in the oligotrophic central waters of the North Pacific Ocean further decreased to $1 \cdot 7$ mg m^{-3}. Small-sized planktonic and micronektonic fishes represent 40 to 75% of the total biomass, next in rank being euphausiids and decapods. In the ultra-oligotrophic waters off Hawaii, CLARKE (1973) found in the 0- to 1000-m water column, an average biomass of only 1 mg m^{-3}, approximately one-third of the total represented by myctophids.

According to PARIN and NESIS (in VINOGRADOV, 1977), many myctophid fishes (genera *Myctophum, Symbolophoros*, etc.) migrate at night up to most superficial $0 \cdot 15$- to $0 \cdot 20$-cm neuston layer; the biomasses recorded (in mg m^{-3}) would be $2 \cdot 1$, $6 \cdot 0$ to $8 \cdot 3$, $31 \cdot 2$ to $33 \cdot 7$ and $88 \cdot 0$ off the Philippines Islands, at the equator and 155° W, at the equator and from 97° to 122° W, and at the equator and 140° W, respectively.

Nekton in the Intertropical and South Pacific Ocean

Our present knowledge on macroplankton and micronekton assemblages in the intertropical and southern regions of the Pacific Ocean is insufficient for a reliable assessment of the biology of some large-sized nektonic species which represent the terminal link of the food chain in the pelagial. This is especially true for species of high economic value such as tunas and tuna-like fishes, which mainly feed on macroplankton and micronekton.

With regard to the food resources available for tunas and related fishes, GRANDPERRIN (1975) and ROGER and GRANDPERRIN (1976) found that the Coral Sea, the equatorial current system and the south tropical Pacific yielded some interesting results which can be summarized as follows. There are three characteristic groups of macroplankton and micronektonic animals: (i) a surface and subsurface fauna which in spite of circadian migrations of some species always remains in the 0- to 450-m layer; (ii) an intermediate water fauna, with components never ascending above 450 m; (iii) an 'interzonal' fauna, descending during the day below 450 m but rising at night at least into subsurface layers. Fishing tunas by long line and studying their gut contents, ROGER and GRANDPERRIN observed that tunas preyed on animals in the surface and subsurface layers (0–450 m) and that the two most important long-line species, albacore and yellowfin, are day eaters; micronektonic mesopelagic fishes represent 60% of their diet. On the contrary, infrapelagic micronektonic fishes, which enter the 0- to 450-m layer at night, i.e. the biotope of these tunas, are rarely ingested. This suggests that the main food chain leading to both these tunas is: phytoplankton and/or small zooplankton → euphausiids → micronektonic fishes → long-line albacore and yellowfin. Each link of this food chain is restricted to the biomass in the 0- to 450-m layer during daytime. Thus, both these day-feeder tunas cannot benefit from such a chain, because the very important biomass of migrating species aggregate in deeper layers by day and ascend into the 0- to 450-m layer at night. Since migrating species feed on epipelagic species at night, a downward directed energy transfer prevails benefiting the infrapelagic and bathypelagic micronekton with a fairly large biomass. However, cephalopods, which represent nearly 40% of the tunas' diet, probably feed to some extent on the migrating fauna.

From these observations, it may be inferred that there are two different but connected assemblages: a surface and subsurface assemblage, in which tunas and tuna-like fishes participate as top-level predators, and a deeper system from which these fishes seem to be excluded. The two assemblages are quite different from the energy point of view: the surface and subsurface assemblage which includes phytoplankton and small-sized zooplankton has its own source of energy, while the deeper system (except for the potential primary production of rather mysterious autotrophic deep-sea microalgae) receives its energy entirely from the assemblage via the migratory 'interzonal' fauna. Therefore, even tunas living in subsurface waters and feeding during the day, directly depend on food resources from the upper assemblage; however, since some of their prey organisms feed on animals of the deeper system, they are to a small extent indirectly linked with the latter assemblage.

In the Pacific Ocean four tuna species comprise ecologically important components. The yellowfin *Neothunnus albacora* is predominantly a surface and subsurface (0–450 m) species; it is most abundant in the area covering 5° N to 6° S and around the equator, but its distribution also extends southwards into the waters west of the international date line up to 30° S. *N. albacora* is also present on the eastern side of the Pacific Ocean off western Baja California, Ecuador and northern Peru (BLACKBURN, 1968). The big-eye or patudo tuna *Parathunnus obesus* has a distribution similar to the yellowfin but is less restricted to surface and subsurface waters (0–650 m); *P. obesus* usually occurs, as does *N. albacora*, in large shoals. Unlike the yellowfin, however, the albacore *Germo alalunga* is more frequently found in mid-latitude waters; presumably, it entertains separate populations in both hemispheres. For example, the southern population in the Pacific Ocean occurs between 6 to 10° S and 30° S, with a distributional centre between 20 and 30° S. The southern bluefin *Thunnus maccoyi* is the most common tuna species in the southern subzone of the South Pacific central water mass. It mainly occurs between New Zealand and the Australian coasts. Unlike the preceding species, the bluefin often comes very close to the shoreline, sometimes feeding on nektobenthic fishes.

Among the scombriform fishes, another species of high economic value is the skipjack *Katsuwonus pelamys* which is common in the Pacific Ocean; this species, smaller than the yellowfin and patudo, is a warm-loving species (thermal preference: 24–28 °C). However, off Japan in summer, the skipjack extends a little farther to the north than the other two species.

In the South Pacific Ocean the skipper *Scomberesox saurus* is also an important terminal link of the planktonic food chain. This typical zooplankton feeder belongs to the same family as the saury of the Northern Hemisphere (p. 214) and the biology of these two species is somewhat similar. *S. saurus* is a cold-temperate species; its northern limit off the South American coast is 15 to 18° S in summer and shifts towards the north reaching 5 to 7° S in winter. Its southern limit corresponds approximately to the 10 °C isotherm in surface waters. In spite of the rather low present landings, it seems that the annual yield might reach 3·0 to 3·5 million tons.

Another fish which is a very important component of the pelagic assemblage in the South Pacific Ocean is the anchovy of the Peru current, *Engraulis ringens*. This species which formerly supported a huge population of guano-producing birds, later sustained the most important fisheries all over the world (average landings from 1960–1970: 10 million metric tons yr^{-1}), until unfavourable oceanographic conditions occurred in 1973 and 1974. This tremendous production is based on the high primary production in the Peru current upwelling areas and results from the fact that the anchovy is mostly herbivorous—at least as adults—and thus represents the second link of the food chain.

In general, squid populations are poorly investigated in the World Ocean, particularly in the intertropical and South Pacific Ocean. These fast swimmers are difficult to sample. The discrepancy between fishing-boat catches and investigations on gut contents (mainly species of the families Enoploteuthidae and Cranchidae) of top-level predators known as squid eaters, e.g. tunas and sperm whales, emphasizes the gap in our knowledge of squid abundance and distribution. Another source of data on squid populations is the abundance of their rostra on the sea bottom. The number of rostra collected correlates well with rates of primary production and plankton distribution. For example, the upwelling areas linked with the California current containing 500 to 1000 rostra m^{-2} and deep-sea sediments close to the Japanese oceanic coasts, are noted for their large number of rostra. However, there is also a wide belt in the intertropical Pacific (15° N–20° S), mainly eastwards from the international date line, where rostra abundance on abyssal sediments averages at least 100 to 300 m^{-2}. In most equatorial areas and in the New Guinea and Fiji regions, there are as many as 500 rostra m^{-2}. These data (BELAYEV, 1970) emphasize the ecological significance of squids in the pelagial of the intertropical and South Pacific Ocean.

In general in the Southern Hemisphere, the largest squid concentrations of distant-neritic species occur in the western regions of the oceans in the vicinity of the subtropical convergence. For example, *Notodarus sloani* is common in the frontal zone off Tasmania and New Zealand. The large-sized *Dosidicus gigas* is common at low latitudes of the eastern Pacific Ocean 200 to 250 miles from the shoreline wherever macroplankton and micronekton biomass in mesopelagic and infrapelagic layers are high, for example, especially in the upwelling areas of the Peru current. During periods of maximum abundance—for example between 1934 and 1937 and again in 1974—juveniles have been observed in California current upwelling waters. In the western tropical regions of the Pacific Ocean and the Australasian Seas—Sulu Sea, Flores Sea, Banda Sea—the more oceanic species are represented by the rarely abundant *Stenoteuthis oualaniensis*; larger concentrations of *Ommastrephes bartrami* have been observed off Idzu Archipelago and Norfolk Island (PARIN and NESIS, in VINOGRADOV, 1977).

(e) Deep-sea Plankton of the Pacific Ocean

Since I have largely adopted the vertical zonation in the pelagial (p. 11) proposed by BIRSHTEIN and co-authors (1954) for the Kuril-Kamchatka Trench, I can now develop and specify this with regard to the Pacific Ocean. In my opinion, the deep-sea plankton truly begins with the infrapelagic zone (200–250 to 600–650 m). However, the presence of vertical migrants in the epi- and mesopelagic zones requires that their role in infrapelagic assemblages be taken into account, especially the 'upper interzonal' zooplankters (VINOGRADOV, 1977) which perform large-range seasonal migrations, sometimes down to 2000 m.

As mentioned above (p. 207) the three 'upper interzonal' and herbivorous copepods, *Calanus cristatus*, *C. plumchrus* and *Eucalanus bungii*, strongly predominate (82·8%) in the total biomass of the epiplankton of the North Pacific Ocean assemblage: next in rank are chaetognaths (mainly *Sagitta elegans* and related forms), euphausiids and amphipods with 8·7, 3·6 and 2·2%, respectively. In the mesopelagic zone, populations of *C. cristatus* and *C. plumchrus* abruptly decrease to 40 mg m^{-3} and represent only 40% of the total biomass; thus, chaetognaths which maintain approximately the same biomass as in the epiplankton assemblage rise to 45% of the total, while euphausiids and amphipods display only a slight increase.

In the infrapelagic zone (200–250 to 600–650 m)—which corresponds to intermediate waters of higher temperature than mesopelagic waters—the three 'upper interzonal' copepod species become abundant again (60–76% of the total plankton biomass), whereas chaetognaths decrease to 10 to 15%; a slight increase and decrease in biomass occur in this zone for amphipods.

In his 1977 review VINOGRADOV distinguished some changes in plankton composition between the layers 300 to 500 and 500 to 1000 m. The former is inhabited by less abundant, large-sized bathypelagic species, represented mainly by juveniles: the shrimps *Hymenodora frontalis* and *Gennadas borealis*, the hyperiids *Primno macropa*, *Scina borealis* and *S. incerta*, sometimes the mysid *Eucopia grimaldii*, the hydromedusa *Crossota rufobrunnea* and by rather abundant fishes such as *Gonostoma gracile*, *Leuroglossus stilbius*, *Chauliodus macouni*, etc. According to VINOGRADOV, the 500- to 1000-m water column, which approximately corresponds to the lowest layers of cold-temperate waters, is mainly characterized by a decrease in the populations of the three herbivorous 'upper interzonal' copepod species; chaetognaths (small individuals of *Eukrohnia hamata*) represent here approximately 12 to 15% of the total zooplankton biomass; next in rank are mysids (3·8–5·7% of the total zooplankton biomass) and shrimps (1·2–1·6%); amphipods are rather poorly represented.

For the Kuril-Kamchatka area, VINOGRADOV (1977) limits the bathypelagic zone to the range 1000 to 3000 m, while I have proposed the range 2000 to 2500 m (p. 12) for the whole World Ocean. However, the characteristic species mentioned by VINO-GRADOV are almost the same as mine. With regard to assemblage composition, the most striking feature of this bathypelagic zone is the important total biomass percentage of carnivorous species, mainly chaetognaths (30–43% in the 1500- to 3000-m layer; average: 29% in the 1000- to 3000-m layer), decapods and mysids (approximate average percentages: 7% and 4% respectively). Fishes—for example *Cyclothone atraria*, melamphaeids, small macrurids—increase in total biomass percentage from 1% in the 1000- to 1500-m layer to 3% in the 2000- to 2500-m layer; copepods exhibit an abrupt change between 1000 to 1500 m (65·6%) and 1500 to 2000 m (48·9%) and attain only 32 to 33% in the 2000- to 3000-m layer. In the whole bathypelagic zone of this region, the total biomass does not exceed 20 to 30 mg m^{-3}, and carnivorous species are always more abundant than detrivores and herbivores, the latter being mainly represented by interzonal filter feeders.

According to VINOGRADOV (1977), in the abyssopelagic zone (below 3000 m), planktivorous fishes largely disappear except in the 3000- to 4000-m layer of the zone; medusae, siphonophores, chaetognaths and decapod crustaceans also disappear below 4000 while amphipods become abundant (19·3% of the total biomass) in the 5000- to 6000-m layer. The sharp decrease in total biomass which characterizes the abyssopelagic zone in comparison with the bathypelagic zone is mainly related to the decrease in copepod biomass due to: (i) the decrease with depth in populations of copepods endemic from 0·78 mg m^{-3} (58·0% of total biomass) in 3000 to 4000 m to 0·18 mg m^{-3} (28·4%) in the 5000- to 6000-m layer. VINOGRADOV further emphasized the amazingly high abundance (20–30% of the total biomass) in 4000 to 6000 m of mysids (e.g. *Boreomysis incisa*) whose stomach is filled with large-sized diatoms and tintinnids sunk from the upper water layers.

Below 6000 m—in the hadopelagic (ultra-abyssal) zone—the total biomass is 0·02 to 0·03 mg m^{-3}. Copepods always predominate (25–27%); next rank amphipods (10–15%) and polychaetes (average 4·7%). Mysids, which still represent 6·9% of the total biomass in 6000 to 7000 m largely disappear below; medusae, euphausiids and decapods seem to be absent. Among the gammarid amphipods, the species *Paralicella tenuipes*, *Parargissa arquata* and *Vitjaziana gurjanovae* were recorded from abyssopelagic and hadopelagic zones, whereas *Scopelocheirus schellenbergi*, *Hirondellea gigas* and *Halice quarta* occurred only in the hadopelagic zone.

In the Kuril-Kamchatka Trench some hadoplankton species are endemic and gammarid amphipods predominate: *Vitjazania guzjanovae*, *Hirondellea gigas*, *Hyperiopsis laticarpa*. Some of these species also inhabit the Japan and Idzu Bonin Trenches which extend beyond the Kuril-Kamchatka Trench towards the south. However, presumably due to depths of only 4000 to 5000 m between the above-mentioned trenches and the Aleutian Trench, the latter harbours different, if related, species.

In the deep-sea plankton of the Kuril-Kamchatka Trench, carnivores tend, as usual, to predominate in the plankton assemblages, but each group or species (e.g. medusae, chaetognaths and decapods) at different depths. The smaller the food resources, the more severe the exclusion of species with similar ecological requirements and, in turn, the more conspicuous the alternation of layers where a given species or group of closely-related species markedly predominated.

Taking the Kuril-Kamchatka Trench as an example of high latitude regions—'subpolar' *sensu* VINOGRADOV (1977)—one can conclude that surface and subsurface waters (0 to 500–750 m) exhibit large changes in mesozooplankton biomass mainly due to the seasonal rising of 'interzonal species' to upper layers where they feed intensively on phytoplankton and, therefore, accumulate fatty reserves (Fig. 6-20). In intermediate

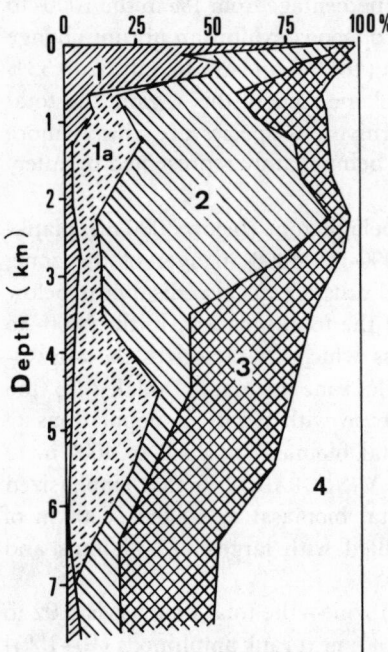

Fig. 6-20: Changes in percentage of total zooplankton for different trophic groups in the north-west areas of the Pacific Ocean. 1: filter feeders, both herbivorous and detritivorous; la: filter feeders which gather fatty reserves in subsurface waters but do not feed in deep layers; 2: carnivores (i.e. predators and scavengers); 3: euryphagous species; 4: radiolarians and other groups. (Based on information provided by VINOGRADOV, 1968.)

depths (about 1000 to 2500 m), 'upper interzonal species' diminish somewhat but still remain relatively high, whereas autochthonous carnivores (both meso- and macroplank-ters) increase. At the upper boundary of the abyssopelagic zone—i.e. below 3000 m—a new sharp decrease in biomass occurs; the carnivores decrease and practically disappear in the deepest abyssopelagic layers where the mesozooplankton biomass falls below 0.5 to 0.25 mg m^{-3} (Fig. 6-21). In these layers and in the hadopelagic zone, small-sized detritivorous species and mixed feeders predominate (VINOGRADOV, 1977). Macro-plankters do not play a significant role in the deep-sea pelagial. In subsurface and intermediate layers—down to 2000 m and sometimes below—herbivorous 'upper inter-zonal species' are mainly preyed upon by mesoplanktonic non-migrant predators. e.g.

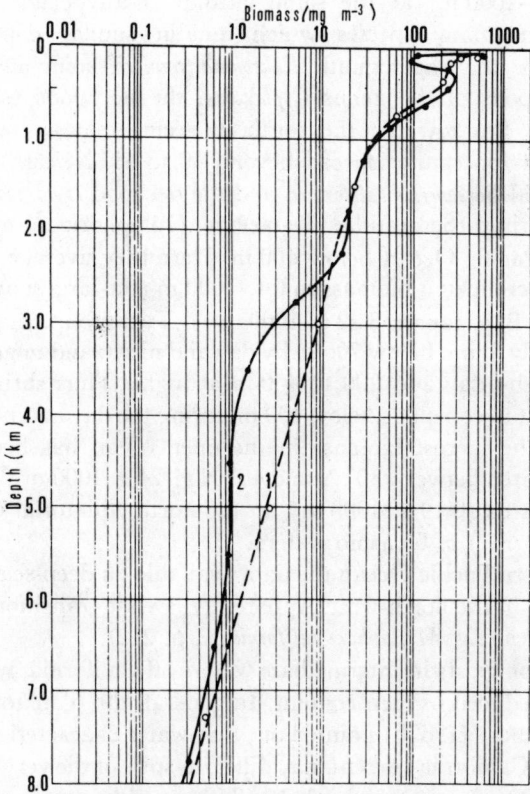

Fig. 6-21: Changes in mesozooplankton biomass (mg m^{-3}) with depth in the Kuril-Kamchatka Trench in spring 1953. 1: average from six stations and summer 1966: 2: average from nine stations. (Based on information provided by VINOGRADOV, 1968.)

carnivorous copepods and the small-sized chaetognath *Eukrohnia fowleri*. Hence, macroplankton and micronekton (mainly decapod crustaceans and fishes) abundance in these layers is less than in tropical regions. In these 'subpolar' regions maximal macroplankton and micronekton biomasses occur in 750 to 1000 m (3·4 mg m^{-3}) and 1000 to 1500 m (5·7 mg m^{-3}), but the percentages of the total zooplankton biomass are only 4·4% and 5·7%, respectively. In 2000 to 3000 m, macroplankton biomass is only 1·3 mg m^{-3} but represents 11·8% of the total zooplankton biomass.

An interesting aspect of the deep-sea plankton in the north-western Pacific Ocean is the sinking of some subarctic and/or boreo-arctic zooplankters in the vicinity of the polar front. At 31° N, the boreo-arctic copepods *Calanus cristatus* and *Eucalanus bungii* may occur between 500 and 1000 m. This is not surprising since they are 'interzonal species'. However, one may wonder whether such low latitude populations should not be considered merely expatriated and thus unable to migrate up to the epipelagic layer. A little deeper (1000–4000 m) at the same latitude, bathypelagic and abyssopelagic assemblages contain many species which inhabit similar depths in the Kuril-Kamchatka Trench: the chaetognath *Eukrohnia fowleri*, some mysids of the genus *Eucopia* and amphipods of the genus *Cyphocaris*, the decapods *Gennadas borealis* and *Hymenodora glacialis*. But towards the south, it seems that a portion of the boreo-arctic deep-sea plankton cannot spread beyond 40 to 45° N, e.g. the medusa *Crossota rufobrunnea*, the mysid *Boreomysis californica* and the decapod *Hymenodora frontalis*.

Micronektonic shrimp (Sergestidae, Penaeidae, Oplophoridae) may be abundant in the north-western Pacific Ocean below 400 m. Here they average about 10% of total zooplankton and micronekton biomass in 0 to 1000 m (the largest average biomass was recorded in Sagami Bay, Japan: $3 \cdot 22$ g. 1000 m^{-3}). Two groups may be distinguished: non-migrants, usually found below 700 m by day and night; and migrants inhabiting the layer 400 to 700 m by day and 200 to 400 m at night. Thus shrimps were the most abundant group next to copepods below 400 m during the day and below 200 m at night (AIZAWA, 1974). The largest biomass in the open ocean has been recorded in the northern Kuroshio area between 30° N and 40° N ($1 \cdot 24$ g. 1000 m^{-3}); the biomass tends to decrease southwards ($0 \cdot 19$ g. 1000 m^{-3} in the north equatorial current) and north-wards ($0 \cdot 44$ g. 1000 m^{-3}) in Oyashio waters.

In the north-eastern Pacific Ocean the data available on deep-sea plankton are much more scattered than in the north-west. Apparently, a wide transition zone prevails with some endemic species (e.g. *Myctophum californiense*, p. 212).

A singular situation exists in intermediate waters off California, which are character-ized by a strongly reduced oxygen content. In general, the plankton of these waters is scanty, but LONGHURST (1967b) pointed out that water characteristics near the south-ern boundary of the California current may differ; in spite of the very low oxygen content ($0 \cdot 5$–$0 \cdot 2$ ml l^{-1} at 400 m) zooplankters below 150 m—although qualitatively poor—are quantitatively rich with a predominance of Copepodite V stages of *Calanus helgolandicus*. This layer may constitute a winter refuge for the species, because its lower temperature reduces the metabolic rate and because there exists a minimum horizontal advection with a net transport towards north-north-east. These local conditions are most con-venient for optimum utilization of phytoplankton blooms (mainly *Coscinodiscus* species). While adult copepods are grazing on phytoplankton in the euphotic layer, the California current progressively transports them towards lower latitudes. However, at the same time the copepods sink deeper and finally reach the undercurrent which carries them back—during the winter resting stage—to their original surface waters in northern areas; here they find again ample food during the subsequent phytoplankton bloom. Some epiplanktonic species with circadian migration can tolerate large changes in oxygen content: from $5 \cdot 0$ ml l^{-1} in surface waters at night to 0.2 ml l^{-1} in intermediate waters during the day.

The deep-sea plankton assemblages in the intertropical and South Pacific Ocean are insufficiently investigated. There are some examples of equatorial submergence, e.g. the chaetognath *Eukrohnia hamata* (ALVARIÑO, 1964). As regards vertical changes in the mesozooplankton biomass in intertropical oceanic areas, VINOGRADOV (1977) sug-gested that the two most marked discontinuities occur at depths of about 100 to 200 m and around 1500 to 2500 m. From a quantitative point of view, VINOGRADOV distinguished surface layers down to 100 to 200 m, and 'intermediate' layers between 100–200 to 1500–2500 m, and abyssal—as well as hadal—layers below 1500–2500 m. In the intermediate layers the mesozooplankton exhibits only slight changes. The percentage of 'upper interzonal species' of the zooplankton biomass is smaller than at higher latitudes, and the vertical range of their migrations—partly related to physical discontinuities—is also smaller. In other words, the changes in mesozooplankton biomass occurring at the boundaries between VINOGRADOV's three zones are primarily controlled by biological factors, and the changes in composition, structure and biomass of the assemblage depend essentially on the trophic resources available in each of the vertical zones.

As regards the vertical distribution of macroplankton and micronekton VINOGRADOV (1977) distinguished two different patterns in the oceanic intertropical areas of the Pacific Ocean in relation to more or less important productivity of surface waters. (Incidentally, the same pattern exists in intertropical areas of the Indian Ocean.) In the oligotrophic tropical regions, the maximum of macroplankton biomass occurs in 500 to 1000 m; macroplankton abundance is lower above 200 to 500 m and below 1000 to 2000 m, and practically non-existent in 0 to 200 m. Even in 500 to 1000 m, macroplankton and micronekton biomasses are low and represents $\frac{1}{3}$ and $\frac{1}{4}$ of the mesoplankton and total zooplankton biomasses, respectively.

In the eutrophic equatorial areas the macroplankton and micronekton biomasses approximately account for 30% and 52% of the mesozooplankton biomass in 200 to 500 m and 500 to 1000 m, respectively. In the 1000- to 2000-m layer, the macroplankton and micronekton biomass is five times that of mesozooplankton, and large-sized organisms are more numerous than in the 500- to 1000-m layer. The large-sized forms, e.g. decapod crustaceans, are more abundant in the upper half of the 1000- to 2000-m layer than in the lower half; they completely disappear from trawl samplings below 2000 m.

In other words, in equatorial regions the 500- to 2000-m layer contains the bulk of large-sized carnivores—decapod crustaceans, cephalopods and fishes—which form a 'living net' or 'filter' below the productive zones that intercept most of the sinking organic material. Consequently, food resources in the deeper layers are very poor and plankters very scarce. In the tropical regions where surface waters are less productive, the layer inhabited by carnivores is much thinner and their biomass much lower than in equatorial regions (VINOGRADOV, 1977) (Fig. 6-22; Table 6-6). Therefore, in depths less than 500 to 800 m of tropical regions where vertical small-range diel migrations prevail, the 'living filter' of large-sized carnivores is located immediately below the productive layer. This results in an increase of the contribution by diel migrants to the food supply in deeper layers.

The most pronounced qualitative differences between boreo-arctic and subtropical deep-sea plankters concern medusae and several families of prawns: Hymenodoridae, Pasiphaeidae, Penaeidae. Probably the specific diversity of the subtropical deep-sea plankton fauna is the highest all over the World Ocean: Russian scientists listed as many as 47 species of mysids, 41 of Gammaridae and about 30 of isopods.

In the south-western Pacific Ocean, the zooplankton attain an abundance peak in the 500- to 1000-m layer at 25° S, possibly because of the influence of the Antarctic intermediate water. Another small abundance maximum occurs below 2000 to 3000 m; this might be related to the upper layers of the Antarctic bottom water flowing northwards.

Table 6-6

Differences in vertical distribution of mesozooplankton and macroplankton plus micronekton biomass (mg m^{-3}) in mesotrophic equatorial waters (12° N–12° S) and oligotrophic tropical and subtropical waters (40° N–12° N and 12° S–40° S) in the Pacific and Indian Oceans
(Based on information provided by VINOGRADOV, 1977)

Depth (m)	Mesozooplankton biomass		Macroplankton plus micronekton biomass			
	12° N–12° S	40° N–12° N and 12° S–40° S	12° N–12° S	Macroplankton (% of total zoo-plankton biomass) 12° N–12° S	40° N–40° S and 12° S–40° S	Macroplankton (% of total zoo-plankton biomass) 40° N–40° S and 12° S–40° S
0–50	63·5	27·6	0	0	0	0
50–100	52·3	25·9	15·2	22·5	0	0
100–200	18·8	14·7	0·3	1·6	0	0
200–500	7·8	6·8	2·2	22·3	0·8	10·5
500–1000	5·2	4·8	2·7	34·2	1·6	25·0
1000–2000	1·2	1·8	5·9	74·0	0·3	14·2
2000–4000	0·23	0·40	0·02	8·0	0	0

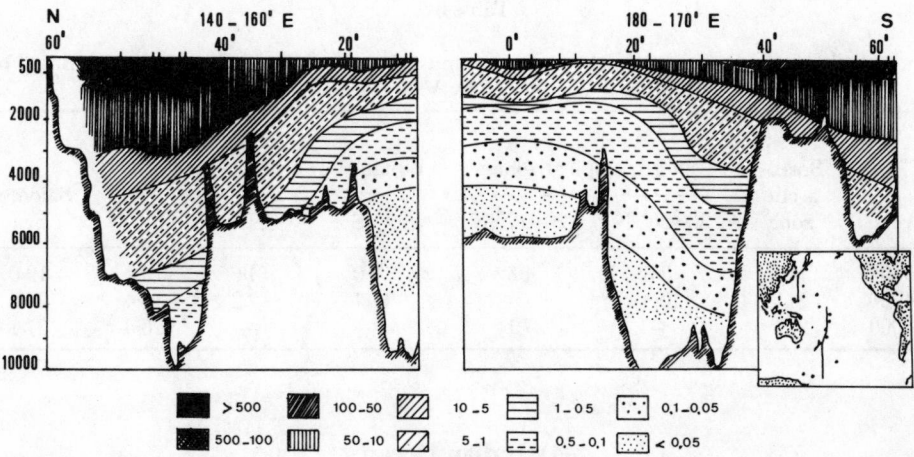

Fig. 6-22: Vertical distribution of zooplankton biomass (mg m^{-3}) in the North Pacific Ocean from 60 to 10° N along a transect 140 to 160° E (left) and in the South Pacific Ocean and Southern Ocean from 5° N to 62° S along a transect 170° to 180° W (Based on information provided by VINOGRADOV, 1968.)

In the latter water body, the Southern Ocean deep-sea amphipods *Hirondellea antarctica*, *Korogo megalops* and *Orchomenella abyssorum* have been collected (VINOGRADOV, 1968).

Of course, the hadoplankton of low-latitude trenches, which are very impoverished, exhibit a strong endemism. VINOGRADOV (1968) pointed out some interesting quantitative relationships between biomasses in surface and subsurface waters in deep layers at different areas. In trenches of rather high latitudes, biomass decrease with increasing depth is slower at intermediate depths (Kuril-Kamachatka 500 to 1000 m, Kermadec, 1000 to 2000 m); this may possibly be ascribed to an input of plankton (advection process) from higher latitudes. Below 3000 m, the curves of biomass versus depth are almost parallel in these two high-latitude trenches and those at low latitudes (MARIANAS and BOUGAINVILLE, Fig. 6-6).

Fig. 6-21 and Table 6-7 reveal a sharp decrease in plankton biomass in the whole water column between 20° N and 20° S with only a very slight increase at the equator; this may be related to the increase in primary production in equatorial divergence areas.

In the Sea of Japan, as in other mediterranean seas, the deep-sea plankton biomass is much less than at corresponding depths of adjacent ocean areas. Whereas the abundance of copepods is of the same order of magnitude in both the Sea of Japan and the ocean from 1500 m down to 3300 m (deepest sampling in the Sea of Japan), euphausiids disappear in the Sea of Japan at about 850 to 900 m, but at 1500 m in the neighbouring ocean; specialized predators (plankton feeders) do not exist in the greatest depths of the Sea of Japan.

Table 6-7

Total plankton (g wet weight) in the water column below 1 m^2 in the Pacific Ocean (Based on information provided by VINOGRADOV, 1960)

Layers	Boreo-arctic zone	North tropical zone		Equatorial zone	South tropical zone		Salomon Sea
		30–20° N	20–12° N	4° S	24° S	30° S	
0–500	140	8·6	4·6	8·0	4·6	10·4	19·0
0–4000	211	20·5	7	11·1	7·2	19·4	26·9
0–8000	217	—	7·1	—	—	20·3	27·1

(4) Indian Ocean

Less extensive than the Pacific and Atlantic Oceans, the Indian Ocean exhibits two striking singularities: (i) it may be considered as 'a half ocean' because it does not stretch northwards beyond 23° N; and (ii) the alternating monsoon winds result in seasonal changes in currents and upwelling presence (or intensity) on several coastal areas, thus modifying physical and chemical water characteristics.

(a) Phytoplankton

Offshore Phytoplankton

As regards its specific composition, the Indian Ocean offshore net phytoplankton displays a predominance of dinoflagellates (55 species), followed by diatoms (35 species); cyanophytes plus other flagellates total more than 100 species. Four species exist everywhere: *Pyrocystis pseudonoctiluca*, *Ceratium carriense*, *C. trichoceros* and *C. massiliense*; they constitute the 'fundamental dinoflagellate quartette' (SUKHANOVA, 1962). However, some species exhibit a more restricted distribution and may be found only in the upper layer of the two main surface and subsurface water masses of the Indian Ocean; for example, several species of the genus *Rhizosolenia* occur only in the regions occupied by the central Indian water mass; the dinoflagellates *Ceratium deflexum*, *Amphisolenia bidentata*, *A. trinax* and the diatom *Planktoniella sol* cannot be found beyond the boundaries of the equatorial Indian water mass. The total list of net phytoplankton species reveal only slight variations with respect to regions and seasons. On the contrary, very important changes in relative cell numbers of the above-mentioned main taxa may be observed in different places or periods.

KREY (1973) recognized different biogeographical units in the Indian Ocean offshore phytoplankton according to the predominance of either of the groups: (i) In central areas of the Arabian Sea and Gulf of Bengal, dinoflagellates predominate together with cyanophytes; these are followed by diatoms and coccolithophorids. (ii) In the north-eastern region of the Somali current, diatoms predominate (dinoflagellates in part), followed by cyanophytes. (iii) In the Mozambique current, diatoms predominate, followed by dino-flagellates and coccolithophorids. (iv) In the equatorial current, dinoflagellates and coccolithophorids predominate, followed by diatoms and cyanophytes. (v) In the southern subtropical gyre, dinoflagellates predominate; next in order are coccolitho-phorids and diatoms; however, there is a strong decrease in dinoflagellate abundance both southwards and eastwards, and this group is very poorly represented between 10 and 30° S and 98 and 108° E (HUMPHREY, 1960).

In order to analyse phytoplankton bottle samples collected in west Indian Ocean offshore waters, approximately between 10° N and 20° S, TORRINGTON-SMITH (1974) used an adaptation of the numerical taxonomic methods which allowed her to classify the samples into groups. In this way she separated groups and related them to the different water masses and currents: (i) equatorial subsurface water; (ii) equatorial undercurrent; (iii) south equatorial current; and (iv) tropical surface water. The author considered the group with its distribution centred in the equatorial subsurface water as the most important; this group contains 50 different species. The main close-clustering species which form the core of this floral element were found in more than 40 of the 59 samples studied in the entire sampling programme: the diatoms *Coscinodiscus lineatus*, *Chaetoceros atlanticum*, *C. danicum*, *Dactyliosolen mediterraneus*, *Thalassionema nitzschioides*, *Nitzschia bicapitata*, members of the *N. seriata* group, a species of *Navicula*; a species of *Coccolithus*, two species of silicoflagellates—*Dictyocha fibula* and *Mesocena polymorpha*—and the xanthophycean *Meringosphaera mediterranea* (TORRINGTON-SMITH, 1974, pp. 376 and 379). This equatorial subsurface group exhibited the highest values of standing crop; it is restricted to waters with a PO_4-P content equal to, or higher than $1·0$ μg at l^{-1}.

Quantitatively, data on regional or seasonal variations are more often expressed in cell number m^{-3} or l^{-1} than in biomass.

In the eutrophic regions, such as some offshore areas located south of Java and west of Australia approximately between 15 and 20° S, SUKHANOVA (1976) found in some places in the 0- to 100-m layer, 1000 to 5000 phytoplankters l^{-1}. In the Andaman Sea, cell numbers as high as 6000 l^{-1} are common; east of Socotora, between 2 to 5° N and 55 to 70° E, cell numbers l^{-1} in the 0- to 100-m layer always exceed 2000; in the Gulf of Aden, 3600 (ZERNOVA, 1962). In these areas, diatoms almost always predominate. All offshore areas mentioned above, as well as some parts of the Arabian Sea (e.g. off the south-eastern Arabian coast and the West Pakistan coast) are located off zones with coastal upwellings; therefore, this may benefit from fertility increase which upwelling generates where a seaward drift occurs.

In oligotrophic regions the average cell number l^{-1} is always below 500 and sometimes below 100, e.g. in central areas of the Gulf of Bengal and all regions occupied by the Indian central water mass, the northern limit of which is about 8 to 10° S during the south-west monsoon, and 18 to 20° S during the north-east monsoon. In these regions the four dinoflagellate species of SUKHANOVA's (1976) 'fundamental quartette' always predominate quantitatively. An intermediate situation with average cell numbers $(500-1000$ $l^{-1})$ may be observed in transitional areas between rich and poor regions and in the central parts of the Arabian Sea. In these areas, diatoms are always associated with the dinoflagellate quartette. Apart from this general scheme of offshore phytoplankton abundance, two peculiarities must be emphasized:

(i) In a given area very important differences may occur between summer and winter. For example, according to SAVICH (1968a), in the north-western parts of the Indian Ocean, dinoflagellates and cyanophytes are almost completely absent during the winter monsoon, while diatoms of the genera *Rhizosolenia*, *Chaetoceros*, *Nitzschia* and *Thalassiosira* largely predominate; in the 0- to 25-m layer, the biomass may be higher than 2 g m^{-3}, but it generally falls to between 500 and 1000 mg m^{-3}. In the Gulf of Aden, which may be considered an offshore area due to its width and depth, the highest phytoplankton biomass (>2000 mg m^{-3} on an average) occurs when the south-westerly, i.e. summer, monsoon blows and induces upwelling. SAVICH (1968b) recorded in a bloom 5400 mg m^{-3}; more than half of the total number of cells was represented by species of the genus *Chaetoceros*, and one-third by species of the genus *Lauderia*. At this time, the discontinuity layer is located at a shallow depth and the phytoplankton, concentrated in the 0- to 50-m layer, grows faster and exhibits its maximum abundance in the north-western parts of the Gulf. Diatoms and cyanophytes are the most important taxa in July and August; the highest number of dinoflagellates is observed in September. During the 'dry' monsoon, the phytoplankton biomass is lower than in late spring and summer, and its distribution is almost homogeneous in the 0- to 100-m layer due to the deeper location of the discontinuity layer (SAVICH, 1971). In other words, the depth of the productive layer exhibits seasonal changes according to the position of the upper boundary of the thermocline; this upper limit itself depends on the presence and intensity—or absence—of the upwelling process.

(ii) The vertical distribution of phytoplankton may be different according to regions and species considered. Where both the water's density gradient and stability is high, the phytoplankton biomass concentrates in the 0- to 30-m layer. Where a sharp thermal discontinuity does not exist, the vertical distribution of the plankton flora becomes more homogeneous. However, the fundamental dinoflagellate quartette and the diatom genus *Thalassiothrix* exhibit their maximum abundance in the 50- to 100-m layer.

The northern parts of the Arabian Sea deserve particular attention because they exemplify the two main sources of fertility—upwelling and land drainage—and the influence of these coastal processes on offshore areas.

According to KUZ'MENKO (1973), primary production in the surface waters of the northern Arabian Sea in August 1969 varied widely: the highest values (10–20 mg C m^{-3} d^{-1}) occurred on both sides (Pakistan shelf and Ras-el-Hadd on the eastern Arabian coast) and gradually dropped with distance from the coast down to the lowest values (0·4–0·6 mg m^{-3} d^{-1}) in the most offshore areas. The anticyclonic gyre which occurs in summer in the northern region of the Arabian Sea results in two strong upwellings: (i) On the western side, at the boundary with the Gulf of Oman east of Ras-el-Hadd, where the phytoplankton biomass is very important (2400 mg m^{-3}) and diatoms largely predominate, especially with species of the genera *Rhizosolenia*, *Bacteriastrum* and *Nitzschia* (Fig. 6-23). The Arabian upwelling zone is wider than most others in the World Ocean because the south-west monsoon is particularly strong, blowing nearly parallel to the coast and increasing in intensity with distance offshore; this results in a combination of coastal upwelling and open-ocean upwelling (SMITH and BOTTERO, 1977). (ii) On the eastern side, i.e. on the Pakistan shelf from Cape Sahid to Karachi, another upwell-

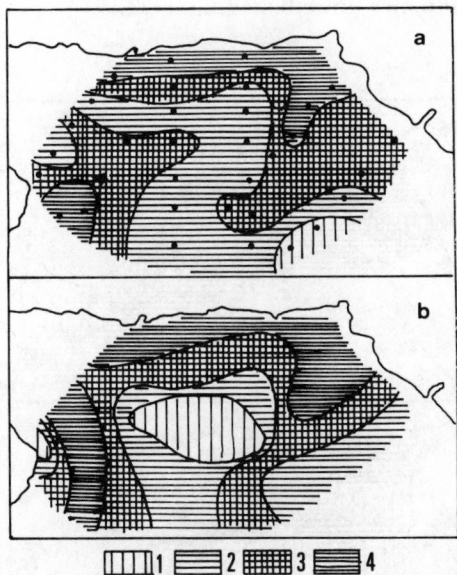

Fig. 6-23: Distribution of primary production
(a) and phytoplankton biomass (b) in the
northern Arabian Sea; August 1969. (a): 1,
<1; 2, 1–5; 3, 5–10; 4, >10 (mg C m^{-3} d^{-1})
(b): 1, <100; 2, 100–500; 3, 500–1000; 4,
>1000 mg m^{-3}. (After KUZ'MENKO, 1973;
reproduced by permission of American
Geophysical Union, Washington, D.C.)

ing exists; here the primary production averaged 13·8 mg C m^{-3} d^{-1} and the average
biomass 1660 mg m^{-3}; *Rhizosolenia* species always predominated together with some
species of the genera *Nitzschia*, *Chaetoceros* and *Thalassiothrix*. As on the north-eastern
Arabian coast, phytoplankton and photosynthesis rate decrease towards the central and
southern regions, probably due to subsidence of surface waters.

In the transition period between summer and winter monsoons (September–
November), although water circulation remains generally the same, both primary
production and phytoplankton biomass decrease on the western side and various
species of dinoflagellates represent a more important part of the phytoplankton assemblage. On the contrary, a different situation occurs on the Pakistan shelf, due to discharge
from the Indus River and general land drainage following the summer rainy season;
the highest primary production (50–60 mg C m^{-3} d^{-1}) and the highest biomass
(6766 mg m^{-3}—3 billion cells m^{-3}) occur in the most inshore areas; the most abundant
phytoplankters are the diatoms *Skeletonema costatum*, *Nitzschia* spp., and *Rhizosolenia* spp.,
but the dinoflagellate *Gyrodinium fusiforme* is also abundant. However, primary produc-

tion rapidly decreases both southwards and towards the open sea. The fertilizing influence of fresh-water input in autumn is thus less in distant-neritic and, especially true
offshore areas, than in summer upwellings (Fig. 6-24).

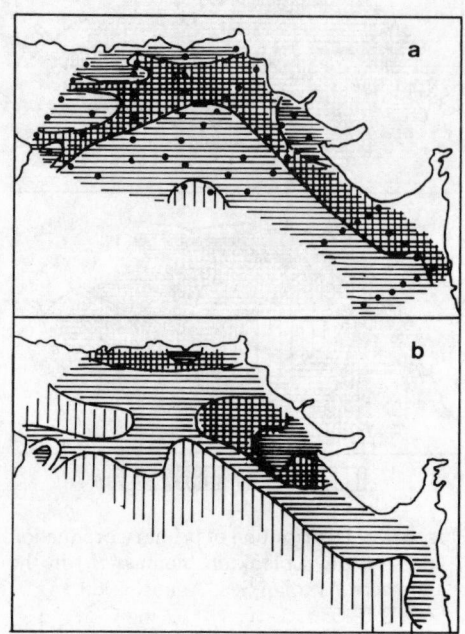

Fig. 6-24: Distribution of primary production
(a) and phytoplankton biomass (b) in the
northern   Arabian   Sea;   September–
November 1969. Same symbols as in
Fig.   6-23.   (After   KUZ'MENKO,   1973;
reproduced by permission of American
Geophysical Union, Washington, D.C.)

In the whole area studied by KUZ'MENKO (1973) maximum photosynthesis occurs
near the surface; the lower boundary of the euphotic layer seems to be 10- to 15-m deep
at the coastal stations and not deeper than 20 to 30 m in the open sea. KABANOVA
(1968) suggests that the differences in the primary productivity rate in different areas of
the Arabian Sea mainly arises from differences in the renewal rate of inorganic nitrogen
originating from subsurface or intermediate waters; he assumes productivity control by
phospate content to be less important. However, KUZ'MENKO (1973) found very high
PO_4–P values both in the summer upwelling off Ras-el-Hadd (30–40 mg m^{-3}) and in
the autumn Indus River discharge (more than 59 mg m^{-3}).

In a more recent paper KUZ'MENKO (1977) took into account many additional data
and presented an even more complete view on phytoplankton distribution in the Arabian Sea. The most striking feature is possibly the fact that the phytoplankton fluctuates

within very wide limits both in abundance—from 2 to 700×10^6 cells m^{-3}—and in biomass, from 8 to 1600 mg m^{-3} (averages 0–100 m). During the winter monsoon, the highest abundance and biomass (360×10^6 cells m^{-3} and 1566 mg m^{-3}, respectively) were recorded from central areas of the Arabian Sea; the poorest regions were those of the Hindustan shelf. During the summer monsoon, the amount of phytoplankton was lower by a factor of 1·5 to 2 than in winter. Diatom biomass was 2·5 times higher in winter than in summer, but cyanophyte biomass was ten times higher in summer than in winter. In summer and autumn the phytoplankton maximum occurred in waters of the Indo-Pakistan shelf with up to 100×10^6 cells m^{-3} and 1186 mg m^{-3} in the 0- to 25-m layer. KUZ'MENKO (1977, p. 74) also noticed that

'comparison of data on the concentration of phospates in the waters of the Arabian Sea with the amount of phytoplankton in the upper 100 m layer in various seasons of the year, demonstrated the direct relation between them.'

The latter remark agrees with the relationship found by TORRINGTON-SMITH (1974) in offshore waters.

The higher primary productivity in the whole Arabian Sea probably arises from regional advection processes bringing nutrients up to the euphotic layer. In other words, the upwelling in the peripheral regions of the Arabian Sea itself would counterbalance the water loss related to the Arabian Sea deep water which sinks deeper and deeper as it flows towards south and south-east.

Coastal Phytoplankton

Because of the influence exerted by some coastal processes on the open-sea phyto-plankton assemblages and biomass, some data on coastal phytoplankton have been dealt with in the preceding section, e.g. for the north-eastern Arabian coast and for the Pakistan shelf. On many other shores of the Indian Ocean one may observe seasonal blooms induced by seasonal upwelling. For example, on the south-eastern Arabian coast where upwelling occurs in summer, chlorophyll a values tended to be higher close to the coast, but decreased seawards more rapidly than the PO$_4$–P content. The zooplankton biomass followed a similar pattern. Neither chlorophyll a nor zooplankton display an exceptional abundance. Within a solitary bloom of *Gonyaulax polyedra*, cell numbers of the order of 10^7 l^{-1} have been observed offshore. *Streptotheca* sp. was one of the most common diatoms in the coastal belt with other chain-forming species of the genera *Chaetoceros* and *Fragilaria*. Another example is the Somalia coast where upwelling also exists around the latitude of Ras Hafun. However, the south-east to north-west extension of upwelled fertile waters is not as wide as on the south-eastern Arabian coast. Maximum chlorophyll a values are accompanied by an increase in zooplankton abundance (CURRIE and co-authors, 1973).

Only the neritic province of the Indian peninsula has been sufficiently investigated at length, mainly on its western coast. As usual, the specific richness of the flora is higher than offshore; SUBRAHMANYAN and VISWANATHA-SARMA (1960) listed 226 diatoms, 120 dinoflagellates and seven cyanophytes for all Indian coasts. Seasonal changes in the species list seem to be rather small, but most of the species listed are uncommon. In fact, there are only 21 diatoms, 7 dinoflagellates and one cyanophyte which really participate in the blooms. Diatoms always seem to predominate, at least in percentage of biomass, but sometimes cyanophytes may be more abundant than diatoms. However, the role of nanoplankters in the primary production of the most inshore areas may have been overlooked. Since restricted light penetration generally tends to nanoplankton production, it may be assumed that nanoplankton contributes more than net plankton to the total primary production in highly turbid coastal areas, especially during the rainy season.

Two periods of phytoplankton outburst have been recognized on western Indian coasts. The most important one occurs during the summer monsoon. It might be related to vertical mixing processes induced by winds; the mineralization of the abundant organic material deposited on the shallow-water muddy bottom is the main source of nitrates and phospates in these fertile waters. The less important autumn bloom, which occurs especially in northern areas, seems to be induced by the great amount of nutrients from river outflow and land drainage (see also Pakistan shelf, p. 245). According to QASIM and co-authors (1972) the photosynthetic activity of many phytoplankton species in Indian neritic areas is greatly enhanced by a salinity decrease; the authors believe this to be an adaptation which allows the algae to benefit optimally from nutrient input caused by rivers and land drainage. MOVCHAN (1971) studied the seston content in the neritic province of the western Indian coast. He reports two distinct divisions at the latitude of Bombay; to the north of this latitude seston content is high (up to $1000-2000 \ mm^{-3} \ m^{-3}$), to the south seston rarely amounts to more than $500 \ mm^{-3} \ m^{-3}$.

Biological productivity on the western Indian coast, in the 0- to 50-m layer between latitudes $17° 30' \ N$ and $08° \ N$, has been investigated by QASIM and co-authors (1978) from March 23 to April 1, i.e. during the period between the end of the winter monsoon and the beginning of the summer monsoon. The total production varied from 0·85 to 6·75, that of nanoplankton from 0·8 to 2·92 $mg \ C \ m^{-3} \ h^{-1}$, respectively. At stations with a production less than $2 \ mg \ C \ m^{-3} \ h^{-1}$, the nanoplankton contributed 78 to 99%. There was a major production peak at $11° \ N$ and a less marked one at $15° \ N$.

In general, fertility on the eastern coast of the Indian peninsula seems to be lower than on the western coast (SUBRAHMANYAN, 1971) but appears to increase in summer towards the north. According to RADHAKRISHNA and co-authors (1978a), who carried out primary production measurements in the Bay of Bengal between $80°$ and $92° 20' \ E$ and $10°$ and $20° 40' \ N$ during the summer monsoon, the highest average value ($315 \ mg \ C \ m^{-2} \ d^{-1}$) was observed above the slope. In offshore waters with PO_4-P contents lower than in slope waters, the primary production averaged $220 \ mg \ C \ m^{-2} \ d^{-1}$; although shelf waters have the highest PO_4-P content, they are less productive ($295 \ mg \ C \ m^{-2} \ d^{-1}$) than slope waters; this may be attributed to their higher turbidity which restricts photosynthesis to shallower depths. Data from various authors indicate that in spring, i.e. prior to the rainy season, primary production in slope and offshore areas does not exhibit marked changes in comparison with the summer season. On the contrary, in shelf waters which are more transparent during the dry season, winter production is approximately three to five times higher than in summer.

Aberrant N/P ratios or unusual amounts of growth factors frequently occur on Indian coasts and may induce discoloured waters, featuring for example the chloromonadin *Hornellia viridis* or *Noctiluca* sp., the latter harbouring the symbiotic green flagellate

Proteuglaena noctilucae. Noctiluca miliaris red tides, which often induce severe mortalities in plankton and benthos assemblages are not similar to those mentioned above, because *N. miliaris* is phagotrophic feeding on diatoms, and hence must be considered a zooplankter. Heavy discoloured waters with the blue-green alga *Trichodesmium erythraeum* (up to 600,000 filaments ml^{-1}) have also been frequently observed, mainly during March and April on the western coast of India. The blooms are 0·5 to 1·0 m thick with maximum concentrations near the surface. They constitute streaks extending 40 to 50 miles containing the cyanophyte exclusively. Such blooms last for four to nine days (RAMAMURTHY, 1972).

According to MARKINA (1971) large biomasses of neritic phytoplankters may also occur in the Great Australian Bay, the waters of which are very rich in nutrients. SUKHANOVA (1976) recorded phytoplankton abundances exceeding 10,000 cells l^{-1} within a narrow belt close to the Northwest Cape, and between 5000 and 10,000 more offshore. The latter abundances may be also observed south of Java. In all these areas diatoms predominate. A general synopsis on the fertility of the Indian Ocean has been presented by KREY (1973) (Fig. 6-25). His figure shows a rather good approximation of the distribution of primary production, at least with regard to average values. Some information on maximum and minimum values can be found in the preceding pages.

(b) Zooplankton

In the Indian Ocean distant-neritic or oceanic zooplankton—as well as micronekton and nekton—are abundant only off areas with seasonal coastal upwelling, e.g. the coasts of Arabia and Somalia, and to some extent off western Australia and in the area of the Java Dome.

Surface and Subsurface Offshore Zooplankton

The most comprehensive and synthetic study of the Indian Ocean surface and subsurface offshore zooplankton remains that by VINOGRADOV and VORONINA (1962a). These authors have investigated the winter distribution of 40 species in the 0- to 500-m layer: 30 copepods, 1 cladoceran, 1 ostracod, 8 pteropods. They consider these 40 species a homogeneous zooplankton assemblage, which includes other species and is widespread in the whole Indian Ocean up to the subtropical convergence—its southern limit. In spite of the qualitative uniformity of this zooplankton assemblage, important differences may occur in the relative abundance of species pertaining to different trophic groups. In the most productive areas (divergence and upwelling), 'coarse' filter-feeding copepods predominate—such as species of *Eucalanus, Rhincalanus* etc.,—whereas carnivores never represent more than 25 to 35% of the total biomass. The specific diversity index is low. In oligotrophic regions, 'fine' filter feeders make up 25 to 35% of the biomass, but carnivores (copepods of the genus *Euchaeta*, together with macroplanktonic and micronektonic forms) sometimes account for 45% of the total biomass. The specific diversity index is high. In areas with features intermediate between those mentioned above, for instance in the presence of a weaker divergence, thin filter feeders always predominate.

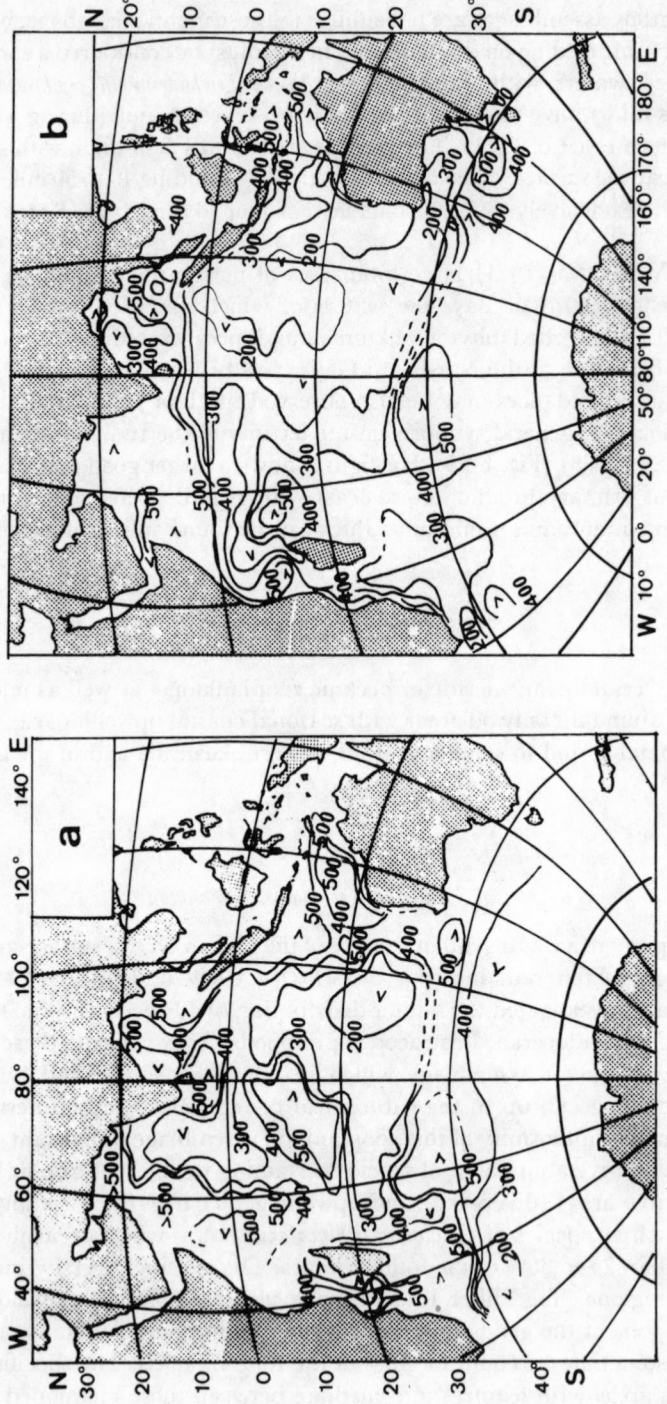

Fig. 6-25: *In situ* primary production (mg C m^{-2} d^{-1}) during SW (a) and NE (b) monsoon. (After KREY, 1973; reproduced by permission of Springer-Verlag, Heidelberg.)

During winter monsoon, VINOGRADOV and VORONINA (1962a) observed a shifting of successive links of the trophic chain from divergence areas, due to the meridian component of equatorial currents. In Fig. 6-26 the areas of maximal abundance in phyto-

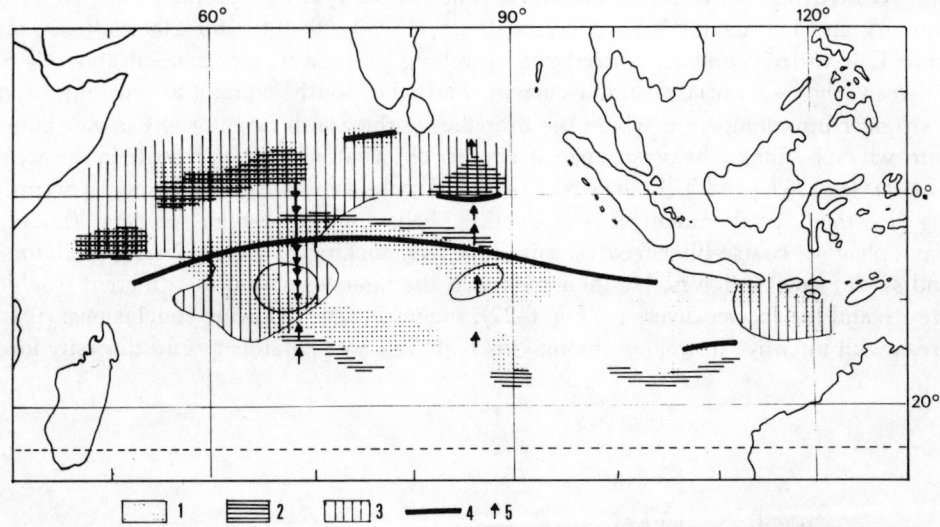

Fig. 6-26: Distribution of maximum abundance of three steps of the trophic pyramid in the equatorial Indian Ocean. 1: herbivorous copepods; 2: predatory copepods; 3: macroplankton; 4: areas of divergence; 5: meridian components of latitudinal currents at 300 m depth. (Based on information provided by VINOGRADOV and VORONINA, 1962a.)

plankton and herbivorous copepods correspond to those where upheaval is most pronounced (lower surface temperature and higher PO_4–P content at 100 m). Examples are the divergence between south equatorial current and equatorial counter-current, the divergence to the west of the Maldive Islands, the cyclonic gyre around Chagos Archipelago and the South Java upwelling. Areas of maximum abundance of the most common carnivorous copepods (*Euchaeta marina* and *E. concinna*) which feed mainly on small herbivores are shifted 60 to 160 miles away from divergence areas; such distance corresponds approximately to drifting over 55 to 65 days. The belt with the maximum abundance of macrozooplankton and micronekton (euphausiids, decapod crustaceans, small fishes) is more distant from the upheaval than is the small-carnivore zone and also wider. Along the convergence line which may exist at some distance of an upheaval zone, the composition of the accumulated plankton assemblage obviously depends on the distance between divergence and convergence.

Such a disjunction of successive links of the trophic chain partly arises from different durations in the average life-cycle and generation time of species participating in each link: a few days for phytoplankters, a few weeks for copepods, longer times for larger macrozooplankters and micronekton. However, it probably also depends on the gradual decrease in the food resources available for each trophic link as the drifting proceeds. Below a given threshold, food shortage causes decreases in population growth while the population concerned begins to be more and more intensively preyed upon by organisms of the next link whose population strength tends to increase.

TIMONIN (1969, 1971, 1973) proposed a more subtle interpretation of the disjunction of successive links of the trophic chain. He collected material on a profile among 76° E in January and February, 1960, in layers (0–25, 25–50, 50–100, 100–200, 200–500 m) from 1° N to 16° S and recognized two upwelling zones: a weaker zone at about 4° S, between the equatorial counter-current and the south equatorial current, and a stronger upwelling zone 'where the branches of the South Equatorial Current bifurcate' which is 'due to the persistence of a center of low atmospheric pressure in the area' (TIMONIN, 1969, p. 687). TIMONIN (1973) tried to discriminate different species according to their predominant food-collecting behaviour: fine-filter feeders (mainly phytophages), coarse-filter feeders, mixed feeders, sucking predators, biting predators, and swallowing predators. He then estimated the biomass within the different trophic groups and the species diversity (Fig. 6-27), suggesting the following conclusions: (i) In areas with intensive upwelling, biomass is high and zooplankton species diversity low,

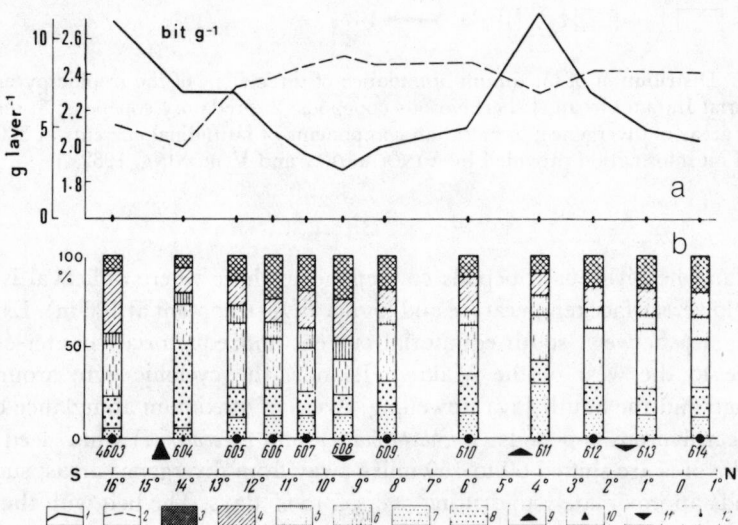

Fig. 6-27: Distribution of zooplankton biomass (solid line) and trophic diversity (broken line) and ratio of trophic groups (b) on a profile between 16° S and 02° N in the Indian Ocean. 1: zooplankton biomass; 2: trophic diversity (H_{tr}); 3: swallowing predators; 4: biting predators; 5: sucking predators; 6: organisms with mixed-feeding type; 7: fine-filter feeders; 8: coarse-filter feeders; 9: weak upwelling; 10: strong upwelling; 11: slight sinking; 12: stable water stratification. (After TIMONIN, 1973; reproduced by permission of American Geophysical Union, Washington, D.C.)

due to increase in total zooplankton quantity caused by the presence of a few largely dominant species; filter feeders (mainly 'coarse') predominate, but other feeding types are poorly represented although predators (biters and swallowers) may amount to 20 to 25% of the total biomass. (ii) In areas with weak upwelling, total zooplankton biomass is slightly lower and species diversity higher than in areas of stronger upwelling; filter feeders always predominate (40%), but the proportion among them of fine filter feeders is higher than 50%; biting and swallowing predators amount to 30 to 35% of the total biomass. (iii) In areas of weak convergence, biomass is low and diversity high; filter feeders (mostly 'fine') are always poorly represented and make up only 20 to 30% of the total biomass, whereas sucking and swallowing predators may represent up to 45%.

In spite of the scarcity of data on the specific composition of surface and subsurface zooplankters, some singularities may be emphasized. For example, according to LOSSE and MERRETT (1971), shoals of *Oratosquilla investigatoris* occur in the western tropical Indian Ocean; this species—the only planktonic hoplocarid—may be fed upon by large fishes such as tunas; however, due to its rather large size (sometimes more than 10 cm) it can also feed on young fishes, e.g. *Sardinella*. The abundance of euphausiids (*Euphausia pacifica*, *E. brevis*, *E. diomedae*, etc.) (Fig. 6-28) in the Arabian Sea, mainly in winter, is

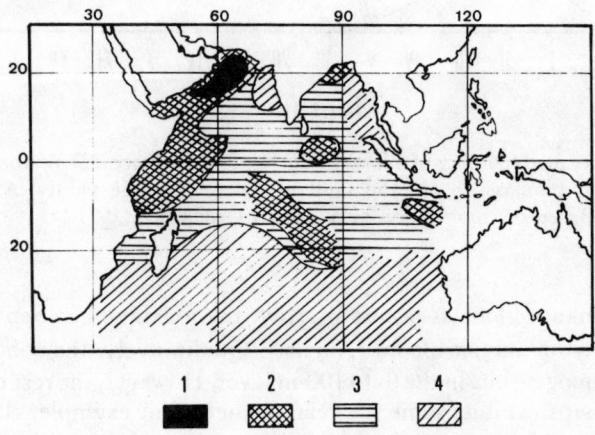

Fig. 6-28: Abundance of euphausiids in the 0- to 100-m layer of the Indian Ocean. 1: >1000; 2: 500–1000; 3: 100–500; 4: <100 (ind. 1000 m^{-3}). (Based on information provided by PONOMAREVA, 1972).

another striking peculiarity. PONOMAREVA (1972) emphasized that these euphausiids, whose presence is quite unusual in a tropical sea, represent a very important food source for many young fishes: tunas, mackerel (*Rastrelliger*), *Coryphaena* and for flying fishes of the family Exocoetidae.

In the Gulf of Aden, GAPISHKO (1971) observed that the monsoon regime controls changes in zooplankton abundance as well as in phytoplankton (Fig. 6-29). The

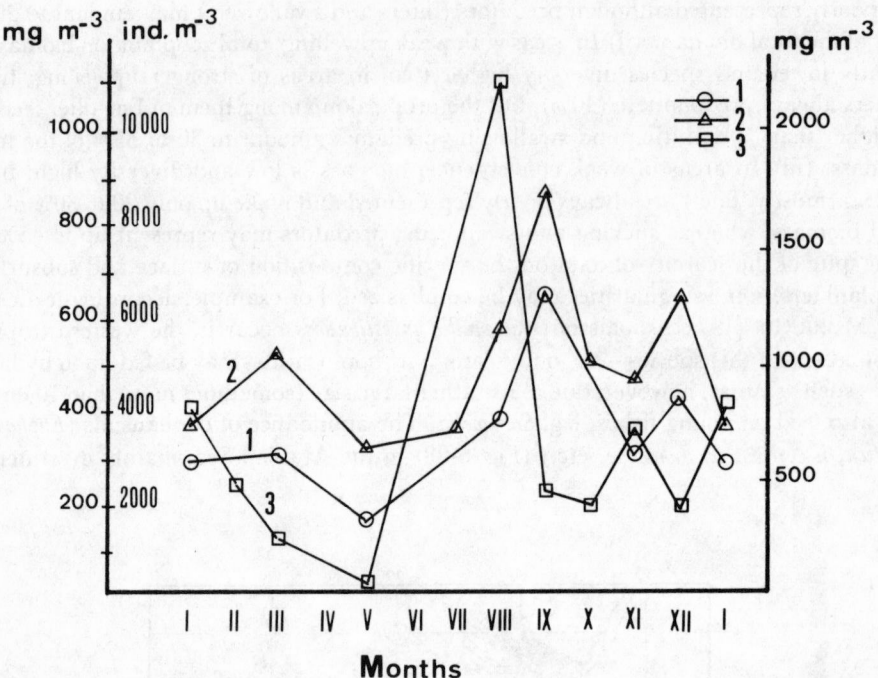

Fig. 6-29: Seasonal dynamics of biomass (1) and abundance (2) of zooplankton and phytoplankton biomass (3) in the Gulf of Aden. Average values over the period 1963–1970. (Based on information provided by GAPISHKO, 1971.)

minimum zooplankton biomass occurs in May, the maximum in September, immediately after the phytoplankton bloom (August). Qualitatively, the zooplankton composition is rather homogeneous in the 0- to 100-m layer. However, the respective abundance of the various groups exhibits some seasonal changes. For example, cladocerans which are abundant in May almost disappear in November when copepods of the genera *Eucalanus*, *Clausocalanus*, *Calocalanus* become very numerous, together with members of the genus *Pseudodiaptomus* (transported from the Arabian Sea by currents generated by the winter monsoon). Copepods always make up the bulk of the zooplankton (49–80% of the total biomass). Next in rank are appendicularians, chaetognaths and molluscs (GAPISHKO, 1971). According to GROBOV (1968), in the 0- to 100-m layer the zooplankton is strongly impoverished during periods between the two monsoons: small-sized species predominate between summer and winter monsoons, larger-sized species between winter and summer monsoon.

Due to its very high primary and secondary production, the Gulf of Aden and the neighbouring areas are very rich in pelagic fishes of economic importance. Seasonal changes in hydrological climate and associated planktonic populations probably result in a large diversity in both temperature conditions for breeding and appropriate food supply for many species, e.g. *Istiophorus gladius* and *Neothunnus albacora* spawn during the south-western monsoon, while *Coryphaena hippurus* spawns in winter (KRAKATITSA, 1970). Gut content and zooplankton studies investigated by GAPISHKO (1968) demonstrate that scombroid fishes and the carangid *Decapterus macarellus* feed on zooplankton in August, whereas in October 80 to 90% of the food supply of *Sardinella longiceps* consists of phytoplankton. ALI KHAN and HEMPEL (1974) concluded from their observations in the Gulf of Aden during October to November 1966 and February to March 1967, that an inverse relationship exists between the number of fish larvae and the accompanying volume of zooplankton. Predatory pressure on fish larvae by zooplankton would be very low in the period of low zooplankton abundance (e.g. October) but very high during the macrozooplankton peak which corresponds to a stronger effect of the north-east monsoon.

From a more general point of view, a correlation between fish eggs and larvae on the one hand and the monsoon regime on the other was clearly demonstrated by PETER (1974) in the Arabian Sea. Both eggs and larvae were two to three times more abundant in the 0- to 200-m layer during the south-west monsoon than during the north-east monsoon, especially in areas such as the Somalia coast, the Arabian coast and the west coast of India. With regard to pelagic fishes, the oceanic areas have large resources of Thunnidae, Scomberomoridae, Trichiuridae and Coryphaenidae.

In other respects, it is amazing that larval stages of young individuals of cephalopods, which are generally very scarce in macroplankton and micronekton samples from the Indian Ocean, may be very abundant in regions of high plankton production: western Arabian Sea, northern and western Bay of Bengal, and off the Somalia Coast. Lesser concentrations have also been recorded in the equatorial zone, in the north-eastern areas of the Arabian Sea, off south-eastern Africa and the western Australian coasts. Large concentrations of adults of the Indopacific *Stenoteuthis oualaniensis* occur on the slope above bottoms 1000 to 2000 m deep in the north-western areas of the Arabian Sea and the Gulf of Aden. On the bottoms of the latter area, the highest numbers of squid beaks in the whole World Ocean have been observed, suggesting the presence of a very large stock. In the eastern Indian Ocean, offshore zooplankton assemblages have been more thoroughly investigated in six cruises carried out between August 1962 and August 1963 on a transect along the meridian $110°$ E from Latitudes $32°$ S to $9°$ S. Samples were taken in the upper 200-m layer with a 5 ft. Isaacs–Kidd mid-water trawl.

From a general point of view, TRANTER and KERR (1977) analysed the numerical abundance of 13 zooplankton taxa in relation to some factors likely to control their distribution. Through regression analysis they demonstrated that season, latitude—and their mutual interaction—as well as time of sampling may be significant sources of variance. For example, coelenterates, amphipods, decapods, fish eggs and sometimes copepods and euphausiids (especially adults: MCWILLIAM, 1977) were more abundant at night than by day. In general, abundances were higher between 9 and $15°$ S and lower from 25 to $32°$ S, except for fish eggs and fish larvae whose centre of abundance was at 24 to $25°$ S.

An analysis on the same transect—restricted to seasonal variations in copepod populations—by GUEREDRAT (1972) supports the following facts: (i) In the northern regions (during the south-east monsoon) upwelling intensity, high primary production rate, phytoplankton maximum and maximum in copepod populations of the first consumer level coincide with a minimum of Shannon's diversity index. (ii) Near the boundary between tropical and subtropical waters, two different copepod assemblages seem to meet, resulting in increased diversity. In this area the chlorophyll maximum occurs from June to August, and copepod proliferation from August to December featuring spawning females, a bulk of copepodite stages and a low diversity index. GUEREDRAT further suggests that the succession of copepod populations observed would be the same for total meso- and macrozooplankton and micronekton, especially for some fish and amphipods of the family Phronimidae.

TRANTER and KERR (1977) distinguished a group of three characteristic amphipod species (*Phronimella elongata*, *Anchylomera blossvillei*, and *Hyperietta stephenseni*) in the low-temperature and high-salinity subtropical water mass and in the high-temperature and low-salinity tropical water mass another three groups: (i) *Vibilia armata*, *Eupronoe armata* and *E. minuta*; (ii) *Platyscelus serratulus* and *Paratyphis*; (iii) *Paraphronima gracilis*, *Eupronoe laticarpa*, *Amphithyrus bispinosus*.

Many pteropod species of the group Euthecosomata reach peak abundance in the tropical water mass, located in winter (May–September) between 9 and 20° S, during the south-east monsoon. Characteristic species in this tropical water mass are *Creseis virgula*, *Diacria quadridentata* and *Cavolinia longirostris*. During other seasons, *Limacina lesueuri*, *Styliola subula*, *Cuvierina columnella* and *Cavolinia inflexa* exhibit their maximum abundance south of 20° S in the subtropical water mass (SAKHTIVEL, 1977).

From a quantitative point of view, VINOGRADOV and NAUMOV (1961) point out a very strong relationship between the depth of the upper thermocline and zooplankton abundance: the shallower the upper thermocline the higher the zooplankton biomass. It is still difficult to draw a satisfactory picture of the quantitative distribution patterns of the surface zooplankton in the Indian Ocean due to the scarcity of data (mainly south of 20° S). Unfortunately, different authors used different devices and sometimes present data in different units (e.g. ml. 1000 m^{-3}, mg m^{-3}, using wet weight or dry weight). I adopted the scheme presented by BOGOROV and VINOGRADOV (1962, 1968) with slight additions and modifications converting all the data into mg m^{-3} wet weight (Fig. 6-30).

In terms of zooplankton abundance in offshore areas, it seems that two main regions may be distinguished in the Indian Ocean: (i) The southern region from approximately 16° S up to the subtropical convergence contains a very poor zooplankton (less than 21 mg m^{-3}) except for some divergence zones located east of 95° E where TRANTER (1962) found about 80 mg m^{-3} and observed rather important populations of tunas and sperm whales. (ii) The northern region, which may be subdivided in three subregions. (a) The north-eastern subregion is rather poor (35–40 mg m^{-3}), except for some up-welling areas between Australia and Java where biomasses may be higher than 100 mg m^{-3}. (b) The central-eastern subregion, which is a little richer than the north-eastern subregion, with an average zooplankton biomass of about 90 mg m^{-3}. (c) The richest areas located on the two opposite sides of the northern Indian Ocean, namely the southern areas off Java, the region off north-eastern African coasts and the Arabian Sea and its dependences. In Java offshore waters, biomasses of about 200 mg m^{-3} (sometimes 300–400 mg m^{-3}) have been observed. The zooplankton bulk is somewhat similar in southern areas of the Arabian Sea, but in its north-western areas biomasses over 600 mg m^{-3} have often been observed.

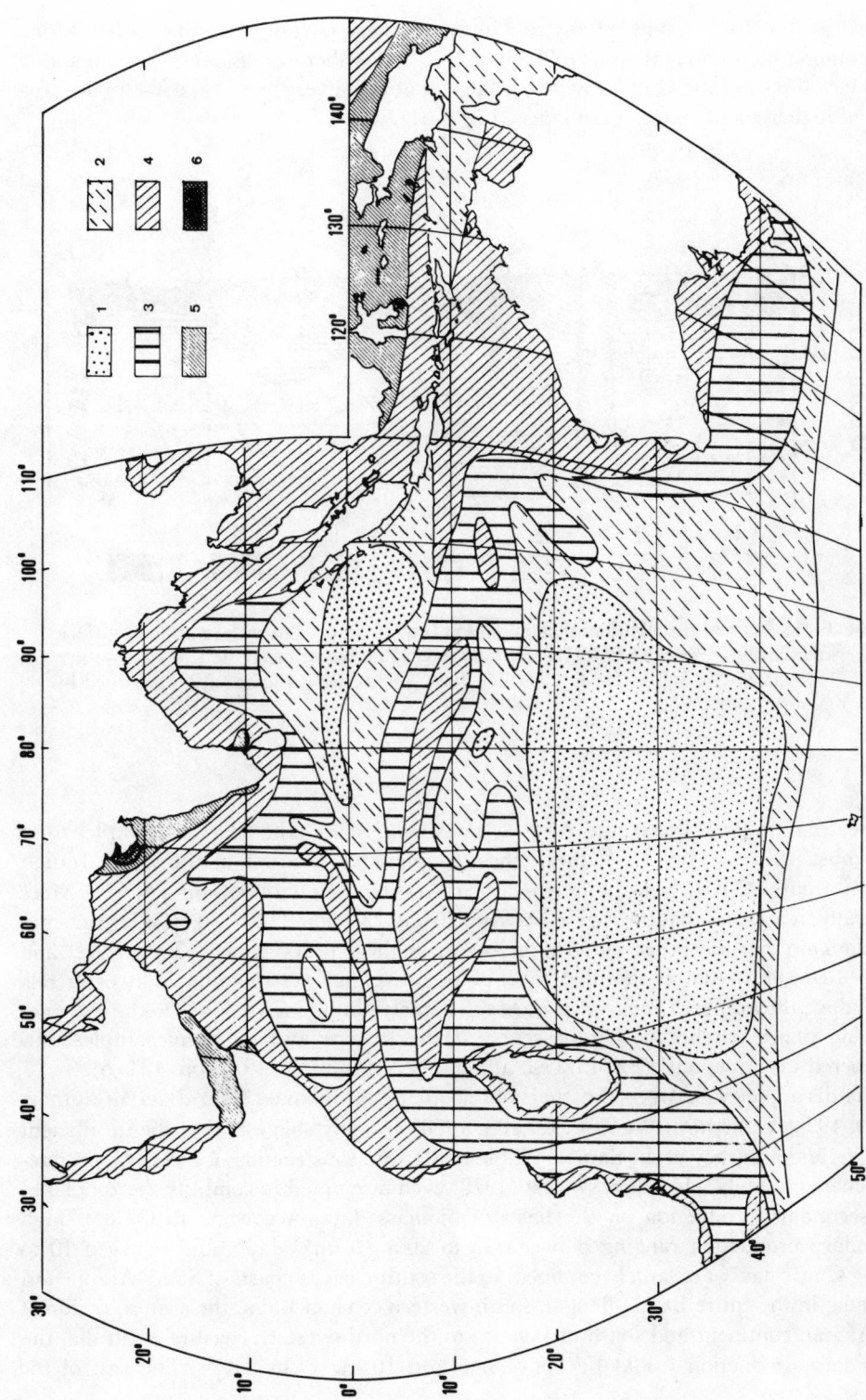

Fig. 6-30: Distribution of net zooplankton biomass (mg m^{-3}) in the 0- to 100-m layer (annual averages). 1: <25; 2: 25–50; 3: 50–100; 4: 100–200; 5: 200–500; 6: >500. (Combined from data by BOGOROV and co-authors, and RAO, 1973.)

GAPISHKO (1971) found 500 mg m^{-3} in the northern part of the Gulf of Aden during the summer monsoon and up to 1080 mg m^{-3} in September near Bab-el-Mandab and in the central part of the Gulf. Obviously, the equatorial divergence also corresponds to a belt of higher zooplankton abundance (Fig. 6-31).

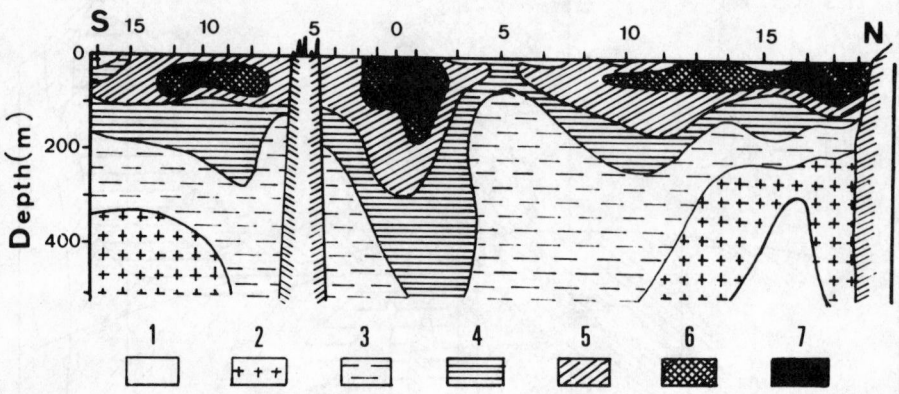

Fig. 6-31: Vertical distribution of zooplankton (ml. 1000 m^{-3}) in surface and subsurface waters in the western Indian Ocean (Tamatave–Bombay transect. 1: <5; 2: 2·5–10; 3: 10–25; 4: 25–50; 5: 50–100; 6: 100–250; 7: >250. (Based on information provided by VINOGRADOV, 1968.)

According to PAULINOSE and ARAVINDAKSHAN (1978), who collected samples from December 1973 to May 1974, the highest zooplankton biomass in the whole Indian Ocean occurs off Kutch in the north-eastern Arabian Sea with 560 ml 200 m^{-3}. More generally, it appears that the region between 20 and 23° N and 66 to 70° E is far richer in zooplankton than any other region of this ocean. In the 0- to 200-m layer, PAULINOSE and ARAVINDAKSHAN mainly sampled 'most of the collections were dominated by ostracods and salps'; the former (*Cypridina* sp.) were extremely abundant at night, 'to the extent of creating dense patches of luminescence at the surface and in many samples outnumbered the copepods' (PAULINOSE and ARAVINDAKSHAN, 1978, pp. 134–5).

The discrepancy between the charts of zooplankton biomass from different authors (RAO, 1973; VINOGRADOV, 1977; QASIM, 1978) strongly suggests that the insufficient number and accuracy of the data available render the construction of a general production chart untimely. However, QASIM (1978) even attempted to compute the zooplankton secondary production on the basis of biomass data. According to QASIM 'high secondary production ranging between 15 to 20 mg C m^{-2} day^{-1} and between 10 to 15 mg C m^{-2} day^{-1} are largely confined' to the south-eastern coasts of Saudi Arabia and Somalia; in the entire Bay of Bengal, south-western coast of India, the eastern region of the African continent and south of Java up to the north-western coast of Australia, the secondary production would be between 5 and 10 mg C m^{-2} d^{-1}. 'The rest of the

Indian Ocean would have a secondary production less than 5 mg C m^{-2} day^{-1} (QASIM, p. 702).

Coastal Zooplankton

Our present knowledge on zooplankton assemblages in Indian Ocean neritic areas is still rather incomplete. Many coastal areas have never been investigated from this point of view. Even along Indian coasts, the data are scattered and sometimes inconsistent. On the western coast of India, during the summer monsoon, maximum zooplankton abundance generally occurs at the same time or a little after the phytoplankton maximum. Where setoid algal species, such as *Chaetoceros*, *Thalassiothrix*, *Nitzschia*, *Ceratium*, are abundant, the zooplankton tends to be distributed in patches in which cladocerans (mainly *Evadne tergestina* and *Penilia avirostris*) are common. Shoals of the Indian mackerel *Rastrelliger kanagurta* feed largely on these zooplankton patches and a direct relationship seems to exist between cladoceran peaks and mackerel fisheries. During the winter monsoon, zooplankton abundance is lower and both direct and inverse relationships between the respective abundance in phytoplankton and zooplankton may be observed. According to GROBOV (1968), on the continental shelf of the western Indian coast the zooplankton biomass would be higher in the northern region—from Kutch Gulf to Bombay—than in the southern region. In the Gulf of Oman the highest biomass and nutritive value for fishes occur at the end of June.

QASIM and co-authors (1978) studied the surface zooplankton assemblage on the west Indian coast between 17° 30' and 8° 30' N. They found the most important taxa in the total zooplankton to be copepods, *Lucifer* sp. and ostracods, representing 29·56%, 26·75% and 21·01% of the total energy available, respectively. Next are crab larvae, decapod larvae other than crabs and *Cavolinia* sp. (6·22%, 3·79% and 3·61%, respectively). The average secondary production was 125 mg C m^{-2} d^{-1}.

MENON and GEORGE (1978) sampled the surface zooplankton on the south-western coast of India (approximately between the Latitudes 17 and 8° N) from 1971 to 1975. They observed a recurring pattern in abundance and distribution directly related to the upwelling process. In general, the peak (approximately 9 ml m^{-3}) occurs in July to September with a fairly uniform plankton concentration beyond nearshore waters all along the coasts. Thereafter, until December, a shoreward shift becomes evident, especially in the south: 'at the same time the continuous plankton distribution breaks up and becomes patchy and its overall abundance is greatly reduced' (about 1 ml m^{-3}) 'and is at its lowest level during January and February' (MENON and GEORGE 1978, p. 209). In the Mandapam area, PRASAD (1954) found a large meroplankton maximum from February to May and a smaller one from August to September.

For the Arabian Sea as a whole—one of the most promising areas of the World Ocean for fisheries development—it must be emphasized that offshore and coastal zooplankton populations should hardly be considered separately. The general scheme of this sea has been described by PRASAD (1969) as follows: zooplankton biomass is larger during the south-west monsoon when the highest concentrations occur along the coast of Somalia, Arabian peninsula, Iran and south-west India. Lower biomass and production values are encountered in the central and north-western region. During the north-east monsoon, a diffuse distribution of a comparatively lower magnitude is observed throughout the Arabian Sea. In the Bay of Bengal there is no obvious seasonal change in zooplankton biomass. On the bay's northern shelf BIDULYA and BABICHEV (1972) found that the zooplankton biomass in the 0- to 100-m layer is almost the same (500–600 mg m^{-3}) in summer, autumn (except for November: 390 mg m^{-3}) and winter. The most productive waters stretch along the northern coasts.

Although the area of the Arabian Sea is only 1·8 times that of the Bay of Bengal, Indian fisheries landings along the nation's west coast are almost three times those of the east coast (PRASAD, 1969). Engraulids and clupeids represent the most abundant fish resource in the coastal areas. However, the Indian mackerel is also important for Indian inshore fisheries. While its reproduction seems to be almost continuous during the whole year, a striking peak occurs during the summer monsoon, especially in the shallowest areas. Young mackerels (up to about 18 cm in length) remain on the shelf but older individuals occur offshore.

On the south-eastern Arabian coast most of the higher fish concentrations appear to be encountered a short distance seaward from the edge of the continental shelf in areas where the depth is 1000 to 2000 m. While the fish shoals exhibited pronounced diurnal depth movements, rising to surface layers at night, the day-time depth remained remarkably constant. The complete range of depth inhabited extended from 110 to 310 m (CURRIE and co-authors, 1973).

The Great Australian Bay has a rather rich zooplankton with biomasses of 500 to 1000 mg m^{-3} (MARKINA, 1971). Copepods predominate, especially species of the genera *Eucalanus, Pleuromamma, Candacia, Labidocera* and *Rhincalanus*. The euphausiid *Nyctiphanes australis*—a characteristic species of the southern transitional zone—is also present. Zooplankton biomasses in the neritic areas of West Australia were always lower than 500 mg m^{-3}.

Intermediate and Deep-water Zooplankton

Zooplankton data available from intermediate and deep waters of the Indian Ocean almost exclusively concern quantitative aspects. Four main zones may be recognized (VINOGRADOV, 1962, 1968): central, eastern and southern zone; Java Trench zone; Bay of Bengal; Arabian Sea.

Central, eastern and southern zone

In this very large zone intermediate water currents are rather well documented. Two subzones may be distinguished: (i) The eastern and south-eastern subzone reveals an important influence of Southern Ocean water masses. Due to submergence of zooplankters with decreasing latitude, rather high biomasses are observed: 4 to 9 mg m^{-3} in the 500- to 1000-m layer (Antarctic intermediate water) and 0·6 mg m^{-3} in the 3000- to 4000-m layer (Antarctic bottom water). (ii) The central subzone supports much lower zooplankton biomasses: the 500- to 1000-m layer with a biomass of 2 to 5 mg m^{-3} probably corresponds to the deeper levels of the south Indian central water mass, while the south Indian deep water (between 3000 and 4000 m) has a biomass of only 0·1 to 0·2 mg m^{-3}. In general, in this whole subzone the biomass decreases with increasing depth; it is always lower in the intermediate layers than in the upper and lower levels, and a slight increase may even be noticed within the intermediate layers.

Java Trench zone

In this zone zooplankton abundance versus depth is quite similar to that in the Bougainville Trench of the Pacific Ocean, although zooplankton biomass in the surface waters of the latter is about ten times lower. At 3000 m depth, the biomass is only 0.40 mg m^{-3} in the Java Trench, but 0.35 mg m^{-3} in the Bougainville Trench.

Bay of Bengal

In the Bay of Bengal the biomasses of intermediate and deep waters are always higher than in the central subzone. A slight increase occurs between 200 and 500 m, but the decrease with increasing depth is quite uniform below 500 m.

Arabian Sea

In the Arabian Sea a very particular situation prevails because intermediate waters from 125 to 150 m down to 800 to 1000 m have a very low oxygen content (sometimes below 0.15 ml l^{-1}) and often contain hydrogen sulphide. The zooplankton biomass exhibits a sharp minimum between 200 and 500 m (1.7–3.4 mg m^{-3}) but increases again in the 500- to 1000–1500-m layer, where biomasses of 5 to 10 mg m^{-3} may occur (VINOGRADOV and VORONINA, 1961, 1962b) (Fig. 6-32).

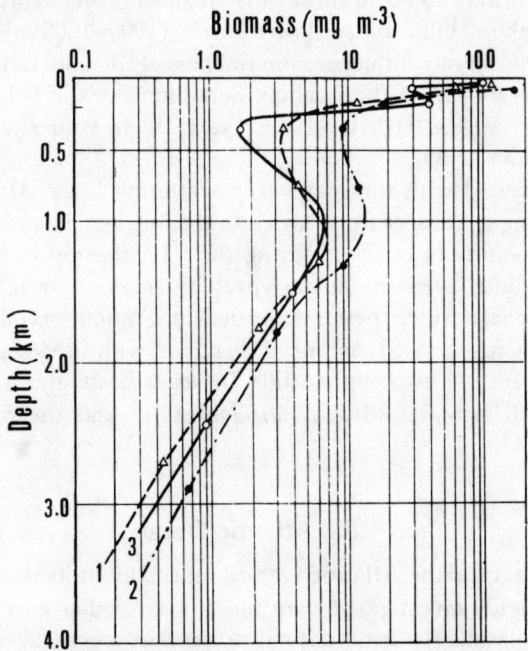

Fig. 6-32: Vertical distribution of plankton biomass (mg m^{-3}) in the Arabian Sea: 'Vitjaz' stations nos 4718 (1), 4721 (2) and 4854 (3). (Based on information provided by VINOGRADOV, 1968.)

In the 200- to 500-m layer the only rather common species are the copepods *Pleuromamma indica* which tolerates very low oxygen content and *Euphausia diomedae*; other copepods, such as *Chirundina streetsi, Euchaeta wolfendeni, Heterostylis longicornis*, are very rare. The vertical migration of *P. indica* and *E. diomedae* is not hindered by the low oxygen content layer. Neritic species, such as the ostracod *Pyrocypris*, the copepod *Centropages*, as well as the epiplanktonic and mesoplanktonic species which usually inhabit the Indian equatorial waters, do not occur there. In the 500- to 1500-m layer the above-mentioned *P. indica, E. diomedae* and *H. longicornis* occur, together with two groups of species: (i) *Rhincalanus nasutus, Calanoides carinatus, Eucalanus bungii, E. attenuatus, Lucicutia maxima*, which are eurybathic (0–50 m down to 1000–2000 m) in other offshore tropical areas, but cannot survive in the 200- to 500-m oxygen-depleted layer; (ii) species which generally cannot be found above 500 m, mainly copepods of the family Augaptilidae.

The difference in the zooplankton bulk between the upper and lower levels of the oxygen minimum layer may be explained as follows. Firstly, in the upper levels the rather high temperatures (18–20 °C) cause higher metabolic rates and oxygen requirements than in deeper-living animals (1000–1500 m) which encounter temperatures of only 9 to 10° C. Secondly, the large amount of sinking organic material which is produced in the surface and subsurface waters cannot be entirely consumed when it passes through the 200- to 500-m layer because the zooplankton populations are very sparse there. Animals which live below the warm-water layer thus receive more food as organic-matter rain than those in most other regions of the World Ocean. The rather important zooplankton bulk at the lower levels (1000–1500 m) of the Arabian Sea intermediate depths, largely influences the richness of the underlying waters: in offshore deep waters (2000–3000 m of the Arabian Sea, biomasses of 0·4 to 1·1 mg m^{-3}) have been recorded by VINOGRADOV (1968); such high values are unusual at such depths in the offshore areas.

The fauna associated with the main deep-scattering layer (DSL) in the equatorial Indian Ocean, along a transect from Africa to the Nicobar Islands has been studied by BRADBURY and co-authors (1971). During the day, the top of the DSL is at 300 to 500 m; an intermediate layer—not always present—may occur at 200 to 250 m. Among the mid-water animals which perform vertical migration, six species exhibit a very strong association with the DSL ascent at dusk and with its descent at dawn: *Abylopsis tetragona, Cymbulia* sp., the two euphausiids *Thysanopoda* sp. and *Nematobrachion* sp., and two fish species, the stomiatoid *Vinciguerria nimbaria* and the myctophid *Notolychnus valdiviae*.

(5) Atlantic Ocean

Pelagos assemblages of the Atlantic Ocean, especially in its northern half, have been much more thoroughly investigated than those in any other part of the World Ocean. However, several reasons render it difficult to present a general outline of the Atlantic Ocean pelagial: (i) The Atlantic Ocean is much narrower than the Pacific and Indian Oceans, and adjacent seas are numerous on both sides of its northern half (Labrador Sea, Caribbean Sea, Norway Sea, North Sea, Mediterranean Sea). This results in more marked exchanges between neritic and offshore plankton assemblages. (ii) Connections and water exchange with the Arctic Basin are much more pronounced than in the North Pacific Ocean. (iii) The South Atlantic Ocean, similar to the South Pacific Ocean, has been insufficiently studied.

(a) North Atlantic Ocean Surface Assemblages

General Aspects

As in the North Pacific Ocean, plankton assemblages in the North Atlantic between the Arctic and north equatorial current comprise three main biogeographical groups: arctic, boreo-arctic and subtropical. In addition, several authors also distinguish a boreal group, whose presence was denied by EKMAN (1953, p. 341) who wrote:

'there are very few, if any, purely boreal epipelagic species. A few were formerly so designated, among the siphonophores, pteropods and euphausiids, but as our knowledge of their distribution increased, they were found to have a more extensive distribution. If one disregards a few species which in reality seem to be neritic, the cold-temperate (boreal) epipelagic region is completely or almost completely inhabited by cosmopolites, by Arctic-North-Atlantic species or more occasionally by pelagic individuals from the plankton community, which is living in the intermediate deep sea zone.'

Perhaps this statement is a little sharp, thus introducing the risk of excessive confusion in the description of assemblages. In the following pages, we shall tentatively distinguish two cold-temperate subassemblages: boreo-arctic (more offshore) and boreal (more neritic).

Arctic Assemblage

The most characteristic phytoplankters which inhabit the areas of the Atlantic Ocean occupied by waters from Arctic origin are the diatom *Chaetoceros decipiens* and the dinoflagellates *Ceratium arcticum*, *C. longipes* and *Dinophysis granulata* (MOVCHAN, 1962; KUZ'MINA, 1962a,b). The species listed on p. 295 for the Arctic Basin itself may occur here too. However, according to many authors the phytoplankters mentioned above are most capable of extending their distributional range into the Arctic areas of the Atlantic Ocean.

Similarly, the list of zooplankton species considered characteristic differs from investigator to investigator. EKMAN (1953, p. 176) distinguished two subgroups: high arctic species which live only in the coldest water and low arctic species which prefer or tolerate waters with temperatures above about 0 to 1 °C (for brief periods up to 4 °C). Among the high arctic species are: the Trachymedusae *Ptychogastria polaris*, *Crossota norvegica* and *Botrynema ellinorae*; the Narcomedusa *Aeginopsis laurentia*; the siphonophore *Stephanomia orthocanna*; the ctenarian *Mertensia ovum*; the copepods *Pareuchaeta glacialis* and *Chiridius obtusifrons*; the amphipod *Themisto libellula*. All these species, except the last one, may also be found at lower latitudes but in deeper layers due to sinking with Arctic waters; these flow downward beyond the Wyville Thomson Ridge or between Iceland

and Greenland. *Dinophysis arctica* and the copepods *Calanus glacialis*, *C. hyperboreus* and possibly *Pareuchaeta glacialis* may be better considered 'low arctic' species. Belonging also to this group are two chaetognaths: *Sagitta maxima*, a subcosmopolite species, and *Eukrohnia hamata*, which exhibits equatorial submergence; it tolerates temperatures from less than 0 °C up to 10 or 15 °C.

Boreo-arctic Assemblage

In the boreo-arctic areas of the North Atlantic Ocean, neritic and offshore phytoplankton assemblages can hardly be distinguished. The following phytoplankters exist in both assemblages: *Rhizosolenia hiemalis*, *R. hebetata* f. *semispina*, *Thalassiosira gravida*, *Chaetoceros atlanticus*, *C. convolutus*, *C. debilis*, *Leptocylindrus danicus*, and *Corethron hystrix*. The proper offshore boreo-arctic subassemblage includes as dominant species *Thalassiosira nordenskiöldii*, *T. longissima*, several species of *Chaetoceros* (the most abundant of which is *C. debilis*), and *Ceratium furca*. The most characteristic species of the neritic subassemblage are *Rhizosolenia alata*, *R. styliformis*, *Chaetoceros atlanticus* and *Ceratium fusus*.

As regards the zooplankton, distinction between boreo-arctic and boreal elements seems to be easier. According to VLADIMIRSKAJA (1962) the main boreo-arctic species are: *Aglantha digitale*, *Metridia longa*, *Oncaea borealis*, *Hyperia medusarum*, *Thysanoessa raschi*, *T. longicauda*, *Themisto abyssorum*, *Limacina helicina* (which has a bipolar distribution), *Oikopleura labradorensis*, and *O. vanhoffeni*: *Calanus hyperboreus* also probably belong to this group. The boreal subassemblage is mainly characterized by the copepods *Calanus finmarchicus*, *Pseudocalanus elongatus Microcalanus pusillus*, *M. pygmaeus*, *Oncaea similis* and the euphausiid *Thysanoessa inermis*.

Subtropical Assemblage

The subtropical assemblage inhabits the waters of the Gulf Stream and the North Atlantic Drift. Its phytoplankton has some species in common with the tropical assemblage (p. 272) which are marked with an asterisk in the following list: the diatoms *Rhizosolenia styliformis* (perhaps only a larger form of *R. hebetata* f. *semispina*), *R. acuminata*, *R. alata**, *Coscinodiscus centralis*, *C. stellaris*, *Chaetoceros densum*, *Bacteriastrum delicatulum**, *B. elongatum*, the dinoflagellates *Ceratium bucephalum*, *C. macroceros*, *C. tripos**, *Dinophysis acutata*, *D. hastata*, *D. homunculus*, the coccolithophorids *Coccolithus pelagicus* and *C. huxleyi*. Among the zooplankters, the most characteristic species are: *Challengeria bidens*, *Aetidus armatus*, *Eucalanus elongatus*, *Rhincalanus nasutus*, *Scolecitricella minor*, *Metridia lucens*, *Pleuromamma robusta*, *Centropages hamatus*, *Spiratella* (*Limacina*) *retroversa*, and *Sagitta serratodentata*.

To consider the zooplankton assemblage of both Gulf Stream and North Atlantic Drift as a single assemblage is probably an oversimplification. As the plankters drift more and more towards the European coasts, the concomitant temperature decrease in the surface layers causes disappearance of the most thermophilic species. Further factors may interfere. As is well documented, forminiferans are sensitive indicators of the physical and chemical characteristics of water masses. CIFELLI and SMITH (1970) observed that in the large region considered, two different foraminiferan assemblages exist: in the assemblage inhabiting the Gulf Stream and Sargasso Sea waters, *Globigerinoides rubra* predominates instead of *Globigerina incompta* in North Atlantic Drift waters. However, this is not caused by temperature—the thermal climate in the North Atlantic Drift is quite suitable to *G. rubra*—but probably suggests that the North Atlantic Drift represents a semi-closed gyre arising from mixing of slope waters and Gulf Stream waters.

In the Gulf Stream and North Atlantic Drift the most important plankton-feeding fish is the Atlantic Saury *Scomberesox saurus*; its maximum concentrations occur between 13 and 70° W. Saury exhibit seasonal migrations: spawning occurs in rather temperate waters, while the coldest parts of the distributional area are used for summer feeding only; eggs and larvae are holo-hyponeustonic. In the central areas of the Atlantic Ocean the biomass of squid, e.g. *Ommastrephes bartrami* and *Todarodes pacificus*, has been estimated at several hundreds kg m^{-2}; most species tend to concentrate at night in subsurface waters, sometimes a little deeper. The general pattern of squid distribution is the same as in the Northern Hemisphere of the Pacific Ocean: the most abundant populations of the distant-neritic species of the family Ommastrephidae occur in the vicinity of the Polar front, e.g. *Illex illecebrosus* which is very common between Newfoundland and Georges Bank in areas where the cold waters of the Labrador Current come into contact with the temperate 'slope waters' preferably inhabited by this squid.

(b) Ecology of North Atlantic Surface and Subsurface Plankton

The circulation of the North Atlantic Ocean is rather complex due to the disposition of continental masses and the existence of many areas where cold and temperate waters come into contact. Hence it is much more difficult than in the Pacific Ocean to seek out general features regarding the distribution and ecology of plankton assemblages.

A very valuable general account on phytoplankton distribution during spring in north-western Atlantic areas located between 35 and 55° N from Cape Finisterre to Newfoundland has been presented by MOVCHAN (1962). For the reasons mentioned above, the phytoplankton population of this large region is somewhat heterogeneous, but boreal species predominate (34%) over arctic (14%), boreo-arctic (14%) and temperate (19%) ones. As expected, the percentage of cold-loving species is greater in the northern areas of the whole region, whereas in southern areas temperate species predominate, with admixtures of some tropical elements. Among the best investigated regions is the eastern coast of North America from the Florida Strait up to Newfoundland. Here, the well-marked differences in total plankton abundance relate closely with water-mass characteristics.

The offshore phytoplankton is everywhere more abundant in cold waters than in Gulf Stream waters (Fig. 6-33). With regard to zooplankton (Fig. 6-34)—according to the data reviewed and complemented by BE and co-authors (1971)—quantitative aspects may be summarized as follows: (i) Along the coasts plankton is generally more abundant in shelf waters due to nutrient input from numerous rivers of the US coast. Examples are 700 to 800 ml. 1000 m^{-3} in summer and 400 in winter from Cape Cod to Chesapeake Bay; in some places up to more than 8000 ml. 1000 m^{-3}. George's Bank, an area characterized by intensive fisheries, is also very rich: 1500 ml. 1000 m^{-3} in summer. (ii) At its starting point in the Florida Strait, the Gulf Stream has a very low plankton content in the 0- to 300-m layer: 20 ml. 1000 m^{-3}. However, towards the north-east its

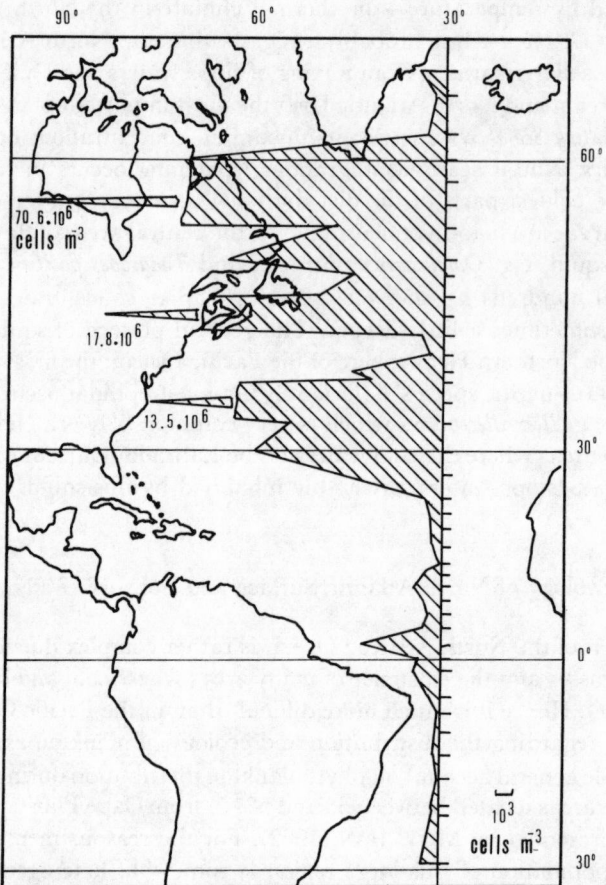

Fig. 6-33: Quantitative distribution of phytoplankton in the
0- to 100-, layer along meridian 30° W in the Atlantic
Ocean. (Based on information provided by SANINA,
1969.)

plankton abundance progressively increases: 50 and 70 ml. 1000 m^{-3} in the 0- to 150-m
layer off Florida and Georgia, respectively; between Cape Hatteras and Chesapeake
Bay (35–37° N): 114 mg m^{-3} (Table 6-4, p. 205) in the 0- to 200-m layer; between 40
and 43° N: 143 mg m^{-3} in the 0- to 200-m layer (p. 295). This gradual enrichment
probably originates from mixing of Gulf Stream waters and more fertile shelf or slope
waters. (iii) More to the north, the Labrador current flows towards the south-west close
to the shoreline and transports the very rich plankton developed during spring and
summer in the Newfoundland area (see below): at 43° N, JASHNOV (1961) observed
more than 1000 mg m^{-3} in autumn. Fig. 6-35 summarizes the relation between primary
production and zooplankton biomass in four water masses of the North Atlantic Ocean

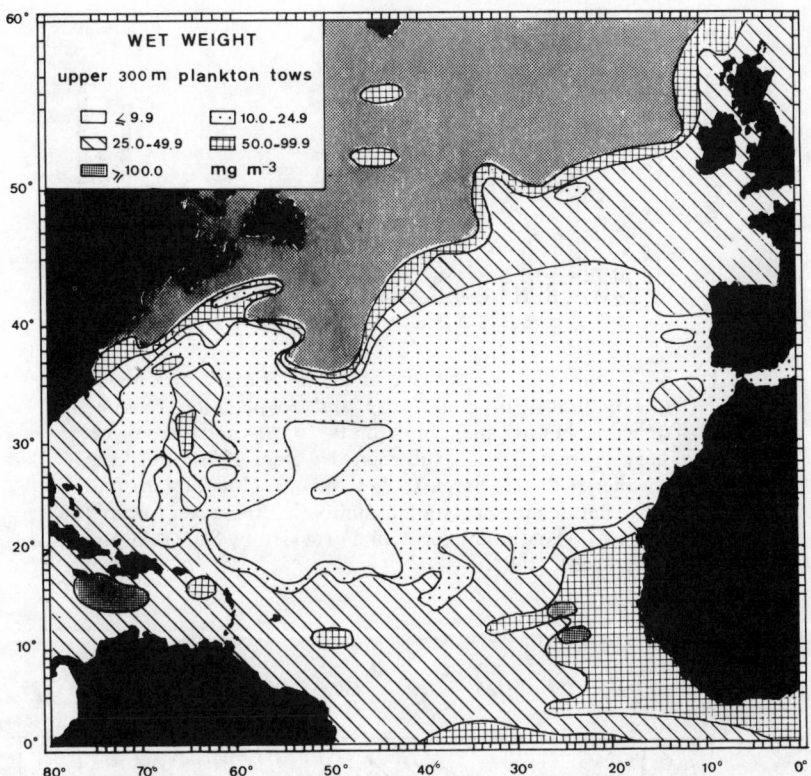

Fig. 6-34: Distribution of zooplankton biomass (wet weight) in the northern half of the Atlantic Ocean based on plankton tows in 0–300 m. (After BE and co-authors, 1971; modified; reproduced by permission of Gordon and Breach, New York.)

according to SOROKIN and KLIACHTORIN (1961). The areas of the eastern and southern Newfoundland coasts and mainly the banks located off the eastern coast are the most productive.

The phytoplankton peak occurs in April, reaching its maximum in the area of Newfoundland Banks with up to 200,000 cells l^{-1} (biomass about 3000 mg m^{-3}). In spring diatoms always predominate and generally account for more than 90% of the total cell number. In the 0- to 100-m layer, the phytoplankton is a little more abundant at 10 m than at any other depth (MOVCHAN, 1967, 1970). According to ELIZAROV and MOVCHAN (1973), the maximum phytoplankton concentrations on the Newfoundland Grand Bank almost always coincide with areas of curls in the vertical plan (Fig. 6-36). Greater phytoplankton abundances are not restricted to the area of upheaval, but may sometimes extend below the euphotic layer and to adjacent areas where population submergence occurs. In spring, when the limiting of nutrient salts is negligible,

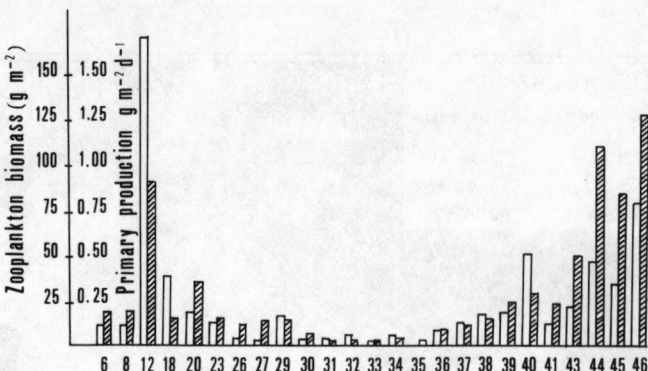

Fig. 6-35: Simultaneous measurements of primary production (white columns) in g m^{-2} d^{-1} and zooplankton biomass (hatched columns) in g m^{-2} for 0- to 200-m layer. 6–12: Canary current; 18–29: subtropical waters of the northern hemisphere; 30–35: Sargasso Sea; 36–46: Gulf Stream and North Atlantic current (station numbering increases towards north-east). (Based on information provided by SOROKIN and KLIACHTORIN, 1961.)

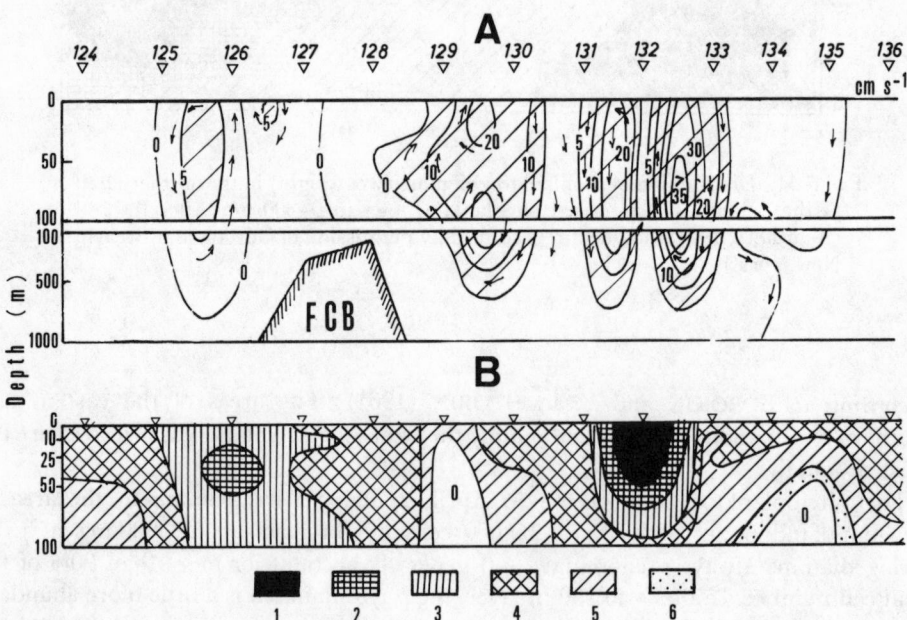

Fig. 6-36: (a) vertical circulation and current speed (cm s^{-1}) off Newfoundland Grand Bank in April 1958. (b) vertical distribution of phytoplankton (10^6 cells m^{-3}) in 0 to 100 m. 1: >100; 2: 10–100; 3: 1–10; 4: 0·1–1; 5: 0·01–0·1; 6: <0·01. FCB: Flemish Cap Bank. (Based on information provided by ELIZAROV and MOVCHAN, 1973.)

phytoplankton development mainly depends on water stratification, whereas in autumn, when water is stratified everywhere, development apparently depends more on the availability of nutrient salts.

In the vicinity of Newfoundland zooplankton biomass in cold waters is 300 to 500 mg m^{-3} in spring and summer, decreasing in autumn to 160 mg m^{-3}. In spring the maximum biomass in the 0- to 200-m layer is found in shallow waters of the north-eastern part of the Grand Bank; in summer on the slope of the same area and on the slope of the Flemish Cap Bank. Maximum biomass occurs again on the north-eastern shelf of the Grand Bank again in autumn. The most important element is *Calanus finmarchicus* representing 40 to 70% of the total biomass in spring in the 0- to 200-m layer, 85 to 90% in summer and 70% in autumn. The population inhabiting the 0- to 200-m layer in autumn is, generally speaking, fairly young, whereas in the 200- to 500-m layer, IV–V copepodites predominate and represent up to 75% of the whole population. This latter fact indicates a population trend towards an overwintering stage. The distribution features of other zooplankton species are rather similar to those of *C. finmarchicus*. Herring that feed on zooplankters are common around the Grand bank and in the St Laurent Estuary. Warmer water masses are inhabited by temperate zooplankton, biomass being low and with no predominant species, as for example, *Calanus finmarchicus* in the boreal assemblage.

Another interesting region is the Greenland Sea. In the southward flowing Arctic waters of the East Greenland current, phytoplankton biomass in the 10- to 25-m layer seems to be the highest in the whole North Atlantic (2220–4370 mg m^{-3}). Phytoplankton here mainly comprises spring and summer species, the most abundant of which is *Chaetoceros decipiens*. In the 20- to 50-m layer the biomass falls to 20 mg m^{-3}. In the northward flowing mixed waters of the West Spitzberg current, the biomass of its surface phytoplankton is low (10–110 mg m^{-3}) and only represented by summer species, the most abundant of which are *Chaetoceros concavicornis* and *Corethron hystrix*. On the 0° meridian, phytoplankton is extremely scarce (KUZ'MINA, 1962a,b). Data on zooplankton from the Greenland Sea are insufficient. It seems, however, that in the Denmark Strait and Greenland Sea, the very rich phytoplankton mentioned above leads to the development of a zooplankton assemblage which includes boreo-arctic and boreal species (among them *Calanus finmarchicus*) with an admixture of some true arctic elements. Composition of the zooplankton may differ as a function of climatic conditions prevailing in spring and summer. For example, a warmer than usual spring and summer may produce more numerous boreal species and their individuals, and perhaps even some temperate species may occur.

Seasonal vertical migration may induce very important changes in the composition of the zooplankton populations, mainly those living in surface waters. For example, in the Norwegian Sea several authors (see review by VINOGRADOV, 1968) have studied the changes in the most abundant species, mainly those comprising food for fishes. In winter (November–January) most *Calanus finmarchicus* inhabit layers below 400 to 700 m, sometimes down to 2000 m, where they overwinter. In April and May the population ascends above 200 m, developing an abundance peak in June after the phytoplankton bloom which provides plenty of food. In the second half of June the IV–V copepodites begin to descend and by August reach depths below 500 m—sometimes 1000 m—before going all the way down to the overwintering layers. The bulk of the winter population of *Calanus hyperboreus* inhabits 1000 to 2000 m, i.e. waters deeper than *C. finmarchicus*. Upward migration begins in March and breeding mainly occurs in 600 to 1000 m; young stages occur in the 0- to 100-m layer only in May to July. By August most of the population is found deeper than 1000 m. Thus, unlike *C. finmarchicus*, almost the whole population of *C. hyperboreus* is restricted to the surface layer from May to July. Both these *Calanus* species may be considered 'interzonal' species on a seasonal scale. Species which are more typically epiplanktonic (*Oithona similis*, *O. atlantica*, *Oikopleura* sp.) generally do not migrate below 600 m in winter.

The hyponeuston layer (0 to 10 cm) studied by GRAVE (1973) at a fixed station in the southern part of the Norwegian Sea exhibited a very low diversity. Whether by day or night, the most abundant species in the uppermost layer were the copepod *Anomalocera patersoni* and larvae of the gadid *Onos* sp. Among the merohyponeustonic forms which occur only at night are the molluscs *Limacina retroversa* and *Clione limacina*, juveniles of the amphipod *Parathemisto* sp. and the euphausiid *Meganyctiphanes norvegica*, larvae of *Sebastes marinus* and *Scomber scombrus*. In the North Sea, larvae of *S. scombrus* inhabit the uppermost layer during the day and night.

Seasonal vertical migrations of zooplankters may result in some succession of zooplankton populations due to reduced competition. For example, in the southern Barents Sea development of the neritic zooplankton assemblage, which mainly consists of small-sized species (genera *Centropages*, *Acartia* and *Temora*) with a low filtration rate (less than 10 ml d^{-1}), is made possible by downward migration of *Calanus finmarchicus* whose filtration rate seems to be about 80 ml d^{-1} (ZELIKMAN and KAMSHILOV, 1960).

Calanus finmarchicus—feeding on particles larger than 10 μm, i.e. mainly diatoms which establish a bloom in April and May in the southern Barents Sea—is later replaced by *Pseudocalanus elongatus* which feeds on smaller algae, e.g. dinoflagellates. These are more abundant between June and October reaching a maximum in August. In general, phytoplankton bloom occurs in spring (April–May) in southern parts of the Barents Sea and in summer in the northern and north-eastern parts. A late summer-to-autumn peak also occurs in the nearshore areas of the North Cape current, where *Rhizosolenia* predominates. An interesting feature of the Barents Sea is the ciliate abundance (mostly *Strombidium strobilus*) in May and June, when the spring bloom ends; tintinnids are also abundant between June and October. The spring spawning of the most important zooplankton species (*C. finmarchicus*, *Thysanoessa inermis*, *T. raschi*) coincides with the beginning of the bloom. However, proliferate development of the 'red' *C. finmarchicus* population generally occurs in areas where the phytoplankton is declining, but ciliates (chiefly of the genus *Strombidium*) are abundant. The latter apparently constitute the main food of adult *Calanus* in this area. Herring do not inhabit areas with abundant phytoplankton, but low-density populations exist in areas where Infusoria predominate.

Special attention has been paid by FRASER (1961) to the Atlantic Ocean and its adjacent seas around the British Isles FRASER distinguished five main zooplankton assemblages but we shall deal here with only three of them, formerly recognized by RUSSELL (1939) (Fig. 6-37): (i) The neritic water assemblage of the North Sea and the Irish Sea characterized mainly by the chaetognath *Sagitta setosa*, the copepods *Labidocera wollastoni* (a holohyponeustonic species) and *Isias clavipes* and the medusae *Tima bairdi*

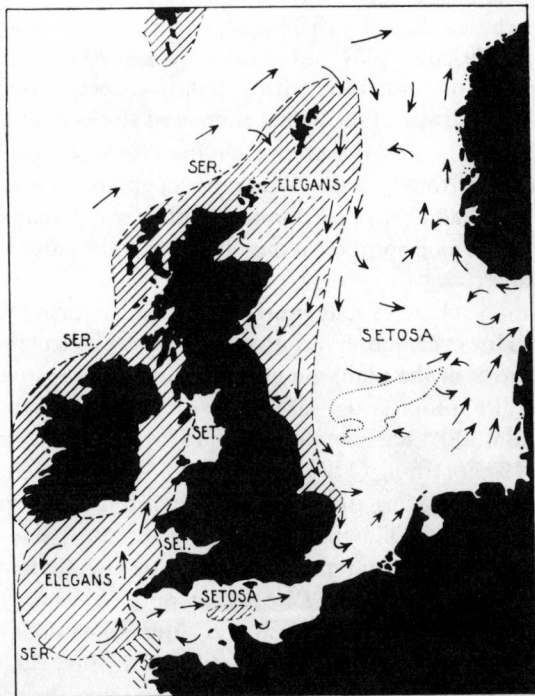

Fig. 6-37: Distribution in surface waters around the British Isles and in the North Sea of the three zooplankton assemblages labelled by three different chaetognaths of the genus *Sagitta* (see text) during a period of strong inflow of Atlantic waters; arrows: main currents. (After RUSSELL, 1939; reproduced by permission of Conseil International pour l'Exploration de la Mer, Charlottenlund.)

and *Eutonia indicans*. (ii) The North Atlantic Drift waters. Its characteristic species have been listed on p. 264, but may be complemented here by the chaetognath *Sagitta serratodentata*, the copepod *Euchaeta hebes* and the siphonophores *Lensia conoidea*, *Physophora hydrostatica*, *Agalma elegans*. (iii) The assemblage of mixed (Atlantic and shelf) waters inhabiting coastal areas on western shores of the British Isles and especially the North Sea, due to the Atlantic inflow entering through the Faroe–Shetland Channel. Some influence of the Lusitanian current (p. 264) may also be observed beyond the shelf edge, but the zooplankton assemblage of this current does not really inhabit the North Sea mixed waters. Obviously some species of the North Atlantic Drift assemblage may be transiently observed in mixed waters; some, like *Sagitta serratodentata* and *Physophora hydrostatica*, may endure the winter but do not breed. Others, such as *Salpa fusiformis* and *Dilioletta gegenbauri*, may breed before the winter cooling but do not overwinter.

The mixed waters are inhabited by a specific organismic assemblage with the indicator species *Sagitta elegans*, generally associated with the copepods *Metridia lucens* and *Candacia armata*, the euphausiid *Thysanoessa inermis* and the mollusc *Clione limacina* and *Spiratella retroversa* (FRASER, 1961). Moreover, *Calanus finmarchicus* is often abundant in the mixed waters; in the North Sea its population seems to originate from both autochthonous stocks which spend the winter there and stocks drifting in from the North Atlantic early in spring. The mixed waters have the largest zooplankton production of all water masses around the British Isles and copepods, euphausiids, amphipods, cladocerans, decapod larvae always predominate other components of the assemblage. This plankton is the main food source not only for herring but also for planktonic stages of many demersal fishes.

Multidecennal climatic changes may markedly influence mixed water formation and thus fish landings. However, observations made near Helgoland in 1975 showed that the principal aspects of the plankton cycle in the North Sea neritic waters did not change, either on a multi-annual scale—as did the mixed waters—or due to an increased eutrophication (HAGMEIER and co-authors, 1975). The average phytoplankton biomass observed in 1975 ($41 \cdot 8$ μg C l^{-1}) lies in the middle of the values observed during the 14 preceding years. In spring diatoms predominate, mainly species of the genera *Coscinodiscus* and *Thalassiosira*, followed in late May to early June by *Phaeocystis*. The characteristic species of the second peak from August to September are *Cerataulina bergonii*, several species of the genus *Rhizosolenia* and *Guinardia flaccida*; *Schroederella schroederi* and *Eucampia zodiacus* predominate in August. Among microzooplankton, *Noctiluca* occurs in July and August and various ciliates in July, August and December.

The zooplankton assemblage of the Lusitanian current and the problem of its sinking to deeper layers of arctic elements is reviewed on p. 292.

(c) Plankton Assemblages in the Intertropical Atlantic Ocean

Offshore Assemblage

A comprehensive review on phytoplankton has been presented by FERGUSON-WOOD (1971) in a paper mainly devoted to the Caribbean region (see below). Here the micro-algal community is controlled by the phytoplankton assemblage of the north equatorial current. According to FERGUSON-WOOD, the phytoplankton populations in the oceanic waters of the north equatorial current—especially those of the subtropical undercurrent—are characterized by the diatoms *Gossleriella tropica*, *Planktoniella sol*, *Synedra superba*, *Mastogloia rostrata* and *Corethron criophilum* and dinoflagellate species of the genera *Histioneis*, *Parahistioneis*, *Amphisolenia* as well as some species of *Ceratium*, such as *C. praelongum*, *C. gravidum*, *C. geniculatum* and *C. lanceolatum*. The coccolithophorids *Discosphaera thomsoni*, *D. tubifer*, *Syracosphaera pulchra*, *S. brasiliensis* and *Calyptrosphaera oblonga* are also related to the north equatorial current, and *Scyphosphaera apsteini* to the subtropical underwaters. FERGUSON-WOOD takes the presence of *Scyphosphaera* in the upper waters as evidence of vertical mixing of the subtropical underwater with the surface water. He further points out that the large outflow of Amazon water and its subsequent dispersal across the north equatorial current sometimes results in *Skeletonema costatum* aggregations in the offshore waters up to 1500 miles from the population centre.

Unfortunately, our present knowledge of zooplankton assemblages is very small, but presumably the assemblages are highly diversified. According to VLADIMIRSKAJA (1962), the most characteristic elements are *Collozoum* sp., *Arachnactis albida*, *Solmundella bitentaculata*, *Physophora hydrostatica*, *Clausocalanus arcuicornis*, *Pareuchaeta barbata*, *Pleuromamma abdominalis*, *Lucicutia simulans*, *Anomalocera patersoni* (hyponeustonic species) and *Corycaeus speciosus*. The plankton assemblages of some marginal areas of the tropical Atlantic Ocean are of special interest. Their main features are summarized below.

Sargasso Sea

The Sargasso Sea has unusual characteristics, most important of which is the existence of two thermoclines: a permanent lower thermocline (400–500 m) and an upper thermocline at about 100 m depth. Primary productivity in general is rather low and does not exhibit large seasonal changes, except when the winter cooling is sufficiently marked for breaking the upper thermocline. When that happens, vertical mixing down to the lower thermocline causes nutrient enrichment of the euphotic layer and thus a small spring bloom. While studying hydrological features and primary productivity on a transect from Nova Scotia to Barbados, HULBURT (1966, 1967) noticed a very obvious decrease in phytoplankton abundance in the 0 to 25-m layer in parts of this transect located in the Sargasso Sea itself, together with an increasing number of 'oceanic' species characteristic of oligotrophic areas, such as coccolithophorids, dinoflagellates (e.g. *Oxytoxum*), the cyanophyte *Trichodesmium thiebautii* and the diatoms *Hemiaulus hauckii* and *Stigmophora rostrata*. In winter when the nutrient-salt content is high, neritic species (chiefly diatoms) may be more abundant, sometimes even more numerous than oceanic species. This seasonal change mainly concerns the northern half of the Sargasso Sea, phytoplankton in the southern half being, on an annual scale, much more uniform.

Although we are presently concerned with surface and subsurface plankton, since the Sargasso Sea is a very particular hydrographic unit, it seems desirable to elucidate here the main features of its infrapelagic and bathypelagic assemblages as well. The structure of zooplankton populations may be summarized according to DEEVEY (1971) for the 0- to 500-m layer, and to DEEVEY and BROOKS (1971) for the 0- to 2000-m water layer. The dominant organisms in the 0- to 500-m layer are small copepods, ostracods and appendicularians, followed by coelenterates, chaetognaths, foraminiferans, pteropods and various larval forms. The highest number (over 200 ind. m^{-3}) are found in April and October; the maximum displacement volume (0·04 ml m^{-3}) in April does not last until October because an important part of the assemblage consists of small copepods. Below 500 m, crustaceans increase in number and diversity: for example, copepods which represent 70% of the plankton in the 0- to 500-m layer, rise up to more than 85% in the deepest waters sampled by DEEVEY and BROOKS. Ostracods are next in importance, but relatively less numerous below 1500 m. Maximum numbers of calanoid genera and ostracod species have been recorded between 500 and 1500 m. Euphausiids were most numerous between 1500 and 2000 m exhibiting a November maximum. Other taxa listed above occur in small numbers below 500 m. Due to increase in body size below 500 m—although the average total number of microzooplankton between 500 and 2000 m is only $\frac{1}{4}$ of that in the upper 500 m—the displacement column is 61 to 88% of the mean volume obtained from the 0- to 500-m layer (DEEVEY and BROOKS, 1971, 1977).

The annual cycle of zooplankton abundance mainly depends on the intensity of vertical mixing which controls primary production (DEEVEY, 1971). Equitability of copepod populations in the Sargasso Sea seems to be inversely related to primary productivity. The low level of primary production suggests that food shortage may occur for exclusively herbivorous species, e.g. the three copepod species of the genus *Pleuromamma*, whose average size is correlated with phytoplankton abundance. No correlation exists between phytoplankton abundance and average size of *Lucicutia flavicornis* and *Haloptilus longicornis*. It would seem therefore that these filter feeders appear to live on another food source, possibly organic aggregates. RILEY and co-authors (1964, 1965) observed that the amount of organic aggregates, which is relatively low, decreases from 0 to 50 m in the Sargasso Sea—with small seasonal changes related to those in phytoplankton abundance—but remains almost the same from 50 to 3000 m. It thus could provide some food for filter feeders.

Investigating the annual qualitative and quantitative distribution of zooplankton over four depth zones (0–500 m, 500–1000 m, 1000–1500 m and 1500–2000 m) at station S (32° 10′ N, 64° 30′ W) DEEVEY and BROOKS (1971, 1977) observed that total numbers varied widely during the year. Below 500 m the numbers tended to be greater from spring to autumn, but minimal in late autumn and/or early winter. DEEVEY and BROOKS stressed the high diversity of Sargasso Sea zooplankton (326 copepod species) attributing this to processes of lateral advection and mixing of water masses and water currents from widely differing sources.

Large changes in relative strength and direction of water flow cause variations in the number of species recorded. Examples of submerged perennial species whose abundance fluctuates irregularly are *Calanus finmarchicus* and *Metridia lucens*. These are constantly recruited by southward depth transport from more northern regions. Examples are bipolar species such as *Temorites brevis*, *Spinocalanus magnus* or *Heterorhabdus compactus* which could have been transported north in Antarctic intermediate waters or south in North Atlantic deep or intermediate water. 'Meanwhile, the annual cycles of the epipelagic species continue consistently from year to year, regardless of events in the deeper waters' (DEEVEY and BROOKS, 1977, p. 288).

American Mediterranean

The Caribbean Sea and Gulf of Mexico constitute the American Mediterranean, which in contrast to the European Mediterranean intensively communicates with the Atlantic Ocean, participating in its current system. It is also more fertile, its intermediate waters having a higher PO_4–P content than the Mediterranean Sea and an average primary production which is two to four times higher. Roughly, the surface circulation is an anticyclonic gyre. Although such a typical gyre exists only in the western half of the Gulf of Mexico, the central areas of both the Gulf and the Caribbean Sea are stratified and highly oligotrophic.

In the Gulf of Mexico the total neritic plankton is very abundant in the northern and north-eastern regions, especially on the northern part of the west coast of Florida and near the mouth of the Mississippi west of the estuary (KHROMOV, 1965a; BOGDANOV and co-authors 1968; Fig. 6-38). In winter the total plankton biomass in the 0- to 100-m layers exceeds 1000 mg m^{-3} and diatoms predominate. Zooplankton is plentiful as well

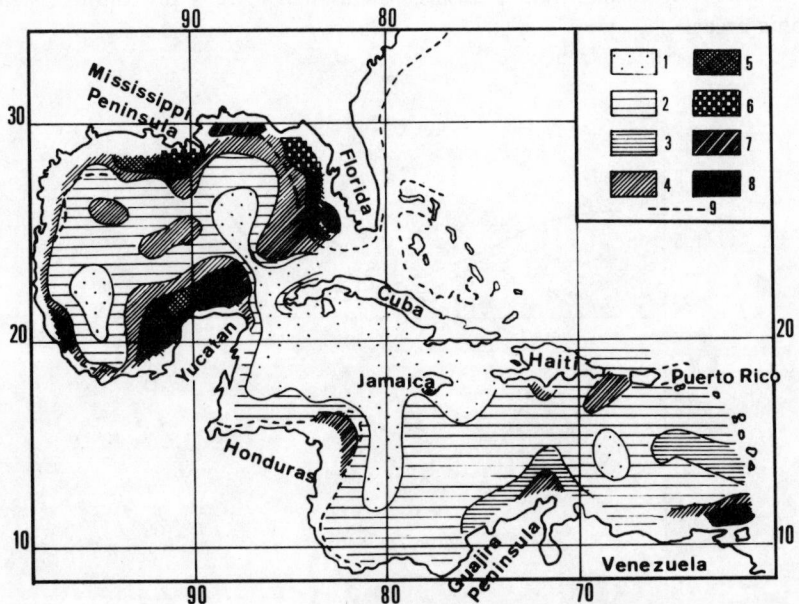

Fig. 6-38: Distribution of total plankton biomass (mg m^{-3}) in Gulf of Mexico and Caribbean Sea in 0–100 m (from data collected 1962–1966). 1: 30–100; 2: 50–150; 3: 100–200; 4: 100–300; 5: 200–600; 6: 200–1000; 7: 100–3000; 8: 300–1000; 9: shelf-edge. (Based on information provided by BOGDANOV and co-authors, 1968.)

as shoals of fishes which feed upon it. The fertility of this area depends mainly on run-off from the Mississippi. This brings nutrient salts to the sea (especially phospates), decaying organic matter and freshwater diatoms (over 1000 *Melosira* cells l^{-1}) (Fig. 6-39b). However, fertility also depends on winter vertical mixing and the north-north-east wind regime. More to the south, the shelf plankton of the west coast of Florida is still plentiful, only displaying a small decrease in summer (Fig. 6-39).

The northern Yucatan coast and the adjacent Campeche Bank are further highly productive areas due to a divergence which almost permanently provides the bank with intermediate waters from the Yucatan current (KHROMOV, 1965b, 1967). Primary production is very high (500–1400 mg C m^{-2} d^{-1}) and diatoms predominate (KABANOVA and LOPEZ BALUJA, 1970). The total plankton biomass on the inner shelf is 500 to 1000 mg m^{-3} (sometimes more than 1000 mg m^{-3} in the most inshore areas) and 300 to 500 mg m^{-3} on the shelf edge. The zooplankton maximum occurs in late summer and is associated with a concentration of fish shoals. Most of the fish species seem to spawn at this time, suggesting that their larval stages can benefit from the abundant food supply.

In central areas of the Gulf of Mexico the total plankton biomass is about 100 mg m^{-3}, sometimes less. The cyanophyte *Trichodesmium* almost always predominates in these oligotrophic areas.

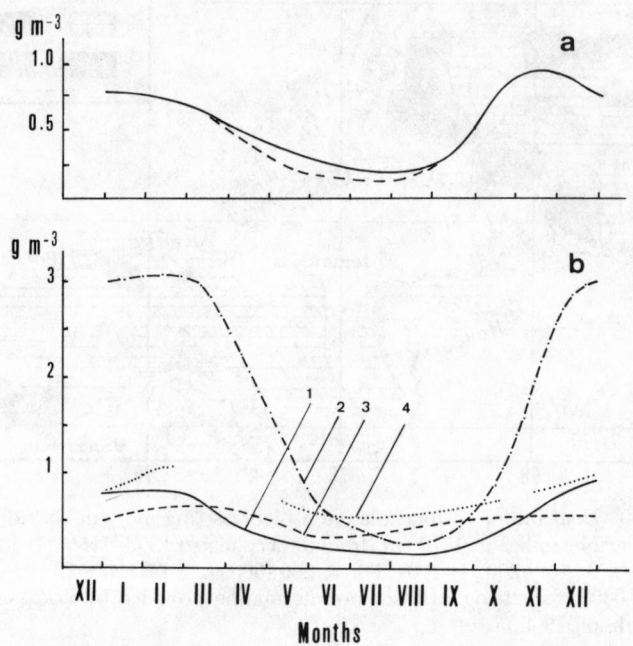

Fig. 6-39: Seasonal dynamics of total zooplankton biomass (g m^{-3}). (a) North-western Florida shelf, average depth 30–100 m; broken line in the period March–August possibly corresponds to samplings in June with a non-standard plankton net. (b) Northern shelf of Gulf of Mexico; 1 and 2: average and maximum values in areas east of Mississippi mouth; 3 and 4: average and maximum values in areas west of Mississippi mouth. (Based on information provided by KHROMOV, 1962.)

In the pelagic ecosystem of the Caribbean Sea MARGALEF (1971) distinguished three factors which augment enrichment of the euphotic layer: (i) river discharge; (ii) islands, which represent sites of fertilization both by land drainage and because they induce a vertical mixing by surface and internal waves; (iii) peripheral upwelling, the most active one located on Venezuela's north-eastern coast. Fig. 6-40 illustrates the striking differences between the most eutrophic areas east of Trinidad Island (I) and Margarita Island (III) and the relative oligotrophic offshore area (V). With respect to the taxonomic composition, an oversimplification of MARGALEF's detailed account allows a distinction of two main phytoplankton communities: (i) In upwelled waters the most common species are *Chaetoceros socialis*, *C. curvisetus*, *Thalassiosira subtilis*, *Asterionella*

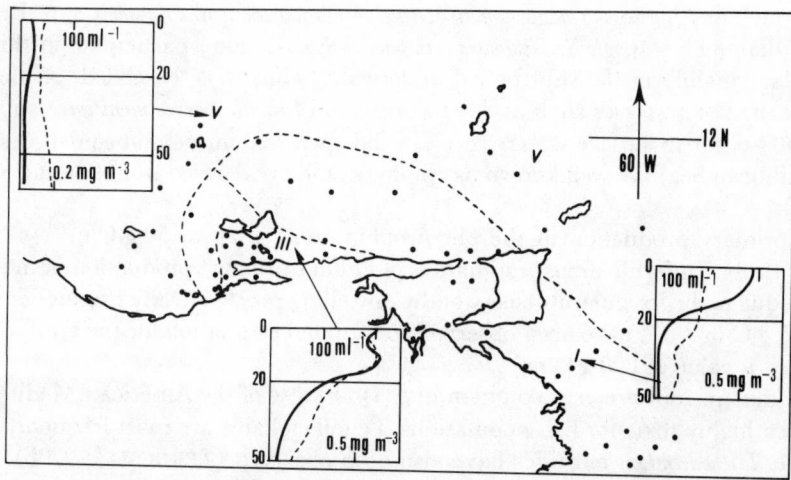

Fig. 6-40: Vertical distribution of phytoplankton, averaged over the year in surface waters (0 to 50–70 m) of north-eastern Venezuelan coasts, in eutrophic (I and III) and oligotrophic (V) areas. Solid and broken lines correspond to abundance (cells 100 ml^{-1}) and chlorophyll a content (mg m^{-3}), respectively. (After MARGALEF, 1971, simplified; reproduced by permission of UNESCO, Paris.)

japonica and *Stephanopyxis* (= *Skeletonema*) *costata* which are characteristic of the fertile period in the temperate seas. Other abundant species are *Cyclotella* cf. *caspia* and *Nitzschia seriata*. Total cell numbers in surface and subsurface waters may be over 200 ml^{-1}. (ii) In more stable situations, cell numbers range between 5 and 15 ml^{-1}. Diatoms—mainly associated with ciliates—are much more rare than in upwelled waters. However, *Chaetoceros coarctatus*, *Rhizosolenia cylindrus*, *Hemiaulus indicus* and *H. membranaceus* occur; more abundant are coccolithophorids, dinoflagellates (various species of the genera *Peridinium*, *Ceratium* and *Pyrocystis*) and also the cyanophyte *Trichodesmium*.

In neritic waters, diatoms (mainly *Thalassiosira subtilis*) predominate both in cell numbers and percentage of the total biomass, whereas in offshore areas *Trichodesmium* predominates in abundance and dinoflagellates in biomass (ZERNOVA, 1970a).

MARGALEF (1971) also noticed, at the boundary between the eastern more or less eutrophic areas and the western highly oligotrophic areas, a belt where *Noctiluca* is very common in the surface layer; this suggests, according to MARGALEF, possible coincidence with a moderate convergence or sinking.

The more detailed data of FERGUSON-WOOD (1971) on phytoplankton distribution in the Caribbean region are generally in agreement with those of MARGALEF (1971) and ZERNOVA (1970a). However, FERGUSON-WOOD distinguished a neritic phytoplankton assemblage in the warmer waters of the South American shelf which exists both inside and outside the Caribbean. This community includes *Bacteriastrum* sp., *Lauderia annulata*,

Stephanopyxis spp., *Leptocylindrus danicus*, many species of *Chaetoceros* as well as benthic species such as *Pleurosigma distortum*, *Navicula membranacea* and *Diploneis* spp. Benthic or epontic diatoms (*Mastogloia*, *Diploneis*, *Amphora*, *Synedra*, etc.) participate in this neritic assemblage mainly in the subtropical underwater which has 'cascaded' off the banks. Some of the latter species such as *Synedra* spp., and *Climacosphenia moniligera* spp., which frequently occur in surface waters (even in the open sea, inside and sometimes outside the Caribbean Sea) are well known as epiphytic and are derived from floating *Sargassum* (p. 353).

The primary production in the oligotrophic areas is 25 to 50 g C m^{-2} yr^{-1} (MAR-GALEF, 1971). In fertile areas, estimation of annual primary production is much more difficult due to the irregular intensity of the upwelling process. Daily productions as high as 1 to 5 g C m^{-2} d^{-1} have been observed. An annual estimation for the Gulf of Cariaco suggested a value of 800 g C m^{-2}.

Red tides (p. 360) are very common in coastal areas of the American Mediterranean and often highly toxic for fish populations. Dinoflagellates are most frequently respon-sible but *Trichodesmium* red tides have also been observed (ZERNOVA, 1970b).

At the time of phytoplankton maxima, or later, phytoplankton populations are domi-nated by copepods—in general small-sized and herbivorous—such as *Paracalanus aculeatus*, *P. parvus*, *Oncaea venusta*, *Nannocalanus minor*, *Clausocalanus arcuicornis* (MARGALEF, 1971). Cladocerans (*Evadne*, *Penilia*) are present and medusae were also identified: *Aglaura*, *Liriope*, *Rhopalonema*. Most of the species common in places of pulsating produc-tivity are cosmopolitan or of wide geographical distribution. The same species are also present near the coasts or in bays of other Caribbean areas (Cuba, Puerto Rico, etc.), in short, everywhere fertility is high. Farther offshore, in deeper water or during the season of decreased productivity, there is an obvious increase in species belonging to higher trophic levels forming low-density populations. This change in faunal composition is clearly pronounced as far as copepods are concerned. Many of the additional species (e.g. the copepod genera *Labidocera*, *Calanopia*, *Scolecithricella* and many deep-water inhabitants) exhibit a restricted geographical distribution. Quantitatively, MARGALEF (1971, p. 496) has estimated the zooplankton biomass at 0·03 to 0·1 g m^{-3} in oligo-trophic regions, at 0·2 to 0·3 g m^{-3} in fertile upwelling and coastal spots and at up to 1 g m^{-3} and even more in exceptionally rich places, such as coastal lagoons or areas close to a river discharge. The zooplankton fraction is important as fish food in general; for example, around Margarita Island, on average 100 to 600 copepods m^{-3} plus 10 to 100 m^{-3} other zooplankters are present. Along the coasts of Cuba, densities of 4000 copepods m^{-3} were observed.

The very rich plankton in some areas of the American Mediterranean provides pelagic fishes of the third and fourth stages of the trophic pyramid with abundant food. Shoals of clupeoid fishes (*Sardinella*, *Brevoortia*, etc.) are numerous in most fertile areas—mainly in places where the shelf is not too narrow and receives some fresh-water input—e.g. *Brevoortia patronus* in the Gulf of Mexico and *Sardinella anchovia* in the south-eastern region of the Caribbean Sea. In the vicinity of the productive areas, large populations of Scombridae (*Pneumatophorus japonicus* and *Scomber scombrus*) and Carangidae (*Trachurus lathami*) and other medium-sized fishes may be found. Big pelagic fishes like tunas and similar species occur more frequently in the upper layers of the boundary between the peripheral centres of higher productivity and the central oligotrophic areas. Most common are *Neothunnus albacora*, *Makaira albida*, *M. ampla*, *Histiophorus americanus*, *Coryphaena hippurus*, *Parathunnus obesus*, *Thunnus atlanticus*, *Germo alalunga* and *Xiphias gladius*. The bluefin *Thunnus thynnus* rarely enters the Gulf of Mexico and even more rarely the Caribbean Sea. Demersal fishes are also plentiful in some coastal areas, especially north of the Yucatan Peninsula and in front of the Mississippi mouth (BOGDANOV and co-authors, 1968). In view of the abundance and diversity of its resources, the pelagic ecosystem of the American Mediterranean presently appears underexploited.

Central and Eastern Intertropical Atlantic Ocean

In the eastern regions of the tropical Atlantic Ocean in inshore and distant-neritic areas, nutrient replenishment in the surface waters and the resulting blooms and zoo-plankton peaks depend either on river discharge or on the position of fronts between water masses and also on coastal upwellings. Offshore, enrichment in the equatorial region is associated with the fact that the thermocline occurs at a shallower depth.

Coastal areas

On all intertropical coasts of Africa the 'normal' aspect of the plankton assemblage corresponds to the pattern typical for tropical oligotrophic areas: low abundance and biomass and high specific diversity. Regular climatic changes induce conspicuous seasonal changes in the whole pelagos system and its plankton assemblages (THIRIOT, 1977).

Off the west coasts of Africa the hydrological features of shelf and distant-neritic areas are controlled by seasonal movements of thermal fronts. Their position at any time is recognizable by the 25 °C isotherm in surface waters (in fact by marked tightening of 23–27 °C isotherms). There are two main homologous fronts: the front of the Southern Hemisphere moves from Cabo Frio, Angola, in boreal winter to Cape Lopez, Gabon, in boreal summer; simultaneously, the front of the Northern Hemisphere, which lies in March at its southern limit off Cape Verga, Guinea (approximately 10° N), moves northward and reaches its northern limit off Mauritania (20–25° N), especially near Cape Blanc. During the boreal winter (November–March), the whole region between the northern and southern limits of the thermal front shifting is occupied by the warm waters of the equatorial counter-current with a poor content of nutrient salts. Primary production is low and the poor zooplankton assemblage mainly consists of tropical species. Later, biological enrichment takes place along the moving front and in areas of coastal upwelling (CORCORAN and MAHNKEN, 1969); this results from changes in the wind regime, for example when south-east trade winds pass the equator and become the south-west monsoon. Nutrient input by rivers during the summer rainy season also plays some part in the enrichment processes. These hydrographic features result in seasonal changes of the plankton assemblages which consist alternately of temperate species (often called 'cold') and tropical, i.e. 'warm', species. The following examples help to illustrate this alternation.

Off the Mauritania coast a very important upwelling occurs in the vicinity of Cape Blanc (Fig. 6-41); here LLOYD (1971) discovered a high primary production in spring (May–June): $1 \cdot 12$ to $3 \cdot 35$ g C m^{-2} d^{-1}. Such production would be too high to be con-

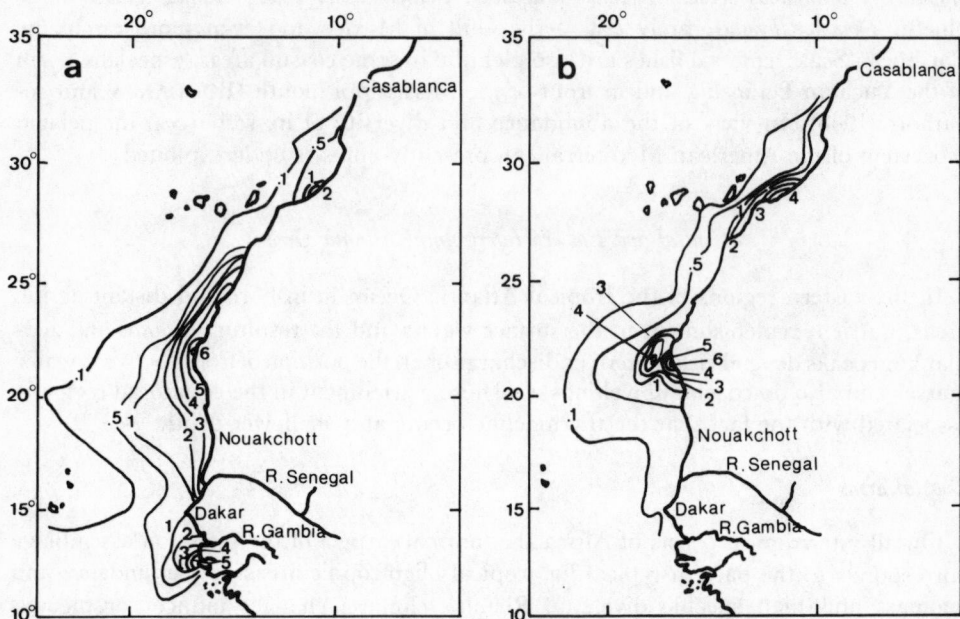

Fig. 6-41: Composite of chlorophyll concentrations (mg m⁻³) for January–June (a) and July–September (b). (After SZEKIELDA and co-authors, 1977; reproduced by permission of Conseil International pour l'Exploration de la Mer, Charlottenlund.)

sumed by grazers. LLOYD assumes that the phytoplankton might be transported more offshore where zooplankton is more abundant. These values of primary production have been corroborated by HERBLAND and co-authors (1973) who observed over an 8- to 15-day period a net production of $10\cdot5$ g C m⁻² and a total organic production (net production and organic excretion) of $19\cdot5$ g C m⁻²; the maximum biomass occurred on the fifth day (377 mg chlorophyll a m⁻²). REYSSAC (1974, 1975) observed 2 upwelling pulses during the period from November 1972 to June 1973. During the first pulse (December–January), intermediate waters reached the surface and the phytoplankton bloom was small (always below 1000 cells ml⁻¹) probably because water renewal at the surface was too quick to permit algal development. During the second pulse (March–May) intermediate waters did not rise above 20 m; this resulted in a very striking bloom (up to 7500 cells ml⁻¹ in March). The highest biomass values were observed at the surface in peripheral areas of the upwelling itself; here phytoplankton populations include almost exclusively diatoms, mainly temperate or cosmopolite: *Thalassiosira partheneia* largely predominated, but *Chaetoceros sociale*, *C. compressum* and *Skeletonema costatum* were also abundant.

As regards the zooplankton, the observations by PAVLOV (1968) do not agree very well with LLOYD's (1971) assumption. The greatest zooplankton biomass (500–1000 mg m⁻³) encountered by PAVLOV was on the shelf. Farther offshore, the 0- to

50-m layer was less rich in zooplankton (200 mg m^{-3}) and the true oceanic surface waters (0–50 m) beyond 50 to 60 miles from the shore contained less than 100 mg m^{-3}. The zooplankters sampled here belonged to the temperate assemblage with the characteristic copepods *Calanoides carinatus* and *Paracalanus parvus*; they largely predominated in shelf waters together with *Nyctiphanes capensis*; *Acartia danae* and *Metridia lucens* were the most abundant species in offshore waters. In the latter, more oceanic euphausiids— mainly *Euphausia krohnii*, associated with *Nematoscelis megalops* and *Thysanoessa gregaria*— predominated in the total biomass, whereas *Calanoides carinatus* mainly occurred in deeper layers (sometimes down to 500–1000 m). According to an assessment by HERB-LAND and co-authors (1973), net mesozooplankton production would be 1·0 to 4·2 g C m^{-2} for six days in the coastal areas. The authors also conclude that zooplankton excretion plays an important part in nitrogen and phosphorus regeneration and induces bacterial activity. The latter increases due to phytoplankton decay at the end of the bloom. In autumn, thermophilic species become somewhat more abundant; then the most abundant phytoplankter is *Thalassiosira subtilis* in the 0- to 100-m layer at both shelf and offshore stations. However, *Calanoides carinatus* (mainly Copepodites IV and V) is the most abundant zooplankter and features a much higher lipid content than in spring.

Off Dakar, the Canary current, strengthened by the trade winds from October to May, induces an upwelling which supports rather abundant phytoplankton. Diatoms predominate: Chaetoceridae are the most numerous in February and March, and *Stephanopyxis palmeriana* in April and May (Fig. 6-41b). The rather cold (17–21 °C) surface waters are inhabited by 'temperate' zooplankters; among these the copepod *Calanoides carinatus* is the most common species, associated with *Eucalanus attenuatus*, *E. crassus*, *Euchirella rostrata*, *Euchaeta hebes*, *Candacia bipinnata* and the chaetognath *Sagitta serratodentata*. The warmer period from June to October corresponds to the north-west monsoon. At the beginning, the surface waters (22–27 °C; 35·8–36·0⁰/₀₀ S) are inhabited by a subtropical assemblage with the copepod *Calanus minor* and *Undinula vulgaris*, the molluscs *Atlanta inclinata*, *A. lesieurei*, *Diacria quadridentata*, the chondrophorid *Physalia physalis*, and the medusa *Phialidium hemisphaericum*. Salinity does not change in July and August but temperatures rise to 29 °C. Pontellid copepods then become very abundant, together with some siphonophores (*Muggiaea atlantica*, *Lensia subtilis*, *L. conoidea*, *Chelophyes appendiculata* and *Abylopsis tetragona*), the ctenarian *Beroe forskali*, the salps *Thalia democratica*, *Salpa maxima*, *S. fusiformis* and *Iasis zonaria*. Finally, in September and October the temperature remains high (27–29 °C), but the salinity drops below 35⁰/₀₀ S owing to run-off following the rainy season. In this period the predominant zooplankters are siphonophores (*Diphyes bojani*) and Thaliacea (*Thalia democratica* and *Doliolum gegenbauri*); fish eggs and larvae are also abundant. During the whole warmer period phytoplankton is scarce, but some small blooms may be observed in autumn—generally restricted to the vicinity (20–25 miles) of the shoreline. In spite of strong seasonal fluctuations in phytoplankton populations, the zooplankton biomass does not change very much (200–400 mg m^{-3}) on an annual scale, possibly due to the high consumption by *Sardinella* (KHROMOV, 1962).

Off the Ivory Coast, the seasonal monsoon (west-south-west) pushes coastal warm waters towards the open sea. This is replaced by colder water from the 200- to 300-m layer. Blowing from January to February and stronger from July to October, the seasonal monsoon induces two upwellings corresponding to the short cold season (surface temperature 24–26 °C) and the 'main cold season' (surface temperature below 25 °C). There are two warm seasons with surface temperatures of 27 to 28 °C: the long one (March–June) is characterized by an increased stability of the water masses with a thermocline at about 20 m; and the short one occurs in November and December. The changes in the hydrological climate result in two main phytoplankton assemblages (REYSSAC, 1970; REYSSAC and ROUX, 1972; DANDONNEAU, 1972). In the main cold season, diatoms constitute at least 50%, sometimes up to 100%, of the algal population: most characteristic are species of the genera *Chaetoceros* (*C. lorenzianum, C. affinis, C. compressum, C. diversum*) and *Rhizosolenia* (*R. hyalina, R. stolterfothii*), several species of the genus *Coscinodiscus, Leptocylindrus danicus, Hemiaulus indicus* (which also produces short and pronounced blooms in April–May) and *Guinardia flaccida*; the dinoflagellate *Ceratium tripos* is also present. As the upwelled waters are transported offshore some substitutions of species occur, chiefly in the genus *Rhizosolenia*; the dinoflagellate *Gymnodinium splendens* appears and sometimes may develop into red tides. During the short cold season, the blooms are weaker and dinoflagellates always predominate. When the main warm season starts, an important development occurs in the 'ivorian waters' of the cyanophyte *Trichodesmium thiebautii*, followed by that of *Hemialus indicus*. Later, dinoflagellates predominate, mainly *Ceratium furca, C. trichoceros, Ceratocorys horrida* and *Dinophysis caudata*. During the short warm season, one observes mainly the same species: *Ceratium* and *Dinophysis*. At maximum run-off, just before the start of the cold season, a small phytoplankton peak occurs which is mainly characterized by increased diversity of *Chaetoceros* spp. and the presence of *Skeletonema tropicum, Bellerochea malleus* and *Cerataulina pelagica*.

The zooplankton of the Ivory Coast has been poorly investigated. However, it seems likely that it has the same characteristics as that of the Congo coast, the hydrological features of which are similar to the Ivory Coast. On the Ivory Coast, BINET and DESSIER (1971) distinguished two different copepod assemblages. During the main warm season (January–March) the 'Guinean waters' (temperature over 24 °C salinity below 34‰ S) are inhabited by tropical species. Typical of these are the copepods *Undinula vulgaris, Eucalanus pileatus, E. subtenuis, Paracalanus aculeatus, Clausocalanus furcatus, Euchaeta paraconcinna, E. marina* and *Centropages furcatus*. The crossing of the cold front and the upwelling modify the shelf waters which become colder (less than 23 °C) and more saline (more than 35‰ S). A more temperate assemblage now prevails with the copepods *Calanoides carinatus, Eucalanus crassus, E. monachus, Oncaea mediterrania* f. *major, Sapphirina migromaculata* and *Corycaeus africanus*. In September, at the end of the main cold season, *Temora turbinata* and *Oncaea venusta* reveal a population peak. During the short cold and warm seasons, one observes an increase in the two above-mentioned assemblages—even more marked for neritic species inhabiting the warm surface waters. Obviously, the sharp thermocline in the main cold season causes a marked zooplankton discontinuity. Some species are restricted to the upper tropical surface waters and others to the underlying South Atlantic central water. However, some species also exhibit diel migrations between the two layers (BAINBRIDGE, 1972). The same author also emphasized that during the upwelling, the zooplankton diversity decreased and the percentage of carnivorous species in the total zooplankton declined below the usual 20 to 40% level. The two copepods *Calanoides carinatus* and *Eucalanus monachus*, both rare over the shelf during stable conditions, become abundant in the upwelling period. *C. carinatus*, one of the most efficient exploiters of the diatom bloom in coastal zones with maximum productivity can accumulate fatty reserves—an unusual phenomenon in subtropical species. It seems to play the same role in the trophic system as *Calanus finmarchicus* in boreal regions. During the warm seasons, *C. carinatus* has been recorded from deeper layers more offshore. According to BINET (1977), the shrimp *Lucifer faxoni* also exhibits ontogenic migration: older individuals inhabit deeper layers than younger ones. Possibly, members of other taxa display the same seasonal vertical migration—inhabiting the surface water during the colder seasons and lowering themselves during the warm seasons.

Due to the high fertility of the west African coasts, pelagic fishes occupying the third and fourth steps of the trophic pyramid often exhibit abundant populations. Among the plankton feeders, the most important are: *Sardinella eba* (flat sardine), mainly living in low-salinity waters between the south of Senegal and Sierra Leone; *S. aurita* (round sardine), inhabiting more saline waters mainly at the margin of upwelling areas where it mainly feeds on *Calanoides carinatus*, and *Ethmalosa fimbriata*, which is restricted to shallow brackish waters. Medium-sized predatory fishes—such as the scombrid *Pneumatophorus colias*, the carangids *Trachurus* sp., *Decapterus* sp. and *Caranx rhonchus*—are common mainly to the north of Sierra Leone, in waters above the deeper parts of the shelf.

Large-sized predators mainly inhabit offshore waters. Seasonal changes in the hydrological climate modify their seasonal distribution. POSTEL (1968) summarizes the situation as follows: The temperate waters are characterized by two tuna-like species *Orcynopsis unicolor* and *Sarda sarda* and the carangid *Pomatomus saltatrix*, whereas the warmer waters are inhabited by the yellowfin *Neothunnus albacora* and the skipjack *Katsuwonus pelamys* which feed on fishes, squids and mainly crustaceans: larval stages of decapods and hoplocarids, hyperiids (e.g. *Phronima* spp.) and *Euphausia hanseni*. The region located to the east of the line from Cape Blanc to the Canary Islands is always occupied by temperate waters, whereas tropical waters permanently predominate in the region located south of the line Cape Roxo–Cape Verde Islands. Between these two regions is a wide transitional area, with seasonal temperature changes related to the movements of the front. The yellowfin moves northwards and disperses in the transitional area and farther offshore, following the summer movement of the warm front and avoiding the areas where upwelling occurs. In winter it moves back behind the line Cape Roxo–Cape Verde Islands. The temperate species mentioned above—restricted to the area north of Cape Blanc in summer, with maximum concentrations in the Bay of Agadir—move southward in winter down to the latitude of Gambia, and are sometimes off the Ivory Coast and Ghana, probably in areas influenced by upwelling.

VOSS (1969) emphasized that the mid-water fauna of the Gulf of Guinea is exceptionally rich both in species and number of individuals, due to the strong upwelling which occurs, especially in the eastern part of this gulf. This peculiarity leads to a very important food supply for surface predatory fishes such as tunas, and perhaps also opens up possibilities for them to exploit directly small-sized mid-water fishes (myctophids, *Cyclothone*).

Offshore areas

In offshore areas, biological enrichment originates from a rising of subsurface and intermediate layers rich in phosphate. For instance, during the cold season the 25 °C isotherm reaches the surface between the Latitudes 2 and 4° S, whereas during the warmer season this isotherm is at a depth of 20 to 50 m (CORCORAN and MAHNKEN, 1969).

The offshore phytoplankton of the intertropical Atlantic Ocean is poorly investigated. Quantitative aspects are summarized in Fig. 6-42. ZERNOVA (1974) observed in

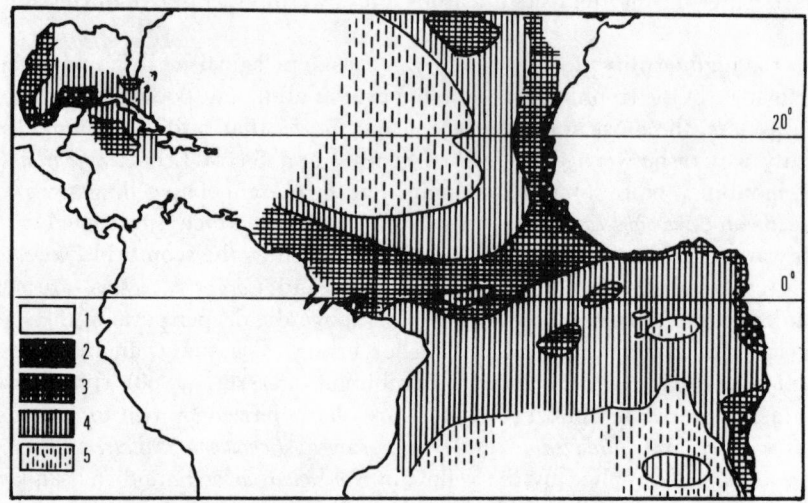

Fig. 6-42: Distribution of phytoplankton biomass (mg m^{-3}) in tropical and equatorial waters of the Atlantic Ocean. 1: >500; 2: 500–100; 3: 100–10; 4: 10–1; 5: <1. (Based on information provided by ZERNOVA, 1974.)

almost all offshore equatorial regions a phytoplankton biomass of 10 to 100 mg m^{-3}. On the meridian 30° W, the respective percentages of the most important taxonomic phytoplankton groups seem to differ largely over a small scale of latitude: at least 12° N, diatoms make up 45%, dinoflagellates 50%; at about 8° N, diatoms 20%, dinoflagellates 65%, coccolithophorids 10%; at about 2° S, diatoms 30%, dinoflagellates 55%, silicoflagellates 7%, cyanophytes 5% (SANINA, 1969).

The integrated primary production almost everywhere exceeds 0·2 g C m^{-2} d^{-1} between 2° N and 8° S (up to 0·375 between 0 and 2° S) during the cold season; it averages 0·1 g C m^{-2} d^{-1} between 6° N and 4° S during the warm season (CORCORAN and MAHNKEN, 1969). According to MINAS (personal communication) production ranged between 0·3 and 1·4 g C m^{-2} d^{-1} in a belt 0 to 3° 30′ S.

The specific composition of the zooplankton assemblage in central areas of the inter-tropical central Atlantic Ocean is probably uniform and hitherto insufficiently investigated. According to GREZE and co-authors (1969) some species, such as *Euchaeta gladiofera* and *Pleuromamma borealis*, seem to exist only in the eastern part of the area. The major part of the biomass consists of copepods, with chaetognaths, euphausiids and siphonophores next in rank. Fig. 6-43 and Table 6-8 show that the highest biomass in the 0- to 200-m layer is encountered not too far from western and eastern sides of the Ocean, whereas the most central regions, particularly those further south, are much

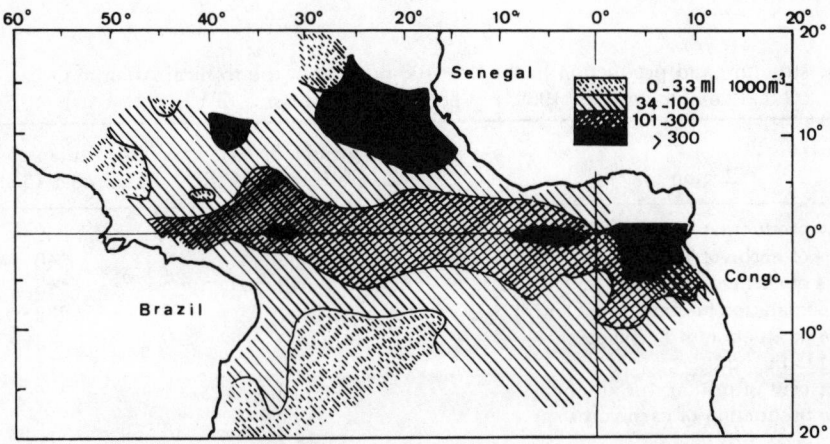

Fig. 6-43: Distribution of zooplankton displacement volumes from stations occupied during Equalant II during the cold season. (After CORCORAN and MAHNKEN, 1969; reproduced by permission of UNESCO-FAO, Paris.)

Table 6-8

Composition of zooplankton biomass (%) in various regions of the tropical Atlantic Ocean (After GREZE and co-authors, 1969; reproduced by permission of UNESCO–FAO)

Regions	$G = 15° W$ $\varphi = 5° N–4° S$		$G = 35° W$ $\varphi = 1·5° N–1·5° S$		$G = 15–22° W$ $\varphi = 8°–14° S$		$G = 17–35° W$ $\varphi = 15° S$	
Layer (m)	0–100	100–200	0–100	100–200	0–100	100–200	0–100	100–200
Siphonophores	8·6	6·2	3·8	0·8	9·0	9·2	1·7	0·2
Copepods	59·2	70·1	49·0	57·0	60·0	41·5	57·6	56·5
Euphausiids	4·7	3·5	12·9	10·1	2·6	1·2	3·3	10·1
Chaetognathes	10·5	3·3	15·8	23·6	12·3	10·5	11·5	21·0
Others	17·0	17·9	18·5	8·5	16·1	37·6*	25·9	12·4
Total (%)	100·0	100·0	100·0	100·0	100·0	100·0	100·0	100·0
Biomass (mg m^{-3})	87·9	53·9	102·0	25·8	28·6	14·9	24·0	13·8

Molluscs: 12·4%.

poorer. GREZE and co-authors assume that the ratio carnivores/herbivores in the whole zooplankton is similar in different areas of the region explored. The 0- to 100-m layer contains 45 to 50% of carnivorous species. Comparison between productivity at the first, second and third trophic levels suggests a fairly efficient energy transfer (Table 6-9).

As regards the quantitative point of view, Table 6-8 clearly demonstrates that the zooplankton biomass in the equatorial region decreases offshore and, at the same

Table 6-9

Trophic structure and production in the 0 to 100-m layer of the tropical Atlantic Ocean (After GREZE and co-authors, 1969; reproduced by permission of UNESCO–FAO)

Region	Longitude, 35° W Latitude, 5° N–4° S	Longitude, 17–35° N Latitude, 15° S
Primary production (mg C $m^{-2} d^{-1}$)	480	252
Biomass of herbivores (mg C m^{-2})	658	140
Biomass of carnivores (mg C m^{-2})	572	148
Total zooplankton biomass (mg C m^{-2})	1230	288
Possible production of herbivores		
in mg C $m^{-2} d^{-1}$	65	14
in per cent of primary production	13	5·5
Possible production of carnivores		
in mg C $m^{-2} d^{-1}$	57	14
in per cent of herbivore production	88	100

longitude in the open sea, also decreases southwards. The figure from GREZE and co-authors (1966, Fig. 2, p. 87) which shows biomasses in mg m^{-3} and Fig. 6-43 (CORCORAN and MAHNKEN, 1969) in which zooplankton abundances are expressed in ml 1000 m^{-3} are rather similar.

Western Regions of the Intertropical Atlantic Ocean

The western areas of the Intertropical Atlantic Ocean has received much less attention than the African coast.

Areas influenced by Amazon and Orinoco discharges

The large area influenced by Amazon River run-off has been investigated by RYTHER and co-authors (1967) and CALEF and GRICE (1967). A large lens of appreciably freshened surface water (200–300 miles in diameter) exists, several hundred miles to the north and east of the Amazon estuary, separated by waters of higher salinity. This freshened surface water, which is transported by the Guiana current towards the north-west, affects an area about one million square miles during the wet season but somewhat more restricted during the dry season. In comparison with the surrounding saline waters, the freshened waters have a higher content of silicate, but nitrate, phosphate and phytoplankton contents are rather low. Therefore it seems that the river outflow causes a decrease in fertility of oceanic waters. However, a more productive region exists between the freshened lens and the coast of the Guianas. This fact is supported by ARZHANOVA (1974) who observed that the photic layer in the Amazon and Orinoco discharge areas is enriched with biogenic elements. In summer favourable conditions occur in inshore waters where the primary production varies between 1·4 and 4·04 g C $m^{-2} d^{-1}$ although the euphotic layer is only 8 to 20 m thick. Further offshore both the content in biogenic substances and the primary production decrease. Throughout the year, STEVEN (1971) observed a constant primary production of about 105 g C m^{-2} yr^{-1} in the Guianas current. This suggests that the increase in fertility originating from the Orinoco (possibly also the Amazon) input may contribute to enriching the most south-eastern region of the American Mediterranean Sea. As regards the total zooplankton, the average displacement volume in the freshened area is almost three times higher during the wet season than during the dry season. The copepod fauna is more diverse during the wet season, but several of the copepods identified in the area are found everywhere. In the offshore freshened lenses, two neritic species are very common: the cladoceran *Evadne tergestina* and the decapod *Lucifer faxoni* (CALEF and GRICE, 1967).

Upwelling area at Cabo Frio, Brazil

J. VALENTIN (personal communication) is currently carrying out an intensive study of zooplankton populations. He kindly made some information available about the plankton assemblage in the vicinity of Cabo Frio, located north-east of Rio de Janeiro. In this area the Brazil current tends to turn left, i.e. offshore towards the south-east. This results in coastal upwelling sporadically enforced in summer by easterlies and north-easterlies. Between September and February, the temperature of the subsurface (42 m depth)—which may rise up to the surface when the wind becomes stronger—is 14 to 20 °C (sometimes lower if deeper waters ascend), instead of 17 to 24 °C from March to August. Therefore, at the same station a subtropical planktonic assemblage alternately occurs in the Brazil current waters (20–24 °C) and a cold (12–16 °C) water plankton assemblage during upwelling. Obviously, some influence of neritic waters may also be involved.

In the Brazil current the most frequent phytoplankters are *Rhizosolenia setigera* and *R. stolterforthii*, next being *Hemiaulus sinensis* and *Coscinodiscus* spp. (among them *C. excentricus*). Opportunists, such as several *Nitzschia* (*N. delicatula*, *N. seriata*, *N. closterium*) and *Leptocylindrus danicus*, also occur. Sometimes *Pleurosigma* sp. may contribute about 90% to the phytoplankton assemblage. Dinoflagellates are rather scarce: less than 20% of the total population. In upwelled waters, which come into contact with the bottom before reaching the euphotic layer, benthic diatoms largely predominate: species of the genera *Melosira* (mainly *M. sulcata*), *Schroederella*, *Diploneis* and often some cyanophytes. Two diatoms (*Asterionella japonica* and *Skeletonema costatum*) which are very common almost all over the World Ocean on shelf areas, exist in both plankton assemblages mentioned.

From the quantitative point of view, phytoplankton abundance averages 3500 cells l^{-1}. In the upwelling season, blooms (60,000 cells l^{-1}) of the opportunist *Asterionella japonica* may be observed. Red tides may occur in semi-enclosed areas; the most important hitherto observed lasted about two weeks; the phytoplankter responsible was a cocco-lithophorid (*Calyptrosphaera* sp.) with an abundance of 3,200,000 cells l^{-1}.

In the zooplankton, discrimination of two assemblages is less obvious than in the phytoplankton due to fast mixing of upwelled waters with neritic or Brazil current waters. J. VALENTIN (personal communication) suggests that the most characteristic species of the two main organismic assemblages—corresponding to those of phyto-plankters—may be the following. (i) Brazil current: *Clausocalanus furcatus**, *Temora stylifera**, *Oncaea media*, *Udinula vulgaris*, *Nannocalanus minor*, the planktonic shrimp *Lucifer* *faxoni* and the chaetognath *Sagitta enflata*. (ii) Upwelled waters: the copepods *Cteno-calanus vanus*, *Microsetella* sp., *Oithona* sp. *Rhincalanus cornutus*, *Scolecithrix bradyi* and the ostracod *Conchoecia* sp.; another copepod, *Calanoides carinatus*, is sometimes also very abundant, similar to the situation in the eastern tropical Atlantic Ocean.

In shelf waters the copepods marked above with an asterisk are also common, together with *Paracalanus quasimodo*, *Corycaeus* sp., *Calanopia americana*, the cladoceran *Penilia avirostris* and the chaetognath *Sagitta fridericii*.

Zooplankton abundance (in mg organic matter m^{-3}) is 40 to 70 in both upwelling and Brazil current waters. Mixed waters are generally a little richer (60–90). There are some slight seasonal differences in abundance in unmixed waters: the bulk of zooplankters is larger in summer (average: 60 mg m^{-3} organic matter) than in winter (average: 40).

(d) The Subtropical Pelagos of the South Atlantic Ocean

The plankton assemblages of the South Atlantic Ocean between the south equatorial current and the subtropical convergence have received little attention. Qualitative data on the specific composition are scattered and possibly defective. Hence, a synthetic overview is almost impossible.

Offshore Areas

In offshore areas the general distribution scheme of the main assemblage in subtropical South Atlantic Ocean waters might correspond to that established by BOLTOVSKOY (1962, 1970) who discussed foraminiferan distribution in different water masses. Because several species in this group are valuable indicators of hydrological characteristics, BOLTOVSKOY distinguished two offshore water masses in the subtropical South Atlantic Ocean: (i) true subtropical waters (37·07–36·11⁰/₀₀ S; 24·4–22·2 °C) with well-grown specimens of *Globorotalia menardi* (f. *typica*), *Globigerinoides ruber* (f. *typica*) and *G. glutinata*; (ii) subtropical waters with some admixture of subantarctic surface waters (35·58–35·75⁰/₀₀ salt; 23–22 °C); here the same species are represented by smaller individuals and *G. ruber*, but not by f. *typica*. Beyond the southern limit of the subtropical convergence, the position of which seasonally changes, the subantarctic surface waters are inhabited by *Globigerina bulloides* (f. *typica* and *conglomerata*) and *G. inflata*; full-grown individuals of the latter only occur in areas not influenced by an admixture of subtropical waters. Intrusion of these warmer waters results in a decrease in number and size of *G. inflata* and the appearance of *G. ruber*.

The true tropical offshore waters are highly oligotrophic, especially between 10° S and 30 to 32° S. The phytoplankton biomass is below 1 mg m^{-3}. Between about 0 and 40° W, cell numbers are always lower than 100 l^{-1}, sometimes below 50 l^{-1} between 10 and 15° S. Primary production seems to be lower than 50 g C m^{-2} yr^{-1} in the whole central area between 10° S and 30 to 32° S. Dinoflagellates always largely predominate (60–75%), whereas the diatoms are poorly represented: sometimes only 10%, never more than 25%. South of 20° S, coccolithophorids may account for about 10% of the total cell number, but never attain their high percentages (often more than 20%) in the North Atlantic Ocean gyre between about 15 and 34° N. Between 32 and 36° S, even in the most open sea areas, fertility slightly increases due to seasonal movement of the subtropical convergence with productivity values of 50 to 100 g C m^{-2} yr^{-1}.

In the 0- to 300-m layer, the zooplankton biomass is less than 50 mg m^{-3} for central areas of the South Atlantic gyre from 10° S to 35 to 36° S. A slight increase (50–100 mg m^{-3}) occurs in the most northern region of the transitional zone to where the subtropical convergence seasonally moves (but seems to be smaller than that of the phytoplankton); abundance expressed by displacement volume is always less than 33 ml. 1000 m^{-3}.

The most marginal western and eastern areas of the South Atlantic Ocean gyre benefit by some peripheral enrichment arising from the higher fertility of neritic areas. Because the zooplankter development and life cycle are longer than those of phytoplankters and because of the direction of the main currents, zooplankton belts extend more in latitude than those of phytoplankton. For example, zooplankton biomass values of 100 to 150 mg m^{-3} occur on the western side between 20° S and the subtropical convergence as far offshore as 30° W; on the eastern side, the same biomass prevails approximately between 15 and 30° S as far as 10° W, and biomasses of 150 to 300 mg m^{-3} have been observed between 18 and 22° S as far offshore as the 0° meridian.

Neritic Pelagos on the Western Side of the South Atlantic Gyre

The pelagos in the neritic province along the coast of Brazil, Uruguay and Argentina is probably the least well-investigated area in the Atlantic Ocean, except for the Cabo Frio upwelling. The Brazilian neritic waters located between the shore and the Brazil current (p. 286) are very poor: phytoplankton biomass values between 3 and 20° S are between 1 and 10 mg m^{-3}, and primary production is lower than 150 mg C m^{-2} d^{-1} (less than 100 mg m^{-3} for the 0- to 300-m layer).

South of 20° S, the Brazil current tends to turn to the south-east then to the east and therefore induces a divergence in the Cabo Frio area where this dynamic process is especially strong. Average primary production probably increases to more than 100 g C m^{-2} yr^{-1}; zooplankton biomass increases to 200 mg m^{-3} in the 0- to 300-m layer along the coasts of southern Brazil and Uruguay, and to higher values (250 mg m^{-3}) along the Argentinian coast from the Rio de la Plata (Parana) estuary in the south to the latitude of the subtropical convergence, the northern boundary of the subantarctic region. Beyond the convergence the neritic primary production exceeds 200 g C m^{-2} yr^{-1} and the zooplankton biomass is 200 to 400 mg m^{-3} in the 0- to 300-m layer.

Pelagic fishes are rare in the northern part (3–20° S) of the neritic province. From 20 to about 30° S *Sardinella* sp. is fairly abundant, its largest populations occurring at the periphery of the divergence area. *Scomber* sp. is also rather common; it migrates seasonally along the coasts of Brazil and Uruguay between 15 and 35° S. More to the south, the most abundant pelagic fish is the anchovy *Engraulis anchoita*; its main spawning area is in front of the Rio de la Plata mouth (below 32⁰/₀₀ S). In general, an inverse relationship may be observed between the respective abundance of zooplankton and eggs and larvae of anchovy, probably because anchovy feeds on zooplankton. Although *E. anchoita* may be observed up to 30° S, its largest populations occur between 35 and 42° S, i.e. in areas where subantarctic influences related to the Falklands current are more marked.

Many squid species display a large geographical distribution in the intertropical and subtropical regions of the Atlantic Ocean. As in the Pacific Ocean, distant neritic species of the Southern Hemisphere exhibit their largest aggregations on the western side of the Ocean and in the vicinity of the subtropical convergence. For example, populations of *Illex argentinus* concentrate on the Patagonian shelf and slope in the front between Brazil current and Falklands current. Among the more oceanic species *Stenoteuthis pteropus* has been recorded in many areas where a tropical current tends to turn offshore and induce upwelling—e.g. Canary current and Benguela current—in the productive areas of the Gulf of Guinea, off Mauritania, Angola and Namibia, as well as near the tropical fronts, the tropical convergences and the equatorial divergence; *S. pteropus* also occurs in the Mediterranean Sea, the Caribbean Sea and the Gulf of Mexico. *Ommastrephes bartrami* exhibits a similar distribution but also occurs near the Azore Islands off the south-west African coasts and in the vicinity of the front between Brazil current and Falklands current. In the latter region the populations extend along the subtropical convergence farther from the shore than elsewhere. It appears that the distribution of both squids depends on increased production in the surface layers, whatever the cause of such increase: divergence, upwelling or 'island effect'. Where the increase is not permanent—e.g. seasonal—abundance of the squid populations reveal parallel annual variation. In contrast, squid populations are very scarce in central water masses.

Neritic Pelagos on the Eastern Side of the South Atlantic Gyre

The side of the South Atlantic Ocean gyre along the west African coast is influenced by the Benguela current which flows northwards almost parallel to the coast, approximately above the shelf edge up to 10° S. At this latitude the current turns gradually to the north-west, then to the west and becomes the south equatorial current. Like the Peru current, the Benguela current induces upwelling of subsurface and upper intermediate waters.

The very high PO_4–P content (up to $3 \cdot 0$ mg at. m^{-3} in the euphotic layer) in areas with strong upwelling, results in striking fertility in the surface layers of the distant shelf. Diatoms always predominate, mainly Chaetoceridae and *Planktoniella sol* in the waters which, once upwelled become stable during a few days. The phytoplankton biomass above the distant shelf exceeds 500 mg m^{-3} between about 1 and 8° S and between 13 and 27° S. A little more offshore in front of these most fertile areas the vegetal biomass amounts to between 250 and 500 mg m^{-3}. Primary production exceeds 500 mg C m^{-2} d^{-1} in the most fertile areas mentioned above and 250 to 500 mg C m^{-2} d^{-1} in peripheric areas. The zooplankton biomass in the 0- to 300-m layer is over 300 and sometimes over 500 mg m^{-3} in the most productive zones.

The diatoms *Planktoniella sol* and *Thalassiothrix longissima* may be considered indicator species of oceanic waters. In inshore waters species of the genus *Chaetoceros* predominate and sometimes constitute up to 90% of the total phytoplankton, but the neritic *Eucampia zodiacus* and *Asterionella* spp. may also be common in upwelled coastal waters.

Between the Benguela current and the shoreline, the inshore areas are occupied by coastal waters. The zooplankton which inhabits these waters is rather scarce and exhibits tropical characteristic species such as the chaetognaths *Sagitta fridericii*; cladocerans seem to be rather common too. In the contact zone with upwelled waters, the zooplankton is rather rich, especially serving as fish food for euphausiids, copepods and pteropods. The waters of the 100- to 400-m layer which flow southwards and correspond to the compensation of the upwelling are inhabited by *Sagitta serratodentata*.

In areas of upheaval the phytoplankton seems to be incompletely consumed by the grazers and a bulk of decaying organic material sinks to the bottom and tends to deplete the oxygen content in bottom waters. Sometimes dynamic processes transport either fertile surface waters with their organic contents or the subsurface waters with low oxygen content close to the shore; then one can observe, in the most superficial layers not completely deprived of oxygen, many benthic species forced up from the bottom by anoxia, e.g. the cumacean *Iphinoe fagei* together with post-larval polychaetes, pelecypods and ophiuroids. In some cases bacterial decomposition of organic material results in a total depletion of oxygen and in hydrogen sulphide formation inducing massive fish-kills.

Among the nekton and, especially, the plankton feeders, *Sardinella* sp. occurs only between 5 and 15° S because it avoids the cooler waters originating from the upwelling. More to the south, anchovy and sardine are abundant, feeding on the rich zooplankton. Eggs and larvae of the South African sardine *Sardinops sagax ocellata* are very common in the 50- to 150-m layer (13–14·5 °C). Presumably, *S. sagax ocellata* is mainly herbivorous. In and around the area of highest fertility, the organismic abundance not only of fishes but also of marine birds and mammals is very striking in comparison with the scarcity of shallow-water marine life.

The best investigated region on south-west African coasts is probably that of Walvis Bay (23° S), i.e. north of the area of maximum upheaval. Here, upwelling is permanent, but stronger in winter and spring; phytoplankton is very abundant almost the whole year, except from December to March. The zooplankton maximum is delayed about two months compared to the time of minimum temperature in the 0- to 50-m layer. According to KOLLMER (1963), three different zooplankton assemblages may be recognized off this coast between 21° 30' and 24° S: (i) In the most inshore assemblage *Acartia* spp. and *Podon polyphemoides* predominate, associated with *Evadne nordmanni*, *Noctiluca* sp. and larval stages. (ii) Beyond the shelf edge the abundance in the zooplankton assemblage is higher; its most characteristic species are *Calanoides carinatus*, *Centropages brachiatus* and *Metridia lucens*. Farther from the shoreline one also observes in this assemblage the copepods *Rhincalanus nasutus* and *Eucalanus elongatus* together with some euphausiids and hyperiids. Large concentrations of radiolarians of the genus *Aulosphaera* may be observed in this zone which mainly control the changes in abundance of the whole assemblage. (iii) In the most offshore zone, jelly-like organisms such as *Liriope*, *Pleurobrachia*, *Beroe*, *Salpa fusiformis*, etc. predominate together with pteropods. In the Walvis Bay region the sardine fishing season almost exactly corresponds (March–November) to the period of high primary production, mainly in the most inshore zone (i) although the highest phytoplankton abundance occurs in zone (ii). In the neritic zone the hyperiid *Parathemisto gaudichaudi*—its younger stages are herbivorous, whereas the adults feed on copepods—and euphausiids are fed by jack-mackerels *Trachurus trachurus* and young individuals of *Dentex macrophthalmus*. Euphausiids constitute the main food source for young hake *Merluccius vulgaris* and *M. capensis*.

(e) Deep-sea Plankton of the Atlantic Ocean

The faunistic composition of the Atlantic Ocean deep-sea plankton has been sufficiently investigated only in the north-eastern regions (FRASER, 1961). He paid particular attention to the fate of the arctic zooplankters which are transported towards the North Atlantic Ocean deep layers. The North Atlantic deep water is mainly formed by waters from the Norwegian Sea surface, intermediate and deep layers which flow over the Wyville Thomson ridge and sink along the sill slope on its Atlantic side. The North Atlantic deep water moves very slowly towards the south-south-west. Its fauna includes two components: (i) Epiplanktonic cold-loving species which are able to subsist in these deeper waters due to the low temperature. Some of them may have a bipolar distribution. The most characteristic species of this group are: the siphonophore *Dimophyes arctica*, the copepods *Calanus hyperboreus*, *Metridia longa*, *Pareuchaeta norvegica* and *P. barbata*, the chaetognaths *Sagitta maxima* and *Eukrohnia hamata* (the latter bipolar), and the shrimp *Sergestes arcticus*. Some epiplankton psychrophilic species, such as the amphipod *Themisto libellula* and the appendicularian *Oikopleura vanhoffeni*, are not eurybathic enough to survive in deep waters. (ii) True deep-water species which inhabit the Norwegian Sea infra- and bathypelagic zones, such as many copepods, the chaetognath *Sagitta macrocephala*, *S. setezios* and *Eukrohnia fowleri*, the nemertine *Nectonemertes mirabilis*, the mysid *Boreomysis microps*, the shrimps *Amalopenaeus elegans* and *Hymenodora glacialis*, the squid *Histioteuthis bonelliana*, the fishes *Argyropelecus hemigymnus*, *Stomias boa*, *Cyclothone braueri*, *C. microdon*, *Paralepis coregonoides* and *Bathylagus* sp..

FRASER (1961) also proposed an interesting analysis of the plankton of the Lusitanian current. This current flows northwards along the western coasts of Spain and Portugal, and farther up to at least the southern part of the Irish Sea. It corresponds to the Mediterranean intermediate (and subsurface) water outflow through the Gibraltar Strait. Its flow is somewhat irregular, but off Lisbon, for example, it occupies approximately the 500- to 1500-m layer. As they move northwards, the Mediterranean waters mix up more with North Atlantic Ocean central water. According to FRASER, the plankton of the Lusitanian current is mainly of Mediterranean origin with some admixture of species which inhabit the infra- and bathypelagic zones between the Azores and the Bay of Biscay. Among the most characteristic species one may quote the following: the siphonophores *Hippopodius hippopus*, *Vogtia* sp. and *Stephanomia bijuga*; the medusae *Rhopalonema velatum* and *Pelagia noctiluca*; the chaetognaths *Sagitta lyra* and *S. serratodentata* f. *atlantica*; the pteropod *Euclio polita*; the crustaceans *Sapphirina* sp. and *Nematoscelis megalops*; the Thaliaceae *Thalia democratica*, *Salpa maxima*, *Doliolina muelleri* and *Doliolum nationalis*. However, some Mediterranean zooplankters do not spread into the Atlantic Ocean due to hydrological changes which arise from the progressive mixing of the waters they formerly inhabited with the central North Atlantic waters. The latter harbour a very characteristic assemblage with *Sagitta bipunctata*; this assemblage may be recognized at least up to Morocco and possibly farther to the south where it would come into contact with the shelf waters characterized by *S. fridericii*. The presence of colourless flagellates in the Atlantic Ocean deep layers has been discussed on p. 39ff.

The main features of quantitative distribution of Atlantic deep-sea plankton have been summarized by JASHNOV (1961, 1962), who assumes that this distribution corresponds to three different patterns: (i) Ultraoligotrophic areas such as the Sargasso Sea, where the surface biomass is low (a few tens mg m^{-3}). The most pronounced decrease in biomass is observed between 500 and 1000 m; below 2000 m, the very small biomass does not exceed a few mg m^{-3}. (ii) Oligotrophic areas, such as the Gulf Stream, the surface biomass of which is larger. In these areas—due to more intensive mixing, mainly in zones of convergence—the most marked decrease in biomass occurs below 1000 m. (iii) Eutrophic areas, e.g. the Newfoundland region—where the zooplankton biomass is high, not only in surface waters but also down to the 200- to 500-m layer. Still deeper, the biomass decrease is almost the same as in other areas, a few tens of mg m^{-3} in 500 to 1000 m and a few mg m^{-3} between 3000 and 4000 m (Table 6-10). Plankton assemblages in the infrapelagic and bathypelagic zones of the Sargasso Sea were discussed on p. 273ff.

During their zooplankton investigation in the tropical Atlantic, GREZE and co-authors (1969) noticed no important changes in plankton composition in the 200- to 500-m layer, compared to the surface assemblage. However, the mesoplankton sometimes exhibits an increase in copepods and ostracods (% total biomass), whereas other groups, such as polychaetes and appendicularians, generally decrease. GREZE and co-authors further pointed out that carnivores are more important in the 200- to 500-m layer than in surface waters making up 60 to 65% of the zooplankton biomass.

As regards the micronekton, it may be assumed from the quantitative distribution of myctophids in the 0- to 1000-m layer (KASHKIN, 1969) that the abundance depends on biomass and production of the euphotic layer. In the central parts of the large ultra-oligotrophic anticyclonic gyres, the biomass of myctophids may be as low as $0 \cdot 06$ g m^{-2} (10^{-4} ind. m^{-2}); in the most productive places the average weight of catches in the 0- to 1000-m layer may rise to $7 \cdot 6$ g m^{-2}. These data are supported by observations in the Gulf of Guinea, where the thermocline is well above compensation depth and permits heavy phytoplankton production in the lower part of the euphotic layer. Table 6-11 clearly indicates that macroplankton and micronekton are abundant down to 3000 m.

In general, abundance and biomass of myctophids are high at mid- and low latitudes—approximately between 40 to 50° N and 40 to 50° S—especially off north-western and south-western African coasts and in the equatorial areas, i.e. wherever the production in the surface layers is sufficient. In contrast, myctophids are less abundant and represented in small-sized—possibly more ubiquitous—species in central waters.

In the Atlantic Ocean, as in the Pacific Ocean, the abundance of mesopelagic and infrapelagic decapod crustaceans is closely related to the production level in surface waters. The maximum biomass occurs in subarctic waters, sharply decreases in north central waters and increases again in equatorial waters. Further decrease occurs in south central waters, whereas a third peak is observed near the subtropical convergence (VINOGRADOV, 1977). Larval stages and young individuals of cephalopods seem to display a rather similar distribution pattern.

Table 6-10

Zooplankton biomass (mg m⁻³) in relation to water depths in various areas of the Atlantic Ocean (Based on information provided by JASHNOV, 1961)

Layer (m)	Gulf Stream 40–43° N	Labrador current 43° N	North American shelf and slope 38–41° N	Gulf Stream 35–37° N	SW part of the Sargasso Sea 20–37° N	North equatorial current 16–19° N	West Africa shelf and slope 15–22° N	Canary current 27–36° N
0–50	74.5	562.0	199.0	54.9	64.7	89.6	1115.0	78.6
50–100	42.3	270.0	94.0	35.9	59.9	80.7	214.0	55.0
100–200	26.2	240.0	35.1	23.3	32.1	30.6	94.7	27.0
200–500	25.4	61.0	30.8	11.1	10.2	15.9	59.9	15.6
500–1000	15.7	—	19.0	6.0	4.1	5.4	27.2	6.5
1000–2000	5.0	—	7.8	2.8	1.2	1.6	—	2.1
2000–3000	—	—	—	2.1	0.6	1.0	—	0.8
3000–4000	—	—	—	—	0.3	0.3	—	0.2
4000–5000	—	—	—	—	0.1	—	—	—

Table 6-11

Number of individuals of different macroplankton and micronekton taxa in the Gulf of Guinea. The four layers sampled approximately correspond to the epi-, meso-, infra- and bathypelagic zones (After VOSS, 1969, p. 98; reproduced by permission of UNESCO–FAO)

Layer (m)	0–50	51–300	301–1000	1001–3000
Coelenterates	0	21	4	3
Molluscs (other than decapods)*	5	24	16	7
Crustaceans (other than decapods)†	231	366	142	177
Chaetognath	2	9	4	7
Chordates	3	30	18	16

*Squid (especially *Ommastrephes*) are very abundant.
†Shrimp of the genus *Nematocarcinus* are very abundant.

(6) Arctic Basin

(a) General Aspects

The very severe environmental conditions in the Arctic Basin (Volume I: COLLIER, 1970, pp. 2 ff.)—very low temperature, very low illumination due to both the higher latitude and the frozen (and snow-covered) surface layer, large salinity fluctuations due to seasonal ice melting or river discharge and so on—explain the scarcity of its plankton populations. However, in the higher Antarctic areas where the plankton is rather rich, at least in spring and summer, the same abiotic factors may reach just as extreme values as in the Arctic Basin. This discrepancy may be related to the fact that unlike the Antarctic region which readily participates in the general dynamics of the World Ocean, the Arctic Basin has only narrow and generally shallow connections with the northern parts of the Pacific and Atlantic Oceans, except through the strait between Greenland and Spitzberg. Moreover, it seems that the Arctic Basin has been quite separated twice from the World Ocean during the last 40,000 to 50,000 years.

The plankton of the Arctic Basin consists of about 70 phytoplankton species (of which 50 are diatoms and 80 zooplankton, most of the latter being copepods and mysids).

(b) Phytoplankton

In general, phytoplankton may be observed only in the 0- to 50-m layer. In the true marine areas, the salinity of which in the Arctic surface water is only 30⁰/₀₀ to 32⁰/₀₀ S, the most important species are the diatoms *Thalassiosira baltica*, *Coscinodiscus marginatus*, *Chaetoceros gracilis*, *C. whigami*, *Caloneis brevis* as well as the dinoflagellates *Dinophysis arctica*, *Peridinium breve* and *P. pellucidium*. In the neritic areas off northern Siberia (Laptev Sea, Eastern Siberia Sea, Chukchi Sea) where pronounced fresh-water input from large rivers (Lena, Ienissei, Ob) prevails in spring, brackish waters are inhabited by cyanophytes (*Anabaena*, *Aphanizomenon flosaquae*) and diatoms (several species of *Melosira, Asterionella gracillima*); wherever the salinity is lower than $5^0/oo$ S, fresh-water diatoms mainly belonging to the genera *Diploneis* and *Navicula* may be observed. In the eastern areas of the Chukchi Sea, a small inflow of Bering Sea surface water results in an important phytoplankton development in the spring. Diatoms always predominate, such as *Thalassiosira gravida, Fragilaria islandica, F. oceanica* and some species of the genus *Chaetoceros*. This diatom spring bloom quickly reduces the nutrient salt content and the summer phytoplankton mainly predominated by the species of the genus *Chaetoceros* in the colder areas, and dinoflagellates in the surface waters with somewhat higher temperatures. The spring melting of sea ice seems to stimulate the development of the phytoplankton in the Chukchi Sea.

The primary production rate is insufficiently known, and the bloom period seems to be very brief. Nutrient–salt content appear to be generally higher than the requirements of the phytoplankton populations so that light intensity would be the limiting factor. The bloom begins during the second half of June and suddenly decreases at the end of July. At this time the chlorophyll *a* content is about ten times that in early June. Nevertheless, primary productivity seems to be rather low. The photosynthetic activity: chlorophyll *a* ratio (average normal value: 3·7) is as low as 0·6 at 6 m depth, possibly due to the lower temperature of the Arctic surface water.

However, nanoplankton abundance in Arctic waters has probably been underestimated. According to THRONDSEN (1970), some small green flagellates might be very abundant in Arctic waters (up to 10^5 to 10^6 cells l^{-1}), mainly *Micromonas pusilla* (Prasinophyceae) and several Haptophyceae (for example *Chrysochromulina*). This finding was recently corroborated by REYNOLDS (1974) who points out that after elimination of diatoms and the largest dinoflagellates, the chlorophyll *a* content of the samples is not substantially reduced, due to the abundance of photosynthetic organisms smaller than 25 μm. Among these, REYNOLDS observed very few small diatoms (e.g. *Nitzschia closterium*), an occasional *Gymnodinium* and larger numbers (up to 10 to 15 $\times 10^6$ cells l^{-1}) flagellates (mostly Prasinophyceae) 10 μm or less in diameter; the motile stages of *Phaeocystis* are also common. Obviously, flagellates dominate the arctic phytoplankton even during the short period when diatoms are fairly abundant.

As in all areas covered with sea ice in winter, unicellular algae on the undersurface of ice and in the ice mass may contribute significantly to the total primary production in spring; this applies to the whole arctic marine ecosystem. According to GRANT and HORNER (1976) who studied sea-ice samples from the Beaufort and Chukchi Seas, extensive blooms of microalgae grow on the underside of sea ice resulting in a brownish layer which penetrates approximately 2 cm. Experiments conducted at 5 °C under constant illumination demonstrate that these species (e.g. the diatoms *Melosira jurgensii, Porosira glacialis, Navicula transitans* var. *derasa, Coscinodiscus lacustris*) can grow in a very broad range of salinities—sometimes from 5 to $55^0/oo$ S. This ability allows them to tolerate the salinities prevailing in the brine pockets. Higher salinities deeper than 2 cm with the sea ice may limit upward penetration of ice algae.

It was formerly assumed that these algae would be able to utilize dissolved organic carbon during the 'polar night'. HORNER and ALEXANDER (1972) who carried out experiments with labelled organic compounds (^{14}C glycine, ^{14}C glucose, ^{3}H acetate) demonstrated that bacteria and very small flagellates are mostly responsible for the dissolved organic carbon assimilation observed. According to BUNT and LEE (1972) the survival of algae confined into the sea ice does not correspond to a heterotrophic process but to a mechanism of overwintering which involves a striking decrease in metabolic rate and a very slow consumption of the reserves gathered during spring and summer. The production of benthic microalgae is high: *in situ* measurements in the Chukchi Sea gave values of less than 0.5 mg C m^{-2} h^{-1} in winter up to about 0.57 mg C m^{-2} h^{-1} in August, i.e. about eight times that of the algae living within the sea ice and twice the phytoplankton production. Probably in arctic coastal areas, phytoplankton, microphytobenthos and sea-ice algae constitute a 'pool' of primary producers, which provides food to both plankton grazers and benthic filter feeders.

The above analysis of Arctic Basin phytoplankton is highly schematic. In reality the distributions and dynamics of phytoplankton populations are much more intricate, especially in coastal areas—some eight to ten miles from the shoreline. This was demonstrated by BURSA (1963) in the vicinity of Point Barrow, Alaska. The sea-ice flora can modify close-to-shore phytoplankton assemblages not only due to summer ice-melting but also due to meteorological conditions that induce large changes in temperature and salinity. BURSA observed, for example, that in a station located more than three miles from the beach, *Chaetoceros* species were common as long as the weather was calm, but were replaced after a few days of northerly winds by populations of *Gymnodinium lohmanni Gyrodinium* sp., *Gonyaulax monospina*, *Polykrikos* sp., and other shoreward drifting species.

Average primary production and chlorophyll *a* concentration in the Chukchi Sea during July and August 1960 were 1.5 mg C m^{-3} h^{-1} and 2.1 mg m^{-3}, respectively (HAMEEDI, 1978). Nutrient supply (nitrogen and silicon) into the photic zone was considered the dominant factor determining these low values in permanently stratified waters; HAMEEDI further observed a pronounced subsurface accumulation of chlorophyll *a* at stations near the ice edge relative to those in open water. Since zooplankton is rather abundant near the ice edge (695 to 3176 ind. m^{-3}; average biomass: 5 to 35 mg n^{-3} in the 0- to 40-m layer) HAMEEDI concludes that the subsurface chlorophyll *a* maximum is not related to the sinking of algae due to low grazing pressure but appears to be contributed by the extensive flora near the bottom of solid, unbroken ice. Nevertheless, the contribution of sea-ice algae to the total chlorophyll *a* content in the Chukchi Sea seems to be substantial. During the ice-free period, however, these algae may fail to stay in the water column for any length of time so that the subsurface chlorophyll *a* maximum completely disappears.

Coastal leads in which phytoplankton begins to develop while the rest of the water, still covered by ice, remains cut off from sunlight energy may modify local conditions. This results in areas in which plankton populations—at various stages of the ecological succession—are more or less dense and separated from each other by poorer areas. For example, according to BURSA (1963), a single lead opened close to the shore at Barrow on June 18 revealed a very dense zooplankton population, whereas its phytoplankton contained 12 diatom species together with the dinoflagellate *Gonyaulax tamarensis* (2900 cells l^{-1}). On July 8 another lead—two to five miles from the coast—contained a diatom population of 77 500 cells l^{-1} with a large dominance of *Nitzschia closterium* (60 500 cells l^{-1}). On the same day the *G. tamarensis* population rose to a maximum of 28 000 cells l^{-1} at more inshore stations. The part played in the assemblages by neritic opportunists especially during a bloom can be disturbed by excessive river run-off or pollution; for example *Nitzschia closterium* and *Leptocylindrus danicus* may reach in the area investigated by BURSA maximum values of 78 000 and 45 000 cells l^{-1} respectively.

(c) Zooplankton

The zooplankton as a whole is rather similar to that of most northern areas of the Atlantic Ocean where the Arctic assemblage occurs (p. 263). However, in the eastern areas of the Chukchi Sea, some North Pacific zooplankters may also be present. The information at hand is still insufficient for generalizations concerning zooplankton distribution in the whole Arctic Basin, but the observations by GRAINGER (1962, 1965) and JOHNSON (1963) in the Beaufort Sea provide some clues for a general framework. These authors distinguish three assemblages:

(i) The offshore oceanic assemblage (temperature $< 1\,°C$; salinity $> 32\%_0\,S$) includes many copepod species—such as *Scaphocalanus magnus, Temorites brevis, Gaidius* sp., *Spinocalanus* spp.—the ostracod *Conchoecia borealis maxima* (which exists in the American Basin down to 1500 m) and the chaetognath *Eukrohnia hamata*.

(ii) The neritic but euryhaline assemblage inhabits waters where the salinity falls between $20\%_0$ and $32\%_0\,S$. It is the most diversified and its fauna is rather similar to those of the truly Arctic areas of the Pacific and Atlantic Oceans. The most abundant species are: the medusae *Aglantha digitale* and *Aeginopsis laurentia*; the chaetognath *Sagitta elegans*; the pteropods *Limacina helicina* and *Clione limacina*; the copepods *Calanus hyperboreus, C. glacialis, Pseudocalanus minutus, Microcalanus pygmaeus, Pareuchaeta glacialis, Metridia longa, Oithona similis* and *Oncaea borealis*; the appendicularians *Oikopleura vanhoffeni* and *Fritillaria borealis*.

(iii) The brackish water assemblage which takes place in inshore areas where the salinity is between $8\%_0$ and $25\%_0\,S$ and fluctuates on an annual scale. Its most characteristic species are the copepods *Eurytemora herdmani, Limnocalanus grimaldii* and *Acartia clausi*, as well as the amphipod *Hyperoche medusarum*, together with rather abundant Hydromedusae.

Except in the brackish water assemblage, meroplanktonic elements are rather scarce, because the arctic benthos is very poor and very few of its species were able to adjust their spawning period to the short phytoplankton bloom. Presumably, the rather abundant tintinnid populations (1400 ind. m^{-3}; biomass 0.8 mg m^{-3}) in the 0- to 25-m layer observed by BURKOVSKY (1976) in the north-eastern part of the Kara Sea, belong to this assemblage; the salinity is approximately $20\%_0$ to $25\%_0\,S$.

In general, copepods predominate in zooplankton assemblages and represent more than 80% of the total biomass. *Calanus* species alone average half the biomass in the upper 1500 m; below 1500 m their relative importance declines. Copepods are also the most numerous (90%) of the metazoan plankters, although they are outnumbered by small radiolarians in the 0- to 100-m layer from July to early September (HOPKINS, 1969). Of all copepods recorded, the eight copepod species listed above for the neritic-euryhaline assemblage (ii) make up 99% in the 0- to 50-m layer and 95% in the 0- to 300-m layer. Presumably, they breed in the central areas of the Arctic Basin.

In the eastern areas of Chukchi Sea some mixing occurs between species endemic to the Arctic Basin—such as *Aeginopsis laurentii, Calanus hyperboreus*, and *Euchaeta glacialis*—and boreo-arctic invaders which enter with the Bering Sea surface water.

Among the latter are some tintinnids *(Tintinnopsis japonica, T. kofoidi, Tintinnus rectus)* and the copepods *Calanus cristatus, C. plumchrus, Eucalanus bungii*. According to WING (1974) this general scheme might have to be slightly modified. He collected zooplankton samples in the eastern Chukchi Sea between September 26 and October 17, 1970—apparently mainly in the above-mentioned assemblage (ii) and compared his results with those JOHNSON (1963) obtained in summer of 1947. The comparison revealed several discrepancies. While some appear to be related to differences between years, others seem to be due to the time of the year.

The lesser richness and abundance recorded in autumn 1970 in comparison with summer 1947 for cladocerans, *Acartia, Oithona*, appendicularians and planktonic larvae of benthic invertebrates may be ascribed to a seasonal cycle. This appears rather clear, especially for pelecypod and echinoderm larval stages. For example, because most echinoderms reveal a short spawning period in July or August, their larvae, which spend less than eight weeks in plankton, presumably had settled before the period of WING's (1974) investigations. Some physical and chemical parameters seems also to control abundance and distribution of plankters: (i) abundance decreases in waters where the temperature is below 0 °C; (ii) areas in which a sharp temperature gradient from 1 to 3 °C occurs reveal changes in abundance of many species in a parallel fashion; (iii) no correlation can be demonstrated between salinity and plankton abundance; (iv) a tendency was recorded for an inverse relation between abundance and water content in oxygen. Presumably, however, year to year differences in plankton assemblages may be greater than seasonal ones.

The zooplankton peak occurs a little later (July, August) than the phytoplankton bloom. According to HOPKINS (1969), biomass values are always below 1 mg m^{-3} (d.w.): 0·62 for the Arctic surface water (0–200 m); 0·14 for the Atlantic water (200–900 m); 0·04 for the Arctic deep water. In the Laptev and East Siberian Seas the neritic-euryhaline (ii) and the brackish water (iii) assemblages seem to exhibit larger biomasses: 24 to 200 mg m^{-3} (w.w.). HOPKINS assumes that the primary production cannot meet the metabolic requirements of the zooplankters in the central Arctic basin, and that the particulate organic carbon (POC) brought in by rivers or by surface inflow from the Bering Sea might contribute to their food supply. However, the POC content is not significantly higher in the Arctic Basin than elsewhere in the World Ocean (KINNEY and co-authors, 1971). Dissolved organic carbon (DOC) is very abundant not only in the surface layer but also in intermediate and deep waters. This suggests that some still unknown decomposition processes might take place in the depths of the Arctic Basin. Unfortunately, our present knowledge on local microbial activities is very poor. The surface layer contains ten times less bacteria than the North Pacific surface water. Most bacteria are cocci with a very thick membrane. Bacterial abundance peaks have been recorded at 200 to 250, 900 to 1000 and 2000 m; they correspond—at least the first two—to pycnoclines.

The problem of zooplankton food supply remains unsolved. Certainly, *Calanus hyperboreus, C. glacialis* and many other zooplankters build up reserves during the brief phytoplankton bloom from which they may partly satisfy their metabolic needs during the 'polar night'. The significance of bacterial or POC food resources still remains doubtful. DOC utilization has not yet been proved unequivocally for copepods. It seems that none of these sources can account for the discrepancy (HOPKINS, 1969) between the very low level of the primary production and the relatively high zooplankton abundance, except if one assumes that primary productivity has been greatly underestimated, possibly because smaller flagellates have not sufficiently been taken into account.

Literature Cited (Chapter 6)

AIZAWA, Y. (1974). Ecological studies of micronektonic shrimps (Crustacea, Decapoda) in the western North Pacific. *Bull. Ocean. Res. Inst. Univ. Tokyo*, **6**, 84.

ALI KHAN, J. and HEMPEL, G. (1974). Relation of fish larvae and zooplankton biomass in the Gulf of Aden. *Mar. Biol.*, **28**, 311–16.

ALVARIÑO, A. (1964). Bathymetric distribution of Chaetognaths. *Pacif. Sci.*, **18**, 64–82.

ARZHANOVA, N. V. (1974). Hydrochemical characteristics and primary production in the Brazil-Guiana waters in the summer of 1969. (Russ.) *Trudȳ vses. nauchno-issled. Inst. morsk. rȳb. Khoz. Okeanogr.*, **98**, 70–6.

BAINBRIDGE, V. (1972). The zooplankton of the Gulf of Guinea. *Bull. mar. Ecol.*, **8**, 61–97.

BALECH, E. (1973). Segunda contribucion al conocimiento del microplancton del mar de Bellingshausen. *Boln. Inst. antart. argent.*, **107**, 1–63.

BANSE, K. (1974). On the role of bacterioplankton in the Tropical Ocean. *Mar. Biol.* **24**, 1–5.

BARKHATOV, V. A., VOLKOV, A. F., DOLZHENKOV, V. H. and KAREDIN, E. P. (1973). Macroplankton biomass and composition in some regions of the South Pacific. *Oceanology*, **13**, 677–82.*

BE, A. W. H., FORNS, J. M. and ROELS, O. A. (1971). Plankton abundance in the North Atlantic Ocean. In J. D. Costlow (Ed.), *Fertility of the Sea*, Vol. I. Gordon and Breach, New York. pp. 17–50.

BEERS, J. R. and STEWART, G. L. (1970). The ecology of the plankton off La Jolla, California, in the period April through September 1967. VI. Numerical abundance and estimated biomass of microzooplankton. *Bull. Scripps Instn Oceanogr. non-tech. Ser.*, **17**, 67–87.

BEERS, J. R. and STEWART, G. L. (1971). Micro-zooplankters in the plankton communities of the upper waters of the eastern tropical Pacific. *Deep Sea Res.*, **18**, 861–83.

BEKLEMISHEV, K. V. (1961). Zooplankton of northeastern parts of the Pacific Ocean in winter 1958–59. (Russ.) *Trudȳ Inst. Okeǎnol.*, **45**, 142–71.

BELAYEV, G. M. (1970). Squid beaks in sediments of the Pacific Ocean. (Russ.) *Trudȳ Inst. Okeanol.*, **88**, 236–51.

BIDULYA, O. G. and BABICHEV, E. N. (1972). Zooplankton quantitative distribution on the northern shelf of the Bay of Bengal. (Russ.) *Okeanologiya*, **12**, 1072–7.

BINET, D. (1977). Grands traits de l'écologie des principaux taxons due zooplancton ivoirien. *Cah. ORSTOM, Océanogr.*, **15**, 89–109.

BINET, D. and DESSIER, A. (1971). Premières données sur les Copépodes Pélagiques de la région congolaise. *Cah. ORSTOM, Océanogr.*, **9**, 411–57.

BIRSHTEIN, Ya. A., VINOGRADOV, M. E. and CHINDONOVA, Yu. G. (1954). Plankton vertical zonation in the Kuril-Kamchatka Trench. (Russ.) *Dokl. Akad. Nauk SSSR*, **95**, 389–92.

BLACKBURN, M. (1968). Micronekton of the eastern tropical Pacific Ocean: family composition, distribution, abundance, and relations to tuna. *Fish. Bull. U.S.*, **67**, 71–115.

BOGDANOV, D. V., SOKOLOV, V. A. and KHROMOV, N. S. (1968). Regions of high biologic and commercial productivity in the Gulf of Mexico and the Caribbean Sea. *Okeanologiya*, **8**, 466–78.

BOGOROV, B. G. (1958). Biogeographical regions of the plankton of the northwestern Pacific Ocean and their influence on the deep sea. *Deep Sea Res.*, **5**, 149–61.

Oceanology is the English translation for the Soviet journal *Okeanologiya*, published by the American Geophysical Union, Washington, D.C. It follows the same dating and numbering as the original.

BOGOROV, B. G. (1960). Geographical zones in the pelagial of central parts of the Pacific Ocean (Russ.) *Trudy Inst. Okeanol.*, **41**, 8–16.

BOGOROV, B. G. and BEKLEMISHEV, C. W. (1955). On the plankton production in the north-western Pacific Ocean. (Russ.) *Dokl. Acad. Nauk SSSR.*, **104**, 141–3.

BOGOROV, B. G. and VINOGRADOV, M. E. (1955). Some essential features of zooplankton distribution in the north-western Pacific Ocean. (Russ.) *Trudy Inst. Okeanol.*, **18**, 113–23.

BOGOROV, V. G. and VINOGRADOV, M. E. (1960). Zooplankton repartition in the Kuril-Kamchatka region of the Pacific Ocean. (Russ.) *Trudy Inst. Okeanol.*, **34**, 60–84.

BOGOROV, V. G. and VINOGRADOV, M. E. (1961). Some features of zooplankton biomass repartition in the surface waters of the Indian Ocean in the winter 1959–60 (Russ.). *Mezhduved. Geophys. Komitet Akad. Nauk SSSR.*, Izdat Akad., Nauk, Moskva. **4**, 72–75.

BOLTOVSKOY, E. (1962). Planktonic foraminifera as indicators of different water masses in the South Atlantic. *Micropaleontology*, **8**, 403–8.

BOLTOVSKOY, E. (1970). Masas de agua (caracteristica, distribucion, movimientos) en la superficie del atlantico sudoeste, segun indicatores biologicos: foraminiferos. *Servicio Hidrografia Naval (Armada Argentina)*, **H643**, 1–101.

BRADBURY, M. G., ABBOTT, D. P., BOVBJERG, R. V., MARISCAL, R. N., FIELDING, W. C., BARBER, R. T., PEARSE, V. B., PROCTOR, S. J., OGDEN, J. C., WOURMS, J. P., TAYLOR, L. R., Jr., CHRISTOFFERSON, J. G., CHRISTOFFERSON, J. P., MCPHEARSON, R. M., WYNNE, M. J. and STROMBORG, P. M., Jr. (1971). Studies on the fauna associated with the deep scattering layers in the equatorial Indian Ocean, conducted on RV *Te vega* during October and November 1964. In G. Brook-Farquhar (Ed.), *Proc. Int. Symp. on Biological Sound Scattering in the Ocean*. Maury Center for Ocean Science, Washington, D.C.

BUNT, J. S. (1964). Primary productivity under sea ice in Antarctic waters. 2. Influence of light and other factors on photosynthetic activities of antarctic marine microalgae. In M. O. Lee (Ed.), *Biology of the Antarctic Seas*. (*Antarct. Res. Ser.*, **1**, 13–26.)

BUNT, J. S. and LEE, C. C. (1972). Data on the composition and dark survival of four sea-ice microalgae. *Limnol. Oceanogr.*, **71**, 458–61.

BURKOVSKY, I. V. (1976). New data on Arctic Tintinnida (Ciliata) and revision of their fauna. (Russ.) *Zool. Zh.*, **55**, 325–36.

BURSA, A. (1963). Phytoplankton in coastal waters of the Arctic Ocean at Point Barrow, Alaska. *Arctic*, **16**, 239–62.

CALEF, G. W. and GRICE, G. D. (1967). Influence on the Amazon River Outflow on the ecology of the western tropical Atlantic. II. Zooplankton abundance, copepod distribution, with remarks on the fauna of low salinity areas. *J. mar. Res.*, **25**, 84–94.

CHEKUNOVA, V. I. and RYKOVA, T. I. (1974). Energy requirements of the Antarctic Krill, Euphausia superba. (Russ.) *Okeanologiya*, **14**, 526–32.

CIFELLI, R. and SMITH, R. K. (1970). Distribution of planktonic Foraminifera in the vicinity of the North Atlantic current. *Smithson. Contrib. Paleobiol.*, **4**, 1–51.

CLARKE, T. A., (1973). Some aspects of the ecology of lanternfishes (Myctophidae) in the Pacific Ocean near Hawaii. *Fish. Bull. Calif.*, **71**, 401–34.

COLLIER, A. W. (1970). Oceans and coastal waters as life-supporting environments. In O. Kinne (Ed.), *Marine Ecology*, Vol. I, Environmental Factors, Part 1. Wiley, London. pp. 1–93.

CORCORAN, E. F. and MAHNKEN, C. V. W. (1969). Productivity of the Tropical Atlantic-Ocean. In *Proc. Symp. on Oceanography and Fishery Resources in the Tropical Atlantic*. UNESCO-FAO, Paris. pp. 57–67.

CRADDOCK, J. E. and MEAD, G. W. (1970). Midwater fishes from the eastern south Pacific Ocean. *Scientific Results of the Southeastern Pacific Ocean Expedition, Anton Bruun Report*, **3**, 1–46.

CURRIE, R. I. (1964). Environmental features in the ecology of Antarctic seas. In R. Carrick, M. W. Holdgate and J. Prévost (Eds), *Biologie Antarctique*. Hermann, Paris. pp. 87–94.

CURRIE, R. I., FISHER, A. E. and HARGREAVES, P. M. (1973). Arabian Sea upwelling. In B. Zeitzschel (Ed.), *Ecological Studies. Analysis and Synthesis*, Vol. 3. Springer-Verlag, Heidelberg. pp. 37–53.

DANDONNEAU, Y. (1972). Aspects principaux des variations du phytoplancton sur le plateau continental ivoirien. *Bull. Centre Recherches Océanogr. Abidjan (Ivory Coast)*, **3**, 32–59.

DEEVEY, G. B. (1971). The annual cycle in quantity and composition of the zooplankton of the Sargasso Sea off Bermuda. I. The upper 500 m. *Limnol. Oceanogr.*, **16**, 219–40.

DEEVEY, G. B. and BROOKS, A. L. (1971). The annual cycle in quantity and composition of the zooplankton of the Sargasso Sea off Bermuda. II. The surface to 1.000 m. *Limnol. Oceanogr.*, **16**, 927–43.

DEEVEY, G. B. and BROOKS, A. L. (1977). Copepods of the Sargasso Sea off Bermuda: species composition, and vertical and seasonal distribution between the surface and 2000 m. *Bull. mar. Sci.*, **27**, 256–91.

EKMAN, S. (1953). *Zoogeography of the Sea*, Sidgwick and Jackson Ltd, London.

ELIZAROV, A. A. and MOVCHAN, O. A. (1973). Peculiarities of the waters vertical circulation and phytoplankton distribution in the northwestern part of the Atlantic Ocean (region of the Great Newfoundland Bank). (Russ.) *Okeanologiya*, **13**, 662–8.

EL-SAYED, S. Z. (1970). On the productivity of the Southern Ocean (Atlantic and Pacific sectors). In M. W. Holdgate (Ed.), *Antarctic Ecology*, Vol. 1. Academic Press, London. pp. 119–35.

EL-SAYED, S. Z. (1971). Observations on phytoplankton bloom in the Weddell Sea. In G. A. Llano and I. E. Wallen (Eds), *Biology of the Antarctic Seas*, Vol. 4. (*Antarc. Res. Ser.*, **17**, 301–12.)

EL-SAYED, S. Z., MANDELLI, E. F. and SUGIMURA, Y. (1964). Primary organic production in the Drake Passage and Bransfield Strait. *Antarc. Res. Ser.*, **1**, 1–11.

FAGETTI, E. G. (1968). Quetognatos de la expedicion "Marchile l" con observaciones acerca del posible valor de algunas especies como indicadoras de las masas de agua frente a Chile. *Revta. Biol. mar.*, **13**, 1–155.

FAGETTI, E. G. (1972). Bathymetric distribution of Chaetognaths in southeastern Pacific Ocean. *Mar. Biol.*, **17**, 7–29.

FAGETTI, E. G. and FISCHER, W. (1964). Resultados cuantitativos del zooplancton colectado frente a la costa Chilena por la expedicion "Marchile 1". *Montemar*, **4**, 137–94.

FAY, R. R. (1973). Significance of nannoplankton in primary production of the Ross Sea, Antarctica, during the 1972 austral summer. (Abstract of unpublished Ph.D. thesis, University of Texas) *Rep. in Contr. Oceanogr. Met. agric. mech. Coll. Tex.* (1974–75), **17**, 104.

FERGUSON-WOOD, E. J. (1971). Phytoplankton distribution in the Caribbean Region. In *Proc. Symp. Investigating the Resources in the Caribbean Sea and Adjacent Regions.* UNESCO, Paris. pp. 339–410.

FINENKO, Z. Z. (1978). Production in plant populations. In O. Kinne (Ed.), *Marine Ecology*, Vol. IV, Dynamics. Wiley, Chichester. pp. 13–88.

FOXTON, P. (1964). Seasonal variations in the plankton of Antarctic waters. In R. Carrick, M. W. Holdgate and J. Prévost (Eds), *Biologie Antarctique, Proc. S.C.A.R. Symp., Paris, 1962.* Hermann, Paris. pp. 311–18.

FRASER, J. H. (1961). The oceanic plankton of the northeast Atlantic. H.M. Stationery Office, Edinburgh. *Mar. Res.*, **4**, 48.

FUKUSHIMA, H. and MEGURO, H. (1966). The plankton ice as a basic factor of the primary production in the Antarctic Ocean. *Antarctic Rec..*, **27**, 99–101.

GAPISHKO, A. I. (1968). On the food of pelagic fishes in the Gulf of Aden. (Russ.) *Trudy VNIRO*, **64**, 278–81.

GAPISHKO, A. I. (1971). Zooplankton seasonal changes in the Gulf of Aden in the years 1965–1966. (Russ.) *Okeanologiya*, **11**, 475–81.

GRAINGER, E. H. (1962). Zooplankton of Fox Basin in the Canadian Arctic. *J. Fish. Res. Bd Can.*, **19**, 377–400.

GRAINGER, E. H. (1965). Zooplankton from the Arctic Ocean and adjacent Canadian waters. *J. Fish. Res. Bd Can.*, **22**, 377–400.

GRANDPERRIN, R. (1975). *Etude des Structures Trophiques Aboutissant aux Thons de Longue Ligne dans le Pacific Sud-Ouest Tropical*, Thèse Doct., Université Aix-Marseille (II).

GRANT, W. S. and HORNER, R. A. (1976). Growth responses to salinity variation in four Arctic ice diatoms. *J. Phycol*, **12**, 180–5.

GRAVE, H. (1973). Kurzzeitige Schwankungen im Neuston der Norwegischen See. *'Meteor'* *Forschungsergeb.*, D, **14**, 67–86.

GREZE, V. N., GORDEJAVA, K. T. and SHMELEVA, A. A. (1969). Distribution of zooplankton and biological structure in the tropical Atlantic. In *Proc. Symp. on Oceanography and Fishery Resources in the Tropical Atlantic.* UNESCO-FAO, Paris. pp. 85–90.

GROBOV, A. G. (1968). Zooplankton quantitative repartition in the north-western part of the Indian Ocean. *Trudÿ VNIRO*, **64**, 260–70.

GUEREDRAT, J. A. (1972). Seasonal variations of the specific diversity of the copepods in the Indian Ocean at longitude 100° E. *J. mar. biol. Ass. India*, **14**, 148–59.

GUEREDRAT, J. A., GRANDPERRIN, R. and ROGER, C. (1972). Diversité spécifique dans le Pacifique équatorial: évolution de l'écosystème. *Cah. ORSTOM*, (*Océanogr.*), **10**, 57–69.

GUILLARD, R. R. L. and HELLEBUST, J. A. (1971). Growth and the production of extracellular substances by two strains of *Phaeocystis poucheti*. *J. Phycol.*, **7**, 330–8.

GUILLEN, O. G. and IZAGUIRRE DE RONDAN, R. (1968). Production primaria de las aguas costeras del Peru en el ano 1964. *Bol. Inst. Mar Peru*, **1**, 349–76.

GUNTHER, E. R. (1936). A report on oceanographical investigations in the Peru coastal current. *'Discovery' Rep.*, **13**, 107–276.

HADA, Y. (1970). The protozoan plankton of the Antarctic and Sub-Antarctic seas. *JARE Scientific Reports, Biology (Ser. E)*, **31**, 1–51.

HAGMEIER, E., KANJE, M. and TREUTNER, K. (1975). Unpublished results summarized in *Jahresbericht* 1975. Biologische Anstalt Helgoland, Hamburg. pp. 21–7.

HAMEEDI, M. J. (1978). Aspects of water column primary productivity in the Chukchi Sea during summer. *Mar. Biol.*, **48**, 37–46.

HASLE, G. R. (1969). An analysis of the phytoplankton of the Pacific Southern Ocean: abundance, composition and distribution during the Brategg Expédition, 1947–1948. *Hvalråd. Skr.*, **52**, 1–168.

HEINRICH, A. K. (1962). Main peculiarities of the Pacific Ocean. (Russ.) *Trudÿ Inst. Okeanol.*, **58**, 114–34.

HEINRICH, A. K. (1971). On the near-surface plankton of the eastern South Pacific Ocean. *Mar. Biol.*, **10**, 290–4.

HEINRICH, A. K. (1973). Horizontal distribution of copepods in the Peru Current region. *Oceanology*, **13**, 94–103.

HENDEY, N. I. (1937). The plankton diatoms of the southern seas. *'Discovery' Rep.*, **16**, 151–364.

HERBLAND, A., LE BORGNE, R. and VOITURIEZ, B. (1973). Production primaire, secondaire et régénération des sels nutritifs dans l'upwelling de Mauritanie. *Bull. Centre Recherches Oceanogr. Abidjan (Ivory Coast)*, **4**, 1–75.

HOPKINS, T. L. (1969). Zooplankton standing crop in the Arctic Basin. *Limnol. Oceanogr.*, **14**, 80–5.

HOPKINS, T. L. (1971). Zooplankton standing crop in the Pacific sector of the Antarctic. In G. A. Llamo and I. E. Warren (Eds), *Biology of the Antarctic Seas.* (*Antarct. Res. Ser.*, **17**, 347–62.)

HORNE, A. J., FOGG, G. E. and EAGLE, D. J. (1969). Studies *in situ* of the primary production of an area of inshore Antarctic sea. *J. mar. biol. Ass. U.K.*, **49**, 393–405.

HORNER, R. A. and ALEXANDER, V. (1972). Algal populations in Arctic sea ice: an investigation of heterotrophy. *Limnol. Oceanogr.*, **17**, 454–8.

HULBURT, E. M. (1966). The distribution of phytoplankton, and its relationship to hydrography, between southern New England and Venezuela. *J. mar. Res.*, **24**, 67–81.

HULBURT, E. M. (1967). Some notes on the phytoplankton off the southeastern coast of the United States. *Bull. mar. Sci.*, **17**, 330–7.

HUMPHREY, G. F. (1960). The concentration of Plankton Pigments in Australian Waters *CSIRO—Division of Fisheries and Oceanography. Technical Paper*, **9**, 27.

IKEDA, T. and MOTODA, S. (1973). An approach to the estimation of zooplankton production in the Kuroshio and adjacent regions. In B. Morton (Ed.), *Proc. Special Symp. on Marine Science.* Pacific Science Association, Hong Kong. pp. 24–8.

JASHNOV, V. A. (1961). Vertical repartition of zooplankton bulk of tropical areas of the Atlantic Ocean. (Russ.) *Dokl. Acad. Nauk SSSR.*, **136**, 705.

JASHNOV, V. A. (1962). Plankton of the tropical regions of the Atlantic Ocean. (Russ.) *Trudȳ. morsk. Gidrof. Inst.*, **25**, 195–207.

JEFFREY, S. V. and CARPENTER, S. M. (1974). Seasonal succession of phytoplankton at a coastal station off Sydney. *Aust. J. mar. Freshwat. Res.*, **25**, 361–9.

JITTS, H. R. (1965). The summer characteristics of primary productivity in the Tasman and Coral seas. *Aust. J. mar. Freshwat. Res.*, **16**, 151–62.

JOHANNES, R. E. (1967). Ecology of organic aggregates in the vicinity of a coral reef. *Limnol. Oceanogr.*, **12**, 189–95.

JOHNSON, M. W. (1963). Arctic Ocean plankton. In *Proc. Arctic Basin Symp., Hershey, 1962*. (Rep. in 1964, *Contr. Scripps Instn Oceangr.*, **33**, 1127–37.)

KABANOVA, Yu. G. (1968). Primary production in the northern Indian Ocean. (Russ.; Engl. abtr.) *Okeanologiya*, **8**, 270–8.

KABANOVA, Yu. G. and LOPEZ BALUJA, A. (1970). Primary production in the southern parts of the Gulf of Mexico and the northwestern coast of Cuba. (Russ.) *Mezhduved. Geophys. Komitet Akad. Nauk SSSR, Okeanol. Issled*, **20**, 46–68. Izdat. Akad. Nauk, Moskva.

KASHKIN, N. I. (1969). Main features of the biological productivity in the southern Atlantic. (Russ.) *Trudȳ VNIRO*, **66**, 128–59.

KHROMOV, N. S. (1962). *Sardinella* repartition, dynamics and trophic resources in the cooperation region of western African coast. (Russ.) *Trudȳ VNIRO*, **46**, 214–35.

KHROMOV, N. S. (1965a). Plankton repartition in the Gulf of Mexico and some features of its seasonal dynamics. (Russ.) *Sov. Kub. Ryb. Issled. VNIRO*, **1**, 47–70. Izdat. Pischeraja Promyshlennost, Moskva.

KHROMOV, N. S. (1965b). Plankton repartition in the Caribbean Sea. (Russ.) *Sov. Kub. Ryb. Issledov. VNIRO*, **1**, 71–6. Izdat Pischeraja Promyshlennost, Moskva.

KHROMOV, N. S. (1967). Plankton investigations in the Gulf of Mexico and the Caribbean Sea. (Russ.) *Sov. Kub. Ryb. Issled. VNIRO*, **2**, 39–57. Izdat. Pischeraja Promyshlennost, Moskva.

KINNEY, P. J., LODER, T. C. and GROVES, C. (1971). Particulate and dissolved organic matter in the Amerasian basin of the Arctic Ocean. *Limnol. Oceanogr.*, **16**, 132–6.

KLIACHTORIN, L. B. (1964). Investigations on the primary production in the Antarctic region. (Russ.) *Okeanologiya*, **3**, 458–461.

KOBLENTZ-MISHKE, O. I. (1961). Phytoplankton specific composition in the northeastern parts of the Pacific Ocean in winter 1958–59. (Russ.) *Trudȳ Inst. Okeanol.*, **45**, 172–89.

KOLLMER, W. E. (1963). Notes on zooplankton and phytoplankton collections made off Walvis Bay. *Investl. Rep. mar. Res. Lab. S.W. Afr.*, **8**, 1–78.

KRAKATITSA, V. V. (1970). The ichthyoplankton distribution in the area of the Aden gulf. In M. Uda (Ed.), *The Ocean World: Proceedings of the Joint Oceanography Assembly*. Publication Society for the Promotion of Science, Tokyo. pp. 436–7.

KREY, J. (1973). Primary production in the Indian Ocean. In B. Zeitzschel (Ed.), *The Biology of the Indian Ocean*. Springer-Verlag, Heidelberg. pp. 115–26.

KUZ'MENKO, L. V. (1973). Primary production of the northern Arabian sea. *Oceanology*, **3**, 251–6.

KUZ'MENKO, L. V. (1977). Distribution of phytoplankton in the Arabian Sea. *Oceanology*, **17**, 70–4.

KUZ'MINA, A. I. (1962a). Some data on the spring phytoplankton in the North Atlantic. (Russ.), *Dokl. Akad. Nauk SSSR*, **144**, 1156–9.

KUZ'MINA, A. I. (1962b). Phytoplankton quantitative dynamics and repartition in the northern areas of the Greenland Sea. (Russ.) *Trudȳ VNIRO*, **46**, 287–96.

LE BRASSEUR, R. J. and KENNEDY, O. D. (1972). Microzooplankton in coastal and oceanic areas of the Pacific subarctic water mass: a preliminary report. In A. Y. Takenouti (Ed.), *Biological Oceanography of the North Pacific Ocean*. Idemitsu Shoten Publ., Japan. pp. 355–65.

LLOYD, I. J. (1971). Primary production off the coast of northwest Africa. *J. Cons. int. Explor. Mer*, **33**, 312–23.

LONGHURST, A. R. (1967a). Diversity and trophic structure of zooplankton communities in the California Current. *Deep Sea Res.*, **14**, 393–408.

LONGHURST, A. R. (1967b). Vertical distribution of zooplankton in relation to the eastern Pacific oxygen minimum. *Deep Sea Res.*, **14**, 51–63.

LOSSE, G. F. and MERRETT, N. R. (1971). The occurrence of *Oratosquilla investigatoris* (Crustacea: Stomatopoda) in the pelagic zone of the Gulf of Aden and the Equatorial western Indian Ocean. *Mar. Biol.*, **10**, 244–53.

McROY, C. P. and GOERING, J. J. (1974). The influence of ice on the primary productivity of the Bering Sea. In D. W. Hood and E. J. Kelley (Eds), *Oceanography of the Bering Sea with Emphasis on Renewable Resources*. Institute for Marine Science, Univ. Alaska, College. pp. 403–21.

McWILLIAM, P. S. (1977). Further studies of plankton ecosystems in the eastern Indian Ocean. 6. Ecology of the Euphausiacea. *Aust. J. mar. Freshwat. Res.*, **28**, 627–44.

MALONE, T. C. (1971a). The relative importance of nannoplankton and netplankton as primary producers in tropical oceanic and neritic phytoplankton communities. *Limnol. Oceanogr.*, **16**, 633–9.

MALONE, T. C. (1971b). Diurnal rhythms in netplankton and nannoplankton assimilation ratios. *Mar. Biol.*, **10**, 285–9.

MARGALEF, R. (1971). The pelagic ecosystem of the Caribbean Sea. In *Proc. Symp. Investigating the Resources in the Caribbean Sea and Adjacent Regions*. UNESCO, Paris. pp. 483–98.

MARKINA, N. P. (1971). Composition and the distribution of plankton near the Australian coast. In *Proc. Symp. on the Indian Ocean and Adjacent Seas*. Marine Biological Association, Cochin, India. p. 35.

MARLOWE, C. J. and MILLER, C. B. (1975). Patterns of vertical distribution and migration of zooplankton at Ocean Station 'P'. *Limnol. Oceanogr.*, **20**, 824–44.

MARUMO, R. (1975). *Trichodesmium* bloom in the Pacific Ocean. (Abstr.) In *Proc. 13th Pacific Science Congr.* Pacific Science Association, University of British Columbia, Vancouver. p. 57.

MENON, M. D. and GEORGE, K. C. (1978). On the abundance of zooplankton along the southwest coast of India during the years 1971–75. In *Proc. Symp. on Warm Water Zooplankton*. National Institute of Oceanography, Goa, India. pp. 205–13.

MICHEL, A., COLIN, C., DESROSIERES, R. and OUDOT, C. (1971). Observations sur l'hydrologie et le plancton des abords et de la zone des passes de l'atoll de Rangiroa (Archipel des Tuamotu, Océan Pacifique Central). *Cah. ORSTOM*, (*Océanogr.*), **9**, 375–405.

MIKHEYEV, V. N. (1977). Structural characteristics of the zoocoenosis in the Peruvian coastal upwelling region. *Oceanology*, **17**, 462–5.

MORITA, R. Y., GILLESPIE, P. A. and JONES, L. P. (1971). Microbiology of Antarctic sea water. *Antarct. J. U.S.*, **6**, 956.

MOTODA, S. and MARUMO, R. (1963). Plankton of the Kuroshio Water. In *Proc. Symp. on Kuroshio*. UNESCO and the Oceanographic Society of Japan, Tokyo. pp. 40–61.

MOTODA, S. and TAKASHI, M. (1974). Plankton of the Bering Sea. In D. W. Hood and E. H. Kelley (Eds), *Oceanography of the Bering Sea with Emphasis on Renewable Resources*. Institute of Marine Science, Univ. Alaska, College. pp. 207–41.

MOTODA, S., TANIGUCHI, A. and IKEDA, T. (1974). Plankton ecology in the western North Pacific Ocean: primary and secondary productivities. In *Proc. FAO Indo-Pacific Fisheries Council, 15th Session*. FAO, Rome. pp. 86–110.

MOVCHAN, O. A. (1962). Spring phytoplankton in the western areas of the North Atlantic. (Russ.) *Trudy VNIRO*, **46**, 315–23.

MOVCHAN, O. A. (1967). Phytoplankton distribution and development in the Newfoundland area in relation to seasonal variations of some abiotic factors. *Oceanology*, **7**, 820–31.

MOVCHAN, O. A. (1970). Phytoplankton qualitative composition in the Newfoundland Region. (Russ.) *Okeanologiya*, **10**, 496–504.

MOVCHAN, O. A. (1971). Plankton investigations in coastal areas of the northern parts of the Indian Ocean. (Russ.) *Trudy VNIRO*, **72**, 65–9.

NESIS, K. N. (1974). Oceanic cephalopods of the southwestern Atlantic Ocean. (Russ.) *Trudy Inst. Okeanol.*, **98**, 51–75.

NUÑEZ, A. L. and NOVARINI, J. C. (1976). Observations of deep scattering layers in the southern Atlantic and the Antarctic ocean. *Deep Sea Res.*, **23**, 475–7.

PARIN, N. V. (1968). *Ichthyofauna of the Oceanic Epipelagic Zone*. (Russ.) Izdat. Nauka, Moscow. (Engl. transl., 1970, U.S. Dept. of Interior and the National Science Foundation, Washington, D.C.)

PARIN, N. V. (1971). Some features on the distribution of deep-sea fishes in the Peru Current zone. (Russ.) *Trudy Inst. Okeanol.*, **89**, 81–95.

PARIN, N. V. (1972). Distribution of oceanic pelagic fishes in the Pacific Ocean. In *Proc. 12th Pacific Science Congr.* Australian Academy of Science, Canberra. p. 162.

PARIN, N. V. (1975). Changes in the pelagic ichthyocoenosis on an equatorial transect in the Pacific Ocean between 96 and 155° W. (Russ.) *Trudy Inst. Okeanol.*, **102**, 313–4.

PARIN, N. V., ANDRIASHEV, A. P., BORODULINA, O. D. and CHUVASOV, V. M. (1974). Deep-water pelagic fishes in the southwestern parts of the Atlantic Ocean. (Russ.) *Trudy Inst. Okeanol.*, **98**, 76–140.

PAULINOSE, V. T. and ARAVINDAKSHAN, P. N. (1978). Zooplankton biomass, abundance and distribution in the north and northeastern Arabian Sea. In *Proc. Symp. on Warm Water Zooplankton*. National Institute of Oceanography, Goa, India. pp. 132–6.

PAVLOV, V. Ya. (1968). Plankton distribution in the Cap Blanc region. *Oceanology*, **8**, 381–7.

PAVLOVA, E. V., PETIPA, T. S. and SOROKIN, Yu. I. (1971). Role of bacterioplankton in the nutrition marine planktonic animals. (Russ.) In M. E. Vinogradov (Ed.), *Functioning of Pelagic Communities in the Tropical Regions of the Ocean*. Nauka, Moscow. pp. 142–151.

PEARCY, W. G. and LAURS, R. M. (1966). Vertical migration and distribution of mesopelagic fishes off Oregon. *Deep Sea Res.*, **13**, 153–65.

PETER, K. J. (1974). Seasonal variation of ichthyoplankton in the Arabian Sea in relation to monsoons. In J. H. S. Blaxter (Ed.), *The Early Life History of Fish*. Springer-Verlag, Heidelberg. pp. 263–4.

PETERSON, W. T. (1976). Zooplankton population maintenance in the coastal upwelling zone off Oregon, U.S.A. In *Book of Abstracts of Papers Presented at Joint Oceanographic Assembly, Edinburgh*. FAO, Rome. p. 149.

PETERSON, W. T. and MILLER, C. B. (1977). Seasonal cycle of zooplankton abundance and species composition along the central Oregon coast. *Fish. Bull. U.S.*, **75**, 717–24.

PETIPA, T. S., MONAKOV, A. V. and PAVLUTIN, A. M. (1974). Food and energy balance in tropical copepods. In *Biological Productivity in Southern Seas*. (Russ.) Kiev, Naukova Dumka. pp. 136–52.

PETIPA, T. S., MONAKOV, A. V., SOROKIN, Yu. I. and KUKINA, I. V. (1975a). In *Energetic Aspects of Growth and Exchanges in Aquatic Organisms*. (Russ.) Kiev, Naukova Dumka.

PETIPA, T. S., MONAKOV, A. V., SOROKIN, Yu. I., VOLOSHINA, G. V. and KUKINA, I. V. (1975b). Balance of matter and energy in copepods from tropical upwelling areas. (Russ.) *Trudy Inst. Okeanol.*, **102**, 335–50.

PONOMAREVA, L. A. (1960). On the plankton biomass distribution in tropical waters of the western areas of the Pacific Ocean. (Russ.) *Trudy Inst. Okeanol.*, **41**, 48–54.

PONOMAREVA, L. A. (1972). Quantitative repartition of euphausiids in the Indian Ocean. (Russ.) *Okeanologiya*, **12**, 689–94.

POSTEL, E. (1968). Marine hydrology and biogeography in western Africa. In *West African International Atlas*. Organization for African Unity, L'Institute francais d'Afrique noire (IFAN), Dakar. pp. 13–16.

PRASAD, R. R. (1954). The characteristics of marine plankton at an inshore station in the Gulf of Mannar near Mandapam. *Indian J. Fish.*, **1–2**, 1–36.

PRASAD, R. R. (1969). Zooplankton biomass in the Arabian Sea and the Bay of Bengal, with a discussion on the fisheries of the regions. *Proc. natn Inst. Sci. India*, **35** B, 399–437.

QASIM, S. Z. (1978). Contribution of zooplankton in the food chains of some warm water environments. In *Proc. Symp. on Warm Water Zooplankton*. National Institute of Oceanography, Goa, India. pp. 700–8.

QASIM, S. Z., BHATTATHIRI, P. M. A. and DEVASSY, V. P. (1972). The influence of salinity on the rate of photosynthesis and abundance of some tropical phytoplankton. *Mar. Biol.*, **12**, 200–6.

QASIM, S. Z., WAFAR, M. V. M., VIJAYARAGHAVAN, S., ROYAN, J. P. and KRISHNA KUMARI, L. (1978). Biological productivity of coastal waters of India, from Dabhol to Tuticorin. *Ind. J. mar. Sci.*, **7**, 84–93.

RADHAKRISHNA, K. (1978). Primary productivity of the Bay of Bengal during March–April 1975. *Ind. J. mar. Sci.*, **7**, 58–60.

RADHAKRISHNA, K., BATTATHIRI, M. A. and DEVASSY, V. P. (1978a). Primary productivity of the Bay of Bengal during August–September 1976. *Ind. J. mar. Sci.*, **7**, 94–8.

RADHAKRISHNA, K., DEVASSY, V. P., BATTATHIRI, M. A. and BHARGAVA, R. M. (1978b). *Ind. J. mar. Sci.*, **7**, 137–9.

RAMAMURTHY, V. D. (1972). Studies on the blooms of *Trichodesmium erythraeum* (E.M.R.) in the waters of the central west coast of India. *Curr. Sci.*, **41**, 803–4.

RAO, T. S. S. (1973). Zooplankton studies in the Indian Ocean. In B. Zeitschel (Ed.), *The Biology of the Indian Ocean.* Springer-Verlag, Heidelberg. pp. 243–55.

RATKOVA, T. N. (1974). Phytoplankton repartition in the Peru Current Region on the parallel 8° S. (Russ.) *Okeanologiya*, **14**, 1077–81.

REYNOLDS, N. (1974). What matters in Arctic phytoplankton. *Br. Phycol. J.*, **9**, 222–3.

REYSSAC, J. (1970). Phytoplancton et production primaire au large de la Côte d'Ivoire. *Bull. Inst. fr. Afr. noire*, (Ser. A), **32**, 869–981.

REYSSAC, J. (1974). Observations sur le phytoplancton et la production primaire de la région du banc d'Arguin (Mauritanie) en avril et mai 1972. *Bull. Inst. fr. Afr. noire*, (Ser. A), **36**, 51–61.

REYSSAC, J. (1975). Evolution quantitative du phytoplancton de la baie du Lévrier de septembre à novembre 1973. *Bull. Mus. Hist. nat. Paris*, (Ser. 3), **328**, 69–79.

REYSSAC, J. and ROUX, M. (1972). Communautés phytoplanctoniques dans les eaux de Côte d'Ivoire. Groupes d'espèces associées. *Mar. Biol.*, **13**, 14–33.

RILEY, G. A., WANGERSKY, P. J. and HEMERT, D. V. (1964). Organic Aggregates in tropical and subtropical surface waters of the North Atlantic ocean. *Limnol. Oceanogr.*, **9**, 546–50.

RILEY, G. A., VAN HEMERT, D. and WANGERSKY, P. J. (1965). Organic aggregates in surface and deep waters of the Sargasso sea. *Limnol. Oceanogr.*, **10**, 354–63.

ROGER, C. and GRANDPERRIN, R. (1976). Pelagic food webs in the tropical Pacific. *Limnol. Oceanogr.*, **21**, 731–5.

ROUND, F. E. (1971). Benthic marine diatoms. *Oceanogr. mar. Biol. A. Rev.*, **9**, 83–129.

RUSSELL, F. S. (1939). Hydrographical and biological conditions in the North Sea as indicated by plankton organisms. *J. Cons. perm. int. Explor. Mer*, **14**, 171–92.

RYTHER, J. H., MENZEL, D. W. and CORWIN, N. (1967). Influence of the Amazon River outflow on the ecology of the western Tropical Atlantic. I. Hydrography and Nutrient Chemistry. *J. mar. Res.*, **25**, 69–83.

RYTHER, J. H., MENZEL, D. W., HULBURT, E. M., LORENZEN, C. J. and CORWIN, N. (1970). Production and utilization of organic matter in the Peru coastal current. *Scientific Results of the Southeastern Pacific Ocean Expedition, Anton Bruun Report*, **4**, 4–12.

SAKHTIVEL, M. (1977). Further studies of plankton ecosystems in the eastern Indian ocean. 8. Seasonal, diurnal and latitudinal variations in abundance of Euthecosomata along the 110 °E meridian. *Aust. J. mar. Freshwat. Res.*, **28**, 663–71.

SANINA, L. V. (1969). Phytoplankton composition and distribution in the Atlantic Ocean along the meridian 30° W. (Russ). *Trudy VNIRO*, **65**, 148–63.

SAVICH, M. S. (1968a). Phytoplankton of the Gulf of Aden and Arabian Sea in the winter-monsoon period. (Russ.) *Trudy VNIRO*, **64**, 243–51.

SAVICH, M. S. (1968b). Phytoplankton of the Gulf of Aden in the summer-monsoon period. (Russ.) *Trudy VNIRO*, **64**, 252–9.

SAVICH, M. S. (1971). Some patterns of vertical phytoplankton distribution in the Gulf of Aden as related to oceanographic conditions. *Oceanology*, **11**, 396–8.

SEKI, H. and KENNEDY, O. D. (1969). Marine bacteria and other heterotrophs as food for zooplankton in the Strait of Georgia. *J. Fish. Res. Bd Can.*, **26**, 165–73.

SEMINA, G. I. (1960). Phytoplankton of the parts of the Pacific Ocean along the meridian 174° W. (Russ.) *Trudy Inst. Okeanol.*, **41**, 17–30.

SEMINA, G. I.(1962). Phytoplankton of the central parts of the Pacific Ocean along the meridian 174° W. (Russ.) *Trudy Inst. Okeanol.*, **58**, 3–26.

SEMINA, G. I. (1971). Distribution of plankton in the south-eastern Pacific. (Russ.) *Trudy Inst. Okeanol.*, **89**, 43–59.

SHEVTSOV, V. V. and MAKAROV, R. R. (1969). On the biology of the Antarctic krill. (Russ.) *Trudy vses. nauchno-issled Inst. morsk. ryb. Khoz. Okeanogr.*, **66**, 177–206. (Engl. transl. N.S. 91 of the Ministry of Agriculture, Fishery and Food, Fisheries Lab. Lowestoft, U.K., 1970.)

SIEBURTH, J. McN. (1959). Antibacterial activity of antarctic marine phytoplankton. *Limnol. Oceanogr.*, **4**, 419–24.

SIEBURTH, J. McN. (1960a). Antibiotic properties of acrylic acid, a factor in the gastrointestinal antibiosis of polar marine animals. *J. Bact.*, **82**, 72–9.

SIEBURTH, J. McN. (1960b) Acrylic acid, an antibiotic principle in *Phaeocystis* blooms in Antarctic waters. *Science, N.Y.*, **132**, 676–7.

SMIRNOVA, L. I. (1956). Phytoplankton in the northwestern areas of the Pacific Ocean. (Russ.) *Dokl. Akad. Nauk SSSR*, **109**, 649–52.

SMITH, R. L. and BOTTERO, J. S. (1977). On upwelling in the Arabian Sea. In M. Angel (Ed.), *A Voyage of Discovery*, George Deacon 70th Anniversary Volume. Pergamon Press, Oxford, pp. 291–304.

SOROKIN, Yu. I. (1970). Some data on primary production in the central Pacific. *Oceanology*, **10**, 538–42.

SOROKIN, Yu. I. (1971). Quantitative evaluation of the role of bacterioplankton in the biological productivity of World Ocean tropical waters (Russ.). In M. E. Vinogradov (Ed.), *Functioning of the Pelagic Community in the Tropical Regions of the Ocean*. Izdat. Nauka, Moskva. p. 272.

SOROKIN, Yu. I. (1973a). Productivity of coastal tropical waters of the western Pacific. *Oceanology*, **13**, 551–8.

SOROKIN, Yu. I. (1973b). Data on biological productivity of the western tropical Pacific Ocean. *Mar. Biol.*, **20**, 177–96.

SOROKIN, Yu. I. (1978). Decomposition of organic matter and nutrient regeneration. In O. Kinne (Ed.), *Marine Ecology*, Vol. IV, Dynamics. Wiley, Chichester. pp. 501–616.

SOROKIN, Yu. I. and KLIACHTORIN, L. B. (1961). Primary production in the Atlantic Ocean. (Russ.) *Trud. vsesoj. gidrobiol. Obshchest. SSSR*, **11**, 262–84.

SOROKIN, Yu. I., PAVEL'EVA, E. B. and VASIL'EVA, M. I. (1975). Productivity and trophic role of bacterioplankton in the equatorial divergence area. (Russ.) *Trudy Inst. Okeanol.*, **102**, 184–98.

SOURNIA, A. and RICARD, M. (1975). Phytoplankton and primary productivity in Takapoto Atoll, Tuamotu Islands. *Micronesica*, **11**, 159–66.

SOURNIA, A. and RICARD, M. (1976). Phytoplankton and its contribution to primary productivity in two coral reef areas of French Polynesia. *J. exp. mar. Biol. Ecol.*, **21**, 129–40.

STEVEN, D. M. (1971). Primary productivity of the tropical western Atlantic Ocean near Barbados. *Mar. Biol.*, **10**, 261–4.

SUBRAHMANYAN, R. (1971). Phytoplankton of the Indian Ocean. Some ecological problems. In *Proc. Symp. on the Indian Ocean and Adjacent Seas*. Marine Biological Association, Cochin, India. p. 26.

SUBRAHMANYAN, R. and VISWANATHA-SARMA, A. H. (1960). Studies on the phytoplankton of the west coast of India. *Indian J. Fish.*, **7**, 307–36.

SUKHANOVA, I. N. (1962). Phytoplankton specific composition and distribution in the northern areas of the Indian Ocean. (Russ.) *Trudy Inst. Okeanol.*, **58**, 27–39.

SUKHANOVA, I. N. (1976). The qualitative composition and quantitative distribution of the phytoplankton in the north eastern Indian Ocean. (Russ.) *Trudy Inst. Okeanol.*, **105**, 55–82.

SZEKIELDA, K. H., SUSZKOWSKI, D. J. and TABOR, P. S. (1977). Skylab investigation of the upwelling off the northwest coast of Africa. *J. Cons. int. Explor. Mer*, **37**, 205–13.

TAKEDA-AKIYAMA, A. and ARIA, R. (1968). Plankton. *J. Tokyo Univ. Fish., Spec. edn.*, **9**, 27–51.

TANIGUCHI, A., SAITO, K., KOYAMA, A. and FUKUCHI, M. (1976). Phytoplankton communities in the Bering Sea and adjacent seas. I. Communities in early warming season in Southern areas. *J. oceanogr. Soc. Japan*, **32**, 99–106.

THIRIOT, A. (1977). Peuplements zooplantoniques dans les régions de remontées d'eau du littoral africain. *Bull. Centre Recherches Oceanogr. Abidjan (Ivory Coast)*, **8**, 1–72.

TIMONIN, A. G. (1969). The structure of pelagic associations. The quantiative relationship between different trophic groups of plankton in frontal zones of the Tropical Ocean. *Oceanology*, **9**, 686–95.

TIMONIN, A. G. (1971). The structure of plankton communities of the Indian Ocean. *Mar. Biol.*, **9**, 281–9.

TIMONIN, A. G. (1973). Structure of pelagic communities. Trophic structure of zooplankton communities in the northern part of the Indian Ocean. *Oceanology*, **13**, 114–24.

TORRINGTON–SMITH, M. (1974). The distribution of abundance, species diversity and phytohydrographic regions in west Indian ocean phytoplankton. *J. mar. biol. Ass. India.*, **16**, 371–80.

TRANTER, D. H. (1962). Zooplankton abundance in Australasian Waters. *Aust. J. mar. Freshwat. Res.*, **13**, 106–42.

TRANTER, D. J. and KERR, J. D. (1977). Further studies of plankton ecosystems in the eastern Indian Ocean. 3. Numerical abundance and biomass. *Aust. J. mar. Freshwat. Res.*, **28**, 557–83.

THRONDSEN, J. (1970). Flagellates from Arctic waters. *Nytt. Mag. Bot.*, **17**, 49–57.

TSYBAN, A. V. (1973). Microbial investigations in the northeastern Pacific Ocean. (Russ.) *Trudỹ Inst. Okeanol.*, **91**, 52–66.

TSYBAN, A. V. and TEPLINSKAYA, N. G. (1972). Microbial population of the northwestern Pacific waters. In A. Y. Takenouti (Ed.), *Biological Oceanography of the North Pacific Ocean.* Idemitsu Shoten, Tokyo. pp. 541–57.

VEDERNIKOV, V. I. and STARODUBSTEV, E. G. (1971). Primary production and chlorophyll in the south eastern Pacific. (Russ.) *Trudỹ Inst. Okeanol.*, **89**, 33–42.

VENRICK, E. L. (1974). The distribution and significance of *Richelia intracellularis* Schmidt in the north Pacific Central gyre. *Limnol. Oceanogr.*, **19**, 437–45.

VINOGRADOV, M. E. (1960). Quantitative repartition of the deep-sea plankton in the western and central regions of the Pacific Ocean. (Russ.) *Trudỹ Inst. Okeanol.*, **41**, 55–84.

VINOGRADOV, M. E. (1962). Quantitative repartition of the deep-sea plankton in the northern regions of the Indian Ocean. (Russ.) *Okeanologiya*, **2**, 577–92.

VINOGRADOV, M. E. (1968). *Vertical Distribution of Oceanic Zooplankton.* (Russ.) Izdat. Nauka, Moskva.

VINOGRADOV, M. E. (1970). Vertical distribution of zooplankton in the Kuril-Kamchatka region of the Pacific Ocean. *Trudỹ Inst. Okeanol.*, **86**, 99–117.

VINOGRADOV, M. E. (1976). Equatorial upwelling: its physical and biological peculiarities. In *Book of Abstracts of Papers Presented at Joint Oceanographic Assembly.* Edinburgh Publ., F.A.O., Rome. 35 pp.

VINOGRADOV, M. E. (1977). *Biology of the Ocean*, Vols 1, 2. (Russ.) Izdat Nauka, Moskva.

VINOGRADOV, M. E., KRAPIVIN, V. F., MENSHUTKIN, V. V., FLEYSHMAN, B. S. and SHUSHKINA, E. A. (1973). Mathematical model of the function of the pelagial ecosystem in tropical regions (from the 50th voyage of the R/V. VITYAZ). *Oceanology*, **13**, 704–17.

VINOGRADOV, M. E., KUKINA, I. V., LEBEDEVA, L. P. and SHUSHKINA, E. A. (1977). Variation with depth of the functional characteristics of planktonic community in the equatorial upwelling of the Pacific Ocean. *Oceanology*, **17**, 342–50.

VINOGRADOV, M. E., MENSHUTKIN, V. V. and SHUSHKINA, E. A. (1972). On mathematical simulation of a pelagic ecosystem in tropical waters of the ocean. *Mar. Biol.*, **16**, 261–8.

VINOGRADOV, M. E. and NAUMOV, A. G. (1961). Plankton quantitative repartition in the antarctic waters of the Indian and Pacific oceans. (Russ.) *Okeanol. Issled. M.G.G.*, **3**, 172–6.

VINOGRADOV, M. E. and SEMENOVA, T. N. (1977). Trophic characteristics of the pelagic communities of the equatorial upwelling. *Pol. Arch. Hydrobiol.*, **24** (Suppl.), 307–14.

VINOGRADOV, M. E. and SHUSHKINA, E. A. (1977). Some characteristics of the vertical structure of a planktonic community in the Equatorial Pacific region. (Russ.) *Oceanology*, **16**, 389–93.

VINOGRADOV, M. E. and SHUSHKINA, E. A. (1978). Some development patterns of plankton communities of the upwelling areas of the Pacific Ocean. *Mar. Biol.*, **48**, 357–66.

VINOGRADOV, M. E., SHUSHKINA, E. A. and KUKINA, I. N. (1969). Functional characteristics of a planktonic community in an equatorial upwelling region. *Oceanology*, **16**, 67–75.

VINOGRADOV, M. E., SHUSHKINA, E. A. and KUKINA, I. N. (1975). Functional characteristics of a planktonic community in an equatorial upwelling region. *Oceanology*, **16**, 67–75.

VINOGRADOV, M. E. and VORONINA, N. M. (1961). Influence of the oxygen deficit on the zooplankton repartition in the Arabian Sea. (Russ.) *Okeanologiya*, **1**, 670–8.

VINOGRADOV, M. E. and VORONINA, N. M. (1962a). Some features of the zooplankton repartition in the northern areas of the Indian ocean. (Russ.) *Trudy Inst. Okeanol.*, **58**, 80–113.

VINOGRADOV, M. E. and VORONINA, N. M. (1962b). Influence of the oxygen deficit on the distribution of plankton in the Arabian Sea. *Deep Sea Res.*, **9**, 523–30.

VINOGRADOV, M. E. and VORONINA, N. M. (1963). Plankton distribution in the Equatorial Currents water of the Pacific Ocean. (Russ.) *Trudy Inst. Okeanol.*, **1963**, 22–59.

VINOGRADOV, M. E. and VORONINA, N. M. (1964). Plankton distribution in the Equatorial Currents water of the Pacific Ocean. (Russ.) *Trudy Inst. Okeanol.*, **65**, 58–76.

VINOGRADOV, M. E. and VORONINA, N. M. (1974). Vertical repartition of the zooplankton biomass in the South-Sandwich Islands region. (Russ.) *Trudy Inst. Okeanol.*, **98**, 38–42.

VINOGRADOV, M. E., VORONINA, N. M. and SUKHANOVA, I. N. (1961). Spatial distribution of the tropical plankton and its relation with some structural peculiarities of the offshore waters. (Russ.) *Okeanologiya*, **1**, 283–93.

VLADIMIRSKAJA, E. V. (1962). Zooplankton distribution and seasonal changes in the Newfoundland region. (Russ.) *Trudy VNIRO*, **46**, 296–315.

VOLKOV, A. F., DOLZHENKOV, V. H. and KAREDIN, E. P. (1972). Macroplankton of the southern Pacific ocean. In *Proc. 12th Pacific Science Congr*. Australian Academy of Science, Canberra. p. 155.

VORONINA, N. M. (1962). On the distribution of the macroplankton in the northern half of the Indian Ocean. *Okeanologiya*, **2**, 118–25.

VORONINA, N. M. (1963). Dependence of situation and characteristics of the biogeographic boundaries on the hydrometerological conditions in the southern Ocean. 1. Boundary between the Antarctic and Subantarctic Regions. (Russ.) *Okeanologiya*, **2**, 205–10.

VORONINA, N. M. (1964a). Distribution of subsurface zooplankton in the Equatorial Currents water of the Pacific Ocean. (Russ.) *Trudy Inst. Okeanol.*, **65**, 95–106.

VORONINA, N. M. (1964b). Macroplankton repartition in Equatorial Currents water of the Pacific ocean. (Russ.) *Okeanologiya*, **4**, 884–95.

VORONINA, N. M. (1966a). On the distribution of zooplankton in the Southern Ocean. *2nd Int. Oceanography Congr. (Abstr. papers)*, **2**, 387.

VORONINA, N. M. (1966b). Distributions of the zooplankton biomass in the Southern Ocean. (Russ.) *Okeanologiya*, **6**, 1041–54.

VORONINA, N. M. (1972). Vertical structure of the pelagic community in the Antarctic. (Russ.) *Okeanologiya*, **12**, 492–6.

VORONINA, N. M. (1975). On ecology and biogeography of the Southern Ocean plankton. (Russ.) *Trudy Inst. Okeanol.*, **103**, 60–87.

VORONINA, N. M. and NAUMOV, A. G. (1968). Qualitative repartition and composition of the mesoplankton of the Southern Ocean. (Russ.) *Okeanologiya*, **8**, 1059–965.

VORONINA, N. M. and ZADORINA, L. A. (1974). Plankton quantitative distribution in the South Antarctic in November–December 1971. (Russ.) *Trudy Inst. Okeanol.*, **98**, 30–7.

VOSS, G. L. (1969). The pelagic mid-water fauna of the eastern tropical Atlantic with special reference to the Gulf of Guinea. In *Proc. Symp. on Oceanography and Fishery Resources Tropical Atlantic*. UNESCO-FAO-OAU, Paris. pp. 91–9.

WALSH, J. J. (1969). Vertical distribution of antarctic phytoplankton. II. A comparison of phytoplankton standing crops in the Southern Ocean with that of the Florida Strait. *Limnol. Oceanogr.*, **14**, 86–94.

WALSH, J. J. (1971). Relative importance of habitat variables in predicting the distribution of phytoplankton at the ecotone of the antarctic upwelling ecosystem. *Ecol. Monogr.*, **41**, 291–309.

WALSH, J. J., KELLEY, J. C., WHITLEDGE, T. E. and MCISAAC, J. J. (1974). Spin-up of the Baja California upwelling ecosystem. *Limnol. Oceanogr.*, **19**, 553–72.

WICKSTEAD, J. H. (1958). A survey of the larger zooplankton of Singapore straits. (*J. Cons. perm. int. Explor. Mer*, **23**, 340–53.

WING, B. L. (1974). Kinds and abundance of zooplankton collected by the USCG icebreaker glacier in the eastern Chukchi Sea, September–October 1970. *U.S. National Oceanic and Atmospheric Administration Technical Reports*, **670**, 1–18. National Marine Fisheries Service.

YAMANAKA, N. (1976). The distribution of some copepods (Crustacea) in the Southern Ocean and adjacent regions from 40° to 81° W long. *Bol. Zool. Univ. S. Paulo*, **1**, 161–96.

YASHNOV, V. A. (1961). Vertical repartition of zooplankton bulk of tropical areas of the Atlantic Ocean. (Russ.) *Dokl. Akad. Nauk SSSR*, **136**, 705–8.

YASHNOV, V. A. (1962). Plankton of the tropical regions of the Atlantic Okean. (Russ.) *Trudy morsk. Gidrof. Inst.*, **25**, 195–207.

ZELIKMAN, E. A. and KAMSHILOV, M. M. (1960). Dynamics on several years of the zooplankton biomass in the southern Barents Sea and its controlling factors. (Russ.) *Trudy murmansk biol. Inst.*, (2), 68–113.

ZERNOVA, V. V. (1962). Phytoplankton qualitative distribution in the northern areas of the Indian Ocean. (Russ.) *Trudy Inst. Okeanol.*, **58**, 45–53.

ZERNOVA, V. V. (1964). Net-phytoplankton distribution in the tropical region of the western Pacific Ocean. (Russ.) *Trudy Inst. Okeanol.*, **65**, 32–48.

ZERNOVA, V. V. (1970a). On planktonic algae of the Gulf of Mexico and Caribbean Sea. (Russ.) *Okeanol. Issledov. M.G.G.*, **20**, 69–104.

ZERNOVA, A. A. (1970b). On colour changes of Gulf of Mexico waters originated from planktonic algae. (Russ.) *Okeanol. Issledov. M.G.G.*, **20**, 105–9.

ZERNOVA, A. A. (1974). Phytoplankton biomass distribution in the Atlantic Ocean tropical waters. (Russ.) *Okeanologiya*, **6**, 1070–6.

Marine Ecology Vol. 5, Part 1
Edited by Otto Kinne

7. SPECIFIC PELAGIC ASSEMBLAGES

J. M. PERES

(1) Assemblages at the Air–Ocean Interface

(a) General Aspects

Organisms in the immediate vicinity (uppermost 0–5 or 0–10 cm water layer) of the air–ocean interface are usually classified into two different assemblages: neuston and pleuston. ZAITSEV (1971) reviewed and discussed the definitions by different authors of these two terms, the meaning of which has gradually widened. This led to a rather confused situation which the synthesis by CHENG (1975) also did not clarify.

Aiming at increased accuracy, I have slightly modified the definitions provided by ZAITSEV (1971, p. 57). The term 'neuston' is defined here as 'microorganisms, plants and animals of small-to-medium size, involving both hydrobionts and aerobionts, which live on the aquatic (hyponeuston) or aerial (epineuston) sides of the water's surface film.' 'Pleuston' comprises plants or animals whose bodies are normally situated simultaneously both in aqueous and aerial media, or who are linked with flotsam.

(b) Neuston

Physical and Chemical Characteristics of the Uppermost Water Layer

The physico-chemical properties of the uppermost water layer are often very difficult to investigate because water bottles can not usually be employed there. HARVEY (1975, p. 104) distinguishes the microlayer as a special entity of the uppermost layer, i.e. the structurally organized water layer at the air–water interface, ca 60 μm thick, and made coherent by a surface active film (usually one molecule layer thick) consisting of material adsorbed at the surface. Assuming the monolayer to be one or a few nm thick, and the microlayer to be 30,000 to 60,000 nm thick, HARVEY postulates the microlayer to be some 30,000 times as thick as the monolayer.

Illumination

In the uppermost water layer, light penetration and absorption are most intense. Off the southern coast of the Crimea, the 0–1 cm layer absorbs 20% of the solar radiation, and the 0–5 cm layer, possibly 40%. Long (infrared) and short (ultraviolet) wavelengths are the most intensively absorbed; both are practically absent below 10 cm.

Temperature

Light absorption and temperature are obviously intimately related. When the sea is quite smooth, the uppermost microlayer—plus a thin layer of adjacent water—is 0·1–0·2 °C warmer than water 10 cm below (HARVEY, 1975) due to absorption of infrared. During cloudy weather the uppermost layer may possibly be a little colder due to evaporation. However, even very slight waves may induce vertical mixing which disturbs this microregime.

Salinity and nutrients

Except for cases of either massive rainfall—which are frequent in intertropical regions—or massive river run-off during calm weather, significant salinity differences between the uppermost 1-cm thick layer and the underlying waters have not yet been demonstrated. Differences in the concentration of basic salts were assumed, but not yet demonstrated. Nutrient salts tend to be much more abundant immediately below the interface: DAUMAS (1974), for instance, observed in the 0–100 μm layer a PO_4–P content 15 times higher than below, and also higher NO_3–N concentrations.

Dissolved gases

In the Gulf of Marseilles, France, the oxygen content in the 0–10 cm layer is almost always above saturation level. Amazingly, warmer waters are more supersaturated than colder waters (CHAMPALBERT, 1975). Possibly, phytoneuston and phytoplankton photosynthetic activities play some part in this supersaturation. However, taking into account meteorological sequences, CHAMPALBERT noticed that the highest supersaturation level always occurs just after periods of strong wave motion, and further observed that maximum oxygen supersaturation occurs at 2 p.m., i.e. the time of a slight temperature maximum. She assumes that the oxygen maximum could correspond to the maximum photosynthetic oxygen production cumulated since dawn. ZAITSEV (1971) did not consider dissolved gases.

Wind stress

At 7 m s^{-1} or less, wind compresses surface films or incipient film materials and transports these rapidly downward (HARVEY, 1975, p. 105).

'Water associated with the surface film (monolayer) is dragged along with it to an extent depending upon locally prevailing conditions. At speeds exceeding 7 m s−1, vertical mixing progress rapidly; the film then mixes with deeper water. When the materials adsorbing at the surface have accumulated sufficiently to form a monolayer on the water, nipple damping commences and slicks appear.'

Extensive film thickening, e.g. in the presence of sufficient reservoir material, causes the appearance of interference colours.

Non-living organic material

The uppermost water layer always contains large amounts of non-living organic material: particulate, colloidal and dissolved. Particles in the hyponeuston may originate from atmosphere, land and from the underlying pelagos.

Terrigenous particles may be brought in by river runoff or by winds. Even large particles such as land insects—not only flying but also wingless forms—were observed above even 4500 m in the atmosphere, especially representatives of Isoptera, Hymenoptera and Diptera. These may be transported hundreds of km away from shore. The insects soon die after falling on the sea surface, but do not sink immediately due to tracheal and air sacs which make them highly buoyant. Sinking prevails only after the body has become water-soaked and partly disintegrated. Airborne land insects may be consumed by hyponeuston animals.

Smaller airborne organic particles transported from land to the sea include scales of mosquitoes and butterflies, and vegetal particles such as spores, pollen, small pieces of leaves, etc. All these remain for some time on the interface since they are small in size, light and nonwettable. They may be used as a food resource by small phagotrophs such as *Noctiluca*. The percentage of the total organic particulate matter of the neuston layer represented by the allochthonous—i.e. terrestrial—material is not exactly known, but possibly never exceeds 20–30% . This percentage obviously decreases towards the open sea. Progressive decay of this material results in colloidal and dissolved organic substances.

Nevertheless, especially in areas furthest offshore, the bulk of particulate organic matter arises from marine organisms: dead bodies, crustacean moults, excreta, etc. 'Organic rain' from non-living plankton particles is important as a food source for many benthic animals. However, 'antirain' also exists and results in the concentration of dead bodies, fragments and excreta in the uppermost water layer. For example, it is well established that the ratio of dead/alive crustaceans (e.g. cladocerans, copepods) is always higher in the 0–5 cm layer than at any other level; the same is true for phytoplankton. Moreover, the dead/alive ratio is much higher for phytoplankton than for zooplankton; in general, the shorter the life span, the higher this ratio. The concentration of dead plants and animals is greater below the surface film, where these organisms decompose and thus supply the biotope with 'young' detritus (ZAITSEV, 1971) of great nutritional value, which may also be fed on by neuston animals.

'Antirain' may be due to several processes: (i) an increase in buoyancy due to gas bubbles appearing in dead bodies at the beginning of decomposition; (ii) convection cells (Vol. 1, p. 1528) arising from wind stress; (iii) phytoplankton photosynthesis; (iv) ultraviolet-induced aggregation of proteins or lipids (WHEELER, 1972); (v) bacterial aggregate formation on bubbles. The latter demands further comment. It is generally admitted that the initial nucleus of an aggregate may either be a bubble generated by turbulence or a very small organic debris particle. On the bubble surface, epiphytic bacteria settle and develop at the expense of dissolved organic substances and (in the second case) also at the expense of organic debris supporting the bubble. Aggregates may also exhibit an additional increment by adsorption of colloidal organic substances. Aggregates represent an important food supply for many neuston animals (for example, calyptopsis larvae of euphausiids, some pteropods).

Dissolved organic matter also exists in the uppermost layer. It originates from: (i) *in situ* excreta of neuston organisms; (ii) degradation of dead bodies or organismic fragments; (iii) release and subsequent degradation of organic substances from aggregates and the organic envelope of bubbles which burst when reaching the interface.

As regards the surface-active material in the organic surface film, it was found long ago that phytoplankters can produce such a material, which modifies the physical properties of the water surface. Experiments by HARVEY (1975) revealed that *Dunaliella salina* and *Tetraselmis tetrahele* did not produce detectable amounts of these materials during their logarithmic growth phase. However, after rupturing the cells by ultrasonic vibration, surface-active material was immediately released. This result corresponds with phytoplankton cell break-up from environmental stress. In microlayer samples cell disintegration occurs within seconds, resulting in the escape of tiny granules into the surrounding sea water, as well as cytoplasm droplets which immediately form a delicate skin upon contact with the water. While there are initially no bacteria on or around the cell, dead cells were usually populated by small non-motile bacteria within a day. Table 7-1 summarizes results from surface–skin studies in the Pacific Ocean (at the equator,

Table 7-1

Parameters related to primary production in upper water layers of the Pacific Ocean (equator, 155° W) (After NISHIZAWA, 1971a; reproduced by permission of University of Tokyo, Ocean Research Institute)

Parameter	Surface skin (ca. 150 μm)	Water depth		
		5 cm	10 cm	15 cm
POC (μg l^{-1})	418	39·8	43·4	55·1
PON (μg l^{-1})	38·5	5·6	7·4	6·8
Chlorophyll (*a* μg l^{-1})	0·173	0·091	0·094	0·087
Pheopigments (μg l^{-1})	0·242	0·119	0·088	0·082
DOC (mg l^{-1})	1·42	0·83	0·95	0·97
NH$_4$ (μg at. N l^{-1})	0·65	0·14	0·12	0·14
NO$_2$ (μg at. N l^{-1})	0·12	0·12	0·12	0·14
NO$_3$ (μg at. N l^{-1})	3·2	2·6	3·0	2·8

155° W) listing important chemical parameters, including particulate organic carbon (POC), nitrogen (PON) and dissolved organic carbon (DOC).

The organic matter content in the 0–150 μm layer may be briefly characterized as follows: (i) DOC is almost 3·5 times higher than POC; (ii) the concentration factor in the skin sample, relative to the bulk water, is about 10·5 for POC and 1·7 for DOC; (iii) the C/N ratio is higher in the skin sample. An extremely sharp negative carbon gradient occurs within the immediate surface skin (ca 100 μm thick). This layer appears to be a locus where an intensive formation of particulate material occurs (NISHIZAWA, 1971b, p. 268).

The dissolved organic material of the surface film mainly consists of glycoproteins and various lipids: triglycerides, free fatty acids (mostly C_{11} to C_{22}) and wax esters, together with alcohols (C_{16}–C_{22}) (LARSSON and co-authors, 1974). Most of these hydrophobic substances originate from marine organisms. For example, it was demonstrated that the mucous layer of fish skin contains appreciable amounts of lipids, the fatty acid of which may be discriminated in the surface film. Diatoms also release lipids. All these interfacial organics are exposed to high-energy ultraviolet radiation, which may induce photochemical reactions such as hydrocarbon photo-oxidation or diene and triene conjugation of unconjugated linoleic acid (WHEELER, 1975).

All manmade lipophilic pollutants are also concentrated and stabilized in the upper-most water layer, e.g. petroleum and chlorinated hydrocarbons, pesticides and PCBs.

'Organic acids, proteinaceous material and other surface-active organics may pro-vide complexing sites for many heavy metals, resulting in the concentration of these metals at the water surface' (MORRIS, 1974, p. 109).

This was demonstrated in the Black Sea 0–10 cm layer for Fe, Cu, Mn, V, Co, Ni, Ti, Al, Mo, Sn, Pb and Ag (ZAITSEV, 1971). Concentration factors in the surface skin for many of these pollutants, relative to the underlying waters, commonly range between 10^3 and 10^4, and sometimes higher. This makes the pollutants readily available to neuston organisms and, due to predator–prey relationships between plankton and neus-ton, also to plankters. PIOTROWICZ and co-authors (1972), for example, found a higher particulate lead content in the microlayer than in subsurface water and determined that the enrichment was greater in sea slicks. Larvae and fry of many edible fish species participate in the merohyponeuston (p. 331) and thus possibly might accumulate pol-lutants. Adult specimens of the Hawaiian grey mullet which—as other *Mugil* species—consume mainly benthic microflora and detritus, also feed at the surface after the wind has accumulated a sufficient amount of film and microlayer materials (HAR-VEY, 1975).

Particular Features of the Interface: Sea Slicks and Foam

Sea slicks

Sea slicks (also called calm streaks) develop, when the sea is quite smooth, due to the ripple-damping action of naturally occurring surface films. The attenuation of capillary waves by surface film and the resulting radiation stress breaks up the film layer—or patches—into streaks and produces secondary flows (PARKER and BARSOM, 1970). The most prominent slicks occur in near-shore areas, possibly due to their higher organic production. Petroleum slicks could be distinguished by PARKER and BARSOM (1970) from natural slicks by their higher order of interference colors. According to LAFOND and LAFOND (1972) slicks are related to convergence circulation caused by internal wave motion: the rough band is over the crest and the smooth band $\frac{1}{4}$ wavelength behind. Slicks are more visible when the internal wave period is large and its height great. As wind speed increases, one observes a narrowing of the streaks and a decrease in their width. Winds over 13 knots break up slicks and induce Langmuir circulation. According to GARRETT (1967) the sea surface temperature may increase in the slicks due to accumulation of surface-active material. In contrast, LAFOND and LAFOND (1972) observed that slicks are normally colder than adjacent waters by an average of 0·2 °C; they attributed this to greater evaporation of a thin film over organic particles.

BROCKMANN and co-authors (1976) studied natural slicks near the Island of Sylt in the North Sea. They concluded that dissolved compounds of fatty acids alone could not account for slick formation during calm weather. BROCKMANN and co-authors assume that the slick appears to be caused by active accumulation of the dinoflagellate *Prorocen-*

trum micans at the surface and an increase in bacterial numbers associated with the accumulation. They observed a decrease in dissolved surface-active substances compared with the concentration usually found in normal surface waters.

Seafoam

Seafoam mainly originates from non-living organic matter rising to the surface from the bottom and intermediate waters (ZAITSEV, 1971), together with terrigenic eolian organic material. According to KREY (cited by ZAITSEV, 1971, p. 18), fresh batches of dissolved organic matter are constantly being adsorbed on the surface of detritus particles formed by dead hydrobionts or aggregates. This supports a multiplication of heterotrophic bacteria which are consequently much more abundant in foam than outside. It even seems that this results in a special bacterial flora which differs from that at the interface in the absence of foam. As living organisms, foam also contains fungi, phytoplankters (cyanophytes, naked flagellates, coccolithophorids, dinoflagellates, diatoms) and ciliates. As a whole it seems that foam contains 25% of solid matter, most of which consists of live and dead organisms.

The composition of natural foam has been insufficiently investigated, since it is very complex and variable. However, exometabolites which may inhibit or stimulate many physiological processes in plants and animals have been detected. For example, Russian scientists (see review by ZAITSEV, 1971, pp. 20–2) demonstrated a stimulatory effect of 80% of the tested foam samples upon: (i) root system, length and weight of some cereals; (ii) growth of the cyanophyte *Spirulina tenuissima*; (iii) egg hatching and lifetime of the goby *Pomatoschistus* sp. In 20% of the experiments, an inhibitory effect occurred which may be attributable to an overdose (higher than 3–5%) in the culture medium (ZAITSEV, 1971), but possibly also to chemical peculiarities of the foam. Experiments on artificial foam formed by bubbling in fish farming tanks are a marginal example of the latter case. Attempts to cultivate phytoplankters on media mixed with foam were unsuccessful with fresh foam; however, when using 15–20 day old foam with a chemical composition probably modified by bacterial activities, very dense cultures of *Chlorella* sp. (up to 52×10^6 cells l^{-1}) were obtained which could be used as food for ciliates (e.g. *Stylonychia* sp.) or rotifers of the genus *Brachionus* (DIVANACH, pers. commun).

Recently, HILLIER (unpublished) investigated natural seafoam from five Texas coastal stations: all foams analysed turned out to be enriched in organic matter and several trace metals, and the inorganic and mineral compositions far exceeded their organic constituents. Supernatant or soluble portions of the foams were enriched in organic matter 24–66 times that of nearshore Gulf of Mexico surface waters, but only by 13–26 times that of the underlying waters. Organic-matter enrichment in these foams was much less than that previously reported by other authors. Particulate portions were enriched in organic matter 885–5541 times over Gulf of Mexico surface waters; in trace metals 100–100,000 times over Gulf of Mexico particulates ($\mu g\ l^{-1}$). In a harbour, a lead content of 500 ppm (dry weight of foam) revealed an enrichment of 40,000 over ocean water particulates. Carbohydrates plus lipids constituted on an average 26% of the total organic matter, with 12% carbohydrates and 14% lipids. The protein content was low, with an average of 2%, found primarily in the particulate portions. In all foams studied the major group of organic compounds was a suite of polar, complex aromatic carboxylates of phenolic, basic and reducing character. The next organic group was a medium suite of carboxylic acids, usually associated with phenolics, ammonium ions and salts. The third most prominent group was aliphatic and aromatic esters and acids found in the non-polar fractions. Amides may occur in polar fractions associated with aromatic/phenolic organics of sulphates. According to HILLIER, the major organic classes seem to be humate precursors; if so, these would be relatively refractory to decomposition.

When foam is cast up on shore by wind, its composition is modified mainly by an increase of organisms or debris of terrigenous origin and by addition of living and decomposed psammic organisms. SCHLICHTING (1972) found in foam samples collected from oceanic beaches in North Carolina, Washington State (USA) and Ireland, up to 70,000 algal cells ml^{-1} (59 taxa) and many protozoans (25 taxa). Most of these micro-organisms originated from sediments along the intertidal zone and can be classified as euryhaline freshwater forms; species of the genera *Chlorella*, *Chromulina*, *Phormidium* and *Stichococcus* grow well in both freshwater and sea water media. Obviously, this change in the assemblage of stranded foam is due to input of less saline waters of phreatic origin.

Seafoam may be dispersed and lifted into the atmosphere by strong winds. Airborne individuals of *Pyramimonas* sp. and *Coccolithus huxleyi* and of several other phytoplankters such as dinoflagellates of the genera *Ceratium*, *Peridinium*, *Gymnodinium*, and sometimes diatoms, were collected up to an altitude of 3000 m. This may explain some similarities in the phytoplankton assemblages of both sides of a continental barrier, such as the Panama isthmus (MAYNARD, 1968).

Neuston Assemblages

The usual subdivision of the neuston into epineuston and hyponeuston does not take the form into account. This is closely related to hyponeuston—mainly with regard to the water surface film—but forms above the interface.

Bacterioneuston

In the 0–2 cm layer, bacteria are usually about 10–100 times more abundant than in underlying waters. Bacteria are always much more abundant in the seafoam than in the most superficial water layer. Experiments suggest that bacteria aggregate in a ca 10 μm thick surface layer (McINTYRE, 1968; BEZDEK and CARLUCCI, 1972). Unfortunately, seafoam and the most superficial layer were seldom separately analysed (except in the Caspian and Black Seas, see below), and most devices employed usually sample the 0–1 to 0–2 cm layer. However, bacterial abundance probably decreases sharply in the uppermost 1 or 2 cm.

Table 7-2 illustrates the general scheme of bacterial distribution according to MARUMO and co-authors (1971) who studied bacteria and phytoplankton distribution in the Pacific Ocean at the equator (155° E) during the day. Free bacteria (F) predominate much more in the surface film than at a depth of 10 cm. Viable heterotrophic bacteria (H) and nitrogen-fixing bacteria are very rare in the surface film. Both these peculiarities were attributed by MARUMO and co-authors (1971) to the harmful effect of strong solar radiation; in these authors' opinion, the high bacterial abundance in the film is caused by physical accumulation rather than by *in situ* multiplication. However, this matter requires further investigation because SKOPINTSEV (1939) observed that Cas-

Table 7-2

Abundance of bacteria and detritus in the uppermost layers of the Pacific Ocean (equator, 155° E). All values are expressed in numbers ml^{-1} (After MARUMO and co-authors, 1971; reproduced by permission of University of Tokyo, Ocean Research Institute)

Depth	Total bacteria (T)	Free bacteria (F)	Attached bacteria (A)	Detritus (D)	Viable counts of heterotrophic bacteria (H)	Ratio F/A	A/D	T/H
Surface film (sampled by screen) (S)	$5·5 \times 10^7$	$5·36 \times 10^7$	$2·03 \times 10^6$	$1·1 \times 10^5$	—	26·4	18	—
10 cm (B)	$2·5 \times 10^6$	$2·3 \times 10^4$	$1·5 \times 10^5$	$0·8 \times 10^4$	—	15·3	17·3	—
50 cm (C)	$1·7 \times 10^6$	$1·2 \times 10^6$	$4·4 \times 10^5$	$0·7 \times 10^4$	$7·2 \times 10^2$	2·95	62·9	2·30
Ratio S/B	22	23	13·5	13·7				
S/C	32	44·6	4·6	15·7				

pian Sea foam and sea water samples grown on fish peptone agar, contained 14,000 and 440 colonies ml^{-1} immediately after sampling, but after three days' incubation, the colony number was 2,350,000 and 820 ml^{-1} respectively.

In the Kaneohe Bay, Oahu, Hawaii, HARVEY (1975) recorded within the microlayer bacterial abundances of $0·15–5·10 \times 10^5$ cells ml^{-1} in fresh samples—values much lower than those reported by MARUMO and co-authors (1971). However, HARVEY's fixed and stained samples fall in the same order of magnitude as those of MARUMO and co-authors in the offshore equatorial Pacific Ocean. In the Honolulu area, polluted by sewage outfalls, HARVEY (1975) also found telluric bacteria (Table 7-3) and emphasized that

Table 7-3

Bacteria (cells 100 ml^{-1}) in microlayer and 10 cm depth samples near Honolulu, determined by culture methods (After HARVEY, 1975; reproduced by permission of Pacific Science Association, Hong Kong)

Date	Microlayer Coliforms	Salmonella	Faecal streptoc.	10 cm depth Coliforms	Salmonella	Faecal streptoc.
10 July 1970	380	90	230	88	120	140
3 Aug. 1970	459	9	—	260	13	—

winds blowing shorewards may endanger human coastal populations due to contamination from airborne bacteria and viruses.

According to TSYBAN (1971) the abundance of the bacterioneuston as a whole (i.e. sea foam plus the 0–150 μm layer) is one to four orders of magnitude higher than the general bacterioplankton abundance. The bacterioneuston seems to be extremely stable even when the sea is rough; this suggests that seafoam plays an important part in its formation and maintenance. The specific diversity is very high and pigmented species are rather common. Species of *Bacterium* and *Pseudomonas* are always present; the latter predominates in seafoam assemblages where growth stimulating substances are particularly abundant. Germs in the 0–150 μm layer exhibit strong proteolytic and lipolytic activities. TSYBAN also noticed a strong positive correlation between bacterioneuston density and zooneuston abundance and that some bacteria species seem to be intimately linked with zooneuston species, such as young copepod stages. In the Black Sea bacterioneuston, ammonifying, denitrifying and sulfuroxidizing bacteria are particularly abundant, and the sulphate-reducing microorganisms also exist as in demersal waters.

Phytohyponeuston and protozoohyponeuston

Algae and protozoa may also occur in seafoam; for example they may cross the interface when thrown up by bubbles. Seafoam populations of these groups are therefore probably rather similar to those in the uppermost layer, at least from a qualitative point of view.

No plant can presently be considered restricted to the hyponeuston. All the algae thus far recorded from the uppermost layer also exist in true planktonic surface assemblages.

The information available on phytoneuston and protozooneuston is rather poor. I shall briefly review it here, before attempting to summarize their general features.

Phytohyponeuston. PARKER and BARSOM (1970, p. 88) counted samples off the coast of southern California (6 May 1964) per litre 'in the surface skin': 5000 colorless flagellates, 31,270 dinoflagellates, 330 ciliates and 930 diatoms in contrast to 0, 3, 900, 370 and 3770 cells of the respective groups at 10 cm depth.

At an equatorial station in the Pacific Ocean, MARUMO and co-authors (1971) observed the phytoneuston and zoohyponeuston composition summarized in Table 7-4. Diatoms largely predominate (especially *Nitzschia* spp.), followed by several *Chaetoceros* species and *Thalassiothrix delicatula*. The authors further noticed the absence of minute naked flagellates. *Trichodesmium thiebautii* occurred in the film water not as colony but in the form of solitary filaments, possibly indicating that this alga was not in a good physiological condition.

In three Japanese bays studied by TAGUCHI and NAKAJIMA (1971), diatoms also largely predominated, especially those of the *Nitzschia* and *Navicula* groups followed by *Bacteriastrum delicatulum* (during the May–June bloom). However, the authors had stored their samples in formalin, which destroyed minute naked flagellates. They generally observed higher abundances (100–700 $\times 10^6$ cells m^{-3}) in the surface skin than in the subsurface (10–50 cm) layer when the sea surface was calm, but the inverse relation prevailed when the wind was sufficiently strong.

The investigations of HARDY (1973), who studied phytoneuston and phytoplankton assemblages in and near Griffin Bay (San Juan Archipelago, Washington) were more accurate. Like TAGUCHI and NAKAJIMA (1971), he reports that diatoms (e.g. *Cylindrotheca* sp. and *Navicula* spp.) predominate in the phytoneuston. However, he also noticed that brief blooms may occur, both in the bay itself (mainly due to *Chroomonas* sp., the silicoflagellate *Dictyocha fibula*, and the ciliate *Mesodinium rubrum*), and in the lagoon where *Prorocentrum minimum* is very abundant in August. In general, phytoneuston abundance is maximal between spring and summer. The average abundance increases, while

Table 7-4

Number of plankton organisms l^{-1} in the Pacific Ocean (equator, 155° N), afternoon 30 September 1969 (After MARUMO and co-authors, 1971, p. 37; slightly modified; reproduced by permission of University of Tokyo, Ocean Research Institute)

Depth	Diatoms	*Trichodesmium thiebautii* (filaments)	Dino-flagellates	Silico-flagellates	Radiolaria	Heliozoa	Foraminifera	Tintinnoidea	Copepoda
Surface skin	9988	124	16	2	—	—		34	2
5 cm	2056	—	2	—	—	12		10	6
10 cm	948	—	4	—	—	12	2	6	8
10 m	74	—	—	2	—	—			

the specific diversity decreases from the bay areas towards the most enclosed part of the lagoon. As usual, the higher the abundance, the lower the specific diversity. The phytoneuston was more abundant and less diversified than the phytoplankton. According to HARVEY (1975, p. 107) a given species may not always be present in the microlayer, but may appear in very large numbers at intervals. A characteristic of Kaneohe Bay, Oahu, Hawaii, investigated by him was the wide variation of species and individuals number found at different times and different locations (Table 7-5).

Table 7-5

Living organisms in fresh microlayer (mic.) and 10 cm-depth samples from Kaneohe Bay, Oahu (Hawaii). All volumes expressed in ml × 10^{-3} (After HARVEY, 1975; reproduced by permission of Pacific Science Association, Hong Kong)

Station	Date	Diatoms		Dinoflagel-lates		Colourless flagellates		Ciliates		Detritus	
		mic.	10 cm	mic.	10 cm	mic.	10 cm	mic.	10 cm	mic.	10 cm
A	30 Nov. 1969	0	0	0	21	236	20	0	21	9	4
B	22 Jan. 1970	2983	0	316	86	3471	318	579	86	221	10
D	7 Dec. 1969	848	434	1697	553	2700	163	77	72	31	17
E	16 May 1970	1060	2362	82	0	41	1593	0	204	23	26

With regards to the abundance of large-sized algae both at Kaneohe Bay and at La Jolla HARVEY (1975, p. 107) noticed that only occasionally did diatoms outnumber large-sized flagellates. Among the diatoms, the genera *Nitzschia* and *Chaetoceros*, followed by *Navicula* and *Coscinodiscus*, are most frequently represented. *Prorocentrum* species, which have been repeatedly responsible for red tides at La Jolla and Kaneohe Bay, often migrate into the microlayer in enormous numbers, assisted by their high swimming ability. Next to *Prorocentrum*, species of *Gymnodinium* were most frequently encountered in microlayer samples, followed by species of *Ceratium*, *Goniaulax*, *Cochlodinium* and *Dinophysis*. *Noctiluca* and *Polykrikos* were occasionally present in large numbers. Bacteria were the most numerous organisms; they usually prevailed in high numbers within and around aggregates. Small colourless flagellates (bodonids and related groups) were also invariably present and were the next most numerous. They were usually found in association with particles of detritus and with bacteria upon which they feed.

Our present knowledge regarding the phytoneuston assemblage may be summarized as follows with regard to marine and slightly brackish water areas (Baltic Sea phytoneuston reveal special characteristics due to its many freshwater species): (i) The assemblage attains higher organismic densities than phytoplankton except for periods of rough seas. (ii) It is probably almost invariably less diversified than the phytoplankton. (iii) Its composition and abundance are highly variable: seasonable fluctuations are very pronounced; mean range (several days) fluctuations are large; circadian (day/night) fluctuations are less important than in the 10–20 cm layer and below. The dinoflagellates *Prorocentrum micans* and *Goniaulax polyedra* come nearer to the interface during the day, while *Ceratium fusus* and *C. tripos* exhibit an inverse migration. (iv) Diatoms often seem to be more abundant than other large-sized unicells. Some scattered observations suggest that cyanophytes (where they exist) might be more abundant in the superficial film (0–150 μm), i.e. together with the bacterial maximum. (v) The saturated photosynthetic rate in the surface skin tends to be lower than in the subsurface layer (TAGUCHI and NAKAJIMA, 1971, p. 28); possibly, chlorophyll a is depressed in the surface skin by strong solar radiation.

At this point I would like to emphasize the striking physiological similarities between phytoneuston algae and those which induce discoloured waters (p. 356): increased buoyancy resulting in life restricted to the upper and uppermost water layers; increased requirements for nutrient salts and possibly also for growth factors; high multiplication rates resulting in very high population densities; and high tolerance to strong solar radiation. As is well known, a 'red-tide' generally, if not always, begins in the most superficial water layer; it might therefore be considered to be a prominent phenomenon of the phytoneuston assemblage which subsequently tends to become more and more oligospecific (possibly almost always due to predominant dinoflagellates) and which occupies a somewhat thicker water layer.

Protozoohyponeuston. Hyponeuston protozoa have been insufficiently investigated. Ciliates seem to be rather common in the hyponeuston. Radiolaria of the genus *Spumellaria* are more abundant in the uppermost layer than in other levels of the water column, both in the Gulf of Mexico and in the Pacific Ocean. Similarly, some Foraminifera species occur more frequently in hyponeuston assemblages than in plankton assemblages. The phagotrophic dinoflagellate *Noctiluca miliaris* exhibits its maximum density in the 0–5 cm layer. Tintinnids seem to be the most diversified and abundant protozoans in the hyponeuston all over the World Ocean. For example, in the 0–5 cm layer of the Black Sea, MOROZOVSKAYA (1966, 1969) recorded more than 20 species, several with summer abundance peaks of 5000 to more than 12,000 individuals m^{-3}. All these protozoa probably mainly consume bacteria and play a major role in hyponeuston energy transfer; several, especially the tintinnids, serve as major food source for numerous invertebrates and fish larvae (ZAITSEV, 1971, p. 66). Possibly, this compensates for the fact that (in the Black Sea 0–5 cm layer) relations between phytophagous metazoans and main primary producers (diatoms, small dinoflagellates) are rather insignificant (NESTEROVA and POLISHCHUK, 1975).

Zoohyponeuston (Protozoa Excluded)

General aspects. Except for protozoans, most members of the zoohyponeuston reveal great differences with regard to the periods of their life cycle spent in the uppermost layer. Hence, generalized taxonomy based on an account of hyponeuston metazoans would be premature. It seems better to consider the zoohyponeuston under the following two aspects: (i) permanent or transitory presence in the uppermost layer; (ii) the time spent by a transitory species in the uppermost layer and its original biotope (ZAITSEV, 1971; CHAMPALBERT, 1975). The major difference is between permanent and transitory zoohyponeuston. The permanent zoohyponeuston, known as holohyponeuston, comprises animals which complete their whole life cycle within the uppermost layer (0–10 cm) under normal environmental conditions. The transitory zoohyponeuston is much more diversified and may be subdivided into four units, the names of which are simplified here by omitting the prefixes zoo- and hypo-, since it is obvious that this section only deals with hyponeuston animals: (i) The epiplanktoneuston comprises species which usually participate in epiplankton (p. 11) assemblages, but under certain circumstances may rise up to the surface (0–100 cm) and there establish a stratification up to the uppermost layer. This stratification is almost invariably more marked for males than for females and juveniles. The ascent generally occurs at night. (ii) The bathyplanktoneuston comprises plankton species which exhibit the same behaviour as epiplanktoneuston species, but usually inhabit deeper layers (infrapelagic and even bathypelagic zones, p. 12). Obviously, such species participate in the hyponeuston assemblages only in areas where depth is sufficiently important. (iii) The benthoneuston, i.e. species which live within the bottom during the day and migrate upwards near the interface at night. (iv) The merohyponeuston (ZAITSEV, 1971), i.e. a rather heterogeneous group including all larval stages which seasonally (during the adults spawning periods) inhabit the uppermost layer; some occur there at night, e.g. many crustaceans (decapods, euphausiids), molluscs and polychaete larvae; others are more permanently hyponeustonic, e.g. fish eggs and sometimes fish larvae and fry.

Holohyponeuston. Pontellid copepods largely predominate in the holohyponeuston. Apparently all genera of this highly specialized family contain holohyponeustonic species. Even *Anomalocera patersoni*, which extends far northward into the Atlantic Ocean, and *Epilabidocera amphitrites* from the Bering Sea, retain their hyponeustonic nature (ZAITSEV, 1971, p. 77). Species of *Pontella*, *Labidocera* and *Pontellopsis* constitute the bulk of the 'pontellid hyponeuston' in tropical and temperate waters of the World Ocean. In the Gulf of Marseilles, three species may be considered truly holohyponeustonic: *Anomalocera patersoni*, *Pontella mediterranea* and *Labidocera wollastoni* (CHAMPALBERT, 1975). Several, but not all, species of the copepod genus *Sapphirina* may also belong to the holohyponeuston. ZAITSEV further includes in the hyponeuston assemblage crabs of the genus *Planes*, which live on floating substrates, and the nudibranch *Glaucus atlanticus*, which floats due to an air bubble. In my opinion, both should be considered pleuston species (p. 346).

Transitory zoohyponeuston (merohyponeuston excluded). Circadian rhythms play an important part in the occurrence of many epiplanktoneuston, bathyplanktoneuston and benthohyponeuston species in the uppermost layer. While this phenomenon is better documented in the Black Sea (ZAITSEV, 1971), I have preferred to select examples from areas with salinities more typical of the marine environment.

(i) Epiplanktoneuston. In the subtropical north-eastern areas of the Atlantic, WEIKERT (1972) established that the total neuston abundance in the 0–10 cm layer was much higher at night than during the day (1396 vs 104 ind. 100 m^{-3}); the maximum abundance was observed at sunset (3744 ind. 100^{-3}). There was a corresponding fluctuation in the abundance of species: from the 70 species investigated, 46 invaded the uppermost layer during dusk and at night, and 42 at dawn. During the day the number decreased to 25, further diminishing to a minimum of about ten species at noon. Most species that rise to the surface seem to belong to epiplankton assemblages. WEIKERT further noticed two different types of rising: species that aggregate twice at the surface itself, and species showing a single abundance at the surface. In the Gulf of Marseilles CHAMPALBERT (1975) noticed that the two sexes and different development stages of a given epiplanktoneuston species may exhibit different migration patterns. For example,

at night the ascent of the copepod *Isias clavipes* to the 0–10 cm layer is sharply pronounced in males, but less markedly so in females; juveniles ascend during the day (Fig.
7-1). Calyptopsis of *Meganyctiphanes norvegica* rise mainly during the day, furcilia stages
at night (Fig. 7-2) (CHAMPALBERT, 1975).

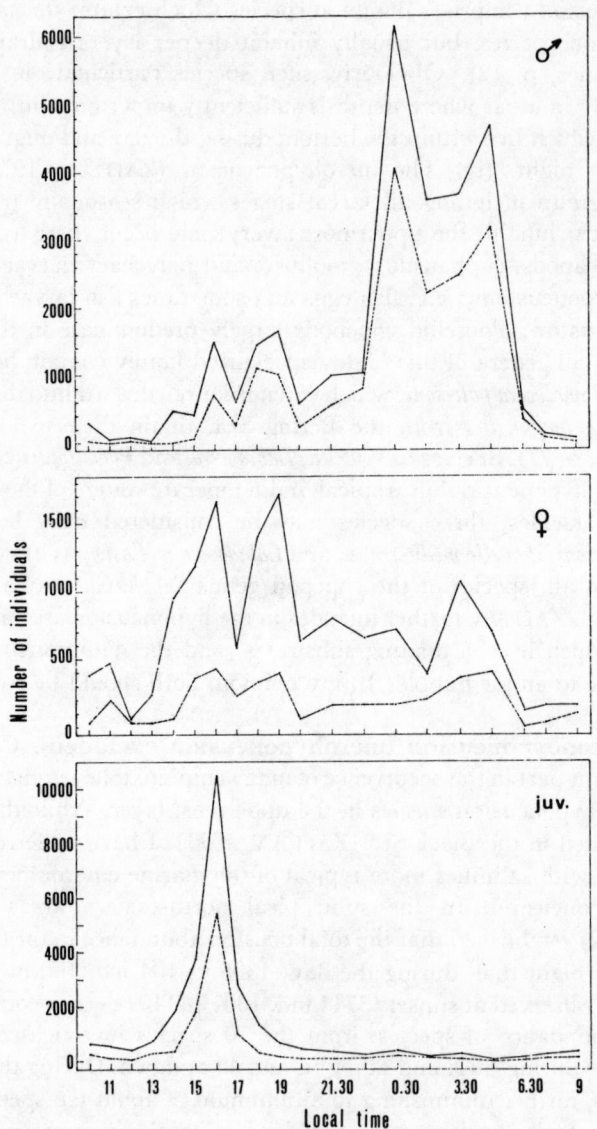

Fig. 7-1: *Isias clavipes*. Changes in abundance in 0–100 cm
(solid line) and 0–10 cm layers (broken line). Number of
individuals correspond to a 15 min haul. Gulf of Marseilles, 13–14 November 1967. (After CHAMPALBERT,
1975; reproduced by permission of the author.)

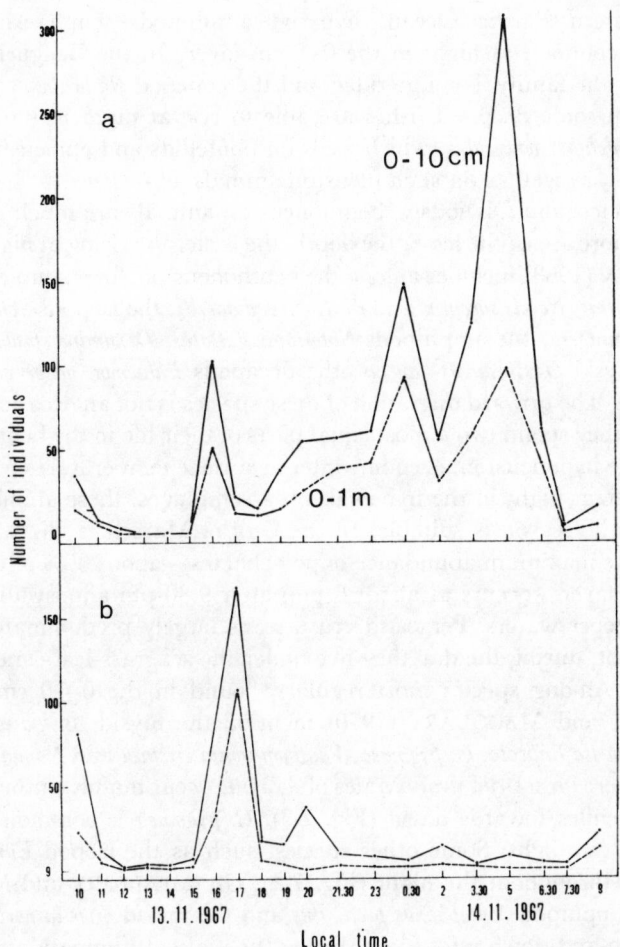

Fig. 7-2: *Meganyctiphanes norvegica*. Nycthemeral fluctuations
in larval abundance (number of individuals per 15 min
haul). Gulf of Marseilles, 13–14 November 1967. a: furcilia;
b: calyptopsis. (After CHAMPALBERT, 1975, reproduced by
permission of the author.)

Halocyprid ostracods (which are completely absent from the neuston at high
latitudes, but common in these regions below 100 m) may migrate into the uppermost
layer in tropical and subtropical latitudes (MOGUILEVSKY and ANGEL, 1975) In the
Atlantic Ocean, males of *Conchoecia spinirostris* sometimes swarm at dusk in the 0–10 cm
layer.

(ii) Bathyplanktoneuston has been insufficiently investigated. Migrant species rising
up to the most superficial layer from depths below 200–250 m seem to be very scarce at

low and middle latitudes. However, at higher latitudes they play an important role in the hyponeuston assemblage, while holohyponeuston species are rare or even absent. In the north-western Pacific Ocean, hyperiid amphipods, euphausiids and calanid copepods are common at night in the 0–5 cm layer. In the Benguela current region polychaetes of the family Tomopteridae and the copepod *Rhincalanus nasutus* were also recorded. Even some deep-sea fishes are able to rise at night to the surface, e.g. the myctophid *Gonichthys tenuiculus*, which feeds on pontellids and epineuston species of the genus *Halobates*, as well as on such pleuston animals as *Janthina*.

(iii) Benthoneuston. Obviously, benthoneuston animals are much more common in shelf than offshore areas; the lesser the depth, the easier the rising at night. On Black Sea coasts, ZAITSEV (1968) includes among the benthoneuston, for example, the polychaetes *Nephthys longicornis*, *Nereis succinea* and *Platynereis dumerili*: the isopods *Sphaeroma pulchellum* and *Eurydice spinigera*; the amphipods *Nototropis guttatus*, *Dexamine spinosa* and *Gammarus crangon*; the mysid *Gastrosaccus sanctus*; the decapods *Palaemon adspersus*, *P. squilla* and *Crangon crangon*. The upward migration of these species is not an occasional but a regular phenomenon; they spend two almost equal parts of their life in the benthos or nektobenthos, and in the hyponeuston, even in winter at surface temperatures below 0 °C. When they participate at night in the hyponeuston assemblages, these animals feed and also breed there if the season is suitable. In the Gulf of Marseilles CHAMPALBERT (1975) noticed that the maximum abundance of polychaetes—about 80% of the population in the 0–100 cm layer—occurs at about 6 pm; after 9.30 pm almost all individuals had returned to deeper waters. Peracarid crustaceans largely predominate in the benthoneuston at night; during the day they live under the seagrass-leaf canopy or burrow in the substrate. Among species most regularly found in the 0–10 cm layer at night, CHAMPALBERT and MACQUART (1970) mention the mysid *Anchialina agilis* and the cumaceans *Cumella limicola*, *C. pygmaea*, *Vauthompsonia cristata* and *Nannastacus unguiculatus* which occur there for a brief time. Males of *A. agilis* occur mainly at the beginning of the night, and juveniles towards dawn (Fig. 7-3). *C. pygmaea* is common only during the second half of the night. Some other species, such as the isopod *Eurydice inermis*, are hyponeustonic throughout the night (Fig. 7-4). CHAMPALBERT and MACQUART even observed the amphipod *Metaphoxus pectinatus* and the mysid *Siriella norvegica* performing two or three up-and-down migrations during the night. Illumination at full moon may be sufficient to induce a decrease in the ascending activities (*Siriella norvegica*, *S. clausi*, *Vauthompsonia cristata*) or even to completely inhibit them (*Cumella pygmaea*), mainly in rather shallow-water areas.

Merohyponeuston. (i) Invertebrates. Merohyponeuston invertebrates exhibit a striking stratification in the Black Sea neuston layer (Table 7-6). Obviously, larval stages of plankton animals whose adults inhabit mostly the epi- and mesopelagic zones (p. 11) may occur in the neuston in both neritic and oceanic provinces. For example, the Black Sea copepod *Centropages ponticus* is hyponeustonic during the nauplius stage, but less so during other stages, although all age groups exhibit a certain predilection for the 0–5 cm layer (ZAITSEV, 1971). The nauplius stages of *Acartia clausi* establish stable concentrations in the hyponeuston, whereas their affinity (at copepodit stages) to the 0–5 cm layer decreases with time, despite the fact that females are always more abundant there. *Paracalanus parvus* participates in the hyponeuston 'during the nauplius stage but later spreads more or less uniformly in the water column' (ZAITSEV, 1971, p. 75). In

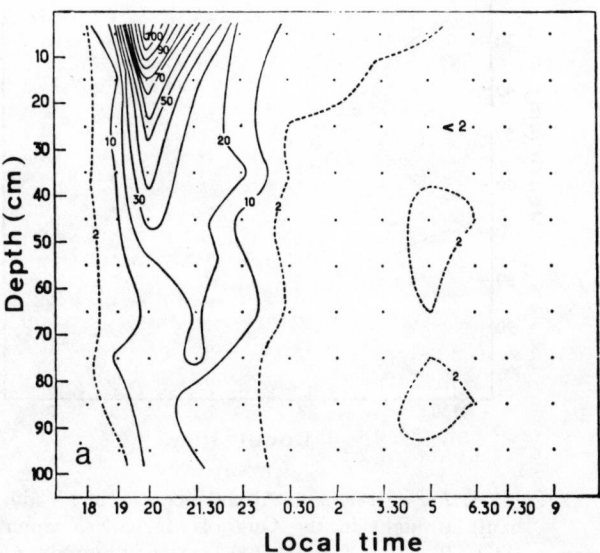

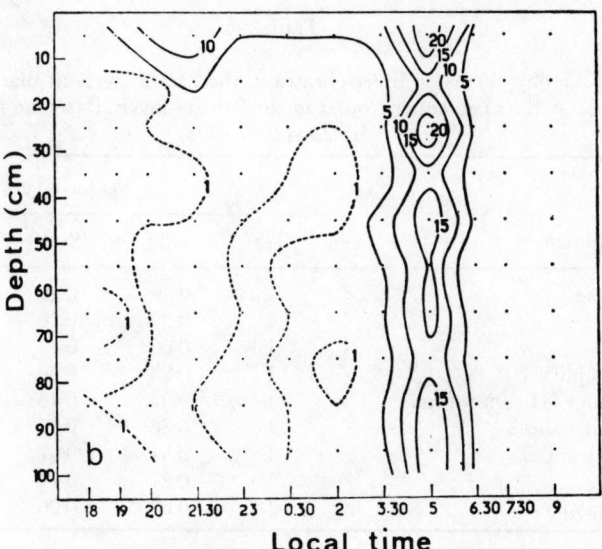

Fig. 7-3: *Anchialina agilis*. Number of individuals (in a 15 min haul). Males (a) and juveniles (b) at night in the Gulf of Marseilles, winter 1968–69. (After CHAMPALBERT, 1975; reproduced by permission of the author.)

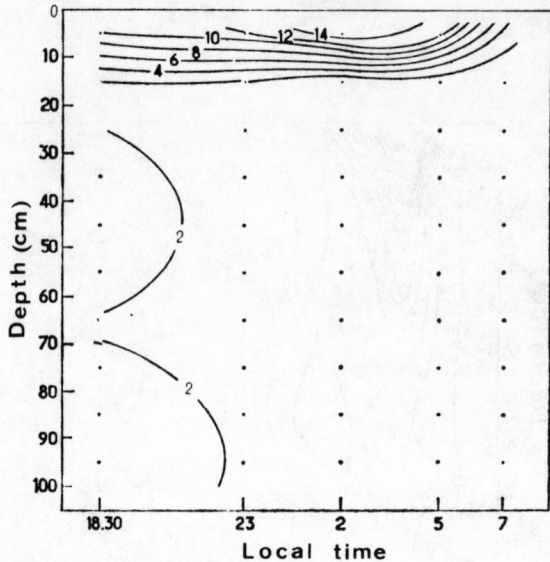

Fig. 7-4: *Eurydice inermis*. Abundance (in a 15 min haul) at night in the Gulf of Marseilles, winter 1968–69. (After CHAMPALBERT, 1975; reproduced by permission of the author.)

Table 7-6

Stratification of merohyponeuston invertebrates in the upper layers of Black Sea waters. Larval abundancies expressed as percentage found in the 0–5 cm layer (Based on information compiled by ZAITSEV, 1968)

Animals	Water layers (cm)				
	0–5	5–25	25–45	45–65	65–85
Polychaeta larvae	1	0·20	0·22	0·24	—
Pelecypoda larvae	1	0·37	0·56	0·56	—
Balanus nauplii	1	0·06	0·73	0·04	—
Acartia clausi nauplii	1	0·28	0·60	0·60	—
Acartia clausi (Stage III copepodits)	1	0·52	0·43	0·53	—
Centropages kroyeri nauplii	1	0·36	0·47	0·62	—
Oithona minuta juveniles	1	0·54	0·61	0·65	—
Brachyura zoeae	1	0·24	0·24	0·22	0·20
Brachyura megalopa	1	0·03	0·06	—	—

the Bering Sea, young individuals (15–20 mm long) of the pelagic squid *Ommastrephes sloani pacificus* often concentrate in the uppermost layer, possibly for feeding. These examples represent particular cases of the general trend of many larval or young planktonic or pelagic animals to exploit waters shallower than those inhabited by their adult populations. The abundance of larval benthic invertebrates is higher in the hyponeuston in neritic than in offshore provinces, due to shallower water on the shelf and its more abundant and more diversified benthos assemblages.

In the Gulf of Marseilles, CHAMPALBERT (1975) used a multiple 200 μm mesh net yielding ten samples in each of the 10 cm layers from 0 to 100 cm (filtering volume about 80–90 m^3 per 15 min haul). He observed two categories of migrant species, very similar to those mentioned above for the epiplanktoneuston (p. 325). For example, gastropod larvae rise up in the uppermost layer before dusk (4 p.m.); after a peak at dusk (6 p.m.) the abundance decreases slowly towards midnight and, more quickly, later. Among the decapod crustaceans, *Ebalia* sp. larvae (Fig. 7-5) exhibit a single peak corresponding to species of WEIKERT's (1972) second group (p. 325), whereas *Macropipus* sp. larvae, exhibiting two peaks, correspond to his first group. CHAMPALBERT also noticed that some decapod species may exhibit three abundance peaks over a 24-h period, e.g. *Maia* sp. larvae (Fig. 7-5). In general, pelecypod larvae are more abundant in the hyponeuston than gastropod larvae. Polychaetes are always well represented, especially species of the families Spionidae and Magelonidae.

(ii) Fish eggs, larvae and fry, although part of the merohyponeuston, are often considered separately as ichthyoneuston; the most thoroughly investigated ichthyoneuston is that of the Black Sea. Due to their high buoyancy, fish eggs are very common in the uppermost water layer. The main families involved are Engraulidae, Pomatomidae, Carangidae, Mugilidae, Mullidae, Callionymidae, Bothidae, Pleuronectidae and Soleidae. Fish eggs concentrated at the sea surface float passively, but are dispersed at wave amplitudes exceeding 1–2 m. Most of them may be destroyed if the sea becomes rougher, but as soon as extensive wave action ceases, the surviving eggs again accumulate below the surface film (ZAITSEV, 1971).

Elsewhere, the 0–5 cm layer harbours large numbers of larvae and fry of the same fish families as well as larvae of such other families as Belonidae, Scomberesocidae, Coryphaenidae, Exocoetidae, Atherinidae, Centriscidae, Berycidae, Labridae, Blenniidae, Gobiidae, Ammodytidae, Balistidae, Syngnathidae. Larvae and fry can withstand rougher seas than eggs, and can remain close to the surface with wave amplitudes of up to 3–4 m.

The possibility of adverse wave motion effects on eggs can be reduced by an acceleration of embryonic development; in the Black Sea, neustonic eggs of anchovy, grey mullet, red mullet and horse-mackerel hatch after 30–40 h, instead of after several weeks, as do demersal eggs such as those of gobies, atherinids and garfish. Accelerated embryogenesis allows hyponeustonic eggs to attain the hatching stage quickly, thus, reducing the risk of being damaged by critical wave forces (ZAITSEV, 1971). Table 7-7 summarizes the spring and summer distribution of eggs and larvae of the most abundant fish species in the Black Sea ichthyoneuston. In April the Black Sea ichthyoneuston consists mainly of cold-water species common during autumn, winter and spring but not in summer (e.g. flounder eggs), or of species which spend the spring season in deeper water layers (sprat eggs) (ZAITSEV, 1971). The ichthyoneuston is much more abundant and diversified in summer than in spring; only the most common species are listed in Table 7-7.

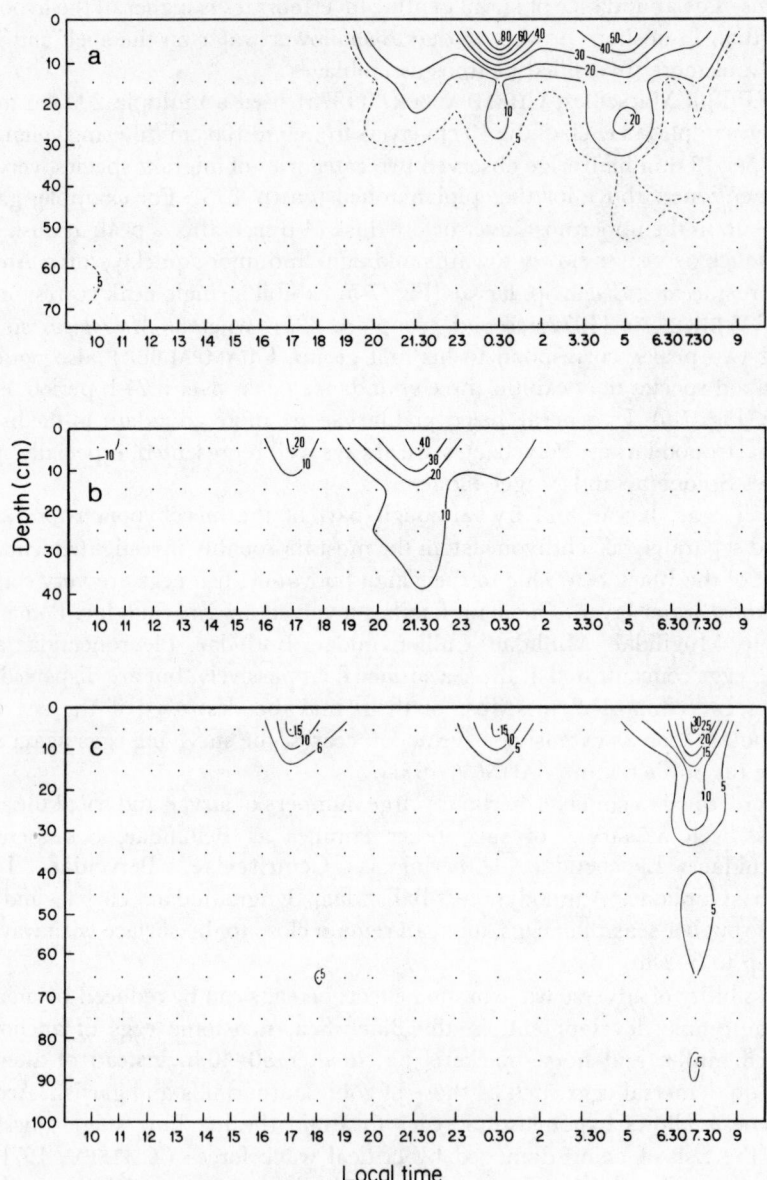

Fig. 7-5: (a) *Macropipus* sp., (b) *Ebalia* sp., (c) *Maia* sp. Nycthemeral changes
in stratification of larval stages. Gulf of Marseilles, 13–14 November 1967.
Number of individuals correspond to a 15 min haul. (After CHAMPALBERT,
1975; reproduced by permission of the author.)

Table 7-7

Composition and average density (individuals m^{-3}) of fish eggs and larvae in the north-western Black Sea, 1970 (based on information provided by ZAITSEV, 1970)

	Species	Water layer (cm)				
		0–5	5–25	25–45	45–65	65–85
Spring (April)	*Platichthys flesus luscus*, eggs	11·40	4·10	5·20	3·80	2·50
	Platichthys flesus luscus, larvae	15·0	6·10	4·10	5·20	3·10
	Scophthalmus maeoticus maeoticus, eggs	16·40	4·80	3·70	3·70	0·40
	Clupea sprattus phalericus, eggs	13·80	9·10	7·40	6·50	7·10
Summer (July)	*Engraulis enchrasicolus ponticus*, eggs	17·60	9·25	6·36	6·42	6·21
	Engraulis enchrasicolus ponticus, larvae	11·0	2·50	1·11	2·40	1·20
	Trachurus mediterraneus ponticus, eggs	15·71	4·31	3·69	2·68	2·17
	Trachusus mediterraneus ponticus, larvae	6·58	1·0	1·22	1·50	0·96
	Blennius sp.	3·04	0·45	0·08	0·18	0·23

In the north-western Pacific Ocean, larvae of the saury *Cololabis saira* represent the most abundant constituent of the ichthyoneuston; in the 0–5 cm layer their density is 20 to 30 times greater than in the 5–25 cm layer. Saury larvae of about 20 mm in length feed mainly on hyperiid amphipods, but larger ones (220–270 mm) also feed on smaller saury larvae.

In the north-eastern Atlantic Ocean, JOHN (1973) observed notable differences in ichthyoneuston composition between neritic (up to about 130 nautical miles from the coast) and oceanic areas. In the neritic areas (27 samples) the average concentration was 190 individuals 100 m^{-3}; 35 species (total number 51) were present. Offshore (32 samples) only 22 taxa were found and the mean abundance was 112 individuals 100 m^{-3}. According to JOHN, the larger number of species near the coast might be attributed to the higher temperature of neritic surface waters and to the fact that, in addition to young stages of bottom-living and epipelagic fishes, mesopelagic species were also caught. Only six species samples in neritic areas were not collected offshore.

Ichthyoneuston, as well as other zooneuston, may exhibit two different types of behaviour: (i) permanent day and night presence (or maximum abundance) in the uppermost layer ('obligate neuston' or 'euneuston'); (ii) occurrence only during certain times of the 24-h period ('facultative neuston'). In the area he investigated, JOHN (1973) includes all stages of *Scomberesox*, Mugilidae, Sparidae and the juveniles Carangidae in the euneuston. The daytime facultative ichthyoneuston is mainly represented by juveniles of *Macrorhamphosus* and *Scombridae*, together with larvae of *Ceratoscopelus*. The night-time facultative ichthyoneuston consisted of young larvae of *Macrorhamphosus* and several genera of Myctophidae (adults). The behaviour may change according to age; for example, early larval stages (up to 3 mm) of *Macrorhamphosus* and small *Belone* belong to the night-time facultative neuston, whereas older stages should be considered as 'day-time facultative neuston' or 'euneuston' in the case of *Belone*.

Adaptations of Hyponeuston Organisms

Adaptation to life just below the interface

Devices for passive buoyancy. Adaptations enabling neuston organisms to remain just below the surface film, some of which are also present in plankters, include means for decreasing specific weight: higher water content and oil droplets are common in hyponeuston and plankton organisms; changes in ionic composition were demonstrated in several plankters which substitute lighter ions (K^+, Na^+, Cl^-) for heavier ones (Ca^{2+}, Mg^{2+}, SO_4^{2-}) without changing isotony. This mechanism probably also exists in some hyponeuston organisms. Increased buoyancy due to air (gas) bubbles is rather rare in plankters such as siphonophores Physophorida, but is common in pleuston animals (p. 346) also exists in such hyponeustons as grey mullet fry 15–20 mm long. The air sac which develops near the dorsal fins (Fig. 7-6) and keeps the carnivorous fry in the

Fig. 7-6: Grey mullet. Position of air sac on dorsal side of young hyponeustonic stage. (After ZAITSEV, 1970; reproduced by permission of the author.)

nutrient-rich 0–5 cm layer, has been described by ZAITSEV (1971). Also observed in atherinid fry, the large bubble has a protective function against predatory birds. In fact, almost all fish larvae in the hyponeuston seem to have a gas bladder, although this may disappear in adults. Mechanisms such as the hygrofuge capacity of mullet egg membranes (Fig. 7-7) or of the spines on the rostrum and posterior expansions of decapod zoeae are less common in hyponeuston organisms (Fig. 7-8).

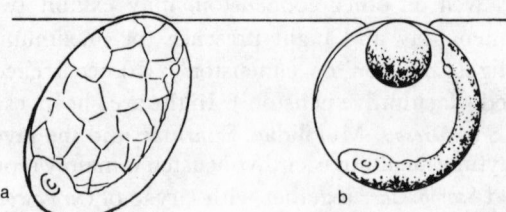

a   b

Fig. 7-7: Position of hyponeustonic fish eggs at sea surface. (a) anchovy, wettable membrane; (b) greymullet, hydrofuge membrane. (After ZAITSEV, 1970; reproduced by permission of the author.)

Fig. 7-8: *Pisidia (Porcellana) longicornis*. Position of metazoea
larval stage below surface film. (After ZAITSEV, 1970;
reproduced by permission of the author.)

Devices for active position maintenance. Mechanisms which enable motile organisms actively to maintain this position in the most superficial water layer are much more sophisticated. Since both light intensity and spectrum width attain maximum values at the water surface, photoactic behaviour requires special attention. *Anomalocera patersoni*, for example, is positively phototactic under strong illumination, but reveals negative phototaxis at low light intensities (CHAMPALBERT, 1975): parallel to this, its photokinesis decreases. Both reactions contribute to dispersion of individuals. Employing illuminations from 4000 to 4 lux, i.e. at dusk and dawn, *A. patersoni* is acutely sensitive both to light intensity and its fluctuations; its swimming activity significantly controls its diurnal activity (Fig. 7-9).

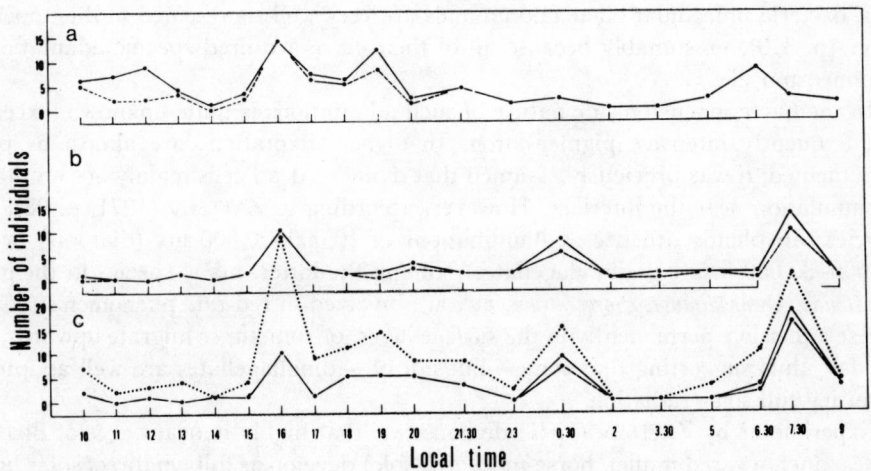

Fig. 7-9: *Anomalocera patersoni*. Nycthemeral abundance fluctuations in the Gulf of
Marseilles, 13–14 November 1967. Number of individuals correspond to a 15 min
haul; (a) juveniles; (b) females; (c) males. Solid line, 0–100 cm layer; broken line,
0–10 cm layer; dotted line, total numbers in the 0–10 cm layer. (After CHAMPALBERT, 1975; reproduced by permission of the author.)

While pontellids are highly sensitive to light, some variations in vertical migration seem to be unrelated to this fact, since migration prevails in the presence of small light fluctuations in the natural environment, as well as under experimental conditions of permanent light or darkness. The control mechanism is presumably based on an endogenous rhythm, modifiable by natural light variations, but possibly also by other environmental factors such as temperature, gravity and hydrostatic pressure.

Changes in hydrostatic pressure can affect activity rhythms and consequently modify the patterns of vertical migration and distribution (Volume I: KINNE, 1972; FLÜGEL, 1972; Volume II: SCHÖNE, 1975; CREUTZBERG, 1975; TESCH, 1975). In *Anomalocera patersoni* and *Pontella mediterranea*, CHAMPALBERT (1975) recorded sensitive responses to pressure changes (threshold of immediate response ca 0·5–0·6 bar). The physiological clock of *A. patersoni* (mainly in males) is modified by prolonged overpressure; this weakens the endogenous rhythm or even abolishes it. Overpressure enhances the effect of light and increases photosensitivity. CHAMPALBERT's results offer an explanation of pontellid behaviour. In the morning, the negatively phototactic individuals exposed to increasing illumination actively leave the uppermost layer and thus become exposed to progressively decreasing light intensity, while the pressure gradient increases. Finally, pressure becomes the dominating stimulus causing upward swimming. In general, for active position maintenance of neustonts, endogenous rhythm, light and hydrostatic pressure appear to be the major denominators.

Adaptation to solar radiation

Full sunlight, particularly short-wavelength (ultraviolet) radiation, reduce bacterial activities and inhibit photosynthesis in phytoplankters (Volume I: GUNKEL, 1970). Similar adverse effects were also demonstrated in many zooplankton species. Nevertheless, bacteria unicellular algae and animals are very well represented in the uppermost layer (p. 319), presumably because all of them have acquired specific adaptations to hyponeuston life.

In the bacterioneuston, the nature of such adaptations is quite unknown, except for the frequently intensive pigmentation. In algae, adaptations are almost as poorly documented; it was previously assumed that dying or dead cells mainly account for the accumulation near the interface. However, according to ZAITSEV (1971, p. 98), some species can photosynthetize at illuminations of 10,000–20,000 lux (diatoms) or even 25,000–30,000 lux (e.g. dinoflagellates); among the latter, many species in the genera *Gonyaulax*, *Gymnodinium*, *Prorocentrum*, etc, are involved in red-tide phenomena (p. 356). These either live permanently in the surface layer, or sometimes migrate upwards, even by day, thus suggesting that some—but not all—dinoflagellates are well adapted for enduring full solar radiation.

Experiments by ZAITSEV (1971) demonstrate that highly buoyant eggs of Black Sea fishes (anchovy, red mullet, horse-mackerel, sole) develop as fully in direct solar light as in shade. 'This indicates that hyponeustonic fish eggs are euryphotic, an essential property for hyponeustonic organisms which do not perform vertical migrations' (ZAITSEV, 1971, p. 95). Totally transparent (anchovy) or poorly pigmented (the Black Sea fish *Ctenolabrus rupestris*) eggs are rather rare in the hyponeuston. In the embryo, adaptation to intensive solar radiation involves various physico-chemical and physiological means, the most obvious of which is pigmentation. Grey mullet eggs, for example, are strongly pigmented. Where oil droplets exist, they are always located at the top of the egg attached to the film (Fig. 7-7) and tend to focus incident light. A pigmented cell layer forms a screen below the droplets, thus protecting the embryo (ZAITSEV, 1971).

Many hyponeuston animals are blue or sometimes blue-green. This colouration may also protect them against excessive solar radiation, although the exact protective mechanism is not known. The entire body of pontellid copepods which are permanent hyponeuston members is blue; the blue is particularly deep in species such as *Anomalocera patersoni, Pontella mediterranea, P. lobiancoi, Pontellopsis regalis* and *P. villosa*. Plankton species which migrate into the 0–10 cm layer only at night are slightly pigmented (ZAITSEV, 1971). Such pigmentation may also protect against predators, especially birds, mainly in species with a blue dorsal side, whereas a silvery ventral side tends to protect against predators from lower water layers.

Adaptation to wave motion

The influence of surface-water motion on the hyponeuston is of considerable interest. The smoother the sea surface, the more marked are organic stratification and abundance. Storm and rough sea periods cause a decrease in both stratification and abundance. The longer the storm period, the more delayed the re-attainment of typical assemblage structures, at least as regards the zoohyponeuston. The restitution dynamics of bacterio- and phytoneuston are less well documented.

The high density of common inhabitants of the 0–5 cm layer remains largely unchanged up to wave heights of 2·5 m, compared with underlying horizons (ZAITSEV, 1971, p. 89). When hyponeustonts move downwards only 1–2 m, this involves a drastic change in their normal living conditions, especially in illumination. Motile hyponeuston organisms, such as young fishes and large invertebrates, therefore obviously attempt to remain near the surface as long as possible when the sea becomes increasingly rough. Where they have been forced downwards, they return to their normal habitat as soon as the wave height decreases below the critical level. However, heavy storms tend to kill a significant percentage of the population. Such non-motile organisms as bacteria, algae, protozoans, fish eggs, which depend on buoyancy, are more easily dispersed. Fish eggs may be seriously injured by rough seas.

Adaptation to variations in temperature and salinity

Adaptations of marine organisms to variations in temperature and salinity have received detailed attention in Volume I. Infrared rays of the solar spectrum are very rich in thermal energy. They are absorbed in the upper centimetres below the interface, and extreme infrared rays in the upper millimetres. However, owing to convection, the thermal energy also reaches deeper levels, and thus the thermal regime of the 0–10 cm layer is not intrinsically different from that of the immediate underlying waters (p. 313).

Temperature effects on growth are particularly marked in pontellids. They often attain different sizes in successive generations born and grown under different thermal conditions (CHAMPALBERT, 1975). Many pontellid species are somewhat stenothermal. In a given area, pontellids tend to exhibit rather strict thermal requirements, although this does not preclude capacities for non-genetic adaptation to thermal fluctuations.

As regards salinity, pontellid copepods, at least in areas with rarely diluted surface waters, seem to be highly stenohaline. An experimental salinity decrease of only 3⁰/₀₀ is sufficient for inducing a marked depression in respiratory rates (CHAMPALBERT, 1975). This peculiarity is striking because most marine animals exhibit increased metabolic rates in diluted waters (Volume I: KINNE, 1971; HOLLIDAY, 1971). In the natural environment, a pronounced salinity decrease—unless restricted to a very brief period—tends to have an adverse effect on hyponeuston populations and to cause alterations in their vertical and horizontal distributions. In summary, meteorological and transient or local hydrographical conditions play an important part in the distribution and abundance of hyponeuston animals.

In the immediate vicinity of the Rhône estuary, a decrease in salinity in the uppermost layer due to rainfall or river runoff is associated with an increase in suspended mineral particles. In this area, CHAMPALBERT (1975) observed that the stratification of the three typical holohyponeuston pontellid copepods (p. 325) was less marked and their abundance in the uppermost layer to decrease sharply. Likewise, most merohyponeuston animals—such as zoeae of the genus *Ebalia*, larval stages of the family Alpheidae, megalopae of various Brachyura and larval stages of gastropods—are scattered in the diluted surface water and attain higher abundances only just below the halocline. Fish eggs, which usually aggregate in the uppermost layer due to their high buoyancy, also occur below the freshened layer.

CHAMPALBERT (1975), further noted that epiplankton species which are sufficiently eurythermal and euryhaline may inhabit the hyponeuston layer; most common are the copepods *Centropages typicus*, *Paracalanus parvus*, *Clausocalanus* sp., *Temora stylifera*, and the appendicularian *Oikopleura* sp.

In the most diluted parts of the area investigated, CHAMPALBERT collected the chaetognath *Sagitta elegans* and the mysids *Mesopodopsis slabberi* and *Neomysis integer* in the hyponeuston layer. These species which, under certain environmental conditions, may substitute for hyponeustonts are sometimes called 'pseudoneuston'. In summary, surface water dilution—except when very brief—causes dispersion of the permanent zoohyponeuston and the occurrence of species which usually do not participate in the neuston assemblage. Due to their rather high stenothermy and stenohalinity, hyponeuston animals—especially pontellids—are highly sensitive to seasonal changes in temperature and salinity, the combined effects of which seem to be cumulative. In addition, seasonal changes in zoohyponeuston assemblages may occur together with fluctuations in these two factors. In the pontellid copepods of the Gulf of Marseilles, for example, *Anomalocera patersoni* is rather psychrophilic and exhibits its maximum abundance during the cold season; the thermophilic *Pontella mediterranea* mainly occurs in summer and the eurythermal *Labidocera wollastoni* always inhabits the uppermost layer but exhibits abundance peaks at the beginning of July, October, November and January (CHAMPALBERT, 1975). Such gaps in the occurrence of holohyponeuston species possibly require the production of resting eggs which remain viable either in the pelagial or, more probably, on the sediments, and facilitate repopulation of the uppermost layer when the temperature becomes suitable again. Such winter, or summer, dormancy was demonstrated for several calanoid copepods. Black Sea females of *Pontella mediterranea* produce two types of eggs which differ in ornamentation, colour, size and number laid (SAZHINA, 1968). One type is produced in summer, the other in September following seasonal decrease in water temperature. However, SAZHINA could not demonstrate that the latter type is a resting egg. The pontellid *Labidocera aestiva*, which is fairly common in Woods Hole Harbour and adjacent areas in summer and autumn, produce resting eggs (GRICE and GIBSON, 1975). The eggs are present on the bottom and remain viable for six months. Hatching occurs in May at water temperatures between 13 and 14 °C.

Significant fluctuations in merohyponeuston composition and abundance are due to seasonal dynamics; most plankton and benthos species exhibit more or less time-limited breeding periods. The changes largely depend on the respective percentages of thermophilic, psychrophilic and more eurythermal species within the fauna considered and correspond to regional or local peculiarities of the prevailing hydrological climate.

Zooepineuston

In limnic waters zooepineuston is much more abundant and diversified than on the surface of marine waters. The latter are inhabited only by insects of the order Hemiptera. Two families are represented: Veliidae and Gerridae. Nearshore areas such as lagoons, bays and coral reefs may be inhabited by veliids (mainly of the genera *Halovelina* and *Trochopus*) and, among the gerriids, members of the genera *Rheumatobates* and *Hermatobates* (CHENG, 1975). The latter family also includes the offshore genus *Halobates* which contains ca 35–40 species. The population density varies between 1 individual 100 m^{-2} and several individuals m^{-2}. Supported by surface tension due to air bubbles confined between setae of the body which makes it nonwettable, *Halobates* lays its eggs upon floating things. The typical species of the genus are wingless and live strictly on the aerial side of the surface film without any connection with the land; they are almost completely restricted to intertropical regions. However, some species are less thermophilic, such as *H. sericeus* which inhabits the subtropical northern and southern gyres in the Pacific Ocean. The most common species, *H. micans*, has a circumtropical distribution; however, it may sometimes be observed in temperate oceanic areas if displaced by strong winds. *Halobates* is carnivorous and mainly feeds on pleuston organisms.

Trophic Relationships

Trophic relationships in the neuston have been insufficiently investigated. As mentioned on p. 321, the primary level of the trophic pyramid may mostly consist of dead or dying algal cells. In subtropical waters of the north-eastern Atlantic Ocean (HEMPEL and WEIKERT, 1972, p. 81), noticed that 'the abundance of phytoplankton seems to be extremely low in the immediate vicinity of the surface. Everywhere, including upwelling areas, more phytoplankton occurred in the lower nets than in the upper ones.' In the 0–5 cm layer of the Black Sea, NESTEROVA and POLISHCHUK (1975) failed to find any correlation between phytoneuston and zooneuston. They assume that small invertebrates utilize other food sources. HEMPEL and WEIKERT emphasize that most larger neuston invertebrates and fishes in the 0–10 cm layer are omnivorous or carnivorous; at noon, about 75% of the invertebrates recorded are macrophages.

All pontellid copepods are predators: *Labidocera wollastoni* feeds on small copepods, *Pontella mediterranea* on larger copepods and fish larvae. According to LILLELUND and

LASKER (1971), pontellids in general require a very abundant food supply. *Labidocera jollae* and *L. trispinosa* cannot endure more than three days' starvation. Pontellid predation pressure on young fish larvae may be very pronounced. LILLELUND and LASKER experimented with *Labidocera jollae*, *L. trispinosa* and *Pontellopsis occidentalis*, all very common in the hyponeuston of the California Current together with larvae of the anchovy *Engraulis mordax*. It seems that the copepods are attracted by vibrations of the tail beat and react by capturing or biting the fish larvae. Due to continuous swimming and high cruising speeds (up to four body lengths s^{-1}), the copepods killed or fatally injured within 24 h all larvae contained in a 3·5 litre jar (ratio of fish larvae to *L. jollae* females: $< 10 : 1$). When the larvae become older, predation diminishes due to increasing swimming and escape abilities. In another experiment, LILLELUND and LASKER demonstrated that all *Artemia* nauplii ($< 11–14$ individuals l^{-1} in a 3·5 litre jar) were also killed within 24 h by *L. trispinosa*. When the density of either anchovy larvae or *Artemia* nauplii increases, more are killed but never all of them. According to WEIKERT (1973), who studied pontellids in the subtropical north eastern Atlantic Ocean, these copepods—especially the most common *Pontella atlantica*—do not play an important role in the hyponeuston community except as predators in the uppermost water layer during the hours of maximum solar radiation.

On the other hand, it is well known that many fish larvae (*Atherina mochon* 35 mm in length, *Belone belone*, etc) feed on pontellids. Halocyprid ostracods, some of which are abundant in the hyponeuston only at dawn or dusk, are of little importance in the neustonic food web, because migrant fish carnivores reach the surface too late and leave it too early to allow extensive predation. However, halocyprid ostracods may represent an important part of the diet for non-migrant neuston carnivores such as young (>30 mm) individuals of *Scomberesox saurus*. According to WEIKERT (1972, p. 82) the smallest *S. saurus* larvae feed on neuston protozoans and gastropods as well as on cladocerans and calanoid copepods. For larger individuals (ca 40 mm in length) ostracods and cyclopoid copepods constitute dominant food items. At this stage, the food supplied in the neuston is no longer sufficient. The fish must increasingly exploit deeper waters during the day, but they return to the surface for further feeding at night.

The trophic structure of the hyponeuston summarized above makes it clear that predation plays a major part within this organismic assemblage. Apparently in response to this, many hyponeustonts have evolved adaptations for reducing the pressure exerted upon them by aerial and aquatic predators. Numerous examples of such adaptations have been discussed by ZAITSEV (1971, p. 100 ff.). I have selected only a few of what I consider to be the most striking examples, which also seem least tinged by anthropocentric bias. ZAITSEV mainly distinguishes cryptism, i.e. protective colouration 'enabling the animal to blend into and become indistinguishable from the background of its normal environment' and mimesis, i.e. 'imitation of various environmental details to which the predator is indifferent' viz. inedible items. Mimicry, i.e. imitation of environmental elements which are poisonous or dangerous to the predator, does not seem to exist in hyponeuston but only in pleuston assemblages (p. 350). Adaptations leading to cryptism are:

(i) High body transparency, e.g. fish eggs, anchovy larvae (up to 30–35 mm), young blenny larvae, many decapod crustacean larvae. While the sparse, finely branched chromatophores do not interfere with transparency, only the strongly pigmented eye tapetum betrays the presence of these hyponeustonts.

(ii) Silvery colour of the ventral side which tends to protect against hydrobiont predators because the interface seen from below appears as a mirror.

(iii) Blue colour is protective against aerobiont and hydrobiont predators. This is widespread in hyponeuston and pleuston organisms and well known among pontellid copepods. ZAITSEV (1971) demonstrated that blue may correspond to local sea water conditions and thus varies, for instance, from deep blue in Black Sea offshore areas to greenish-blue in neritic waters in *Pontella mediterranea* and *Anomalocera patersoni*. Individuals transferred into an aquarium rapidly pale. The blue pigment appears to be a chromoprotein complex of carotenoid and protein which exhibits a wide absorption range with a peak at 640 nm in *P. fera* (ZAITSEV, 1971; p. 101). The protective function of the blue pigment against excessive illumination (p. 313) has been questioned by HERRING (1967), because the wavelengths 625–650 nm which correspond to the absorption peak seem to be dimly noxious.

(iv) Disguising alternation of strongly pigmented and colourless areas tend to mask true body contours and thus to interface with prey recognition. Disguising is characteristic of *Solea nasuta lascaris* larvae (Fig. 7-10) which, although clearly discernible on the sea

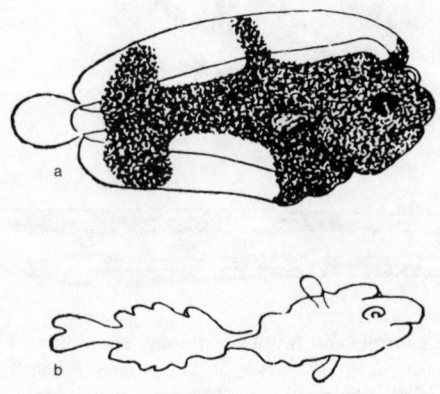

Fig. 7-10: *Solea lascaris nasuta*. (a) disruptive coloration of larva; (b) shape assumed by the same larva in case of danger. (After ZAITSEV, 1970; reproduced by permission of the author.)

surface by the human eye, also exhibit mimesis: they 'freeze' in an unnatural pose and thus resemble floating objects (ZAITSEV, 1971; p. 104).

Mimesis. The air bubble located on the dorsal side of mullet fry, in addition to its buoyancy function (p. 334), also has a protective value against birds. ZAITSEV (1964) observed that grey mullet fry, whose back is normally dark, changes to a silvery colour in rough seas. When caught and placed in a vessel with sea water, the fry change back to their normal greenish or bluish colour. Such mimesis seems to be common in many young fishes; an aggregated shoal looks similar to a swarm of air bubbles. Hyponeuston organisms also often tend to imitate various small flotsams. Fry of *Sphyraena barracuda* (18–22 mm in length) simulate twigs or rods in colour and shape, and drift in a more or less vertical position with the head touching the surface film. Larvae and fry of the Black Sea fish *Callionymus belenus* exhibit a dense network of dark brown melanophores, making them very similar to thallus pieces of the phaeophyte *Cystoseira barbata*. Larvae of *Belone belone euxini* copy this algae even more efficiently: 'their elongated body bears a dense cover of brown and brownish melanophores together with a regular pattern of light spots composed of guanocytes, resulting in an imitation of the alternating widenings (air bladders) and constrictions on the blades of the alga' (ZAITSEV, 1971; p. 107). Young *B. belone euxini* (10–15 cm long) resemble small pieces of *Zostera* leaf and often remain closely parallel to a leaf when they hunt for pontellids (Fig. 7-11).

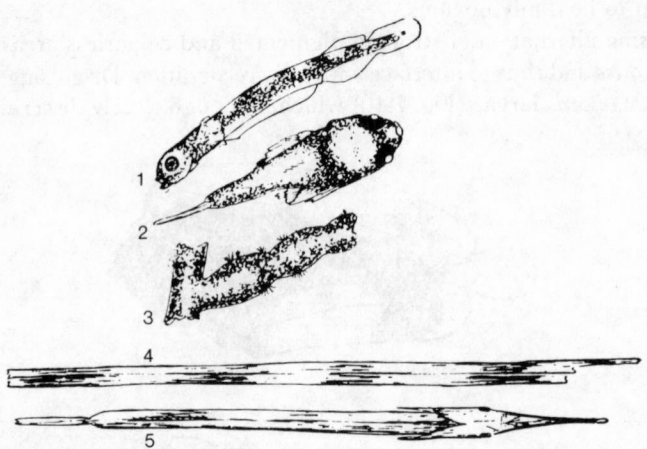

Fig. 7-11: Examples of mimesis among neustonts: 1 : larva of *Belone belone euxini*; 2 : larva of *Callionymus belenus*; 3 : thallus fragment of the phaeophyte *Cystoseira barbata*; 4: leaf fragment of *Zostera marina*; 4 : young *Belone belone euxini*. (After ZAITSEV, 1970; reproduced by permission of the author.)

Jumping ability. A very important adaptation of pontellid copepods for escaping predators is their ability to jump out of the water up to a height of 15 cm, covering distances of 15 to 20 cm in one or several jumps (Fig. 7-12). Such jumping ability—which seems to be lacking in young individuals—probably affords efficient protection for pontellid species whose whole body is dark blue and so makes them highly visible to predators swimming in the 0–5 cm layer or just below. The ability to jump is also known from young stages of *Belone belone*.

A general feature of pelagic organismic assemblages is that the ratio carnivores total zooplankton increases with depth and from temperate towards warmer surface waters. In both cases, the gradient is inversely related to primary production rate; carnivores

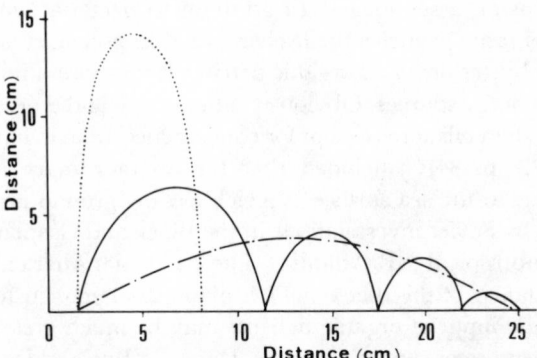

Fig. 7-12: *Pontella mediterranea*. Different types of tra-
jectory in aerial jumps of this Black Sea pontellid.
(After ZAITSEV, 1970; reproduced by permission of
the author.)

thus have a better chance of surviving than herbivores. The neuston assemblage seems
to be characterized by a very low level of phytoneuston production. In the Pacific Ocean
equatorial current system, equatorial divergence areas are characterized by extreme
scarcity in hyponeuston animals, chiefly pontellids. As the meridional components of the
equatorial currents transport upwelled waters further and further away, the succession
phytoplankton → herbivores → small carnivores → macrozooplankton and micronekton proceeds up to a convergence zone; simultaneously, the hyponeuston assemblage
increases in abundance and diversity.

'The fate of hydrobionts reaching a convergence zone depends on a number of
factors, notably their specific weight or buoyancy. Organisms with neutral or slight
positive buoyancy (planktonts) sink easily with the descending "surface layers"
into the water column. Those having high positive buoyancy (neustonts) resist the
sinking currents and remain in the surface biotope to which they are mor-
phophysiologically and behaviorally adapted. Owing to the continuous nature of
this process, large concentrations of neuston, floatage and stable foam develop in
convergence zones. The abundant foam, with its rich organic detritus stimulates
proliferation of bacterioneuston. . .' (ZAITSEV, 1971, p. 127).

Since zooneuston abundance does not seem to be directly dependent on primary
producers, what is the origin of the energy within the neuston ecosystem? It is possible
that plankton assemblages below the uppermost layer may represent an energy source;
neuston abundance increases periodically, mainly in concert with the nycthemeral
rhythm. WEIKERT (1972) assumes that such periodical increase is indicative of
exchanges between neuston and plankton assemblages, but it is unknown which of the
two assemblages obtains a gain. ZAITSEV (1971) believes that bacterioneuston prolifera-

tion supports subsequent links of the neuston trophic chain. While this might be true, utilization of bacteria as food by ciliates, tintinnids and small larvae has not yet been demonstrated for neuston assemblages. In addition to bacteria which live mainly on epiphytic particles, organic particles themselves (such as pollen, spores, cysts, scales of mosquitoes, flies and butterflies) and organic detritus from shores and river run-off, may represent important energy sources. Obviously, abundance and diversity of particles are much higher inshore than offshore (except for convergence areas). This fact may explain why WEIKERT (1972, p. 84) concluded that the 'extraordinary importance of the uppermost centimeters at the sea surface as a rich feeding ground for invertebrates and fish—as emphasized by Soviet investigations in the Black and Caspian Seas—could not be verified for the subtropical NE Atlantic.' The particular characteristics, especially with regard to abundance, of the Black and Caspian Seas hyponeuston are not surprising: it is likely that the input of organic detritus may be much greater in the northern areas of both seas, due to river run-off (Danube, Dniester, Bug and Dnieper for the Black Sea; Volga and Ural Jaïk for the Caspian Sea).

However, one may agree with ZAITSEV (1971) that the significance of neuston cannot be sufficiently evaluated in terms of biomass or production. The role it plays seems to be more significant in the ecology of the pelagic domain via chemical and biochemical transformations, partly due to micro-organisms in the uppermost water layer.

Horizontal Distribution of Neuston—Biogeography

For evaluating the horizontal distribution of neuston it is useful to consider two different scales: (i) topographic distribution inside a biogeographical unit; and (ii) biogeography *sensu stricto*.

Topographic distribution

Various factors affect the topographic distribution and abundance of neuston organisms. Among them, the input of dead and decaying organic material to the uppermost water layer is of paramount importance. For example, on the shelf, terrigenous material from land drainage, river run-off or winds, as well as organic substances lifted from the sea bottom by bubbles, constitute most of the trophic basis that allows an abundant neuston to develop. The specific richness may also be rather high due to bentho-hyponeuston animals and larval stages of benthic invertebrates and fishes. Further, it was mentioned above that convergence areas where detritus accumulates, together with drifting species, harbour a more abundant neuston than divergence or upwelling areas.

Salinity also plays some part in the distribution of neuston, especially as regards holohyponeuston, since most of its members are rather stenohaline. However, some pontellids may be restricted to highly diluted areas, e.g. *Labidocera brunescens* which attains maximum abundances in the Azov Sea, but decreases with distance from the shore in the Black Sea. Most pontellids are highly sensitive to salinity fluctuations. Some species seem to be characteristic of distant neritic areas where the salinity fluctuates more than in open-ocean areas. In the North Pacific Ocean, *Pontellopsis occidentalis* and *Labidocera trispinosa* must be considered tropical neritic species (HEINRICH, 1960), as is *Pontellopsis lubbockii* in the southeastern Pacific Ocean (HEINRICH, 1971).

Biogeography

As far as the biogeography of the neuston is concerned, pontellid copepods—the most characteristic and best adapted group in the holohyponeuston—probably represent the most suitable key organisms for delimiting biogeographical units.

In general, the family Pontellidae is warmth-loving. The specific richness is particularly marked between the tropics in the Pacific and Indian Oceans (HEINRICH, 1960, 1969), as well as in the Atlantic Ocean (HEINRICH, 1974). North of 40° N and south of 35° S—i.e. beyond the north and south subtropical convergences—pontellid richness decreases sharply. Cold-loving species are few. *Epilabidocera amphitrites*, the northernmost pontellid, occurs in subarctic and Arctic waters of the Pacific Ocean and was even recorded in the Chukchi Sea (Arctic Basin); *Epilabidocera longipedata* inhabits Oyashio Current waters. *Anomalocera patersoni*, sometimes classified as cold-loving, should be considered, rather, as cold-tolerant.

Specific thermal requirements of pontellids are also obvious in areas with strong hydrological contrasts. For example, HEINRICH (1971) reported four oceanic species in equatorial waters of the eastern Pacific Ocean (*Labidocera detruncata, L. acutifrons, Pontella tenuiremis, Pontellopsis regalis*), and two distinct neritic species (*Pontella danae* and *Labidocera acuta*), whereas to the south of the boundary of equatorial waters, more influenced by upwelling and higher-latitude waters, he observed only two species: *Pontellopsis regalis* and *Labidocera acutifrons*.

Salinity seems to affect the geographical distribution of most offshore pontellids less markedly than temperature. As mentioned above, some neritic species tolerate diluted waters. *Pontella fera* inhabits equatorial waters of the western Pacific Ocean where the salinity may be lower than 34·5⁰/oo. In the South Atlantic Ocean the 35⁰/oo isohaline seems to separate the group of temperate species *Anomalocera patersoni, Labidocera wollastoni, Pontella mediterranea* and *Labidocera fluviatilis* from the two groups of oceanic species: the subtropical group *Pontellopsis regalis* and *P. villosa* and the tropical group *Labidocera acutifrons, Pontella atlantica* and *Pontellopsis brevis* (WEIKERT, 1975).

Biogeographical data thus confirm that most offshore pontellids are somewhat stenotherm and stenohaline. However, some species are more tolerant to both temperature and salinity. For example, the cold-tolerant *Anomalocera patersoni* inhabits the central North Atlantic Ocean between 29° N and 37° N in winter, the Norwegian Sea in summer, the Labrador Sea in winter, and the St Lawrence Gulf where the temperature fluctuates from 3·8° to 15 °C and salinity from 25⁰/oo to 33⁰/oo. *Pontellina plumata* is widespread in the Pacific, Atlantic and Indian Oceans in salinities of ca 38·0⁰/oo. *Anomalocera patersoni* and *Pontella mediterranea*, which are common in the Mediterranean Sea and Atlantic Ocean, establish very dense populations in uppermost water layers of the Black Sea (ca 16–22⁰/oo).

In an extensive study of pontellids in the Indian Ocean during the winter monsoon, some species turned out to be restricted to a given current, or to be much more abundant there than in adjacent waters (VORONINA, 1962). *Pontella novae-zelandiae*, for example, exists only in the West Australia Current; *P. denticaudata* in waters flowing from the Java Sea; *P. fera* largely predominates in the Monsoon current and the equatorial countercurrent; whereas the more barotolerant *Labidocera detruncata* is the most abundant species in the South Equatorial Current and also in the undercurrent.

Finally, local productivity can affect distributional patterns. The most fertile waters probably carry a poor hyponeuston; it seems less likely, however, that the most oligotrophic areas harbour abundant hyponeuston. WEIKERT (1975) assumes that, in the South Atlantic Ocean, the highest concentrations of pontellids are associated with waters of intermediate productivity (150–250 mg C m^{-3} d^{-1}) and in central regions of the Pacific Ocean. SHERMAN (1963, 1964) recorded high abundances of *Labidocera detruncata*, *L. acutifrons*, *Pontellopsis villosa* and *Pontellina plumata*, east of 140° W, i.e. along the latitudinal gradient of increasing productivity. Further quantitative investigations are necessary before we can definitely answer the question whether or not a correlation exists between water fertility and pontellid abundance.

(c) Pleuston

General Aspects

The definition of the term 'pleuston' (p. 313) requires additional qualification here. I consider the most important difference between pleuston (also called pleiston by Russian authors) on the one hand and plankton plus neuston on the other, to be that pleuston displacements result from combined actions of water currents, and winds, while plankton and neuston only drift with the water current. Pleuston assemblages not only include animals whose body is partly submerged but also species associated with floating objects, either the animals just mentioned or such inanimate flotsams as pieces of wood, lumps of tar and plastics, empty bottles and cans, etc., which are displaced by both water currents and winds.

No multicellular algal species has a truly semisubmerged thallus. Due to their large size, floating *Sargassum* species (p. 350) are both hyponeustonic and planktonic. Nonliving floating objects may support unicellular algae and sessile as well as motile animals (Volume I: BACESCU, 1972; GERLACH, 1972). Moreover, several swimming animals usually remain in the immediate vicinity of flotsams. In my opinion all these species participate in the pleuston assemblage.

Composition of the Pleuston Assemblage

Free floating animals

The physophorid siphonophore *Physalia* ('Portuguese man-of-war') is the largest species in the interface assemblage. Its blue, air-filled float (20–30 cm in length) constitutes a goffered sail; dactylozoïds which capture, strangle and kill prey (mainly fishes) thanks to the nematocyst toxin, may be as long as 9 m, and by retracting they bring the prey to the gastrozoïds. Owing to the chitinoid nature of its walls, the float persists long after the siphonophore has died and may thus support other organisms, such as eggs of surface animals (*Fiona*, *Halobates*) and *Lepas* larvae which settle and develop into adults. Young larvae of *Janthina* settle temporarily before metamorphosing into the free-living stage. *Physalia* species are common only between Latitudes 40° N and 40° S, but may drift with strong winds beyond these boundaries. The *Physalia* sail is set obliquely to the float's long horizontal axis so that the body tends to sail at an angle of about 45° to the wind (CHENG, 1975); there are right-sailing and left-sailing forms. Several authors have suggested that under moderate wind conditions (ca 3–4 Beaufort) drifting of *Physalia* and *Velella* may be influenced by Langmuir circulation, resulting in a predominance of left-sailing forms in the Northern Hemisphere and of right-sailing forms in the Southern Hemisphere. Fig. 7-13 lends some support to the first assumption, but additional data seem necessary for a more definitive assessment.

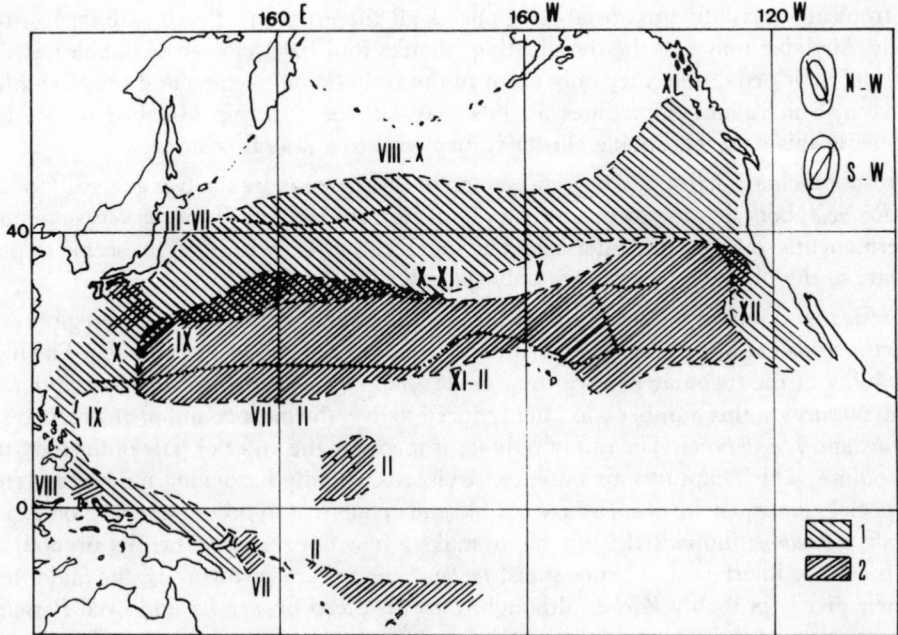

Fig. 7-13: *Velella lata*. General distribution of the two ecological forms in the northern half of the Pacific Ocean. 1 and 2, massive clustering of NW and SW forms, respectively. Dotted lines: winter location of subpolar and subtropical convergences. Roman numerals denote the month in which the boundary of each zone was observed. (After SAVILOV, 1961, reproduced by permission of the author.)

Although *Velella* is a carnivore, feeding on copepods, crustacean larvae, but mostly on fish and euphausiid eggs, etc, it also contains zooxanthellae, like hermatypic corals. *Velella* is one of the most important and common animals in the pleuston. It not only plays a significant role in trophic nets of pleuston and neuston, but also provides a substratum for many organisms in the assemblage.

In the family Porpitidae the colonial structure is simpler than in *Velella*; there is only one gastrozoid, obviously associated with gonozoids and numerous long and tentacle-like dactylozoids moving synchronously for food collection. As in *Velella*, buoyancy is ensured by a large aboral (zenithal) raft, chitinoid, septate and air-filled, but without sail. *Porpita* mainly feeds on motile prey such as copepods and crustacean larvae. Two genera were described, *Porpita* and *Porpema*, each of them with two species: *Porpita umbella* and *Porpema globosa* in the Atlantic Ocean; *Porpita pacifica* and *Porpema prunella* in the Pacific Ocean. The systematic status of the genus *Porpema* is still questionable (CHENG, 1975); it was even assumed that the Porpitidae might contain only one species (*Porpita umbella*). Porpitidae seem to be more restricted to intertropical regions than *Physalia* and *Velella*.

Real free-floating Actiniaria are represented in the pleuston by the family Minyadidae with the sole genus *Minyas*, which may include five different species. These comparatively rare animals, whose discoid pedal sole is modified to form a float, seem to be restricted to the tropical—possibly equatorial—regions of all three oceans. Some actinians of the family Abylidae may also be free-floating, thanks to a hard cluster of bubbles at the centre of their pedal sole; they only occur in the Indo—Pacific region and are often blue or greenish in colour. Sometimes abylids may also be observed attached to floating objects; in this case the bubble cluster is displaced to a lateral position.

Prosobranchial gastropods are represented in the pleuston by the two genera, *Janthina* and *Recluzia*, both floating by means of a raft of air bubbles. The former is very common; it permanently inhabits the interface. The much rarer yellowish *Recluzia* seems to participate in the pleuston assemblage only sporadically.

Janthina is commonly observed in all three oceans, mainly in intertropical regions but sometimes also at higher latitudes, drifted by winds of anomalous persistence. The high variability of the turbinated, very thin, violet shell of *Janthina* led to the description of about 60 species; this number was later reduced to five, the most common of which are *J. janthina* and *J. umbilicata*. The raft of bubbles is made by the anterior part of the foot, the propodium, which captures air bubbles. A viscous, rapidly hardening mucus, secreted by special glands, encloses each new bubble and cements it to preceding ones; on losing its raft, a *Janthina* individual is unable to make a new one, except when its propodium can break the interface and hence sinks. *Janthina* species are carnivorous; the main item of their diet is probably *Velella*, although, they also feed on *Porpita* and even *Physalia*. Some species are viviparous, whereas others lay encapsulated eggs which are attached to the undersurface of the raft.

Nudibranchial gastropods of the family Glaucidae, which comprises the two genera *Glaucus* and *Glaucilla*, are often rather common in the pleuston. Glaucids are usually found at the interface with their ventral side up (zenith). They also float on air bubbles, assisted by their exceptionally flattened body and its numerous and flattened appendages, the 'cerata'. The air bubbles are in the stomach and are not secreted but swallowed.

The dark blue *Glaucus atlanticus* (about 30 mm in length), the only species in the genus, has a circumtropical distribution. Like *Janthina*, *G. atlanticus* feeds mainly on chondrophorids and *Physalia*. It is hermaphroditic and lays its eggs in strings of 12 to 15 either free in the surface waters or attached to the prey. The much rarer genus *Glaucilla* seems to include two valid species, *G. briarea* and *G. marginata*.

Cephalopoda are represented in the pleuston assemblage only by the circumtropical genus *Argonauta*. It elaborates a light, almost paper-thin shell, in the inner chambers of which air is retained for buoyancy. Since they keep their eggs within the shell, females can become insufficiently buoyant and then may attach themselves with their arm suckers to floating objects.

Sessile species on floating objects

Floating objects such as wood pieces, empty bottles or cans, tar balls, plastic chips, dead floats of *Physalia* or chondrophorids, or turtles, may support sessile species on their submerged surfaces (Volume I: BACESCU, 1972; GERLACH, 1972). Unicellular algae, both cyanophytes and diatoms, are common. Animals are represented only by species of the pedunculate cirriped genus *Lepas*. These attach themselves to floating objects at the cypris stage. As other non-parasitic cirripeds, *Lepas* species feed on suspended particles collected by rhythmic beating of their thoracic appendages (cirri). Three species are common: *Lepas anatifera*, *L. pectinata* and *L. fascicularis*. The latter is cosmopolitan and is sometimes found in large numbers. Drifting from its substratum, *Lepas* may occasionally be observed at rather high latitudes. *L. fascicularis* is free-living, i.e. not fixed to a floating object; it has a low-weight skeleton and develops a float which may be shared by up to 20–25 individuals. *L. fascicularis* groups drift quite passively, whereas solitary individuals can swim slowly by stalk movements. The amphinomid polychaete *Hipponoe gaudichaudi*, which may reach 20 mm in length, is a common commensal of *L. fascicularis*, inhabiting its pallial cavity.

Motile animals on floating objects

The most common motile animals living on the various floating objects listed above (inclusive *Sargassum*) is a small (3–19 mm) crab of the genus *Planes*. There are two species: *P. minutus* in the Atlantic and Indian Oceans, and *P. cyaneus* in the Pacific Ocean. The crab's colouration depends on that of its substratum, and so does its abundance. *Planes* appears to be an omnivore and utilizes detritus.

The isopod *Idothea stephenseni* often clings to floating objects. Like *Planes* its colouration tends to parallel that of its substratum. It feeds on algae and small crustaceans and probably on detritus.

The nudibranch *Fiona pinnata*, which is less common than the crustaceans mentioned above, sometimes crawls on flotsams and exhibits the same adaptive homochromy as *Planes* and *I. stephenseni*.

Animals swimming in the immediate vicinity of floating objects

This group, of which the appurtenance to pleuston is less obvious, comprises only the small shrimp *Parapenaeus longipes* and the little fish *Nomeus gronovii*; the latter is often found among the tentacles of *Physalia*.

The occurrence of flatworms at the air–sea interface requires special attention. Sampling the interface near La Paz, Baja California, CHENG and LEWIN (1975, p. 518) observed 'reddish brown streaks extending for several hundred metres'. These consisted of high densities of *Noctiluca* with an admixture of other dinoflagellates and a polyclad turbellarian identified as *Stylochoplana sargassicola*. Such parallel streaks are well known as a result of convection cells generated by wind. However, it was surprising that several of the brown patches consisted almost exclusively of *S. sargassicola*, associated with species of *Porpita*, *Physalia*, *Glaucus* and *Janthina*. All sizes from 2 to 12 mm were camouflaged with associated *Sargassum* sp. leaflets. Behaviour and movement make it difficult to place this flatworm in the neuston or pleuston assemblage. It can glide on solid substrates and even on a relatively undisturbed air–water interface; such behaviour is characteristic of pleuston organisms. However, it can also swim, either 'slowly, in a horizontal position, by undulations of the lateral margins of the anterior two-thirds of the body' or 'much more rapidly by turning on one side. . . with serpentine undulations'. Such behaviour is characteristic of a hyponeustont. *S. sargassicola* was first described, associated with *Sargassum bacciferum*, from the Sargasso Sea, but was later also recorded off the coast of Sierra Leone in the absence of *Sargassum*.

Pleuston Ecology

Adaptations to life in the pleuston are similar to those described for neuston species (p. 334). They mainly involve buoyancy and protection against excessive solar radiation. Blue pigmentation occurs as frequently as in neustonts. Its true significance, except for the protection it provides against aerial predators, remains insufficiently documented. Wind affects pleuston distribution more than that of neuston.

The pleuston assemblage is almost devoid of primary producers. The only exceptions are sessile unicellular algae attached to floating objects and zooxanthellae in some coelenterates. Therefore, almost all energy available at the basis of the assemblage appears to be supplied by the neuston or plankton assemblage. This energy presumably consists mainly of dead and decaying organic material and bacteria multiplying at its expense. Pleuston mainly comprises carnivorous species. Its trophic relationships may be classified as follows: (i) Pleustonts called 'lift predators' (CHENG, 1975) such as *Physalia* and chondrophorids, which feed on hyponeuston animals—most of them also carnivorous—or on plankton animals with the help of their long tentacles (especially *Physalia*) in water below 0–5 cm. The energy requirements of this group are met by food resources from layers below interface. (ii) Pleustonts which feed mainly, and possibly exclusively, on other pleuston species. This group seems to include only gastropods which prey mainly on *Physalia* and chondrophorids. (iii) Pleustonts walking or gliding on floating objects which are probably more or less omnivorous and consume both pleuston and neuston materials.

The feeding habits of the swimming species are poorly documented but it may be assumed that they feed mainly if not exclusively on neuston animals. Therefore, despite the fact that they are usually classified as pleuston species because they are always associated with floating things, it might be better to consider them as occupying an intermediate position between pleuston and neuston.

(d) Floating *Sargassum* Assemblage

Floating Macrobenthic Algae

As is well known, in waters such as coastal lagoons and semi-enclosed bays, which remain sufficiently smooth for lengthy periods, several macrobenthic algae may become free-living at the air–sea interface. Most of them are phaeophytes (*Ascophyllum nodosum, Chorda filum, Desmarestia aculeata, Saccorhiza polyschides, Laminaria saccharina, L. digitata, L. hyperborea*), but this phenomenon also exists in some rhodophytes, such as *Spermothamnion* and *Furcellaria fastigiata* f. *aegagropila*. The free-living condition results in modifications of morphological and physiological characteristics: reduction of the thallus basis; disparition of sexual reproduction; increase in growth rate inducing asexual multiplication; and the thallus may become more slender and more ramified (p. 409).

The ability of benthic algae to develop free-living, more or less modified forms reaches its height in floating *Sargassum* species.

The Sargasso Sea

The Sargasso Sea, located in the Atlantic Ocean approximately between 20 and 40° N and from 40 up to 75° W, has a surface area of about 4,400,000 km^2. It corresponds to a lateral anticyclonic gyre of the Gulf Stream. Its western (Gulf Stream) and northern (North Atlantic Drift) boundaries are well defined due to high velocity gradients, whereas the eastern (Canary current) and southern (north equatorial current) boundaries are less well defined. The average depth is about 4000 m. Despite their very high temperature (February 20–25 °C; August 25–28 °C) Sargasso Sea surface waters have a rather high density, due to their high salinity (37·0–37·3⁰/₀₀) which results from strong evaporation, and thus tend to sink. The sinking in turn transports waters of high temperatures to subsurface (18 °C at 200 m) and upper intermediate layers (15–17 °C at 400 m). Between 400 and 500 m a strong permanent thermocline prevails which blocks upward diffusion of mineral nutrients from deeper layers. Since a seasonal thermocline exists at about 100 m, summer fertility is very low; in winter, the shallower thermocline breaks up and fertility increases, mainly in the northern areas of the Sargasso Sea, but sometimes in all its surface waters when the winter is colder.

Floating Species of the Genus Sargassum

The genus *Sargassum* is widespread on warm-temperate and intertropical hard shelf bottoms. Up to eight species of floating *Sargassum* have been described in the Sargasso Sea, but most of them exist mainly in Gulf Stream surface waters and must be considered as flotsams which decay relatively quickly. The two species wholly adapted to pelagic life are *Sargassum natans* and *S. hystrix* var. *fluitans*. However, it is generally admitted that floating *Sargassum* originate from benthic algae ripped off the rocky shores of Florida, Honduras, Jamaica, etc., during the hurricane season. This assumption is supported by the fact that, in western areas of the Sargasso Sea, some of these algae still exhibit conceptacles including sexual, but apparently not fertile, products. It has been postulated that *S. natans* originates from the benthic *S. vulgare* or *S. filipendula*. The ripped off benthic algae drift first with the Florida Current, then with the Gulf Stream. A small percentage is transported to shelf waters where their standing crop is about 0·06 tonnes per square nautical mile off Carolina (HOWARD and MENZIES, 1969); they may include a high percentage of individuals from species with a low adaptative potential, which therefore cannot survive for long.

Like the other floating multicellular algae mentioned above, floating *Sargassum* fail to reproduce sexually, but have a high capability for vegetative multiplication. The rapid growth takes place at the apex of the branches; the most basal parts progressively degenerate. Nevertheless, growth much exceeds basal necrosis, so that the oldest floating algae attain giant dimensions.

Most of the *Sargassum* thalli at depths of 0–1·5 m must be considered hyponeustonic. When the sea surface is quite smooth, some parts of the phylloids may be exposed to air above the interface; in such a case they rapidly (7–10 min) dry and degenerate. Frequent wetting with sea water (approximately 1-min intervals) keeps the exposed parts alive. Under natural conditions, exposed thallome parts die only in a very calm sea (ZAITSEV, 1971). This fact demonstrates that floating *Sargassum* species really belong to the hyponeuston, not to the pleuston, because the aerial parts of pleustonts can withstand solar radiation and dessication for indefinite periods, even in calm weather when not wetted.

In *Sargassum natans* buoyancy results from both the very low specific weight (0·785–0·788 without overgrowths) and numerous air bladders. Without air bladders the specific weight of *S. natans* thalli is 1·09. Small to medium-sized (2–3·5 mm) air bladders create a higher buoyancy effect than larger ones, possibly due to mineral deposition on the walls of old air bladders (ZAITSEV, 1971). About one-fifth of the bladder buoyancy is required to facilitate buoyancy of the weed itself. About four-fifths of the air bladder buoyancy constitutes a 'safety reserve' enabling the weed to endure the drying of some of its air bladders as well as the additional weight of the attached flora and fauna (ZAITSEV, 1971). The total biomass of floating *Sargassum* in the Sargasso Sea is estimated to range from 4 to 11×10^6 t. This corresponds to an average biomass of about 0·9–2·5 g m^{-2} (wet weight) or 75–225 mg C m^{-2} (lower than the phytoplankton biomass). An increase in biomass seems to occur in summer, possibly due to faster growth and increased vegetative multiplication, but possibly also due to the arrival of new floating weeds recently detached from the shores. In fact, assessments of the global, or average, *Sargassum* abundance do not correspond in the slightest to the actual distribution of floating weeds, which usually concentrate in more or less compact rafts separated by areas of free water.

Several authors have assumed that vertical convection cells generated by winds stronger than 3 on the Beaufort scale would be able to induce the weeds to sink and remain at a certain depth as long as the descending current balanced their ascending tendency; as soon as the wind decreases the thallomes would ascend again and concentrate at the surface layer. Despite experiments which demonstrate that *Sargassum natans* withstand a pressure of 10 bar (i.e. 100 m depth) and retain a positive buoyancy, this assumption remains questionable. On the other hand, the distribution of floating *Sargassum* in spaced streaks, which is sometimes observed, corresponds to lines of weak convergence generated by light winds.

Floating *Sargassum* Assemblage

All species which participate in the floating *Sargassum* assemblage exhibit, like the weed itself, a very high tolerance to solar radiation.

Algae

Floating *Sargassum* support many unicellular epiphytes, such as cyanophytes (*Calothrix*, *Dichothrix*) and diatoms (*Fragilaria*, *Synedra*, *Licmophora*, *Cocconeis*), as well as such small multicellular algae as the chlorophyte *Monostroma* and the rhodophyte *Ceramium*.

The nutrient supply for *Sargassum* and its associated flora has been poorly investigated. As regards *Sargassum* itself, CARPENTER and COX (1974) observed that north of 30° N, its primary production averages about twice that to the south, possibly because vertical winter mixing occurs more regularly in the former region. The authors calculated a production of about 1.0 mg C m^{-2} d^{-1} which approximately corresponds to 0.5% of the total primary phytoplankton production (200 mg C m^{-2} d^{-1}). They did not find clear relationships between production rates and nutrient concentrations. However, since *Sargassum* is mostly restricted to the upper metre of the water column, it may contribute as much as 60% of the total net production in this layer. No pelagic animal seems to feed directly on fresh organic *Sargassum* which probably enters the pelagic food chain only in the form of dissolved exudates or via decaying processes.

The epiphytic cyanophyte *Dichothrix fucicola*, a heterocyst-bearing species, can fix elementary nitrogen (CARPENTER, 1972). In dense aggregations of floating seaweeds, this may result in pronounced enrichment of surface waters with combined nitrogen. While several authors have suggested that *Sargassum* epibionts may depend on exogenous metabolites released by this seaweed, CARPENTER and COX (1974) believe that nutrient availability regulates epiphyte abundance, as it apparently regulates the growth of *Sargassum* itself. They suggest that iron is the limiting nutrient in the south-western Sargasso Sea. Nutrient supply dynamics for vegetal components of the *Sargassum* assemblage require further research, which should also include the release of reactive phosphorus by *Sargassum* patches increasing the PO_4–P concentration in immediately surrounding waters (CULLINEY, 1970).

Associated fauna

Within the *Sargassum*-associated fauna, two groups can be distinguished: (i) species characteristic of the *Sargassum* assemblage and fully adapted to live there; and (ii) species which spend only part of their life cycle in the *Sargassum* assemblage, or which may also be found elsewhere.

The characteristic fauna of floating seaweeds is a benthic fauna closely related to that on infralittoral rocky bottoms where algae predominate, but are greatly impoverished, because apparently only a few are able to adapt to the 'floating bottom'. The characteristic fauna may be divided into three subunits comprising sessile, sedentary and motile animals.

(i) The sessile fauna includes about 20 species. About ten hydroid species, mostly of the genera *Clytia*, and *Laomedea*; one actinian, *Anemonia sargassensis*; two serpulid polychaetes of the genus *Spirorbis*; one bryozoan (*Membranipora*); one didemnid ascidian (*Diplosoma*); two pedunculate cirripeds (*Lepas ansifera* and *L. anatifera*). Several other hydroids and bryozoans were observed mainly on the basal (i.e. the oldest) parts of the thallome and must be regarded as remnants of the sessile fauna which existed on the alga before it was ripped off from the rock; these species constitute the *Sargassum* 'subfauna'. The characteristic species listed above may be easily distinguished from the subfauna, because they are more abundant on the younger parts of the thallome.

Antifouling substances, possibly elaborated by *Sargassum* (CONOVER and SIEBURTH, 1964), could inhibit the surface dwelling microflora including bacteria of the genus *Vibrio*. RYLAND (1974) assumes that tannins in young thallus parts can delay the settlement of larger epibionts such as hydroids, bryozoans and multicellular algae. This assumption is supported by RYLAND's observations that only the hydroid *Clytia noliformis* occurs on the frond tips, spreading from more basal parts of the alga towards the meristem by means of stolons. Species with planktonic stages, such as the cyanophyte *Calothrix crustacea*, the spirorbid polychaete *Janua formosa* and the bryozoan *Membranipora tuberculata*, settle well away from the growing tip (instead of settling near the meristem as is usual on benthic algae).

(ii) The sedentary fauna consists of only a few species: the small nudibranch *Scyllaea pelagica*, the pycnogonid *Tanystylum orbiculare*, and the isopod *Janira minuta*. The latter is closely related to the benthic species *J. maculosa*, whose size is about 10 mm, while *J. minuta* measures only 2 mm in length. The small sedentary crab *Planes minutus* is common on many floating objects (p. 349) and therefore cannot be considered characteristic of the *Sargassum* assemblage. Several small-sized sedentary animals were observed on *Sargassum* sampled between Bermuda and the Azores by YEATMAN (1962), e.g. five species of harpacticoid copopods. However, these harpacticoids must presumably be considered to belong to the 'subfauna'; they occur on European and North African coasts as well as on several areas of the North American shelf (New England, Florida) and the shores of Bermuda. The same seems to pertain to the polyclad turbellarian *Stylochoplana sargassicola* (p. 349). *Sargassum*, which constitutes a 'floating benthic biotope', apparently plays a role in the geographical dispersion of these species.

It has been suggested that *Sargassum* thallomes are coated with a thin mucous film which facilitates attachment of sedentary species. This view requires further investigations. It is well known, for example, that if the nudibranch *Scyllaea pelagica* is placed 10 cm away from the alga, it usually cannot return, and so drowns. According to FINE (1970) who studied the non-sessile *Sargassum* fauna along a transect from the Gulf Stream to 71° W, the qualitative composition always remains the same, while the respective abundance of different species may vary with season and region.

(iii) The motile fauna in the *Sargassum* assemblage may to some extent be compared with that swimming under and around floating objects (p. 349). However, the dense and sometimes thick canopy of floating *Sargassum* streaks or patches provides a much better shelter than a tar ball, a piece of wood, or a *Physalia* float. Hence, the motile fauna inhabiting *Sargassum* rafts is much more abundant and diversified. This fauna consists of two different components: permanent inhabitants which spend their whole life cycle within the *Sargassum* masses, and temporary inhabitants.

Among the permanent inhabitants which are characteristic of, i.e. endemic to, the *Sargassum* assemblage, I should mention the shrimps *Latreutes ensiferus*, *Palaemon tenuicornis* and species of *Hippolyte*, and the two fishes *Syngnathus pelagicus* and *Pterophryne (Histrio) histrio*. Both fishes exhibit protective adaptation for diminishing predation pressure through cryptism (body colour, brown with white spots resembling an algal thallus with its air bladders) and mimesis (elongate body shape in *S. pelagicus* and foliaceous skin appendages copying small *Sargassum* phylloids in *P. histrio*). The latter may be the best adapted species of the whole *Sargassum* fauna; it seems that individuals spend their whole life in the immediate vicinity of the same algal fragment. Their buoyancy is well pronounced thanks to the existence of large subdermal cavities filled by vitellus reserves in larval stages and acting as a float in adults. *P. histrio* is carnivorous and feeds mainly on the shrimps *Palaemon tenuicornis* and *Latreutes ensiferus*. According to SMITH (1973) its assimilation efficiency is very high (72–82%), making it well adapted to the food-limited environment of the Sargasso Sea.

Temporary inhabitants mainly comprise adult or larval fishes seeking shelter against predators. Examples are 'flying-fishes' of the family Exocoetidae which spawn within *Sargassum* rafts and eel larvae (leptocephali). In 3·2 t of *Sargassum* clumps in the Florida current, DOOLEY (1972) found about 8400 individuals belonging to 23 families and 54 species; about 90% of the fishes collected were representatives of the families Carangidae, Monacanthidae, Balistidae and Antennariidae with 14, 10, 4 and 1 species, respectively. DOOLEY further noticed that the fishes were fewer and larger in summer and fall, when floating weeds are highly abundant, than in winter and spring.

Finally, it should be noted that pleuston animals, mainly *Physalia*, *Velella* and *Janthina*, sometimes occur in *Sargassum* rafts.

The perfection of adaptations characterizing floating *Sargassum* plants as well as many animals participating in the *Sargassum* assemblage—together with the speciation phenomena mentioned above—suggest an ancient origin for this interspecific assemblage. Most benthos species attached to floating *Sargassum* thalli only exist in western Sargasso Sea areas; they are apparently unable to adapt and extend to central and eastern areas. Since endemic species are—qualitatively and quantitatively—better represented in the central and eastern areas of the Sargasso Sea, which are most remote from the rocky areas where benthic *Sargassum* exist, it seems justified to postulate that the *Sargassum* assemblage constitutes a benthic biocoenosis isolated in the pelagic domain.

Trophic net

The trophic net of the *Sargassum* assemblage is little known. Primary producers are probably more abundant here than in other interface ecosystems, even though photosynthetic productivity seems to be rather low. At first glance, herbivores appear to be poorly represented.

Respiration measurements of the *Sargassum* community suggest that micro-organisms are energetically the most important component of the community, possibly due to 'the extensive surface area presented by *Sargassum* which is conducive to adsorption of macromolecular products and to chemical processes' (SMITH and co-authors, 1973, p. 216). The brownish exudate released by *Sargassum* in standing sea water contains phenolic products (as in many benthic phaeophytes) which rapidly combine with protein and carbohydrate materials to produce organic aggregates. Presumably such aggregates complement the true primary production and provide additional food for microphagous animals which are in turn consumed by carnivores. However, the microphagous trophic link (ciliates?) remains unknown, as do the nutritional relationships between *Sargassum*, and hyponeuston assemblages. Floating patches of *Sargassum* sometimes mixed with other algal species (mostly phaeophytes), may be found offshore in many regions of the World Ocean. MAKKAVEEVA (1965), for example, observed floating plant masses in the Red Sea, consisting of *Sargassum vulgare* and *Turbinaria* sp., the associated fauna including some pleuston as well as benthic species which had obviously attached themselves to the algae before these were ripped off the rocky substrate. Some sessile species occurred on the thalli but motile animals such as polychaetes, amphipods and prawns largely predominated. The associated fauna must be considered as a 'surviving' benthic assemblage; no endemic species could be recognized.

In summary, the Sargasso assemblage, as a consequence of its very ancient origin, is an ecological unit which consists both of a 'subfauna' (i.e. 'surviving' species) and an endemic species.

The fact that carnivorous species predominate in all interface assemblages (neuston, *Sargassum*, pleuston) reveals a gap in our knowledge of interface ecological dynamics. Possibly exudates, decaying material—with its biochemical transformations—and bacteria play an important part in the interface ecosystems and constitute the energy basis for the first animal (microphage) level. More comprehensive knowledge of the second trophic level must be obtained by further investigations to elucidate energy pathways and establish reasons for the predominance of carnivores in interface ecosystems.

(2) Discoloured Waters

General Aspects

Discoloured waters, also known as 'red tides', occur most frequently in intertropical regions, mainly in shallow water areas and semi-enclosed bays. However, discoloured waters may also be observed in temperate areas (e.g. coasts of eastern North America, the North Sea, etc) and even in colder seas such as coasts of south-eastern Alaska, southern Chile and southern Argentina. In temperate shallow waters, the discoloured waters seem to occur more and more frequently in coastal areas polluted by organic wastes and/or the excessive release of thermal energy.

Discoloured waters are always based on local and temporary (2–3 days to several weeks) alterations in the surface plankton assemblage. The alterations consist either in explosive *ad hoc* multiplication or in the concentration of a single or, more rarely, a few plankters (several million cells l^{-1}). This induces a change in the colour of sea water and is often associated with bioluminescence. The surface area concerned varies from a few to many hundred square miles (in larger bays or offshore), but the layer of discoloured water is always very thin, mostly a few dm, sometimes up to 1·5–2 m. Discoloured waters may form either a continuous sheet or patches (sometimes stripes) which alternate with normal sea water areas. Discoloured waters are paralleled by sharp selection leading to oligospecific, sometimes unispecific, populations. This is due to both multiplication or concentration of the responsible organism and the death or escape of other species to uncontaminated areas.

Organisms Involved

In the majority of cases, discoloured waters are due to dinoflagellates. Many species of the following genera may be concerned: *Gonyaulax* (*G. tamarensis*, *G. breve*, etc.), *Gymnodinium*, *Glenodinium*, *Peridinium*, *Polykrikos*, *Exuviella*, *Prorocentrum* (e.g. *P. micans*), all of which may be considered to be phytoplankters. Some discoloured waters are due to the phagotrophic *Noctiluca*. *Noctiluca* blooms presumably have a quite different ecological background from other dinoflagellate blooms.

Discoloured waters due to other phytoplankters are rarer. In these cases, the organisms involved may be the coccolithophorid *Coccolithus huxleyi* (up to $115 \times 10^6\,l^{-1}$) on the southwestern coasts of Norway (BERGE, 1962) or the euglenoid *Hemieutreptia antiqua*

$(2500–10,000$ cells $ml^{-1})$ in summer in Hiroshima Bay, Japan, which led to massive fish and mollusc kills (KIMURA and co-authors, 1973). Some discoloured waters are due to chlorophytes. For example, along the Gulf of Patti coast (Sicily), GANGEMI (1973) observed two different populations of a species of the family Chlamydomonadaceae: one $(40 \times 10^6$ cells $l^{-1})$ with a blue-green pigment and a negative phototactic reaction, the other, twice as abundant, with yellow-green pigment and a positive phototactic reaction. Both populations follow diurnal cycles with maxima at about 2 p.m., and two minima at about 10 a.m. and 7 p.m.

Next to dinoflagellates, cyanophytes of the genus *Trichodesmium* are a common cause of discoloured waters: SATO and co-authors (1966) observed more than $2 \times 85 \ 10^6$ cells l^{-1} *T. erythraeum* off Recife (Brazil) during periods of high water temperatures and very high salinities, assuming that the latter caused diatom disappearance. *Trichodesmium* species may thrive in highly oligotrophic waters because they are able to fix molecular nitrogen. According to MARUMO (unpublished), two types of *Trichodesmium* bloom may occur in the tropical and subtropical Pacific Ocean: (i) *T. erythraeum* blooms limited to a surface layer of less than 1 m and arising from both *in situ* multiplication and mechanical accumulation by surface currents and wind (p. 365); (ii) *T. thiebautii*, inhabiting a water layer of several metres thick in inshore areas and originating exclusively from *in situ* multiplication. Usually associated with other organisms such as *Oscillatoria*, diatoms and even copepods, *T. thiebautii* must be considered a rather significant primary producer. The buoyancy of *Trichodesmium* species is probably due to gas vacuoles in their filaments.

Diatom blooms rarely reach the very high cell numbers characteristic of true discoloured waters; however, the abundance of some opportunist species (such as *Skeletonema costatum*) may rise up to 75×10^6 cells l^{-1}, for example, in the dilution area off the Rhône mouth (BLANC and LEVEAU, 1973). In general, such massive diatom blooms do not modify the sea water colour and thus cannot be classified as true discoloured waters. However, diatoms may participate in red tides. For example, the red tide off Ostend (Belgium) in May 1972, where the predominant organism was *Cryptomonas profunda* $(41 \times 10^6$ cells $l^{-1})$, followed by the diatom *Chaetoceros gracilis* $(6·7 \times 10^6$ cells $l^{-1})$, *Pseudopedinella pyriforme* $(5·5 \times 10^6$ cells $l^{-1})$, *Chrysochromulina* sp. and *Pyramimonas* sp. $(500,000$ cells $l^{-1})$ (MOMMAERTS and MOMMAERTS-BILLIET, 1972). In some discoloured waters off the southwestern African coast, diatoms of the genus *Thalassiosira* may sometimes be almost as abundant as dinoflagellates (PIETERSE and VAN DER POST, 1967).

A rather special type of red tide is due to the holotrich ciliate *Cyclotrichium meunieri* ($=$ *Mesodinium rubrum*) which rather frequently occurs on the coasts of New Zealand and Peru. The chlorophyll content of the ciliate is about 0·5%, i.e. very similar to that of many plankton algae and is due to a symbiotic alga (RYTHER, 1967). The latter is delimited from the ciliate cytoplasm by only a single membrane; the chloroplast structure, the presence of a compartment containing starch grains, ribosomes and a nucleomorph, as well as the mitochondrial structure, unequivocally confirm cryptophycean affinities of the symbiont (HIBBERD, 1977). During a red tide on Maine coasts (USA) MCALICE (1968) observed an abundance of 2500 ciliates l^{-1}. The ciliate was associated with an abundant population of silicoflagellates (*Distephanus speculum*) and some diatoms, dinoflagellates, copepods and cladocerans. The abundance of *C. meunieri* can attain much higher values, e.g. on Ghanian coasts during maximum summer upwelling (ANANG and co-authors, 1976) with vertical extension from 0 to 20 m of chlorophyll *a*, diatoms plus ciliates, and ciliates alone. Fig. 7-14 demonstrates that *C. meunieri* plays the role of a

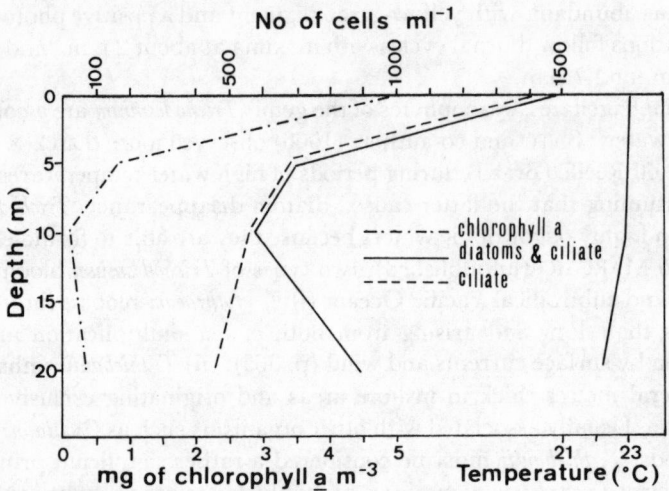

Fig. 7-14: Changes in chlorophyll *a*, temperature, numbers of *Mesodinium rubrum* cells and total cell counts from near the water surface to a depth of 20 m. (After ANANG and co-authors, 1976; reproduced by permission of the authors.)

primary producer and is positively phototactic. The red water appeared for about five days (the ciliate was not observed before and after the bloom). The bloom was preceded by a peak in the numbers of diatom cells and was followed by a second and smaller diatom peak. Changes in the nutrient salts suggest that *C. meunieri* competes with diatoms. As illustrated in Fig. 7-15, 'the lowest temperature (21·5 °C) was recorded just a few days before the development of the *Cyclotrichium* bloom' (ANANG and co-authors, 1976, p. 237 and 238). HOLM-HANSEN and co-authors (1970) assume that essential microelements or organic growth factors may be implicated in the ciliate peak. Since ANANG and co-authors recorded similar water temperatures from the surface to 20 m during and just prior to the red water, 'vertical stability of the water column might be a necessary prerequisite for the vertical migration of the ciliate.'

Lastly, some red tides must be ascribed to bacteria, mainly of the family Athiorhodaceae (*Rhodopseudomonas*), but also to *Thiopolycoccus ruber*, *Chromatium*, *Chlorobium phaeobacterioides*, *Thiocystis*, etc. Bacterial red tides essentially differ from all other phenomena classified under this term and will be considered separately (p. 367).

It must be emphasized that dinoflagellate discoloured waters are by far the most widespread, and hence most of the following pages will be devoted to them.

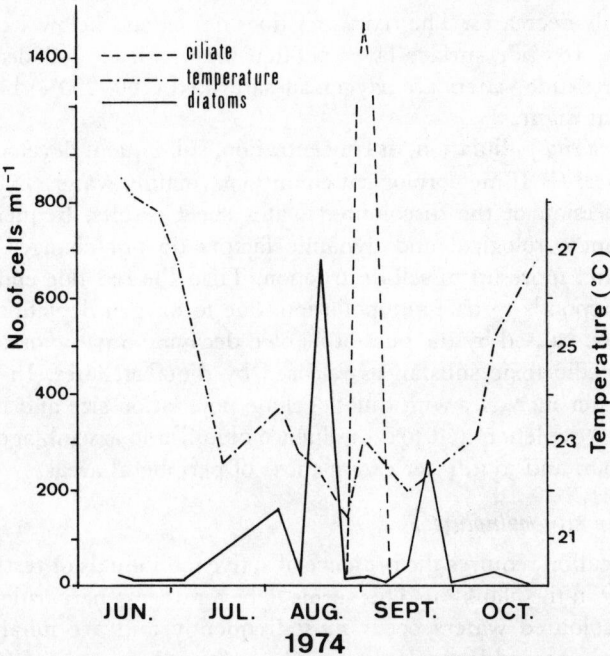

Fig. 7-15: Variations in cells counts of *Mesodinium rubrum* and
diatoms, between June and October 1974. (After ANANG
and co-authors, 1976; reproduced by permission of the
authors.)

Discoloured Waters: Dinoflagellates

Description of the phenomenon

In general discoloured waters in bays and coastal lagoons appear suddenly. Very
occasionally, some scattered patches occur before the continuous sheet becomes established. Along an open shore or offshore, discoloured waters tend to occur in streaks
parallel to the coast or to the wind direction. The streaks may be from a few decimetres
to several metres wide but often extend over many miles or even tens of miles in length.
In brackish water lagoons or semi-enclosed bays, the streaky form is uncommon. The
duration is highly variable and depends on meteorological conditions, mainly the wind
regime. When different meteorological events follow one another at short intervals, the
intensity and quality of the red tide may be more or less modified due to either partial
dispersion of the sheet or aggregation of the organism(s) responsible.

Population abundance is very variable. For example, DE SYLVA (1962) observed a
Prorocentrum micans red tide off the coast of Chile with only 20,000 cells l^{-1}. In another, off
the coast of Jamaica (STEVEN, 1966), the abundance was 29,000 cells l^{-1} (95% *Exuviella*
sp., 4% *Euglena* sp., 1% others). However, in most cases there are several million,
sometimes tens of million cells per litre. The highest populations density is restricted to

the uppermost 2–3 cm; at a depth of 50 cm the abundance generally is only about one-quarter of the average density in the 15–20 cm surface layer. Still deeper, the abundance rapidly decreases. The red water does not extend below 1 or 2 m. Primary production in the red-tide surface layer is often very high (150–200 mg C m^{-3} d^{-1}). During the day red-tide waters are oxygen sursaturated (150–200%) but the oxygen is totally depleted at night.

From massive *in situ* pullulation, or concentration, subsequent development may take one of two courses: (i) If meteorological conditions (mainly water circulation) change extensively, dispersion of the discoloured water sheet results, frequently after some sinking. (ii) If meteorological and dynamic factors do not change, the abundance increases more and more up to self-destruction. Then the red tide ends in a terminal state which corresponds to true autopollution, due to oxygen depletion and hydrogen sulphide formation caused by the bulk of its own decaying organic material and sometimes also to specific toxic substances released by dinoflagellates. In the latter case, toxicity of the water increases with dinoflagellate population size and bloom duration. Of course, oxygen depletion and toxin influence organismic assemblages in underlying waters and bottom, and to a lesser extent those of peripheral areas.

Possible causes of in situ *multiplication*

In situ multiplication requires the presence of active individuals or resting stages of the species concerned in the plankton. This seems to be a rather general rule in intertropical areas where discoloured waters occur most frequently and are inhabited by highly diversified plankton assemblages. However, *Gymnodinium breve* which frequently induces red tides on Florida coasts does not seem to be a usual component of the phytoplankton in this region; possibly *G. breve* is a dimorphic, or even polymorphic, species described under two different names in red tide and in normal plankton assemblages. Observations by HARTWELL (1975) on *Gonyaulax tamarensis* blooms in the Gulf of Maine suggest that red tides may originate from upwelling processes, which bring up dormant cysts together with nutrients into the euphotic layer. The outburst starts as soon as weather conditions sufficiently stratify and stabilize the water column. Hence, since Florida *Gymnodinium breve* red tides are initiated in distant neritic areas (10–40 nautical miles from the shore), it has been assumed that this species might have a benthic resting stage which would constitute the 'seed' for discoloured waters (STEIDINGER, 1975). In any case, initiation of discoloured water requires a 'seed' of the causative agent.

Among the factors which may promote massive multiplication of dinoflagellates, the three most frequently suspected are temperature, salinity and chemical peculiarities of the sea water with regard to non-conservative dissolved organic and inorganic (e.g. an abnormal N/P ratio) substances. We must thus distinguish between red tides (arising from *in situ* multiplication) in lagoons or semi-enclosed bays, and those along open shores.

The higher the temperature, the higher the probability of discoloured waters. Even in intertropical areas where dinoflagellate red tides are particularly common, they occur most frequently in summer. In temperate or cold areas, red tides always occur in periods of seasonal maximum temperature or in areas warmed by thermal release. High air temperatures tend to warm the most superficial sea water layer, especially if the sea is smooth, and thus increase stratification.

Effects of salinity decrease on red-tide initiation have received detailed attention. As has been known for a long time, red tides often succeed heavy rain periods. This relationship is supported by the following considerations: (i) salinities in discoloured waters are often lower than in areas with a normal plankton assemblage; (ii) many dinoflagellates exhibit salinity optima between 15 and 25%₀ S, i.e. much below the normal salinity, and are highly euryhaline.

Ecologically, temperature and salinity cannot be separately considered (Volume I: KINNE, 1970a,b, 1971; ALDERDICE, 1972). Both temperature rise and salinity decrease cause a decrease in density which tends to induce heterogeneous hydrological structure.

The most superficial water layer, and also some lenses resulting from successive peaks of river discharge or land drainage, carry both warmer and less saline waters than ambient water bodies. The most superficial layer or the lenses represent environments which are 'closed' from a hydrological point of view. In bays and lagoons, these may be considered as the initial focal point of discoloured waters. Subsequently, a red tide extends gradually, at the expense of neighbouring waters, providing the meteorological and dynamic conditions remain stable. While temperature and salinity may combine in increasing stratification or hydrological heterogeneity, if the weather keeps calm turbulence may also directly affect dinoflagellate development. Experiments by WHITE (1976) demonstrate that low-speed shaking of laboratory cultures of the red-tide-forming species *Gonyaulax excavata* causes growth inhibition compared to non-shaken cultures. Intermittent shaking, even for as little as 30 min d^{-1}, results in growth inhibition. Small-scale water turbulences, possibly even when insufficient for markedly disturbing hydrological structures, may therefore inhibit or reduce growth in some dinoflagellates.

For *in situ* multiplication, nutritional conditions are of basic importance. It is generally admitted that nutrient salt requirements are less for dinoflagellates than for diatoms. High temperatures may enhance the bacterial mineralization of decaying organic material from the bloom itself. Dinoflagellates seem to have particularly high requirements for nitrates. If we again consider the red tide observed near Kingston Harbor, Jamaica, STEVEN (1966) reports that prior to the bloom, the concentration of all nutrients (nitrates, phosphates, silica) had increased and had an extremely high N/P ratio; towards the end of the red tide, the mineral nitrogen content was very low, and so was the N/P ratio. This was recently observed in the Berre Lagoon (west of Marseilles): at the end of a small red tide, phosphates were abundant, but nitrates were completely absent. Thus it may be assumed that in bays or lagoons a high N/P ratio and massive nitrogen intake tend to trigger the red tide, whereas nitrogen depletion—even if phosphates are still available—tends to induce the red-tide fall.

Enrichment in organic matter from sewage, as well as other organic wastes, greatly increases the probability of red-tide development. For example, in the Los Angeles–Long Beach Fish Harbor, OGURI and co-authors (1975) observed red tides as the ultimate stage of organic pollution, mainly caused by waste discharge from a cannery. At the first stage, oxygen depletion in the polluted waters leads to sulphide production by bacteria. This leads to the second stage, a 'white-tide' consisting of a mixture of bacteria and colloidal sulphur. A little later, red tide begins to appear in waters peripheral to white-tide areas. Subsequently, the white tide begins to disappear while the red tide expands. In the *Gonyaulax polyedra* red-tide, the oxygen content (depleted in the white-tide water) rapidly exceeds saturation (up to 14–16 ppm) during the day and, of course, decreases at night due to the respiration of phytoplankters.

Oligodynamic substances, mainly vitamins—for instance, cyanocobalamine (vitamin B_{12})—can stimulate *Gymnodinium* growth. INGLE and DE SYLVA (1955) assume that Florida red tides may be related to vitamin B_{12} input due to bacteria from continental soils. The role of vitamins in triggering red tides requires further investigation. EVANS (1973) suggests that insufficient attention has been devoted to bacteria associated with *Gymnodinium breve*, the principal cause of Florida red tides. Since many marine bacteria produce vitamins and growth factors, EVANS postulates that the *G. breve* bloom is stimulated by prior or simultaneous growth of bacteria and that the red colour is mainly due to the presence of saprophytic bacteria supported by organic matter from fishes and invertebrates that are killed by the *G. breve* bloom. EVANS' assumption is supported by observations of OGURI and co-authors (1975) on *Gonyaulax polyedra* red tides in the Los Angeles–Long Beach Fish Harbor, where a bacterial 'white tide' precedes the beginning of the red-tide. Experiments by OGURI and co-authors revealed that the delay in the growth phase of *Gymnodinium breve* is largely reduced after the addition of naturally occurring marine bacteria to its axenic culture. Several other observations and experiments suggest that a substantial increase in the population of some marine bacteria may provide an environment which stimulates dinoflagellate growth, possibly due to the release of chemical substances.

In addition to stimulating substances brought in by fresh water, such substances may also be released from the sea bottom. In some bays, dissolution of anaerobically decomposed material from bottom mud might be one of the causes of red tides. Experiments by HONJO and HANAOKA (1972) demonstrate that the flagellate *Heterosigma* sp. cannot develop without the addition of 'supernatant mud'. In Ise Bay (Japan), the growth-promoting effect of mud collected in October turned out to be about twice as high as that of mud collected from May to August (UYENO and NAGAI, 1973). In general, it appears that anoxic bottom water may be closely related to red-tide occurrence.

According to IWASAKI (1973) and HONJO (1974) growth promoting substances may be classified in two main groups: (i) heavy metals such as Fe and Mn, mainly in species of the genera *Eutreptiella* and *Exuviella*; and (ii) organic substances such as purines and pyrimidines and, more generally, organic products from protein decomposition (amines, aminoacids), for instance, in *Heterosigma inlandica, Peridinium hangoei, Gymnodinium nelsoni, Polykrikos schwartzi* and *Exuviella* sp. . Red tides may occur due to a leakage of substances through discontinuity layers or to the direct supply of such substances from the bottom mud by turbulent processes.

In summary, all the factors considered above may play some part in the development of dinoflagellate discoloured waters in bays and lagoons. Our present knowledge is insufficient for a detailed analysis of their respective effects and influences. These presumably depend on area, season and stage of bloom development.

In situ multiplication combined with concentration by dynamic processes may also be observed outside lagoons and semi-enclosed bays, i.e. at some distance from the shore. In the southwestern approaches to the English Channel (Fig. 7-16) PINGREE and co-authors (1975) observed surface temperature discontinuities (about 1·5 °C) in summer, which could be followed up to 100 nautical miles. They deliminated the transition between well stratified shelf water and well mixed coastal water. In the stratified region

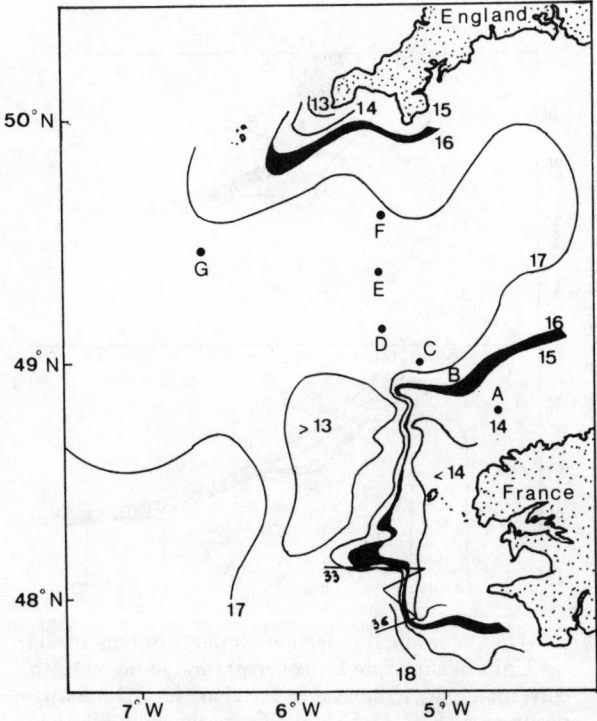

Fig. 7-16: Sea-surface temperature contours between July 26 and July 30, 1976. The 15–16 °C contours define shelf front or outcropping of seasonal thermocline at the sea surface. (After PINGREE and co-authors, 1975; reproduced by permission of Macmillan Journals Ltd., London.)

where primary production in the surface layer was restricted by low levels of inorganic nutrients, PINGREE and co-authors identified three different mixing regimes: (i) turbulence in the top mixed layer, derived from wind stress and night-time convection; (ii) turbulence in the bottom layer, derived mainly from tides but also from swell near the shore; and (iii) the thermocline prevailing between the two mixed layers when the wind above the tide can no longer stir the excess surface heating downwards (PINGREE and co-authors, 1975, p. 674). In the frontal region of recently stabilized mixed water, the combination of high nutrients and a shallow upper mixed layer creates conditions suitable for rapid phytoplankton growth. Counts of a representative sample (34 mg m^{-3} chlorophyll a) yielded the following values (1 × 10^6 cells l^{-1}): *Gyrodinium aureolum*, 1·7; *Prorocentrum micans*, 0·01; flagellates, 1·4; all other photosynthetic cells, 0·005. Chlorophyll a contents of up to 100 mg m^{-3} were observed (Fig. 7-17). Such a composition and abundance correspond to a red-tide phenomenon. A week later, PINGREE and co-authors observed the typical streaks. Continual series of blooms may result from interaction of weather and tide. Boundaries are sites of phytoplankton blooms which, if environmental conditions are suitable, may develop into red tides.

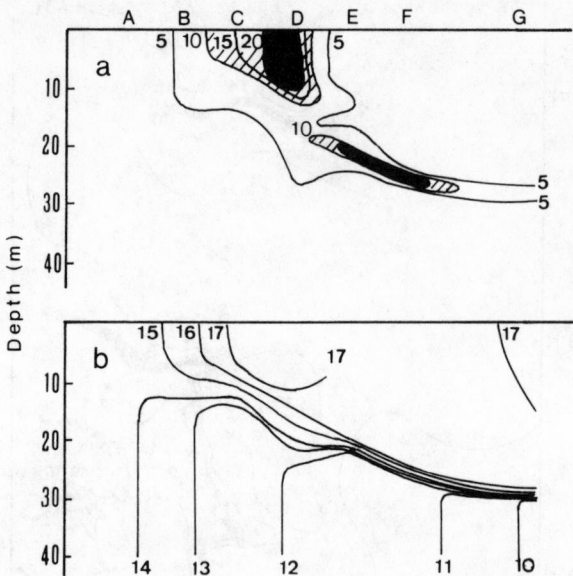

Fig. 7-17: Chlorophyll *a* section through frontal region. (a) Units of chlorophyll *a* concentration in mg m^{-3}; (b) corresponding temperature section in °C. Station positions A, B, C, D, E, F and G are shown in Fig. 7-16. (After PINGREE and co-authors, 1975; reproduced by permission of Macmillan Journals Ltd., London.)

The oligospecific and sometimes almost unispecific composition of many dinoflagellate red tides remains unexplained. Since dinoflagellates exhibit greater swimming speeds than most other phytoplankters, diurnal vertical migration was proposed as a mechanism for buildup of a near unialgal population of *Gonyaulax tamarensis* in some estuaries. Several authors assumed that reduction in vertical mixing rates occurring in areas where discoloured waters start and develop quickly may induce such motile organisms as dinoflagellates to concentrate in preferred layers with optimum light conditions for multiplication. Population maxima in the most superficial layer receiving maximum illumination is surprising. In general, optimum illumination for photosynthesis corresponds to about one-third of the light energy passing across the atmosphere–sea interface. Red-tide dinoflagellates, like hyponeuston algae (p. 321), may require, or at least tolerate, very strong illumination.

In *Gonyaulax polyedra* red tides along the coast of California, red patches generally appear in mid-morning. This suggests morning migration or aggregation of the algae (SMITH, 1975). This observation, together with the fact that *G. polyedra* red tides are associated with a stratified hydrological structure (a nutrient-depleted, shallow mixed layer above a shallow thermocline and nutrient-rich waters below) suggests that diurnal vertical migration of *G. polyedra* may play some part in the red-tide development. Locomotion allows dinoflagellates to outcompete such immotile organisms as diatoms (BLASCO, 1975). Experiments indicate that *G. polyedra* is able to assimilate nutrients at night and does not exhibit repression of nitrate reductase by ammonium. Species which do not have the same physiological peculiarities—circadian migration and permanent nitrate reductase activity—might be eliminated (EPPLEY and HARRISON, 1975). It has also been assumed that some dinoflagellates do not require nitrate but can utilize ammonium or urea as nitrogen source.

Other factors may contribute to oligo- or even unispecificity. (i) Nitrate exhaustion by dinoflagellates. Many dinoflagellates require extremely high amounts of nitrogen nutrients and may prefer organic compounds or those available from recycling. Hence it may be hypothesized that dinoflagellates which develop slowly in a rather large range of nutrient conditions exhibit sharp growth peaks only under a small range of conditions and then rapidly exhaust the N supply to the detriment of other co-existing species. (ii) Release of stimulating or inhibiting circumstances. (iii) Oxygen depletion at night in the discoloured waters, or permanently in underlying layers. Factors (ii) and (iii) may contribute to zooplankton elimination from discoloured waters. This results in decreased grazing pressure and progressive red-tide algal predominance, either because these algae are not consumed as food or because their density is far too high (depression of zooplankton filtering activity).

Grazing suppression will enhance algal abundance and reduce zooplankters and microphages as well as carnivores which feed upon the latter.

In situ *concentration by dynamic processes*

RYTHER (1955) was the first author who suggested that some red tides may originate either from changes in the phytoplankters' physiology or from dynamic processes. RYTHER supported his hypothesis by the following arguments. (i) Since the phytoplankters generally require moderate illumination, their abundance in the uppermost layer involves a deep change in their photosynthetic mechanism or puts them in a 'self-defence state'. (ii) At some stages of their life cycle, a variety of dinoflagellates can store lipid inclusions—often yellow or red in colour—which increase the buoyancy. (iii) Massive concentration of discoloured water organisms in narrow streaks may originate from a discontinuity between upwelled and coastal waters. Casual transport shorewards of such streaks by meteorological events sometimes results in massive kills of fish and benthos assemblages.

Offshore discoloured waters in parallel streaks—the responsible organisms of which are generally cyanophytes (p. 357)—arise from wind stress. Under gentle to fresh breezes (3–5 on the Beaufort scale, i.e. $3.4–10.7$ m s^{-1}), windrows oriented in the direction of the wind appear; these wind-generated convection cells—'Langmuir circulation'—force lighter plankters as well as foam, lighter debris, and *Sargassum* when they exist to concentrate along the convergence line between two adjacent vortices. Spacing of windrows is proportional to wind speed, varying from about 20 m for 4 m s^{-1} wind to about 50 m for 12 m s^{-1}; it was also demonstrated that the shallower the thermocline, the larger the windrow spacing.

Some exceptional meteorological events may sometimes result in discoloured waters. HARTWELL (1975), for example, noticed that in the Gulf of Maine in 1972 a *Gonyaulax tamarensis* red tide arose from a bloom of this species pushed across the gulf from eastern Canadian coasts by northern winds preceding the hurricane 'Carrie'. Offshore winds and subsequent upwelling then brought the dinoflagellates into coastal waters where the bloom ensued. FUNG and TROTT (1973) observed a *Noctiluca scintillans* red tide off Hong Kong, which immediately followed the passing of a typhoon. They identified the cause as convergence associated with high-velocity typhoon winds.

Influence of Dinoflagellate Discoloured Waters on Adjacent Assemblages

Providing the population density of red-tide organisms is not too high, it may enhance zooplankton production. Off the Chilean coasts, DE SYLVA (1962) noticed that many zooplankters, such as meroplanktonic larvae, copepods, euphausiids and medusae, fed largely on a *Prorocentrum micans* red tide (20,000 cells l^{-1}). However, as soon as the population density increases, mainly in lagoons, bays and coastal areas, the harmful influence of discoloured waters becomes conspicuous. As mentioned above, plankters inhabiting waters in which a red tide develops are eliminated. In adjacent areas, the plankton assemblage may retain some diversity and abundance as long as the putrefaction arising from the bulk of decaying organic matter does not spread too far. As the red tide develops in population density and occupied area, the consequences rapidly become catastrophic.

Oxygen depletion and sometimes hydrogen sulphide formation due to microbial decomposition of excessive amounts of organic material from red-tide organisms induce massive kills of benthos invertebrates and fishes. Decaying albuminoid substances make the water increasingly viscous, sometimes attaining the consistency of a jelly which, together with the dinoflagellate cells, clogs the gills of fishes and crustaceans as well as the filtering devices of such filter feeders as pelecypods, ascidians, etc.

Since neither pelecypods of the genus *Donax* which feed on red-tide dinoflagellates nor marine birds which feed on *Donax* were injured, PIETERSE and VAN DER POST (1967), who studied red tides at Walfish Bay, South Africa, assumed that only anoxia and gill clogging lead to the massive fish kills commonly observed in this area. Consequently, they attributed the death of some marine birds to bacteria in stranded putrefied marine animals, but denied any effect related to dinoflagellate toxin. This interpretation is no longer tenable. In May and June 1958 off the coast of Northumberland (Great Britain) a *Gonyaulax tamarensis* (associated with *Prorocentrum micans*) red tide was studied in detail. As soon as the first warnings of danger were recorded—for example, poisoning of people after mussel consumption and increasing numbers of dead marine birds (cormorants and sterns)—ecological investigations revealed important mortalities in the populations of many pelecypods (*Venus striatula, Cardium edule, Macoma baltica, Lucina borealis*) as well as in the sand eel (*Ammodytes*). However, neither *Mytilus edulis* and *Pecten maximus* which contained large amounts of dinoflagellate toxin nor predatory gastropods which feed on bivalves died. The death of piscivorous marine birds may have been due to sand eel consumption. The dinoflagellate toxin can apparently be transmitted along a food chain: dinoflagellate → herbivorous copepod → *Ammodytes* → birds. Of course, since most edible pelecypods—mainly mussels, oysters and various clams—may accumulate the 'paralytic shellfish poison' without any damage to themselves, they can be dangerous for seafood consumers. Cases of toxic plankton-feeding fishes (clupeids and engraulids) were observed in tropical regions. Even carnivorous fishes which feed on these filter feeders may be poisoned.

It is clear that red tides may cause severe damage to shallow-water fisheries and marine farming projects. The effects on fisheries have already been mentioned, especially the escape of fishes from the contaminated areas. However, where red-tide areas are intermixed with uncontaminated areas, some pelagic fishes may aggregate in the latter, thus becoming more available to fishermen. When the putrid jelly formed at the end of a red tide falls to the bottom, the normally benthic Indian shelf flatfish *Cynoglossus fasciatus* swims above the bottom and thus may be fished with floating nets instead of trawls. When a red tide is dispersed by winds blowing towards the shore, aerosols containing the toxin may provoke bronchial and lung diseases in coastal human populations.

Not all dinoflagellates responsible for red tides elaborate paralytic shellfish poison. However, toxic species are numerous and the chemical nature of the paralytic shellfish poison—still insufficiently investigated—may be different in different species. Several species (*Gymnodinium breve*, for example) elaborate two different toxins. However, it seems that all dinoflagellate poisons are neurotoxins.

Up to the time of writing, it has been neither possible to prevent the start of a red tide nor to stop it. The possibility of biocontrol was recently suggested by KUTT and MARTIN (1975) who noticed that the cyanophyte *Gomphosphaeria aponina* elaborates a toxin which is lethal to the Florida red-tide organisms *Gymnodinium breve*.

Discoloured waters: Bacteria

A few examples of bacteria responsible for discoloured waters have been mentioned on p. 358. Bacterial discoloured waters have been less thoroughly investigated than those caused by dinoflagellates. Their development differs in two main features: (i) The phenomenon always begins near the bottom or at least in subsurface layers (never in the most superficial layers). (ii) Oxygen depletion and hydrogen sulphide formation precede (not follow) discoloured waters. The characteristics common to dinoflagellate and bacterial discoloured waters are: (i) both develop in stratified and dystrophic areas (lagoons or shallow water areas with very poor circulation), and (ii) both occur mainly in summer. This latter parallelism arises only from the fact that the higher the temperature, the stronger the bacterial activity which depletes the oxygen content.

Bacterial discoloured waters are in fact the ultimate stage of a progressive dystrophic process arising from the excessive accumulation of decaying matter and subsequent oxygen depletion. The bulk of organic material may originate either from a red-tide process of the dinoflagellate type or from exogenous—usually man-made—massive input. The two successive discoloured water processes described (p. 361) by OGURI and co-authors (1975) demonstrates that under the same conditions of temperature, salinity, circulation, etc, bacterial discoloured waters arise from organic matter accumulation and concomitant oxygen depletion, whereas the subsequent dinoflagellate red tides arise from enrichment of nutrients and growth substances, mostly originating from the preceding stage, and resulting finally in a new phase of organic matter accumulation and oxygen depletion.

Literature Cited (Chapter 7)

ALDERDICE, D. F. (1972). Responses of marine poikilotherms to environmental factors acting in concert. In O. Kinne (Ed.), *Marine Ecology*, Vol. I, Environmental Factors, Part 3. Wiley, London. pp. 1659–722.

ANANG, E. R., OBENG-ASAMOA, E. K. and JOHN, D. M. (1976). Observations on the red-water ciliate *Mesodinium rubrum* Lohmann in Ghanaian coastal water. *Bull. Inst. fr. Afr. noire (Ser. A)*, **38**, 235–40.

BACESCU, M. (1972). Substratum: animals. In O. Kinne (Ed.), *Marine Ecology*, Vol. I, Environmental Factors, Part 3. Wiley, London. pp. 1291–313.

BERGE, K. (1962). Discoloration of the sea due to *Coccalithus huxleyi* bloom. *Sarsia*, **6**, 27-40.

BEZDEK, H. F. and CARLUCCI, A. F. (1972). Surface concentration of marine bacteria. *Limnol. Oceanogr.*, **17**, 566–9.

BLANC, F. and LEVEAU, M. (1973). *Plancton et Eutrophie: Aire d'Épandage Rhodanienne et Golfe de Fos*, Thèse Doct., Université Aix-Marseille (II).

BLASCO, D. (1975). Red tides in the upwelling regions. In V. R. Lo Cicero (Ed.), *Proc. 1st Int. Conf. on Toxic Dinoflagellate Blooms*. Massachussetts Science and Technology Foundation, Wakefield. pp. 113–19.

BROCKMANN, U. H., KATTNER, G., HENTSCHEL, G., WANDSCHNEIDER, K., JUNGE, H. D. and HÜHNERFUSS, H. (1976). Natürliche Oberflächenfilme im Seegebiet vor Sylt. *Mar. Biol.*, **36**, 135–46.

CARPENTER, E. J. (1972). Nitrogen fixation by a blue-green epiphyte on a pelagic *Sargassum*. *Science, N.Y.*, **178**, 1207–9.

CARPENTER, E. J. and COX, J. L. (1974). Production of pelagic *Sargassum* and a blue-green epiphyte in the western Sargasso Sea. *Limnol. Oceanogr.*, **19**, 429–35.

CHAMPALBERT, G. (1975). *Répartition du Peuplement Animal de l'Hyponeuston Étude Expérimentale de la Physiologie et du Comportement des Pontellidés*, Thèse Doct., Université Aix-Marseille (II).

CHAMPALBERT, G. and MACQUART, C. (1970. Les Péracarides de l'hyponeuston nocturne du Golfe de Marseille. *Cah. Biol. mar.*, **11**, 1–29.

CHENG, L. (1975). Marine pleuston animals at the sea–air interface. *Oceanogr. mar. Biol. A. Rev.*, **13**, 181–212.

CHENG, L. and LEWIN, R. A. (1975). Flatworms afloat. *Nature, Lond.*, **288**, 518–19.

CONOVER, J. T. and SIEBURTH, J. McN. (1964). Slicks associated with *Trichodesmium* blooms in the Sargasso Sea. *Nature, Lond.*, **205**, 830–1.

CREUTZBERG, F. (1975). Orientation in space: animals. Invertebrates. In O. Kinne (Ed.), *Marine Ecology*, Vol. II, Physiological Mechanisms, Part 2. Wiley, London. pp. 555–655.

CULLINEY, J. L. (1970). Measurements of reactive phosporus associated with Pelagic *Sargassum* in the northwest Sargasso Sea. *Limnol. Oceanogr.*, **15**, 304–6.

DAUMAS, R. A. (1974). Influence de la température et du développement planctonique sur le mécanisme d'accumulation de la matière organique à la surface de la mer. *Mar. Biol.*, **26**, 111–16.

DE SYLVA, D. P. (1962). Red-water blooms of northern Chile, April–May 1956, with reference to the ecology of the swordfish and the striped marlin. *Pacif. Sci.*, **16**, 271–9.

DOOLEY, J. K. (1972). Fishes associated with the pelagic *Sargassum* complex, with a discussion of the Sargassum community. *Contrib. mar. Sci.*, **16**, 1–32.

EPPLEY, E. W. and HARRISON, W. G. (1975). Physiological ecology of *Gonyaulax polyedra*, a red-water dinoflagellate of southern California. In V. R. Lo Cicero (Ed.), *Proc. 1st Int. Conf. on Toxic Dinoflagellate Blooms*. Massachussetts Science and Technology Foundation, Wakefield. pp. 113–19.

EVANS, E. E. (1973). The role of bacteria in the Florida red tide. *Environ. Lett.*, **5**, 37–44.

FINE, M. L. (1970). Faunal variation on pelagic *Sargassum*. *Mar. Biol.*, **7**, 112–22.

FLÜGEL, H. (1972). Pressure: animals. In O. Kinne (Ed.), *Marine Ecology*, Vol. I, Environmental Factors, Part 3. Wiley, London. pp. 1407–50.

FUNG, Y. C. and TROTT, L. R. (1973). The occurrence of a *Noctiluca scintillans* (Mc Cartney) induced red tide in Hong Kong. *Limnol. Oceanogr.*, **18**, 472–6.

GANGEMI, G. (1973). Appearance of 'red waters' caused by Volvocales along the coast of the Gulf of Patti (Messina). Preliminary note. In S. Genovese (Ed.), *Proc. 5th Int. Colloq. on Medical Oceanography.* Ellebi Publications, Messina, Sicilia. pp. 475–86.

GARRETT, W. D. (1967). Dumping of capillary waves at the air–sea surface by oceanic surface-active material. *J. mar. Res.*, **25**, 279–91.

GERLACH, M. (1972). Substratum: introduction. In O. Kinne (Ed.), *Marine Ecology*, Vol. I, Environmental Factors, Part 3. Wiley, London. pp. 1245–50.

GRICE, G. D. and GIBSON, V. R. (1975). Occurrence, viability and significance of resting eggs of the calanoid copepod *Labidocera aestiva. Mar. Biol.*, **31**, 335–337.

GUNKEL, W. (1970). Light: bacteria, fungi and blue-green algae. In O. Kinne (Ed.), *Marine Ecology*, Vol. I, Environmental Factor, Part 1. Wiley, London. pp. 103–24.

HARDY, J. T. (1973). Phytoneuston ecology of a temperate marine lagoon. *Limnol. Oceanogr.*, **18**, 525–33.

HARTWELL, A. D. (1975). Hydrographic factors affecting the distribution and movement of toxic dinoflagellates in the western Gulf of Maine. In V. R. Lo Cicero (Ed.), *Proc. 1st Int. Conf. on Toxic Dinoflagellate Blooms.* Massachussetts Science and Technology Foundation, Wakefield. pp. 47–68.

HARVEY, G. W. (1975). Marine microbial ecology of the sea surface microlayer and nearshore waters. In B. Morton (Ed.), *Proc. Spec. Symp. on Marine Science* (1973). Pacific Science Association, Hong Kong. pp. 104–10.

HEINRICH, A. K. (1960). The plankton of surface waters in the central part of the Pacific Ocean. (Russ.) *Trudy Inst. Okeanol.*, **41**, 42–7.

HEINRICH, A. K. (1971). On the near suface plankton of the eastern South Pacific Ocean. *Mar. Biol.*, **10**, 290–4.

HEINRICH, A. K. (1974). On neuston pontellids (Pontellidae, Copepoda) of the southern Atlantic. (Russ.) *Trudy Inst. Okeanol.*, **98**, 43–50.

HEMPEL, G. and WEIKERT, H. (1972). The neuston of the subtropical and boreal northeastern Atlantic Ocean. A review. *Mar. Biol.*, **13**, 70–88.

HERRING, P. J. (1967). The pigments of plankton at the sea surface. *Symp. zool. Soc. Lond.*, **19**, 215–235.

HIBBERD, J. D. (1977). Observations on the ultrastructure of the cryptomonad endosymbiont of the red water ciliate *Mesodinium rubrum. J. mar. biol. Ass. U.K.*, **57**, 45–61.

HOLLIDAY, F. G. T. (1971). Salinity: animals. Fishes. In O. Kinne (Ed.), *Marine Ecology*, Vol. I, Environmental Factors, Part 2. Wiley, London. pp. 997–1033.

HOLM-HANSEN, O., TAYLOR, F. J. R. and BARSDATE, R. J. (1970). A ciliate red tide at Barrow, Alaska. *Mar. Biol.*, **7**, 37–46.

HONJO, T. (1974). Studies on the mechanism of red tide occurrence in Hakata Bay. 4: Environmental conditions during the blooming season and essential factors of red tide occurrence. *Bull. Hokkaido reg. Fish. Res. Lab.*, **79**, 77–84.

HONJO, T. and HANAOKA, T. (1972). Studies on the mechanism of red tide occurrences in Hakata Bay. I. On the regional distribution of organic matter in bottom mud. *Sci. Bull. Fac. Agric. Kyushu Univ.*, **26**, 191–216.

HOWARD, K. L. and MENZIES, R. J. (1969). Distribution and production of *Sargassum* in the waters off the Carolina coast. *Botanica Mar.*, **12**, 244–54.

INGLE, R. M. and DE SYLVA, D. P. (1955). *The Red Tide*, Education Series 1, Florida Board of Conservation.

IWASAKI, H. (1973). The physiological characteristics of neritic red tide dinoflagellates. *Bull. Plankton Soc. Japan*, **19**, 104–14.

JOHN, H. Ch. (1973). Surface living ichthyoplankton in the Canary Current. *'Meteor' Forschungsergeb.*, D, **15**, 36–50.

KIMURA, T., MIZOKAMI, A. and HASHIMOTO, T. (1973). The red tide that caused severe damage to the fishery resource in Hiroshima Bay: outline of its occurrence and the environmental conditions. *Bull. Plankton Soc. Japan*, **19**, 82–112.

KINNE, O. (1970a). Temperature: general Introduction. In O. Kinne (Ed.), *Marine Ecology*, Vol. I, Environmental Factors, Part 1. Wiley, London. pp. 321–46.

KINNE, O. (1970b). Temperature: animals. Invertebrates. In O. Kinne (Ed.), *Marine Ecology*, Vol. I, Environmental Factors, Part 1. Wiley, London. pp. 407–514.

KINNE, O. (1971). Salinity: animals. Invertebrates. In O. Kinne (Ed.), *Marine Ecology*, Vol. I, Environmental Factors, Part 2. Wiley, London. pp. 821–95.

KINNE, O. (1972). Pressure: general introduction. In O. Kinne (Ed.), *Marine Ecology*, Vol. I, Environmental Factors, Part 3. Wiley London. pp. 1323–60.

KREY, J. (1961). The balance between living and dead matter in the oceans. In *Oceanography: Invited Lectures Presented at the International Oceanographic Congress* (New York, 1959). American Association for Advancement of Science, Washington, D.C. pp. 539 ff.

KUTT, E. C. and MARTN, D. F. (1975). Report on a biochemical red tide repressive agent. *Environ. Lett.*, **9**, 195–208.

LAFOND, E. C. and LAFOND, K. G. (1972). Sea surface features. *J. mar. biol. Ass. India*, **14**, 1–14.

LARSSON, K., ODHAM, G. and SÖDERGREN, A. (1974). On lipid surface films on the sea. I. A simple method for sampling and studies of composition. *Mar. Chem.*, **2**, 49–57.

LILLELUND, K. and LASKER, R. (1971). Laboratory studies of predation by marine copepods on fish larvae. *Fish. Bull. U.S.*, **69**, 655–67.

MCALICE, B. J. (1968). An occurrence of a ciliate Red Water in the Gulf of Maine. *J. Fish. Res. Bd Can.*, **25**, 1749–51.

MCINTYRE, F. J. (1968). Bubbles. A boundary-layer 'microtome' for micron-thick samples of a liquid surface. *J. phys. Chem., Wash.*, **72**, 589–92.

MAKKAVEEVA, E. D. (1965). Biocoenosis of the Sargasso 'algae' in the Red Sea. (Russ.) *Benthos K.*, **1965**, 81–3.

MARUMO, R., TAGA, N. and NAKAI, T. (1971). Neustonic bacteria and phytoplankton in surface microlayers of the equatorial waters. *Bull. Plankton Soc. Japan*, **18**, 36–41.

MAYNARD, N. G. (1968). Aquatic foams as an ecological habitat. *Z. allg. Mikrobiol.*, **8**, 119–26.

MOGUILEVSKY, A. and ANGEL, M. V. (1975). Halocyprid ostracods in Atlantic neuston. *Mar. Biol.*, **32**, 295–302.

MOMMAERTS, J. P. and MOMMAERTS BILLIET, F. (1972). Observation d'une eau rouge dan le port d'Ostende. *Bull. Inst. r. Sci. nat. Belg.*, **48**, 1–3.

MOROZOVSKAYA, O. I. (1966). Zoogeographical affinities of Tintinnids of the surface layer of the Black Sea. (Russ.) In *Tezisy dokl. Chetvertoi mezhvuzovskoi zoogeograficheskoi konferentsii*. Odessa.

MOROZOVSKAYA, O. I. (1969). Composition and distribution of infusorians of the Tintinnoidea in the Black Sea. (Russ.) In *Biologicheski problemy okeanografii yuzhnykmorei*. Naukova Dumka, Kiev.

MORRIS, R. J. (1974). Lipid composition of surface films and zooplankton from the eastern Mediterranean. *Mar. Pollut. Bull., N.S.*, **5**, 105–9.

NESTEROVA, D. A. and POLISHCHUK, L. N. (1975). Phyto- and zooplankton relationship in the surface microlayer of the Black Sea. *Gidrobiol. Zh.*, **11**, 20–5.

NISHIZAWA, S. (1971a). Concentration of organic and inorganic material in the surface skin at the equator, 155° W. *Bull. Plankton Soc. Japan*, **18**, 42–4.

NISHIZAWA, S. (1971b). Concentration of particulate and dissolved organic material at the sea surface skin. In *The World Ocean: Proc. Joint Oceanographic Ass.* Tokyo. pp. 267–70.

OGURI, M., SOULE, D., JUGE, D. M. and ABBOTT, B. C. (1975). In V. R. Lo Cicero (Ed.), *Proc. 1st Int. Conf. on Toxic Dinoflagellate Blooms*. Massachussetts Science and Technology Foundation, Wakefield. pp. 41–6.

PARKER, B. and BARSOM, G. (1970). Biological and chemical significance of surface microlayers in aquatic ecosystems. *BioScience*, **20**, 87–93.

PIETERSE, F. and VAN DER POST, D. C. (1967). The pilchard of southwest Africa (*Sardinops ocellata*). Oceanographical conditions associated with red tides and fish mortalities in the Walvis Bay region. *Investl Rep. mar. Res. Lab. S.W. Afr.*, **14**, 125.

PINGREE, R. D., PUGH, P. R., HOLLIGAN, P. M. and FORSTER, G. R. (1975). Summer phytoplankton blooms and red tides along tidal fronts in the approaches to the English Channel. *Nature, Lond.*, **258**, 672–7.

PIOTROWICZ, S. R., RAY, B. J., HOFFMAN, G. L. and DUCE, R. A. (1972). Trace metal enrichment in the sea-surface microlayer. *J. geophys. Res.*, **77**, 5243–54.

RYLAND, J. S. (1974). Observations on some epibionts of gulf-weed *Sargassum natans* (L.) Meyen. *J. exp. mar. Biol. Ecol.*, **14**, 17–25.

RYTHER, J. H. (1955). Ecology of autotrophic marine dinoflagellates with references to red water conditions. In F. H. Johnson (Ed.), *The Luminescence of Biological Systems*. American Association for the Advancement of Science, Washington, D.C. pp. 387–414.

RYTHER, J. H. (1967). Occurrence of red water off Peru. *Nature, Lond.*, **214**, 1318–19.

SATO, S., NOGUEIRA, M., PARANAGUA, E. and ESKINAZI, E. (1966). On the mechanism of red tide of *Trichodesmium* in Recife, northeastern Brazil, with some considerations of the relation to the human disease 'Tamandare fever'. *Trav. Inst. Oceanogr. Univ. Recife*, **1966**, 7–50.

SAVILOV, A. I. (1958a). Pleuston of the western part of the Pacific Ocean. (Russ.) *Dokl. Akad. Nauk SSSR.*, **122**, 1014–1017.

SAVILOV, A. I. (1958b). Pleuston biocoenosis of the Pacific Ocean. In *Abstract Papers from the 1st Int. Oceanographic Congr.*, New York. p. 322.

SAVILOV, A. I. (1961). Extension of the ecological forms of *Velella lata*. Cham. and Eysen and *Physalia utriculus* (La Martinière) in the northern part of the Pacific Ocean. (Russ.) *Trudy Inst. Okeanol.*, **45**, 223–238.

SAVILOV, A. I. (1967). Pleuston of the Pacific Ocean. *2nd Int. oceanogr. Congr.* (Abstr. papers), **2**, 316.

SAZHINA, L. I. (1968). On hibernating eggs of marine Calanoida. (Russ.) *Zool. Zh.*, **47**, 1554–6.

SCHLICHTING, H. E., Jr. (1972). Seafoam algae and Protozoa. *J. Elisha Mitchell scient. Soc.*, **88**, 186–7.

SCHÖNE, H. (1975). Orientation in space: animals. General Introduction. In O. Kinne (Ed.), *Marine Ecology*, Vol. II, Physiological Mechanisms, Part 2. Wiley, London. pp. 449–553.

SHERMAN, K. (1963). Pontellid copepod distribution in relation to surface water type in the central North Pacific. *Limnol. Oceanogr.*, **8**, 214–27.

SHERMAN, K. (1964). Pontellid copepod occurrence in the central South Pacific. *Limnol. Oceanogr.*, **9**, 476–484.

SKOPINTSEV, B. A. (1939). Organic surface-active compounds in sea water. (Russ.) *Meteorologiya Gidrol.*, **2**, 75.

SMITH, G. B. (1975). The 1971 red tide and its impact on certain reef communities in the mid-eastern Gulf of Mexico. *Environ. Lett.*, **9**, 141–52.

SMITH, K. L., Jr. (1973). Energy transformations by the *Sargassum* fish *Histrio-histrio* (L.), *J. exp. mar. Biol. Ecol.*, **12**, 219–27.

SMITH, K. L., BURNS, K. A. and CARPENTER, E. J. (1973). Respiration of the pelagic *Sargassum* community. *Deep Sea Res.*, **20**, 213–17.

STEIDINGER, K. A. (1975). Implications of dinoflagellate life cycles on initiation of *Gymnodinium breve* red tide. *Environ. Lett.*, **9**, 129–39.

STEVEN, D. M. (1966). Characteristics of a red-water bloom in Kingston Harbor, Jamaica, W. I. *J. mar. Res.*, **24**, 113–23.

TAGUCHI, S. and NAKAJIMA, K. (1971). Plankton and seston in the sea surface of three inlets of Japan. *Bull. Plankton Soc. Japan*, **18**, 20–36.

TESCH, F.-W. (1975). Orientation in space: animals. Fishes. In O. Kinne (Ed.), *Marine Ecology*, Vol. II, Physiological Mechanisms, Part 2. Wiley, London. pp. 657–707.

TSYBAN, A. V. (1971). Foam as an ecological niche for bacteria. (Russ.) *Gidrobiol. Zh.*, **7**, 14–24.

UYENO, F. and NAGAI, K. (1973). Seasonal change of growth promoting effect of mud extracts and sea water collected during various seasons at the Ise Bay on a red tide flagellate *Heterosigma inlandica* Hada. *Bull. Plankton Soc. Japan*, **19**, 97–102.

VORONINA, N. M. (1962). Plankton of the surface waters of the Indian Ocean. (Russ.) *Trudy Inst. Okeanol.*, **58**, 67–79.

WEIKERT, H. (1972). Verteilung und Tagesperiodik des Evertebratenneuston im subtropischen Nordostatlantik während der Atlantischen Kuppenfahrten 1967 von F. S. 'Meteor'. *'Meteor' Forschungsergeb*. D, **11**, 29–87.

WEIKERT, H. (1973). Zur Ökologie der Pontellidae (Copepoda, Calanoida) im subtropischen Nordostatlantik. *'Meteor' Forschungsergeb*. D, **16**, 42–59.

WEIKERT, H. (1975). Distribution and occurrence of pontellids (Copepoda, Calanoida) in the central and South Atlantic Ocean. *Ber. dt. wiss. Kommn Meerefors*, **24**, 134–50.

WHEELER, J. R. (1972). Some effects of solar levels of ultraviolet radiation on lipids in artificial sea water. *J. geophys. Res.*, **77**, 5302–6.

WHEELER, J. R. (1975). Formation and collapse of surface films. *Limnol. Oceanogr.*, **20**, 338–42.

WHITE, A. W. (1976). Growth inhibition caused by turbulence in the toxic marine dinoflagellate *Gonyaulax excavata. J. Fish. Res. BdCan.*, **33**, 2598–2602.

YEATMAN, H. C. (1962). The problem of dispersal of marine littoral copepods in the Atlantic Ocean, including some redescriptions of species. *Crustaceana*, **4**, 253–72.

ZAITSEV, Yu. P. (1964). Documents on the Black Sea hyponeuston. (Russ.) *Odessk. gosud. Univ. Odessa, Autoreferat.*, **1964**, 1–22.

ZAITSEV, Yu. P. (1968). La neustologie marine: objet, méthodes, réalisations principales et problèmes. *Pelagos*, **8**, 1–47.

ZAITSEV, Yu. P. (1970). *Morskaya Neustonologiya*, Naukova Dumka, Kiev.

ZAITSEV, Yu. P. (1971). *Marine neustonology*, Israel Program for Scientific Translations, Jerusalem.

Marine Ecology Vol. 5, Part 1
Edited by Otto Kinne
© 1982 John Wiley & Sons Ltd

8. MAJOR BENTHIC ASSEMBLAGES

J. M. PERES

Following the general concept proposed on p. 11, this chapter considers the vertical zonation of benthic assemblages beginning with the shallowest and ending with the deepest waters investigated. Within each vertical zone, I first deal with hard and then with soft substrate assemblages. My treatment is mainly based on the benthal of the northeastern Atlantic Ocean and the Mediterranean Sea, which have been far better investigated than other ocean regions. Fortunately, in each vertical zone the assemblages of different biogeographic regions of the world ocean are similar, providing the major substrate characteristics are the same. This fact facilitates the worldwide classification of most biocoenoses and communities, employing the concept of assemblage and isoassemblages. Whether the authors used qualitative or quantitative methods for assessing the samples they considered representative of an ecological entity, I have attempted to combine the biocoenosis or community considered into a single organismic assemblage wherever I found this sufficiently backed up by the information available.

(1) Supralittoral Zone

(a) Hard Substrates

On rocky bottoms of the supralittoral zone, organismic assemblages exhibit a rather striking similarity on all coasts of the world ocean, except for polar shores. The vertical range of these assemblages depends on the degree of wave exposure: the stronger the degree—and thus the moistening by spray—the higher the upper boundary of the supralittoral zone, although humidity may also exert some influence. Both seasonal fluctuations and differences in tidal regime exert only little influence on supralittoral rocky-bottom assemblages.

The following organismic components are almost invariably present: (i) Unicellular algae, chlorophytes and cyanophytes; the most common genus of the latter is *Calothrix*, often with species of *Plectonema* and *Entophysalis*.

(ii) Highly desiccation tolerant prosobranchs of the family Littorinidae which feed on the algae mentioned above, among them the European *Melaraphe neritoides* and, on other shores, species of *Tectarius*, *Nodilittorina*, *Risella*, *Peasiella*, etc. These littorinids move slowly upwards or downwards according to changes in the degree of moistening (spring or neap tides, storms). However, they are generally more abundant at the lower level of the supralittoral zone. Sometimes two or more species of these littorinids may inhabit the same area, but have different ecological requirements. For example, on the coast of

Margarita Island (Venezuela) *Tectarius muricatus* may be found in drier places than *Littorina ziczac*, and males of the latter species tend to inhabit higher levels than females (ROGRIGUEZ, 1959). On strongly exposed shores some of the supralittoral littorinids may go down to the higher level of the midlittoral zone.

(iii) Isopods of the family Ligiidae: e.g. species of *Ligia* and *Megaligia*; these detritus eaters, which quickly move over rocks, sometimes go down to the midlittoral zone; during sunny weather they seek shelter in crevices where they huddle.

Almost as common as the above-mentioned components are lichens (species of *Verrucaria*), some halophilic Diptera and, in tropical and subtropical areas, often some crabs of the family Grapsidae. These crabs are not characteristic of the supralittoral zone, but move up and down with the sea level, e.g. *Cyclograpsus natalensis* (South Africa), *Brachynotus sanguineus* and *Doclea bidentata* (Sea of Japan).

Our present information about the biomass of these communities is very poor and restricted to littorinids: 131 g m^{-2} for *Littorina sitchana subtenebrosa* on north-western coasts of the Sea of Japan with a small tidal amplitude (MOKIEVSKY, 1960a, b, 1962); 190 g m^{-2} for *Littorina brevicula* on the Chantung peninsula, Yellow Sea, with an average tidal amplitude of ca 4 m (GURJANOVA and co-authors, 1958).

On the rocky coasts of polar seas which are ice-covered for at least a part of the year, supralittoral assemblages seem to be missing, owing to both ice abrasion and melting fresh water (KNOX, 1960). KNOX found supralittoral rocky shores in the subantarctic region to be inhabited only by *Verrucaria* and cyanophytes, and sometimes some insects. There are no littorinids, except on the southernmost shores of New Zealand.

On islands with abundant guano-producing birds, such as, in many areas of the Peruvian and Chilean coasts, the whole typical assemblage may be missing (GUILER, 1959a). On the coast of Puerto Pardelas (Argentina) OLIVIER and co-authors (1966a) mention a community consisting only of nitrophilic blue-green algae, with *Lyngbya aestuarii* predominating and with species of *Microcalcus, Oscillatoria, Phormidium* and *Calothrix*.

On shorelines bordering coastal deserts, e.g. the south-western coast of Africa influenced by the Benguela Current (PENRITH and KENSLEY, 1970a,b), the supralittoral assemblage is strongly impoverished due to exceptional air dryness; *Littorina punctata* is comparatively rare here, but is more abundant below, i.e. in the midlittoral zone.

(b) Soft Substrates

Sand and Muddy Sand

On supralittoral beaches, jumping amphipods represent the most prominent component of the assemblage except on tropical shores. Two groups of communities may be distinguished.

(i) On dry, well drained sands, species of the digging *Talitrus* usually predominate, feeding on small detritus particles in the sand.

(iii) On less dry sands (more or less covered with organic debris, mainly of phanerogams or large seaweeds), the amphipod genera *Orchestia, Talorchestia*, etc., prevail, feeding on decaying organic material. While less fitted for digging, they are protected from excessive desiccation by stranded flotsams (mainly benthic seaweeds). Together with these amphipods, several other community components occur, which, according to our present state of knowledge, seem to be less constantly present: some isopod species of the genus *Tylos* on most of the European beaches, e.g. *T. europaeus* on dry sands, *T. sardous* and *Halophiloscia couchi* on wet sands (Mediterranean Sea), *Excirolana natalensis* on the south-western coast of Madagascar; numerous Coleoptera, some feeding on small organic particles (*Bledius*)—similar to amphipods—others feeding on detrivorous insects and crustaceans; pseudoscorpions; myriapods, etc.

In tropical and equatorial regions the supralittoral beach community also includes crabs, sometimes together with amphipods, but substituting for them when the sands become very dry (coarse sands) and the climate is warmer. For example, in the Chan-tung Peninsula (Yellow Sea) GURJANOVA and co-authors (1958) found a supralittoral community including only crabs: *Sesarma* sp., *Scopimera globosa longidactyla* and *Helice tridens*, with a biomass of about 25 g m^{-2}. Most of these crabs are detritus feeders like amphipods or graze on the microphyte cover of sand grains (*Scopimera*, *Dotilla*). How-ever, some species of *Ocypode* are carnivorous.

The substitution of crabs for amphipods on tropical beaches may be interpreted in terms of varying ability to withstand the stress of terrigenous life in warm climates (DAHL, 1953). Amphipods have a very thin exoskeleton and generally release their young from brood pouches into the biotope; crabs have a more calcified exoskeleton and breed on lower, immersed beach levels. This interpretation is supported indirectly by several observations. On some beaches amphipods exist at levels lower than those occupied by Ocypodidae, for example, on the north-eastern Brazilian coasts (*Orchestoidea tuberculata*), in the Congo estuary (*Talorchestia tricornuta*), at Inhaca Island Indian Ocean (*Talorchestia malayensis* and *T. australis*); the two latter species occur only in stranded masses of dead sea-grass, whereas *Ocypode ceratophtalmus* and *O. kuhli* inhabit dry sand.

Anomurans, such as species of *Coenobita* and some prosobranchs of the genus *Bullia*, sometimes exist in the supralittoral zone of tropical beaches. However, they cannot be considered characteristic of this because they are also found in non-marine habitats.

According to MOKIEVSKY (1953, 1960b, 1962) the biomass of the sandy-bottom supralittoral assemblage amounts only to a few tens of g m^{-2} on estuarine beaches of the Okhotsk Sea but may rise to about 460 g m^{-2} in very sheltered areas, where the com-munity is more diversified: *Orchestia ochotensis*, oligochaetes, Muscidae, a species of the mite genus *Molgus*; the *Orchestia* biomass tends to be somewhat lower at low tide than at high tide, because some individuals migrate downwards at ebb tide. On tropical shores, the biomass seems to be less: REYS and REYS (1965) found on the south-western coast of Madagascar 1–2 g m^{-2} (d.w.) for the *Talorchestia–Excirolana natalensis*–tenebrionids assemblage (irrespective of the crabs' biomass).

Mud Flats

Supralittoral assemblages on mud or sandy mud have not been sufficiently investi-gated, except on the European coasts where such bottoms are named 'schorres'. On the mud itself, the higher stratum of terrestrial plants (*Salicornia*, *Arthrocnemum*) is associated with a lower stratum of marine blue-green algae. The fauna includes numerous insects: Diptera, Coleoptera (namely *Bledius*), the pulmonate *Alexia myosotis* and the isopod *Halophiloscia couchii*. Due to the high water-retaining ability of mud, its infauna does not suffer excessive desiccation. Consequently, although living in the supralittoral zone, infaunal species must be considered as midlittoral from an ecological point of view.

An assemblage similar to that of schorres sometimes exists above the mangrove swamp complex (p. 391), such as on the south-western coast of Madagascar, but with an impoverished insect fauna (*Bledius hasticeps*, *Pogonus gillipes*) and the large-sized crab *Cardisoma carnifex*, which digs very deep burrows.

(c) Other Assemblages of the Supralittoral Zone

Many intermediates may exist between true hard and soft bottoms, e.g. boulders, pebbles, gravels, all sometimes more or less mixed with sand or even mud. The composition of assemblages inhabiting such mixed substrates depends on morphometric features and the degree of air exposure. For example, a very sheltered area with boulders tends to harbour an assemblage rather similar to that of an uninterrupted rocky substrate but with some particularities due to the spaces between boulders. Here, on French Mediterranean Sea coasts, COSTA and PICARD (1956) found littorinids and idoteids and sometimes the small-sized pulmonate *Truncatella subcylindrica*. Between the boulders, sand and vegetal detritus can accumulate, resulting in a higher degree of moistening. The resulting assemblage is similar to that mentioned above for sand with decaying vegetal matter (*Orchestia* assemblage). The two biocoenoses described by SCARLATO and co-authors (1967) and GOLIKOV and SCARLATO (1967) in the Possjet Bay (Sea of Japan) on pebbles mixed with gravel and sand very rich in decaying marine vegetals (*Zostera*, *Phyllospadix*, *Punctaria*) probably belong to this group; their rather poor fauna (biomass $11-27 \text{ g m}^{-2}$) mainly consists of different species of *Orchestia* and *Talorchestia* and Dermaptera.

In more exposed areas where boulders are frequently stirred by waves the assemblage become impoverished. Pebbles in the supralittoral zone are usually frequently stirred, thus diminishing their cover of epilithic unicellular algae.

(2) Midlittoral Zone

(a) Hard Substrates

Shores with Small Tidal Ranges

Rocky shores with a tidal range smaller than 1·5–1·8 at spring tides support assemblages whose vertical distributions reveal a very striking similarity. There are two separate subzones: (i) the upper subzone where thoracic cirripeds predominate; and (ii) the lower subzone, the most characteristic components of which are crustose Corallinaceae, often called 'Melobesieae'.

Upper midlittoral subzone

In general, the plant component of the assemblage includes only epilithic and endolithic unicellular algae, mainly cyanophytes. The thoracic cirripeds which occur in the upper subzone are most tolerant to emergence stress: desiccation, temperature and salinity fluctuations, shortening of the time span available for respiration (except for animals with aerial respiration) and for food collection. Some cirripeds of this subzone belong to the genus *Balanus (B. trigonus* on the tropical Atlantic American coasts) or to *Elminius*, but more often to *Chthamalus: C. stellatus* in the Mediterranean Sea, *C. dentatus* on the Ghanaian coast (GAULD and BUCHANAN, 1959), *C. dalli* in the Possjet Bay, Sea of Japan, (SCARLATO and co-authors, 1967; GOLIKOV and SCARLATO, 1967). The third important component of the upper subzone consists of herbivorous prosobranch gastropods, sometimes species of *Nerita* or *Littorina*, but more frequently limpets, such as *Patella lusitanica* in the Mediterranean Sea, some species of *Cellana*, or limpet-like pulmonates such as *Siphonaria* spp. (Fig. 8-1).

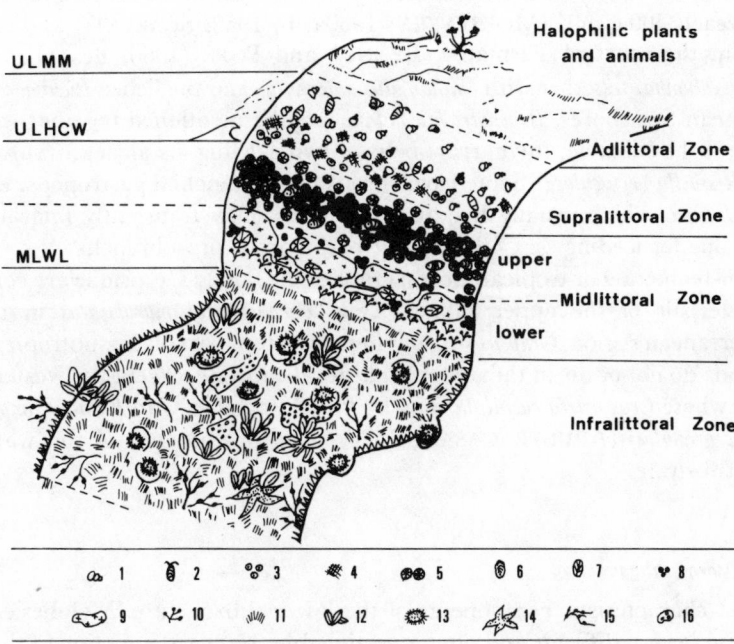

Fig. 8-1: Shallow-water assemblages on rocky substrates (mean exposure) of the French Mediterranean Sea coast. 1: *Melaraphe neritoides*; 2: *Ligia italica*; 3: *Chthamalus depresus*; 4: *Verrucaria*; 5: *Chthamalus stellatus*; 6: *Patella lusitanica*; 7: *P. aspera*; 8: *Rivularia atra*; 9: *Neogoniolithon notarisi*; 10: *Rissoella verruculosa*; 11: articulate Corallinaceae; 12: *Mytilus galloprovincialis*; 13: *Paracentrotus lividus*; 14: *Marthasterias glacialis*; 15: *Cystoseira*; 16: *Lithophyllum incrustans*. ULMM: upper level of marine moistening; ULHCW: upper level of high calm waters; MLWL: mean low-water level. (Original.)

The abundance of *Chthamalus* species depends on the degree of wave exposure. They are scattered or missing on both very sheltered areas where herbivorous gastropods predominate and on strongly exposed rocks where their cypris larvae probably cannot settle. The slope of the rock also plays a role; for example, on the Ghanian coast, small *C. dentatus* only cover 10% of the whole surface of horizontal rocks where *Nerita senegalensis* is more common, whereas on vertical rocks the bigger *C. dentatus* cover up to 90% of the surface (GAULD and BUCHANAN, 1959).

Apart from these three most important and almost invariable components of the assemblage several others may be observed. Most multicellular algae are unable to withstand the rather severe environmental conditions of the upper subzone. However, species of the genus *Bostrychia* (and more rarely of the genera *Porphyra* and *Bangia*) occur—sometimes only in crevices or overhangs, sometimes only during the coldest or most stormy seasons—on many atidal shores of the world ocean. The phaeophyte *Gloiopeltis capillaris*, which is rather common on northern shores of the Sea of Japan in this upper subzone (together with *Chthamalus dalli* and some species of *Littorina*) also exists in the lower subzone where it seems to be more abundant, and down to the upper fringe of the infralittoral zone. Its biomass varies between 100 and 300 g m^{-2}; that of *C. dalli* may reach 500 g m^{-2} (MOKIEVSKY, 1960a, b, 1962; SCARLATO and co-authors, 1967). From the coast of Tasmania, BENNET and POPE (1960) described a belt of *Chamaesipho columna*, together with *Chthamalus antennatus* and the lichen *Lichina confinis*. On Mediterranean Sea shores, *Brachytrichia balani* may be mentioned together with, at the lower part of the subzone, the narrow belts of two calcifugous algae: *Mesospora mediterranea* and *Rissoella verruculosa*. Some carnivorous prosobranchial gastropods, e.g. *Cronia*, *Thais* and *Drupa*, which usually live at lower levels, may transiently migrate into the upper subzone for feeding on *Chthamalus* or herbivorous prosobranchs.

On warm-temperate or tropical shores, crabs of the family Grapsidae are common but not characteristic of the upper subzone; e.g. *Pachygrapsus marmoratus* in the Atlantic–Mediterranean region, *Grapsus strigosus* in the western part of the subtropical Pacific.

Pelecypods do not occur in the upper midlittoral subzone, except the western coast of Sri Lanka, where *Crassostrea cucullata* may be found, together with *Balanus amphitrite* and *Nodilittorina granularis*(ARUDPRAGASAM, 1970), possibly because of the water-soaked sandstone substrate.

Lower midlittoral subzone

The most characteristic components of the lower subzone are 'Melobesieae', whose identification is very difficult and often questionable. Hence, distinction of this subzone from the upper fringe of the infralittoral zone is sometimes difficult, particularly on exposed shores where this fringe also supports some species of encrusting rhodophytes which probably always belong to other species; for example, in the Mediterranean Sea, the most common infralittoral species are *Lithophyllum incrustans*, *L. trochanter* and *Tenarea undulosa*.

The crustose Corallinacea of the lower midlittoral subzone may exhibit two main extreme aspects: (i) a thin crust covering the rock, such as the 'veneering' of *Neogoniolithon notarisi* in the Mediterranean Sea; (ii) reef-like cornices superimposed on the rock itself, such as the *Lithophyllum tortuosum* 'trottoir' (Fig. 8-2). This is rather common on the northern coasts of the western Mediterranean Sea. The so-called 'algal ridge' (*Porolithon*) of tropical coral reefs belongs to the latter group of assemblages (p. 414).

The flora and fauna associated with the crustose Melobesieae are richer than those of the upper subzone. Apart from the unicellular, epilithic and endolithic chlorophytes and

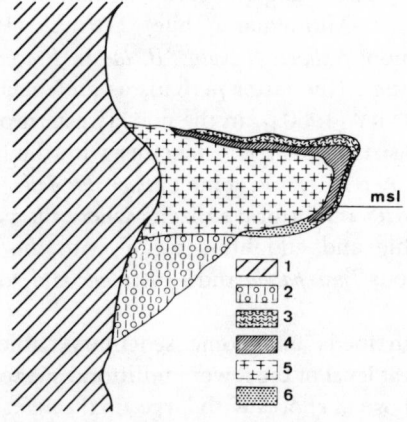

Fig. 8-2: *Lithophyllum tortuosum*. Reef-like cornice on Mediterranean Sea rocky shores. msl: mean sea level; 1: rocky substrate; 2: precoralligenous assemblage; 3: live *L. tortuosum*; 4: dead, but unconsolidated *L. tortuosum*; 5: dead, consolidated *L. tortuosum*; 6: uncrystallized biogenous limestone. (After PERES and PICARD, 1952, reproduced by permission of the authors.)

cyanophytes, the multicellular soft algae are much more common there than in the upper subzone; for example, species of the genera *Ralfsia, Chaetomorpha, Bryopsis, Caulacanthus*, and some Ceramiaceae and Polysiphoniae. Among these, some are permanent, while others are only seasonal, depending on their particular ecological requirements. The herbivorous gastropods are fairly diversified: limpets, such as *Patella aspera* (in the Mediterranean Sea), *Cellana, Acmaea*, and limpet-like forms such as *Siphonaria* and *Fissurella*, and also the trochids *Gibbula, Monodonta* and *Tegula*. Carnivorous prosobranches (*Drupa* and *Thais*) are permanent inhabitants of this subzone, feeding mainly on the herbivorous snails.

Crabs are also present (Grapsideae, *Eriphia*) on warm-temperate or tropical shores. Where the Melobesieae crust is thick enough to provide sufficient shelter, small mussels of the genus *Brachidontes* are present, such as *B. rostratus* in *Lithothamnium* sp. and *Lithophyllum hyperellum* on the Tasmanian coast (BENNET and POPE, 1960).

On coasts where Corallinaceae build very important cornices (Fig. 8-2), both the irregular growth of different thalli and the activity of boring organisms (cyanophytes, pelecypods, etc.) result in a very intricate system of small holes, crevices and pipes, supporting a very rich cryptic fauna. Most of its components really belong to the fauna of the infralittoral zone, enclaved in the midlittoral. These species are able to live there because of the high degree of moisture in this calcareous formation, which is more or less porous and has its lower part permanently immersed. In the Mediterranean 'trottoir' one finds, for example, the actinian *Phellia*, some hydroids, the polyplacophore *Acanthochiton fascicularis*, the gastropods *Fossarus* and *Oncidiella*, the small pelecypods *Lasaea rubra* and *Brachidontes minimus*, the sipunculid *Physcosoma granulatum*, many polychaetes (mostly syllids), isopods and amphipods and the small compound ascidian *Diplosoma*. In addition to these marine species, forms of terrigenous origin also occur: collemboles, myriapods, chernetes, mites (Acari), and even the spider *Desidiopsis racovitzai*, closely allied to the genus *Desis* which lives in the Indo–Pacific coral reefs (PERES and PICARD, 1952, 1964).

The higher the latitude, the lesser the role the Melobesieae play in the assemblage; they may totally disappear on polar sea shores. Then, limpets, barnacles and sometimes soft algae substitute for them. For example, at Montemar (Chile), GUILER (1959b) found the following species to be most common: *Balanus flosculus*, *B. laevis*, *Brachidontes purpuratus*, some *Ulva* and also *Iridaea laminarioides* (the latter perhaps as an 'enclave' of the infralittoral zone). According to MOKIEVSKY (1960a), in the north-western part of the Sea of Japan, most of the assemblage consists of *Gloiopeltis* together with the limpet *Acmaea testudinalis*, whose biomass may reach 56 g m^{-2}, and some *Chthamalus dalli*. From Barbados, LEWIS (1960, 1965) describes a lower midlittoral subzone where the vegetal component includes only unicellular epilithic and endolithic algae, with the poly-placophore *Acanthopleura granulata*, the gastropods *Thais patula* and *T. floridana* and, locally, the barnacle *Tetraclita squamosa*.

Zonations concerning the pedunculate cirripeds and some sedentary tubicolous polychaetes which sometimes exist at the lowest level of the lower midlittoral subzone on shores with small tidal range are similar to those of shores with large tidal range. They will be treated at the end of the next section

Shores with Large Tidal Ranges

The subdivision of the whole midlittoral into two subzones is generally less obvious on tidal shores—firstly because the tidal flow and ebb regularly dampens the whole zone, and secondly, because the vertical distribution of sessile or sedentary organisms is often obscured by belts of large-sized soft algae. At high or mean latitudes these algae are mainly phaeophytes, but at lower latitudes chlorophytes and rhodophytes coexist and sometimes substitute for the brown algae (FELDMANN, 1951).

The vertical zonation of algal belts depends on the ability of the different species to withstand emergence effects (Fig. 8-3). The species at the highest levels are most tolerant to (or possibly require) long-term emergence, e.g. some species of *Pelvetia* (*P. canaliculata* on the north-eastern Atlantic coast), *Gloiopeltis furcata* and *G. complanata* on southern Sea of Japan shores, and some species of *Hildenbrandtia*. Those at the lowest levels are less able to withstand emergence: e.g. *Fucus vesiculosus* (western Europe), *F. evanescens* (northern Sea of Japan at lower levels), *Adenocystis* and *Durvillea antarctica* (Antarctic shores).

Species which predominate near the mean tide level would be most tolerant to breaker effects or require more pure waters. The horizontal distribution mainly depends on the degree of wave exposure. Some species, such as *Ascophyllum nodosum* and *Adenocystis* sp. only exist in rather sheltered areas, whereas others, such as *Durvillea antarctica*, require a high degree of exposure. In brackish waters and in coastal marine areas rich in organic wastes (especially sewage) from temperate and cold-temperate shores, green algae such as *Ulva* and *Enteromorpha* substitute for phaeophytes. The algal standing crop often does not exceed a few hundred g m^{-2} (e.g. *Gloiopeltis* or green algae), but may rise up to several kg m^{-2} (ca 5 kg m^{-2}, for example, for *Halosaccion* on the lower levels of the western Kamchatkan coast) (VOZZHINSKAJA, 1964).

The belts of large-sized soft algae result in the occurrence of a particular assemblage which mainly includes sessile or sedentary species but also some motile forms. Among the sessile forms are included epiphytic diatoms and small multicellular algae (Polysiphoniae, Ectocarpaceae, and some small patches of crustose Corallinacea), many

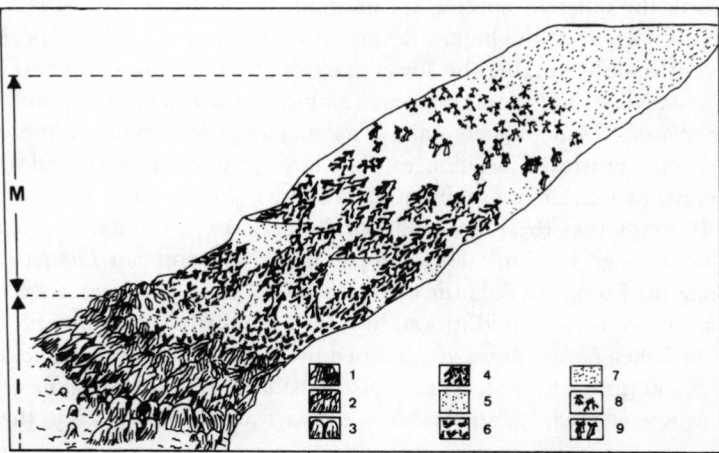

Fig. 8-3: Vertical zonation of intertidal assemblages on a rather sheltered rocky substrate in the English Channel. 1: *Fucus serratus*; 2:
Laminaria digitata; 3: *L. hyperborea*; 4: *F. vesiculosus*; 5: *Balanus* and *Patella*;
6: *Laurencia*; 7: *Verrucaria* and *Melaraphe neritoides* (supralittoral zone); 8:
Pelvetia canaliculata; 9: *F. spiralis*; M: midlittoral zone; I: infralittoral
zone. (After LEWIS, 1964; reproduced by permission of The English
Universities Press, London.)

hydroids and polyzoans, some serpulid polychaetes (chiefly Spirorbinae) and small
compound ascidians (*Botryllus*, Didemnidae). Small gastropods, polychaetes, nematodes
and small crustaceans (isopods, amphipods, crabs) are among the sedentary forms. At
high tide, numerous crustaceans (amphipods, prawns, shrimps) and some fishes swim
between the floating leafage of the algae. At low tide, the motile species either escape
with the ebb or find refuge in small pools, while the sessile and sedentary forms remain
under the algal cover, which retains some water. Gastropods, hydroids, polyzoans, serpulids and ascidians close up their shells, hydrothecae, zoeciae, tubes or siphons thus
retaining as much water as possible. Some nematodes temporarily exist under anoxic
conditions. It seems that the whole flora and fauna directly associated with these algal
belts constitutes a single organismic assemblage, although certain sessile forms may
exhibit some vertical microzonation.

Below the algal canopy, i.e. on the rock itself, the assemblage components are not very
different from those on atidal rocky shores. Algae (mainly diatoms and small multicellular soft species) and sponges (halichondrines) are more common due to higher and more
permanent moistening, and one also observes Melobesieae, barnacles (mainly *Balanus*),
limpets, Trochidae, Littorinidae, etc. The floristic and faunistic composition of this
assemblage is almost the same over the whole midlittoral zone, with only some changes
in the abundance of the different species.

In general, algal belts occur neither on ice-covered shorelines nor in the presence
of ice abrasion, nor on strongly wave-exposed shores. At middle latitudes they are
replaced by an assemblage in which barnacles predominate. Unlike on shores of small

tidal range, the delimitation of two subzones is rather indistinct and vertical changes in populations of the different species are gradual. Usually the most euryplastic species (e.g. *Chthamalus*) occur at the highest levels; other barnacles (such as species of *Balanus* and *Tetraclita*) mainly inhabit the lower part of the midlittoral zone. Together with barnacles one always finds some limpets (mainly *Acmaea*, *Notacmaea*, etc), carnivorous gastropods (*Nucella*, *Thais*), which mainly feed on cirripeds, and often, at the lower levels of the zone, some crustose Corallinacea. The latter, small and scattered on the cold or cold-temperate shores, increase in abundance as the average annual temperature increases. It seems that these crustose Corallinacea are more developed on shores of moderate tidal range (3–5 m): for example the Mediterranean *Lithophyllum tortuosum* extends along the European Atlantic coasts to 46° 30′ N. Sometimes in exposed areas of western European coasts, midlittoral brown algae belts are replaced by scattered patches of the lichen *Lichina pygmaea* inhabited by the small pelecypod *Lasaea rubra*. Some sessile pelecypod (mussels, oysters) may also exist in the midlittoral zone on shores with large tidal ranges whereas on atidal shores they are always restricted to the infralittoral zone.

At the lowest level of very wave-exposed rocky shores, one often finds, sometimes together with mussels, populations of pedunculate cirripeds such as *Mitella pollicipes* on the West African Atlantic coasts or *Pollicipes polymerus* on the North American Pacific coast. It seems that the narrow belts of some serpulid polychaetes (*Pomatoleios*) in some tropical areas belong to the infralittoral zone (p. 405).

In general, on tropical and equatorial rocky shores, the midlittoral assemblage tends to exhibit two kinds of change. (i) The percentage of the rocky surface covered by soft algae decreases to the advantage of sessile and sedentary animals, mainly at the upper levels. (ii) A more marked subdivision of the midlittoral zone may be observed as a consequence of the climate which is more severe in the upper part of the zone: solar irradiation and drying effects, washing by tropical rains, etc.

The following examples illustrate both of these tendencies.

Coast of São Paulo State, Brazil (spring tide range 1·8–2 m): upper subzone with *Chthamalus* (except in very sheltered areas) and, on exposed rocks, *Ectocarpus breviaculeatus*; lower subzone inhabited by *Crassostrea* and *Hildenbrandtia prototypus* in sheltered areas, *Tetraclita squamosa* in exposed places, and *Acmaea subrugosa* on strongly exposed rocks (NONATO and PERES, 1961).

Coast in the vicinity of Bombay, India (mean tidal range 3–6 m): upper subzone with either *Balanus amphitrite* (sheltered), or *Chthamalus challengeri* and *C. whitersi* (exposed); lower subzone with *Crassostrea cucullata* everywhere but mixed with *Balanus* on exposed rocks. The whole midlittoral zone features the alga *Caloglossa leprieurei* and many herbivorous gastropods, such as *Nerita polita*, *Cerithium morus* and *Planaxis sulcatus* (BAL and BHATT, 1959).

Chantung Peninsula, Yellow Sea (mean tidal range 4 m): GURJANOVA and co-authors (1958) found an upper subzone inhabited by *Chthamalus dalli*, *Balanus amphitrite albirostrata* and *Volsella atrata* (closely related to *Modiolus*); the biomass may exceed 4·5 kg m^{-2}. In the lower subzone (biomass up to 6 kg m^{-2}) *Crassostrea cucullata* predominates with gastropods both herbivorous (*Patelloida schrenki*, *Turbo coronatus*, *Monodonta* spp.) and carnivorous (*Thais*), polychaetes, isopods, crabs (*Heteropanope makiana*, *Hemigrapsus penicillatus*) and pagurids.

South-western coast of Madagascar (mean tidal range 3 m): although the species are different from those on Yellow Sea shores, the general scheme observed by PLANTE (1964) is rather similar, but three differences may be pointed out. (i) In very sheltered places, the cyanophyte cover eliminates *Chthamalus*, whereas an important population of the prosobranchial gastropod *Cerithidea decollata* feeds on these algae. (ii) Barnacles of the genus *Tetraclita* and crabs mix with oysters, and species are different for each group in sheltered and exposed areas. (iii) The soft alga *Caulacanthus* sp. is common on exposed rocks.

Konakry (Guinea): while the vegetal component is rather uniform over the whole midlittoral zone (*Bostrychia binderi*, *Gelidiella tenuissima*), the upper subzone supports *Chthamalus dentatus*, *Nerita senegalensis*, *Siphonaria grisea*, *Brachidontes puniceus*; the lower subzone, *Crassostrea tulipa*, *Morula nodulosa*, *Pachygrapsus transversus*, etc. The standing crop is very low: 20 g m^{-2} for the upper and 54 g m^{-2} for the lower subzone (USHAKOV, 1965, 1970).

Western coasts of South Africa: Both the low sea-water temperature (due to the Benguela Current upwelling) and the very high air temperature and dryness (coastal desert conditions (cause changes in intertidal assemblages. At Lüderitz Bay (26° 36′ S; water temperature generally below 14 °C; spring tidal range 3 m), the supralittoral *Littorina punctata* is thriving in the midlittoral zone, together with *Porphyra capensis*, *Siphonaria capensis* and, locally, *Patella granularis*. Most limpets (3–4 species) are in the infralittoral fringe, together with *Fissurella mutabilis*, mussels, some Lithothamnieae and soft algae (*Gymnogongrus*, *Gigartina*). Barnacles are rare or absent, and where mussels (*Aulacomya*, *Choromytilus*) exist, the highest-placed individuals occur in the upper fringe of the infralittoral zone, in spite of the rather large tidal range. At Rocky Point (18° 59′ S; water temperature 14–17 °C; tidal range: 1·8–2·1 m) the infralittoral fringe is only slightly modified due to the addition of some more thermophilic species. The animals of the supralittoral and midlittoral zones, except for the pulmonate gastropod *Siphonaria capensis*, exhibit a general descent; *L. punctata* attains its maximum abundance in the midlittoral, together with *Chthamalus dentatus* and *P. granularis*; the latter two also exist in the upper fringe of the infralittoral zone (PENRITH and KINSLEY, 1970a,b).

Finally, it must be emphasized that on the most wave-exposed shores the vertical distribution of the main characteristic species of the midlittoral zone can be completely obscured: *Chthamalus* often extends down to the lowest level of the zone and occupies the vacant place of *Balanus* or *Tetraclita* (whose cypris larvae are unable to settle there). Soft algae of the lower midlittoral and even of the upper fringe of the infralittoral zone (e.g. species of *Laurencia*), may often rise up to the highest levels of the midlittoral zone supported by permanent wave-moistening (PERES, 1967a,b).

Mussel and Oyster Beds

Mussels and oysters occur permanently in the midlittoral zone only on shores with large tidal ranges (>1·5–1·8 m), although some individuals of these typical intertidal species may also be found below spring tide low-water level. On shores with small tidal amplitudes the same species are restricted to the infralittoral zone.

The causes of the distribution of mussel beds remained long unknown until SEED (1969) observed that the planktonic larvae do not settle directly on the rock itself but at first, at the 'plantigrade' (pediveliger) stage, on filamentous algae. The heterogeneous distribution of such algae in the vicinity of midlittoral rocky shores partly explains why places very similar in terms of wave exposure, organic matter content, rock characteristics, presence of predators, etc., may or may not be inhabited by mussel beds.

The biomass of mytilid communities in the midlittoral zone is among the highest of all benthos assemblages: values of 20–25 kg m^{-2} are common, but sometimes may rise up to 80 kg m^{-2}. However, about 70–75% of the standing crop consists of calcareous matter. Obviously, the mytilid species vary between different geographic areas: *Mytilus edulis* in the North Atlantic Ocean; *Mytilus californicus* on the American Pacific coast; *Perna perna* on the western coasts of Africa; and *Mytilus chilensis* and *Aulacomya magellanica* in Patagonia, etc.

In the clumps formed by mytilids, a very rich associated fauna occurs: sponges, hydroids, polyzoans, polychaetes, small crustaceans and species of *Balanus* and *Tetraclita*. Predators are gastropods (*Nucella*, *Thais*) and Asteroidea (*Asterias*, *Pisaster*), feeding on both barnacles and mussels, and thus controlling their abundances.

On shores with large tidal ranges, mussel beds always develop in the lower half of the midlittoral zone, except on the shores influenced by the Benguela Current. Most of the associated fauna, e.g. barnacles, the small pelecypod *Lasaea rubra* and carnivorous gastropods, are the same as those outside the mussel beds. Nevertheless, the network of byssus filaments forms a felt-like layer, which retains sea water. Hence, one also finds some species in the mussel beds which usually inhabit the infralittoral zone; e.g. in the byssus felt of the European *Mytilus edulis*, many polychaetes—the most common of which is *Cirratulus cirratus*—small molluscs, amphipods (e.g. *Gammarus marinus*, which also inhabits clumps of *Balanus balanoides*, and *Hyale nilsoni*), *Amphipholis squamata*, and even two fish species, *Blennius gattorugine* and *B. galerita*.

At low tide, mytilids require an adequate degree of moisture, although in areas where large seasonal changes in wave movement and/or in air temperature and humidity occur, young mussels may be found at levels higher than normal, e.g. on the English Channel coasts, young *Mytilus edulis* in winter and early spring at the upper level of the midlittoral, intermixed with *Chthamalus*. At Ubatuba, Brazil, where fully grown mussels inhabit the infralittoral zone, young individuals also exist in winter in the lower part of the midlittoral zone. All these young mussels—not to be confused with *Brachidontes*—disappear early in summer when the sea becomes smoother and warmer.

As one proceeds to more tropical coasts, oyster beds tend to substitute for mussel beds. In general, when oysters and mussels exist together, the latter are found at a lower level (Fig. 8-4).

Southern Ocean Assemblages

The Antarctic rocky shores seem to be all but lifeless down to 2 m at Haswell Islands (PROPP, 1970) and 7–8 m at Molodezhnaya and Mirny Stations (GRUZOV and PUSHKIN, 1970), due to permanent ice and its covering or erosion effects. However, according to KNOX (1960) some algae (*Ulothrix*, *Bangia*, *Monostroma*, *Enteromorpha*, *Cladophora*) may exist in sheltered places of other areas that are ice-free during the summer. There are no barnacles or midlittoral mussels: *Aulacomya* only occurs in the infralittoral zone; some limpets (*Nucella*, *Patinigera*) which usually live in the infralittoral zone may rise up to the

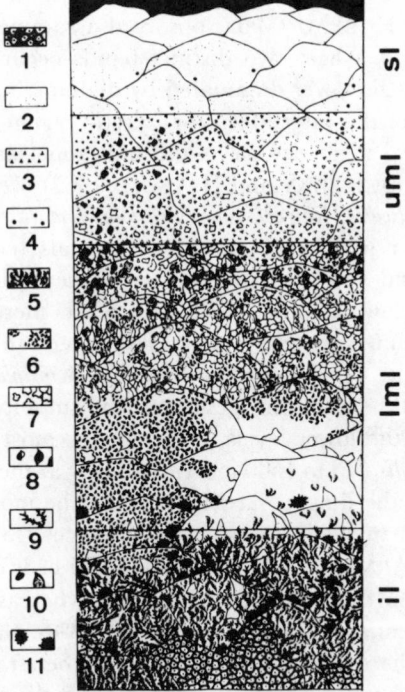

Fig. 8-4: Rocky-shore intertidal assemblages at Beaufort, North Carolina (USA). sl: supralittoral zone; uml and lml: upper and lower subzones of the midlittoral zone, respectively; il: infralittoral zone. 1: *Aiptasia pallida*; 2: Cyanophytes; 3: *Chthamalus fragilis*; 4: *Littorina irrorata*; 5: mixed algal turf; 6: *Mytilus exustus*; 7: *Crassostrea virginica*; 8: *Porphyra leucosticta*; 9: *Sargassum filipendula*; 10: *Enteromorpha* and *Ulva*; 11: *Arbacia punctulata*. (After STEPHENSON and STEPHENSON, 1952, modified; reproduced by permission of Blackwell Scientific Publications.)

midlittoral zone for short periods (GAIN, 1912). This has been corroborated by HEDGPETH (1969) at Palmer Station (64° 45′ S, tidal range ca 1·5 m), who found the upper half of the midlittoral quite bare except for some scattered *Patinigera polaris*, which are much more abundant at depths of 2–3 m. Nevertheless, HEDGPETH found in the lower third of the whole intertidal (probably corresponding to both the lower midlittoral and the upper fringe of the infralittoral) a thick felt of diatoms, mixed with some chlorophytes (*Enteromorpha, Ulothrix*, etc) and several species of polychaetes and amphipods. The latitude of Palmer Station, which is near the boundary between Antarctic and subantarctic regions, may well explain these peculiar characteristics. However, DELEPINE and HUREAU (1963) observed a summer midlittoral zonation at Petrel Island, Adélie Land, where two distinct bands occur between tidemarks: the upper of *Ulothrix australis*, the lower dominated by diatoms.

The aberrant features of the intertidal subantarctic region have already been mentioned for the supralittoral zone. In the midlittoral zone, at the Macquarie Islands (spring tidal range: 1·45 m), KENNY and HAYSOM (1962) observed a rather rich algal population (*Porphyra umbilicalis, Rhizoclonium, Prasiola, Iridaea boryana*, etc) in the upper third of the zone together with some terrestrial animals (collemboles), oligochaetes (*Lumbricillus, Marionia*) and scattered individuals of the herbivorous gastropod *Macquariella hamiltoni*. Just below, at about mean sea level, there is a belt of bare rock, possibly related to maximum water movement. The lower half of the midlittoral zone is much richer, with the algae *Chaetangium fustigium* and *Acrosiphonia pacifica* at the upper level and *Rhodymenia* sp. at the lower; in these algae the amphipod *Hyale novae-zelandiae* is very common, mainly at the lower levels, although the most abundant species is the pulmonate *Siphonaria lateralis* (up to 1500 individuals m^{-2} at the lower level), mixed with scatterd *M. hamiltoni* and the limpet *Nacella delesserii*. The most striking feature of the midlittoral zone in the subantarctic region is the absence of barnacles.

At Kerguelen Islands, ARNAUD (1971) studied mixed beds of *Mytilus edulis desolationis* and *Aulacomya ater regia*. The first is mainly midlittoral, whereas *Aulacomya* appears at the low water level and extends down to about 30 m. ARNAUD noticed that the beds which may exist on both hard and soft bottoms, thrive better in rather sheltered areas, especially on shallow and narrow sills between bays and the open sea; such a position, which corresponds to a very important and regular water transit due to tides, provides the mussels with an abundant food supply.

(b) Soft Substrates

For several reasons, it is much more difficult to delimit the midlittoral zone and its assemblages on soft than on hard substrates: (i) the more or less pronounced capacity of sediment to retain sea water depending on granulometry, porosity, beach profile, etc) when emerged; (ii) the digging ability of many macrobenthic species (as regards the micro- and meiobenthos see p. 313), which depends on the local characteristics of the sediment and also on those of the fauna itself (biogeographical and ethological). The first-mentioned peculiarity is of much more general value and always results in an altitudinal narrowing of the 'true' midlittoral zone, i.e. the belt of the beach where the consequences of emergence can decisively affect the species. In other words, on soft substrates the upper and lower boundaries of the midlittoral zone never correspond to those recognized on neighbouring rocky substrates and the vertical species zonation is less dependent on tidal range. Moreover, the general concept of a whole 'intertidal zone' by many authors often makes it difficult to interpret literature data and to incorporate them into the general scheme of vertical zonation adopted here.

Sand and Muddy Sand

The beaches of the Atlanto–Mediterranean region exemplify the main characteristics of the midlittoral zone with sand and muddy sand substrates. At first glance, there is a single assemblage, the main components of which are the polychaetes *Nerine cirratulus* and *Ophelia bicornis*, the isopod *Eurydice affinis* and the pelecypod *Mesodesma corneum*. The possible disjunction of some species of the assemblage during periods of low water level (high atmospheric pressure and smooth sea) is illustrated in Figure 8-5.

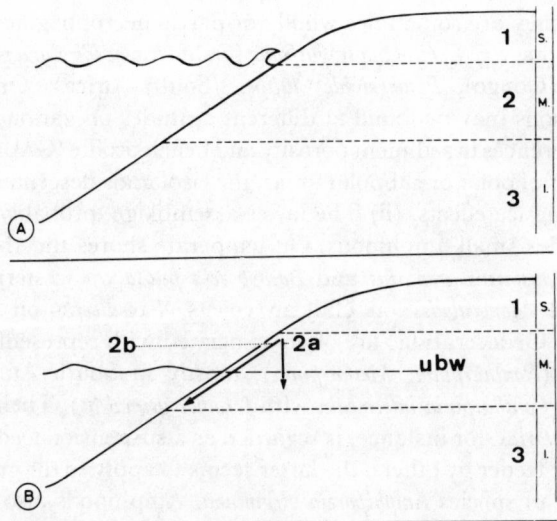

Fig. 8-5: Migrations of species of the midlittoral sand assemblage on Mediterranean Sea shores related to changes in sea level. (A) high level of calm waters or rough sea. (B) low level of calm water. S, M, I: supra-, mid- and infralittoral zone, respectively. 1: *Talitrus* assemblage; 2 (A): homogeneous midlittoral assemblage, with *Ophelia bicornis. Nerine cirratulus, mesodesma corneum* and *Eurydice affinis*. 2 (B): disjunction of the midlittoral assemblage; arrows: migration of *O. bicornis* (2a), and *N. cirratulus, M. corneum* and *E. affinis* (2b) at low level of calm water; ubw: highest level of imbibition by interstitial sea water; 3: upper fringe infralittoral assemblage. (After PERES and PICARD, 1964; reproduced by permission of the authors.)

Similar assemblages seem to exist on many shores, obviously with different species but pertaining to the same ethological groups. Unfortunately, the authors often took into account only some elements of the fauna: the presence of *Thoracophelia* on Californian beaches; assemblages of *Nerine agilis* and *Haustorius* with *Callianassa* and *Lysiosquilla* on North Carolinan beaches; *Excirolana* and *Donacilla* on southern coasts of Japan (HABE, 1959); *Buzonus articus*, *Spio filicornis* and other spionids, with juveniles of *Mactra sulcataria* on north-western coasts of the Sea of Japan (MOKIEVSKY, 1960a); *Nerine cirratulus* and *Excirolana latipes* (LAWSON, 1966) on Ghanaian beaches; *Nerine cirratulus, N. lefbvrei, Perinereis nuntia*, *Excirolana orientalis* and two species of *Mesodesma* on the south-western coast of Madagascar (PICHON, Mir, 1965).

According to DAHL (1953) who accurately studied the distribution of small peracarids, the midlittoral zone on sands or muddy sands always features two main different assemblages, except for beaches at very high latitudes. (i) The upper assemblage is characterized by cirolanid isopods only superficially buried at low tide but very active when immersed; they are sometimes wholly or partly necrophagous or detritophagous, but mostly predators, e.g. *Excirolana linguifrons* (California), *Cirolana mayana* (Galapagos), *Eurydice carangis* (Congo), *Pontegeloides latipes* (South Africa). On Ghanaian coasts cirolanid populations may be found at different altitudes on various beaches, probably depending on differences in sediment porosity and beach profile (GAULD and BUCHANAN, 1959). On beaches of polar or subpolar areas, the cirolanids descend into the infralittoral zone, thus avoiding ice effects. (ii) The lower assemblage, probably only in part really midlittoral, includes small amphipods. On temperate shores they belong to the family Oedoceratidae: *Haustorius arenarius* and *Bathyporeia pilosa* on western European coasts, *Bathyporeiopus* and *Monoculodes* on Chilean coasts, *Exoediceros* on eastern Australian shores, etc. These Oedeceratidae are often associated with representatives of the family Phoxocephalidae (*Pontharpinia, Metharpinia*), mostly in South America, but also on Ghanaian coasts (*Pontharpinia intermedia* with *Urothoe grimaldii*). Their feeding habits are still debated: *Haustorius*, for instance, is regarded as a suspension feeder by some authors but as an epistrate feeder by others; the latter seems to apply to the genus *Bathyporeia* and the North American species *Amphiporeia virginiana*. Amphipods also exist at the lowest level of subarctic and Arctic beaches, often belonging to the family Lysianassidae (e.g. *Pseudolibotrus*). They are fairly tolerant to mud content of the sediment and feeding on plant detritus, and probably inhabit the infralittoral zone.

On tropical and equatorial beaches, as an addition to the typical assemblage (or assemblages) of the midlittoral zone, some species of three main groups occur: crabs, Anomura of the family Hippidae and pelecypods of the genus *Donax* (whose distribution will be further discussed on p. 389).

While the true crabs (Brachyura) are very common in the mangrove swamps (p. 390), they also may inhabit beaches. Species of the carnivorous genus *Ocypode*, for example, exist on the south-western coast of Madagascar and on Ghanaian beaches. On the latter, they are associated with other fauna components (*Nerine, Excirolana*) only if the beach profile has its concavity upwards (LAWSON, 1966). On the midlittoral muddy sand of Chantung Peninsula, GURJANOVA and co-authors (1958) and GURJANOVA (1959) found the crabs *Helice tridens* and *Scopimera globosa*, together with the molluscs *Cerithidea sinensis* and *Glaucomya* sp. (standing crop 400 g m^{-2}).

Like true crabs, hippids (Anomura) are strictly intertropical. The different species of the genus *Emerita* of the American coasts probably belong to the midlittoral zone. Partly buried in sand, they follow the tidal shifts of the surf zone, thus always selecting optimum conditions for collecting suspended organic particles. On the Gold Coast of Ghana, a *Hippa* species exhibits the same behaviour, moving up during neap tides and down during spring tides.

These truly midlittoral hippid species must not be confused with hippids in the permanently immersed infralittoral zone, such as *Blepharipoda occidentalis*, which lives just below the *Emerita* populations or with *Hippa cubensis*, which is common on the West African shores. All these infralittoral hippids seem to be mostly carnivores or scavengers. The infralittoral *Corystes cassivelaunus*, which is common at the lowest beach levels of the North Atlantic Ocean and the English Channel, is a brachyuran in spite of its 'hippid-like' body shape; its feeding habits rather similar to those of *Blepharipoda*.

The zonation of *Donax* populations is very complex. On temperate beaches the species of this genus always inhabit the infralittoral zone. In tropical areas, non-migrant infralittoral species also exist. However, others permanently migrate with the tides. These migrations, which often involve hundreds or even thousands of individuals m^{-2}, have been well investigated in Japan and in the Gulf of Mexico. In general, the population maximum occurs near the mean sea level: a little above it at ebb, a little below at flow tide. Unfortunately, the systematics of *Donax* has been rather insufficiently investigated to date, and hence discrimination between species is sometimes difficult. However, it is indisputable that some migrating *Donax* species really participate in a midlittoral assemblage, e.g. *Donax faba* on the south-western coast of Madagascar (PICHON, Mir., 1965) and *Donax pulchellus* on West African coasts. The population density of the latter may rise up to 20 000 individuals m^{-2}; the bivalve is associated with the amphipods mentioned above in the lower midlittoral subzone.

Some species of *Mesodesma* may also migrate along the beach, such as *Donax*, e.g. *M. mactroides* in the Mar del Plata area, Argentina (OLIVIER and co-authors, 1966b, 1971). The latter genus probably also includes both midlittoral and infralittoral species. In general, species of *Mesodesma* seem to prefer medium-sized sands, especially those with a rather high percentage of calcareous remains.

Owing to occasionally overcrowded populations of pelecypods, some carnivorous gastropods may occur in the midlittoral on intertropical shores. However, these species of *Terebra*, *Oliva*, *Olivancillaria*, *Buccinanops*, etc, are only transient intruders from the infralittoral zone. Sometimes mysids have been mentioned in the lower part of the midlittoral zone, e.g. *Gastrosaccus sanctus* in the Mediterranean Sea, *G. spinifer* on West African shores, *Archaeomysis grebnitzkii* on the north-western coasts of the Sea of Japan. I believe that these swimming peracarids participate in the upper infralittoral assemblage rather than in the midlittoral one.

The above synthesis should not obscure the fact that the cohesion of some of the sandy-bottom assemblages may be uncertain. They cannot invariably be considered as true biocoenoses due to the lack of—or insufficiently strong—mutual member interdependence. Moreover, seasonal changes in surf intensity modify the granulometry and thus sometimes the slope of the beach as well. On Atlantic coasts of France, for example, when the average grain size becomes coarser, *Nerine cirratulus* and *Eurydice affinis* disappear, while *Ophelia bicornis*, *Haustorius arenarius* (and the supralittoral species *Talitrus saltator* and *Tylos europaeus*) move to other areas with better moisture conditions. Where strong surf disturbs the beach profile, only *Haustorius arenarius* can endure such stress (LAGARDERE, 1966; MCINTYRE, 1971). Thus it appears that specific responses of many midlittoral sandy-bottom animals to environmental fluctuations may induce consider-able dynamics and hence some uncertainty regarding the composition and definition of the assemblages concerned (LAGARDERE, 1966).

Mud

Owing to the higher capacity of mud to retain water, the vertical range of muddy strands which may be considered as corresponding to the midlittoral zone is much more restricted than on sandy beaches. The species number is always smaller owing to both the low oxygen content of interstitial water, and the very fine silt or clay particles which may clog respiratory or food-collecting mechanisms. The 'slikke' assemblage of west European coasts should be mentioned here, which is only inhabited by the two polychaetes *Nereis diversicolor* and *Streblospio shrubsoli*, the crab *Carcinus maenas* (an almost ubiquitous species in intertidal and shallow-water areas) and, in brackish water areas, the amphipod *Corophium saltator*. On the cliffs of consolidated mud of estuaries in Britanny, France, MAGNE (1957) found populations of eight species of Chrysophyceae, the vertical distribution of which corresponds to the macrophytes zonation on rocky shores.

On tropical and equatorial shores where pulmonate gastropods and crabs in particu-lar are very numerous, the midlittoral mud assemblage is richer and more diversified. From an extensive study by DERIJARD (1965) on the south-western coast of Madagascar (tidal amptiude 3–4 m), it may be inferred that all epifaunal species exhibit a rather narrow distribution related to the presence or absence as well as to the nature and level of a canopy of mangrove trees (p. 391). The upper midlittoral assemblage is character-ized by the pulmonate *Melampus lividus* and the crabs *Sesarma meinerti* and *S. eulymene*. On very consolidated mud, the crab *Uca inversa* is more common. The lower midlittoral assemblage is much richer, including the pulmonates *Cassidula labrella* and *Onchidium verruculatum* and the crabs *Uca chlorophthalmus*, *U. annulipes*, *U. urvillei*, *Sarmatium crassum*, *Sesarma longipes*, *S. smithi*, *Ilyograpsus rhizophorae*, *Tylodiplax derijardi* and *Euplax boscii*. This epifaunal assemblage of the lower midlittoral consists of at least five different facies, depending on granulometric characteristics of the sediment and shading from mangrove trees (abstract in PERES, 1967a). The infauna of mangrove muddy grounds is discussed on p. 391).

Other Substrates

Gravels

The assemblage of Mediterranean Sea midlittoral gravels—which is always more or less mixed with sand, and sometimes mud—is rather similar to that of the sand itself: *Ophelia radiata*, *Eurydice* and *Mesodesma corneum* are present, but not *Nerine cirratulus*, which requires find sand. On the western coast of France, very similar gravels support the alga *Hildenbrandtia prototypus*. The pelecypod *Venerupis aureus*, which inhabits the underlying sediment, is certainly an infralittoral species (COSTA and PICARD, 1956). In the Eilat region of the Red Sea, POR and LERNER-SEGGEV (1966) described a midlittoral coarse gravel with *Hippa* and *Mesodesma glabratum*; in the underlying sediment *Polygordius* sp. and a polyclad turbellarian exist, which must probably be considered as infralittoral species.

Pebbles

On Mediterranean shores, COSTA and PICARD (1956) found in midlittoral pebbles the isopod *Sphaeroma serratus*—more abundant among larger pebbles—and the amphipod *Gammarus olivii*, which predominates between small pebbles. In a rather sheltered area of the Possjet Bay (Sea of Japan) SCARLATO and co-authors (1967) observed a similar assemblage, with *Gnorimosphaeroma noblei* and *Anisogammarus posseticus*.

Boulders

The assemblage of boulders is not very different from that of the rocky shore, but it is impoverished. However, when the boulders are more or less piled up, sciaphilic species, mainly animals like those on overhangs and in crevices, may occur.

Mangrove Swamps

Mangrove swamps develop on tropical and equatorial shores of small or medium tidal range (e.g. 3–4 m), in sheltered areas where sedimentation rate of silt and clay is sufficiently high (e.g. in or near an estuary and sometimes in parts of large coral reef lagoons). On the muddy ground erect individuals of some tree species thrive, which may reach more than 40 m in height, but sometimes form big bushes of only 2–3 m or less. Detailed biological, ecological and physiological comment on mangrove trees would be out of place here. From the assemblage point of view, it is sufficient to list two important facts. (i) A mangrove swamp frequently exhibits both horizontal and vertical zonation, the different species forming large belts parallel to the shoreline on the gentle slope of the midlittoral zone. This zonation is less obvious in estuaries, owing to the alternate flows of sea water, brackish or fresh water, and also behind coral reefs where substrate diversity influences the distribution of the different tree species. (ii) Mangrove tree leafage and branches are above the sea level even at high tide, but trunks and piling roots and pneumatophores (where the latter exist) are transiently or permanently immersed. Hence, the mangrove assemblage comprises a complex of hard (wood) and soft (mud) substrates. The latter supports both an epifauna, which has been partly considered above with the midlittoral muddy bottoms in general, and an infauna. A part of the latter does not really belong to the midlittoral zone owing to water saturation in the sediment, but participating animals exhibit a special ability to endure frequent anoxic conditions in the interstitial water and low dissolved oxygen values in the upper-lying waters, which usually prevail during ebb tide.

From the biogeographical point of view, two entities may be recognized: the western mangrove which includes all coasts of subtropical and tropical American and West Africa, and the eastern mangrove on all coasts of the Indian and western Pacific Oceans. The two entities are separated by the immense width of the central Pacific Ocean and by the low temperatures of the Benguela Current on the south-west coast of Africa. Only a few genera (no species) are present in both entities. The eastern mangrove (about 45 tree species) is much richer than the western mangrove (ten species). The most widespread tree genus is *Rhizophora* which, as a 'pioneer', probably promotes the settlement of other species. Another widespread genus is *Avicennia*. As regards the associated fauna, the difference between the two entities is less marked.

Paradoxically, the primary production of the occasionally dense mangrove forest was generally considered to be rather low; GERLACH (1958), for example, estimated about $150 \text{ g C m}^{-2} \text{ y}^{-1}$. In fact, the role played by mangrove trees in primary production is still unclear. On the eastern coast of the United States, the amount of dead leaves accumulating on the bottom was estimated at approximately $7 \cdot 5$ metric tons (d.w.) $\text{ha}^{-1} \text{ yr}^{-1}$; fresh leaves only comprise 6% proteins, whereas their decay results in detritus which comprises 22% proteins after one year, probably owing to microbial activities. It seems that these very rich detritus sources provide an abundant food supply, not only for invertebrates but also for juvenile fishes.

All species of mangrove swamps are highly eurythermal, euryhaline, euryionic and tolerate large changes in dissolved oxygen. In the most schematic way the bios of mangrove swamps may be subdivided into three assemblages: hard-substrate, soft-substrate, and swimming-species assemblages.

(i) **The hard-substrate assemblage** has been the most thoroughly investigated, particularly on the southwestern coast of Madagascar (PLANTE, 1964). In general, trunks, pile-roots and pneumatophores bear assemblages similar to those of the supralittoral, midlittoral and upper fringe of the infralittoral zones of the neighbouring rocky shores. For example, at Tulear (Madagascar) the following zones may be distinguished: (*a*) A supralittoral zone without cyanophytes due to the substrate nature; *Littorina scabra* substitutes for *L. glabrata*, which is typical on the neighbouring rocky substrates. (*b*) A midlittoral zone supporting some algae (*Bostrychia*, *Catenella*) with an upper *Chthamalus* subzone and a lower *Elminius–Balanus–Crassostrea* subzone. In both these zones, the peeling conditions of bark, which differ with tree species, largely control the assemblage composition. (*c*) An upper fringe of the infralittoral zone with an assemblage, although to a limited extent more diversified than in the two preceding zones, is impoverished in comparison with that of rocky shores. The algal component includes some colonial diatoms, chlorophytes (*Cladophora*) and rare rhodophytes (e.g. Gelidiaceae and *Caloglossa*). Among the animals, the most abundant are the wood-boring pelecypods (Teredinidae) and the xylophagous isopod *Sphaeroma terebrans*. However, there are also sponges (e.g. *Prostylissa foetida*), alcyonarians (*Telesto*), and sometimes oysters as well as ascidians (Fig. 8-6). The motile fauna of the wood substrate, which may occur in all zones depending on sea level, mainly includes true amphibious genera, e.g. some crabs (*Aratus*, *Cardisoma*) and the air-breathing fish *Periophthalmus*. All of these may also be observed on the mud in the vicinity of the trees.

The brief description by LAWSON (1966) of the bios on trees of the western mangrove (West Africa) is very similar, with a *Littorina cingulifera* supralittoral zone, an upper midlittoral where *Chthamalus rhizophorae* predominates and a lower midlittoral with *Ostrea tulipa* and *Thais callifera*.

(ii) **The soft-substrate assemblage** has been less thoroughly investigated than the hard-substrate assemblage. Some aspects of the assemblage on the mangrove ground itself have been discussed above (p. 390). Herbivorous gastropods which cannot bury seem to be active mainly at low tide; they are either pulmonates (*Melampus*, *Cassidula*) or prosobranchiates (Cerithiidae). The most striking component of the assemblages is

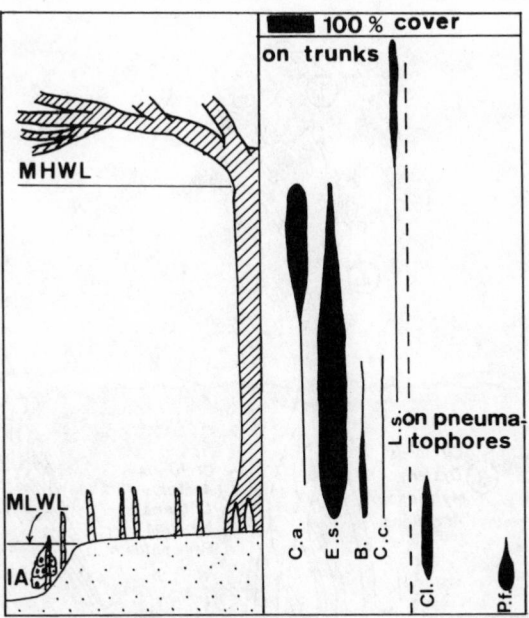

Fig. 8-6: Zonation of prevailing species on trunk and
pneumatophores of the mangrove-swamp tree
Avicennia nitida at Tulear (Madagascar); C.a.:
Chthamalus antennatus; E.s.: *Elminius sinuatus*; B.:
Balanus sp.; C.c.: *Crassostrea cucullata*; L.s.: *Littorina
scabra*; Cl.: Chlorophytes; P.f.: *Prostylissa foetida*
(sponge); IA: infralittoral, permanently immersed
assemblage with predominance of sponges and
Isognomon sp.; MHWL: mean high-water level;
MLWL: mean low-water level. (After PLANTE,
1964; reproduced by permission of the author.)

represented by crabs; all are burrowers and also seem to be active during emergence
periods. Their distribution depends on their individual ability for amphibious life; the
genera *Cardisoma* and *Sesarma* occupy the upper levels, whereas *Uca*, *Ucides*, *Goniopsis*,
Dotilla, etc., require wetter sediments (Fig. 8-7). In West Africa, LAWSON (1966)
noticed the gastropod *Tympanonotus fuscatus* and the pagurids *Clibanarius cooki* and *C.
africanus*.

At lower levels, in spite of transient emergences, the mud is water-saturated. The
infauna thus includes animals which belong to the upper fringe assemblage of the infralittoral zone. Among the most characteristic forms are the polychaetes *Marphysa* and
Dendronereis, many nematodes, some holothurioids, some deep burrowing decapod crustaceans (e.g. Alpheidae and *Upogebia*) and such fishes as *Gobinellus* and *Boleophthalmus*. All
are active during high tide and this sustains their infralittoral appurtenance (Fig. 8-7).

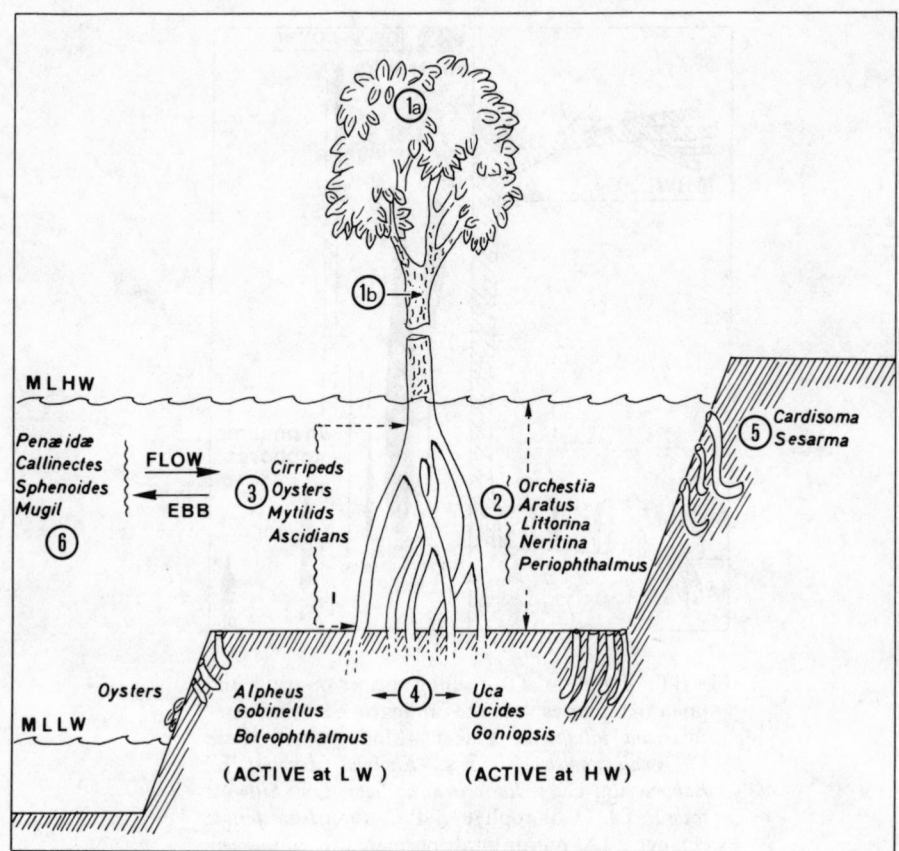

Fig. 8.7: Schematic outline of mangrove-swamp macrofauna (Cananeia, Brazil). 1a: terrestrial fauna of the tree's leafage; 1b: terrestrial fauna of wood and bark; 2 and 3: motile
and sessile species of the supralittoral and midlittoral zones, respectively; 4: burrowing
species of the mud flat; 5: upper-level burrowing crabs; 6: species migrating with flow
and ebb; MLHW: mean level high-water; MLLW: mean level low-water. (After
GERLACH, 1958; modified; reproduced by permission of Springer-Verlag, Heidelberg.)

(iii) **The swimming-species assemblage** contains species which migrate with the
tide; at high tide they find both food and shelter under the tree canopy. The important
assemblage components are fishes such as grey mullet (Mugilidae), *Scarichthys* and
Sphenoides, crabs (*Callinectes*), and, in particular, penaeid shrimps, which are sometimes
of very high commercial value.

Beyond the last mangrove trees, the outer muddly slope supports assemblages which
are no longer part of the mangrove swamp complex and may differ widely depending on
local conditions. Rather frequently a cover of phanerogams (*Cymodocea, Thalassia,
Halophila, Halodule*) occurs, sometimes mixed with chlorophytes. In other places there
are no macrophytes, at least in the upper parts of the muddy slope, and the assemblage
is dominated by crabs (*Uca, Dotilla, Macrophthalmus*) and lucinid pelecypods.

(3) Infralittoral Zone

In spite of their apparent diversity, in terms of both biogeographic differences and heterogeneity of environmental conditions, the numerous organismic assemblages of the infralittoral zone—certainly the best investigated in the whole marine benthal—may be fairly easily incorporated into a few large ecological entities.

First, we must emphasize that the upper level of the infralittoral zone almost always exhibits particular features. The so-called infralittoral fringe (STEPHENSON and STEPHENSON, 1949) on rocky shores with mean or large tidal ranges probably has a wider meaning than that implied by the authors. Obviously, this fringe includes all assemblages able to tolerate transient emergence, but also many other permanently immersed assemblages. In fact, emergence is only one of the environmental stress factors in very shallow waters, in addition to excessive illumination, changes in temperature and salinity, wave motion, etc.

Considering water movement, RIEDL (1964 and Volume I, 1971, pp. 1123 ff.) described, multidirectional flow patterns down to a depth of about $2\cdot5 \times h$ immediately beneath the surface, where h is the wave amplitude. It is striking coincidence that descriptions of the infralittoral zone by many authors point to more or less important changes in species distribution at a critical depth of about $2\cdot5$ to 5 m. RIEDL suggested that another critical level exists at a depth between λ and $\lambda/2$, where λ is wavelength. Above this level there is a layer where water movements are bidirectional, going alternatively up and down; below this layer, the movements are always unidirectional for a given direction of the waves down to the deepest level, where water may be moved by wave action. In fact, the second critical depth usually does not occur above a depth of 15 m and thus can influence the structure and composition of infralittoral assemblages only on very exposed shores. In the circalittoral zone, it might play some part in the distribution of some filter feeders whose bodies or colonies are at least partly fan-shaped.

(a) Hard Substrates

Species inhabiting the infralittoral zone are photophilic, except for those in places where illumination diminishes because of substrate topography. On hard substrates, two major groups of assemblages may be distinguished, either with predominant photophilic algae, or with predominant corals (coral reefs). In fact, this opposition is more apparent than substantial, because the reef-building hermatypic corals require sufficient illumination for their symbiotic zooxanthellae.

Assemblages with Predominant Photophilic Soft Algae

A comprehensive review of assemblages where photophilic soft algae predominate may be found in PERES (1967a). Hence, we can now restrict ourselves to more general, synthetic aspects. The basic assemblage component is algal turf, consisting mainly of articulate Corallinaceae (e.g. *Jania, Corallina, Arthrocladia*) together with *Hypnea, Bryocladia, Botryocladia, Gracilaria, Grateloupia, Gelidium, Gelidiella, Laurencia*, etc. Rhodophytes always predominate in this lower stratum. Obviously, genera and species differ depending on biogeography, wave exposure, water quality, etc. For example, in the Mediterranean, *Corallina officinalis* and *C. mediterranea* substitute for *Jania rubens* in slightly polluted areas (BELLAN-SANTINI, 1961). The algae of the lower stratum, especially when they are turfy, not only retain their own detritus (small thallus pieces) as well as detritus from coexisting animals, but also collect some suspended particles. This may result in the formation of something like a 'soil' with humic substances. The thickness of the algal turf depends on both its constitutive species and its location; exposure to strong wave action is generally adverse to the development of algal turf; it also seems that its thickness tends to decrease with increasing water depth, possibly because the turf-forming species are among the most photophilic species. The biomass of the algal turf, including the sessile flora and fauna which cannot be separated, varies between ten and a few hundred g (d.w.). BELLAN-SANTINI (1966), for example, found $234 \cdot 6$ g m^{-2} (d.w. without calcareous material) for the *C. mediterranea* algal turf near Marseilles. In polluted areas (especially when they are sheltered) most rhodophytes disappear and some tolerant chlorophytes (*Ulva*, *Enteromorpha*) tend to replace them. On tropical shores, rhodophytes are partly and sometimes almost completely replaced by such chlorophytes as *Caulerpa* (p. 406).

In many cases, large-sized algae and phaeophytes (e.g. species of *Cystoseira* and *Sargassum*) and, more rarely, rhodophytes, are associated with algal turf and form a higher stratum; for example, in the western Mediterranean Sea, *Cystoseira abrotanifolia*, *C. crinita*, *C. stricta*. The standing crop is much higher than in places where only the algal turf exists, mainly owing to the presence of these large algae: In the Marseilles area, BELLAN-SANTINI (1966) found a biomass of $2217 \cdot 4$ and $1256 \cdot 8$ g m^{-2} for the *Cystoseira stricta* and *C. crinita* assemblages (d.w. without calcareous material), respectively. In the Possjet Bay (Sea of Japan), SCARLATO and co-authors (1967) recorded a biomass of $1800-200$ g m^{-2} for assemblages dominated by species of *Sargassum*.

On tidal shores, a particular aspect of the higher stratum is represented by infralittoral belts of large-sized algae very similar to those of the midlittoral zone mentioned above. For example, on the shores of the Macquarie Islands, the higher stratum in the infralittoral fringe consists of *Durvillea antarctica*, which successfully withstands the surf effects and protects the lower stratum; on western European rocky shores in the infralittoral fringe, the belt of *Fucus serratus* and sometimes, a little lower, those of *Himantalia elongata* and *Bifurcaria rotunda*; on the coasts of northern New Zealand, from top to the beginning of the Laminarian zone (p. 406) here represented by *Ecklonia*, a belt of the brown alga *Carpophyllum* and another one of rhodophytes (*Pterocladia*, *Vidalia*, *Melanthalia*) (BERGQUIST, 1960); in the Okhotsk Sea, an upper belt of *Fucus inflatus* and a lower belt of rhodophytes (*Rhodymenia*, *Halosaccion*) with the shallowest-living individuals of the king crab *Paralithodes brevipes* (USHAKOV, 1951). In False Bay (South Africa), MORGANS (1962) pointed out different aspects of the infralittoral upper fringe depending on wave exposure (in sheltered areas, algal belts are replaced by a belt of *Vermetus natalensis*, p. 403); in very exposed areas, a belt of the brown alga *Bifurcaria brassicaeformis* prevails with the barnacle *Octomerus angulosus* just above the first kelps (*Ecklonia*). For details see PERES (1967a).

When the canopy of large algae of the higher stratum is sufficiently dense, the underlying algal turf of smaller species may be greatly impoverished. The same occurs under the leafage of the kelps, which must be considered as a special aspect of the higher stratum.

In tropical seas the higher stratum of large-sized algae is always less important and often in fact absent.

If we now consider the faunistic component of the photophilic algae assemblage from the same synthetic point of view, it is possible to subdivide it into three units: (i) fauna of the algal turf itself; (ii) fauna of higher-stratum algae; (iii) fauna of the rocky substrate itself. Of course, this classification is somewhat arbitrary, because many species belong to more than one unit or are able to move from one unit to another. Moreover, the nature of the animals which are associated with a given algal species largely depends on the characteristics of the alga itself: animals which settle (or crawl) on calcified Corallinaceae and soft algae are not always the same. Other differences in associated animals depend on thallus shape (filamentous or wide and flattened) and nature of its surface (smooth or rough), etc. While the fauna alone is considered in this paragraph, there are also many epiphytic small algae (diatoms and small multicellular species) which sometimes constitute an important part of the sessile cover of larger algae; some species of the algal turf, e.g. *Jania rubens*, may be epiphytic on species of the higher stratum.

(i) **The algal turf fauna.** The interstices of the algal turf are inhabited by a large number of small-sized species. Most of them are nematodes, polychaetes (chiefly syllids, juveniles of nereids, small sabellids), gastropods (mainly rissoids), harpacticoid copepods and many small isopods and amphipods. The genus *Caprella* often predominates, e.g. in the assemblage described by SCARLATO and co-authors (1967) in the infralittoral fringe on fairly exposed rocks of Possjet Bay (Sea of Japan) where the algal turf of *Grateloupia filicina*, *Sphaerotrichia dissessa* and *Laurencia nipponica* is inhabited by four species of *Caprella*; the most abundant (*C. cristibrachium*) attains a density of up to 12 750 individuals m^{-2}. Although rather scarce on Corallinaceae, sessile animals are more common on soft algae, i.e. hydroids, bryozoans, ascidians of the family Didemnidae, *Spirorbis*, etc.

A few species of the algal turf are a little larger than those mentioned in the preceding paragraph; in Mediterranean Sea Corallinaceae, the most common are prosobranchial gastropods (*Cerithium rupestre*, *Gibbula adansoni*, *Columbella* spp.) and decapods: the pagurid *Clibanarius misanthropus* and the crabs *Pirimela denticulata* and *Acanthonyx lunulatus*. Even some fishes (*Blennius* and *Gobius*) may be found there, but pelecypods are rare (*Cardita calyculata*).

The flattened basal part of the articulate Corallinaceae may produce, together with other calcified organisms (e.g., in the Mediterranean Sea, the gastropod *Vermetus triqueter* and the polychaete *Pomatostegus polytrema*) some calcareous concretions lying between the algal turf and the rock itself; the holes and crevices of these very irregular concretions are inhabited by many animals, the most common of which is the polycheate *Perinereis cultrifera*.

(ii) **Higher-stratum fauna.** The assemblage of higher-stratum algae comprises approximately the same taxa as the algal turf. However, three peculiarities must be pointed out. (*a*) The composition and abundance of sessile and sedentary fauna depends markedly on both shape and surface characteristics (smooth or rough, mucoid or not) of the algal thallus (see also 'phytal water' fauna, p. 143). (*b*) In general, except for algae which are filamentous or have a mucus-producing surface, the sessile fauna is more abundant on higher-stratum species than in the algal turf. For instance, on the Mediterranean, *Cystoseira stricta*, the foraminiferan *Miniacina miniacea*, the hydroids *Coryne muscoides* and *Sertularella ellisi f. lagenoides*, and the bryozoan *Schismopora armata* are common. Some polychaetes of the genus *Spirorbis* (often highly selective with regard to the nature of the substrate) also occur frequently, as do small compound ascidians (*Botryllus*, didemnids, some polyclinids). Small nudibranchial and prosobranchial gastropods are common on large-sized algae in sufficiently sheltered areas. Sometimes they are more or less specific for a given algal species. In Possjet Bay (Sea of Japan), for example, SCARLATO and co-authors (1967) found large numbers of *Thapsiella plicosa* on *Sargassum kjelmanianum* at a depth of 1·5 m, 10 000 individuals m^{-2} of *Epheria turrita*, and 1 m deeper, on *S. pallidum*. The two most common prosobranchial gastropods on the Mediterranean Sea *Cystoseira stricta* are *Rissoa guerini* and *Persicula clandestina*. (c) The higher algal stratum supports more motile species. Most of them, in particular amphipods and isopods, can both walk on the thalli and swim about. Some other crustaceans, generally large in size, are more effective swimmers and rarely clutch to the algae for resting; thus, in Possjet Bay *Sargassum* populations, *Neomysis mirabilis* is very common (up to 750 individuals m^{-2} in *S. kjelmanianum*), whereas the shrimp *Pandalus borealis* (11 individuals and 80 g m^{-2}) joins the mysid in the deeper *S. pallidum* population.

(iii) **Fauna of the rocky substrate itself.** On the rocky substrate itself, vacant areas are scarce, particularly if the algal turf is dense and made up of bushy species. In such a case this faunal compartment mainly consists of medium-sized species of the turf fauna able to weave their way through the turf. Prosobranchial gastropods, pagurids and crabs may be found under as well as within the turf. Large polychaetes (mainly nereids, eunicids, large-sized syllids, and sometimes aphroditids), together with nemertines, find more suitable conditions for moving about there than in the turf itself (where juveniles and smaller-sized species predominate). Only the largest-sized species of the rocky substrate fauna, e.g. actinians and regular echinoids, are able to compete for space with the turf-forming algae.

Among the herbivores, prosobranchial gastropods are the most numerous and diversified, e.g. species of limpets and littorinids, *Gibbula*, *Monodonta*, *Calliostoma*, *Tegula*, *Fissurella*, haliotids, etc, which are different from those in the midlittoral zone. The opisthobranchial gastropods present are often specialized for feeding on a single prey species (frequently a chlorophyte). The Polyplacophora (*Chiton*, *Acanthochites*, *Poneroplax*) which, like most prosobranchial gastropods, feed on epiphytic unicellular or small multicellular algae, are never common. The species of regular echinoids are few, but their individual numbers are often very high; thus, they may largely control the populations of small algae. On shores washed by cold waters, *Strongylocentrotus* is the most important genus, whereas *Arbacia* and *Echinometra* predominate in warm water areas. In temperate waters of western Europe, *Paracentrotus lividus* and *Psammechinus* spp. are common (e.g. ZAVODNIK, 1965).

The particular feeding behaviour of many carnivores common in this fauna, such as nemertines, grazers on sponges or colonial invertebrates (e.g. some opisthobranchs and asteroids), has been discussed on p. 137 ff. It therefore suffices to mention here that the carnivores which appear to play the major role in the system are: (i) crabs and pagurids, which are often also scavengers; (ii) prosobranchial gastropods, e.g. species of *Nucella*, *Thais*, *Ocinebra* and *Drupa*, which usually feed on herbivorous prosobranchial gastropods and barnacles. Their abundance increases when pelecypods and barnacles make up an important part in the assemblage. The same is true for such sea stars as *Asterias*, *Coscinasterias* and *Pisaster*.

Among the suspension feeders, one should also mention holothurioids of the order Dendrochirota, which are never common, and the vermetid gastropods which may become so abundant as almost to exclude the algae.

In some subtropical or tropical areas where coral reefs could not develop, some cnidarians may exist together with photophilic algae. One of the most common is an alcyonid of the genus *Telesto*, but there are also some hermatypic and ahermatypic scleractinians. For example, on the west coast of Kyu-Shu (Japan) where the tidal range is 3·7–6 m, KIKUCHI and ARAGA (1968) described an assemblage below an upper fringe of the infralittoral zone with a pelecypod of the genus *Septiger* and *Sargassum thunbergi*, where the algae (*Corallina*, *Galaxaura*, *Sargassum*) are mixed with *Balanus tintinnabulum*, some echinoids (*Echinometra*, *Pseudocentrotus*), the alcyonid *Dendronephthya gigantea* and some stony corals (*Favia*, *Platygyra*, etc).

Some suspension-feeding polychaetes also represent a significant part of the fauna of the rocky substrate itself wherever the water content in organic particles is sufficient for meeting their food requirements. Such sabellids as *Sabella*, *Spirographis* and *Dasychone* often coexist with algae. While serpulids may participate by their tube-building activities in the building of the basal concretion of some photophilic algae, they tend to exclude the latter where the environmental conditions are particularly suitable for them (p. 405). Sabellariids compete with algae in occupying the rocky substrate in areas washed by waters with high contents of suspended sediment; they can also completely substitute for them (p. 402).

Barnacles are sometimes present, mainly at shallow-water levels of the infralittoral zone in rather exposed places; they always belong to species different from those found in the midlittoral zone, e.g. *Balanus perforatus* in western Europe, *Octomerus angulosus* and, somewhat lower, *Balanus maxillaris* at False bay, South Africa, etc.

Pelecypods may be another important component of this assemblage. As has already been mentioned, mussel beds only exist in the infralittoral zone either on atidal shores or on tidal ones with special environmental conditions (p. 383). According to BELLAN-SANTINI (1962, 1966), Mediterranean Sea mussel beds are only a facies of the photophilic algae assemblage and their organic matter content (1473·1 g m^{-2}, d.w. after decalcification) is not as large as previously assumed. Arcidae are also sometimes encountered, e.g. *Arca barbata* in the Mediterranean Sea or *Arca boucardi* with up to more than 100 individuals m^{-2} and a standing crop of 250 g m^{-2} and in the Possject Bay *Sargassum pallidum* assemblage (SCARLATO and co-authors, 1967). On tropical shores, *Spondylus Chama* and Aviculidae (e.g. *Isognomon*) may be found.

Ascidians are rarely a predominant element of the assemblage, except for some special areas (p. 405). However, small Aplousobranchiata, mainly didemnids and polyclinids, and species of *Botryllus*, are fairly common everywhere; species of the families Polycitoridae and Polystyelidae become more important in warm water areas.

The photophilic soft-algae assemblage, with its three subassemblages (algal turf, higher algal stratum, rock itself) includes so many different ecological niches and has thus such a rich and diversified flora and fauna that a sufficiently marked change in one edaphic factor can result in facies characterized by a tremendous abundance of one or a few species. These facies are further discussed on p. 402.

The 'Laminarian Zone'

Phaeophytes of the laminarian group are a striking feature of rocky substrates on cold and cold-temperate shores. Some species may even inhabit some soft bottoms owing to the ability of their holdfasts to form an anchor-like spheroid enclosing a sediment mass. However, it appears that the ecological requirements of these phaeophytes are too diverse, and in consequence their vertical distribution as well, for the term 'laminarian zone' to be adequate. For example, on southern California coasts EMERY (1960) found three different successive levels: (i) From spring tide low water down to 3 m, a belt of *Egregia laevigata* exists with some rhodophytes (*Gigartina spinosa*, *Gelidium cartilaginum*), the phanerogam *Phyllospadix aorryi* (with epiphytic *Melobesia mediocris*) and a rich associated fauna: the echinoids *Strongylocentrotus franciscanus* and *S. purpuratus*, the large actinian *Anthopleura xanthogrammica*, some Pholadidae, the sea star *Pisaster ochraceus* which feeds on *Mytilus californianus*, etc. (ii) From 3 to 10 m the very euryplastic *Pterygophora californica* predominates. (iii) Deeper, down to at least 30 m, the giant kelp *Macrocystis pyrifera* occurs (together with the smaller species *Pelagophycus porra*); the distal thallus parts of *Macrocystis* float at the sea surface, and thus act as a breakwater.

It seems that, depending on their specific requirements and tolerances with respect to illumination, laminarians may be subdivided into three groups.

(i) Photophilic species, such as, on European coasts, *Saccorhiza polyschides*, *Laminaria saccharina*, *L. digitata*, *L. hyperborea*, or, on Californian shores, all species of *Laminaria*, *Costaria*, *Dictyoneurum*, *Dictyopsis*, *Postelsia*, *Lessoniopsis*, *Alaria*, *Pterygophora* and *Egregia*. The Antarctic species *Lessonia frutescens* and *Adenocystis lessoni* also belong to this group.

(ii) Sciaphilic species, for example, the mediterranean *Laminaria rodriguezei* which exists in areas of the circalittoral zone where the bottom water is pure and the water currents rather strong. The Antarctic *Lessonia nigrescens* belongs to this group, as may *Phyllogigas grandifolius* (although some individuals may be found down to 4 m depth).

(iii) Euryphotic species which may thrive both in shallow well illuminated and in deeper waters. Thus, *Laminaria ochroleuca*, which may be found on western European Atlantic coasts almost up to spring tide low-water level, exists in the Mediterranean Sea (e.g. in the Alboran Sea) at a depth of 40 m. Since young plants may be observed below the canopy there or larger individuals where the illumination is almost the same as at a depth of about 70 m, this species (the largest individuals of which may reach 8 m in length) may be considered as euryphotic. The same physiological peculiarity probably exists in other kelps; for example, in *Macrocystis angustifolia* of the Fuegian Province, in *Nereocystis luetkana* which may be found as deep as 55 m off California, and in some species of *Phyllaria*. In the eastern Atlantic Ocean and the Mediterranean Sea, *P. reinformis* has been observed at depths of 1–2 m down to 130 m. The most thoroughly investigated species from this point of view is the 'giant kelp' *Macrocystis pyrifera*, which is very common on Californian shores and in the subantarctic region; its thallus may reach several tens of metres in length. GRUA (1964a,b) found it at the Kerguelen Islands between 2 and 12 m in turbid waters and between 8 m and 30–40 m in clearer waters. Despite the fact that the giant kelp's full photosynthetic activity has been demonstrated to be about one-tenth of the incident light, this alga may be considered euryphotic because (i) young plants develop under the canopy of the older ones together with sciaphilic algae (Fig. 8-8, 7 and 8), and (ii) the photosynthetic rate decreases only slightly under full light (NORTH, 1959, 1971). In other respects floating thalli make giant kelp banks important as breakwaters, allowing the occurrence of assemblages characteristic of sheltered areas behind them.

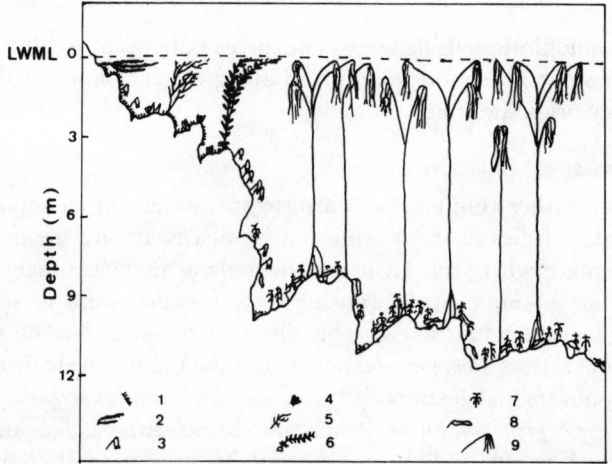

Fig. 8-8: Vertical zonation of the most abundant large-sized algae within and near a giant kelp (*Nereocystis*) belt at Carmel, California (USA). 1: *Calliarthron*; 2: *Egregia*; 3: *Laminaria*; 4: *Dictyoneurum*; 5: *Cystoseira*; 6: *Macrocystis*; 7: *Pterygophora*; 8: *Costaria*; 9: *Nereocystis*. LWML: low-water mean level. (After MCLEAN, 1962; reproduced by permission of Marine Biological Laboratory, Woods Hole.)

The kelp's standing crop is generally high but decreases with increasing depth in photophilic species. On Britanny coasts ERNST (1966) found 4 kg and 30 individuals m^{-2} of *Laminaria digitata* and *L. hyperborea* at a depth of 3–4 m, but only three individuals and 0·75 kg m^{-2} at 15 m. KAIN (1966) observed that the growth rate is very low at the maximal depth tolerated by *L. hyperborea* where light saturation rarely occurs. On the Kerguelen Islands GRUA (1964a,b) estimated the standing crop of *Macrocystis pyrifera* as high as 40 kg m^{-2} under optimal light conditions.

The fauna associated with European laminarians has been insufficiently ecologically investigated. In the Sea of Japan kelp beds DERJUGIN and SOMOVA (1940) distinguished three subassemblages (thallus, holdfast, substrate itself), which are rather similar for the three extant species: *Laminaria saccharina*, *L. bullata* and *L. japonica*. In general, it seems that most animal species associated with kelps are not very different from those found with other large-sized algae or on rocky surfaces and in crevices; only the species living on the thallus, for example, some hydroids and the limpet-like *Helcion*(= *Patina*) *pellucidus*, might be considered characteristic. The species which live on the thallus are far less numerous than, for example, those living on eel-grass blades, probably due to the mucoid nature of the kelp surface. For the sessile and sedentary associated fauna of *Macrocystis pyrifera*, GRUA (1964a,b) found a mean biomass of 3–4 kg m^{-2} and a maximum of 15 kg m^{-2}. On the other hand, fishes are rare and usually live in the marginal parts of kelp beds; these kelps may release a repellent substance.

Facies of the Photophilic Soft Algae Assemblage

Facies of the photophilic soft algae assemblage usually occupy rather restricted surface areas but are sometimes more widespread in a given region under the environmental conditions they originate from.

Facies of crustose calcareous rhodophytes

These facies are rather common on warm-temperate and intertropical shores. The *Lithophyllum incrustans* facies of the Mediterranean Sea is among the most widespread, thriving in every place where soft algae cannot settle or maintain their populations for various reasons, for example, rock abrasion by suspended sand in stirred waters or intensive grazing by invertebrates, e.g. echinoids, which mainly feed on soft algae. Then the whitish crusts of *Lithophyllum incrustans* cover most of the substrate together with some sponges. The prosobranchial gastropods *Patella caerulea*, *Vermetus gregarius* and *V. triqueter*, the pelecypod *Chama gryphina*, the scleractinian *Balanophyllia italica* and the echinoid *Arbacia lixula*—which is able to feed on *Lithophyllum*—are the most common associated species. In some places on crustose Corallinacea thrive sufficiently well for building some cornice-like formations by successive piled-up layers (MOLINIER, 1955). The *L. incrustans* facies also exists in lower rock pools of the Atlantic coasts of Europe, the water of which is renewed at each tidal cycle. In these rock pools many boring species (*Clione celata*, *Polydora* and pelecypods) produce numerous holes and cracks which are inhabited by a rich endolithon: polychaetes, nematodes, harpacticoids, ostracods, amphipods, etc.

The assemblage of *Arbacia*, sponges and bryozoans described by HEDGPETH (1954) on man-made hard substrates in the Gulf of Mexico could be interpreted to be a similar assemblage with a predominating animal component. In the Mediterranean Sea, on the north-western coast of Corsica, a *Lithophyllum byssoides* facies was observed. At the higher levels of wave-exposed shores of the eastern Mediterranean Basin, a facies with *Tenarea undulosa* and *Lithophyllum trochanter* is common (HUVÉ, 1957).

Facies of sabellariids

On all temperate and tropical rocky shores washed by water with a high content of suspended sand, polychaetes of the family Sabellariidae (*Sabellaria*, *Phragmatopoma*, etc) build their sandy tubes on the hard substrate. In places with serpulid or vermetid facies, the sabellariids live just below these. It seems that the associated fauna does not display particular characteristics.

Isoassemblages

On subtropical and intertropical shores without true coral reefs, one often observes assemblages with species characteristic of, or related to, those of photophilic soft algae assemblages occupying the same levels on rocky substrates of temperate coasts. In my opinion, they may be considered as isoassemblages of the former. These isoassemblages reveal two fairly common, sometimes associated, trends: (i) substitution of chlorophytes for rhodophytes and (ii) a marked increase in animal species at the expense of the algal component. Four main animal groups can predominate in these isoassemblages: vermetid gastropods, serpulid polychaetes, zoantharians and ascidians. Sometimes some scattered individuals of hermatypic corals also occur.

Vermetid isoassemblages

An upper fringe infralittoral zone occupied by aggregated tubes of vermetid gastropods was observed by CROSSLAND (1905) at Cabo Verde Islands, with a rich associated fauna in fissures and anfractuosities of the aggregate: e.g. polychaetes, sipunculids, nemertines, small crustaceans (mostly amphipods). On western African coasts SOURIE (1954) found *Vermetus adansoni*, together with typical infralittoral algal turf species, e.g. of *Jania* and *Amphiroa*.

The coexistence of Corallinaceae and vermetids (*Spiroglyphus irregularis*) was observed by LEWIS (1960, 1965) at Barbados; the associated fauna includes *Chiton marmoratus*, the prosobranchial gastropods *Fissurella barbadensis*, *Leucozonia ocellata*, *Thais floridana*, etc, as well as some actinians such as *Bunodactis stelloides* and *Bunodosoma* spp. At the lowest level, vermetids coexist with the zoantharian *Palythoa variabilis* and the sabellariid polychaete *Phragmatopoma californica*. The serpulid *Spirobranchus giganteus* sometimes substitutes for the vermetid.

In the West Indies, the Bermuda 'boilers' are the most popular formations of this group. These are funnel-shaped reefs of pleistocene calcareous sandstone with their summits surrounded by a mixture of two vermetid species, *Spiroglyphus irregularis* and *Petaloconchus nigricans*, together with infralittoral soft algae (*Gelidium pusillum*, *Dilophus*, *Sargassum*), *Fissurella barbadensis*, some serpulids, etc. On the coasts of Ghana, GAULD and BUCHANAN (1959) also observed vermetids, coexisting with *Palythoa* at their lowest level; further down a belt of the brown alga *Dictyopteris delicatula* occurs.

In areas of the Mediterranean Sea with surface temperatures that are not too low in winter (Algeria, Sicily, Lebanon), the vermetid *Spiroglyphus cristatus* often thrives. Depending on geomorphological and mineralogical characteristics of the shoreline, the vermetid formations can exhibit different aspects (Fig. 8-9). The most interesting aspect comprises wide flats named 'trottoirs' by DE QUATRE-FAGES (1854) who described them from Sicilian coasts. MOLINIER and PICARD (1953a,b) demonstrated that vermetids form a crust which is only 5–10 cm thick on calcareous sandstone. Owing to physico-chemical and biological (endolithic algae) erosion in the spray zone, the shore line slowly moves backwards; as soon as the rock is immersed the vermetids settle and thus protect it. At the outer edge of the flat, the most active water stirring induces quicker growth of the vermetids, resulting in a small ridge. The upward increment of vermetids sometimes oversteps the upper boundary of the infralittoral zone, allowing the settlement of midlittoral lower subzone (p. 378) species, such as *Neogoniolithon notarisi*, *Lithophyllum tortuosum*, *Rivularia atra*, the chiton *Middendorfia caprearum*, *Patella aspera* and *Brachidontes minimus* (Fig. 8-9).

Owing to the existence of the outer ridge, a very thin water sheet may be retained on the flat; heating by sun when the sea is smooth makes the flat assemblage very poor. For

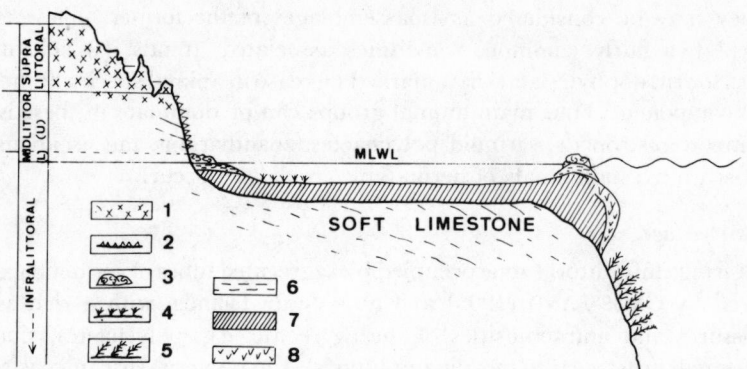

Fig. 8-9: Vermetid 'trottoir' on soft calcareous rocky substrate of Mediterranean Sea shores. 1: supralittoral zone with endolithic (decaying)
cyanophytes and rockpools; 2: *Chthamalus stellatus*; 3: small cushions of
Lithophyllum tortuosum; 4: very shallow basin with scarce rhodophytes (e.g.
Laurencia) and fauna (see text); 5: *Cystoseira stricta*; 6: *Neogoniolithon notarisi*;
7: dead vermetid 'veneering'; 8: living vermetids; (L) and (U): lower and
upper subzone of the midlittoral zone; MLWL: mean low-water level.
(After MOLINIER and PICARD, 1953a,b; reproduced by permission of the
authors.)

example, at Beirut (Lebanon) it only includes some scattered small rhodophytes, a
polychaete of the genus *Nereis* and some amphipods. On Algerian shores, in more
exposed places, water renewal is more frequent and consequently the assemblage is
richer, comprising species usually found in the typical photophilic soft algae assemblage;
apart from the vermetids, some subtropical elements occur, such as the pagurid *Calcinus
ornatus* in empty vermetid tubes, the echinoid *Arbaciella elegans*, and the boring pelecypod
Lithophaga aristata.

It must be emphasized that vermetid 'trottoir' and *Lithophyllum* 'trottoir', although
they have the same name in French, are quite different from each other: while both are
flats near the mean sea level, the first one consists only of rock with a thin crust of
agglomerated mollusc tubes and belongs to the infralittoral zone; the second is in fact a
biogenic building and belongs to the midlittoral zone.

Serpulid isoassemblages

In tropical and subtropical areas, the most widespread serpulid genus is *Pomatoleios*,
which forms cushions or linear pads at the highest level of the upper fringe of the
infralittoral zone; for example, at Tokara Islands, Kyu-Shu Archipelago, where these
cushions are inhabited by dense populations of the sipunculid *Phascolosoma albolineatum*
(TOKIOKA, 1953); on the western coast of Sri Lanka (ARUDPRAGASAM, 1970); in many
places on the South African coast (EYRE and STEPHENSON, 1938; MORGANS, 1957,
1959, 1962) where the serpulid is associated with some infralittoral soft algae such as
Gelidium. On southern Queensland coasts, another serpulid of the genus *Galeolaria*

replaces *Pomatoleios* and co-exists with typical infralittoral species, such as *Corallina*, *Jania fastigiata* and *Actinia tenebrosa*, some polychaetes (*Nereis*, *Phyllodoce*) and mytilids. A little deeper, *Galeolaria* is associated with an ascidian of the genus *Pyura*. In the highest crevices or empty tubes of the calcareous anfractuous mass built by *Galeolaria*, some midlittoral species are present, such as the pulmonate *Onchidium* and the spider *Desis* sp.

In some areas of the Corsican coasts, PERES and PICARD (1964) observed corbelled structures built by serpulids in the upper infralittoral fringe of rather sheltered places. Such a structure may reach about 1 m in height with a thickness of about 0·65 m at its lower third. The animals which build this very anfractuous and fragile cornice are mainly serpulids: *Protula* sp., *Serpula vermicularis*, *S. concharum*, *Vermiliopsis multicristata* and *Pomatostegus polytrema* (the last one predominates) with some participation of *Vermetus*. The associated very rich fauna mainly includes numerous individuals of the sipunculid *Physcosoma granulatum*, many polychaetes (mostly syllids and eunicids), and some pelecypods (e.g. *Arca barbata*, *Cardita calyculata*, *Modiolaria costulata*), etc.

Zoantharian isoassemblages

Infralittoral species of zoantharians are fairly common in some assemblages in the coral reef complex (p. 415). However, they also frequently occur in the upper infralittoral fringe. For example, in Barbados, LEWIS (1960, 1965) observed in some places a belt of *Palythoa variabilis* between the highest vermetid level and the belt of *Phragmatopoma californica* a little below. On some sheltered shores of South Africa where the tidal range is small (1·2–1·8 m), soft algae are replaced by *Palythoa nelliae* and the less common *Zoanthus durbanensis*. Most of these zoantharians contain symbiotic zooxanthellae, as well as hermatypic scleractinians.

Ascidian isoassemblages

On the same South African coasts which support zoantharians in sheltered areas, one finds an isoassemblage in more exposed places where the ascidian *Pyura stolonifera* excludes all algae except *Plocamium coccineum*; in the ascidian clumps the fauna is rich with *Bunodosoma capensis*, *Dasychone violacea*, etc. In the assemblage of *Corallina chilensis* and *Cynthia praeputialis* of the Sydney area and on the Tasmanian coast (tidal amplitude 1·6 m) an infralittoral fringe occurs, the highest level of which is typical with predominant algae (*Corallina officinalis*, *C. cuvieri*, *Jania fastigiata*, *Pterocladia*, *Laurencia*); the ascidian *Pyura stolonifera* lives a little below, together with a very rich associated fauna: sponges, polychaetes, bryozoans, many gastropods, etc. (BENNET and POPE, 1960). At Montemar (Chile) *Pyura chilensis* is associated with the algae *Corallina chilensis*, *Gelidium filicinum*, *Centroceros clavulatum*, *Colpomenia sinuosa*, some chitons, *Fissurella*, etc (GUILER, 1959b).

Filter feeders such as vermetids, serpulids (with the exception of those building the Mediterranean corbelled structure) and ascidians, inhabit rather exposed areas. On the contrary, zoantharians, whose metabolic requirements are partly met by organic matter released from their zooxanthellae, may occupy more sheltered places.

Chlorophyte isoassemblages on tropical shores

Because of the very large areas they cover on tropical shores, chlorophyte assemblages deserve special attention, even though they have been insufficiently investigated.

Assemblages with predominant chlorophytes may live on both soft (p. 448) and hard (see below) substrates. On well-sheltered hard substrates of the Andros Island Lagoon (Bahamas), for example, a large variety of chlorophytes occur, such as species of *Caulerpa*, *Acetabularia*, *Halimeda*, *Dasycladus* and *Penicillus*, together with some *Sargassum*, *Padina*, *Laurencia* and the crustose Corallinaceae *Goniolithon strictum*. Some chlorophytes have a more or less calcified thallus and their remains, together with those from *Goniolithon*, largely contribute to the accumulation of neighbouring soft bottoms. The fauna includes many boring animals, e.g. *Cliona*, *Eunice*, some sipunculids, and also a large variety of sessile pelecypods, e.g. *Pinctada radiata*, *Mytilus exustus*, *Pteria*, *Codakia*, *Asaphis* and *Chama*; the echinoid *Centrechinus antillarum* inhabits small holes. As usual, in the West Indies, gorgonians are fairly common (p. 420); some corals are also present, the most common of which are *Siderastrea radians*.

Very similar assemblages with a mixture of chlorophytes and corals inhabit immersed flats of many coral reefs (WELLS, 1957).

Polar Sea Shores

Immersed algal assemblages on rocky bottoms in the Arctic have rarely been described, possibly because ice-free upper rocky surfaces are rather scarce, possibly too, because at depths where wave and ice effects are not too severe, many algae cannot obtain sufficient light for photosynthesis. However, near Point Barrow, Alaska (Latitude 71° N) MOHR and co-authors (1957) observed two different assemblages at 8–12 m deep: (i) an upper assemblage of phaeophytes with very common *Phyllaria dermatodes*, *Laminaria saccharina* and *Desmarestia viridis*; and (ii) a lower assemblage of predominating rhodophytes with *Turnerella pennyi*, *Phyllophora interrupta*, *Antithamnion americanum*, *Phycodrys sinuosa*, *Polysiphonia arctica*, *Odonthalia dentata* and *Rhodomela lycopodioides* f. *flagellaris*. The invertebrate fauna, which is rather similar in both algal assemblages, includes some polychaetes—mainly of the genus *Spirorbis* (common) and a species of Polynoinae—gammarids and *Caprella* (numerous), as well as some decapods (*Hyas coarctatus aleuticus*), pagurids and shrimps. Among the latter, *Sclerocrangon* predominates, but species of *Agris*, *Enhalus* and *Spirontocaris* also occur. The fishes, mainly represented by *Boreogadus saida* and four species of the family Cottidae, seem to feed on crustaceans, mostly amphipods.

For Antarctic coasts, new information which largely complements the review by KNOX (1960) has been provided by GRUZOV and co-authors (1967) for Enderby Land (Davis Sea), PROPP (1970) for eastern Antarctic and Haswell Islands (1970), GRUZOV and PUSHKIN (1970) for Molodeshnaya and Mirny Stations, and DAYTON and co-authors (1970) for Cape Armitage and Hut Point areas (McMurdo Sound).

Before attempting a brief synthesis of our present knowledge of Antarctic assemblages on infralittoral hard substrates, I wish to make four general statements. (i) A vertical succession of different assemblages can be recognized everywhere. These assemblages exhibit some similarities on all Antarctic shores; the boundaries between adjacent assemblages, mainly depend on points (ii) and (iii) below. (ii) The upper level of the infralittoral zone is invariably bare, but the vertical range of this 'lifeless' belt tends to differ from place to places, according to differential direct effects of anchor ice shelves (see (iii)). (iii) GRUZOV and PUSHKIN (1970) distinguished two types of rocky upper infralittoral surface: (a) on steep coasts with ice shelves with snowdrifts which absorb most of the solar radiation, the bare zone extends down to 7 or 8 m; (b) on low coasts, no snowdrifts occur and the ice is fairly transparent; the bare zone extends only down to about 2 m; at this depth diatom populations begin; macrophytes have been observed at Molodezhnaya Station only below 5 or 7 m. In general, one may assume the lower boundary of the infralittoral zone to be at ca 30 m depth. (iv) The trophic network (ARNAUD, 1970; DAYTON and co-authors, 1970) is highly specific owing to large seasonal changes in primary production. If alcyonarians and hydroids, for example, permanently feed on suspended particles, some benthic invertebrates must change their diets completely twice a year. For instance, the asteroid *Odontaster validus* is planktonophagous or feeds on benthic diatoms and multicellular algae in summer, but is a scavenger in winter. The echinoid *Sterechinus neumayeri* is also a scavenger or a detritus feeder in winter, but feeds on algae or the superficial film of soft bottoms in summer. ARNAUD assumes that necrophagy represents a substituting diet when lack of light diminishes primary production. The fact that the same species are necrophagous throughout the year in areas which are not ice-free during the summer supports this view.

Synthetizing the information provided by the authors quoted above, the general scheme of infralittoral-zone hard bottoms in the Antarctic region may be described as follows:

(i) From the surface down to 2 m (sometimes 7 or 8 m): 'lifeless' zone.

(ii) Diatom assemblage: Diatoms belonging to *Pleurosigma, Amphiroa, Achnantes, Nitzschia*, etc, form a layer on the bottom in spring and in summer, which may be several cm thick. For the rest of the year this shallow-water rock is coated with a thick cover of ice crystals formed when the sea freezes early in winter. The diatom flora extends down to 5 to 10 m and sometimes overlaps the macrophyte assemblage which occurs deeper. The most characteristic species of the associated fauna—for example, in the Davis Sea—is *Odontaster validus* which, together with the hydroid *Tubularia ralphy*, make up more than 90% of the standing crop; the prosobranchial gastropod *Falsimargarita iris* is also present. The total biomass is 20–25 g m^{-2}, but only about half that in places not under snow-covered ice; in the latter case, diatoms are ten times less abundant, but *F. iris* is two or three times more abundant than in ice-free or transparent fast-ice areas, and many bryozoans are present.

(iii) Macrophyte assemblage: from 6–10 m down to 25–30 m. The plant fraction of this assemblage seems to be rather variable. In the Davis Sea only the rhodophyte *Phyllophora antarctica* (standing crop 35 g m^{-2}) and some calcareous rhodophytes occur. In other places the flora is more diversified. In general, it seems that in some areas two different vegetation levels can be distinguished: the upper level with predominant rhodophytes (*Lithothamnium, Lithophyllum, Phyllophora, Phycodrys, Plocamium*) and some brown algae such as *Desmarestia* and *Ascoseira mirabilis*; the lower level where phaeophytes predominate, mostly the large species *Phyllogigas grandifolius* and *Cystosphaera jacquinotti*. The most conspicuous animal in this assemblage is the echinoid *Sterechinus neumayeri*, whose standing crop may exceed 400 g m^{-2} in the Davis Sea. In the same area, there are also some *O. validus, T. ralphy, Lineus* sp. and *F. iris*, already observed in the diatom assemblage. The lower illumination results in the presence of some alcyonarians (*Eunephthya*) and bryozoans. On bottom areas below snow-covered fast ice, *Phyllophora* is less abundant (less than 1 g m^{-2}), as is *S. neumayeri*; in contrast, the bryozoans and especially the *Eunephthya* (biomass 45 g m^{-2}) thrive. *Eunephthya* and the bryozoans mentioned are intruders from the circalittoral assemblage.

According to ARNAUD (1974) the lower depth limit of the infralittoral zone is ca 30 m. The only multicellular algae in the circalittoral zone are *Phyllophora antarctica* and *Physodrys antarctica*, which disappear at a depth of ca 60 m. However, the vertical algae zonation may be strongly modified below a cover of sea ice; in the latter case, sciaphilic (circalittoral) species may be encountered at shallower depths, whereas the photophilic species are eliminated due to the insufficient illumination.

A very particular Antarctic assemblage has been observed by ANDRIASHEV (1967) on the under surface of fast ice, with very important diatom populations, small polychaetes, copepods and amphipods (mostly *Orchomenopsis*). Fingerlings of the nektonic fish (family Nototheniidae) *Trematomus borchgrevinski* are also common. ANDRIASHEV assumes that a particular food chain exists: diatoms → herbivorous invertebrates (mainly amphipods) → fish. This food chain would produce some fall-out of detrital organic material for the benefit of true benthic assemblage members.

DAYTON and co-authors (1970, p. 246) described anchor ice effects on benthic animals thus:

'The mats of ice crystals grow to be as much as 0·5 m thick and entangle benthic animals present on the substratum. Mobile animals crawl on the surface of the ice, where they also become entrapped. Portions of this anchor ice mat become detached from the substratum and float to the undersurface of the annual sea ice, where the animals are effectively trapped. . . . [The authors] frequently observed individuals of the fish genus *Trematomus*, the echinoid *Sterechinus neumayeri*, the asteroid *Odontaster validus*, the nemertine *Lineus corrugatus*, the isopod *Glyptonotus antarcticus* and various pycnogonids frozen into the undersurface of the sea ice. In one instance 12 *S. neumayeri* and 14 *O. validus* were frozen into the undersurface of 1 sq. m of the sea ice. Entrapped animals are not necessarily killed by the large ice crystals.'

DAYTON and co-authors further recorded many individuals especially of *Odontaster validus* and *Lineus corrugatus*, which escaped and fell, apparently unharmed, to the bottom. Those that fail to escape are transported out to sea each summer when the ice breaks up.

The decrease in the specific richness of marine life that is well documented on polar shores may be partly ascribed to the severe climatic conditions prevailing at present. However, the degree of impoverishment much more probably depends on the recent (in geological terms) history of each polar or subpolar region. The benthos of the Antarctic shelf, formerly surrounded by uninterrupted Southern Ocean, is two or three times richer in species numbers than the Arctic Ocean, which was probably wholly frozen and/or landlocked at least at different times of the Quaternary Era and at present has only rather poor communications with the Atlantic and Pacific Oceans.

Free-Living Multicellular Algae

As a response to some particular and transient environmental changes, several multi-cellular algae which are usually sessile, may separate from their substrate and continue to live and grow unattached for a more or less lengthy period. Two different types of this phenomenon may be distinguished in relation to both the initial cause of the release and the location of the free-living individuals in the water column.

Non-floating, free-living algae

In many sheltered coastal areas where—mainly in summer—long periods of very calm weather prevail, individuals of several benthic multicellular algal species separate from the bottom and often constitute a fairly thick mattress just above the bottom. In Port Erin Bay, Isle of Man, Great Britain (BURROWS, 1958) observed (i) permanent populations of *Laminaria digitata*, *L. saccharina* and *L. hyperborea* on the rocky substrates, but (ii) free-living populations, predominantly *L. saccharina*, *Saccorhiza polyschides*, *Chorda filum* and *Desmarestia aculeata*, on non-rocky bottoms (stones, pebbles, sand mixed with broken shells). The free-living populations become dispersed by autumn storms and subsequently died. In the brackish water of the Vaïne Lagoon near Marseilles (France), HUVE (1960) observed the same phenomenon in *Spermothamnion* sp., whose free-living forms exhibited very intensive growth and asexual multiplication.

GIBB (1957), who analysed populations of free-living *Ascophyllum nodosum*, suggested that environmental peculiarities which enhance development are (i) shelter and lack of strong currents (tidal or from river run-off); (ii) low sedimentation rates, i.e. rather firm bottoms (sand, gravel or rock); (iii) horizontal or gently inclined bottoms; and (iv) large salinity fluctuations. Free-living *A. nodosum* thalli exhibit rather striking modifications compared with sessile conspecifics: the thallus changes its colour and consistency and becomes more ramified, gas-bladder formation ceases, and receptable numbers increase. AUSTIN (1960), who studied *Furcellaria fastigiata* f. *aegagropila* on Danish coasts, mentions peculiarities which seem to be fairly common to all free-living multicellular algae: a disappearance of sexual reproduction; an increase in asexual multiplication; the disappearance and rhizoids; and adaptation to buoyancy due to an increase in ramification and a decrease in thallus dimensions.

Floating free-living algae

The adaptive trends which make some benthic algae able to survive when separated from their substrate reach their highest level in the floating *Sargassum* species. These have been dealt with in detail on p. 350.

Assemblages with Predominant Hermatypic Corals

General aspects

The term 'coral reef' should denote only those assemblages where true 'coral', i.e. Scleractinaria predominate. In reality, every 'coral reef' includes several different assemblages: (i) some in which corals (*sensu lato*) predominate in the biomass or in the percentage of bottom cover; (ii) others with or without corals (e.g. hard substrates of dead corals or other calcareous organisms, or sedimentary deposits originated from reef tract destruction), whose structure and dynamics are strongly controlled by group (i) assemblages. A 'coral reef' and its surrounding areas influenced by corals is a complex of assemblages.

'Coral' is not a scientific but a common term comprising all cnidarians which belong to two main taxa: Scleractinaria and Hydrocorallia, all exhibiting a coherent calcareous skelton; such a skeleton also exists in some Octocorallia (*Tubipora*, *Heliopora*). Most of these cnidarians contain symbiotic zooxanthellae and are named 'hermatypic' corals. The unicellular algae also exist in the soft tissues of many other invertebrates—sometimes also in those of protozoans—in the coral reef complex. Zooxanthellae are specialized dinoflagellates which live almost completely intracellularly; they are of paramount importance in coral reef ecology. The symbiotic relationship between corals and zooxanthellae may constitute one of the main causes for the development of coral reefs which cover about 19 million km^{-2} on intertropical shores at present.

The coral reef complex is the richest and most diversified floristic and faunistic assemblage in the whole World Ocean, owing to both the general increase in species richness with increasing temperature and the diversity of ecological niches.

According to age, prevailing environmental conditions and shape of the pre-existing rocky substrates, coral reef structures have been classified into four well-known categories: the fringing reef, on the shoreline itself; the barrier reef, separated from the shoreline by a back-reef channel; the atoll, where coral reef formations encircle a lagoon generally connected with the open sea by one or several openings; and the platform reef with an almost horizontal upper surface, located farther away from the shorline.

Ecological requirements of the coral reef complex

Before describing the different parts of the coral reef complex, we should recapitulate here the essential ecological requirements of corals (see review in YONGE, 1963).

(i) Temperature. Coral reefs thrive only in areas with average annual temperatures above 23 °C; they are never found below an annual average of 18 °C. Some species can withstand, but not build true reefs at, lower temperatures. Coral reefs thus exist only between Latitudes 30° N and 30° S, except for areas where warm currents extend beyond these parallels. The scarcity of coral reefs on the eastern side of the oceans has been attributed to upwellings which are more common and active there. Some authors assume that the relationship between temperature and coral reef building ability is due to CO_3Ca fixation conditions; calcareous elaboration is very slow at temperatures below 15 °C, but exhibits a logarithmic increase between 15 and 28 °C and is inhibited above 30 °C.

(ii) Salinity. In general, corals are sensitive to diminished salinities. For example, salinity decreases due to rain or estuarine discharge exert depressing effects on corals, differing according to species and duration. Many species can endure salinities as low as 75% of its normal value, but only for a few hours (HIATT, 1958). However, heavy and long-lasting rains associated with hurricanes and typhoons adversely affect corals, killing or at least 'whitening' them, owing to the release of zooxanthellae (GOREAU, 1964). The regular and heavy rains that occur in some regions are certainly an indirect but important factor in the geographical distribution of coral reefs. Conversely, rather high salinities such as those of the Red Sea and Persian Gulf coastal waters $(41^0/_{00}\text{ S})$ are tolerated by many species. Nevertheless, in general 'corals' cannot withstand salinities over $45^0/_{00}\text{ S}$, although *Porites* species may tolerate up to $48^0/_{00}\text{ S}$ (for details on temperature and salinity effects on corals, see KINSMANN, 1964).

(iii) Wave exposure. Mean wave exposure offers optimum conditions for coral growth: it maintains water purity, renews the zooplankton food supply, and prevents sedimentation of mineral particles on polyps. Although polyps can eliminate sand or silt particles by ciliary movements, this requires additional energy (see also Volume I: RIEDL, 1971, p. 1125).

In the most exposed places, the substrate is primarily covered by calcareous crustose algae, but water movement also directly affects coral distributions. The usual assumption that strong, massive forms occur at the highest levels, and more delicate ones (erect, cup-like or branched) at calmer, deeper levels requires qualification. In fact, corals at the highest levels must be sufficiently resistant to withstand wave stress, but they are mainly bush-shaped. They live epibiontically on calcareous dead coral masses and do not in fact participate in reef building. The most massive forms, which are real builders, live somewhat deeper, probably because they are more sensitive to fluctuations in environmental factors (temperature and salinity changes, excessive illumination, emergence) than the shallower forms. Because the massive forms flourish at these levels, vacant places for other species are very rare. However, where a place is available, colonies of more fragile coral species may be observed wherever increasing water depth results in a sufficient decrease in turbulence.

(iv) Emergence. Most corals cannot endure long-term emergence at low tide; the most tolerant species belong to the genus *Porites*. In general, the more porous the skeleton, the higher the resistance to emergence. Porosity probably allows the retention of sufficient moisture in the highest colony parts due to the ascent of capillary water from wet basal parts.

(v) Light. Due to the light requirements of their symbiotic zooxanthellae, hermatypic corals usually inhabit the infralittoral zone. However, some species may be found down to 60 m. According to GOREAU (1959) maximum species diversity occurs between 5 and 20 m; at 60 m the species number decreases to 25% of the subsurface value. Vertical species distribution greatly facilitates assemblage recognition on the outer slope and in the deepest atoll lagoons. The influence of light on hermatypic coral shape, metabolism and growth has been extensively reviewed in Volume I: SEGAL, 1970, pp. 162 ff.

(vi) Turbidity. Increased turbidity diminishes water transparency and hence does not suit hermatypic corals. Mineral particle deposition on polyps is also harmful despite the sweeping ability of polyps by ciliary currents; even if efficient, this consuming energy weakens the colonies. According to LOYA (1972), increased sedimentation rates favour branched species to the detriment of more massive ones, providing the latter lack sufficient sweeping mechanisms.

Physiognomy of the Indo-Pacific coral reefs

Indo-Pacific reefs are the most typical, except for those in marginal regions, e.g. some areas of the Persian Gulf and the western coast of Central America. Otherwise, fringing reefs tend to be atypical due to land influences. Barrier reefs and atolls which are sufficiently similar in structure and evolution in many areas of the Pacific and Indian

Oceans may both be taken as a common model of coral reef organization (PERES, 1967a), following the general scheme proposed by PICARD (1967) for the Tulear 'Great Reef' on the southwestern coast of Madagascar. The terminology used here has been suggested by CLAUSADE and co-authors (1971).

In every fringing reef or atoll three large units* may be distinguished (PICARD, 1967): front-reef, on-reef and back-reef† (Fig. 8-10).

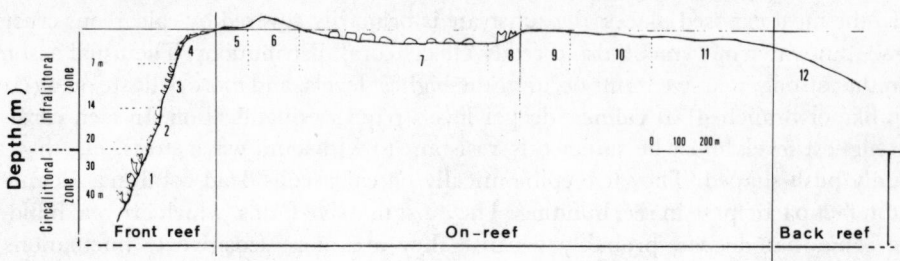

Fig. 8-10: Schematic transect of the Tulear (Madagascar) 'Great Reef', at 0–40 m depth, with its main assemblages. 1: Front-reef circalittoral assemblage with predominant sponges, soft algae and scattered scleractinians (Agariciidae, Pectiniidae); 2: Front-reef infralittoral assemblage in which massive scleractinians predominate; 3: assemblage of scleractinians, Corallinaceae and alcyonarians; 4: predominant *Acropora* spp. with Corallinaceae and *Millepora platyphylla*; 5: outer reef-flat (red algae) *Phyllospongia* assemblage; 6: boulder tract; 7: inner reef-flat assemblages with coral patches on dead reef, intermixed with sandy couloirs; 8: *Porites* microatolls; 9: sparse sea-grass beds (*Thalassia, Cymodocea, Halodule*) with large-sized actinians; 10: dense sea-grass beds of the same phanerogams with a more abundant and diversified fauna; 11: sea-grass bed of *Thalassodendron ciliatum* with scattered fauna; 12: back-reef assemblage: scleractinians and *Millepora* mixed with algae and sea-grass beds (mainly *Cymodocea serrulata* and *Halophilia stipulacea*). (Original.)

Front-reef. Front-reef is the part of the reef which faces the open sea. It includes two main parts: the outer slope and the outer reef flat.

(i) The outer slope, the declivity of which is generally steep, consists of coral-constructed formations and sedimentary deposits, mostly of skeletal origin. The upper part of this slope often exhibits coral-built crests called spurs, roughly perpendicular to the reef front and alternating with grooves, the floors of which are bare or largely filled with biogenic coarse sediments—sometimes mixed with gravel, pebbles or even boulders. Grooves and spurs also exist on the outer reef flat.

On the outer slope, Scleractinaria usually cover most of the substrate surface. Different levels may be distinguished depending on changes in dominant hard-substrate species. For example, on the windward outer slopes of the Marshall Island atolls, WELLS (1954, 1957) observed three main assemblages: (*a*) 0–15 m, with spurs and

*In French: *Ensembles*.
†In French: *fronto-récifal, épirécifal, postrécifal*.

grooves, where the most abundant corals (which never cover more than 25% of the whole hard substrate surface) are *Acropora cuneata*, *Pocillopora meandrium* and some species of *Millepora*, together with abundant algae; (*b*) 15–45 m, where branched and foliated forms, mainly *Echinophyllia*, predominate and (*c*) a lower level where cup-shaped and branched, partly hermatypic, forms such as *Leptoseris* are most abundant and extend down somewhat below 100 m. On the leeward outer slope of the same reefs, the two lowest subzones are similar, but the upper zone does not exhibit spurs and grooves but an assemblage where Scleractinaria predominate, both with branched forms (*Acropora reticulata*) and massive species (*Favia*, *Favites*, *Porites*). Soft-corals, such as alcyonarians of the genera *Sarcophytum*, *Lobophytum* and *Sinularia*, some Antipatharia and Gorgonacea, holothurioids, molluscs, etc., coexist.

On the upper part of the outer slopes of the Tulear Great Reef down to 19–24 m, Scleractinaria, alcyonarians and Melobesieae are the most important groups in the assemblages, one of them predominating depending on different microhabitats. Soft algae cover only a small percentage of the whole hard substrate. According to PICHON (1973, 1974), this part of the outer slope is the only one which may be considered as belonging to the infralittoral zone. At its upper levels, encrusting corals and Corallinaceae (*Porolithon*) predominate; a little deeper, branched but rather robust corals (*Acropora*) cover most of the reef surface. Massive forms (*Porites* spp., *Montastrea* spp.), the most active reef builders, predominate at deeper levels (Fig. 8-11).

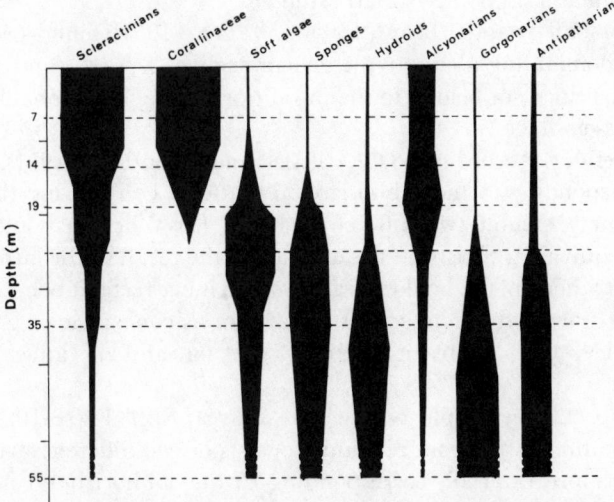

Fig. 8.11: Changes with depth in the substrate cover by different groups on the front-reef—i.e. outer slope—of the Tulear Great-Reef; the most marked species renewal takes place between 20 and 25 m depth. (After PICHON, 1973; reproduced by permission of the author.)

On the Tulear Great Reef, the outer slope below 19–24 m consists of coral flagstone with an abundant sessile flora and fauna. According to PICHON (1973), the assemblage on this flagstone belongs to the circalittoral zone and should be considered in the group of 'coralligenous assemblages' (p. 453). However, I prefer to mention it at this juncture, thus stressing the unity of the whole outer slope reef. This circalittoral outer slope assemblage includes only a few hermatypic corals of the Agariciidae and Pectiniidae families (*Leptoseris*, *Blastomussa*, *Cynarina lacrymalis*, etc), which play an insignificant role in reef building. Soft corals of the family Nephthyidae substitute for shallow-water Alcyoniidae. Soft algae are abundant, mainly at the upper levels of the flagstone, whereas the invertebrate fauna is rich and diversified mainly at the lower levels. Sponges are common (*Clathria* sp., etc), as are hydroids (*Aglaophenia*, *Halicornaria*), antipatharians, bryozoans, the holothurioid *Pseudocolochirus* sp. and others (Fig. 8-11). This assemblage probably corresponds to WELLS' (1957) '*Leptoseris* zone'.

Outer slope assemblages on soft substrates have been described only from the Tulear Great Reef. In the grooves, THOMASSIN (1972, 1978) found the following communities: (*a*) A very sparse macrofauna, including some *Conus* and crabs of the family Leucosidae, inhabits the upper-part (0–4 m) gravel bottom. (*b*) From 4 m down to 18–20 m, fine sands covering the grooves 'floor are permanently stirred up by swell and closely resemble those of hydraulic dunes. The polychaete *Ophelia peresi*, some isopods (*Eurydice*, *Exosphaeroma*), and amphipods (Haustoriidae) and decapods of the family Hippidae (*Hippa*, *Albunea*) are their most important inhabitants. (*c*) Deeper, the sedimentary trails, spots and hollows of the coral flagstone (GRAVIER and co-authors, 1970) are inhabited by a rich fauna of amphipods and cumaceans, together with the pelecypod *Venus lamellaris* and lancelets (*Asymmetron lucayanum*, *Branchiostoma* sp.). At the lowest level, the large gastropod *Lambis digitata* may be observed. In some hollows Melobesieae 'nodules' are common on poorly sorted sediment.

Near the Marshall Islands, below 150 m, WELLS (1954) found organismic assemblages with predominating ahermatypic genera such as *Sclerhelia* and *Dendrophyllia*. Of course, this fauna does not belong to the infralittoral zone, but probably to the offshore rocky-bottom assemblage (p. 463).

(ii) Outer reef flat. According to CLAUSADE and co-authors (1971), this part of the front-reef, corresponding to the subhorizontal platform constituting the top of a coral reef structure, may exhibit two different aspects: (*a*) Where a boulder tract (or its morphological equivalent) exists on the forepart of the reef flat, the latter is divided into an outer reef flat, ahead of the boulder tract, and an inner reef flat behind. The outer reef flat, the boulder tract and the inner reef flat appear as concentric or parallel strips. (*b*) Where the boulder tract is absent, the outer reef flat and the inner reef flat are not separated.

In the first aspect, for example, on the Tulear Great Reef (Fig. 8-10: 4) the part of the reef flat participating in the front reef unit consists of two different structures. The fore platform is the more external, corresponding to the subhorizontal upper part of the spurs. Immediately beyond there is a slight depression, the outer moat, parallel to the reef front, which lies between the fore platform and the reef glacis. The glacis is the first belt of the on-reef unit (p. 415). On the fore platform, scleractinians are abundant (*Acropora* spp., *Goniastraea retiformis*, *Pocillopora damicornis*, etc), coexisting with zoantharians and above all with crustose Corallinaceae. The exuberant growth of these calcareous algae, which is due to rather strong water movement, makes the reef platform sufficiently solid to withstand adverse effects. In the outer moat, reef builders (corals and 'Melobesieae'), which are more scarce and stunted, elaborate less calcareous material, whereas reef destroying organisms are more abundant. The preponderance of borers over builders explains the depressed profile of the outer moat and the dented surface of its floor.

Since increased water movement increases 'Melobesieae' growth, these algae may produce, on the windward side of atolls, for example, a structure whose highest part exceeds the mean low-water level and the reef platform by about 0·6 m (WELLS, 1957; LEWIS and TAYLOR, 1966; TAYLOR, 1968). Most of the bulk of this craggy 'algal ridge' consists of calcareous crustose rhodophytes (*Porolithon*, *Lithophyllum*, etc). Other components of the organismic assemblage are insignificant. However, at Mahé (Seychelles) the algal ridge supports some phaeophytes (*Sargassum*, *Turbinaria*), and—in holes and crevices—the echinoids *Diadema* and *Echinometra*, as well as various prosobranchs: *Turbo*, *Trochus*, *Drupa*, *Conus*, etc.

Additional evidence of the direct relation between wave exposure and 'Melobesieae' growth is provided by an increase in coral covering percentage on leeward atoll outer reefs. At Mururoa Atoll, for example, corals cover up to 60% of the whole surface area. In general, species diversity is higher leeward than windward.

On-reef. As emphasized by PICARD (1967), transportation and sedimentation predominate on the on-reef. Both of these processes tend to utilize materials originating from the reef itself: boulders, broken pieces of corals, pebbles, gravel and sand (resulting from reef erosion) as well as finer particles (Silt or clay), mainly in the vicinity of continents or islands. The transported materials sediment in different areas of the on-reef unit, depending on water movement (swells, storms, tidal currents), bottom profile and the specific characteristics of the material concerned (size, density, etc). Nevertheless, once deposited, the materials may be moved again, at least in the outer part of the on-reef unit, by unusual intensities of water movement, e.g. spring tide currents or hurricanes. In coral reef areas frequently exposed to hurricanes or typhoons, the distribution of the materials of different sizes may be deeply disturbed: sandy bottoms formed since the last typhoon may be moved out and big boulders pushed further landwards. Substrate stability is higher in less exposed coral reef areas: transport processes predominate only in on-reef outer parts (usually more or less recurrent, e.g. alternations of spring and neap tides or calm and stormy weather), whereas sedimentation processes predominate in on-reef inner zones. The latter situation is best exemplified by the Tulear Great Reef.

(i) The reef glacis, the most external zone of the on-reef, (CLAUSADE and co-authors, 1971) is a layer of organic concretions gently inclined seaward, the thickness of which depends on the location; it is always thinner in the front part (Fig. 8-10: 5 and 6). The assemblage of the reef glacis is very different from that of the reef platform. In the latter, 'Melobesieae' and corals always predominate, whereas these are very rare on the reef glacis; soft algae (*Sargassum*, *Turbinaria*, various chlorophytes, etc) predominate, associated with zoantharians and sometimes with sponges. Vermetid gastropods (*Dendropoma*) are represented by different species on the fore platform and reef glacis (and boulder tract). Sometimes the reef glacis exhibits belts, for example, on the Tulear Great Reef, from the outer moat towards the boulder tract: (*a*) a belt where the stony octocoral

Tubipora musica is common; (*b*) a belt where the large cup-shaped sponge *Phyllospongia madagascariensis* occurs; (*c*) a belt where chlorophytes (*Cladophora*, Ulvaceae, *Penicillus*, etc) predominate.

(ii) The boulder tract is that part of the reef flat where detrital and predominantly angular particles are deposited; among these, centimetric or decimetric sizes predominate. Two extreme aspects of this structure may be observed: (*a*) Oblong crags (domes) stretched backwards with the long axis roughly perpendicular to the reef front; between two crags or domes, 'tidal couloirs' occur with strong tidal currents; in the hydrodynamic lee of the crags there are 'gravel toils', tapering away at the back of the boulder tract. (*b*) Block ramparts, i.e. continuous deposits of detrital particles, with a marked relief, the crest of which roughly parallels the reef front; its profile is asymmetrical, with a steep slope at the back. The assemblage of the boulder tract has still been insufficiently investigated. Corals are absent, but 'Melobesieae' are abundant, together with foraminiferans (*Homotrema*, *Carpenteria*), sponges, zoantharians, cirripeds and pelecypods (*Arca*, *Chama*) (THOMASSIN and GALENON, 1977). Most of these sessile species inhabit the under surface of the boulders. Sedentary and motile species occur in the crevices and hollows in or between the blocks: sipunculids, small polychaetes, small gastropods, grapsid crabs, the apod fish *Lycodontis annulatus*, etc. The top of some boulders, if frequently emerged at each tidal cycle for a sufficient length of time and may be occupied by some typical midlittoral species such as the alga *Ralfsia* and the barnacle *Tetraclita serrata*.

(iii) The inner reef flat begins just behind the boulder tract. As on the outer reef flat, its assemblages comprise both corals (mainly Scleractinaria) and algae; the percentage of the hard substrate covered by the former generally decreases inwardly: even at low-water spring tide a water layer of 0·10–1·50 m always overlays the bottom. On the Tulear Great Reef, PICHON (1964) and CLAUSADE and co-authors (1971) distinguish four types of coral-built formations: the compact reef flat; the reef flat with coral alignments and sandy couloirs filled with sediments from skeletal origin; the reef flat with scattered coral growth; and the reef flat with annular-shaped microatolls. All these formations have a levelled flat top. In general, they succeed one another inwardly in the order listed.

The compact reef flat consists of a dead coral flagstone covered by living colonies of Scleractinaria, chiefly *Echinopora gemmacea*, *Goniastraea retiformis* and *Galaxea fascicularis*; the echinoid *Echinometra mathaei* is common. In the reef flat with scattered coral growth, *Acropora palifera* predominates.

The assemblage of the sandy bottoms ('couloirs' and various depressions) includes two different strates (THOMASSIN, 1969, 1978). The upper one has coarse sand and well-oxygenated interstitial water, caused by intensive stirring. Some phaeophytes may be observed, such as *Turbinaria* and *Sargassum*. The fauna is rather diversified: pelecypods (*Venus lamellaris*, *Donax veneriformis*, *Angulus vestalides*, etc) are less numerous than gastropods: *Cerithium hawaiensis*, *Otopleura auricasti*, *Nassarius albescens* and different species of the genera *Natica*, *Conus*, *Terebra* and *Pyramidella*. There is also a large diversity of crustaceans such as crabs (*Portunus*, *Leucosia*), the pagurid *Diogenes senex*, many small amphipods, etc. Polychaetes (*Euthalenessa djiboutiensis*, *Onuphis holobranchiata* and *Glycera* spp.) are common as are echinoderms such as ophiuroids, the echinoid *Metalia spatangus*, the holothurioids *Halodeima atra* and *Theloneta ananas*, which are eurytropic. The lancelet

Asymmetron lucayanum and the fish *Aserragodes filifer* are also present. In the lower sand layer, not moved by currents, interstices between coarse particles tend to become clogged by finer particles, and the oxygen content is lower. Polychaetes (*Pista cristata, Metaphoxus fultoni, Owenia fusiformis* and some tubicolous chaetopterids) predominate, while true burrowing species (scaphopods, pelecypods) are less abundant. The upper layer assemblage is similar to that of the 'Amphioxus sands' (p. 450, Fig. 8-10: 7).

Microatolls consist of one colony of a sceleractinian coral, mainly massive species such as *Porites* (here, *P. somaliensis*), more rearly, hydrocorals (*Millepora*), or alcyonarians. While the colony can develop in diameter, its vertical growth is restricted by the thin water sheet which remains at low water of spring tides. Living polyps only occur at the periphery of the microatoll and at the sidewalls of the enclosed basin or 'microlagoon' (Fig. 8-10: 8).

(iv) Sea-grass beds. The most internal part of the on-reef acts as a retention area for the suspended material, mainly for the lightest and smallest particles. Finer soft bottoms and more sheltered conditions combine to promote sea-grass bed settlement. For example, moving inwards, with slowly increasing depth on the Tulear Great Reef, one finds the following series: a scatterd sea-grass bed with mixed *Thalassia, Cymodocea, Halodule*, few Scleractinaria and a considerable number of large actinians (*Actinodendron plumosum, Cryptodendron adhesivum*); a sea-grass bed supporting more dense populations of the same species but with a much richer fauna; a sea-grass bed of *Syringodium* and *Thalassodendron ciliatum* with an impoverished fauna (Fig. 8-10: 9, 10 and 11).

In spite of small discrepancies between these three different sea-grass beds, they are inhabited by a common fauna (THOMASSIN, 1971, 1973). In this assemblage the same subdivisions as in Mediterranean *Posidonia* beds may be recognized (p. 441). A diatom felt, foraminiferans, many hydroids, copepods, amphipods and many small gastropods, such as *Phasianella, Smaragdia*, Cerithiidae can be found on living leaves. Amphipods, isopods (*Paracilicaea, Synisoma*), carid prawns, small cephalopods and fishes, swim beneath the leaf canopy. The most important components of the assemblage inhabit the sediment itself: numerous polychaetes and echinoderms; among the latter, asteroids are the most diversified, with species of *Pentaceros, Pentaceraster, Protoreaster, Linckia*, the cushion-like *Culcita* (THOMASSIN, 1976), etc. Regular echinoids (*Tripneustes, Toxopneustes, Astropyga*), holothurioids (*Holothuria, Halodeima, Synapta*) and endopsammic ophiuroids (*Amphioplus integer*) are also abundant. Some of the gastropods feed on 'detritus' or small algae, such as species of *Cypraea* and *Strombus*. Others are predators: *Mitra, Pusia, Vexillum, Conus, Naticarius, Polinices, Fasciolaria, Murex*; the large crabs *Thalamita, Portunus, Matuta* and *Calappa* are predators or scavengers. Suspension feeding is mainly represented by pelecypods: Lucinidae (*Codakia*) and Arcidae predominate in *Syringodium*—*Thalassodendron* meadows, whereas the large Pinnidae (*Pinna, Atrina*) are more common in the mixed sea-grass beds.

(v) On-reef unit in fringing reefs. From the Mahé (Seychelles) fringing reef, TAYLOR (1968) describes the general scheme of on-reef organismic assemblages similar to that occurring in the Tulear Great Reef. Immediately behind the algal ridge, the outer reef flat is depressed and a water sheet 1–2 m deep remains at low tide. Most of the bottom is sandy, but with scattered boulders and pebbles. The algal component of the assemblage is rich: *Sargassum* and *Turbinaria* predominate and are associated with *Padina, Jania, Halimeda*, etc. Corals are sparse and represented by different species on sand and boulders. Gastropods are the most abundant molluscs. Holothurioids are uncommon, but ophiuroids abound, while echinoids are common (*Tripneustes* and *Toxopneustes* on sand, *Echinometra* on hard substrates). The inner part of the reef flat, with a water sheet of only a few decimetres at low tide, features a sea-grass bed of predominant *Thalassia*, associated with other phanerogams, in particular *Cymodocea*. Scleractinians are rare. Crustaceans (hoplocarids, *Alpheus*, *Calappa*) and molluscs are more abundant. Among the latter, the percentage of gastropods (ca 55%) is lower than in the outer reef flat; pelecypods (45%) are mainly represented by suspension feeders of the family Lucinidae together with some Veneridae.

The rocky shoreline is fairly steep and exhibits only trivial assemblages of thoracic cirripeds in the midlittoral zone and of small littorinids in the supralittoral zone.

(vi) On-reef unit in atolls. On the windward side of Eniwetok Atoll (Marshall Islands) WELLS (1957) observed a sill separating front-reef and on-reef at the inner limit of the *algocorallian* belt belonging to the front-reef unit. Just beyond the sill, the most external part of the on-reef forms a basin which is always immersed, even if at low tide only by a very thin water sheet. The covering by scleractinians—the commonest is *Acropora digitifera*—may exceed 50%. Soft algae, mainly chlorophytes (*Halimeda*, *Caulerpa*), but sometimes rhodophytes (*Jania*), are abundant, together with foraminiferans (*Calcarina*, *Marginopora*), many zoantharians, large actiniarians (*Stoichactis*, *Discosoma*), the echinoid *Echinometra*, etc.

Behind a slight raising of the reef flat approximately up to the small spring tide low-water level, WELLS (1957) distinguished three successive assemblages. (*a*) The most striking feature of the first one, always immersed, is the presence of numerous scleractinian microatolls (mainly built by *Acropora palifera*) scattered on a coarse coral sand bottom. Both the vicinity of the microatolls and the sand couloirs and patches are inhabited by a very rich and diversified fauna: polychaetes, crustaceans, molluscs (*Tridacna*, *Lima*, *Conus*, *Cypraea*, Nudibranchiata, etc), echinoids, holothurioids and fishes. At the shallowest places a facies of mat-forming chlorophytes may be observed: *Cladophora*, *Cladophoropsis*, *Dictyosphaeria*, etc. (*b*) Inwardly, both the water sheet at low tide and the mean water motion diminish. The stony octocoral *Heliopora caerulea* builds micratolls, the coalescing of which may produce a flagstone: the sand couloirs between these constructions support some scleractinians with fragile skeleton and many small crabs and fishes. (*c*) The most internal assemblage mainly includes coral heads of the highly emergence-tolerant *Porites lutea*, with a very poor associated fauna.

On the leeward side of Eniwetok Atoll, the outer-reef assemblages are similar to those on the on-reef and the inner part of the reef flat of the windward side. However, the coral diversity is much lower than on the windward reef flat; large *Acropora haebes* bushes and colonies of branched forms of the genus *Montipora* are common. The most internal belt of the leeward reef flat is inhabited by *Porites lutea*, as is the similar level of the windward reef flat.

Back-Reef. The 'back-reef' is formed by the assemblages located behind on on-reef, and is directly influenced by on-reef characteristics, particularly the input of suspended organic material and the intrusion of some species from neighbouring on-reef assemblages. This means that wherever trival infralittoral and midlittoral assemblages occur, both on hard substrates (see above: Mahé reef; TAYLOR, 1968) and soft substrates (Hainan Island Beach; NAUMOV, 1959), the back-reef must be considered to be absent.

Unlike on-reef formations, which are always supported by a dead coral mass, back-reef structures may exist either on sediments (whether of skeletal origin or not) or on rock beds. In fact, the back-reef comprises all atoll lagoons and back-reef channels of barrier reefs, providing the latter are markedly influenced by the neighbouring barrier reef.

(i) Back-reef of the Tulear Great Reef. The channel behind the Tulear Great Reef (PICARD, 1967) is a good example of a back-reef unit because its assemblages are only partly controlled by the neighbouring reef complex.

On a substrate made of dead corals mixed with sands, the inner slope of the Tulear Great Reef exhibits a succession of three different levels with increasing depth: (a) upper level predominated by *Acropora pharaonis*; (b) middle level inhabited by *A. formosa*, *Tridacna crocea*, and the large pelecypod *Atrina vexillum* in sandy patches mixed with small boulders; (c) subhorizontal lower level where colonies of *Millepora intricata* and *M. dichotoma* predominate, together with scattered colonies of massive scleractinians and big boulders of dead corals covered by a very rich epifauna of anthozoans, sponges, etc. A little deeper, some soft-bottom assemblages occur on which coral reef influences still predominate: coarse clean sands with a rather poor fauna including lancelets, pelecypods and crustaceans; sea-grass beds of *Thalassodendron ciliatum* on poorly sorted sediments or of *Syringodium isoetifolium* on finer and better sorted sediments. Some knolls rise from the bottom up to spring tide low-water level. They vary from pinnacle-like structures to flat-topped (sometimes tabular), more massive forms (Fig. 8-10: 10–12).

In other respects, terrigenous influences from the adjacent river Fiheranana predominate in the deepest areas of the back-reef channel. Where the sediment contains both silt and coral sand, one finds lancelets, which usually occur in the coarser and current-stirred soft bottoms, together with more interesting free-living, cup-shaped species of two usually sessile taxa: the corals *Heterocyathus* and *Heteropsammia*, and the bryozoan *Anotesopora magnicapitata*. On more muddy sands the assemblage mainly consists of pelecypods (*Tellina*, *Solen*, *Ensiculus*), with some brachiopods (*Lingula*), many polychaetes and crustaceans (mainly amphipods of the genus *Ampelisca*), whereas the mud filling bottom hollows are inhabited by pelecypods (*Macoma*, *Paphia*) and the flattened echinoid *Lovenia*.

On the island of Mauritius the very narrow barrier reef consists of a thin pad made of algae and corals on a basaltic rock; the on-reef unit is absent. With increasing depth, the back-reef reveals a succession of three assemblages: an upper boulder level covered with *Sargassum* and *Turbinaria* with the echinoid *Echinometra mathaei*; a middle level with predominating coral patches (mostly *Acropora* and *Pavona*) separated by sandy couloirs; and a lower level with sea-grass beds (*Thalassodendron ciliatum*, *Syringodium*, *Halodule*) similar to those described from the Tulear back-reef channel. Farther from the reef there are sands covered with chlorophytes not influenced by the reef complex (PICHON, 1967).

(ii) Back-reef of atoll lagoons. Coral reef structures in atoll lagoons largely depend on lagoon size and depth as well as on number, and on the depth and width of channels with the open sea. In large lagoons, water circulation and stirring are usually sufficiently intense to meet the requirements of most reef complex organisms, except for Melobesieae, which require a rather high degree of water movement and can be observed only on windward shores of very large lagoons. Coral growth in the lagoon is strongly influenced by water turbidity and sedimentation rate. The most important coral constructions tend to occur in the leeward part of the lagoon, where the higher mean water movement intensity reduces the sedimentation rate.

The Eniwetok Atoll lagoon (Marshall Islands), which has been analysed by WELLS (1975), is used here as example. There are three main zones:

(a) Leeward lagoon reef flat. Owing to similar environmental conditions, the lagoon leeward reef flat assemblage is similar to that of the outer leeward reef flat. Corals, with the most characteristic species *Porites andrewsi*, cover most of the hard substrate surface. The fauna is rich and diversified, with many polychaetes, crustaceans, molluscs (*Spondylus, Tridacna, Hippopus, Lambis, Cypraea, Conus*, nudibranchs), some actinians (*Stoichactis*) with their commensal shrimps and fishes, soft alcyonarians (*Sarcophyton, Sinularia*), and many sponges and echinoderms. The calcareous chlorophytes *Halimeda*, whose thallus remnants are an important source of lagoon sediments, is also common. In the sandy couloirs between coral heads, the fauna is rich in polychaetes and crustaceans; molluscs and echinoderms predominate. Among the pelecypods, species of the genera *Tellina, Mactra, Codakia, Pinna* are common; the gastropods mainly belong to *Strombus, Polinices, Voluta, Terebra, Oliva* and *Melo*. Echinoids are rare (except for *Centrechinus antillarum*); holothurioids and above all asteroids (*Acanthaster, Nardoa, Echinaster, Culcita, Oreaster*, etc) abound. The fish fauna is also rich. In the vicinity of the channels between the lagoon and open sea, stronger currents occur; this results in increased turbidity, dampening coral growth and benefiting the algal component of the assemblage.

(b) Windward lagoon reef flat. The upper level of the windward lagoon reef flat supports the *Porites lutea* assemblage, which usually inhabits other reef flats with different exposures. Below this belt, the reef flat surface is irregular, and coral growth seems to be more or less diminished by the higher turbidity. Coral heads are scarce, mixed with microatolls and sandy couloirs and banks, the surface of which is sometimes densely covered by chlorophytes. When the lagoon is wide, a small algal ridge may be observed at the outer edge of the reef flat.

(c) Lagoon bottom and slopes. These are occupied by a coarse sand consisting of *Halimeda*, coral and foraminiferan remnants. *Halimeda* is by far the most important plant of the assemblage, but one may also observe other chlorophytes and sometimes mixed sea-grass lawns (*Thalassia, Cymodocea*) and *Halimeda*. The sand infauna does not present peculiar characteristics: the best represented taxa are crustaceans, molluscs and echinoderms. Corals are widely scattered, the most characteristic species being *Acropora formosa*. According to WELLS (1957), two different levels may be distinguished: an upper level (3–25 m) in which *A. formosa* is associated with *A. reticulata* and some gorgonians; and a lower level, below 25 m, where *A. formosa* occurs, together with *A. rayneri*.

In some places there are coral heads documenting successful colonization of the sandy substrate by corals. In general, the assemblage of these coral heads in the Eniwetok Atoll is similar to that of the leeward lagoon reef flat. Lagoon coral formations rising from the bottom may exhibit very different sizes and shapes (see review in CLAUSADE and co-authors, 1971). The composition of coral head assemblages is related to water depth.

Closed or semi-closed atoll lagoons. Benthos assemblages of closed lagoons are insufficiently investigated. An example of a wholly closed lagoon is Fangataufa Atoll (French Polynesia), which probably became isolated about 1000 years ago. An artificial channel was opened in 1966. Corals are poorly developed; the lagoon is mainly inhabited by pelecypods with a low biomass; the most common species are *Pinctada maculata, Chama imbricata, Arca ventricosa* and *Tridacna maxima*. The latter, probably because it contains symbiotic zooxanthellae, is so abundant that empty shells from dead individuals increasingly clog the lagoon.

The semi-closed lagoon of Maturei Vavao Atoll (Tuamotus) (depth: 45 m), is only incidently connected to the sea through very shallow channels (< 2 m deep) at high tide; amount and strength of the water inflow depend on wind and swell. According to RENAUD-MORNANT and co-authors (1971), the fauna of this lagoon is currently undergoing a process of impoverishment. Its specific richness has already markedly decreased, mainly with regard to corals, of which there are only three species: *Porites solida*, *Acropora pulchra* and *Fungia repanda*. Echinoderms are still common, mainly the holothurioid *Halodeima atra*, with others, such as *Holothuria illa* and *Microthele difficilis*, and the echinoids *Echinometra mathaei* and *Laganum depressum* (which may be found together with an enteropneust) on hard and sandy bottoms respectively. Molluscs seem to be less influenced by the isolation: *Tridacna maxima*, *Pinctada maculata*, *Crassostrea cucullata*, *Arca plicata*, *Chama* sp., *Isognomon* sp. and *Cypraea moneta* are common. The macrobenthos biomass is fairly high, the meiobenthos highly diversified in shallow-water soft bottoms; abundance increases with depth and also in finer sediments. In the deepest parts of the lagoon, nematodes and mollusc larvae predominate and probably feed mainly on detritus. In coarse sediments, the percentage of predators increases.

Biogeography of Indo-Pacific coral reefs

The global description of Indo-Pacific coral reefs, based on few examples, must not obscure their very large variability—a consequence both of the rather narrow ecological requirements of hermatypic corals, and of drastic changes in surface temperature during the Quaternary glacial periods.

The geographical distribution and population abundance of many benthic taxa, such as foraminiferans, sponges, alcyonarians, gastropods, several crab and fish families, are closely correlated with the distribution and development of hermatypic corals. Hence, changes in environmental conditions preventing corals developing or, at least, maintaining their reefs results in the disappearance of most of the associated fauna. Moreover, like all highly mature ecosystems, the coral reef complex, if destroyed or damaged, requires a very long time for recovery.

According to STODDART (1969), the maximum luxuriance of the coral reef complex, for both hermatypic corals (about 50 genera and 700 species) and their associated fauna, occurs in the western part of the Central Pacific Ocean, from the Ryu-Kyu Islands up to Indonesia. This focal area comprises many archipelagos (Philippines, Marshall, Mariana, etc.). With increasing distance from this area, the centrifugal impoverishment in species numbers of hermatypic corals and their associated fauna is evident. For example, the Australian Great Barrier reefs comprise only 150 species of hermatypic corals and the Tuamotus reefs only 20 genera. The reasons for this centrifugal impoverishment are still being debated, the two most frequently cited being eastwards decrease in surface temperature and geographical disjunction by wide oceanic areas.

The coral formations of the California Gulf are sufficiently atypical to deserve particular mention. They do not in fact constitute true reefs. For example, at El Pulmo there are only two species of *Pavona*, two of *Pocillopora* and some of *Porites*—all in patches on Quaternary limestone beds. There are also some *Gorgonia* and soft algae. Plants and animals usually occur on crests of vertical parts of the blocks, because the horizontal parts are sediment-covered. In San Gabriel Bay and near Espiritu Santo Island, the coral formations consist of clumps of *Pocillopora elegans*, sometimes up to 12 m in diameter with filamentous soft algae, some molluscs and rare colonies of *Porites californicus*. The percentage of covering by corals is high only on top of the blocks. Algae predominate on the lateral faces. In the Panama Gulf where temperatures as low as 16 °C can fleetingly occur owing to casual water inflow from upwelling or doming areas, corals are largely absent. It is generally admitted that the coral reef fauna of the Pacific coasts of Central America is Indo-Pacific in nature, but highly impoverished. However, EKMAN (1953) noticed some similarities between coastal assemblages of Atlantic and Pacific Central America in spite of their present isolation by the impassable barrier of the Panama isthmus. While there are very few warm-water stenothermal species common to both coasts, the similarity is very high at the generic level of many zoological groups. EKMAN (1953, p. 44) concluded

'that the Pacific coast of Central America, despite its geographical position, shows a lesser degree of affinity than the Atlantic coast to the Indo-West-Pacific. This is doubtless due to the fact that both the West Indian and the Indo-West-Pacific fauna were formerly parts of the same original fauna, namely that of the Tethys Sea'.

In the Indian Ocean the specific diversity of corals and their associated animals also decreases with the distance from the focus of maximum luxuriance postulated by PICHON (1973) and located in the Seychelles–Mauritius region. The most marked impoverishment occurs on the southern coasts of the Mozambique Channel, probably due to the lower average temperature, and in the Persian Gulf, presumably because of both increased temperature and salinity.

Tropical Atlantic Ocean coral reefs

True coral reefs exist only on the western side of the Atlantic Ocean (WALTON-SMITH, 1954). On African coasts (mainly in the Gulf of Guinea) only 20 Scleractinaria species are known, five of which are hermatypic. The most typical coral reef assemblages occur in the West Indies. Here, GLYNN (1973) listed 62 Scleractinaria species (48 of which are hermatypic) belonging to only 25 genera. The associated fauna is also much poorer than in the Indo-Pacific. Some animal groups or genera which are widespread in that region are completely absent, e.g. the pelecypods *Tridacna* and *Hippopus*, the large actinians and their commensal fishes and the stony octocorals. The specific diversity of many other taxa is strongly diminished, e.g. 'Melobesieae', molluscs (mainly Conidae and Cypraeidae) and soft alcyonarians. On the other hand, gorgonians are more diversified and often abundant; the genus *Rhipidogorgia* is endemic.

A rather well-developed fringing reef occurs on the northern coasts of Jamaica; a 240 km long and 12–40 km wide barrier reef exists on British Honduras coasts. Some coral reef banks occur on the southern coasts of Jamaica and Puerto Rico and elsewhere in the Caribbean Sea and Gulf of Mexico. Nevertheless, even the most developed Atlantic coral reef cannot be compared to the impressive coral constructions in many Pacific Ocean areas.

The main features of the Atlantic reefs have been very well analysed by YONGE (1963), who lists five main differences from Indo-Pacific reefs: Atlantic reefs (i) rarely occur on the edge of submarine platforms, probably due to erosion processes in dead reefs during the Quaternary period; (ii) are restricted to the most suitable windward shorelines; (iii) often do not reach the water surface; (iv) always lack an algal ridge; and (v) have very poor associated fauna at frequently emerged upper intertidal levels. YONGE assumed that the surface-water cooling during the glacial Quaternary episodes killed the majority of older reefs and that those existing at present are very young (3000–6000 years). According to GLYNN (1973a), the frequent hurricanes in this area may partly inhibit coral constructions.

Jamaican reefs are the most complete and typical in the West Indies. They have been very well analysed by GOREAU (1959). According to him, hermatypic corals cannot grow near Jamaica below 30 m; from that depth up to ca 15 m, coral colonies are not very numerous and cover less than 50% of the substrate; number and size of the colonies sharply decrease below 20 m. *Montastrea annularis* predominates. Betwee 15 and 8 m a 30–100 m wide assemblage with *Acropora cervicornis* as dominant species occurs, as well as some colonies of *M. annularis* and *Porites* sp.

The highest percentage (about 90%) of substrate covering by corals is reached in the spur and groove zone. Spurs steeply rise from depths of 6–7 m up to 1 m; the very robust *Agaricia agaricites* predominates, associated with various species of the genera *Montastrea* and *Porites*. Sandy grooves between the spurs are very narrow. In the 0–1 m breaker zone, small bushes of *Acropora palmata* predominate, sometimes mixed with *Millepora alcicornis*. The specific diversity of corals decreases with decreasing depth.

Near the mean level of low-water spring tides, the reef flat mainly consists of dead and unconsolidated colonies of *Acropora palmata*; this reef flat is 40 m wide on an average and the water sheet at low-water spring tides does not exceed 0·5 m. Living scleractinians are scarce except for encrusting colonies of *Diploria clivosa* (mainly in pools and ditches); *Millepora, Lithothamnium, Halimeda* and especially *Zoanthus sociatus* are common. Behind the reef flat is the back-reef channel 10–300 m wide and 2–15 m deep; its bottom consists more of *Halimeda* and foraminiferan remains than of coral fragments. Coral colonies are rare, except for the free-living *Manicina areolata*, but gorgonians, molluscs and echinoderms are numerous. In some places, small patches of the turtle grass *Thalassia testudinum* may be observed.

The two slopes of the back-reef channel are different. Close to the reef flat is GOREAU's (1959) 'rear zone' where colonies of massive corals (*Montastrea* and *Porites*) predominate. The assemblage on the other side (i.e. the shoreline) down to depths of 2–3 m exhibits a higher coral diversity: *Acropora palmata, Montastrea annularis, M. cavernosa, Diploria strigosa, Porites astreoides, P. porites* and *Siderastrea siderea* are most common; *Manicina areolata* lives on sandy patches. In places with stronger wave motion, some Lithothamnieae may occur.

Jamaican coral reefs fit rather well into the general scheme adopted for Indo-Pacific coral reefs. The front-reef unit includes four of GOREAU's assemblages: *M. annularis* zone, *A. cervicornis* zone, spur and groove zone, *Acropora palmata* zone. The latter corresponds to the algo-corallian zone, which is an almost invariable component of all coral reefs. The back-reef unit is also easy to recognize with the back channel and the assemblages of both its slopes. The homology of the *Acropora palmata–Zoanthus* reef-flat assemblage is less obvious. However, investigations by PICHON (1967·) on some reef flats off Mauritius (Indian Ocean) suggest that the Jamaican reef flat belongs to the front-reef unit. The on-reef unit seems to be missing in West Indian coral reefs, possibly due to both insufficient abundances and growth rates of building corals and the fairly recent recovery of construction processes.

Jamaican reefs appear to be built mainly by *Montastrea annularis*; *Acropora palmata* which predominates in shallower areas participates at a later stage of reef construction.

Crustose Corallinaceae microatolls, near Cozumel Island, Yucatan, (BOYD and co-authors, 1963) form the upper rims of cylindrical structures with vertical to overhanging sides. Calcareous rhodophytes are abundant and diverse (*Goniolithon*, *Archaeolithothamnium*, *Lithothamnium*, *Lithophyllum* and *Lithoporella*) whereas coral colonies (*Porites*, *Diploria*, *Agaricia*, *Favia*, *Colpophyllia*) cover less than 5% of the substrate. The origin of these microatolls remains speculative.

The West Indies region, at least with regard to the corals and their associated fauna, seems to constitute a single biogeographical unit, with a roughly centrifugal impoverishment away from the West Indies. In upwelling areas, such as on north-eastern Venezuelan coasts, coral growth is greatly diminished. The biogeography of corals and coral reefs seems to be more intricate there than in the Indo-Pacific region. During the Quaternary glacial periods, some corals appear to have been protected from low temperatures in refuge areas with less pronounced decreases in surface temperature. Jamaica was such a refuge, representing the focus of the true Caribbean coral fauna. As an example of centrifugal impoverishment from the focus, it should be mentioned that Bermuda reefs contain only 20 coral species. The larvae of several species may have too short a planktonic life to be transported from the Bahamas to Bermuda. The lower temperature in Bermudan surface waters probably inhibits the growth of some species such as *Agaricia agaricites*. Bermudan reefs, whose thickness never exceeds 10 m, are mainly built by *Diploria* and *Montastraea annularis* and at the highest levels (probably midlittoral) by vermetid gastropods (p. 403), which are less important in Antillean and Caribbean reefs (LABOREL, 1966). To the south, from the mouth of the Orinoco River up to São Luiz do Maranhão, the surface waters of the Amazonian region, with their higher content of suspended mineral matter and low salinity, are adverse to hermatypic coral life. Hence, the coral species (not more than 25) and their associated fauna in the Brazilian tropical region differ somewhat from those of the Antillean–Caribbean region. The rather significant percentage of endemic species in Brazilian reefs, many showing affinities with Indo-Pacific or Miocene forms, strongly suggests a very ancient separation between the two regions.

However, according to LABOREL (1969a,b) the general structure of Brazilian coral reefs is similar to that of Jamaican reefs, although the following differences must be noticed. (i) *Acropora* species are absent (as on Bermuda reefs), and the most abundant scleractinians in the outer slope down to 15 m are species of *Mussismilia*. (ii) The most important West Indian reef builder, *Montipora annularis*, is replaced by *M. cavernosa*. (iii) The reef flat lies at a higher level than in West Indies (where it is always covered at low tide with a water sheet of at least a few decimetres) and is inhabited by many algae; zoantharians (*Palythoa* and *Zoanthus*) are also common. (iv) The reef flat's highest levels which emerge at low tide bear vermetid-built formations (as in Bermudan reefs) and 'Melobesieae'.

It seems that the tropical Brazilian region, the southern limit of which corresponds to the Cabo Frio upwelling area (near Rio de Janeiro), has two foci of coral reef activity which probably functioned as refuges during the Quaternary surface-water cooling. The southern and most important focus lies in the Bahia–Abrolhos Archipelago area; the northern focus is encompassed by line through Cape São Roque, Recife, Fernando de Noronha and Das Rocas Island (LABOREL, 1969b).

Trophic net in the coral reef complex

A detailed account of the trophic net in coral reefs is beyond the scope of the present chapter. However, this net is of considerable ecological interest and exhibits such distinctive features that a brief account appears necessary (see also Volume V, part 2). In view of the general oligotrophy of tropical waters the luxuriant richness and high specific diversity of coral reef assemblages appear paradoxical, seeming to challenge evaluations of standing crop and production. However, for Eniwetok Atoll (Marshall Island ODUM and ODUM (1955) found a biomass of about 850 g organic matter (d.w.) m^{-2}; this seems a reasonable average value. The authors suggest a production rate of ca 12·5 times the standing crop. Such a global production estimate for all trophic levels is of little ecological significance. In fact, reliable data on production in coral reef assemblages are currently available only at the primary (plant) level. The multiple relationships between the numerous assemblages of the coral reef complex preclude dealing only with assemblages in which hermatypic corals predominate. It is necessary to consider the trophic net in the reef complex as a whole.

One could cite the cyanophytes among the classic primary producers; their production has been estimated at 1–2 g C m^{-2} d^{-1} (BAKUS, 1967). The production of benthic diatoms is probably of the same order of magnitude. For soft multicellular algae, a value of 5–10 g C m^{-2} d^{-1} seems to be reasonable. Recently, WANDERS (1976) estimated the yearly gross and net production of the algal 'community' in the shallow-water reef of Curaçao (Netherlands Antilles) 3840 and 2550 g C m^{-2}, respectively. According to MARSH (1970), the primary production of calcareous crustose rhodophytes is about ten times lower (0·66 g C m^{-2} d^{-1}). Measurements of *Thalassia* bed primary production by ODUM (1957, 1963) gave a value of 4650 g C m^{-2} yr^{-1}, i.e. of the same order of magnitude as that for soft algae.

JOHANNES (1967) and JOHANNES and GERBER (1974) estimated input and output of net plankton (seston would be a more appropriate term) in a coral 'community' of the windward reef at Eniwetok Atoll. They noticed that fragments of benthic algae which occupy the algal flat comprised more than 90% (w.w.) of the net input by the coral community over two days and almost 60% during a single night. Faecal pellets accounted for a minor input during the day, but for 36% of the net input at night (JOHANNES and GERBER, 1974). Suspended algal fragments may be taken up by certain herbivorous benthic fishes (Pomacentridae, Acanthuridae) and also by plankton-feeding fishes.

The problem of symbiotic zooxanthellae (*Microsymbiodinium adriaticum*) in hermatypic corals (Volume I: SEGAL, 1970) and many other animals of reef assemblages, e.g. some xeniid alcyonarians, zoantharians (*Zoanthus sandwichiensis*; REIMER, 1971), the giant clams of the genus *Tridacna*, etc, is much more intricate. Measurements by many authors of zooxanthellae primary production yielded values of 100–2700 g C m^{-2} yr^{-1}. Such variability may be partly related to the different species used in the experiments. It is now well documented that organic substances synthetized by zooxanthellae may provide an energy source for the host animal. It is further well documented that many hermatypic corals cannot be considered as truly photo-autotrophic organisms, as they require exogenous food.

It is difficult to determine the pathways of zooxanthellae primary production to the subsequent link of the food chain. Certainly part of this energy is directly used by the host animal, with a high efficiency because the producer lives within the consumer. Experiments on ^{14}C fixation by *Pocillopora damicornis* revealed that the fraction of fixed carbon found in host tissues amounts to 35–50% (MUSCATINE and CERNICHIARI, 1969). More recent experiments by MUSCATINE and co-authors (1972) showed that release of soluble products (glycerol, glucose, alanine) by cultivated zooxanthellae is enhanced by an homogenate of host tissue compared to a culture in sea-water-enriched medium.

Another pathway of energy transfer from zooxanthellae to the secondary trophic level is cellular release under particular environmental conditions. It was well documented long ago that heavy rains during hurricanes or typhoons induce a massive release of zooxanthellae by corals, which become 'whitened'. In Tulear Great Reef it has also been observed that, at low-water spring tides, the sea water in reef pools turns dark brownish due to suspended zooxanthellae. In *Palythoa* and *Zoanthus* populations it has been observed that their released zooxanthellae, which are·embedded in amorphous bodies or in spiral structures like a mesenteric filament, seem to be in very good physiological condition. The periodic release of zooxanthellae may thus occur in response to environmental stress. The life cycle of *Microsymbiodinium adraticum* (FREUDENTHAL, 1962) suggests periodic release of the gymnodinioid zoospore which probably represents a free-living stage. Unfortunately, a detailed evaluation of primary production by zooxanthellae which enter the trophic chain via cellular release by host animals is currently not possible.

We also must take into account the 'paraprimary' production (PERES and PICARD, 1969), at the lowest level of the trophic pyramid. Its existence, although sometimes questioned, has been demonstrated by JOHANNES (1967). He noticed a marked increase in the concentration of suspended particulate organic aggregates in oceanic waters as they were crossing the windward coral reef at Eniwetok Atoll (Marshall Islands). The aggregates mainly consist of mucus released by corals, zoantharians and other animals. Micro-organisms, chiefly bacteria, but also ciliates, small flagellates and diatoms are embedded in—or clustered on the surface of—the aggregates. An additional source of aggregates may be adsorption of dissolved organic matter onto bubbles produced during the turbulent crossing of water over the reef. QASIM and SANKARANARAYAN (1970) suggested that very small calcareous particles may function as initial nucleus for aggregate formation. The suspended material may be caught not only by zooplankters (JOHANNES, 1967) but also by various other animals of coral reef assemblages, among them scleractinians (*Pocillopora, Fungia*) and xanthid crabs. This has been proved by DISALVO (1971), using ^{35}S-labelled bacteria. SOROKIN (1972; see also Volume IV: SOROKIN, 1978) demonstrated that polyps of several coral species actively consume and quickly assimilate ^{14}C from labelled bacterioplankton and dissolved organic substances (hydrolysate of algal protein at a rather dilute concentration of 1 mg l^{-1}); dinoflagellates and flagellates are consumed and assimilated to a much lesser extent. Entrapping by filtering structures such as the 'reef mesh' or the sandy bottoms probably facilitates utilization of such material by the infauna.

Even without considering the food supply from aggregates and dissolved organic matter as data available are insufficiently reliable, it seems likely that the primary production of the coral reef complex as a whole amounts to ca 1200–2500 g C m^{-2} yr^{-1}. The very low production level at the highest levels of the trophic pyramid, especially with regard to fishes of commercial value, is all the more surprising as the production at the basic level of the pyramid is very high. For instance, on the 10 000 square miles of the Seychelles Plateau, the fish standing crop, including migratory species, is as low as 2200 tons during optimum seasonal conditions. According to BARDACH (1959) the mid-summer standing crop of fishes in Bermuda coral reefs amounts to only 490 kg ha^{-1}.

Thus, it may be assumed that part of the primary (and paraprimary) production is exported from the reef complex towards neighbouring pelagic and benthic assemblages. In the pelagos of the Bikini Atoll, zooplankton concentrations average four times those in oceanic waters on the windward reef (JOHNSON, 1949, 1954), but this enrichment mainly benefits lagoon assemblages; downstream from the atoll zooplankton concentrations are only twice those in offshore oceanic waters. In general, enrichment of bottom areas leeward of coral atolls seems to be rather insignificant, except in areas close to the reef complex, for instance, interisland atoll reef benthos assemblages. This means that the coral reef complex may possibly be an almost autarchic ecosystem retaining and accumulating almost all the energy it receives. The low level of production at upper trophic levels (mainly fishes) might be explained by energy leakage to blind alleys of the trophic chain, related to the high diversity of ecological niches or by blockings.

The relatively hypothetical causes for the discrepancy between lowest and highest levels of the trophic pyramid may be classified into three groups.

(i) **Energy loss by carbonate production.** According to CHAVE and co-authors (1972) carbonate gross production, i.e. the amount of CO_3Ca produced by the reef 'community' per unit area of sea floor, ranges from 4×10^2 to 6×10^4 g CO_3Ca m^{-2} y^{-1}. For the entire reef system, the carbonate average gross production is approximately 10^4 g CO_3Ca m^{-2} y^{-1}.

(ii) **Blocking mechanism.** A rather important part of the primary and paraprimary production caused by decaying fragments of metaphytes, released zooxanthellae, organic aggregates, unicellular algae ripped away from the bottom by waves or tidal currents, etc, all suspended in sea water running over the reef surface, may be entrapped by two main filtering structures; the 'reef mesh' and the coarse sand banks of the inner reef flat.

The 'reef mesh' (PERES and PICARD, 1969) comprises a porous system, i.e. a network of small holes (0·5–5 cm in diameter), cracks and crevices, originated from both failures in the building process and activities of boring organisms. It is inhabited by a rich fauna of microphagous species feeding on the suspended organic particles transported by the sea water which flows through the mesh; there are also some small carnivores. Most marine ecologists concerned with coral reef hard substrates have focussed their attention on the most superficial layer of these substrates and neglected the fact that the dead reef parts do not consist of massive and compact limestone. According to CLAUSADE (1970) and PEYROT-CLAUSADE (1977) the main components of the cryptic reef mesh fauna are, in sequence of decreasing importance: nematodes, small polychaetes (mainly Eunicidae and Nereidae), crustaceans (mainly peracarids, hoplocarids, Alpheidae, Xanthidae); molluscs (e.g., Polyplacophora, small gastropods and mytilids; small-sized ophiuroids and aspidochirote holothurioids. Even fishes, such as Syngnathidae and Gobiidae, are present, although less abundant and diversified. The accessibility of cryptic species to predators is probably higher than originally assumed by PERES and PICARD (1969), because many of them leave the reef mesh at night. Moreover, young (small-sized) fishes may enter the network and feed on the small invertebrates. Nevertheless, under-consumption of these microphages and small carnivores by larger-sized predators is quite likely.

Coarse sand banks of the inner reef flat entrap part of the suspended organic material while tidal currents run over the inner reef flat. The material entrapped benefits various microphages such as small amphipods, polychaetes, pagurids, and ophiuroids.

(iii) **Food-web peculiarities adverse to fish production.** The secondary production (i.e. that of first-level consumers inhabiting sandy couloirs and banks) is of little benefit to fishes because it is markedly utilized by large-sized, fast-growing predatory invertebrates, mainly prosobranchial gastropods (Terebridae, Harpidae, Conidae, Naticidae and Olividae) and, to a lesser extent, by asteroids and crabs. Although these invertebrates are less abundant in sea-grass beds, the food available for fishes is also restricted. Among the pelecypods, *Atrina* and *Pinna* species are too large to be eaten by fishes (except the biggest which rarely enter these shallow biotopes), while lucinids are buried too deeply. Holothurioids, asteroids and large gastropods are rarely consumed (for the same reason as the largest bivales). Infaunal polychaetes and sipunculids are protected from predation by the dense network of phanerogam rhizomes and roots. Therefore, fishes can feed only on small epifauna components, e.g. nuculid pelecypods, amphipods and carid prawns.

In temperate waters, BAKUS (1966, 1967, 1969) pointed out that species which feed on unicellular algae are mainly invertebrates, later consumed by carnivorous fishes and invertebrates, whereas in coral reefs many herbivorous fishes—e.g. Scaridae—occur which feed on sessile microalgae, mainly cyanophytes. Most of these fishes are both herbivorous and corallivorous, but represent a blind alley of the food chain because they are active almost exclusively by day, whereas most piscivorous fishes are nocturnal. For these two fish groups, feeding behaviour determines the periods of major activity; during inactive periods, security requirements dominate behavioural patterns. Predator–prey relations between these two fish categories seem rather weak and to occur almost only during transition periods from day to night and night to day, probably mainly at twilight (HOBSON, 1972; VIVIEN, 1973a,b, 1974a,b; HARMELIN-VIVIEN, 1979).

The respective importance of zooxanthellae and external food sources for the nutrition of hermatypic corals remains controversial. Probably the capability of hermatypic coral animals or other animals harbouring zooxanthellae for meeting their energy requirements either from internal substances released by their symbiotic algae or from external sources (e.g. dissolved organic substances, aggregates, bacterioplankton, phytoplankton, zooplankton) differs widely in the different species. QASIM and co-authors (1972) found that production (P) and consumption (R)—deduced from respiration measurements—are almost equal for *Pocillopora damicornis*, whereas *P. verrucosa* and *Acropora indica* exhibit excess production, i.e. a $P : R$ ratio > 1. Differences in the $P : R$ ratio have also been recorded in other animals with zooxanthellae. While a burrowing actinian was found to meet half its energy requirements from external sources ($P : R$ = ca 0·50), *Tridacna* ($P \pm R$ = 1·75) meets its energy requirements from zooxanthellae (JOHANNES and co-authors, 1972).

Ecologically, the energy budget of the whole reef complex is more interesting than the autecological peculiarities of distinct species. In a Puerto Rican coral reef (GLYNN, 1973b), the total energy intake from planktonic populations amounts to 65·4 g C m^{-2} y^{-1}; the filtering efficiency of all animals of the reef communities (91% for diatoms, 60% for zooplankton) might meet 6–13% of the energy requirements of the whole reef. According to JOHANNES and GERBER (1974, p. 100) on Eniwetok Atoll, zooplankters constitute 'an average of only 3% of the net input of net plankton into the coral community'. The meroplankton biomass exceeds the holoplankton biomass both upstream and downstream of the coral community and is consumed in greater amounts by the reef community. Considering the tremendous overall increase in coral reefs, at least since the mid-Tertiary era, and the general low value of the plankton standing-crop and production in tropical waters, coral reefs appear to represent an energy-storing ecosystem.

Symbiotic relationships between zooxanthellae and various animals may have largely contributed to the coral reefs' success. At the beginning, external organic materials were presumably necessary not only from point of view of energy but also with regard to the specific requirements of some components (possibly phosphorus, vitamins, growth factors, etc). As the reef structures increased in size, the flora and fauna became *more and more* luxuriant and diversified. As a consequence, the abundance and diversity of living food and of organic dissolved or suspended materials increasingly met the requirements of all species involved. Thus the coral reef complex, whose degree of maturity is probably the highest of all marine ecosystems, gradually became an almost self-sustaining (autarchic) system.

Other Hard-Substrate Assemblage

Apart from the two main groups characterized by predominance of algae or hermatypic corals, the infralittoral zone comprises several other assemblages, which are treated here under the subheadings, serpulid and barnacle reefs, pearl oyster beds, beach-rock assemblages, and fouling and hard-substrate harbour assemblages.

Serpulid and barnacle reefs

In the Gulf of Mexico, in areas subject to large salinity changes, HEDGPETH (1953, 1954) briefly described very specific reefs of serpulid polychaetes and barnacles with some amphipods. Further research is required for assessing their possible analogy to the serpulid facies of photophilic algae assemblages (p. 404).

Pearl oyster beds

Pearl oyster beds are common in Indo-Pacific regions at depths of 10–20 m. Pearl oysters (*Pinctada*) attach themselves to rocks or biogenic concretionments, together with many other sessile animals such as sponges, alcyonarians, bryozoans, ascidians etc. There are also predatory asteroids which feed on the pelecypods. To a certain extent pearl oysters can invade the sandy couloirs between the rocky flagstone, providing turbulent wave motion not too heavy. Stony corals sometimes compete with oysters for substrate colonization. Pearl oyster beds which exist only in areas of undiminished salinity may be closely related to coral reef assemblages.

Beach-rock assemblages

A beach rock is a hard substrate of sand and various mineral remnants consolidated by calcareous cement. Although it may extend over the three upper vertical zones of the benthal, I prefer to deal with it here as a whole. According to STEPHENSON and SEARLES (1960), who studied this assemblage at Heron Island, the two upper zones are poorly populated: very few *Verrucaria* in the supralittoral and some *Rivularia* in the upper midlittoral subzone. A *Calothrix* belt, grazed by a chiton of the genus *Acanthozostera*, overlaps the lower midlittoral subzone and the upper infralittoral fringe. Below the low-water spring tide level, the beach-rock is covered by chlorophytes of the genus *Blidingia* grazed by *Nerita* species and fishes (*Chaetodon*, Siganidae, Scaridae); the latter are saved from shark predation because of their location in very shallow waters.

Fouling and hard-substrate harbour assemblages

Hitherto, 'fouling'—i.e. the growth of plants and animals on surfaces of submerged objects—has been almost exclusively studied in polluted harbour areas. According to BELLAN-SANTINI (1970a,b), artificial substrates immersed offshore support an assemblage rather different from that usually considered as 'fouling', especially after 6–9 months. It thus appears that two different assemblages exist: (i) the 'pioneer' species assemblage, which includes *Hydroides norvegica*, *Balanus amphitrite*, *Caprella aequilibra*, *Bougainvillia ramosa*, *Ciona intestinalis* and some chlorophytes (*Ulva*, *Chaetomorpha*); (ii) the assemblage of species whose luxuriant growth is characteristic of relatively polluted waters: e.g. *Cardium exiguum*, *Amycla corniclum*, *Ventroma halecioides*, *Bugula neritina*, *Ascidiella pellucida*, *Styela plicata*. In harbour areas, on ship hulls, etc, both assemblages mix, but on artificial structures immersed in pure water the second assemblage is absent. In the latter case, species of the first assemblage gradually disappear and are replaced by species of hard-substrate assemblages characteristic of neighbouring bottoms at the same depth. Some of them (e.g. in the Mediterranean Sea *Ostrea edulis*, *Musculus marmoratus*, *Mytilus edulis*, *Phallusia mamillata*) are much more abundant on offshore immersed structures due to better conditions (well-oxygenated water, abundant food supply, etc).

Both assemblages comprise euryhaline and eurythermal species and are almost cosmopolitan. In certain biogeographical areas they may, however, acquire additional species of the local flora and fauna that are sufficiently euryplastic.

(b) Soft Substrates

Infralittoral organismic assemblages on soft bottoms of the infralittoral zone are highly diversified. Since they provide important fishing grounds and nursery areas for various commercial species, they have been investigated in detail. It is difficult to incorporate the information available into a general concept.

Obviously, the photophilic (mainly vegetal) species whose presence features the infralittoral zone are all epibiontic and thus directly submitted to illumination effects.

Microphytobenthos always exists, but its abundance and production vary in different areas as a function of granulometric characteristics, wave movements and depth. Seasonal fluctuations in assemblages composition frequently occur (p. 161).

Metaphytes sometimes occur but their role in the assemblage of a given area may differ. Some metaphytes are only superimposed on a soft-bottom assemblage, i.e. they do not modify its composition in comparison with those inhabiting the same, but unvegetated, sediments. In other cases, particular morphological and physiological characteristics of the metaphyte facilitate the addition of floristic and faunistic components resulting in a very important structural change of the assemblage. Sedimentation conditions are often modified by the metaphyte, and the organisms which live on or in the bottom tend to be different from those in unvegetated bottoms.

Unvegetated Soft Bottoms

Shallow-water Macoma assemblages

PETERSEN's (1914a, 1915, 1918) *Macoma baltica* community, first described from Danish coasts, may be considered typical of this group. It occurs in rather sheltered shallow waters, either at permanently immersed levels or at levels which merge only at low tide, providing the water-retaining ability of the sediment is sufficiently high. Pelecypods always predominate with both detritus feeders (*Macoma*) and suspension feeders (*Cardium*) and a polychaete of the genus *Arenicola*. All species are predominately euryhaline. The assemblage may exhibit various facies depending on differences in salinity and water current (mainly tidal) intensity. The latter controls both sediment characteristics and suspended or deposited food resources. While the biomass is usually high, the populations of the commonest species include several year classes and productivity is rather low (THORSON, 1957). Predatory echinoderms never appear to participate in assemblages of this group.

On sandy bottoms with stronger currents, Cardiidae (for example *Cerastoderma edule* on European coasts) predominate and the biomass may be as high as $2 \cdot 5$ kg m^{-2}; *Macoma baltica* and *Arenicola marina* are more common on muddy-sand bottoms. In brackish water areas, highly euryhaline species prevail, such as *Mya arenaria* (biomass up to $3 \cdot 3$ kg m^{-2}) on firm sandy muds and *Corophium* and *Scrobicularia* (biomass up to $1 \cdot 2$ kg m^{-2}) in very soft muds. Within the same group, the following three communities may be mentioned: (i) north-eastern Pacific shores: *Macoma nasuta–M. secta–Arenicola claparedei* community in which the venerid *Paphia straminea* substitutes for the cardiid; *Zostera marina* may be superimposed (MCGINITIE, 1939). (ii) *Macoma incongrua–Cardium hungerfordi–Dentalium octangulatum* community of north-western Pacific temperate shores. (iii) On tropical shores, THORSON (1957) observed *Macoma* communities in the Saloum estuarine area (Senegal) and near Mombasa (Kenya). At Tulear (Madagascar), PICHON, Mir. (1965) also observed sands and muddy sands inhabited by *Macoma dubia*, together with other pelecypods, such as *Tellina palatam*, *Solen corneus*, *Mysella* sp., the gastropod *Galeodes parasidica*, the polychaete *Nephthys* cf. *dibranchis* and the amphipod *Grandidierella mahafalensis*; the d.w. biomass averages 10 g m^{-2} (REYS and REYS, 1965).

It seems that the *Macoma* community does not exist in its typical form in the Mediterranean Sea.

Upper clean-sand assemblages

The assemblages of upper clean sands may be observed on almost all shores of the world ocean. They probably exhibit the most diverse features in the infralittoral zone on soft bottoms, owing to biogeographical appurtenance, tidal regime and degree of air exposure.

The main features of upper clean-sand assemblages may be summarized as follows. (i) They are always, but episodically, subjected to more wave exposure than those of the preceding group. (ii) They invariably exhibit pelecypod populations of the genera *Donax* and/or *Tellina*. (iii) Predatory species which feed on young stages or fully grown pelecypods are always present. On temperate shores they are mainly represented by irregular echinoids but on tropical shores prosobranchial gastropods also occur. (iv) On tropical shores, *Anomura* of the family Hippidae are present and sometimes form large populations. (v) Due to seasonal differences in wave action, large fluctuations in assemblage composition may occur.

Since tidal range and flow velocity differences may induce changes in the distribution of some species, we shall take an assemblage from the Mediterranean coast of France at depths of 0–3 m described by PICARD (1965) to exemplify this group. The most common inhabitants of the fine and rather poorly sorted sand are the pelecypods *Donax semistriatus*, *D. trunculus*, *Tellina tenuis*, *Lentidium mediterraneum*, the scavenger gastropod *Cyclonassa donovani*, the isopods *Idothea baltica basteri* and *Iphinoe inermis*, the echinoid *Echinocardium mediterraneum*, etc. The biomass is large and so is the production in all likelihood, because most of the species have short life cycle (MASSE, 1971). However, the accessibility of this production for consumers at higher trophic levels (e.g. fish) seems limited. This assemblage is widespread in the northern parts of the Adriatic Sea (*Lentidium mediterraneum* zoocenosis; VATOVA, 1961). *Corbulomya maeotica* community in the Black Sea is rather homologous. (BACESCU, 1961); it extends down to 20 m.

In brackish water sandy areas of the Mediterranean Sea, *Lentidium mediterraneum* predominate; *Cyclonassa neritea* substitutes for *C. donovani*, and *Echinocardium* is absent. When the silt content in the substrate increases—i.e. in more sheltered places—populations of *Scrobicularia cottardi* develop, which might also be considered as a facies of the *Macoma* assemblage group.

On the lower beaches of the European western Atlantic and on North Sea shores, the rather clean sands of the surf zone inhabited by *Donax vittatus*, *Tellina tenuis*, *Sipunculus nudus*, *Echinocardium cordatum*, etc, correspond to the above-mentioned Mediterranean Sea assemblage. Pelecypods of the family Solenidae may be abundant in some areas, e.g. *Ensis siliqua* and *Pharus legumen* in well-exposed and finer sands, and *Ensis ensis* and *Solen* in slightly muddier, sheltered sands.

Apart from the general characteristics of tropical shores upper clean-sand assemblages—mentioned above under (ii) and (iv)—it seems that polychaetes are more abundant and diversified there and *Tellina* more tolerant to a slight admixture of silt than on temperate coasts: *Tellina*, *Donax rugosus*, *Terebra*, Olividae, *Spio*, *Cirolana* on the western Africa coasts (LONGHURST, 1958; COLLIGNON, 1960); *Tellina*, *Donax*, *Dosinia*, *Terebra*, *Astropecten* and the sand-dollar *Mellita* on Gulf of Mexico shores (HEDGPETH, 1954); various species of *Tellina*, and polychaete *Ophiophragmus* and the sand-dollars *Mellita* and *Encope* in the Miami area (KNEELAND and MCNULTY, 1962); *Tellina*, *Solen*, the scaphopod *Cadulus*, many polychaetes (*Nothria, Magelona, Prionospio*), the cumacean *Diastylopsis tenuis*, etc., of the 'grey-sands' on Californian shores. This latter assemblage is probably similar to the *Ensis ensis* and *Solen marginatus* facies of European coasts, etc.

Areas with patchiness in their sediment characteristics reveal concomitant differences in their assemblages. For example, at Tulear (Madagascar) on rather exposed beaches, PICHON, Mir. (1967) found fine sands inhabited by the migrants *Donax elegans* and *D. aemulus*, together with the pelecypod *Iacra petiti* and some polychaetes (*Gonadopsis incerta, Spio magnus, Lumbriconereis* sp. etc), whereas in coarser sand patches, the predominant species of the community were *Terebra caerulescens* and *Albunea*. At levels which are most exposed to wave effects—i.e. the lower beach—significant seasonal changes in assemblage composition may occur. For example, on the south-western Atlantic coasts of France, overthrow of sand due to heavy storms forces all animals except *Gastrosaccus sanctus* and *Eurydice pulchra* to move to deeper levels. Similar phenomena have been observed by ANSELL and co-authors (1972) in the Cochin area (India) where changes in the beach profile (e.g. erosion) and salinity occur as a consequence of the southwest monsoon. The tidal migrants *Donax incarnatus, Emerita holthuisi* and *Bullia melanoides* keep pace with the erosion and form a permanent element of the beach fauna, whereas other species, mainly polychaetes, largely disappear.

Upper muddy-sand assemblages in sheltered areas

Living in sheltered areas down to a depth of about 3 m, the Mediterranean Sea biocoenosis (PERES and PICARD, 1964; HARMELIN and SCHLENZ, 1964; TRUE-SCHLENZ, 1965) of muddy sands, sometimes mixed with gravel, is an example of upper muddy-sand assemblage. The abundant infauna comprises the following characteristic species: the polychaetes *Aricia foetida, Paraonis lyra* and *Heteromastus filicornis*, the pelecypods *Loripes lacteus* and *Venerupis (Tapes) decussata*, and the burrowing crustacean *Upogebbia pusilla*. There is also a very rich epifauna of species crawling or walking on the sediment surface, e.g. the holothurioids *Holothuria polii* and *H. tubulosa*, the prosobranchs *Cerithium vulgatum* and *C. rupestre*, and the crustaceans *Clibanarius misanthropus* and *Carcinus mediterraneus*.

Among the numerous facies in the Mediterranean Sea (PERES, 1967a,b) the most striking are those with superimposed metaphytes: *Cymodocea nodosa, Caulerpa prolifera* (only in warmer areas), *Zostera nana* (in more silty sediments and sometimes in slightly brackish waters), *Penicillus mediterraneus* (only in warmer areas and sometimes mixed with *Caulerpa ollivieri*). Few sessile species live on the metaphytes of the two first-mentioned facies: the large foraminiferans *Sorites variabilis* and *Peneroplis* sp., the hydroid *Laomedea angulata*, the actinian *Bunodeopsis strumosa*, the bryozoan *Electra pilosa* and the compound ascidian *Trididenmum fallax*.

On the Atlantic coasts of western Europe, from Britanny to Portugal and Spain, this assemblage has its homologue in the *Venerupis (Tapes) decussata* assemblage and also in the *V. T. aurea* (sometimes mixed with *Zostera nana*) assemblage, the latter occurring in more sheltered areas.

Assemblages which belong to this group also exist on tropical shores; they comprise pelecypods (mainly Lucinidae and Veneridae, but sometimes also *Macoma* species), polychaetes, burrowing crustaceans of the family Thalassinidae. Enteropneusts also occur frequently and, sometimes, superimposed metaphytes. The muddy fine sands at

Tulear (Madagascar) for example, are inhabited by *Macoma dubia*, the polychaetes *Grandidierella multiannulata*, *Owenia fusiformis*, *Nepththys tulearensis* and *Scoloplos chevalieri*, actinians of the family Edwardsiidae, the enteropneusts *Ptychodera* and *Glossobalanus*, etc (PICHON, Mir., 1967); the biomass is about $4 \cdot 6$ g m^{-2} (d.w.) (REYS and REYS, 1965). Other examples are:

(i) Muddy coarse sands and gravels, 1–3 m deep, at Eilat (Red Sea) with *Mactra olorina*, *Mesodesma glabrum*, *Polynices pyriformis*, many polychaetes (*Perinereis nuntia*, *Notomastus latericeus*, *Dasybranchus caducus*, *Pectinaria* sp., *Eunice australis*), some decapods (e.g. species of *Thalamita*, *Calappa*, *Portunus*), *Holothuria arenicola*, the echinoids *Echinodiscus auritus* and *Lovenia elongata*, some individuals of *Cerianthus* (FISHELSON, 1971). The facies of finer calcareous muddy sand is inhabited by a similar fauna, but *Ptychodera flava* is more common with the burrowing actinian *Radianthus koseirensis*; the plant component consists of *Thalassodendron ciliatum* instead of the above-mentioned algae.

(ii) Assemblages inhabiting the sandy mud slope beyond the most peripheral mangrove trees. On South African coasts, BROEKHUYSEN and TAYLOR (1959) found numerous polychaetes (*Dasybranchus*, *Dendronereis*, *Perinereis*, *Scolelepis*), the sipunculid *Siphostoma australe*, the lucinid pelecypod *Loripes clausus*, the crustaceans *Clibanarius longitarsus*, *Penaeus japonicus*, etc.

(iii) In the Chantung Peninsula, Yellow Sea, GURJANOVA and co-authors (1958) and GURJANOVA (1959) observed an assemblage with biomasses varying from 50 to 230 g m^{-2} on muddy sands. In spite of a common faunistic background of polychaetes (*Glycera*, *Amphitrite*, *Marphysa*, *Potamilla*), together with the ophiuroid *Amphiura vasicola* and the brachiopod *Lingula*, there are differences at different levels: at the upper level *Aloidis* is the predominant pelecypod, together with opisthobranchs of the genera *Hima* and *Bullacta*, some crabs (*Hemigrapsus*, *Macrophthalmus*) and the burrowing decapods *Alpheus brevicristatus* and *Callianassa japonica*; at lower levels, pelecypods are more diverse (*Dosinia japonica*, *Solen*, *Mactra*), some *Balanoglossus* appear, *Diogenes* and *Palaemon* substitute for crabs and the burrowing decapod is *Upogebia whulstenweni*.

(iv) At outer Kingston Harbour (Jamaïca) WADE (1972) observed, on fine sands mixed with silt and clay, a very interesting assemblage closely related to this group but living at greater depths (5–16 m), probably owing to better shelter. Polychaetes (mainly *Lumbrinereis* and *Armandia*) predominate, whereas only two crustacean species are common: *Periclimenes americanus* and *Upogebia affinis*; the pelecypods *Diplodonta rotunda* and *Phacoides muricatus* (Lucinidae) may be considered as the best indicators of the appurtenance of this assemblage to the group of upper muddy sand assemblages.

Fine, well-sorted sand assemblages

These assemblages live between the lower limit of the upper clean sands (above) to a depth of about 20–40 m, on terrigenous, hard, well-sorted fine sands. The upperlying water masses are always of undiminished salinity; any influence of brackish waters, even episodic, causes a sharp decrease in specific diversity, which deforms the assemblage composition. The fine well-sorted sands (sometimes slightly muddy) exhibit very similar features all over the world. The rich and diversified fauna includes polychaetes and many swimming crustaceans. The most important and general characteristics are the following:

(i) There are many pelecypods, both suspension feeders (*Venus* or other venerids and mactrids) and detritus feeders (e.g. tellinids), together with gastropods of the family Naticidae drilling the bivalves. As usual, other predatory gastropods are more common in tropical regions.

(ii) Among the echinoderms there are always species of *Astropecten*, which feed on testaceous molluscs, mainly pelecypods, but also often on juveniles of irregular echinoids (*Echinocardium*, *Spatangus*, etc).

(iii) The permanent fish fauna almost invariably comprises flatfishes and gobiids; young stages of species whose adults inhabit the deep shelf are also common in some seasons. For resource management (Volume V, Part 3) it is of paramount importance to protect these nursery grounds from excessive fishing.

(iv) Biomass and production rate are extremely diverse, owing to differences in life span of the species constituting the assemblage. Year-to-year fluctuations in biomass and production are frequent due to recruitment irregularities in species with a long planktonic phase. The production of the terrigenous fine-sand organismic assemblages largely depends on the production in upperlying water masses (MASSÉ, 1971).

PETERSEN's (1918) '*Venus gallina* community', which is widespread in northeastern areas of the Atlantic Ocean and adjacent seas is typical of this group; in addition to 'pilot' species, it includes *Tellina fabula*, *Solen pellucidus*, *Philine aperta*, many polychaetes, *Echinocardium cordatum*, *Ophiura albida*, etc. According to HAGMEIER (1951), two main groups of facies may be distinguished: (i) in rather clean sand areas, *Venus gallina* predominates and biomass is low; (ii) in more muddy sand, *Spisula subtruncata* is the most abundant species and the biomass may be as high as 5 kg m^{-2}.

The faunistic diversity of this assemblage is higher in the Mediterranean Sea. According to PICARD (1965), the following belong to the most characteristic species: the pelecypods *Venus gallina*, *Cardium tuberculatum*, *Mactra corallina*, *Spisula subtruncata*, *Donax venustus*, *Tellina fabuloides*, *T. nitida*, *T. pulchella*; the gastropods *Nassa mutabilis*, *N. pygmaea*, *Neverita josephinia*, *Actaeon tornatilis*; the crustaceans *Idothea linearis*, *Eocuma ferox*, *Iphinoe trispinosa*, *Pseudocuma longicornis*, *Macropipus barbarus*; the polychaete *Clymene oerstedi*; the echinoid *Echinocardium mediterraneum*; and the fishes *Gobius microps* and *Callionymus belenus*. Multicellular algae are absent, but eel-grasses *Cymodocea nodosa* and *Halophila stipulacea* (the latter only in the eastern basin) may be superimposed on the assemblage locally; since their rhizomes are usually not crowded, the infauna remains unchanged. The microphytobenthos is abundant.

Biocoenoses and 'communities' of this group are numerous; hence only a few are mentioned.

The boreo-arctic *Astarte borealis* 'community' contains the pelecypods *Mactra corallina*, *Cyprina islandica*, the polychaetes *Thelepus cincinnatus*, *Onuphis conchylega*, etc. On the coasts of Norway and Iceland and in the Barents Sea, it seems to be homologue to the boreal *Venus gallina* 'community'; its biomass may exceed 200 g m^{-2}.

The Arctic *Venus fluctuosa* 'community' (THORSON, 1957) with some other pelecypods e.g. *Cardium groenlandicum*, *Axinopsis orbiculata*, the polychaetes *Scoloplos armiger* and *Euchone analis* and some amphipods (*Ampelisca*) may belong to this group, but might also be considered as impoverished facies of the *Macoma calcarea* community (p. 452), with admixtures of *V. fluctuosa*.

On New Zealand shores, pleistocenous fine sands of the shallow-water shelf (7–40 m) are inhabited by assemblages that contain predominantly suspension-feeding pelecypods: *Scalpomactra*, *Maorimactra*, *Glycimeris*, *Venericardia*, *Diplodonta*, and *Plurigens*. Detritus feeders are rare owing to the generally lower content of organic material in the sediments. Echinoderms also seem to be rare (MCKNIGHT, 1969a).

The *Cytherea* community in the Persian Gulf, with *Spisula*, *Tellina*, *Ensis*, *Cardium* (THORSON, 1957).

The *Cultellus tenuis*–*Macoma cumana**–*Diopatra neapolitana* assemblage in the fine sands (5–14 m deep) of Ghanaian coasts (BUCHANAN, 1958) contains several polychaetes (*Owenia fusiformis**, *Prionospio pinnata*, *Lumbriconereis impatiens*), *Dentalium maltzani** and *D. coarti**, *Actaeon* sp. and some ophiuroids (*Amphioplus*, *Amphipholis*). The epifauna is rich with large prosobranchial gastropods (*Cymbium*, *Tonna*, *Murex*), brachyurans and anomurans (among the latter *Diogenes pugilator*). The assemblage may consist of three facies exhibiting a large predominance of the above species marked with an asterisk (*); the biomass of the *Owenia* facies may be as high as 120 g m^{-2}.

Silty or muddy-sand assemblages in the deeper areas of the infralittoral zone

On Denmark's coasts PETERSEN (1918) described an *Abra alba* 'community', which also comprises *Macoma calcarea*, various species of the genus *Astarte*, *Corbula gibba*, *Nucula tenuis*, *Echinocardium cordatum*, and *Ophiura albida*. HAGMEIER (1951) and THORSON (1957) suggested that a transition may exist, depending on increasing silt or mud contents from the *Macoma* or *Venus gallina* assemblages to the *Abra* (= *Syndosmya*) association. This assumption is supported by STRIPP (1969) who found in Helgoland Bight and near the Wester and Elbe estuaries the *Abra alba* association on silty bottoms, 15–40 m deep, below the *Macoma balthica* association on nearshore sediments of all types (0–15 m depth) and as a lateral facies of offshore *Venus gallina* sands (12–30 m).

In the Mediterranean Sea, *Abra alba* does not in fact inhabit the infralittoral zone but is a silt- or mud-loving species, also occurring in the circalittoral zone.

In the more silty sands of the English Channel, the *Corbula gibba* association (CABIOCH, 1961) also belongs to this group; increasing percentage of silt results in compact mud bottoms, the most abundant species of which are the polychaetes *Ampharete grubei* and *Melinna palmata* (Gulf of Morbihan; GLEMAREC, 1964).

The *Abra alba* assemblage generally seems to have a rather low biomass (<200 g m^{-2}) but the production is probably high due to the short life span of most species. In Kiel Bight the production was estimated to be as high as 98 g m^{-2} yr^{-1} (ARNTZ, 1971), of which only about 43 g represented 'first-class' food for fishes. The biomass of the large bivalve *Cyprina islandica* amounts to 571 g m^{-2}, its production to 530 g m^{-2} yr^{-1}.

Very similar facies which develop as the silt and/or mud components increase in sandy bottoms seem to be rather widespread, e.g. the *Theora* (a genus closely related to *Abra*) assemblages with amphipods of the genus *Ampelisca* and the mud-loving polychaete *Sternaspis* sp., which occur in such areas (sometimes polluted) on Japan's coasts. In the latter case, biomass seems to be rather low (13 g m^{-2}). However, it may be expected that the production would be larger due to a high turnover of amphipods.

Decapod–crustacean assemblages on tropical and subtropical shores

Increased abundance in decapod crustaceans, mainly small crabs and pagurids, is a rather general feature of infralittoral soft bottoms on tropical and subtropical shores. According to THORSON (1957), the following assemblages may be distinguished in this group: (i) *Pinnixa rathbuni* 'community' on muddy sand and gravel bottoms of Japan's coasts (7–37 m depth); this small-sized pinnotherid crab may attain an abundance of 3000 individuals m^{-2}; other species (*Tellina pallidula, Limopsis multistriata, Corbula venusta*) are rare; (ii) *Xenophthalmus pinnotheroides* sand 'community' of the Persian Gulf; the population density of this crab attains 1500 individuals m^{-2}; the associated fauna consists only of some polychaetes; (iii) *Diogenes rectimanus* and *D. costatus* 'community' on muddy bottoms (20–25 m depth) in the Gulf of Madras.

In the Makan River mouth (New Zealand) MCKNIGHT (1969a) found in 37 m depth a rather similar sandy-bottom assemblage with abundant pagurids associated with several gastropods, sponges, hydroids and the sea-star *Coscinasterias calamaris*, but without pelecypods and further echinoderms.

The 'community' observed off Nosy-Bé (Madagascar) at 40 m depth may also belong to the same assemblage group (PLANTE, 1964). Species of the genus *Callianassa* predominate, associated with many other decapods—both brachyurans (*Leucosia, Arcania, Podophtalmus, Cosmonotus, Ceratoplax, Notopus, Oreophorus*) and Natantia (Alphaeidae and *Processa*). Other species are rare: polychaetes of the genera *Scolelepis, Chloeia, Onuphis, Owenia*, and molluscs (*Dentalium, Tellina, Pyrene*). Several suspension feeders (such as sponges), an antipatharian, a caryophyllid stone coral and lancelets (genus *Asymmetron*) indicate a rather high content of suspended organic matter and strong bottom currents; the biomass is 1·88–2·50 g m^{-2} (d.w.).

According to THORSON (1957), such apparently highly productive assemblages can exist only in areas of highly productive upperlying waters, the food supply for the decapods originating as it does from the 'planktonic rain'. Such a contribution of plankton production to the food supply of a benthos assemblage suggests placing the highly specific 'community' observed on western beaches of northern New Zealand near this assemblage group. The trophic basis consists mainly of blooms of diatoms (mostly *Chaetoceros* but sometimes also of *Asterionella japonica*), the primary production of which may be as high as several hundred mg $m^{-3} h^{-1}$. The most abundant species in the assemblage are the polychaete *Aglophamus macrura*, the crustaceans *Callianassa filholi* and *Lysiosquilla spinosa*, the pelecypod *Amphidesma subtriangulatum* and the irregular echinoid *Arachnoides zelandiae*.

Antarctic and subantarctic regions

(a) **Antarctic region.** Soft-bottom assemblages near the Antarctic continent and adjacent islands, at depth corresponding in other regions to the infralittoral zone are little known, not least because soft bottoms are rather rare in that region.

ARNAUD (1974) investigated such bottoms on Adélie Land coasts and reviewed all data obtained from other Antarctic shores. To summarize his conclusions. On Adélie Land coasts the substrate is well-sorted sand mixed with pebbles; its mud content increases below 25–30 m; other regions have more or less muddy sands. The assemblages may be broadly characterized as follows: (i) Polychaetes are abundant and rather diversified (*Haploscoloplos kerguelenensis*, *Capitella perarmata*, *Aglaophamus ornatus*, *Barrukia cristata*, etc); their standing crop is sometimes high (180–250 g m^{-2} in the Davis Sea; GRUZOV and co-authors, 1967). (ii) The tanaid *Nototanais antarcticus* is often very abundant (up to 14 000 individuals m^{-2} in the Ross Sea; DEARBORN, 1967) especially when the sediments contain more debris from metaphytes and phytoplankton. (iii) Irregular echinoids of the genus *Abatus* are always present and sometimes highly abundant, e.g. *A. cavernosus* reaches densities of up to 70 individuals m^{-2}. (iv) The pelecypod *Laternula elliptica* seems to predominate in areas of diminished salinity and may reach a standing crop as high as 5 kg m^{-2} (GRUZOV and co-authors, 1967); it seems to feed on diatoms ingested with the sediment. (v) Among the less common species ARNAUD (1974) mentions the gammarids *Orchomene litoralis* and *Heterophoxus videns*, the burrowing pelecypod *Thracia meridionalis*, the opisthobranch *Philine alata*, holothurioids, asteroids, etc.

Three further generalization seem justified (ARNAUD, 1974): (i) Some species exhibit abundances similar to those recorded from corresponding depths in hard bottom assemblages, e.g. *Ophiura mimaria* (up to 300 individuals m^{-2}). (ii) Some other species are largely eurybathic, e.g. the ophiuroids *Ophionotus victoriae* and *Amphiura belgicae* were recorded down to 752 and 2450 m, respectively (iii) In Adélie Land—and more generally on the eastern Antarctic—the assemblage is less diversified than in areas of the western Antarctic and the subantarctic region; echinoids and ophiuroids tend to substitute for pelecypods; *Yolidia eightsi* and *Mysella charcoti* have been recorded from the South Orkney coasts with abundances of 2700 and over 75 000 individuals m^{-2}, respectively (WHITE and ROBINS, 1972). The standing crop amounts to several kg m^{-2}.

(b) **Subantarctic region.** In the subantarctic region, exemplified by the Kerguelen Islands, ARNAUD (1974) distinguished two main assemblages in the infralittoral zone.

(i) An assemblage on fine muddy sands from the low-water spring tide level down to 25–30 m, with the dominant species *Abatus cordatus* (up to 60 individuals m^{-2}) and *Laternula elliptica*, together with numerous polychaetes and the ophiuroid *Ophionotus hexactis*. (ii) An assemblage on muddy bottoms, from 3–5 m depth in fjords down to more than 200 m; the most abundant species are the pelecypods *Yolidia* sp. and *Malletia gigantea* and some polychaetes; in some places the fauna becomes very impoverished due to the oxygen deficiency.

Soft Bottoms with Metaphytes

The following pages deal with bottoms in which metaphytes significantly modify both the granulometric characteristics of the sediment and the composition of the assemblage concerned. While we here make a distinction, for didactic reasons, between phanerogams and algae, these two types of metaphytes may occur together.

Sea-grass beds

Marine phanerogams, like their continental counterparts, always require true soils. However, the requirements for grain size and humic substances differ between species. In the Mediterranean Sea, for instance, *Posidonia oceanica* only grows on muddy sands with narrowly defined humic substance contents, whereas *Cymodocea nodosa* thrives on relatively clean sands and putrid muds. However, on clean sediment with low organic contents, the rhizomes and roots of *C. nodosa* are poorly developed, while the leafage is very dense. On muddy bottoms with large amounts of organic materials and nutrient salts, the leaves are smaller and scarcer, but the rhizomes and root system are highly developed (MOLINIER and PICARD, 1952). These structural differences may signify differences in nutritional physiology. Labelled isotopes of phosphorus and nitrogen indicate that sea-grasses are able to absorb nutrients across the surface of their leaves or roots (MCROY and MCMILLAN, 1973). Nevertheless, the sediments appear to be the principal sites of nutrient absorption.

Sea-grass beds, while composed of various species, are widely distributed in shallow waters and exhibit common features of considerable ecological significance.

(i) Sea-grass as habitat: Sea-grass increases the surface area available for epiphytic algae and associated fauna. For *Zostera marina* in south-western Japan the ratio of leaf area: bottom area is 2–2·7 in winter and about 4–5 in June; it may reach 18 in *Thalassia hemprichi*. Leaves, stems and rhizomes increase the diversity of microhabitats and ecological niches.

(ii) Sea-grass as a shelter: The dense canopy of leaves attenuates turbulence and diminishes illumination. In sea-grass beds between tidemarks, adverse effects on the associated biota from desiccation and fluctuations in temperature and salinity are reduced. The dense leafage also provides some protection from predators to its inhabitants.

(iii) Sea-grass and sedimentation: Sea-grass traps sediment, in some cases leading to underwater terraces (*Posidonia* and *Thalassia*, pp. 440 and 447).

(iv) Sea-grass as an oxygen producer: Due to their high photosynthetic activity, sea-grasses and their epiphytic algae produce large amounts of excess oxygen during the day, while at night they consume oxygen, as do their associated animals, plants and microorganisms. This leads to significant diurnal and seasonal fluctuations of O_2 and CO_2 contents.

(v) Sea-grass as nutrient resource: Blue-green algae are an important epiphytic component of sea-grasses (*Thalassia testudinum, Halodule baudettei, Syringodium filiforme, Ruppia maritima*). They can fix large amounts of nitrogen and thus play an important role in the nitrogen economy of sea-grass assemblages (GOERING and PARKER, 1972). While phosphorus is absorbed by *Zostera marina* (more intensively in the light) through leaves and roots and rapidly transported to all parts of the plant, it may be returned in part to the surrounding water through the leaves (MCROY and BARSDATE, 1970). In nature, sea-grasses may act either as sink or source for dissolved phosphorus. The release of dissolved organic substances, well documented in large soft algae (especially phaeophytes), may exist in phanerogams, but this requires investigation. It can probably be taken to be of relatively insufficient ecological value (PENHALE and SMITH, 1977). Consumption of living sea-grasses by animals seem to be negligible, except in some echinoids and fishes (*Sparisoma*, some Scaridae and Acanthuridae). Most grazers in sea-grass beds consume associated algae (Opisthobranchial gastropods, *Strombus*, etc), or epiphytic algae mixed with detritus. Most plant material is utilized by animals in a semi-decomposed state. Different consumers specialize on different stages of decay. While some sea urchins or crustaceans feed on rather big pieces of dead leaves, deposit-feeding molluscs and polychaetes utilize more degraded organic matter (KIKUCHI and PERES, 1977).

In the microbial communities living on detrital particles derived from turtle grass *Thalassia testudinum*, the number of organisms and the rate of oxygen consumption are approximately proportional to the total sea-grass surface area (FENCHEL, 1970). In field samples of such detritus FENCHEL found about 3×10^{10} bacteria, 5×10^7 flagellates, 5×10^4 ciliates and 2×10^7 diatoms g^{-1} dry weight. He investigated the consumption of turtle grass detrital particles by the amphipod *Parahyalella whelphleyi* and found that it only uses the associated micro-organisms, while the dead plant fragments remained undigested. The more the amphipods consume detritus, the more the average particle size decrease, leading to an increase in total surface area and thus of microbial activity. Using detritus particles of different sizes (0·18–2 mm) and ages (0–54 d) derived from *Zostera marina* TENORE (1975) recorded increased rates of net incorporation by the polychaete *Capitella capitata* with increasing detritus age at all different particle sizes. However, 'young' particles (0–11 d) were incorporated more readily in the smallest particle size group; with subsequent ageing, greater net incorporation occurred with larger-sized particles.

Because they provide habitat and shelter, as well as abundant and diversified food resources, sea-grass beds are inhabited by an exceedingly rich flora and fauna and function as spawning sites and nursery grounds for fishes or other motile animals, even for those species whose adults inhabit other biotopes. An example of seasonal variation in the motile fauna is presented on p. 126 for *Zostera marina* beds. If the increase in specific diversity and standing crop of epiflora and epifauna (sessile, sedentary and motile species) is obvious in comparison with unvegetated neighbouring bottoms, comparable increases prevail in the infauna. Chesapeake Bay *Zostera* beds, for example, are inhabited by ca 70 infaunal species and 15 000 individuals m^{-2}, while a similar sandy area without *Zostera* contains only 37 species and 6000 individuals m^{-2}. In *Thalassia* beds ORTH (1971) found 55 species at Whalebone Bay (Bermuda), but only 22 species in open sand, the relative abundance in the latter being only one-quarter of that in *Thalassia* beds.

(a) Mediterranean *Posidonia* meadows. In the Mediterranean Sea, *Posidonia* meadows exert the greatest influence on sedimentation conditions and support the most complex organismic assemblage (MOLINIER and PICARD, 1952; PERES and PICARD, 1964). *Posidonia oceanica* is endemic mediterranean; the sole other species of the genus lives on warm-temperate shores of Australia. Three important ecological characteristics of this sea-grass must be emphasized. (i) At the end of autumn most of the oldest leaves fall and the decaying ones accumulate on the shore. The leaf fall is probably enhanced by the summer development of many epiphytes (especially the encrusting rhodophyte *Melobesia lejolisi*), which increasingly restrict gas exchange. (ii) Fruiting is commonly observed in warmer areas of the Mediterranean Sea (the eastern basin and coasts of north Africa), more rarely in the northern half of the western basin and then only after a very hot, early summer. However, even though *P. oceanica* does not appear to be well adapted to the environmental prevailing conditions, in the latter areas it extends its cover by vegetative multiplication by means of shoots and natural cutting. (iii) *P. oceanica* rhizomes grow not only horizontally, as in other sea-grasses, but also vertically. Particles suspended in the water, except the finest ones, are entrapped by the dense leafage; they accumulate on the bottom where they are retained by the rhizome network; most remains of the epiflora and epifauna—especially the calcareous ones—are added to the entrapped suspended particles falling down around the base of the plant. Thus *Posidonia* must counteract this progressive sedimentation by upward growth of its rhizomes; this results in the building of underwater terraces or 'mats' which rise slowly (on average 1 m century^{-1}) but constantly. The mat consists of a strong and crowded network of rhizomes and roots, the interstices of which are completely clogged with a poorly sorted sediment. In this way the top of the terrace progressively nears the water surface. The final state depends on wave exposure.

In very exposed areas, the upward extension stops quickly; the waves wash the upper part of the mat and thus carry away the sediment; finally, the substrate becomes unsuitable for *Posidonia*, and one observes patches of dead plants. In fairly exposed places, erosion occurs when the top of the terrace reaches a depth with sufficiently strong wave effects. In the presence of a boulder or any other hard material on the mat surface, wave action results in a pot hole with fine sand on its bottom. A confluence of several pot holes may produce a channel, whose width may subsequently be enlarged by currents. In sheltered areas, especially in small bays, the wave action is not sufficient to prevent the terraces rising; these are then able to reach such a high level that the *Posidonia* leaves start floating at the sea surface and act as wave-breakers. Behind such a barrier there is a relatively calm area similar to the channel lagoon behind a coral barrier reef. The coarser suspended particles are still trapped by the leaves and the wave breaker strip both rises up and moves towards the bay entrance. The finest particles brought by currents and those coming from the land tend to fill up the lagoon area; in sheltered areas its bottom becomes inhabited by the muddy-sand assemblage. Behind the emerging barrier, *Posidonia* becomes increasingly damaged due to the calm water and so degenerates. Finally, the dead mat becomes covered with very fine sediments and then often supports a *Cymodocea nodosa* lawn (p. 433).

The specific assemblage of the *Posidonia* meadow only comprises those species linked with the leafage and upper parts of living plant rhizomes. Five different subassemblages may be distinguished: algae and sessile animals; micro- and meiofauna of the epiphytic felt; creeping and walking animals on green leaves; swimming animals able to rest on leaves; and animals swimming under the leaf canopy.

(i) Algae and sessile animals. Epiphytic algae of *Posidonia* leaves belong to three different groups (FELDMANN, 1937): (*a*) small species forming a felt-like coating—diatoms and small-sized phaeophytes such as species of the genus *Ectocarpus*, *Ascocyclus orbicularis*, etc; (*b*) encrusting calcareous species, the most important of which is *Melobesia lejolisi*, especially on the older leaves; (*c*) larger and more or less erect soft algae—mainly phaeophytes such as the characteristic species *Castagnea mediterranea*, *C. irregularis*, *C. cylindrica*, *Gyraudia sphacelarioides*, but also rhodophytes (chiefly Nemalionales and Ceramiales)—which are, however, not characteristic. Both qualitative and quantitative maximal development of these algae is observed in shallow-water *Posidonia* meadows. Five characteristic animal species usually represent more than 95% of the sessile invertebrates: the hydroids *Monotheca posidoniae*, *Sertularia perpusilla*, *Campanularia (Orthopyxis) asymetrica* and the bryozoans *Electra posidoniae* and *Microporella johannae*. Many foraminiferans are present, e.g. *Webbinella crassa*, *Iridia serialis* and *Rhizonubecula adhaerans*; the actinian *Parastephanauge paxi* is often common.

(ii) Micro- and meiofauna of the epiphytic felt. This insufficiently documented micro- and meiofauna is exceedingly rich. It includes many protozoans (ciliates, flagellates), free-living nematodes, small polychaetes (*Polyophthalmus pictus*, *Pionosyllis pulligera*, etc), rotifers (*Notommata naias*, *Colurus leptus*), tardigrades, copepods (*Idyaea furcata*, *Laophonte stromi*, *Dactylophusia tisboides*) and the amphipod *Amphithoe rubricata*. Even chironomid larvae sometimes occur among this epiphytic felt.

(iii) Creeping and walking animals. Most of the creeping species consist of small gastropods, both prosobranchial, (e.g. *Rissoa*, *Tricolia*, *Alvania*, *Persicula*, *Bittium*, *Marginella*) and opisthobranchial (*Hancockia uncinata*, *Loviger serradifalci*, *Glossodoris gracilis*, *Polycera quadrilineata* and *Elysia viridis*), the latter being more abundant among algae. Further characteristic creeping species are the medusa *Eleutheria dichotoma*, the small asteroid *Asterina pancerii* and the pelecypod *Propeaumussium hyalinum*. Except for a few halacarid (*Pontarachna globosa*), all animals walking by means of articulate appendages are crustaceans: the harpacticoid *Porcellidium fimbriatum*; the amphipod *Cymadusa crassicornis*; the highly characteristic isopods *Idothea hectica*, *Synisoma appendiculata* and *Astacilla mediterranea* (the latter probably more abundant in *Cymodocea* beds); the common pagurids *Eupagurus anachoraetus* (young individuals), *E. chervreuxi* and *Catapaguroides timidus*.

(iv) Swimming animals able to rest on leaves. Most animals of the *Posidonia* assemblage can cling to the leaves by means of their claws: various copepods and ostracods, many amphipods, such as *Atylus guttatus* and *Apherusa bispinosa*; many mysids, e.g. *Siriella clausi*; many shrimps and prawns, e.g. *Hippolyte inermis*, *H. holthuisi*, *Thoralus cranchi* and *Palaemon ziphias*. Some other animals temporarily attach to *Posidonia* leaves by means of suckers: the Anthomedusa *Cladonema radiatum*, the characteristic Limnomedusae *Olindias phosphorica*, *Gonionemus vertens* and *Scolionema suvaense*, the chaetognath *Spadella cephaloptera*, the small squid *Spiola rondeleti* and the fish *Lepadogaster microcephalus*. Finally, some rest by twisting their body around the leaves, e.g. the sygnathid fishes *Hippocampus guttulatus*, *H. brevirostris*, and the highly characteristic *Nerophis maculatus* and *N. ophidion*.

(v) Animals swimming under the leaf canopy. This group mainly consists of fishes: *Serranus* spp., Labridae (*Ctenolabrus*, *Crenilabrus*, *Julis*, etc), *Anthias*, young *Mullus* and sparids, etc, and of decapod crustaceans, such as *Palaemon serratus*, *Athanas*, *Alpheus*, *Processa*, *Scyllarus arctus* and *Macropipus*. In spring the summer *Sepia officinalis* and *Octopus vulgaris* are also very common. In general, this subassemblage is greatly impoverished in winter due to both temperature decrease and leaf fall diminishing the protective canopy.

Sessile and sedentary plant and animal species, which participate in the coralligenous organismic assemblage (circalittoral zone, p. 453), may also live on rhizomes in the deepest *Posidonia* meadows, providing the leaf canopy is sufficiently dense, thus heavily diminishing the degree of illumination.

Three additional generalizations concerning the typical Mediterranean *Posidonia* assemblage can be made:

(i) The associated fauna exhibits a large variability, depending on depth and wave action. Deeper meadows are the most typical and their fauna is qualitatively and quantitatively the richest, particularly as regards the motile species. In fairly exposed shallow-water beds stenohaline and wave-loving species, e.g. *Rissoa auriscalpium*, *Gibbula umbilicalis*, *Idothea hectica* and *Grubia crassicornis* predominate, whereas in more sheltered areas, more eurytherm and euryhaline species (such as the crab *Macropipus arenatus*) are more abundant (LEDOYER, 1962).

(ii) The ratio day : night abundance in samples from the upper stratum (i.e. leaves), varies from 1 : 3 to 1 : 50, not only for swimming but also for crawling species. The day : night differences in abundance are particularly striking in the following photophilic species: the gastropods *Columbella rustica*, *Mitrella scripta*, *M. gervillei*, *Cerithium vulgatum*, *Phasianella pulla*, *P. speciosa*, *Gibbula ardens* and *G. speciosa*; the amphipods *Apherusa bispinosa*, *Hyale dollfusi* and *Grubia crassicornis*; the mysid *Siriella clausi*; the decapods *Thoralus cranchi*, *Palaemon serratus*, *Clibanarius misanthropus*, *Catapaguroides timidus*, *Macropipus arcuatus* and *Acanthonyx lunulatus*. Even several sciaphilic forms, which during the day remain in the most shaded microenvironments, migrate at night up to the leaves: the shrimps *Eualus occultus* and *Athanas laevirhynchus*, and several species of *Alpheus* and *Processa*; the echinoderms *Sphaerechinus granularis*, *Psammechinus microtuberculatus*, *Genocidaris maculata*, *Antedon mediterranea*, *Astropecten spinulosus* and *Holothuria impatiens*. The more active behaviour and upward migration of these species are stimulated by the increasing carbon dioxide content, which is mainly brought about by photosynthesis cessation at night. The downward migration is related to the direct influence of increasing light; motile species—even the most photophilic ones—do not tolerate the high degree of illumination in the uppermost layer of the leaf canopy (LEDOYER, 1962).

(iii) Mediterranean *Posidonia* meadows are the terminal stage (i.e. the climax) of a succession which begins in well-sorted terrigenous sands, whose humic substance content gradually increases owing to accumulation of organic detritus from sand inhabitants. The second stage consists of a *Cymodocea nodosa* lawn (p. 433); *Cymodocea* seeds first settle in small substrate depressions where the humic substances are somewhat more abundant; it then spreads out progressively and gradually occupies the whole bottom area. The second stage results in an additional increase in humic substances; thus the substratum becomes suitable for *Posidonia* seeds or cuttings. *Posidonia* begins progressively to eliminate *Cymodocea* and finally replaces it completely. This climatic succession may be reversed if the balance of environmental factors becomes disturbed. For example, erosion of the *Posidonia* terrace leads to bare hollows with a bottom of well-sorted fine sand; if the erosion process stops, *Cymodocea* may settle and thus prepare for a new *Posidonia* settling. In contrast, the *Cymodecea* lawn developing behind the *Posidonia* 'barrier reef' (Fig. 8–12) cannot induce *Posidonia* re-installation because the sediment is too fine and contains too much organic matter. Under exceptional circumstances, locally scattered *Posidonia oceanica* may be observed on hard substrate with a sufficiently thick sediment layer. The formation of such soil depends on previous development of tufty algae, mainly *Jania*, which entrap suspended mineral and organic particles.

The organismic assemblage of *Posidonia* terraces (mats) studied by HARMELIN (1964) exhibits very specific features. Only loosely related to such mats, the assemblage may also be found in terraces where *Posidonia* died. The most abundant inhabitants of the mat are polychaetes, especially *Pontogenia chysocoma*, *Nereis irrorata*, *Lumbriconereis paradoxa*, *Clymene lumbricoides*, *Magelona equilamellae* and *Leiocapitella fauveli*. Other characteristic components are the decapods *Upogebia deltaura* and *Callianassa minor*, and the pelecypods *Lima hians*, *L. inflata* and *Venus verrucosa*. The biomass is about 130–150 g m^{-2} in the most superficial layer (20–30 cm) of the terrace, decreasing rapidly in deeper layers.

(b) *Zostera marina* beds in the North Atlantic Ocean. The *Zostera marina* (eel-grass) beds which are common on both the sides of the North Atlantic are similar to the Mediterranean *Posidonia* beds. The *Zostera* leafage has some influence on sedimentation

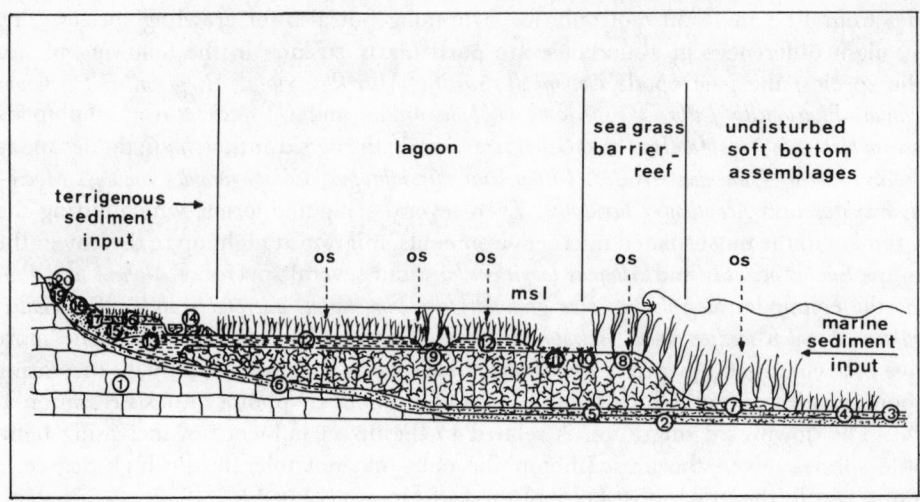

Fig. 8-12: Scheme of a *Posidonia* meadow with emerging 'barrier-reef' and its lagoon. 1: original rocky substrate; 2: original sandy bottom, formerly inhabited by the assemblage of infralittoral well-sorted sands; 3: sediment inhabited by infralittoral assemblage of well-sorted sand or by 'amphioxus' coarse-sand assemblage; 4: sediment increasing due to entrapment of organic detritus from *Cymodocea* lawn; 5: same, covered by *Posidonia* terraces ('mattes'); 6: sediment fastened on rocks by photophilic algae assemblage, formerly humus-enriched by *Cymodocea* lawn facies; 7: *Posidonia* meadow; 8: *Posidonia* barrier reef resulting from terrace rising; 9: *Posidonia* vestigial patches: 10 and 11: *Padina pavonica* and *Jania rubens* facies of photophilic algae assemblage on erect stems of dead *Posidonia*; 12: area of silty sedimentation with *Cymodocea nodosa* lawn superimposed on dead *Posidonia* terrace; 13: silty and muddy sediment inhabited by *Upogebia* facies of the muddy-sand of sheltered-area assemblage; 15: same but *Upogebbia* burrows filled up by later sedimentation; 16: silty sediments with *Zostera nana*; 17: same covered by terrigenous elements; 18: midlittoral beach assemblage; 19: supralittoral beach assemblage; 20: mass of dead and decaying stranded leaves of *Posidonia*; msl: mean sea level; os: predominant input of organogenous material. (After PERES and PICARD, 1964; reproduced by permission of the authors.)

conditions; with in the beds the sediment is always better sorted, finer and muddier than in neighbouring areas. However, there is not a full-scale homology between *Posidonia* and *Zostera* formations, mainly because most *Zostera* rhizomes spread horizontally; *Zostera* therefore never produces mats like *Posidonia*. *Zostera* settlement begins with unraised patches (i.e. at the same level as the neighbouring unvegetated sediment); later, entrapped suspended material causes a slight rising, and small benches appear. However, bench height never surpasses a few decimetres because the beds develop in very shallow water, often in intertidal areas. Where *Zostera* sets are sufficiently crowded, bottom rising results in water depth decrease and consequently in stronger tidal currents, which erode the benches. Where unable to withstand such sediment erosion, *Zostera* disappears and the subsequently reduced sedimentation rate results in the formation of sandy pools or couloirs.

From a biocoenotic point of view, *Zostera* beds are similar to *Posidonia* meadows and may support essentially the same subassemblages, although, of course, the mat assemblage (p. 443) is missing. Two peculiarities probably arise from the intertidal situation of most *Zostera marina* beds. During low tide, the leafage of *Zostera* and its associated algae covers the bottom and thus protects its inhabitants from excessive illumination and temperature and extreme salinity fluctuations. Hence, less euryplastic animals are more numerous here than on neighbouring bottoms without *Zostera*. On the other hand, alternating submersion and emergence probably enhances the release of organic material by stressed organisms. This may be the reason why sponges are more abundant there than in *Posidonia* meadows, e.g. *Halichondria panicea*, and *Hymeniacidon sanguinea* (both also exist beneath brown algae on intertidal rocks) as well as several species of *Mycale*.

The eel-grass beds of the Atlantic American coasts are similar to those of European shores. For example, MARSH (1973, p. 95) who carried out a 14-month scuba study of *Zostera marina* beds in the lower York River (Virginia, USA) noticed that many species are the same on both sides of the Atlantic. The five most abundant non-colonial species (*Bittium varium*, *Paracerceis caudata*, *Crepidula convexa*, *Amphithoe longimana*, *Erichsonella attenuata*) dominate the epifauna throughout most of the year and account for approximately 59% of the total fauna. Minimum total numbers and species counts were recorded in February and early March. Despite *Zostera* exfoliation after June, causing a steady decline in plant biomass, epifauna abundance continued to increase into the summer and fall. Many abundant species (e.g. *Polydora ligni*, *Ercolania fuscata*, *Balanus improvisus*, *Molgula manhattensis*) exhibit short-duration peaks and many epifauna components apparently move into the bottom sediments during winter. In general, gastropods and amphipods followed by polychaetes dominate all *Zostera* areas on the American east coast. THAYER and co-authors (1976) observed a fourfold higher fish biomass in the *Zostera* bed than outside; juvenile-sized classes predominated.

Although primary production in *Zostera marina* beds is not as high as in *Thalassia* beds, it is rather significant in temperate areas. Production in a dense *Z. marina* meadow may be as high as 4–5 g C m^{-2} d^{-1} with a standing crop of about 1 kg m^{-2} (d.w.).

The causes of the so-called 'eel-grass disease' which almost eradicated *Zostera marina* in 1931–33 on both sides of the North Atlantic has been recently discussed by RASMUSSEN (1973). Two fungi, *Ophiobolus halimus* and *Labyrinthula macrocystis*, had been observed in diseased plants especially on Danish coasts, but the latter also exist in healthy *Zostera*. RASMUSSEN believes that medium-range (multi-decennial) climatic changes played the decisive role due to a sequence of warmer summers and particularly mild winters. These adverse climatic effects may have both directly influenced the sea-grass and made it more accessible to parasites. Brackish-water *Zostera* beds were less injured than those in more saline waters, possibly because the salinity optimum for seed germination is between 4·5 and 9·1⁰/₀₀ S. The heavy reduction of *Zostera* beds caused large changes in many shallow-water biotopes and assemblages, but the global benthic production of the Danish shelf seemed to remain unaffected.

(c) North Pacific Ocean *Zostera* beds. A very detailed study of *Zostera* beds was made by SCARLATO and co-authors (1967) in Possjet Bay, Sea of Japan. In the most wave-exposed parts of the bay, on slightly muddy sand 3–4 m deep, *Zostera asiatica* predominates (biomass 4450 g m^{-2}), while *Z. marina* is rare (110 g m^{-2}). Among the infauna, the pelecypod *Spisula sachalinensis* predominates (110 g m^{-2}), together with the echiurid *Urechis* (50 g m^{-2}), *Echinocardium cordatum* (40 g m^{-2}) and some large-sized asteroids such as *Luidia* and *Patiria*. The fauna of the leafage exhibits a striking richness in small gastropods (e.g. *Ephesia turrita* and *Diffalaba vladivostokensis*) and amphipods of the genera

Ischyocerus, Caprella, etc; in spite of its rather small biomass (50–60 g m^{-2}) this fauna probably has a rather high production due to the short life span of its species. In slightly less wave-exposed areas 1·8–2 m deep, only *Zostera marina* is found (biomass 3500 g m^{-2}) with a very similar leafage fauna, but an impoverished infauna where polychaetes tend to replace pelecypods; *Spisula sachalinensis,* however, occurs there (biomass 88 g m^{-2}). The most amazing assemblage lives in sheltered embayments, where the *Zostera marina* biomass reaches its maximum (7300 g m^{-2}). The infauna is very poor, probably due to excessive rhizome abundance and intrication, and polychaetes (biomass 15 g m^{-2}) are almost as abundant as bivalves (20 g m^{-2}). Small gastropods and amphipods of the leafage fauna are both less diversified and more scattered (biomass 14 g m^{-2}). However, on the leaves the small mussel *Musculus senhousia* accounts for up to 3000 g m^{-2}; it is preyed upon by the common sea star *Asterias amurensis.*

In southern areas of the Siberian coast of the Sea of Japan *Zostera* beds are also very common and seem to exhibit the same subassemblages as their European homologues. There are two species of *Zostera*: *Z. pacifica* on relatively clean sands and *Z. marina* on more muddy sands. The fauna associated with their green leaves seem to be richer and more diversified than in Possjet Bay, especially with regard to gastropods (various species of *Rissoa, Lacuna divaricata, Diffalaba vladivostokensis, Gibbula derjugini*), compound ascidians of the genus *Botryllus,* and the prawn *Pandalus latirostris.* Fishes are also more numerous, e.g. the lophobranch *Syngnathus soldatovi.*

Farther to the south, KIKUCHI (1962, 1966)—who studied the fauna, especially fishes and decapod crustaceans of *Zostera marina* beds in Tomioka Bay, Kyu-Shu Islands—demonstrated a striking analogy with Mediterranean *Posidonia* beds. Both groups contain permanent and seasonal residents. About 80% of the fish species are permanent inhabitants (*Rudarius ercodes, Hypodytes rubripinnis,* syngnathids, blennies, gobies, etc); changes in monthly catch mainly arise from their own population fluctuations. As in Mediterranean *Posidonia* beds, seasonal residents (*Sebastes inermis, Lateolabrax japonicus,* etc) mostly occur in summer; maximum and minimum diversity occurs in September and February, respectively. Among the decapods, most of the permanent residents are small-sized species, e.g. Hippolytidae, Palaemonidae, *Alpheus brevicristatus,* whereas in summer various penaeid shrimps, *Processa canaliculata, Sclerocrangon angusticauda* and portunid crabs occur as well. The trophic net analyzed by KIKUCHI also exhibits a striking similarity to that in *Posidonia* beds: among the benthic invertebrates, grazers are very scarce and most of them feed on the superficial sediment film (microphytobenthos and detritus). Permanent fish species mainly feed on sedentary invertebrates, but summer visitors feed mostly on crustaceans (shrimps and crabs). In a more detailed way KIKUCHI (1974) analysed the trophic structure of *Zostera marina* beds revealing dynamic seasonal variations in food-chain relations due to

'(i) the presence or absence of some trophic links by the migration of seasonal residents and transients; (ii) by progressive changeover of food accompanying ontogenetic development of predators; and (iii) by seasonal change of food habits of predators influenced by a seasonal variation of the relative abundance of available foods.' (pp. 152–153).

Zostera beds are also found on American Pacific coasts from Alaska to Baja California. In the north they may often be observed on muddy sands inhabited by the *Macoma–Paphia* assemblage and exhibit the usual green-leaf subassemblage with small gastropods (*Lacuna, Haminea*), as well as many small amphipods and isopods, etc.

(d) *Thalassodendron ciliatum* beds on East African coasts. *Thalassodendron ciliatum*, which participates in assemblages of the on-reef biocoenotic complex (p. 417), is a rather peculiar species due to both the length of its stem (sometimes up to 1 m) and its high ability to support epiphytes (even scleractinians) on stem and leaves. It often constitutes a rather dense canopy (higher stratum) and is very versatile depending on the nature of the substrate, even being able to colonize the ballast of broken stony corals acting as a 'pioneer' species which prepares the substrate for other sea-grasses. *T. ciliatum* may form the largest of all sea-grass beds in the entire Indian Ocean. Its organismic assemblage has been described by MCNAE and KALK (1958) from Inhaca Island (Mozambique) and also the south-western coast of Madagascar (PICHON, 1964).

Thalassodendron ciliatum beds support many species which also live in the shallower *Cymodocea rotunda* and *C. serrulata* beds, e.g. pelecypods of the genus *Atrina*, many echinoderms (*Holothuria*, Synaptidae, *Protoreaster*, *Tripneustes*, *Prionocidaris*, *Temnopleurus* and *Diadema*). However, the *T. ciliatum* assemblage also exhibits three striking peculiarities: (i) abundance and diversity of gastropods (*Trochus, Cypraea, Conus*) are higher than in the shallower *Cymodocea* beds and induce an increase in pagurid abundance; (ii) ophiuroids, which are rare in the shallower beds, are much more numerous; (iii) *T. ciliatum* green leaves exhibit a rich and specific sessile fauna: a species of *Zoanthus*, may characteristic hydroids and bryozoans, the ascidian *Polyandrocarpa*, etc.

(e) *Thalassia* beds. In some places *Thalassia* species, like the preceding sea-grasses, is able to entrap suspended mineral particles and thus create terraces as high as 1 m above the surrounding bottom. *Thalassia* is often mixed with chlorophytes such as *Penicillus*, *Udotea, Avrainvillea*; the only common epiphytes are species of *Fosliella*. According to GESSNER (1971; see also Volume I: GESSNER and SCHRAMM, 1971) the turtle grass *Thalassia testudinum* has a very high tolerance to desiccation; it survives up to 65% water loss and its large leaf surfaces heap up when emerged, the upper leaves protecting the lower ones, which are usually younger and hence less tolerant.

From Bimini Island (Bahamas) VOSS and VOSS (1960) described turtle-grass beds with *Thalassia testudinum, Halodule baudettei* and *Syringodium filiforme*, but the channels between the beds with *Halimeda*. Rather similar to those at Margarita Island (Venezuela), the fauna comprises three subassemblages (RODRIGUEZ, 1959): (i) on the green leaves of some epiphytic algae: *Colpomenia sinuosa, Ulva* and other chlorophytes; (ii) on and in the shallowest sediment: some pelecypods (*Codakia orbiculata, Modiolus tulipus*), gastropods (*Cantharus, Nitidella*) and a holothurian of the genus *Thyone*; (iii) below 0·5–2 m: a very rich fauna of large-sized gastropods (*Strombus gigas, Murex brevifrons, Fasciolaria tulipa, Aplysia protea*) together with some echinoderms, e.g. the asteroid *Oreaster reticulata* and the echinoid *Lytechinus variegatus*. Whether the latter subassemblage is really linked with the turtle grass remains to be investigated.

The assemblage of *Thalassia* and *Halodule* beds in Key Biscayne and Virginia Key (Florida, USA) has been investigated, particularly the infauna (O'GOWER and WACASEY, 1967). Polychaetes are common, their dominant species being *Onuphis magna* and *Loimia medusa*, followed by species of *Nothria, Notomastus, Terebellides*, etc. The most abundant pelecypods are *Codakia orbicularis*, *Cardita floridana*, *Chione cancellata* and the motile *Aequipecten radians*; gastropods of the genera *Columbella*, *Bittium*, *Nassarius*, *Prunum* are also common. The echinoids *Lytechinus variegatus* and *Tripneustes esculentus* feed on sea-grass leaves; in places of maximum density *Lytechinus* may consume 120–140 g d.w. 30 d^{-1}, this figure corresponding to maximal values of *Thalassia* production within the same period. Some fishes are also able to graze on *Thalassia* leaves (RANDALL, 1965, 1967).

Interesting considerations on *Thalassia testudinum* beds in the Caribbean Sea have been developed by WELCH (1965) and may explain the mixing of turtle grass with other metaphytes. According to WELCH, *T. testudinum* beds represent climax conditions, like Mediterranean Sea *Posidonia oceanica* beds. Comparatively mobile sand bottoms are first settled by *Halodule baudettei*, *Halimeda incrassata* and associated species, more steady sediments by *Syringodium filiforme; T. testudinum* settle only on sediments previously stabilized by preceding metaphytes. *T. testudinum* may also exist on hard substrates (like *Posidonia oceanica*, p. 443). In that case the first settlers are *Jania capillacea* and other small tufty species that begin to retain suspended foregin or self-produced detritic particles. Later, other species, such as *Haliméda opuntia* and *Amphiroa fragilissima*, tend to increase the layer of entrapped sediment enriched in humic substances; *Thalassia* may grow on this. In both cases the ultimate post-climax stage appears to be mangrove swamp.

(f) Other sea-grass beds. As mentioned above, intermediate aspects may be recognized between real sea-grass beds, which support an original assemblage, and simple sea-grass facies superimposed on soft-bottom assemblages, which are usually without phanerogams. Assemblages of the first group always exhibit a more diversified and dense subassemblage on living leaves with characteristic species. The major differences in the sessile and sedentary flora and fauna of various phanerogam may depend largely on both leaf morphology and plant physiology (gas exchange).

In tropical areas, phanerogams tend to modify the soft-bottom assemblage because their leaf canopy offers protection from excessive temperature and light intensities at shallow depths. This results in an enrichment of the associated fauna in comparison with that of unvegetated sediments. Such development has been especially noticed on the coasts of East Africa (MCNAE and KALK, 1958) and south-western Madagascar (PICHON, 1964) in three compartments of intertidal sea-grass beds consisting of: (i) burrowing crustaceans, such as alpheid shrimps, but also crabs, e.g. *Lupa* and *Thalamita*; (ii) large-sized invertebrates, such as gastropods (Cypraeidae, Conidae, Strombidae and opisthobranchs) and echinoderms: echinoids, the asteroids *Linckia* and species of the family Oreasteridae, various species of *Holothuria*, etc; (iii) swimming species, mainly fishes and shrimps. A general review of flora and fauna associated with sea-grass beds may be found in KIKUCHI and PERES (1977).

Soft bottoms with algae

(i) Chlorophytes. Organismic assemblages with a dense covering of chlorophytes, some with a calcareous thallus, are very common in intertropical areas both on hard and soft substrates (p. 405). In general, these soft-bottom assemblages have been insufficiently investigated. They should probably be considered as epiflora facies of bottoms without metaphytes.

All soft-bottom species of *Caulerpa* seem to be versatile in regard to substrate nature and do not significantly modify the assemblage, except for their own presence. In the Mediterranean Sea, *C. prolifera* may be observed on the shallowest muddy sands as well as on coarse sands or gravels, and even on terrigenous mud in the circalittoral zone down to at least 100 m in the eastern basin. *C. scalpelliformis*, which is widespread on the coasts of Syria, Lebanon and Israel, also lives on soft bottoms of very diverse characteristics. At False Bay (South Africa), MORGANS (1962) found *C. filiformis*, both on shallow (4–5 m) coarse sands with *Cucumaria insolens* and gobiid fishes, and on deeper sands with the polychaete *Diopatra neapolitana* as the most characteristic inhabitant. The latter assemblage exists as deep as 40 m, but *Caulerpa* cover does not extend below 10–11 m.

A very particular assemblage, with the predominant species *Cladophora prolifera* (biomass 50 g m^{-2} d.w.), was observed by PARENZAN (1969) in the Gulf of Taranto, Ionian Sea, on a 30-m deep soft bottom, forming a belt parallel to the shoreline. Since *C. prolifera* is very euryhaline it may be assumed this assemblage corresponds to a freshwater submarine resurgence.

(ii) Phaeophytes. Some large-sized brown algae, mainly laminarians (p. 400), may exist on soft bottoms. Of course, young plants settle on small hard substrates but fully grown individuals are attached to an anchor-like mass of compacted sediment, enclosed by the holdfasts.

(iii) Rhodophytes. Most of soft bottoms covered to a greater or lesser extent with free-living 'Melobesieae' belong to the circalittoral zone. However, such bottoms may also exist in the infralittoral zone. Near the Cape Verde Islands, CROSSLAND (1905) described a bottom covered with '*Lithothamnium*' nodules and another one with branched or foliaceous Lithothamnieae on coarser sand inhabited by many foraminiferans at depths of 8–18 m; boring sponges and polychaetes were common in all calcareous organisms. This assemblage may represent a facies of the assemblage occurring below on coarse sands and fine gravels under bottom currents (p. 450).

Spongiferous bottoms

On 'spongiferous' bottoms of tropical and subtropical seas live horny sponges of commercial value, ranging from very shallow water down to 30–40 m. Horny sponges grow only on hard substrates, but often on those scattered on sandy bottoms: boulders, gorgonians or dead corals, large-sized algae, rhizomes of dead sea-grasses, etc. Together with their epifauna and infauna, horny sponges seem to constitute a single 'community'. According to WALTON SMITH (1954) the horny sponge assemblage in the Gulf of Mexico includes three ethological groups: (i) Sessile organisms, e.g. algae (mainly the chlorophyte *Acetabularia*, hydroids, actinians e.g. *Aiptasia*), bryozoans and compound ascidians. (iii) Motile epifauna, e.g. harpacticoid copepods, crabs, nudibranchs and asteroids (*Echinaster*). (iii) Infauna, e.g. polychaetes (mainly syllids and terebellids), amphipods (*Leucothoe*, *Colomaestix*), hoplocarids (*Gonodactylus*), decapods (*Synalpheus*, *Coralliocaris*), and some ophiuroids (*Ophiactis savignyi*).

I cannot agree with the rather widespread opinion that the horny sponge assemblage is a 'subcommunity' of the coral reef complex. Spongiferous bottoms often occur in areas near coral reefs, where reef builders are scattered or dead. In fact, I believe that horny sponges and their associated organisms may exist on every vacant hard substrate, provided that bottom currents carry a sufficient amount of suspended (probably small-sized) organic particles. In the Gulf of Gabès (Tunisia) *Hippospongia* ('horse sponges') and *Euspongia* ('baby sponges') are common, for example, in *Cymodocea* or *Caulerpa* lawns, on rhizomes of *Posidonia* and on gravels mixed with broken shells and pebbles (sometimes even in the circalittoral zone).

(4) Organismic Assemblages of the Phytal System Unrelated to the Vertical Zonation

A small number of distinctive and frequently widespread assemblages cannot be considered as belonging to a single vertical zone. Their composition depends on the intensive influence of one (or more) edaphic factor(s) predominating over climatic effects.

(a) Coarse Sands and Fine Gravels under Bottom Currents

Assemblages on coarse sands and fine gravels under bottom currents have been described in many areas of the world ocean. European marine ecologists know them well from the north-eastern Atlantic Ocean and Mediterranean Sea, where they are commonly called 'Amphioxus sand'.

In the Mediterranean Sea the typical 'Amphioxus sand' consists of clean, coarse sands and fine gravel without any admixture of silt or clay, but including many organogenous remains from neighbouring assemblages. Both the presence of these exogenous elements and the lack of fine particles are due to strong bottom currents, which are generally intermittent and linear. WEBB and THEODOR (1969) demonstrated that at shallow depths wave action may have the same effect. Carbon and nitrogen contents of the sediment, which is almost permanently washed at least in its most superficial layer (a few cm), are very low. While micro- and meiobenthos are highly diversified (SWEDMARK, 1956), macrobenthic organisms are rather rare. The most characteristic species are: the calcareous rhodophyte *Lithophyllum racemus*; the free-living coral *Sphenotrochus intermedius*; the polychaetes *Armandia polyophtalma*, *Sigalion squamatum* and *Euthalenessa dendrolepis*; the echinoderms *Ophiopsila annulosa* and *Echinocardium fenauxi*; the scaphopod *Dentalium vulgare*, the pelecypods *Venus casina*, *Diplodonta apicalis*, *Dosinia exoleta*, *Venerupis rhomboïdes*, *Tellina pusilla*, *Arcopagia crassa* and *Donax variegatus*; the crustaceans *Cirolana gallica*, *Anapagurus breviaculeatus*, *Thia polita*, *Macropipus pusillus*, etc; the lancelet *Branchiostoma lanceolatum*; and the sand-eel *Gymnammodytes cicerellus*. One also finds gravel-loving species, such as the pelecypods *Laevicardium crassum* and *Venus fasciata*, the polychaete *Glycera lapidum*, the asteroid *Astropecten aranciacus* and the echinoid *Sphaerechinus granularis*.

In the Mediterranean Sea the most typical aspects of this assemblage occur at 10–20 m. A rather common facies exhibits a peculiar abundance of *Lithophyllum racemus* and admixture of *Acetabularia marina*. In summer, soft algae may grow on gravel or dead *Lithophyllum*, e.g. the phaeophytes *Arthrocladia villosa* and *Sporochnus pedunculatus*, the rhodophytes *Brongniartella byssoides* and *Polysiphonia* sp. (FELDMAN, 1937); *Psammobia costulata* and *Euthalenessa dendrolepis* become rarer, whereas young *Sphaerechinus granularis* is very common. At lesser depths, mainly in the channels between *Posidonia* terraces, some characteristic species (*Arcopagia crassa*, *Sigalion squamatum*, *Thia polita*) disappear, while others, such as *A. polyophtalma* and *D. variegatus*, are much more abundant. The assemblage becomes impoverished, both qualitatively and quantitatively, at greater depths (75 m).

In the north-eastern areas of the Atlantic Ocean this assemblage is very well known since PETERSEN (1918) described it as 'deep-sea, *Venus* community' (V.), later studied by FORD (1923) who named it 'Sp-Vf community' with the characteristic species *Spatangus purpureus, Venus fasciata, Glycimeris glycimeris, Arcopagia crassa* and *Branchiostoma lanceolatum*. THORSON (1975) recognized it as *Venus fasciata–Spisula elliptica–Branchiostoma* 'community' recorded by him between depths of 10 to 300 m. In the English Channel, the *Venus casina* 'community' (CABIOCH, 1961) seems to be similar to the 'Amphioxus sand' and also to the circalittoral assemblage named 'Coastal detritic' on p. 466ff. Because of strong tidal currents, it seems that several species may shift from one assemblage to another, depending on the botom water content of either organic nutritive material or silt particles. In some places an epifauna facies occurs in which the hydroid *Hydrallmania falcata*, the pelecypod *Chlamys varia*, and the solitary form of the styelid ascidian *Dendrodoa grossularia* predominate.

'Amphioxus sands' seem to be rather common wherever bottom currents are sufficiently strong. The following examples may be mentioned here: Persian Gulf and off Madras (Indian Ocean, *Branchiostoma indicum* assemblage with the pelecypods *Venus, Spisula, Glycimeris taylori* and the echinoid *Temnopleurus toreumaticus*, 2–60 m (THORSON, 1957); West African coast, on gravels 20–70 m deep, a *Branchiostoma–Venus declivis–Glycimeris* assemblage with admixture of sessile epifauna on large broken shells (LONGHURST, 1958); at False Bay (South Africa), clean, coarse sand bottoms 10–40 m deep, with *Branchiostoma capense*, many amphipods and isopods, the polychaete *Diopatra neapolitana*, the opisthobranchial gastropod *Philine aperta*, the asteroids *Asteropecten irregularis pontoporaeus* and *Marthasterias glacialis rarispina*, and the hydroid *Aglaophenia pluma dichotoma* (MORGANS, 1962); in Mississipi Sound, an assemblage with *Branchiostoma caribaeum* (73 individuals m^{-2}), together with some pelecypods (mainly *Dosinia*), polychaetes, nemerteans, the flat echinoid *Mellita quinquiesperforata*, etc; on the coast of California (USA) on red and brown sands BARNARD and HARTMAN (1959) observed a *Branchiostoma californiense* assemblage with the hippid *Blepharipoda occidentalis*, the amphipod *Ampelisca cristata*, the echinoid *Dendraster excentricus*, several polychaetes (*Ophelia, Lumbrinereis*), etc; in coarse sands which frequently occur in the vicinity of coral reefs, 'Amphioxus' assemblages are also common (p. 414) and often contain numerous large-sized gastropods (MASSÉ, 1964; SALVAT, 1964; PLANTE, 1967).

(b) Deep Mussel Beds

On coarse sands some mytilids may constitute large beds both in the infralittoral and circalittoral zone. The best example is PETERSEN's (1914a, 1918) *Modiolus modiolus* 'community' (M) in the Danish Straits wherever bottom currents are sufficiently strong to provide abundant food and to prevent fine-particle sedimentation. The typical qualitative and quantitative composition in Øresund at a depth of 28 m, calculated for 0.25 m^{-2} is: *Modiolus modiolus* (40 individuals, biomass 2500 g), *Buccinum undatum* (1 ind., 6.9 g), *Lepidonotus squamatus* (2 ind., 2 g), *Echinocardium cordatum* (1 ind., 10.1 g), *Ophiopholis aculeata* (numerous ind., 16.5 g), *Ophioglypha robusta* (90 ind., 7.7 g), *Strong-*

ylocentrotus droebrachiensis 3 ind., 12·9 g), 'Balanidae' (2 ind., 15 g); the total biomass is 2579 g 0·25 m^{-2}.

ROBERTS (1975), who investigated this assemblage in the Strangford Lough (Northern Ireland), also noticed the pelecypods *Chlamys varia* and *C. opercularis* among the important species, together with the crab *Cancer pagurus*, which seems to be the most voracious predator in the assemblage. He drew the further general conclusions: (i) *Modiolus modiolus*, sometimes epifaunal, often partly buries itself in the substrate with its siphonial ends extruding, and thus must be classed as semi-infaunal; while the exposed part of the shell is heavily incrusted, the buried part remains free of epibionts. (ii) *M. modiolus* is not uniformly distributed but forms clumps of 5 to 30 individuals; first settlers require a small gravel or shell fragment; protection due to clump formation benefits both *Modiolus* itself and its associated fauna. Since the *M. modiolus* assemblage may occur at depths of 5–150 m, it overlaps the infralittoral and circalittoral zones.

Another example of mytilid occurrence in the circalittoral zone was observed by GUILLE and SOYER (1976) in the Gulf of Morbihan (Kerguelen Islands) at depths of 20–50 m, near capes and mouths of fjords; wherever the bottom circulation is active *Aulacomya ater regia* forms large beds on sandy mud bottoms; other filter feeders, e.g. the sabellid *Euchone pallida* and the ophiuroid *Amphiura eugeniae*, are common, together with the detrivorous polychaete *Thelepus setosus*.

Off the northern shores of the Buenos Aires province, south of the Rio de La Plata estuary at depths of 40–50 m, PENCHASZADEH (personal communication) found a much more specific assemblage with the mussel *Mytilus platensis* as the predominant component, supporting a small fishery. All food for the mussel and other filter feeders appears to originate exclusively from the euphotic pelagos layer. The rich associated fauna seems to include many species which prey upon different sizes of mussels: filter feeders on swimming larvae, echinoids and gastropods on young benthic stages, Astropectinidae on mussels less than 6 months old and the gastropod *Trophon* on large-sized mussels.

(c) *Macoma calcarea* Assemblage

The comprehension of the *Macoma calcarea* assemblage, widespread in Arctic and subarctic areas, is still problematic. According to the original description by THORSON (1933, 1934) from muddy sands in fjords of eastern Greenland, its main components—apart from *Macoma calcarea*—are other pelecypods, e.g. *Mya truncata, Cardium (Serripes) groenlandicum, Astarte borealis, A. elliptica* and *A. montagui*, the polychaete *Pectinaria hyperborea*, and the ophiuroid *Ophicoten sericeum*. In this assemblage, which may be observed from very shallow waters down to 300 m, *C. groenlandicum* is more common on cleaner sands, whereas the actinian *Halcampoides* appears where the silt content increases.

The *Macoma calcarea* assemblage was observed in many areas with very diverse facies and aspects: Newfoundland (NESIS, 1965); the northern Kuril Islands and the Gulf of Kamchatka (KUZNETSOV, 1959b, 1961); the Bering Sea (FILATOVA and BARSANOVA, 1964); and Alaska (McGINITIE, 1959). Its biomass may exceed 500 g m^{-2}. The comprehensive study by ELLIS (1959, 1961) suggests that the *Macoma calcarea* assemblage is characteristic of areas permanently or temporarily covered with 'subarctic' waters, i.e. waters with temperatures between 0 and 5 °C. In the most septentrional areas, the assemblage occurs in rather shallow waters (0–50 m) and most of its species breed in summer; at lower latitudes it occurs deeper and its species spawn in winter. The bottom-water temperature is probably the predominant edaphic factor.

(5) Circalittoral Zone

The circalittoral zone is characterized by the presence of the most sciaphilic multicellular algae. Except for some laminarians (p. 400) these are mainly represented by rather small-sized soft species which form an upper stratum and therefore reduce the illumination on the bottom itself, or by incrusting species (calcareous rhodophytes) (PERES, 1967). In general, soft algae are rare or absent on substrates with very fine grain sizes. Unicellular algae, mainly diatoms, are always present but their abundance and, probably, their production decrease at the lowest levels of the continental shelf.

In comparison with the infralittoral zone, the circalittoral one is generally characterized by an increase in the animal components of the populations, especially as regards sessile species on hard substrates, and this occurs at the expense of the algal cover (PERES, 1967). On soft substrates, the vertical distribution of the fauna depends less on illumination than on granulometric characteristics (related to sedimentation rate and bottom currents) and food supply. The latter was considered to depend mainly on the planktonic production of the upperlying waters, but microphytobenthos *in situ* production probably plays a rather important part, at least at the upper levels of the zone.

One of the most striking features of the circalittoral zone is the ability of some assemblages to produce calcareous concretions elaborated by animals, or, more frequently, by rhodophytes (coralligenous assemblage).

(a) Hard Substrates

Mediterranean Coralligenous Assemblage

Although 'coralligenous' is not adequately descriptive (since *Corallium rubrum* belongs to another assemblage) we shall nevertheless retain this established term. The coralligenous assemblage is very well defined (PERES, 1967a,b); it occurs on hard substrates, either rocks or concretions produced by organisms. In its most typical aspect, the important algal fraction mainly comprises the calcareous rhodophytes *Neogoniolithon mamillosum*, *Pseudolithophyllum expansum* and *Mesophyllum lichenoides*, but also soft algae such as *Udotea petiolata*, *Halimeda tuna*, *Cystoseira opuntioides*, *C. spinosa*, *Vidalia volubilis*, *Peyssonnelia rubra* and *P. squamaria*. Sponges are numerous but less characteristic than the gorgonians *Paramuricea clavata* and *Eunicella cavolinii* and the alcyonarians *Alcyonium acaule* and *A. (Parerythropodium) coralloides*. The most frequent polychaetes are *Eunice siciliensis*, *E. schizobranchia* and *Serpula vermicularis*. Bryozoans are numerous, the most characteristic being large-sized species with a ramified or lamellous erect zoarium: *Porella cervicornis*, *P. concinna*, *Hippodiplosia foliacea* and *Myriosoum truncatum*, *Retepora*. The pelecypods *Chlamys pes-felis* and *Lima squamosa*, the decapod crustacean *Lissa chiragra*, the asteroids *Hacelia attenuata* and *Ophidiaster ophidianus* and the ascidian *Rhodosoma verecundum* (in warmer areas) are also characteristic.

The coralligenous assemblage occurs either on rocky substrates wherever the illumination is sufficiently reduced to hinder settlement of species of photophilic assemblages (e.g. lower levels of submarine cliffs, isolated rocks emerging from deep shelf

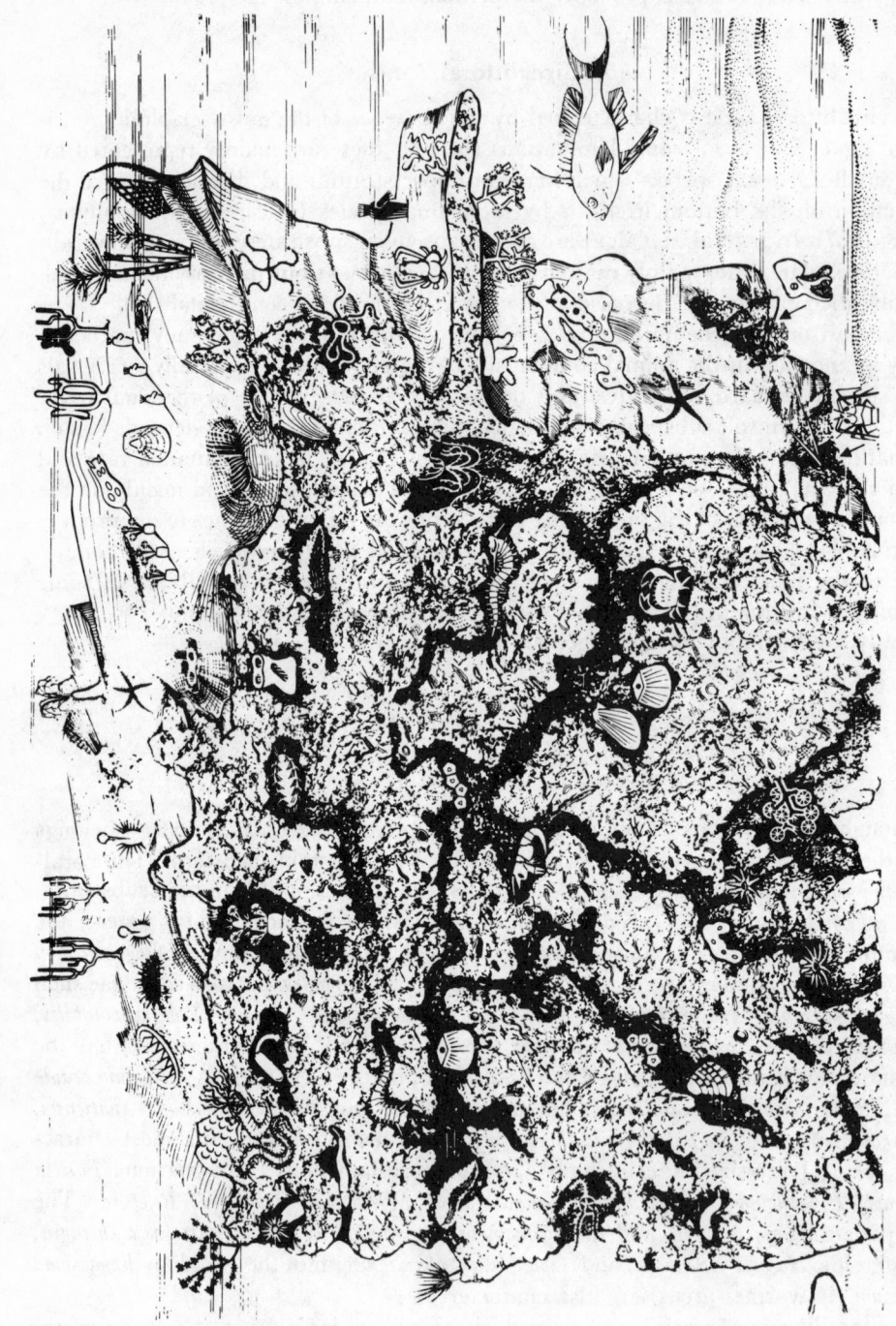

Fig. 8-13: Concretioned hard substrate with motile, epilithic and endolithic animals of the coralligenous assemblage. For explanation see opposite page. (After LAUBIER, 1966; reproduced by permission of the author.)

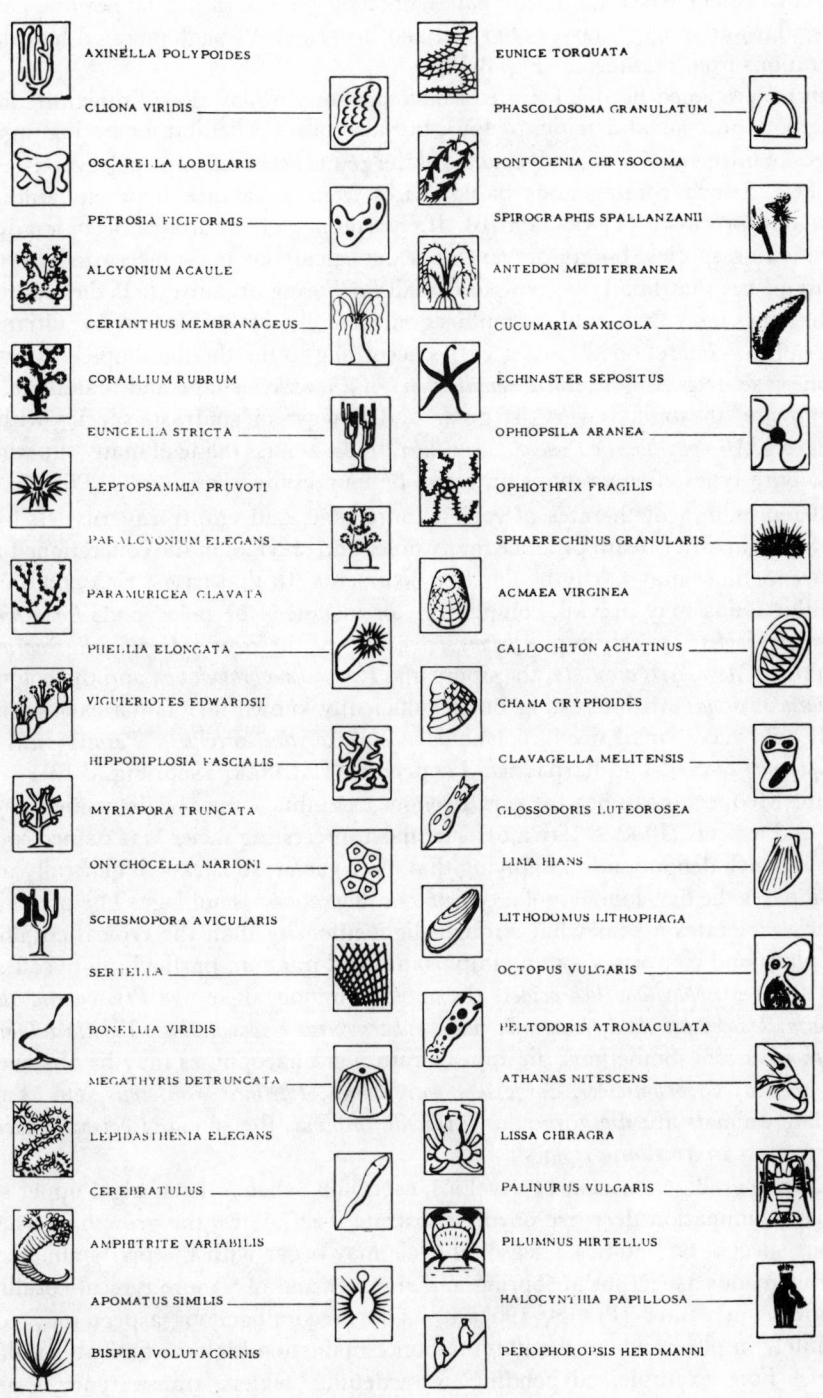

AXINELLA POLYPOIDES

CLIONA VIRIDIS

OSCARELLA LOBULARIS

PETROSIA FICIFORMIS

ALCYONIUM ACAULE

CERIANTHUS MEMBRANACEUS

CORALLIUM RUBRUM

EUNICELLA STRICTA

LEPTOPSAMMIA PRUVOTI

PARALCYONIUM ELEGANS

PARAMURICEA CLAVATA

PHELLIA ELONGATA

VIGUIERIOTES EDWARDSII

HIPPODIPLOSIA FASCIALIS

MYRIAPORA TRUNCATA

ONYCHOCELLA MARIONI

SCHISMOPORA AVICULARIS

SERTELLA

BONELLIA VIRIDIS

MEGATHYRIS DETRUNCATA

LEPIDASTHENIA ELEGANS

CEREBRATULUS

AMPHITRITE VARIABILIS

APOMATUS SIMILIS

BISPIRA VOLUTACORNIS

EUNICE TORQUATA

PHASCOLOSOMA GRANULATUM

PONTOGENIA CHRYSOCOMA

SPIROGRAPHIS SPALLANZANII

ANTEDON MEDITERRANEA

CUCUMARIA SAXICOLA

ECHINASTER SEPOSITUS

OPHIOPSILA ARANEA

OPHIOTHRIX FRAGILIS

SPHAERECHINUS GRANULARIS

ACMAEA VIRGINEA

CALLOCHITON ACHATINUS

CHAMA GRYPHOIDES

CLAVAGELLA MELITENSIS

GLOSSODORIS LUTEOROSEA

LIMA HIANS

LITHODOMUS LITHOPHAGA

OCTOPUS VULGARIS

PELTODORIS ATROMACULATA

ATHANAS NITESCENS

LISSA CHIRAGRA

PALINURUS VULGARIS

PILUMNUS HIRTELLUS

HALOCYNTHIA PAPILLOSA

PEROPHOROPSIS HERDMANNI

sediments, outer parts of submarine caves) or on biogenous, rather flat bottoms consisting of irregular and craggy masses of gravel and broken shells agglomerated by calcareous elaborations from organisms (Fig. 8-13).

This concretioned hard substrate of biological origin may slowly substitute for a soft coarse bottom (coastal detritic, p. 466), but also offers a habitat for boring organisms: sponges of the genus *Cliona*, polychaetes of the genus *Polydora*, some pelecypods, etc. The existence of such coralligenous banks arises from a balance between building and destruction processes (PERES, 1967b). If environmental conditions become adverse to concretioning species, borers prevail; thus the calcareous mass increasingly crumbles, forming pieces that finally become too small for boring organisms. If the environment then becomes more favourable to builders once more, a new cycle may be initiated. The texture of the concretioned masses varies according to the thallus shape of the algae; if 'Melobesieae' (e.g. *Neogoniolithon mamillosum*) of a massive shape and mamillate surface predominate, the bulk is very irregular and craggy; in contrast, species with a flat thallus, e.g. *Mesophyllum* or *Pseudolithophyllum*, build a mass made of many superimposed layers. Both types of concretions may also be found on rocks (PERES, 1967b).

Differences in growth rates of various organisms and empty caverns left by many boring animals after death produce many holes and crevices in the concretioned masses, which sometimes may partly be filled by sediments. In that case a rich and diversified endolithic fauna may prevail, comprising, for instance, the pelecypods *Lima hians* and *Chama gryphoides*, the shrimp *Athanas nitescens*, the bryozoan *Onychocella marioni*, the brachiopod *Megathyris decollata*, the sipunculid *Physcosoma granulatum* and the holothurioid *Cucumaria saxicola*. An interesting but insufficiently known meiofauna exists consisting mainly of very small-sized polychaetes (*Plakosyllis brevipes*, *Paratyposyllis peresi*, *Chrysopetalum caecum*) and harpacticoid copepods (LAUBIER, 1966; Fig. 8-13).

In the Mediterranean Sea the coralligenous assemblage may be diversified into many facies and aspects (PERES, 1967a,b). The most interesting facies was named 'precoralligenous', such denomination implying that, on a vacant substrate, it generally precedes and prepares the development of a typical coralligenous assemblage. The precoralligenous facies tolerates a somewhat stronger light intensity than the typical coralligenous assemblage and comprises a more important algal fraction, particularly of soft species. Apart from *Mesophyllum lichenoides*, the most common algae are *Peyssonnelia rubra*, *P. squamaria*, *Udotea petiolata*, *Halimeda tuna*, *Sphaerococcus coronopifolius*, *Vidalia volubilis* and *Phyllophora nervosa*. Sometimes an upper stratum of phaeophytes may be observed: *Cystoseira spinosa*, *C. opuntioides*, *Sargassum hornschuchii*, *Phyllaria reniformis*, etc. The most abundant animals are the gorgonian *Eunicella cavolinii*, the sponge *Chondrilla nucula* and the bryozoan *Scrupocellaria reptans*.

If the precoralligenous aspect develops, especially when a dense algal upper stratum prevails, illumination decrease on the substrate itself allows the growth of true coralligenous species. Sometimes a seasonal cycle may occur with the predominance of the precoralligenous aspect in late spring and summer and of a more typical coralligenous assemblage in winter (PERES, 1967a,b). The precoralligenous aspect may continue indefinitely in places where the illumination remains too high for typical coralligenous species. For example, depending on depth, water transparency, substrate microtopography, etc., precoralligenous aspect may become permanent on well-shaded vertical rocks or small open crevices, on the console supporting the *Lithophyllum tortuosum* cornice (p. 378) and on *Posidonia* rhizomes emerging from the sediment—both in deeper meadows and in shallow-water beds—providing the phanerogams are sufficiently crowded (PERES, 1967a,b).

On vertical cliffs and in submarine caves with greatly diminished illumination, aspects occur with the higher stratum consisting of gorgonians; *Eunicella cavolinii* in the above-mentioned precoralligenous aspect is more tolerant to light than *Paramuricea clavata*. The stratification of these assemblages has been studied by TRUE (1970).

Other Assemblages which may be referred to the 'coralligenous group'

North Eastern Atlantic Ocean

On rocky substrates on western European coasts—at different depths depending on water transparency—an assemblage may occur which is very similar to the coralligenous assemblage of the Mediterranean Sea. The algal fraction is less important than in the Mediterranean Sea and seems to include only rhodophytes: *Delesseria, Plocamium coccineum, Halarachnion, Phyllophora* and some insufficiently known calcareous crustose species. Sponges are much more abundant: *Axinella** spp. (for the meaning of the asterisks, see next paragraph), *Tethya, Stelligera stuposa, Raspailia hispida, Stelleta grubei, Pachymatisma johnstoni, Aplysilla, Dysidea fragilis*, etc. The characteristic erect bryozoans (*Porella*, Retepora*, Hippodiplosia foliacea**) are present together with the Anthozoa *Corynactis viridis, Alcyonium digitatum* and *Eunicella verrucosa**, many hydroids, some cirripeds (*Scalpellum*, Balanus*) and many echinoderms such as *Cucumaria, Antedon, Ophiothrix fragilis, Ophiopsila aranea*. Ascidians are also common, e.g. *Distomus variolosus, Dendrodoa grossularia* (solitary form), *Stolonica socialis** and *Pyura*. Some aspects of this assemblage which exist on boulders in the English Channel (*Axinella dissimilis–Phakellia ventilabrum–Corynactis viridis*) were described by CABIOCH (1961).

In caves or overhanging rocks which emerge at spring low tides this assemblage is quite recognizable, although slightly modified: (i) species marked with an asterisk above are absent or exceptional; (ii) two species thriving to such an extent that both sometimes cover more than 80% of the rocky surface: *Dendrodoa grossularia* (gregarious form) and *Pachymatisma johnstoni*; (iii) some species absent or very rare in the deeper assemblage are very common: some hydroids (*Sertularia operculata, Aglaophenia septifera*) or bryozoans (*Crisia, Bugula, Bicellaria*), many sponges—both Calcarea (*Clathrina coriacea, Grantia*, some *Leuconia*) and Acalcarea (*Dercitus*)—and many ascidians mainly Polyclinidae). Some species of this assemblage may occur among and on laminarian holdfasts due to the decreased illumination caused by the leaf canopy: small hydroids and bryozoans species and scattered *Dendrodoa* and *Distomus variolosus* are the most common.

However, it must be pointed out that concretioned hard substrates of biological origin (like those of the Mediterranean Sea) have never been observed in the north-eastern Atlantic Ocean. This may possibly be related to the lack of some 'Melobesieae' species, which are particularly able to build such concretions, but is more probably due to the usually higher turbidity that allows these algae to grow only at depths where water movement (waves, tidal currents) is sufficiently strong to hinder their aggregation.

North-western Atlantic Ocean

Near Newfoundland Banks, on mixed bottoms (rocks, boulders, pebbles, broken shells) 45–55 m deep, NESIS (1965) discovered an assemblage belonging to the coralligenous group, with some rhodophytes (Lithothamnieae, *Ptilota*), many hydroids and bryozoans, *Balanus crenatus*, some polychaetes (*Spirorbis, Thelepus cincinnatus*) and echinoderms (*Strongylocentrotus droebrachiensis, Ophiopholis aculeata, Cucumaria frondosa*). The latter species largely predominates and accounts for about $\frac{5}{6}$ of the whole biomass (about 590 g m^{-2}).

North Pacific Ocean

In the north-western areas of the Pacific Ocean, many assemblages are similar to the coralligenous assemblage. The rocky bottoms on the southern coasts of Sakhalin Island harbour many rhodophytes (among them *Ptilota pectinata*) and sponges (e.g. *Tethyum aurantium*) as well as *Asterias amurensis, Cucumaria japonica*, etc., together with numerous hydroids, bryozoans, crustaceans echinoderms and ascidians (USHAKOV, 1951).

Arctic and subarctic seas

Several communities described from Arctic and subarctic areas must also be referred to the coralligenous assemblage group. (i) In the Kamchatka Gulf, KUZNETSOV (1961) observed a rich assemblage on 40–80 m deep hard substrates; its most characteristic species are: the hydroids *Lafoëa grandis* and *Grammaria abietina*; the alcyonarian *Eunephthya*; bryozoans of the genera *Leischara, Membranipora, Cellaria*; species of the echinoid *Stronglylocentrotus* (12 ind., amounting to a biomass of 200 g m^{-2}); the ascidians *Chelyosoma* and *Boltenia*; some shrimps, e.g. *Sclerocrangon intermedia* and *Nectocrangon crassa*, which, swim above the bottom. The total biomass is about 500 g m^{-2}. The biomass of a very similar 'community' in the Bering Sea may be as high as 3 kg m^{-2}, with sponges, hydroids (*Abietinaria, Thuiaria, Sertularella, Lafoëa, Halecium*), *Eunephthya, Balanus crenatus, Hyas coarctatus alutaceus* and the echinoderms *Strongylocentrotus sacchalinicus* and *Ophiopholis aculeata*. Rocky substrates are intermixed here with sandy patches inhabited by polychaetes of the genus *Nephthys* and the pelecypod *Chlamys beringianus*. (ii) In subarctic regions of the North Atlantic Ocean, the *Saxicava arctica* community', with *Balanus, Alcyonidium, Rhynchonella*, down to 100 m deep, its biomass averaging 100 g m^{-2}. (iii) In the Barents Sea DERJUGIN (1915) described a community from rocks and pebbles in which ascidians (*Ascidia, Pyuraarctica*) and Calcarea sponges (*Grantia, Leucosolenia*) predominate, together with some hydroids and many bryozoans (*Porella, Flustra*); its biomass may exceed 500 g m^{-2}. (iv) At Point Barrow (Alaska, USA) gravel, pebbles and broken shells mixed with mud, are inhabited by a rich fauna of bryozoans (McGINITIE, 1955): *Eucratea loricata, Electra crustulenta, Dendrobeania murrayana, Porella compressa*: other common animals are hydroids (*Thuiaria, Lafoëina maxima*), actinians, the octocoral *Eunephthya rubiformis, Balanus crenatus*, some ophiuroids, *Strongylocentrotus droebrachiensis*, the holothurioid *Psolus stimpsoni* and the large pedunculate ascidian *Boltenia ovifera*.

Antarctic shores

The arctic and subarctic assemblages described above—with their main components, sponges, hydroids, *Eunephthya*—are also present on Antarctic shores.

From the Davis Sea GRUZOV and co-authors (1967) described an *Eunephthya* assemblage. There are some exceedingly rare algae, but animals largely predominate. The faunistic richness is markedly higher than in arctic assemblages. Some species of the infralittoral assemblages mentioned above (p. 407), such as *Odontaster validus* and *Sterechinus neumayeri* (standing crop: $88 \mathrm{\ g \ m}^{-2}$) still exist, but *Eunephthya* largely predominates, its biomass amounting to up to 80% of the whole standing crop, which is about $1000 \mathrm{\ g \ m}^{-2}$. There are also numerous hydroids, the most abundant of which is *Oswaldella antarctica*, and polychaetes (e.g. *Potamilla*).

In McMurdo Sound the eastern Antarctic infralittoral and the *Eunepthya* assemblages seem to be somewhat mixed (DAYTON and co-authors, 1969, 1970). Not a single alga was mentioned in the circalittoral zone. However, apart from rather eurytopic species (*Odontaster validus*, *Sterechinus neumayeri*, *Lineus* sp. and some fishes of the genus *Trematomus*), there is a rich cnidarian fauna: the octocorals *Clavularia frankliniana* and *Alcyonium paessleri*, the actinians *Artemidactis victrix*, *Isotealia antarctica*, *Urticinopsis antarctica* and *Hormathia lacunifera*, the hydroids *Tubularia hodgsoni*, *Lampra parvula* and *Halecium arboreum*. Pycnogonids are also common: *Thaumastopycnon striata*, *Colossendeis robusta* and *C. megalonyx*, and the large ascidian *Cnemidocarpa verrucosa* has been recorded.

In a more recent paper on Cape Ermitage, McMurdo Sound (30–60 m deep), DAYTON and co-authors (1974) described an amazing epifaunal assemblage in which sponges and their asteroid and nudibranch predators largely predominate. The sponge *Mycale acerata* appears to be the potential dominant in competition for substrate space, because its growth rate is much higher than that of the other sponges. However, *Mycale* is prevented from dominating the space available by two asteroids: *Perknaster fuscus*, which feeds mainly on *Mycale* and, to a smaller extent, *Acodontaster conspicuus*, the latter asteroid and the nudibranch *Austrodoris mcmurdensis* usually feed on the rossellid sponges *Rossella racovitzae*, *R. nuda* and *Scolymastra joubini*. Despite the rather high consumption of rossellids by the two specialized predators, very large standing crops of these sponges have accumulated, presumably due to the predation on larval stages and juveniles of these predators by the asteroid *Odontaster validus* (both a detritus feeder and filter feeder) and the actinian *Urticinopsis antarcticus*.

Tropical and subtropical seas

At a depth of about 80 m on Australian coasts LEMCHE (1956) observed a bottom on which large-sized species of ramose and highly calcified bryozoans seem to play an important role in the aggregation process; the fauna is rich with *Galathea* sp., gastropods (*Doridunculus*), *Antedon*, ophiuroids, etc. In the Macassar Strait the 80-m deep gravel and pebble bottom, briefly described by the same author, mainly comprises antipatharians, large-sized hydroids (*Nematophanus*), together with crinoids, and ophiuroids of the genus *Euryale*. These organisms seem to belong to the group of coralligenous assemblages.

On Ghanian coasts BUCHANAN (1958) found between 60 and 100 m hard substrates with the following most striking inhabitants: the scleractinian *Dendrophyllia ramea**, *Balanus tulipiformis*, the echinoderms *Antedon hupferi* and *Astrospartus mediterraneus*, the pelecypods *Avicula hirundo* and *Pycnodonta cochlear**; species with an asterisk (*) in the list above indicate admixtures to this coralligenous assemblage by some species characteristic of the offshore rocky-bottom assemblage (p. 463).

The 80-m deep rocky-bottom 'community' described by MORGANS (1962) at False Bay (South Africa) exhibits more similarity to temperate-sea coralligenous assemblages; sessile species are numerous and highly diversified, e.g. sponges, hydroids, encrusting compound ascidians (*Didemnum stilense*) and in particular bryozoans (*Cellepora, Chaperia, Retepora, Stomatopora*). Small-sized motile forms are also abundant (polychaetes, amphipods, isopods), together with the ophiuroid *Dictenophiura anoidea*.

Near Barbados at depths of 50–150 m on a poorly defined bottom with biogenous hard substrates and coarse-sand areas, LEWIS (1965) observed an assemblage which seems to be closely related to the coralligenous group. Unfortunately, algae were not considered by the author. In this assemblage the concretioning process seems to arise mainly from encrusting sponges (*Agelas, Verongia, Geodia, Pachychalina*) and hermatypic corals (*Agaricia, Mussa, Madracis*); alcyonids are numerous, e.g. *Neospongondes portoricensis, Telesto, Nidalia*; most of the polychaetes and crustaceans present are associated with the sponges. Pelecypods are fairly rare and mainly belong to genera usually inhabiting hard or very coarse substrates: *Chama, Lima, Pecten*. However, in soft-bottom patches some *Tellina* and *Laevicardium* may be found. Below 100 m, corals are replaced by anti-patharians (*Antipathes, Parantipathes, Stichopathes*); consequently, the concretioning processes diminish and soft-bottom areas enlarge; they are inhabited by more diversified pelecypods (*Limatula, Limopsis, Pitar, Propeamussium, Pseudomussium*), pagurids (*Polypagurus*), crabs (*Iliacantha, Parthenope*), echinoderms, etc.

The occurrence of some hermatypic corals in the Barbados community described above brings up the problem of the existence of coralligenous assemblage group in the coral reef complex. Unfortunately, data on sciaphilic assemblages of coral reefs are rather scarce. According to YONGE (1963), the *Leptoseris* community (p. 414) observed by WELLS (1954) on the windward slope at the Bikini Atoll (90–150 m) includes only a small percentage of hermatypic corals. WELLS (1957) also assumed that a depth of 50 m corresponds to the lowest level of luxuriant hermatypic corals in the Marshall Islands.

In the global description of the assemblages on the hard substrates of the outer slope of coral reefs, it was emphasized that the deeper parts (19–24 m down to 50 m) of the coral flagstone in the Tulear Great Reef are occupied by an assemblage belonging to the circalittoral zone (p. 414). On the same reef complex, VASSEUR (1964) noticed a very similar assemblage at shallower but more shaded places: overhangs, lower surfaces of boulders, crevices, openings of submarine reef caves, etc. Like that of the deeper flagstone itself, the later assemblage includes a very large algal fraction: e.g. calcareous crustose algae (*Lithothamnium, Neogoniolithon, Dermatolithon*) soft rhodophytes, such as Squamariaceae (*Ethelia, Cruoriella*), *Galaxaura, Phyllophora, Amansia*, and the phaeophyte *Pocockiella variegata*. The predominant animals are foraminiferans (*Carpenteria, Homotrema*), sponges (e.g. *Timea unistellata, Microciona curvichela*), hydroids (*Tyroscyphus, Dynamena, Aglaophenia* and *Solanderia*), some bryozoans (*Margareta levinsoni*), etc. In places of more diminished illumination, the algae are more scattered and less diversified, but the corals *Astroides aurea* and *Dendrophyllia elegans* and the sponge *Iotrochota baculifera* become more abundant; this is a transitional aspect towards the assemblage described on p. 462.

Assemblages in Semi-obscure Caves

The 'biocoenosis' of semi-obscure caves, first described on rocky limestone coasts of the Mediterranean Sea, was distinguished from the coralligenous assemblage group by LABOREL (1960). It may include multicellular algae, always very rare and sometimes of doubtful taxonomic position because they are always in a very poor condition and probably infertile. The luxuriant development of sponges significantly hinders the growth of encrusting algae. Unicellular algae are probably present but have never been studied. Sponges and anthozoans clearly predominate. Among the most characteristic species are the following: the sponges *Verongia aerophoba, Petrosia ficiformis* (with its specific nudibranch *Peltodoris atromaculata*), *Oscarella lobularis, Agelas oroides*; the alcyonarian *Corallium rubrum*; the zoantharian *Parazoanthus axinellae*; the scleractinians

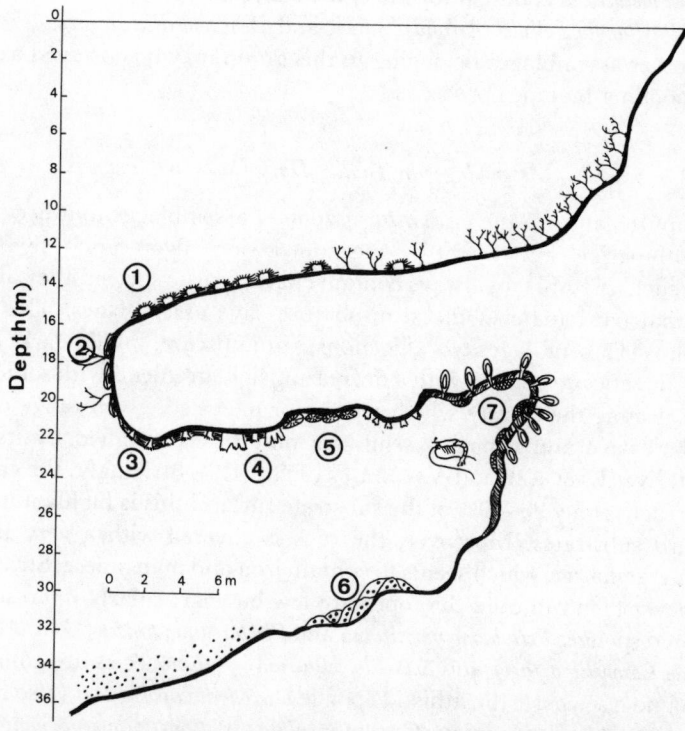

Fig. 8-14: Distribution of different assemblages along a gradient of decreasing illumination in a submarine cave (Paleokastritza, Greece). 1: highly photophilic assemblage (*Arbacia lixula, Sargassum* spp.); 2: *Sargassum* and *Peyssonnelia* assemblage; 3: *Leptopsammia pruvoti* and rhodophyte assemblage; 4: *Hoplangia durothrix*; 5: hollows occupied by *Madracis pharensis*; 6: rock, more or less covered with muddy sand occupied by the sponge *Hymeniacidon caruncula*; 7: dark-cave assemblage with bryozoans, boring pelecypods (*Lithodomus*) and bathyal shrimps (*Stenopus scaber* and *Parapandalus narval*). (After LABOREL, 1960; reproduced by permission of the author.)

Leptopsammia pruvoti, *Caryophyllia smithi*, *Hoplangia durothrix*, *Madracis pharensis* (the latter more common in the eastern basin); the bryozoans *Coastazzia caminata*, *Adonella calveti*, *Schismopora avicularis*; and the ascidian *Pyura vittata* (PERES, 1967a,b; Fig. 8-14). Some different aspects of this assemblage can be observed, depending on changes in light intensity, wave motion, silt deposition, etc. This assemblage may also occur in holes or overhangs of coralligenous concretioned masses where the illumination is sufficiently diminished.

An assemblage corresponding to that above was observed in reef caves by VASSEUR (1964) at Tulear (Madagascar). Here, algae are very scarce and are represented only by a few poorly developed *Cruoriella* sp. and *Peyssonnelia conchicola*. In contrast, sponges are very abundant and diversified: *Placospongia carinata*, two species of *Spirastrella*, *Timea curvistellifera*, *Agelas marmarica*, *Gelliodes* sp., *Strongylophora durissima*, etc. The stylasterid *Distichopora violacea* is common together, with ahermatypic scleractinians, such as *Culicia cuticulata*, *Phyllangia pallida*, *Astroides aurea*, and *Dendrophyllia elegans*.

Many other assemblages belonging to this group may be observed as the exploration of corresponding biotopes proceeds.

Assemblages in Totally Dark Caves and Tunnels

The appurtenance to the circalittoral zone of assemblages that occur in totally dark parts of submarine caves might be questioned, since algae cannot live there. However, although such assemblages always contain characteristic species, they also include many animals that participate in the semi-obscure cave assemblages. This means that the three groups of assemblages (coralligenous, semi-obscure, totally dark) in fact constitute a syngenetic series correlated with a decreasing light gradient, with some thresholds each of them selecting the species which require or tolerate a given range of illumination.

The dark cave and tunnel assemblage may be exemplified by its Mediterranean representative (LABOREL and VACELET, 1958, 1959). Strikingly, the covering by sessile animals reaches only 20–30% of the substrate surface; this is highly unusual on shallow-water hard substrates. Moreover, the rock is covered with a very thin veneering of black mineral matter, which seems to contain iron and manganese. Strictly characteristic species never found in other biotopes are few but particularly interesting, e.g. the two pharetronid sponges *Petrobiona massiliana* and *Plectoroninia hindei* (VACELET, 1964) and the bryozoans *Coronellina fagei* and *Setosella vulnerata*. Some species are common which may also be found elsewhere: the lithistid sponge *Discodermia polydiscus* (also recorded from the bathyal zone); the scleractinians *Gwynia annulata* and *Conotrochus magnanii* (which exist also in semi-obscure caves); the serpulids *Omphalopoma janita* (= *fimbriata*) also present in the coralligenous assemblage) and *Vermiliopsis monodiscus* which also occurs in the bathyal zone); the brachiopod *Megathyris decollata* (also present in coralligenous and semi-obscure assemblages). In totally dark caves, other sessile species also known from the coralligenous assemblage and from semi-obscure cave assemblage occur, always exhibiting a highly reduced vitality (list in VACELET, 1964, and PERES, 1967a,b). LEDOYER (1965) also found a fauna of motile species there; some of them are strictly characteristic, such as the small gastropod *Leptothyra sanguinea*, and mysid *Hemymisis speluncola* (which sometimes forms large shoals) and the fish *Apogon ruber*. Other motile species, which are not characteristic but occur more commonly in this very particular biotope than elsewhere, are the gastropods *Alvania reticulata* and *Rissoa cancellata* (which also exist in coralligenous and semi-obscure cave assemblages), the amphipod *Aristias tumidus* (also in semi-obscure caves), the crab *Herbstia condyliata* (also in coralligenous assemblage), the shrimp *Stenopus scaber* which lives outside the caves only in the bathyal zone, etc. (Fig. 8-14).

A similar assemblage was observed by VASSEUR (1964) in the totally dark tunnels of coral reefs at Tulear (Madagascar) with numerous and sometimes very specific sponges, such as the Silicocalcarea *Astrosclera willeyana*, the lithistids *Macandrewia cavernicola* and *Aciculites tulearensis*, together with the foraminiferan *Carpenteria monticularis*, a hydroid of the family Tubulariidae and the ascidian *Leptoclinides tulearensis*.

Offshore Rocky-bottom Assemblages

In the Mediterranean Sea investigations with Jacques Cousteau's diving saucer (LABOREL and co-authors, 1961) allow us to state categorically that between coralligenous (p. 453) and 'white coral' (p. 480) assemblages a fairly specific assemblage exists which on the upper part of the slope corresponds to an illumination lower th..n that in the coralligenous assemblage but still supports some poorly developed calcareous rhodophytes. The rocky substrate often seems to be covered with a very thin layer of silty sediment and the peduncle or basal part of sessile species may penetrate the sediment layer to attach to the rock.

The most striking inhabitants of offshore rocky bottoms are sponges. Among them, *Poecillastra compressa*, *Rhizaxinella pyrifera*, *Phakellia ventilabrum*, *Suberites carnosus* (f. *ramosus* and f. *typicus*), *Acanthella acuta**, *Axinella polypoides**, *A. verrucosa**, *A. damicornis**, *Ciocalypta penicillus*, *Erylus discophorus*, *Ircinia (Sarcotragus) muscarum*, *I. oros*, some large individuals of *Tylodesma inornata*, *Petrosia ficiformis**, and the pharetronid *Plectroninia hindei**. Among the cnidarians two species are characteristic: the yellow scleractinian *Dendrophyllia cornigera* and the antipatharian *Antipathes fragilis*†. One also finds *Lafoea dumosa* (quite common), *Corallium rubrum*, *Paralcyonium elegans*, *Alcyonium acaule*† (in poor condition) and *Madrepora oculata*†. *Hornera* sp., *Retepora** sp. and *Porella cervicornis*, common brachipods, *Terebratulina caput-serpentis* and *Gryphus vitreus*† are common bryozoans. The rest of the fauna comprises the echiurid *Bonellia viridis**; the decapods *Munida* sp.†, *Palinurus mauritanicus** and *Paromola cuvieri*†; the polychaetes *Serpula vermicularis* var. *echinata*, *Placostegus tridentatus* and *Omphalomopopsis fimbriata*; the pelecypod *Spondylus gussoni*†. The echinoderms are represented by *Ophiacantha setosa* (often in the oscula of large sponges), *Echinaster sepositus*, *Ceramaster placenta**, *Cidaris cidaris*†, and *Antedon mediterranea*. In this list, the rather numerous species marked with an asterisk (*), often quite well represented, inhabit the circalittoral zone, while species marked with a plus (†) are intruders from the underlying bathyal zone. Sestonophagous species usually predominate, apparently owing to the rather strong bottom currents observed during dives.

The offshore rocky-bottom assemblage group, first accurately investigated in the Mediterranean Sea by direct observation, seems to exist in many other areas of the world ocean at similar depths:

(i) In the North Atlantic Ocean ANCELLIN (1957) observed at depths of 100–200 m, on rocky or boulders bottoms, an assemblage predominated by sponges (*Geodia*, *Poecil-*

lastra, Phakellia) and brachiopods of the genus *Waldheimia*. The bathyal scleractinian *Madrepora oculata* is also present.

(ii) From the Barents Sea at depths of 90–180 m DERJUGIN (1915) gave a detailed description of an assemblage that is probably the same as that dredged by ANCELLIN (1957), with the sponges *Geodia baretti, Stryphnus fortis, Phakellia bowerbanki*, many hydroids (*Lafoea, Diphasia, Grammaria, Thuiaria*) and bryozoans (*Retepora, Smittina, Flustra, Caberea*). Brachiopods, such as *Terebratulina caput-serpentis, Rhynchonella psittacea, Terebratella spitzbergensis* and *Waldheimia cranium*, are common and the echinoderms are represented by *Heliometra quadrata, Ophiacantha bidentata* and *Gorgonocephalus eucnemis*. Most inhabitants are sestonophagous, as in the Mediterranean assemblage: the standing crop is about 170 g m^{-2}.

(iii) In Antarctic areas USHAKOV (1963) and BULLIVANT (1967) observed on bottoms of rock or boulders covered with a thin mud layer at depths 100–500 m, a similar assemblage with predominant sponges (*Rosella, Tetilla, Latrunculia, Tedania, Iophon*), gorgonians and a fauna of bryozoans that is particularly diversified and comprises both crustose forms (*Mucronella, Smittina, Schizoporella*) and erect species (*Retepora, Cellepora, Hornera, Tubulipora*). This siliceous sponge assemblage deserves particular attention because of its exceptional abundance and diversity on Antarctic shores. Its upper level overlaps the circalittoral zone, but it extends deeper down to several hundred metres. GRUZOV and co-authors (1967), for instance, observed in the Davis Sea a biomass of about 3 kg m^{-2}, to which the sponges (mainly *Scolymastia joubini* and *Rossella* spp.) contributed 1·5–2 kg; the hydroid *Oswaldella*, 170 g; and the polychaete *Potamilla*, 150 g. The specific diversity of the siliceous sponges is very high: although taxonomically insufficiently investigated, there seem to be at least several tens of different species present. This assemblage has several species in common with the circalittoral assemblage described from MCMURDO Sound's circalittoral zone by DAYTON and co-authors (1970, 1974) and possibly represents its extension towards the depths (ARNAUD, 1965). However, the higher percentage of filter feeders other than sponges, e.g. hydroids and bryozoans, suggests that it represents a distinct unit.

In some areas of McMurdo Sound, the high abundance of siliceous sponges results in a substrate constituted of unsorted rock debris embedded in a sponge-spicule mat sometimes more than 1 m thick. This mat supports a fairly specific assemblage of epibenthic sessile organisms (DAYTON and co-authors, 1970; DELL, 1972). Apart from sponges, the actinians *Stomphia selaginella, Artemidactis victrix*, and a few individuals of *Urticinopsis antarcticus*, the hydroids *Lampra parvula, L. microrhiza* and *Halecium arboreum*, a few sabellid polychaetes, some bryozoans, and numerous molluscs (mainly the pelecypod *Limatula hodgsoni*) are the most conspicuous species.

(iv) The deep-sea barnacle 'community' in the north-western Pacific Ocean at depths 87–115 m on coarse detritic bottoms with rocks and boulders (FILATOVA and BARSANOVA, 1964) is somewhat similar to offshore rocky-bottom assemblages: sessile sestonophagous species predominate, e.g. the sponges *Myxichella ochotensis* and *Phorbos salebrosus*, as well as many ascidians (*Boltenia, Phallusia, Tethyum*), the pelecypod *Chlamys beringianus* and the barnacle *Balanus crenatus*; the standing crop may exceed 1100 g m^{-2}. In the Bering Sea *B. evermanni* and *C. albidus* substitute for *B. crenatus* and *C. beringianus* respectively. However, this assemblage also exhibits some similarity to the coastal detritic group (p. 471), the typical fauna of which may be modified when large-sized and scattered hard substrates are more abundant.

(v) On West African coasts with the shelf often occupied by soft bottoms, most of the sessile fauna which usually inhabits coralligenous areas is forced downwards near the shelf edge (about 100 m deep) where bottom currents are stronger. Here the insufficiently investigated assemblages exhibit features intermediate between those of coralligenous and offshore rocky-bottom assemblages: sponges are numerous, together with gorgonians, scleractinains (among them *Dendrophyllia ramea*), pelecypods (*Pteria hirundo*, *Ostrea cochlear*), the ophiuroid *Ophiothrix tomentosa*, etc. In the Gulf of Guinea, LONGHURST (1958) observed a particular aspect of this assemblage on marle beds near the shelf edge; in this rather soft rock the pelecypod *Pholadidea* sp. bores holes that are also inhabited by the sipunculid *Phascolosoma antillarum*. The remainder of the fauna mainly consists of actinians, large serpulids, the gastropods *Leina excavata* and *Emarginula cancellata*, pedunculate cirripeds (*Ibla*), and many ophiuroids, e.g. *Ophiothrix tomentosa*, *Ophiactis balli*, *Amphiura grandimana*, *Ophioscolex purpureus*.

(vi) The assemblages of ahermatypic corals (*Sclerhelia*, *Dendrophyllia*) below 150 m on coral reef slopes may belong partly to the offshore rocky-bottom assemblage group. They have been insufficiently studied to date.

(b) Soft Substrates

General Aspects

The soft substrates of the circalittoral zone consist of terrigenous sediments and organogenous remnants. The latter mainly originate from benthic organisms and are frequently rather coarse. Both these fundamental components may be mixed.

The distribution of the different assemblages mainly depends on the granulometric characteristics of the sediment which in turn depend on sedimentation rate, water movement and sources of organogenous material. In general, smaller terrigenous particles (silt, clay) are transported further away from the shoreline than are coarser particles, so that the average particle size of soft bottoms should diminish towards offshore areas. In fact, this theoretical expectation does not always correspond to the situation observed. For example, in the vicinity of a river mouth, where a bulk of fine particles is brought to the sea, massive silt and clay sedimentation arises from the mixture of sea and fresh water. In polar seas melting of icebergs from glaciers may sometimes transport boulders or pebbles far from the shoreline. Moreover, it must be pointed out that, at any place of the shelf, many changes in depth occurred in the past due to transgressions and regressions related to interglacial and glacial periods respectively. For example, a bottom with a present depth of 100 m and which currently receives fine sediments, may have been under only 20 m of water during one of the Quaternary regressions, receiving at that time coarser terrigenous sediments or even supporting an organogenous formation. With the most recent transgression, coarse detritic bottoms came nearer and nearer to the shoreline (PERES, 1967a,b). When, at present, a shelf does not receive fine sediments from rivers, one observes from the shelf edge to the present shallowest levels of the circalittoral zone an uninterrupted (both in time and space) strip of coarse soft bottoms called 'detritic bottoms'. Where the present deposition of fine sediments is abundant (e.g. in an estuary), silt, clay and mud may cover all of the older detritic sequence, forming a more or less uniform strip of very soft bottoms down to the beginning of the slope which, in general, is also covered in such areas by fine deposits. Where the supply of terrigenous material is less pronounced or where a sufficiently strong current occurs at some distance from the shore-line, the fine particles do not accumulate as readily and are transported relatively far away from the shoreline. In such a case, one observes a strip of muddy bottom ('shelf terrigenous mud') more or less parallel to the coast and intercalated between two strips of detritic bottoms. Closer to the coast there is the 'coastal detritic', with organic remains (almost always calcareous) mainly from the present or recent benthic biota. Some of these remnants may be produced by the local assemblage, whereas others are brought in by bottom currents from neighbouring assemblages. Their nature depends on existing or recent benthos assemblages: broken shells, coral fragments, calcarous algae, pieces of bryozoan zoaria, etc. On the other side of the intercalated strip of terrigenous shelf mud there is nother strip of detritic bottom, called 'shelf-edge detritic'. The presence of this deeper detritic bottom is correlated with the stronger currents in the vicinity of the shelf edge; calcarous remnants of local benthic organisms often mainly arise from thanatocoenoses. Finally, the coastal detritic may be mixed with some mud. In this case, first distinguished by PICARD (1965), on the Mediterranean shelf of France, a fourth type of circalittoral soft bottom, the 'muddy detritic' assemblage, may be observed.

Coastal Detritic Assemblages

Mediterranean Sea

Once more, we choose a Mediterranean 'biocoenosis' to exemplify the group of coastal detritic assemblages (PICARD, 1965), because it is the best known and also shows some features that facilitate comparisons with homologous assemblages in both colder and warmer regions (Fig. 8-15).

On the French Mediterranean-shelf the coastal detritic assemblage mainly occupies the upper part of the circalittoral zone, just below the last *Posidonia* beds or well-sorted infralittoral sands. In areas with no or very low fine-sediment supply, it may occur as deep as 90–95 m. This maximum depth corresponds to the beginning of the shelf-edge detritic assemblage (p. 000). Where the supply of fine elements is larger, the coastal detritic assemblage disappears in shallower waters in favour of an assemblage belonging to the 'muddy detritic' group or the 'shelf terrigenous mud' group.

The sediment consists of organogenous gravel from present benthos organisms with an admixture of sand; the percentage of silt is low, and muddy particles are generally absent. Many algae and animals of the assemblage show brigher colours (mostly red, yellow, orange) than in any other assemblage on soft substrates in the circalittoral zone. This striking peculiarity indicates a link with the coralligenous assemblage; in fact, several species occur in both assemblages and as was mentioned above (p. 453ff) depending on the predominance of either concretioning or destruction processes; some alternation may occur between these two assemblages.

According to PICARD (1965), the most characteristic species of the coastal detritic biocoenosis are, for algae: the soft rhodophyte *Cryptonemia tunaeformis* and the calcareous species *Lithothamnium calcareum*, *L. coralloides*, *L. fruticulosum*; for sponges: *Suberites domun-*

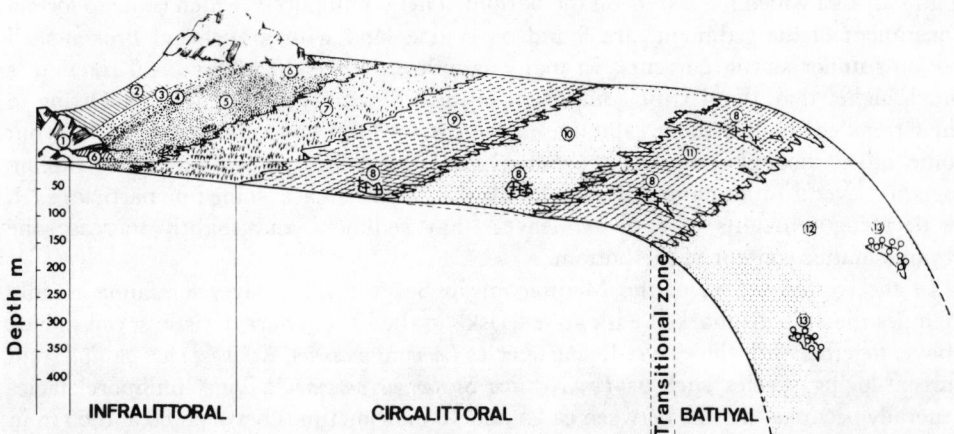

Fig. 8-15: General scheme of the distribution of benthos assemblages on the shelf and the upper part of the slope in the Mediterranean Sea. 1: rocky points; 2: alluvial plain; 3: upper and middle beach (supralittoral and midlittoral zones and upper part of the infralittoral zone); 4: upper clean-sand assemblage; 5: well-sorted fine-sand assemblage; 6: various assemblages on rocky substrates; 7: sea-grass (*Posidonia*) beds; 8: coralligenous assemblage; 9: coastal detritic assemblage; 10: terrigenous mud shelf assemblage; 11: shelf-edge detritic assemblage; 12: bathyal mud assemblage; 12: deep-sea coral assemblage. (After PERES and PICARD, 1964; reproduced by permission of the authors.)

cula, *Basiectyon pilosus**, *Bubaris vermiculata*; for pelecypods: *Modiolus phaseolinus*, *Pecten jacobaeus**, *Lima loscombei**, *L. elliptica*, *Laevicardium oblungum*, *Tellina donacina**, *Psammobia faroense**, etc.; for gastropods: *Turritella triplicata**, *Eulima polita*, *Drillus maravignae*; for decapods: *Paguristes oculatus*, *Anapagurus laevis*, *Ebalia tuberosa*, *E. edwardsi*, etc.; for echinoderms: *Astropecten irregularis*, *Anseropoda placenta*, *Ophioconis forbesi*, *Ophiura grubei*, *Genocidaris maculata*, *Psammechinus microtuberculatus*, *Stereoderma kirchbergi*; for ascidians: *Molgula oculata*, *Ctenicella appendiculata*, *Polycarpa pomaria*, *P. gracilis*. (Asterisks are explained below.) Apart from these highly characteristic species, some others may occur whose presence depends on a certain characteristic of the sediment, e.g. gravel-loving species such as the echinoids *Echinocyamus pusillus* and *Spatangus flavescens*, the decapod *Lambrus massena*, the pelecypods *Astarte fusca* and *Venus fasciata* or—where strong currents often occur—the sand-loving tectibranchail gastropod *Philine aperta*. Small scattered concretioned patches may be occupied by the coralligenous assemblage.

In the Mediterranean Sea the coastal detritic assemblage exhibits several different facies; the most important ones are listed below.

(i) Group of facies in which calcarous rhodophytes play an important role. Among this group, the most interesting and widespread facies is characterized by the abundance of two species of branched Lithothamnieae: *Lithothamnium calcareum* with a rather thick purple-pink thallus and *L. corallioides* with a thinner light pink thallus. The very young algae settle on a sediment particle rapidly enclosed by the growth of the thallus, resulting in an alga which lives 'free' on the bottom. These 'nullipores', which come to form a constituent of the sediment, are found on coarse sand with gravel and broken-shell bottoms under strong currents. In many samples, the weight of the algal fraction is much higher than the detritic component of the sediment (PERES, 1967a,b). Owing to their ramified thallus, these Lithothamnieae cannot form concretioned structures, but some other rhodophytes (*Jania, Gelidium*) sometimes form a felt-like tissue linking together several Lithothamnieae individuals. In the latter case, some fine particles such as decaying remnants of *Posidonia* leaves, may sediment and slightly increase the organic matter content of the bottom.

In the western basin of the Mediterranean Sea the relatively poor fauna mainly includes the animals marked with an asterisk* in the list of characteristic species given above, together with the gravel-loving species *Lambrus massena, Echinocyamus pusillus*, and current-loving species such as *Venus casina Spatangus purpureus*. This 'nullipore' facies generally occurs at depths between ca 25 and 40 m, sometimes between 60 and 65 m in very clear waters. In the eastern basin of the Mediterranean Sea, the nullipore facies is frequently observed as deep as 80–100 m due to the greater incident light intensity, and under very clear waters down to 150–160 m. In these areas the chlorophyte *Palmophyllum crassum* is common and the fauna is impoverished, mainly with regard to the characteristic animal species. Owing to the slight increase in finer particles related to the greater depth, deposit feeders are the most common, among them the pelecypods *Tellina donacina* and *T. balaustina* and the gastropods *Turritella triplicata* and *Aporrhais pespelicani*; the echinoderms *Stylocidaris affinis* and *Ophiura albida* are rare.

These are several other typical 'nullipore' facies. (*a*) On the coasts of Corsica and in the Balearic Island region, bottoms occur with numerous individuals of the ramified and 'free-living' *Lithothamnium valens*, together with the pelecypod *Chlamys commutata*. (*b*) In some areas of the western basin of the Mediterranean Sea *Lithothamnium fruticulosum*, which forms mamillated spheres lying free on the bottom, is very common; and the presence of *Spatangus purpureus* indicates that bottom currents are fairly strong. (*c*) On shelf areas or isolated banks washed by offshore waters in the western basin of the Mediterranean Sea, one may observe a rather fine gravel mixed with many broken shells, inhabited by a typical coastal detritic assemblage, together with many irregular nodules (a few centimetres in diamter) formed by superposed layers of an unidentified encrusting alga belonging to the 'Melobesieae' group. In the central part of each nodule the initial nucleus (small gravel or remnant of a calcareous organism) can often be recognized. Since the alga lives on the whole surface of the nodule, these spheroidal masses seem to be rolled up on the bottom by currents or motile animals. There are also the current-loving *Venus casina* and in particular the sciaphilic large kelp *Laminaria rodriguezei* (p. 400) affixed to nodules; its ribbon-shaped thallus lies on the bottom parallel to the current direction. On both nodules and kelp some epibiotic hydroids and bryozoans may occur. (*d*) While the calcareous algae of the preceding facies are crustose Corallinaceae, another common facies in the Mediterranean Sea is characterized by calcareous rhodophytes of the family Squamariaceae. In these areas the typical coarse sediment of the coastal detritic assemblage is covered by a layer of a very fluid mud which buoys up the algae. *Peyssonnelia polymorpha* largely predominates, but some *P. harveyana* are also present. Both algae have a dark purple thallus, consisting of a thin, strongly twisted plate with a more or less spheroidal shape. The thalli have a single layer and 'float' on the fluid mud. They are probably not immobile and may frequently 'see-saw'. The see-sawing hypothesis not only explains the general spheroidal shape, but also the fact that the parts of the thallus close to the mud surface are generally alive and fertile. The see-sawing movements may arise both from differences in growth rate (displacement of the gravity centre caused by different amounts of illumination received by different thallus parts) and from the movements of animals inhabiting the anfractuous thallus, mainly *Ophiopsila aranea*, which lives inside the alga during the day but moves out at night. The assemblage is rather similar to the typical coastal detritic assemblage described above. However, the holes and crevices of the thallus, containing greater or lesser amounts of muddy sand, are inhabited (in addition to *O. aranea*) by two small pelecypods: *Kellia suborbicularis* and *Mysella bidentata*. On the thallus itself, the bryozoans *Mollia patellaria* and *Chorizopora brongniarti* are common, but not genuinely characteristic, as they occur on every circalittoral rhodophyte with a sufficiently stiff thallus. The current-loving *Venus casina* and *Spatangus purpureus* are also present. It has often been observed that the areas supporting the Squamariaceae facies may change in both space and time; the facies may then sometimes completely disappear, leaving a pure coastal detritic assemblage. The facies only develops at the mouth of open bays, i.e. in areas with alternating periods of storm-swept whirling currents and of sedimentation during smooth-sea conditions (CARPINE, 1958).

(ii) *Ophiura texturata* facies. The ophiuroid *O. texturata* may be abundant (up to 10 inds m^{-2}) on coastal detritic bottoms close to heavily populated well-sorted infralittoral sands, probably because the larvae produced by the rich pelecypod fauna inhabiting the latter provide the ophiuroid with plenty of food.

(iii) *Vidalia volubilis* facies. This facies featuring the soft rhodophyte *Vidalia volubilis*, often accompanied by another red alga (*Rythiphlaea tinctoria*), lives in areas of rather clear water due to the higher light requirements of these algae. The facies is more widespread in the eastern than in the western basin of the Mediterranean Sea.

(iv) Compound ascidian facies. Where coastal detritic bottoms prevail in the vicinity of infralittoral bottoms with very dense meadows of sea-grasses or algae, the higher content in organic detritus and bacteria of by-passing waters can induce a luxuriant development of ascidians, mainly compound species of the families Polycitoridae, Polyclinidae, Didemnidae and Botryllidae.

Coastal detritic assemblages in the north-eastern Atlantic Ocean

Due to the discrimination by PICARD (1965) between two groups of assemblages (coastal detritic and muddy coastal detritic) and to the frequently insufficient data available on either granulometric characteristics of faunistic composition, it is often difficult to classify communities formerly described by many authors. The most useful criteria that allow a community to be included in the group of coastal detritic assemblages are a coarse sediment and the lack (or scarcity) of mud-loving species. Nevertheless, it is quite certain that most of the boreal 'offshore gravel association' recorded by different authors on the West European shelf belong to the coastal detritic group, because they comprise a relatively large number of species that participate in the Mediterranean Sea coastal detritic assemblage, e.g. *Astropecten irregularis*, *Anseropoda placenta*, *Anapagurus laevis*, *Modiolus phaseolinus*, *Psammobia faroense* and *Abra prismatica*.

Facies and aspects are numerous but have been insufficiently investigated. It seems that assemblages which exhibit intermediate features between coastal detritic and muddy coastal detritic are much more common there than in the Mediterranean Sea.

However, in the north-eastern Atlantic Ocean the *Lithothamnium calcareum—L. coral-lioides* facies plays a more important role (p. 450). This facies is well known on the Britanny shelf as 'maërl', because these algae were used long ago as calcareous fertilizer for the acid soils of this region. Although 'maërl' may be found in front of some estuaries up to the low-water spring tide level owing to strong tidal currents, the real maërl bottoms (living at shallower levels than in the Mediterranean Sea), usually occur in depths from a few metres down to about 17 m. In addition to *L. calcareum*, two forms of *L. corallioides* prevail depending on granulometric characteristics of the sediment (CABIOCH, 1968). Reproduction (both asexual and sexual) seems to occur irregularly, possibly following a 5–6 year rhythm, resulting in large-range pluriannual differences in the respective abundance of both forms. The associated flora of soft algae is similar to that formerly observed in the Mediterranean Sea and may similarly tend to agglomerate some *Lithothamnium* thalli. The fauna is impoverished in comparison with that of the homologous assemblage in the Mediterranean Sea, mainly as far as echinoderms (ophiuroids) are concerned. In contrast, crustaceans are more common, especially the decapods *Anapagurus hyndmanni* and *Porcellana longicornis*, and the amphipod *Melita gladiosa*.

In temperate areas of the north-eastern Atlantic Ocean, Lithothamnieae facies are not restricted to the coastal detritic assemblage, but also exist on coarse sands and fine gravels under strong bottom currents (p. 450). In colder areas of the north-eastern Atlantic Ocean, both flora and fauna paradoxically become richer and more diversified than in temperate areas comprising sessile species, probably because *Lithothamnium* (insufficiently documented at the specific level) display a more marked trend to form clumps. On Norwegian coasts SNELI (1968) observed a sandy bottom with broken shells, heavily covered with clumps of ramified Lithothamnieae supporting epiphytic algae such as the chlorophyte *Halicystis* sp. and phaeophytes (*Desmarestia viridis*—upon which *Ophiocomina nigra* is common—and *Chorda tomentosa*). Owing to the mixing of hard and soft substrates one finds both infauna and epifauna there. The polychaete *Nereis pelagica*, the pelecypod *Hiatella arctica*, the gastropod *Acmaea virginea*, the ophiuroid *Ophipholus aculeata* are the predominant species; The serpulid *Pomatoceros triqueter* and the pelecypods *Kellia suborbicularis* and *Modiolus modiolus* are also present.

In the Barents Sea, in addition to typical motile species such as the crustaceans *Hyas araneus*, *Eupagurus pubescens* and shrimps of the genera *Spirontocaris* and *Sclerocrangon*, etc, Lithothamnieae bottoms support many sedentary forms (nemerteans, Polyplacophora, *Phascolosoma*, *Ophiopholis aculeata*, *Cucumaria frondosa*) and even sessile animals: the medusa *Lucernaria*, the actinian *Metridium dianthus*, some sabellids, *Anomia*, the ascidians *Ciona* and *Pyura*, etc. Some components of this fauna (biomass may exceed 300 g m^{-2}) depend on the existence of holes and crevices in algae-built masses (ZENKEVICH, 1963). Near the Ainov Islands, some differences related to depth may be observed in the Lithotham-nieae bottom assemblage (PROPP, 1964): below 25 m, some laminarians may exist, and *Strongylocentrotus droebrachiensis* is very common; patches of *Modiolus modiolus* occur as well; deeper, the pelecypod *Chlamys islandicus* and the gastropod *Neptunea despecta* become abundant.

Wherever the gravel fraction is sufficiently large, a facies of the coastal detritic assemblage may prevail with large-sized pectinids: either *Chlamys opercularis* or *Pecten maximus*.

Other assemblages attributed to the group of coastal detritic assemblages

(i) From 35–80-m deep bottoms of poorly sorted calcareous sands of the Ghanian coasts washed by the Guinea Current and mixed with hard substrate areas, BUCHANAN (1958) described a community in which the following animals predominate: small tectibranchial gastropods (*Acteocina, Ringicula, Cylichna*), prosobranchial gastropods (*Xenophora senegalensis* and *Calliostoma striatum*), pelecypods (*Aloidis* sp. and *Pecten jacobaeus*), the polychaete *Onuphis eremita*, the ophiuroid *Ophiopsila guinensis*, and the small solitary coral *Caryophyllia clavus*. There are also some intruders from the underlying bathyal zone, e.g. the pennatularian *Kophobelemnon stelliferum*. *Caryophyllia clavus* alone amounts to 22 g m^{-2} of the whole biomass there: 28 g m^{-2} (organic matter content less than 5 g m^{-2}).

(ii) Near the mouth of the Rio de la Plata (Argentina) OLIVER and co-authors (1968) found a coarse sand with shell remnants, mainly inhabited by the pelecypods *Sunetta americana* and *Tivela isabelliania*, together with *Pectunculus longior, Darina solenoides*, and isopod *Serolis polaris*, etc.

(iii) In the Gulf of Eilat (Red Sea) one may refer to the coastal detritic assemblage, the shelly gravels and sands at depths of a few tens of metres; these are rich in foraminiferans (mainly *Operculina gaimardi*) and inhabited by antipatharians, the brachiopod *Lingula* sp., *Murex tribulus*, etc.

(iv) Off Tulear (Madagascar) at depths of 40–90 m, the sandy bottom mixed with a very small amount of mud is inhabited by a small *Turritella* species and a free-living, cup-shaped bryozoan. This assemblage may also belong to the coastal detritic group of assemblages. The sediment is sprinkled with broken shells bearing a rich epifauna (sponges, hydroids) and with small mamillated masses formed by 'Melobesieae'.

(v) From the Kamchatka Gulf, KUZNETSOV (1961) described many 'communities', several of which may be considered as belonging to the coastal detritic assemblage group, for example, the *Astarte alaskensis* 'community', which inhabits sands sometimes mixed with gravel and pebbles and has a biomass of about 250 g m^{-2}. Pelecypods predominate (40% of the total) followed by the ascidians *Boltenia* and *Molgula* (20%), the polychaetes *Nichomache, Laonice* and *Nephthys* (9%), crustaceans (*Hyas, Ampelisca*), some hydroids (*Sertularella polyzonias*), etc. The *Astarte rollandi* 'community' was observed by the same author on sands 40–80 m deep in the same gulf; it is rather similar to the preceding assemblage; here, the pelecypods account for 92% (150–200 inds, 225 g m^{-2}) of the total biomass; some polychaetes and bryozoans are associated species. The latter 'community' has also been observed by FILATOVA and BARSANOVA (1964) in the Bering Sea on coarse sand mixed with pebbles 25–68 m deep (biomass 304 g m^{-2}).

(vi) In Arctic areas (0–4 °C) the 'community' observed off the mouth of Gaspé Bay, eastern coast of Canada by BRUNEL (1971) on sand mixed with pebbles, broken shells and a very small silt fraction at depths of 50–90 m seems to parallel the pectinid aspects of the Atlantic Ocean coastal detritic assemblage, owing to the high abundance of *Chlamys islandicus*. *Ophiura robusta, Ophiacantha bidentata, Strongylocentrotus droebrachiensis*, the asteroid *Solaster papposus*, the polychaete *Onuphis conchylega*, and the shrimp *Pandalus montagui* are other common species.

Muddy-Detritic Assemblages

According to PICARD (1965) it is easy to recognize an assemblage on the Mediterranean Sea shelf corresponding to bottom areas in which the characteristic sediment of the coastal detritic exhibits a significant admixture of finer material. The sediment may be a muddy sand, a sandy mud or sometimes a relatively firm mud, but gravels, clinkers, broken shells are always present, and the sedimentation rate is sufficiently low to allow the settlement of sessile species. The muddy fraction always predominates.

Mediterranean Sea

PICARD (1965) lists 12 characteristic species of the muddy detritic assemblages: the sponge *Raspailia viminalis*; the anthozoans *Alcyonium palmatum* and *Anemonactis mazeli*; the polychaetes *Aphrodite aculeata*, *Polyodontes maxillosus*, *Eupanthalis kinbergi*, *Leiocapitella dollfusi* and *Clymene palermitana*; the sipunculid *Golfingia elongata*; the pelecypod *Tellina serrata*; the isopod *Cirolana neglecta* and the holothurioid *Pseudothyone raphanus*. Mud-loving species such as *Nepthys incisa*, *Pectinaria auricoma*, *Amphiura chiajei*, etc., are also present.

Two facies of this assemblage are common. (i) In the *Ophiothrix quinquemaculata* facies, this ophiuroid with its spiny arms may be very abundant (several tens of inds m^{-2} or more); it clings to the substrate or, more often, to broken shells and dead organogenous concretions with two or three of its arms; the others are lifted up for collecting particles suspended in the bottom water layer 10–15 cm above the sediment. (ii) In the *Alcyonium* facies the rate of fine particles sedimentation is particularly low. This allows a significant development of sessile species, in particular *Alcyonium palmatum*, but also some hydroids and bryozoans, the big compound ascidian *Diazona violacea* and several large-sized solitary ascidians: *Ascidia mentula*, *Phallusia mamillata*, *Microcosmus* spp. and *Polycarpa pomaria*.

Other regions of the World Ocean

So many 'communities' (probably more than 100) that might be accommodated in the group of muddy detritic assemblages have been described that we can neither discuss them all here nor evaluate the degree of their similarity to the typical Mediterranean Sea assemblage. I shall attempt to place more emphasis on generalization and on the faunistic criteria for assemblage characterization and to illustrate these by some examples.

Among the animals whose presence may be used for ascribing the relationship of a community to the muddy detritic assemblage group, I believe the first ranked indicators are burrowing ophiuroids (cf. THORSON, 1957) with flexible arms raised above the bottom and collecting suspended particles. The genera *Amphiura*, *Amphiodia* and *Amphioplus* may serve as examples. Dendrochirote holothuroioids (*Cucumaria*, *Thyone*, *Trachythyone*) are also of primary importance, but these are sometimes represented by different species, depending on the percentage of mud in the sediment. The most characteristic mud-loving species are pelecypods (*Thyasira*, *Abra*, *Nucula* or related genera, *Corbula*, some tellinids) and some holothurioids of the order Apoda. Sand loving species always include suspension-feeding pelecypods: *Dosima*, *Astarte*, Carditidae, and, more rarely, Cardiidae; scaphopods are typical and among the gastropods, filter feeders of the genera *Aporrhais* and *Turritella*. Among the sand-loving species, polychaetes of the genera *Nephthys*, *Onuphis*, *Pectinaria* also frequently exist. Irregular echinoids of the genera *Echinocardium* or *Echinarachnius* are also very characteristic of the most sandy substrates.

I agree with THORSON's (1957) idea that increasing abundance of some filter-feeding gastropods may result in a decrease in the amphiurid abundance. For example, in his detailed study of the New Zealand shelf communities, McKNIGHT (1969b) pointed out that in the *Amphiura rosea–Dosinia lambata* community, *Maoricolpus (Turritella) roseus* becomes more abundant as the mean grain size increases. However, a generalization cannot be deduced from these findings, as each species of *Turritella* and allied genera exhibits its own requirements and tolerance as regards sediment grain size, food supply, etc.

Summarizing our present knowledge regarding the group of muddy detritic assemblages, it appears that, despite considerable diversity, they might be categorized into three subgroups depending on the predominant component of the fauna: Amphiuridae, *Turritella* or related genera, and foraminiferans.

(i) Typical amphiurid assemblages. In the north-eastern Atlantic Ocean and adjacent areas, the muddy detritic assemblage group is represented by the PETERSEN (1915, 1918) E. Fil. 'community' with the characteristic species *Echinocardium cordatum*, *Amphiura filiformis*, *Thyasira flexuosa*, *Aporrhais pes pelicani*, *Turritella communis*, etc. BUCHANAN (1963) gave a more detailed study of this community and distinguished three facies: (*a*) hydroids and bryozoans facies where the gravel fraction is high; (*b*) *Cucumaria elongata–Diastylis rathkei* (cumacean) *Turritella communis* facies on fine clean sands and with an admixture of *Melinna palmata* and *Ampelisca tenuicornis* when the silt and mud fractions increase; (*c*) *Astrorhiza limicola* (foraminiferan) facies on medium-sized sands slightly mixed with mud. Some additional data may be found in GLEMAREC (1965, 1969, 1973) and HOLME (1961, 1966); on muddy gravels off Plymouth, the latter author observed a highly specific facies in which burrowing crustaceans predominate (*Upogebia deltaura*, *U. stellata* and more rarely *Squilla desmaresti*), and the burrowing actinian *Mesaemaea (Ilyanthus) mitchelli* occurs. *Turitella communis*, *Nucula nucleus*, *Golfingia elongata*, *G. vulgaris*, *Thyone* sp., etc., are other fairly common species.

In the north-western Atlantic Ocean the *Echinarachnius parma* assemblages which were studied by NESIS (1962a, 1965) in the vicinity of Newfoundland are similar to those of the north-western Pacific Ocean (see below).

In the Pacific Ocean, some *Amphiura* communities have been described on sandy muds, e.g. in south-western regions the *Amphiura rosea* community, with *Echinocardium australe*, *Nucula hartwigiana*, *Dosinia limata*, *Onuphis aucklandensis*, *Pectinaria australis*, etc. (THORSON, 1957), or the *Amphiura norae* community of Milford Sound, New Zealand, where HURLEY (1964) observed *Echinocardium cordatum*, the scaphopod *Cadulus delicatulus*, *Nucula hartwigiana*, *Thyasira peroniana* and many polychaetes.

However, in most of the muddy detritic assemblages in the Pacific Ocean, species of *Amphiodia* and/or *Amphioplus* substitute for *Amphiura* species. For example, one may mention the *Amphiodia craterodmeta–Turritella fortilirata* 'community' (biomass 100–200 g m^{-2}), together with the pelecypods *Axinopsis orbiculata* and *Yoldia johanni* and the polychaete *Magelona longicornis*, which exists in the northern areas of the Sea of Japan (THORSON, 1957).

In the boreo-arctic region of the north-western Pacific Ocean and adjacent areas (Gulf of Kamchatka, Bering Sea) *Amphiodia craterodmeta* participates in another 'community', where the scutellid echinoid *Echinarachnius parma* represents up to 90% of the whole biomass (200–2000 g m^{-2}), together with *Astarte alaskensis* and the polychaetes *Nephthys ciliata* and *Owenia fusiformis* (KUZNETSOV, 1961; FILATOVA and BARSANOVA, 1964). On the Californian coast *Amphiodida–Amphioplus* 'communities' have been recorded by BAR-NARD and HARTMAN (1959) and BARNARD and ZIESENHENNE (1961), e.g. the *Amphiodia urtica–Cardita ventricosa* 'community', which may take different aspects related to differences in sediment characteristics. Among the most amazing is that in which the echiurid *Listriolobus pelodes* predominates (100 inds, i.e. ca 1000 g m^{-2}) associated with *Phoronopsis* sp., some polychaetes (*Marphysa, Ceratocephala, Hesperonoe*), the pelecypod *Saxicavella arctica*; its total biomass 1370 g m^{-2} may reach up to 2000 g m^{-2} in a facies where *Chaetopterus* and *Lima* are also abundant.

Amphiodia–Amphioplus communities also exist outside the Pacific Ocean. For example, on bottoms 20–50 m deep in the Persian Gulf THORSON (1957) observed that typical ophiuroids are associated with *Nucula*, a burrowing Axiidae decapod and the polychaete *Ammotrypane aulogaster*. LONGHURST (1958, 1959) on the West African coast described muddy sands mixed with broken shells 15–70 m deep inhabited by *Amphioplus congensis*, together with sipunculids (*Ochetostoma mercator* and *Sipunculus titubans*) decapods (*Callianassa guineensis, Upogebia* sp., *Alpheus floridanus*) and pelecypods (*Cultellus tenuis, Tellina nymphalis, Aloidis dautzenbergi*).

(ii) Assemblages in which *Turritella* predominate or substitute for amphiuroids. In the Sea of Japan *Turritella fortilirata* occurs together with *Amphiodia* and may prevail over the ophiuroid in some areas such as Peter the Great Bay (DERJUGIN and SOMOVA, 1940). The same was observed by BUCHANAN (1958) on the Ghanian coast, where muddy sands 14–35 m deep are inhabited by a very abundant population of *Turritella annulata* (up to 5000 inds m^{-2}) together with *Amphiodia acutispina, Tellina compressa*, etc. Crustaceans are abundant, e.g. the shrimps *Penaeus duorarum, Penaeopsis miersi* and *Solenocera membranacea*, and species which live in *Turritella* empty shells (*Diogenes, Upogebia*); the sipunculid *Aspidosiphon venabhelum* is present. The biomass may be as high as 540 g m^{-2}, but *Turritella* alone accounts for 500 g m^{-2}. Among the assemblages where a *Turritella* species seems to replace Amphiuridae, one may quote the 'muddy gravel boreal association' (HOLME, 1966) and some *T. communis* assemblages of the English Channel (FORD, 1923). In the Gulf of Eilat (Red Sea) POR and LERNER-SEGGEV (1966) described a sandy mud mixed with gravels inhabited by *Turritella* cf. *columnaris*, with *Dentalium* and some pelecypods (*Aloidis, Chione*), but this assemblage occurs from 100 down to 260 m, i.e. not exclusively in the circalittoral zone.

(iii) Assemblages in which foraminiferans predominate or substitute for amphiurids. In the North Sea E. Fil. 'community' (PETERSEN, 1918) BUCHANAN (1963) found a facies in which the arenaceous foraminiferan *Astrorhiza limicola* predominates. This assemblage was formerly classified as an independent 'community' by THORSON (1957), who also mentions *Saccamina sphaerica* and *Psammosphaera fusca*.

On poorly sorted muddy sands 35–45 m deep off the Ghanian coasts, BUCHANAN (1960) observed a 'community', the predominant species of which are colonial foraminiferans of the genera *Julienella* and *Saccamina* together with an impoverished macrofauna e.g. *Cardium kobelti* and the detritus-feeding ophiuroid *Rhopalodina*;

Amphiodia, Amphioplus and the polychaete *Maldane sarsi* are also present; the biomass is 36 g m^{-2} (foraminiferans only 32 g m^{-2}). The distribution of the different foraminiferan species depends on the sediment granulometric and mineralogical characteristics, as non-calcareous particles (mainly quartz) are the most often used for nest-building: *J. faetida* and *S. arborescens* occur on fine sand (40–50 m) mixed with silt and containing less than 50% of CO$_3$Ca, whereas *S. labyrinthica* and *S. furcata* live deeper on less silty and coarse sand with a more important CO$_3$Ca fraction.

In the Arctic regions of the Atlantic Ocean fine muddy sands are inhabited by an assemblage with the foraminiferans *Rhabdamina cornuta*, *Miliolina bucculenta*, the polychaete *Asychis biceps* and the pelecypod *Axinopsis orbiculata*. ELLIS (1961) on the American coast of the Arctic basin found a similar assemblage with some scattered pelecypods, the polychaete *Onuphis conchylega* and the holothurioid *Myriotrochus heeri*. The Arctic foraminiferan assemblages seem to extend down to ca 700 m, suggesting that their distribution depends mainly on non-climatic factors, e.g. sediment nature, very low temperatures and food scarcity.

Some assemblages related to the muddy detritic group may be transitional between terrigenous shelf mud assemblages and shelf-edge detritic assemblages. An example is the muddy sands described by GLEMAREC (1969) on the deeper shelf parts in northern areas of the Gulf of Biscay where the polychaete *Terebellides stroemi* is the most characteristic inhabitant, together with the foraminiferan *Astrorhiza limicola*, the anthozoans *Pennatula phosphorea* and *Funiculina quadrangularis* and the pelecypod *Bathyarca pectunculoides*. Finally, as pointed out by PICARD (personal communication), ascribing an assemblage to the muddy detritic group demands consideration of both the sediment composition and the ecological requirements of its major inhabitants, in particular their feeding behaviour. When the sediment contains a sufficient percentage of silt or mud, it is easy to assign a community to this group. But sometimes one observes species characteristic of the muddy detritic group in (or on) sediments not containing very fine particles. In this case the gut content of most filter feeders contains silt or mud. This may mean that the finest mineral particles are transported above the bottom by a current strong enough to prevent their sedimentation. Under such conditions a species may be able to establish itself because it is able to collect very small particles (required as intestinal 'ballast') not from the sediment itself but from the bottom water.

Terrigenous Mud-shelf Assemblages

North Atlantic Ocean and Mediterranean Sea shelf

The marine terrigenous sediment is always a relatively mobile (fluid) mud constituted of silt and clay; admixture of sand is rare and always slight. Owing to a high sedimentation rate and very soft sediment consistence, hard bodies tend to be buried quickly. Consequently, sessile species are largely absent.

For the Mediterranean Sea the list of the most characteristic species reviewed by PICARD (1965) may be summarized as follows: the pennatularian *Virgularia mirabilis*; many polychaetes e.g. *Lepidasthenia maculata*, *Phyllodoce lineata*, *Nereis longissima*, *Nephthys hystricis*, *Goniada maculata*, *Sternaspis scutata* and *Pectinaria belgica*; decapod crustaceans, e.g. *Callianassa truncata* and *Goneplax rhomboides*; the pelecypods *Thyasira croulinensis*, *Mysella bidentata, Abra nitida* and *Thracia convexa*; the holothurioid *Oerstergroenia digitata*; the fishes *Caecula imerbis* and *Gobius lesueurei*. Mud-loving species, some of which may occur in other assemblages, providing the sediment contains sufficient mud, are also common, e.g. the polychaetes *Lumbriconereis fragilis* and *Terebellides stroemi*, *Alpheus glaber* and *Amphiura chiajei*. In the Mediterranean Sea shelf, a very specific aspect of the terrigenous mud assemblage exhibits a tremendous abundance of *Turritella tricarinata* (up to 95% of the total number of individuals), while the other inhabitants are mainly molluscs. The flat-shaped holothurioid *Stichopus regalis* is common only where the sediment is sufficiently firm for crawling.

In north-eastern Atlantic Ocean temperate areas, many different associations ('communities') have been described which may easily refer to the two communities B-Ch and B.S. formerly defined by PETERSEN (1918).

B-Ch is characterized by the echinoderms *Brissopsis lyrifera* and *Amphiura chiajei* and corresponds to the '*Amphiura chiajei* subcommunity' of BUCHANAN (1963), who found some species highly characteristic of the terrigenous mud assemblage there, such as *Abra nitida* and the polychaetes *Laonice cerrata* and *Goniada maculata*. According to GLEMAREC (1965), the name *Brissopsis lyrifera–Nucula sulcata* would probably be more convenient for this 'community'. The B.S. 'community' with its leading species *B. lyrifera* and *Ophiura sarsi* corresponds to the assemblage recently redescribed by GLEMAREC (1965), who selected as 'pilot' species *Virgularia tuberculata*, *Sternaspis scutata* and *Amphiura filiformis*, together with *Abra nitida*, *Nucula turgida*, *Lucina spinifera*, *Thyasira flexuosa*, the gastropod *Cylichna cylindracea* and *Oestergroenia digitata*. *Amphiura filiformis* is relatively eurytopic and may be found in almost all circalittoral soft substrates. Later, GLEMAREC (1969) distinguished, on the shelf of the northern Gulf of Biscay, two different subassemblages related to different pelite contents of the sediment; the polychaetes *Ninoe armoricana* and *Scalibregma inflatum* predominate where the pelite fraction exceeds 60%; *Nucula sulcata* is the most abundant species where the pelite percentage is 20–50%; many common species are represented in both these subassemblages.

In my opinion, PETERSEN's B-Ch and B.S. 'communities' represent two subcommunities (or aspects) or a single assemblage. The B-Ch aspect seems to exist on relatively mobile muddy bottoms, either very shallow or, if deeper, exposed to swell; maximum biomass is about 400–450 g m^{-2}. The B.S. aspect seems to prefer deeper or more sheltered areas; the average biomass is below 300 g m^{-2}.

Subarctic and Arctic areas of the North Atlantic and Pacific Oceans

The 'community' observed by SPÄRCK (1929) in Icelandic fjords and in the vicinity of the Faroe Islands features *Amphiura filiformis*, *Virgularia mirabilis* and *Sternaspis fossor* as its most characteristic species; it corresponds to the B.S. aspect of the boreal temperate assemblage.

The *Yoldia hyperborea* (= *Y. limatula*) 'community' is of a more typical subarctic character; it was described from Iceland coasts (except for the southern shelf), the White Sea and the southern Greenland shelf at depths of about 10–70 m. Similar to the preceding assemblage, it also includes *Leda pernula*, *Sternaspis fossor* and *Pectinaria hyperborea* (SPÄRCK, 1937). The 'deep-water muddy-bottom community' identified by LIE and KISKER (1970) off Washington State (USA) coasts at depths of about 150 m is homologous in the north-eastern Pacific Ocean. The polychaetes *Sternaspis fossor, Prionos-*

pio malmgreni, *Ninoa gemma*, the pelecypods *Axinopsida serricata*, *Adotonrhina cyclica* and *Macoma charlottensis*, the amphipod *Heterophoxus oculatus*, the echinoderms *Brisaster latifrons**, *Ophiura lütkeni** and *Amphioplus* sp.* are the most abundant species in the sediment with an average mud percentage above 50. The asterisk refers to the major contributors to the standing crop (ash-free d.w. 3.058 g m^{-2}).

The *Pandora filosa* 'community' with the venerid *Marcia subdiaphana*, the gastropod *Phacoides tenuisculptus*, the shrimp *Crago alaskensis* and some mud-loving polychaetes is also closely related; according to CLEMENTS and SHELFORD (1952) two aspects may exist: (i) on mud very rich in organic matter, *Cucumaria populifera*, *Ophiopholis aculeata* and the polychaete *Scalibregma inflatum*; (ii) on bottoms with a lower organic matter content under rather brackish waters *Yoldia* predominates, together with the polychaete *Clymenella rubrocincta*, *Amphiodia urtica* and the big nudibranchial gastropod *Dendronotus arborescens*.

The 'community' identified by FILATOVA and BARSANOVA (1964) on Bering Sea muddy bottoms (96–200 m, i.e. spreading downwards onto the upper slope) with the pelecypods *Yoldia thraciaeformis* and *Leda pernula*, *Sternaspis scutata* and the echinoderms *Ctenodiscus crispatus* and *Ophiura sarsi* (standing crop: 88 g m^{-2}) probably also belongs to the subarctic group of terrigenous mud shelf assemblages, as well as the 'community' observed on the northern Newfoundland shelf (NESIS, 1962a) on muddy clay bottom (165–280 m) inhabited by *Ctenodiscus crispatus*, some actinians and the polychaete *Stylarioides plumosus* (biomass 80 g m^{-2}). NESIS also found a 'community' in the same area, on mud mixed with some sand, with predominating *Ctenodiscus crispatus*, *Ophiura sarsi*, holothurioids (*Trochostoma turgidum*) *Sternaspis scutata*, the scaphopod *Dentalium entale*, etc.

In high Arctic areas near the mouth of glacial rivers and large glaciers (East Greenland, Svalbard) THORSON (1957) identified a *Yoldia arctica* 'community', with the holothurioid *Myriotrochus rincki* and the ophiuroid *Ophiocten sericeum*, on soft (milky-coloured to light grey) between 10–15 and 50–60 m depth. In the Bering Sea FILATOVA and BARSANOVA (1964) found a *Myriotrochus–Ophiura sarsi* 'community' on 64–67 m deep blackish mud.

It seems necessary at this juncture to discuss the significance of the *Maldane sarsi–Ophiura sarsi* 'community' first described by SPÄRCK (1937) from Icelandic shores. The distribution of this community was accurately reviewed by THORSON (1957, p. 514) who considered it to be a circumpolar community inhabiting 'soft, fine mud bottoms in rather shallow estuaries and at greater depths (about 100–300 m) in the open sea'. THORSON believed it to correspond to PETERSEN's B.S. community. FILATOVA and BARSANOVA (1964), who investigated 50–200 m deep muddy bottoms in the Bering Sea (bottom-water temperature 0·5–2·0 °C), attributed two other characteristic species to this 'community': the pelecypods *Macoma calcarea* and *Nucula tenuis*. However, these two species should not be considered as members of this 'community': the status of the psychrophilic *Macoma calcarea* was already discussed (p. 452); in the Atlanto–Mediterranean region *Nucula tenuis* is characteristic of bathyal muddy bottoms (p. 484); it might inhabit shallower waters in Arctic regions because its vertical distribution is not upwardly limited by temperature there. Let us now turn to the distribution of the two pilot species *O. sarsi* and *M. sarsi*. *O. sarsi* is widespread in northern USSR seas and seems to require both very fine substrates (fine sand, silt, mud) and low temperatures

($-2-+7$ °C); since it has been collected from 10 to 360 m, it largely exceeds the limits of the circalittoral zone. The requirements and tolerance of *M. sarsi* seem to be similar; the two species may occur in both Arctic and subarctic areas.

In my opinion, careful scrutiny of the data available allows us to conclude that the *Ophiura sarsi–Maldane sarsi* 'community' does not exhibit any characteristic species. Its inhabitants may be divided into two stocks: psychrophilic species, e.g. *Macoma calcarea*, and very fine sand or mud-loving species. Hence, this 'community' may occur in very shallow waters as well as in the upper bathyal zone, and seems to be lacking any species which might be considered genuinely characteristic of the circalittoral zone.

Transitional aspects may be observed between terrigenous mud assemblages and those grouped above in the muddy-detritic group. Examples are the *Amphiodia urtica* facies of SHELFORD's *Pandora filosa* community, or the existence of some individuals of *Amphiodia craterodmeta* in the *Ophiura sarsi–Maldane sarsi* 'community' in the Bering Sea (FILATOVA and BARSANOVA, 1964).

Other areas of the World Ocean

I shall restrict myself to four examples:

(i) A False Bay, South Africa, the 76–88 m deep green mud, mixed with faecal pellets and some fine sand, is characterized by a high abundance of the pelecypod *Dosinia limbata*, together with many species that do not exist in other soft bottoms of the same area: *Virgularia mirabilis** f. *pendentula*; the polychaetes *Drilonereis monroi*, *Goniada maculata**, *Diopatra dubia* and *Prionospio pinnata**; the pelecypod *Tellina gilchristi* and the crab *Goneplax rhomboides** (MORGANS, 1962). Although this assemblage lives very far from the northern Atlantic Ocean and Mediterranean Sea, one must emphasize that all specified marked with asterisks also occur in these two areas.

(ii) In northern areas of the Bay of Bengal, muddy bottoms support a holothurioid of the family Molpadidae together with Hoplocarida and swimming crabs as predominant species; these muddy bottoms may correspond to the *Oerstergroenia digitata* mud of northern Atlantic and Mediterranean shelves. Where the sandy sediment fraction increases, the tube-dwelling actinian *Sphenopus marsupialis* is more abundant.

(iii) From Japan coasts a muddy-bottom 'community' has been reported with abundant species such as the polychaete *Sternaspis scutata*, the pelecypod *Theora lubrica*, the gastropod *Cerithium pfefferi*, *Echinocardium cordatum*, etc (MIYADI, 1941). In areas of more turbid waters, this 'community' may occur in shallower places (e.g. 5–10 m in bays).

(vi) Off the La Plata river estuary, beneath turbid waters on the sandy mud bottoms at 12–70 m depth, OLIVIER and co-authors (1968) observed an assemblage with the following predominant species: anthozoans *Renilla* sp. and *Virgularia* sp., pelecypod *Mactra marplatensis*, gastropod *Olivancillaria unetae*, flat-shaped echinoid *Encope marginata* and ophiuroid *Amphiodia planispina* (OLIVIER and co-authors, 1968).

Shelf-edge Detritic Assemblage

Mediterranean Sea

In areas with a belt of terrigenous mud, the shelf-edge detritic bottom assemblage begins at a depth of 90–95 m and extends down to 120–150 m, sometimes deeper

(200–250 m). The sediment consists of a mixture of gravel, limy remnants of animals from the Quaternary thanatocoenoses, sand, silt and mud; the percentage of the two latter fractions is always higher than in coastal detritic bottoms.

At least in its typical aspect, the assemblage does not usually comprise multicellular alga. According to PICARD (1965), there are only 9 characteristic species: the scaphopod *Dentalium panornum*, the pelecypod *Astarte sulcata*, the crustaceans *Haploops dellavallei*, *Lophogaster typicus* and *Ebalia granulosa*, the ophiuroid *Ophiura carnea*, the holothurioids *Thyone gadeana* and *Neocucumis marioni*, the crinoid *Leptometra phalangium*. Owing to strong currents which at the shelf edge flow mainly perpendicularly to the isobathes, the suspension feeder *L. phalangium* is invariably common (often 5 ind. m^{-2}, sometimes up to 10–15). Among the non-characteristic, but fairly abundant species are the following: some hydroids (see (ii) below), the pelecypod *Pitaria rudis*, the polychaetes *Syllis cornuta* and *Lumbriconereis latreillei*, the sipunculids *Phascolion strombi* and *Aspidosiphon mulleri*, the amphipod *Ampelisca diadema* and the ophiuroid *Amphiura filiformis*.

The facies of shelf-edge detritic bottoms are less numerous than those of the coastal detritic. Two of the most important facies are:

(i) Facies of large hydroids. When the silt and mud fraction of the sediment is sufficient, a facies of the large hydroids *Lytocarpia myriophyllum* and *Nemertesia antennina* may be found, whose rhizoidal net strengthens the superficial layer of the finer detritic bottom. The hydroids support a rather diversified epifauna which includes, for example, smaller hydroids of the genus *Lafoea*, the actinian *Gephyra dohrni*, several species of neomenians, the gastropod *Capulus hungaricus* and the pedunculate cirriped *Scalpellum*.

(ii) Impoverished facies. When the silt and mud fraction of the sediment is particularly small due to strong water circulation above the bottom, a very impoverished facies may occur in which the characteristic species of the typical assemblage (even *D. panormum*) are rare; the single characteristic species of this facies is the very small echinoid *Neolampas rostellata*.

In the eastern basin of the Mediterranean Sea, the shelf-edge assemblage is less typical due to greater water transparency which allows calcareous rhodophytes to extend their depth range. The characteristic crinoid *Leptometra phalangium* seems to be missing.

Shelf-edge detritic assemblages outside the Mediterranean Sea

(i) North-eastern Atlantic Ocean and adjacent seas. On the French and Spanish Atlantic coasts the shelf-edge detritic assemblage exhibits a striking similarity to that in the Mediterranean Sea. *Leptometra celtica* substitutes for *L. phalangium*; it is uncertain, however, whether it represents a good characteristic species there; it may also be observed in the bathyal zone (and off the Algerian Mediterranean coast). Moreover, it seems that the two species have often been confused by some authors. The facies of large hydroids (*Polyplumaria*) is also well known there, with a rich fauna of large-sized gastropods *Ranella*, *Morio*, etc.

In the North Sea, sediments of the deeper part of the shelf generally seem to contain more silt and mud; thus the shelf-edge detritic assemblage is replaced by PETERSEN's B.S. 'community' (p. 476).

(ii) On West African coasts of Senegal and Guinea on sand and mud, both mixed with broken shells LONGHURST (1958) found, an assemblage which clearly belongs to the shelf-edge detritic group. The most common species are the hydroid *Lytocarpia myriophyllum*, the pennatularians *Pteroides griseum*, *Pennatula phosphorea* and *Veretillum cynomorium*, the pelecypod *Cuspidaria cuspidata*, the gastropod *Xenophora senegalensis* and the echinoderms *Ophiothrix tomentosa*, *Cidaris cidaris*, *Stichopus regalis*, *Leptometra celtica* (the latter up to 20 inds m^{-2}). However, this assemblage comprises some intruders from other circalittoral soft-bottom assemblages and from the bathyal zone.

(iii) Off Rio de Janeiro, Brazil, TOMMASI (1969) observed at depths of 125–180 m a coarse-sand bottom with numerous limy organogenous remnants (from molluscs, brachiopods, bryozoans), inhabited by many corals (*Madracis*, *Cladocora*, *Trochocyathus*, *Desmosmilia*) and ophiuroids, some with spiny arms (*Ophiothrix*, *Ophiomisidium*), others epibiotic on gorgonians (*Astrocyclus*); due to the high coral abundance, some concretioning processes may occur.

(iv) At the entrance of the Milford Sound (New Zealand), HURLEY (1964) found bottoms of gravel and broken shells beneath strong currents between 100 and 110 m, inhabited by various brachiopods and some pelecypods (*Chlamys*, *Modiolus*, *Cardita*), which seem to belong to the shelf-edge detritic group.

Antarctic and subantarctic regions

In the Antarctic region at depths usually corresponding to those of the circalittoral zone, ARNAUD (1974) considered the sandy mud with an abundant admixture of gravel and sponge spicules as belonging to the group of shelf-edge detritic bottoms. The assemblage of these bottoms has been reviewed on p. 458 because it consists exclusively of epifaunal suspension-feeding species. Obviously, the nature of the substrate is highly adverse to the infauna. Soft bottoms of the lower circalittoral zone in the Antarctic region have been insufficiently investigated.

These assemblages of large-sized and often erect suspension-feeding species that characterize the lower shelf of the Antarctic continent extend deeper to depths usually corresponding to the upper bathyal zone (p. 482). USHAKOV (1963), as well as ARNAUD (1974), could neither recognize a vertical succession of benthos assemblages nor a species which may be considered as genuinely characteristic of a given depth, 50–100 m down to 500–700 m (for details consult discussion in ARNAUD 1974, pp. 537–538).

(6) Bathyal zone

The data available on assemblages of the bathyal zone mainly concern the northern areas of the Atlantic and Pacific Oceans. However, there is evidence for assuming that assemblages in other regions of the world ocean are essentially not very different. We shall consider now hard-substrate and soft-substrate assemblages.

(a) Hard Substrates

North-eastern Atlantic Ocean

On hard substrates the most striking assemblage is that of ahermatypic deep-sea ('white') corals, mainly characterized (e.g. in the north-eastern Atlantic Ocean) by colonies of the ramose species *Lophelia pertusa* (= *L. prolifera*) and *Madrepora (Amphelia) oculata*, which are sometimes very large, as well as solitary species such as *Caryophyllia armata, Desmophyllum cristagalli* (for an extensive list see SQUIRES, 1959; ZIBROWIUS, 1976). *Dendrophyllia* is the only genus represented both in deep-sea coral assemblages and deep-shelf coral reefs.

The fauna is rich. The most important sponge genera are *Iophon, Eurypon, Myxilla, Aphrocallistes* and *Leucopsacus*. There also are many hydroids; among the alcyonarians some species are eurybathic, e.g. *Rolandia coralloides* and *Alcyonium palmatum*, whereas others belong to strictly bathyal genera (e.g. *Bellonella* and *Anthomastus*). The gorgonians are diversified belong to three main families: Isidae (*Isidella, Chelidonisis, Ceratoisis*), Primnoidae (*Stachyodes, Calligorgia*) and Muricaeidae (*Muricea, Paramuricea*). There are also some zoantharians, actinians and antipatharians. Polychaetes are common, of which the most abundant is *Enice floridana*, which lives in the coral masses in a parchment-like tube coated with limestone. *E. floridana* is frequently associated with *E. pennata*, as well as several species of *Harmothoe, Pholoe, Lagisca*, Phyllodocidae, etc. There are some sabellids among the sedentary forms, but serpulids are more common (e.g. species of *Serpula* and *Vermiliopsis*). Bryozoans and brachiopods are also present. Although crustaceans are fairly common, most of them are not characteristic of the bathyal zone, except, except for such cirripeds as some species of *Scalpellum* and, above all, 'barnacles' of the genera *Verruca* and *Hexelasma*. Molluscs are relatively rare, the most characteristic ones being the chiton *Hanleya hanleyi* and some pelecypods: *Arca nodulosa, A. oblicatula, Spondylus gussoni, Chlamys bruei*. Among the echinoderms, crinoids (*Trichometra, Atelecrinus, Actinometra*) and ophiuroids (*Ophiactis corallicola* and species of *Ophiacantha, Ophiomyxa, Astrochema, Gorgonocephalus*) are very common, while asteroids, echinoids and holothurioids are relatively rare. It seems that many of them inhabit soft-bottom patches intermixed with deep-sea coral clumps. The echinoid *Cidaris cidaris*, which is rather common on soft substrates wherever it can find sufficient food, is much more abundant on rocky substrates, and even on vertical cliffs. It seems to feed mainly on sponges. Hemichordata are represented by the pterobranch *Rhabdopleura normanni*, ascidians mainly by some eurybathic species of the genus *Ascidia*. Since trawling among deep-sea coral clumps is difficult, characteristic fishes are insufficiently known; they may not be very different from those living on soft-bottoms at the same depth.

According to SQUIRES (1959), deep-sea corals occur in the central Atlantic Ocean down to 4000 m, but at higher latitudes might also inhabit shallower areas. Shallower water appear to be inhabited where either reduced illumination due to increased water turbidity or lower temperatures due to circulation prevail. STETSON and co-authors (1962) suggested that deep-sea corals can tolerate only a rather narrow temperature range of between 8 and 12 °C. This was corroborated by NESIS (1962b) who studied the distribution of bathyal anthozoans in the northern Atlantic Ocean (on both hard and soft bottoms) and noticed that they provide good indicators for bottom-water temperature. Deep-sea corals are absent in subarctic and Arctic areas. On the Blake Plateau, STETSON and co-authors (1962) found deep-sea corals on a consolidated calcareous mud, not on rocky substrates. Such findings are probably exceptional.

Another hard-substrate assemblage has been desribed by SOUTHWARD and SOUTHWARD (1958) from the south-western slope of the British Isles. Its most prevailing component consists of cirripeds and seems to comprise two different levels. At the upper level (900–1250 m depth) the cirripeds *Verruca recta* and *Hexelasma hirsutum* are abundant, together with a sponge of the genus *Hymedesmia*; some solitary corals are also present, the brachiopods *Hispanirhynchia cornea* and *Dallina septigera*, the holothurioid *Psolus squamalis*, the echinoid *Stereocidaris ingolfiana* and some ophiuroids (*Ophiacantha*) similar to those participating in the deep-sea coral assemblage. At the deeper level (1500–1800 m depth) live—in addition to *V. recta*, *Ophiacantha* and *Ophiactis*—the coral *Anisopsammia rostrata*, a brachiopod of the genus *Platidia*, the decapod *Munida microphthalma*, the echinoderms *Korethraster hispidus* and *Hypsecrinus* sp. Both levels might be considered as impoverished facies of the deep-sea coral assemblage.

Bathyal Zone off the Antarctic Continent

Before dealing with the bathyal assemblages off the Antarctic continent, we must consider the specific geomorphology and the substrate nature of its continental margin. The shelf (0 to 130–200 m) is very narrow (10–15 miles) and is partially ice-covered. At about 15 and 20 miles off shore there are two narrow depressions, often parallel to the shore, which are 1000 m or more deep. Further than 25 miles from the shore a pene-planed area, about 25 miles in width begins, whose depth decreases progressively from 500 to 700 m. At 700 m (50 miles from shore) the true continental slope begins and extends up to 75 miles from the shore, leading to the abyssal plain at about 3100 m.

As mentioned on p. 458, below ca 50 m (30–60 m) the substrate consists of sandy mud with gravels, pebbles, boulders and scattered rocks, always mixed with spicules from silicaceous sponges. The latter may become relatively consolidated and then form a 'sponge-spicule mat'. Hence, discrimination between hard- and soft-substrate areas is difficult in the Antarctic bathyal zone.

At depths of about 60–80 m a very rich epifaunal assemblage of suspension feeders dominated by large-sized sponges and branched bryozoans suddenly occurs. This assemblage, which in fact belongs to the circalittoral zone, extends almost unmodified down to 500–700 m. It seems that the most marked change species composition occurs below 700 m, corresponding to the depth of the outer edge of the submarine peneplain. We may assume therefore that very large areas on the Antarctic continental margin are inhabited from 30–60 m down to about 700 m by an eurybathic assemblage that is rather homogeneous, at least in qualitative composition. The biomass (1200–1400 g m^{-2} from 100–200 m; 240–320 g m^{-2} from 200–500 m) seems to decrease progressively down to 500 m. Deeper, the decrease in biomass with increasing depth is more marked: 20–40 g m^{-2} between 500 and 1000–1500 m, and 1–2 g m^{-2} below 1500 m.

According to ARNAUD (1974), most assemblages described by BULLIVANT (1959, 1961) and BULLIVANT and DEARBORN (1967) at 160–800 m, must be considered to represent facies of the large-sized sponges and branched bryozoans assemblage.

Although ramified 'white corals' do not exist in the Antarctic bathyal zone, in some places solitary forms such as *Flabellum impensum* and *Caryophyllia antarctica* may occur together with some stylasterine corals of the genus *Errina*. For example, from Pennel Bank, Ross Sea, (180–350 m depth), BULLIVANT and DEARBORN (1967) recorded a substrate consisting of cobbles up to several cm in diameter with inter-cobble patches of muddy sand. Such attached animals as calcareous bryozoans, gorgonians, tunicates and stylasterine corals were common. Among the species identified, the author listed the ophiuroids *Ophiacantha antarctica*, *Amphiura belgicae*, *Ophioceres incipiens*, *Ophiurolepis gelida*, the asteroid *Peribolaster powelli*, the pycnogonids *Achelia* (*Pigrocalavatus*) *spicata* and *Colossendeis lillei*. This assemblage seems to exhibit some different facies owing to differences in local abundance of some other species, e.g. the scleractinian coral *Gardinieria*, siliceous sponges and some brachiopods, down to 1300–1500 m.

A barnacle assemblage similar to that in the north-eastern Atlantic Ocean has also been recorded in the Ross Sea (BULLIVANT and DEARBORN, 1967) at 350–500 m depth with strong bottom currents on a substrate of boulders and rocky flagstones. Live *Bathylasma corolliforme* (previously *Hexelasma antarctica*) cover the rocks; the associated fauna includes stylasterid corals, the scleractinian *Flabellum impensum* and some amphipods. Around the rocks, accumulating dead barnacles build up an almost homogeneous substrate of dead barnacle plates (DELL, 1972).

It seems that differences between rocky-substrate assemblages in the bathyal zone are mainly related to the pattern of bottom circulation (which results in different amounts of particulate food) and to mineralogical characteristics of the rocky substrate itself. For example, important formations of deep-sea corals appear to occur only on non-calcareous rocks. Since sampling in a deep-sea coral assemblage is very difficult, detailed data are rather scarce except for the north-eastern Atlantic Ocean. However, except for the Antarctic Region, this assemblage seems to occur on almost all slope areas with suitable rocky bottoms. Among the 'pilot' species of the assemblage, *Madrepora oculata* has been recorded in the North (up to 69° N) and South Atlantic, and in the Pacific and Indian Oceans. *Lophelia pertusa* occurs in the North (up to 70° N) and South Atlantic Ocean, in the western Indian Ocean and possibly on Californian coasts.

(b) Soft Substrate

General Aspects

The boundary between bathyal zone and abyssal zone (p. 21) has been defined partly according to the renewal in species of elasipod holothurioids. However, SIBUET (1977) who investigated echinoderms along two transects in the Bay of Biscay (northern transect 2200–4700 m deep; southern transect 2100–4400 m) observed that 11 out of a total of 35 holothurioid species listed are eurybathic between 2100 and 4700 m. Since the maximum specific richness occurs at about 3000 m, one may assume that this depth corresponds to the boundary between bathyal and abyssal zones. That only one of the 35 asteroid (*Dytaster rigidus*) species recorded by SIBUET exists in both these vertical zones supports this assumption.

While the boundary corresponds to the end of the slope, we must keep in mind that when the lower part of the slope is particularly gentle (abyssal rise), a transitional belt between bathyal and abyssal assemblages may exist.

North Atlantic Ocean and Mediterranean Sea

Based on specialist literature and his own observations, LE DANOIS (1948) attempted to distinguish several levels within the bathyal zone of the north-eastern Atlantic

Ocean, mainly related to epifauna (e.g. hexactinellids and echinoderms) and more motile animals: decapods (Natantia) and fishes. LAGARDÈRE (1977) investigated the distributions of 254 crustacean species (decapods, euphausiids and peracarids) from 200 to 1300 m on the continental slope of the Atlantic Ocean off France. he noticed: (i) a transitional belt (200–400 m) between circalittoral and bathyal zones; (ii) a first critical depth at about 400 m corresponding to a pronounced renewal in mysids and decapods; (iii) a second critical depth with larger changes in faunistic composition occurring at about 1000 m. Consequently, LAGARDÈRE proposed vertical zonation within the bathyal zone. However, on the basis of available information, I do not believe that the 'critical depths' proposed by several authors in various places of the world ocean are necessarily of general significance. Rather, these subdivisions within the bathyal zone appear to be of merely local significance.

On muddy bottoms of the north-eastern Atlantic Ocean slope from the lower boundary of the transitional belt down to the beginning of the abyssal plain (ca 3000 m) a relatively uniform assemblage occurs. The scaphopods *Dentalium agile* and *Siphonodentalium quinquangulare*; the pelecypods *Abra longicallus*, *Propeamussium vitreum*, *Limopsis aurita* and *Nucula tenuis*; the polychaetes *Panthalis oerstedi*, *Ammotrypane aulogaster*, *Leocrates atlanticus*, etc; and crustaceans *Callocaris macandreae*, *Leucon longirostris*, *Diastylis cornuta*, etc, are among its characteristic infaunal species. There is also an important epifauna: the sponge *Thenea muricata*; the sea-pen *Umbellula*; the polychaetes *Hyalinoecia tubicola* and *Aphrodite aculeata* (both eurybathic); the brachiopod *Gryphus vitreus* (on small gravels or pebbles mixed with mud); the decapods *Parapagurus pilosimanus* and *Geryon tridens*; the gastropods *Fusus bocagei* and *Neptunea bernicienesis*; the ophiuroids *Ophiocten abyssicola* and *Amphilepis norvegica*; the asteroid *Brisingella*; holothurioids such as *Mesothuria intestinalis* and various species of the order Elasipoda. Pogonophorans have probably always existed on muddy bathyal bottoms but have often been disregarded by investigators. According to BRATTEGARD (1967), pogonophoran populations as dense as 130–140 individuals m^{-2} exist both on the Norwegian slope (*Siboglinum*, *Scleroclinum*, *Oligobrachia*) and the North American Atlantic slope.

In an attempt to define vertical subzones in the bathyal, LE DANOIS (1948) points out: (i) Natantia decapods which exhibit migrations (seasonal and circadian) can hardly be used as indicators for a given depth. (ii) The distribution of the hexactinellid sponges is mainly related to the availability of suspended food which, in turn, depends on bottom circulation. (iii) The taxonomy and feeding behaviour of most deep-sea echinoderms are too little at present known to allow them to be considered as characteristic of certain subassemblages.

However, in the north-eastern Atlantic Ocean and the Mediterranean Sea it may be possible to distinguish several facies of the muddy-bottom assemblage: (i) In the *Funiculina quadrangularis* facies, the name-giving pennatularian is sometimes very common in the upper levels of the slope, probably in places where the amount of suspended organic material transported by bottom currents is sufficiently high. Filter-feeding gastropods *Aporrhais serresianus* occur and the shrimp *Parapenaeus longirostris* may be sufficiently abundant to support a small fishery. (ii) In the *Isidella elongata* facies, this large-sized articulate gorgonian depends on sufficient suspended food and requires a hard substrate for attachment. Hence, the facies exists either on slopes sufficiently steep to induce a slow downward gliding of the mud layer, the latter remaining sufficiently thin to allow the *Isidella* stem to pass through, or on slopes with an irregular rocky bottom where *Isidella* can attach itself outside the mud-filled depressions. While some epizoics live on *Isidella*, these are rare and not characteristic. The large-sized penaeid shrimps *Aristeus antennatus* and *Aristeomorpha foliacea* are often common in this facies, together with the burrowing crayfish *Nephrops norvegicus* (which also exists in the *Funiculina* facies), and some cephalopods (*Pteroctopus tetracirrhus*, *Bathypolypus sponsalis*), which feed mainly on crustaceans. Owing to their high economic value, large shrimps and crayfish are often overfished and *I. elongata* has been greatly injured by trawling. (iii) In the hexactinellid facies, the abundance of hexactinellids depends on a sufficient supply of very fine particulate suspended organic material. This tends to be available in areas where decantation processes prevail. On the eastern coast of Corsica, for example, I have observed a dense population of *Asconema setubalense* and, a little deeper, of *Hyalonema* and *Pheronema* species. In the Atlantic Ocean, in the vicinity of Azores Islands, I have further noted from the bathyscaphe many patches of hexactinellids on muddy bottoms, where even bottom irregularities induced a faster bottom current, for example, on steps of consolidated mud.

The muddy-bottom bathyal assemblage described above corresponds to the Al. P. community (*Amphilepis norvegica*, *Pecten vitreus*, *Thyasira flexuosa*) described by PETERSEN (1918) in the deepest areas of the Skagerrak and Oslo Fjord.

The abundance of hexactinellids and, more generally, of large species of sponges is one of the most striking features of the Antarctic seas but this does not specifically apply to the bathyal zone (p. 481). Sometimes macrobenthic species are very rare or even absent, and the soft-bottom bathyal assemblage only comprises foraminiferans. This may be observed on the most impoverished muddy bottoms whose upperlying waters have a very low productivity, e.g. in some areas of the Mediterranean Sea (p. 499ff).

Considering briefly the fishes (obviously all carnivores or scavangers) of the bathyal zone in the north-eastern Atlantic Ocean, two facts emerge: (i) The abundance of individuals and species diversity decrease slowly with increasing depth down to ca 2000 m, but faster below that; in general, body size increases with increasing depth. (ii) Several species exhibit a restricted depth distribution. The genus *Chimaera*, which seems to be strictly bathyal, does not extend below 2000 m. Among the elasmobranchs, *Pristiurus melanostomus*, sometimes common down to 1000 m, is replaced in deeper waters by species of the genera *Centrophorus* and *Centroscymnus*, the latter occurring down to 3000 m. The apods *Halosaurus johnsonianus* and *Simenchelys parasiticus* are often common from 1000 to 2500–3000 m. Representatives of the genus *Synaphobranchus*, the vertical distribution of which extends deeper, are much rarer; the genus *Notacanthus* does not seem to extend below 2000 m. *Bathypterois longipes* which mainly inhabits deeper levels of the slope (about 1500–3000 m) may also be observed on the abyssal plain, whereas *B. dubius* occurs from 1200 to 2000 m; *Bathysaurus agassizii* and *Benthosaurus grallator* have been observed mainly between 1000 and 3000–3400 m. Soft-bottom slopes down to 800–1000 m are of particular interest because fishes of commercial value sometimes abound there, e.g. the sparid *Pagellus centrodontus* and in particular the gadiforms *Merluccius merluccius*, *Micromesistius poutassou*, *Molva*, *Mora*, *Phycis*, *Onos*, which do not extend below 1500–2000 m but are replaced at this depth level by species of the genus *Haloporphyrus*. The family Macruridae, whose shallowest-living species may occur up to about

300 m, includes numerous species which seem to replace each other with increasing depths down to the abyssal plain.

While all the assemblages described above occur on muddy bottoms, occasional sandy bottoms may exist on the slope in some areas, but their assemblages are almost unknown and probably rare. However, in the North Atlantic Ocean, the brachiopod *Gryphus vitreus* may be common where pebbles are mixed with sand, and sometimes the echinoid *Cidaris cidaris*, which seems to be highly eurytopic (p. 463).

Pacific Ocean

According to ZENKEVICH and FILATOVA (1958), the upper muddy bottoms (200–800 m) of the Bering Sea exhibit an infauna characterized by the pelecypod *Yoldia beringiana* and an epifauna whose most predominant species are an echinoid of the genus *Brisaster* and the asteroid *Ctenodiscus crispatus*. Deeper, the bathyal zone supports an assemblage with the following most important elements: several hexactinellid sponges, some actinians, large-sized *Bathysiphon*, holothurioids, pogonophorans and, among the echiurids, *Tatjanellia gracilis*.

Some data on 'communities' living on the upper (200–600 m) muddy bottoms of the slope in the Kuril–Kamchatka area may be found in KUZNETSOV (1959a, 1959b, 1961), but it seems impossible to derive any general conclusion from them, except that the biomass is higher when pelecypods predominate than when polychaetes are more abundant. However, in the Kamchatka Gulf KUZNETSOV found a *Brisaster townsendi* assemblage on fine silty sand (250–500 m depth) with strongly oxygen-depleted bottom water (down to 20% saturation). This echinoid made up 30–50% of the total (150–250 m^{-2}) biomass, other inhabitants being polychaetes and amphipods of the genus *Ampelisca*.

An assemblage with predominant ophiuroids *Ophiomusium lymani* and *Amphilepis* sp. and the holothurioid *Scotoplanes* sp., was observed from the bathyscaphe 'Trieste' in the San Diego Trough on a transect at 1243 m depth (BARHAM and co-authors, 1967). Fishes (*Anoplopma fimbria*, *Sebastolobus altivelis* and others) were also present; the biomass seemed to be unusually high (up to about 200 g m^{-2}).

Off Cape Kinkasan, Honshu (Japan), from the bathyscaphe FNRS 3, I observed two highly distinct epifaunal assemblages (PERES, 1959) on the muddy-bottom slope. At a depth of 980–1000 m, the current speed was ca 5–10 cm s^{-1}; actinians were the predominant assemblage elements (*Bolocera longicornis*, *Chondractinia* or *Actinothrix*), together with asteroids (*Ceramaster*, *Luidiaster*, *Pseudarchaster*) and buried amphiuroids (possibly *Amphiodia* or *Amphioplus*). Empty pelecypods shells possibly a result of sea-star predation were abundant on the mud. Fishes were also common and relatively diversified: *Ophichtus*, *Coelorynchus*, *Lepidotrigla*, *Sebastodes*. At a depth of 1650 m, the current was weaker but the sediment apparently the same; dead pelecypods were not observed; asteroids and actinians were very rare. The most abundant elements of the assemblage were the crinoid *Heliometra glacialis*, the pennatularian *Funiculina quadrangularis* (often carrying an ophiuroid of the genus *Asteronyx*) and a dendrochirot holothurioid. As far as fishes were concerned, I only observed macrurids (*Coelorynchus*). Thus it seems that suspension feeders are more numerous at depths below 1000 m.

In the Gulf of Alaska within the bathyal zone FILATOVA (1973) observed a vertical zonation which seems to be better founded than on the eastern Atlantic Ocean slope. She distinguished four vertical subzones.

(i) The *Ophiophthalmus cataleimmodius–Astarte derjugini–Onuphis pallida–Aphrodite talpa* assemblage occurs at depths of 450–570 m on grey clayish mud mixed with pebbles (bottom temperature range: $4·2–3·5$ °C). *O. cataleimmodius* (biomass, $2·5$ g m^{-2}) predominates, together with *O. pallida*, *A. talpa* and *A. derjugini*. The following characteristic species are less abundant: the echinoderms *Ophiura leptoctenia*, *O. sarsi*, *Henricia* sp., *Heliometra glacialis*, *Brisaster latifrons*; the polychaetes *Onuphis conchylega* and *Pectinaria koreni*; the brachiopods *Frielleia palli*, *Terebratulina unguicula* and *Laqueus californicus*; various pelecypods of the genera *Cuspidaria*, *Dermatomya*, *Delectopecten* and *Crenella columbiana*; the pycnogonid *Hedgpethia californica* and the cirriped *Scalpellum columbianum*. Many species in this assemblage are eurybathic, extending far downwards from the shelf. The biomass ranges from $4·7$ to 12 g m^{-2}; ophiuroids and polychaetes largely predominate.

(ii) The *Onuphis pallida–Pavonaria pacifica–Ophiophtalmus nordmani–Ophiura leptoctenia* assemblage inhabits clayish mud (525–1500 m; $4·1–3$ °C). *O. pallida* and the large pennatularian *P. pacifica* largely predominate; these two ophiuroids are endemic in the northern Pacific Ocean. The biomass ranges from $5·1$ to $16·8$ g m^{-2} (average $10·5$), but sponges and hydroids which are sometimes very abundant in this assemblage may induce a very large increase in biomass (up to about 80 g m^{-2}).

(iii) The *Onuphis pallida–Yoldia beringiana–Ophiophtalmus nordmani–Ophiura leptoctenia–Virgularia cystifera* assemblage inhabits fine clayish mud (1090–1930 m; $2·0–1·9$ °C). The percentage of species typical for the bathyal zone clearly increases; among them are the pelecypod *Y. beringiana* (widespread in the whole bathyal zone of the northern Pacific Ocean) and the pennatularians *V. cystifera*, *Protoptilum orientale*, *Pavonaria pacifica* and *Kophobelemnon affine*. In addition to *O. pallida*, polychaetes are numerous: *Brada irenaia*, *Samythella neglecta*, *Travisia forbesi*, etc, as are pelecypods: *Delectopecten randolphi*, *Cuspidaria (Myonera) garetti*, *Vesicomya pacifica*, *Malletia pacifica*, *Poromya* sp., *Lyonsiella*, etc. Large-sized individuals of the decapod crustacean *Chinoecetes angulatus* were also recorded. Among the echinoderms, the most common are the ophiuroids *O. normani*, *O. leptoctenia* and *Ophiolimna bairdi*, as well as holothurioids (Stichopodidae and *Molpadonia*). Some animals that usually inhabit abyssal depths have also been observed: the echinoid *Pourtalesia laguncula beringiana*, the large echiurid *Prometor* sp. and some pogonophorans of the family Siboglinidae. The average biomass is $8·7$ g m^{-2}.

(iv) On the deeper and steeper parts of the slope (1960–2340 m), FILATOVA (1973) distinguished a fourth assemblage, which she named *Prometor grandis–Onuphis pallida–Yoldia beringiana–Pennatularia* biocoenosis. Here the biomass is greatly reduced (about 1 g m^{-2}) and it might thus be better to consider this deepest bathyal assemblage as a qualitatively and quantitatively impoverished aspect of assemblage (iii); in some areas the biomass at 3000 m may be lower than $0·5$ g m^{-2}.

Other areas of the World Ocean

In Gulf of Eilat (Red Sea) at 250–500 m depth, POR and LERNER-SEGGEV (1966) observed an assemblage on muddy sand in which the following animals predominated:

polychaetes of the genus *Hyalinoecia*, the echinoid *Palaeostoma mirabile*, the gastropod *Turris cingulifera* and the pelecypods *Cuspidaria*, *Abra* and *Nucula*.

In the above discussion of the bathyal zone of the Antarctic region I emphasized that the epifaunal assemblage occupies very large areas (p. 482). At this juncture I should like to add some information on three assemblages living mainly in the Ross Sea, where rather large portions of the superflacies are covered by sediments: (i) The 'deep-shelf mixed assemblage' recorded by BULLIVANT and DEARBORN (1967) between 160 and 520 m on fine sediment with boulders comprises tubicolous polychaetes, molluscs (*Limatula hodgsoni*, *Cyclocardia astartoides*, *Thracia meridionalis*, *Cadulus dalli antarcticus*) and various echinoderms: *Ophiacantha antarctica*, *Amphiodia joubini*, *Porania antarctica glabra*. Of course, epifaunal species (bryozoans, gorgonians, etc.) also occur on boulders. (ii) The 'deep-shelf muddy-bottom assemblage' (BULLIVANT and DEARBORN, 1967) lives on a muddy bottom with scattered boulders at depths of 430–750 m. Apart from the sessile fauna (*Flabellum impensum*, crinoids, etc) inhabiting the boulders, the soft-bottom animals seem to differ considerably here from those of the preceding assemblage: tubicolous polychaetes, sipunculids, arenaceous foraminiferans (*Rhabdamina*), holothurioids, asteroids (*Bathybiaster loripes obesus*, *Psilaster charcoti*, *Luidiaster gerlachei*, *Notasterias armata*) and the sea-pen *Umbellula* sp. (iii) Deeper (1200–2200 m), DELL (1972) found a sparse fauna in diatomaceous ooze. It seems that calcareous foraminiferans mostly occur on sediments in association with the corals and the stylasterine assemblage, whereas arenaceous foraminiferans inhabit muddier and deeper bottoms.

Biomass data are scarce and often approximate; except for the last-mentioned assemblage, the standing crop is probably very low. BELAYEV and USHAKOV (1957) emphasized the difference between Barents Sea and eastern Antarctic areas: down to 500 m, the biomass is higher in the latter, but below that it is lower. For example, between 1000 and 3000 m, the biomass in the eastern Antarctic bathyal zone amounts to only one-fifteenth of that of the Barents Sea.

In the Arctic region, particularly in its peripheral areas, an *Arca glacialis–Astarte crenata* assemblage exists, which may belong to the bathyal zone but may extend upwards to the lowest levels of the shelf.

More generally, the bathyal zone in the Arctic is greatly impoverished and its biomass is low; metazoans are very rare (polychaete *Asychis biceps*, pelecypods *Thyasira* and *Axinopsis*), but there are many foraminiferans: *Rhabdammina*, *Saccamina*, *Miliolina*, *Astrorhiza*, *Psammosphaera* (p. 474). Cirromomorph cephalopods have been observed immediately above the bottom in some areas of the Arctic Basin.

According to PAUL and MENZIES (1974), who analysed qualitative and quantitative samples from the Amerasian region in the Arctic Basin bathyal zone (1000–25000 m depth), foraminiferans account for 53%, pelecypods for 27%, tetraxonid sponges for 7%, and polychaetes for 5% of the total biomass, which is only about 0.04 g m^{-2}. This very low biomass is comparable to that at 5000–6000 m on the mid-Pacific Ocean red mud; it is 40 times lower than the biomass at comparable depths of Antarctica and off Peru (PAUL and MENZIES, 1974). The most common metazoans are the sponge *Thenea abyssorum*, the polychaete *Spirorbis granulatus* and the ascidian *Eugyra glutinosa*, which may be considered as 'pilot species' of the assemblage; all three species are suspension feeders. The biomass ratio meiofauna/macrofauna is amazingly low, namely about 1/1. According to PAUL and MENZIES the low average value of the diversity index suggests that the ecosystem is still young. If we exclude the foraminiferans, pelecypods dominate in biomass (58%) and sponges (16%) are more abundant than polychaetes, which is unusual in the bathyal zone.

(c) Heterogeneity in Soft-Bottom Assemblages of the Bathyal Zone

The information presented above was almost exclusively based on samples collected by bottom samplers, dredges and trawls. The resulting picture must be regarded as over-generalized, because underwater-vehicle observations have revealed that the structure of these assemblages is often highly heterogeneous.

As mentioned (Fig. 5-5, p. 137), the relative percentages of animal species which collect their food in different ways (suspension feeders, detritus feeders, limivorous) differ with geomorphological features. However, SOKOLOVA (1959) largely took into account only the most general forms of the submarine relief: shelf, slope, abyssal plain or trenches. Smaller-scale features of the submarine relief may also be important. This may be exemplified by the faunistic differences inside and outside of the submarine canyons that cut across the continental slope in many areas. In the macrobenthic epifauna of a canyon close to Cape Hatteras, ROWE (1971) observed very striking differences between the assemblages in the canyon itself and in the bottoms of the continental slope outside the canyon. According to ROWE one may distinguish three faunistic stocks: (i) a stock inhabiting the canyon, i.e. *Kophobelemnon, Peniagone*, the asteroids *Benthopecten* and *Dysaster* and a cerianthid-like actinian; (ii) a stock whose main element is *Ophiomusium lymani*, together with some other species common on the upper part of the slope; (iii) a stock found both in the canyon and on the slope itself, whose most abundant species is *Hyalinoecia artifex*, together with glass-sponges (*Euplectella suberea* and *Hyalonema boreale*) the anthozoans *Anthomastus grandiflorus* and *Umbellula lindahli*, the decapods *Munida valida* and *Parapagurus pilosimanus* and the echinoderms *Cidaris abyssicola* and *Euphronides depressa*.

ROWE's (1971) observations corroborate my own observations based on five dives with the French bathyscaphe FNRS 3 off Portugal (PERES and co-authors, 1957). On the slope itself, the muddy bottom at 650 m was mainly inhabited by numerous terebellid polychaetes and some brown cerianthids, as well as crustaceans (*Aristeus, Parapandalus, Nephrops norvegicus*) and the cephalopod *Rossia macrosoma*; filter feeders seemed to be absent. On the lower and more gentle part of the slope (2200 m) one could observe a great abundance of cumaceans (*Leucon longirostris*), several holothurioids and terebellid polychaetes and some brown cerianthids, the solitary coral *Flabellum apertum* (rather common), a cerianthid (different from that observed at 650 m). Filter-feeding animals were represented by some sabellids and the crinoid *Rhizocrinus lofotensis*. In the Setubal Canyon at 610 m, the assemblage corresponds to the *Funiculina quadrangularis* facies of the bathyal muddy bottoms, with some brown cerianthids, some sabellids, *Aporrhais serresianus* and *Anapagurus* sp. At 1150–1160 m, on hard substrates almost bare except for some very scattered unidentified sponges, muddy bottom inhabitants were only slightly more numerous: some small sabellids, the glass-sponge *Asconema setubalense*, some individuals of the shrimp genus *Aristeus*. At ca 1690 m I observed rather abundant tubes of sabellids (?), several individuals of a large, violet shrimp (probably *Aristeomorpha*) and numerous (average 1 colony m^{-2}) filter-feeding anthozoans, among them: *Funiculina quadrangularis, Kophobelemnon stelliferum* and *Gyrophyllum* (= *Bathypenna*) *elegans*. Fishes observed during these dives were of little interest (see list by PERES and co-authors, 1957). Only three species were observed on both slope and canyon: the brownish cerianthid, the shrimp *Aristeus* and the pennatularian *Funiculina quadrangularis*. Although the latter usually inhabits the upper level of the bathyal, it was observed in the canyon near the lowest level.

Such faunal differences between the insides and outsides of canyons usually appear to arise from differences in local current systems which transport food particles from the more productive planktonic and benthic shelf assemblages.

Preliminary results of a series of 15 dives with bathyscaphe 'Archimede' in the São Miguel area (Azores) suggest a very striking heterogeneity of the bathyal soft-bottom assemblages. During many dives on gentle bottom slopes, several steps (a few dm high) were observed, consisting of compacted mud with the crest always inhabited by large-sized, erect filter feeders: mainly glass-sponges and sometimes anthozoans. This heterogeneity must be assumed to arise from changes in bottom-water circulation related to the micro-relief. It testifies to smaller-scale distribution patterns of different trophic groups than those reported by SOKOLOVA (1959).

Along many transects on this bathyal mud (always mixed with volcanic ashes) which apparently constitute a quite uniform biotope in terms of declivity, grain size, bottom current and depth, a rather rich assemblage has been observed. Its most abundant elements are echinoderms, mainly *Ophiomusium lymani*, the asteroid *Pedicellaster sexradiatus* and some holothurioids (e.g. and unidentified flat-shaped Synallactidae and several Elasipoda: *Benthodytes typica, B.* cf. *janthina, Euphronides* sp, *Peniagone azorica*) and sponges, mostly hexactinellids (among them several species of *Euplectella*). The small solitary coral *Caryophyllia ambrosia*, many pennatularians (*Kophobelemnon, Scleroptilum grandiflorum, Anthoptilum murrayi*), the large shrimp *Plesiopenaeus edwardsianus* and the echinoid *Asthenosoma hystrix* were also often observed (ARNAUD, 1972; CARPINE, 1972; LAUBIER, 1972b; PERES, 1972b; SALDANHA, 1972; ZIBROWIUS, 1972). During all the dives, which correspond to a transit of several tens of miles over the bottom, all observers were impressed by the very irregular distribution and abundance of the species recorded. Such observations corroborate the opinion of BARHAM and co-authors (1967) that areas of high and low density may exist in close proximity. Although *Ophiomusium lymani* attains speeds of up to $3 \cdot 4$ cm min^{-1} (LAFOND, 1967) and *A. hystrix* is also able to move rather quickly (ARNAUD, 1972), the reasons for the heterogeneous distribution over an apparently uniform bottom remain to be explored both for the fixed and the majority of the sedentary species.

It seems, then, that abundance and food transport by currents, as well as food availability for detritus feeders and limivorous macrobenthic animals, largely control the distribution and abundance of bathyal species (SOKOLOVA, 1976) and thus the distribution and composition of the assemblage concerned. Differences in assemblage composition as a function of depth possibly arise mainly from the influence of depth on food abundance and availability.

We must further keep in mind that data obtained by dredges, trawls, and even bottom samplers are less reliable in quantitative terms than are those obtained more recently by core samplers and anchor dredges. WOLFF (1977, p. 783) refers to detailed analyses of ten box-core samples from two basins (1130–1230 m depth) off southern California (U.S.A.) which showed that polychaetes constitute more than 75% of the total number of individuals. Next in importance are the peracarid crustaceans, with an almost equal share between amphipods, tanaids and isopods. Rather insignificant are bivalves and echinoderms (mainly consisting of brittle stars)'.

In the north-western Atlantic Ocean, epibenthic-sled samples revealed a relative scale of total abundance rather similar to that of box-core and anchor dredge samples; however, only sled samples contained large-sized echinoderms. In 75 quantitative core samples from mainly bathyal depths in the Arctic Ocean, half the total number of individuals (minus nematodes) were polychaetes; next came tetraxonid sponges (13%), crustaceans and bivalves. The latter data differ somewhat from those of PAUL and MENZIES (1974) mentioned above (p. 488).

Finally, WOLFF (1977), who proposed describing as megafauna the larger animals that cannot be collected adequately with a grab or core sampler for quantitative evaluation, reviewed the places where combined use of bottom photographs and trawl allow estimates of megafaunal biomass. As an example, he points out that at about 2000 m off New England the numerical density of the megafauna is three orders of magnitude lower than that of the macrofauna in anchor dredge samples. However, calculations of megafaunal biomass in trawl samples within the same area revealed that macrofaunal and megafaunal biomass are of the same order of magnitude, the most abundant taxa being fishes and echinoderms. The problem of the megafaunal components in the bathyal and abyssal will be discussed further on p. 495.

(7) Abyssal Zone

(a) General Aspects

Data on abyssal assemblages are even more scattered than those on the bathyal zone; they are concerned almost exclusively with soft (almost invariably muddy) bottoms. Hard substrates are rare, insufficiently investigated and generally quite bare. I twice had the opportunity to observe rocky substrates in the abyssal zone from the bathyscaphe 'Archimede': in the Puerto Rico Trench at 3100 m where the numerous flagstones were quite bare (PERES, 1966) and in the Japan Trench (5100 m depth, bottom declivity 15–20 °) where the rocky steps were also bare except for some large gorgonians *Scirpaearia* sp.

However, at a depth near the ca 2600–2800 m, i.e. bathyal–abyssal boundary, an assemblage of sessile and motile species on solid substrate has been recently observed by American scientists along sea-floor spreading centres of the mid-oceanic ridge in the eastern Pacific Ocean. Similar assemblages appear also to exist at genuinely abyssal depths. In these areas, sea water seeps into cracks and picks up minerals from the crystal rock before being rejected; when the heated (up to 350 °C) plume mixes with the cold (only 2 °C) sea water, minerals precipitate. Owing to the high hydrogen sulphide and heavy metal sulphides content of the ejected water, bacteria of the sulphur cycle proliferate in mats and clumps in subsurface spaces of the porous rock; peeling of mats and clumps due to water movement results in high bacterial density (up to 10^6 ml^{-1}) in the

ambient sea water. Hence, chemosynthetic primary production prevails. The first consumer level consists of filter feeders consuming bacterial cells and aggregates, e.g.
tubicolous polychaetes, two pelecypod species (one ca 30 cm long) barnacles and a
galatheid crustacean. A white crab—certainly a scavenger—a siphonophore of the
family Rhodaliidae, some gastropods and a pink-coloured fish were also observed
(BALLARD and GRASSLE, 1979).

At the lower part of the slope whose inclination is more gentle than described above,
between 3000 and 4000 m depth, one observes a complete renewal of the fauna on soft
bottoms in the north-eastern Atlantic. The most characteristic species are the pelecypod
Abra profundorum, the hexactinellid *Hyalonema lusitanicum*, the asteroid *Crenaster semis-
pinosus* and the ophiuroid *Ophiomusium planum*. Among the holothurioids of the order
Elasipoda, species of *Oneirophanta*, *Psychropotes*, *Benthodytes*, etc., substitute for *Laetmogone*,
Benthogone, *Peniagone* and *Deima*. Polychaetes of the typically abyssal genus *Macellicephala*
appear at about 3000 m, whereas the umbrella-like cephalopod *Cirroteuthis* occurs below
5000 m. Crustaceans are represented by *Munidopsis*, the shrimp *Benthesicymus longipes*, etc.

Among the benthic and nektobenthic fishes on the abyssal plain off the coasts of
Europe one often observes the apod *Bythites crassus*, some macrurids (*Nematonurus gigas*,
Chalinura brevibarbis), *Melamphaes crassiceps* (Berycidae); below 5000 m two blind species
may occur; the zoarcid *Leucochlamys cryptophthalmus* and the ophidiid *Alexetenion parfaiti*.

In the vicinity of Madeira and the Azores, observations from the bathyscaphe
'Archimede' (PERES, 1972a, 1972b) provided an opportunity to observe the renewal of
the fauna in the transition between bathyal and abyssal zones. For example, off Madeira
at 3300 m depth the abundance of typical bathyal species (*Euplectella ruberea*, *Rhizocrinus
rawsoni*, *Plesiopenaeus edwardsianus*) suddenly decreases, while the abyssal pagurid *Parapagurus pilosimanus* var. *abyssorum* (with its commensal *Epizoanthus*) and the ophiuroid
Ophiomusium argigerum appear. Near São Miguel (Azores) by comparing two dives in the
bathyal zone (1990 and 2130 m) and a third one in the abyssal zone (3050 m), I
established a rather sharp change in some faunistic elements, chiefly hyalosponges and
holothurioids (PERES, 1972b). From 'Archimede' I could further observe the muddy
bottoms (3100 m) in the abyssal zone of the Puerto Rico Trench (PERES, 1966).
Anthozoa turned out to be extremely rare, but sponges were fairly common, together
with some echinoderms (*Ophiura*, *Zoroaster*, a synallactid holothurioid) and fishes
(*Congermuraena*, *Trachonurus sulcatus*).

In the north-western Pacific Ocean which is from this point of view the best known
area of the world ocean, ZENKEVICH and FILATOVA (1958) found an assemblage at a
depth of 3000–6000 m quite different from that in the deepest parts of the Bering and
Okhotsk Seas. According to these authors, the predominant animals are holothurioids of
the families Elpidiidae and Psychropotidae, asteroids of the families Porcellanasteridae
and Brisingidae, echinoids of the families Echinothuridae and Pourtalesiidae, crinoids of
the genus *Bathycrinus*, some actinians and solitary corals, some polychaetes (mainly
Maldanidae and Ampharetidae) and the pelecypod *Spinula oceanica* (Malletiidae).
Pogonophores may also be observed, but in the north-western Pacific Ocean and adjacent seas they seem to be largely eurybathic; *Siboglinum caulleryi*, for example, occurs from
22 to 8164 m. Echiurids are probably among the most original and characteristic elements of the abyssal (and hadal) fauna of the north-western Pacific Ocean. Such abundance of echiurid species was not recorded from any other region in the world ocean.
The most typical species of the abyssal zone are *Bonellia pacifica*, *B. achaeta*, *Tatjanellia*

gracilis and *T. grandis*; some species may be relatively eurybathic, e.g. *Alomasoma nord-pacifica* (ZENKEVICH, 1958).

The large amount of data collected by USSR expeditions allowed FILATOVA (1959, 1969) to propose a general scheme for northern Pacific abyssal bottom communities living in the 3000–6000 m deep abyssal plain at bottom-water temperatures between 1 and 1·8 °C. It seems that the 150 to 160° W line represents a rather important ecological boundary. On its western side the northern *Spinula oceanica* 'community' lives (biomass 0·2–1 g m^{-2}), to the south a *S. calcar* 'community (biomass 0·1–0·2 g m^{-2}); on the easternside, the northern 'community' includes pelecypods of the family Malletiidae together with ophiuroids (biomass 0·5–0·8 g m^{-2}), whereas the southern 'community' is dominated by sponges, isopods and nematodes (biomass 0·01 g m^{-2}). The largest biomasses always occur in the northern areas with higher planktonic primary production; the average biomass always decreases eastwards.

In the Japan Trench, with a significant bottom declivity of 15–20°, I have observed an assemblage on a muddy substrate at 5100 m with the shrimp *Stylodactylus bimaxillaris*, the holothurioid *Enypniastes eximia* (sometimes swimming), another holothurioid of the family Synallactidae, a glass-sponge (*Hyalonema apertum?*) and fishes (Macruridae and Ceratiadae).

FILATOVA (1973) investigated the upper abyssal zone, from 3200 to 3950 m (1·6–0·9 °C) in the Gulf of Alaska. On clayish mud, usually mixed with pebbles, she observed a very rich and diversified assemblage, the *Abyssaster tarda–Echinocrepis rostrata–Urechinus loveni–Hyalonema* assemblage. Sponges are very common and represented by four species of the genus *Hyalonema*, *Cladorhiza longipinna*, *Abyssocladia bruuni*, *Polymastia sol pacifica* and *Bathydorus laevis*. Small-sized actinians live attached to spicules of *Hyalonema* spp. and Ceriantharia, together with less abundant ascidians and *Stephanoscyphus*. As pennatularians typical of the abyssal zone FILATOVA mentioned *Umbellula thomsoni* and *Kophobelemnon stelliferum* (off Portugal *K. stelliferum* occurs in the bathyal zone; PERES and co-authors, 1957). *Laetmonice* sp., *Melinna cristata*, *Leanira areolata* and the abyssal endemic *Kesun abyssorum* are among the most often observed polychaetes. Species characteristic of the abyssal zone also predominate among the pelecypods: *Neilo fiora*, *Malletia truncata*, *Cuspidaria*, *Tindaria*, etc. Echinoderms are also mainly represented by abyssal species: the asteroids *Abyssaster tarda* and some Porcellanasteridae, the echinoids *Echinocrepis rostrata*, *Urechinus loveni*, *Echinosigra gracilis*, various species of the genus *Pourtalesia* and the cidarid *Aporocidaris fragilis*, the ophiuroid *Amphilepis platytata*, the holothurioids *Psychropotes*, *Myriotrochus*, etc, as the most common species. Small-sized tanaids and isopods abound.

This assemblage clearly exemplifies the change in assemblage composition which occurs at the boundary between the bathyal zone and the abyssal zone: the number of polychaetes, pelecypods and ophiuroids decreases, whereas the number of irregular echinoids, holothurioids and asteroids of the family Porcellanasteridae increases. The biomass is higher than in the deepest levels of the bathyal zone, but this is not likely to be a general characteristic.

At two deeper stations (4190 and 4740 m) FILATOVA (1973) found a decrease in both biomass and faunal diversity. Molluscs were represented by various Malletiidae, *Neilo fiora*, *Tindaria bruunea* and scaphopods. Pogonophores were also recorded but echinoderms predominated, mainly elasipod holothurioids, e.g. *Scotoplanes*, *Peniagone*,

Ophiura bathybia, the small-sized and elongated irregular echinoid *Echinosigra gracilis*, as well as the asteroids *Eremicaster tenebrarius, Vityazaster djakonovi* and *Pteraster* sp. On these bottoms (1·2 °C) the biomass has still 2·2 g m^{-2}, mainly made up of *Ophiura bathybia*, polychaetes and pelecypods.

Although rare, the strictly abyssal monoplacophoran mollusc *Neopilina* is present in the Pacific Ocean; its two species are highly psychrophilic (1–1·8 °C): *N. galatheae* (3591–3718 m) and *N. ewingi* (5611–6324 m). The population density is 0·04 individuals 1000 m^{-2} for *N. galatheae* and 0·7 1000 m^{-2} for *N. ewingi* (MENZIES and co-authors, 1959).

Another abyssal muddy-bottom assemblage is constituted by animals trawled in the Indian ocean in an area located between Madagascar and Mombassa (BRUNN, 1957). Holothurioids and ophiuroids predominate; among the holothurioids, elasipods are most abundant (*Psychropotes, Decina, Benthodytes*) together with molpadids and synallactids; of the ophiuroids, the family Ophiolepididae—which includes a large percentage of abyssal species—is best represented. Other echinoderms are rare: some asteroids (*Stryracaster, Goniaster*) a spatangid echinoid and a crinoid. Polychaetes are also abundant, chiefly of the families Oweniidae, Maldanidae, Glyceridae and Capitellidae. Among the molluscs BRUNN further mentions some scaphopods, prosobranchial and opisthobranchial gastropods, some pelecypods (e.g. *Cuspidaria*) and some decapod crustaceans such as *Parapagurus* and *Ethusa*.

While our present knowledge regarding the abyssal macrobenthos is still incomplete, there can hardly be any doubt that it supports different organismic assemblages. There are presumably some differences of a biogeographical nature and, of course, others depending on food availability, nutritional requirements and appurtenance to a given trophic group (Fig. 5-5; p. 137). However, as in the bathyal zone, heterogeneity in both abundance and distribution of macrobenthos were observed on bottoms with apparently fairly uniform environmental conditions.

According to VINOGRADOVA (1958, 1969) one may assume the abyssal zone to consist of two subzones: an upper subzone from 3000 to 4500 m and a lower one from 4500 to 6000–6500 m. This assumption is based on some rather abrupt changes in species composition at ca 4500 m in the north-western Pacific Ocean. The changes are particularly obvious with regard to sponges (Fig. 8-16) and elasipod holothurioids. Some species of Elipidiidae and Psychropotidae disappear, whereas others of the former family are never found above this depth. VINOGRADOVA's hypothesis requires further investigation.

(b) Recent Advances in the Knowledge of Abyssal Assemblages

The above review of abyssal assemblages, based on dredge samplings, trawls and bottom samplers, is somewhat approximate. More recently, the use of anchor dredges and box-core samplers has provided more detailed and more significant data, mainly with regard to the small-sized infaunal macrobenthos. As in the bathyal zone, polychaetes, crustaceans and pelecypods have generally turned out to be the main components in quantitative samples.

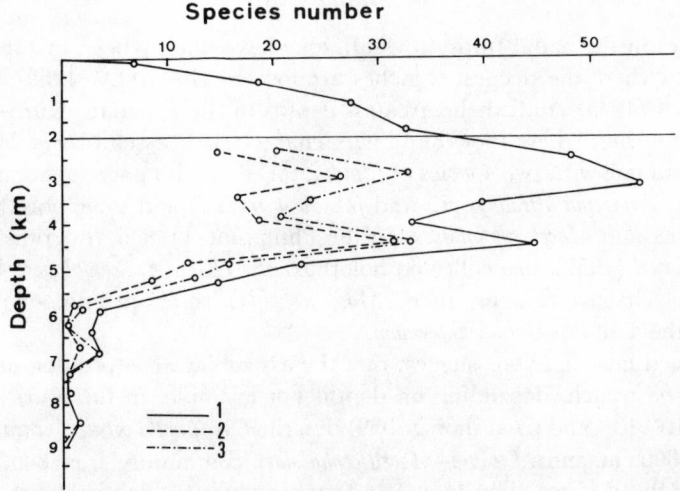

Fig. 8-16: Changes in species number of Porifera as a function of
increasing depth, especially below 2 km. 1: total species number; 2:
number of species which appear at the depth indicated; 3: number
of species which disappear at the depth indicated. (Based on data
from VINOGRADOVA, 1958.)

'Polychaetes dominate, although in the oceanic [oligotrophic] environments they
decrease somewhat in importance. The same applies to bivalves, while crustaceans,
particularly tanaids, become proportionately more abundant. Combined, the three
groups constitute 86–92% of the toal fauna.' (WOLFF, 1977, p. 783).

WOLFF (1977, p. 784) also takes into account data obtained by the epibenthic sled,
which samples both epifauna and infauna. Data obtained with this gear minimize the
role of polychaetes, whereas ophiuroids attain high abundances in more productive
waters closer to the shore, followed by cumaceans. In the deep oligotrophic environment
of the Sargasso Sea bivalves dominate, followed by isopods.

In the abyssal zone, polychaetes constitute the dominant group of the biomass (up to
50%), but their abundance decreases in oligotrophic areas. While crustacean biomass
may rise up to 25% in oligotrophic areas, it is very low in more productive regions.
Echinoderms vary so greatly in relative abundance and are so diversified with regard to
their food requirements and feeding behaviour, that their role cannot be easily interpreted at present. In the Arctic Ocean, bivalves dominate in biomass (58%) and sponges
(16%) are also more abundant than polychaetes (WOLFF, 1977, p. 784).

(8) Hadal Zone

Assemblages in the hadal (ultra-abyssal) zone have mainly been investigated in the Pacific Ocean where the deepest trenches are located (BELAYEV, 1969). ZENKEVICH and co-authors (1955) studied the greatest depths in the Aleutian, Kuril–Kamchatka and Japan Trenches, where they found very characteristic assemblages. Holothurioids largely predominate with two species of *Elpidia*, together with pogonophorans, echiurids (among them: *Vitjazema ultraabyssalis* and *Jakobia biersteini*) and some polychaete species of *Macellicephala* and *Macellicephaloides*. In the Philippine Trench at depths of 9790 and 10210 m, BRUNN (1957) also collected holothurioids (*Elpidia, Scotoplanes, Myriotrochus*), echiurids, polychaetes (among them *Macellicephala*) some pelecypods, amphipods, isopods and the actinian *Galatheanthemum*.

More detailed investigations suggest that the assemblage composition may be different in the same trench, depending on depth. For example, in the Kuril–Kamchatka Trench, ZENKEVICH and co-authors (1959) described a *Spinula vitjazi* 'community' from depths 6000–8000 m, and a *Elpidia–Macellicephaloides* 'community' from 8000 down to the maximum depth (9575 m). The data from trawls at different depths in the Bougainville Trench (BIRSHTEIN and SOKOLOVA, 1960; Table 8-1) also suggest some changes in the

Table 8-1

Percentage of total biomass for different taxa as a function of depth in the Bougainville Trench
(Based on information provided by BIRSHTEIN and SOKOLOVA, 1960)

Taxa	Depth (m)		
	6920–7657	7974–8006	8980–9043
Actinians	6·9	6·7	—
Polychaetes	3·8	0·3	0·2
Molluscs	0·1	0·2	1·6
Amphipods	0·8	0·1	—
Isopods	0·1	0·03	0·1
Asteroids	5·0	—	—
Holothurioids	83·3	92·4	97·7
Total biomass (g m^{-2})	0·014	0·022	0·007

percentages of the main taxa in relation to depth. However, the information at hand is insufficient to decide whether different assemblages are involved or only different aspects of one and the same assemblage.

From one trench to another the predominant element of the total biomass may differ. According to BIRSHTEIN (1959) in the Kuril–Kamchatka Trench 90% of the biomass consists of *Elpidia glacialis*, whereas in the Kermadec Trench pelecypods contribute 70–95% and in the Tonga Trench amphipods and isopods about 50%.

Table 8-2

Occurrence of different macrobenthic taxa in seven trenches of the Pacific Ocean (Based on information provided by BELAYEV and co-authors, 1960)

Taxa	Trench								
	Kuril-Kamch.		Idzu–Bonin 9700 m	Philip. 9000–10,200 m	Boug. 9000 m	Tonga 10,500 m	Kermadec		Marianna 10,700 m
	9000 m	9575 m					9000 m	10,000 m	
Foraminifera	+	+	+	−	−	+	?	+	−
Actiniaria	+	−	+	+	−	−	+	−	+
Nematoda	−	−	−	−	−	+	+	+	−
Echiuroidea	+	+	+	+	−	−	−	−	−
Polychaeta	−	+	+	+	+	+	+	+	+
Crustacea	+	−	−	+	+	+	+	+	+
Pelecypoda	+	−	+	+	+	+	+	+	−
Gastropoda	+	−	+	+	+	+	+	+	−
Holothurioidea	+	+	+	−	+	+	+	+	+
Pogonophora	+	+	+	−	−	−	−	−	−

Table 8-2 lists the presence or absence of several taxa in the seven deepest trenches for the Pacific Ocean. Holothurioids and polychaetes exist in all trenches, whereas echiurids and pogonophores were never found south of the Philippine Trench. Crustaceans seem absent in the three northernmost trenches. However, this requires verification, as, during dives with the bathyscaphe 'Archimede' in the Japan Trench, I have observed an amphipod and a euphausiid species as well as several individuals of an unidentified white shrimp on a muddy bottom more than 7500 m deep, together with some sponges, a rather common holothurioid of the genus *Elpidia* and a liparid fish (probably *Careproctus*); all hard substrates were quite bare.

In the Indian Ocean Java Trench, the pelecypods *Kelliella pacifica* and *Thyasira* (*Axinulus*) sp. largely predominate (up to 20 ind. m^{-2}; 75% of the total biomass), together with the actinian *Paredwardsia lemchei* and the holothurioids *Elpidia glacialis* and *Periamamma naresi* (BELAYEV and VINOGRADOVA, 1961).

In two trawl hauls and one dredge sample from the Romanche Trench in the Atlantic Ocean (7200–7340 m) PASTERNAK (1968) observed more than 100 individuals of about 20 different species representing Foraminifera, Porifera, Actiniaria, Polychaeta, Nematoda, Pelecypoda, Bryozoa, Crustacea (harpacticoids and isopods) and Echinodermata (asteroids, ophiuroids, holothurioids). On the flat bottom of the trench, detritus feeders predominate; the biomass was as high as 0·595 g m^{-2} and thus not only exceeded that in other trenches, but also that usually observed on the abyssal plain.

At 7300 m depth in the Puerto Rico Trench, PERES (1966) observed a relatively rich fauna, whose most striking peculiarity was the high abundance of fishes: several dozen individuals of a liparidid (genus *Careproctus*), one macrurid and one individual of *Melanostigma gelatinosum* (formerly recorded only from the bathyal zone). There were also a fairly common shrimp (*Nematocarcinus*), one isopod, some sabellids, an actinian (probably family Edwardsiidae) and several holothurioids (among them *Myriotrochus*).

In view of the scarcity of data it seems premature to discuss possible common features of hadal zone assemblages. As emphasized by WOLFF (1977, p. 784), 'an evaluation of the composition of the hadal trench fauna cannot yet be based on quantitative data'. In most of the trenches explored so far by grab, trawl and sled, holothurians dominate as much as polychaetes in the bathyal zone. Bivalves, polychaetes and isopods are next in abundance. Other striking characteristics of the hadal zone are discussed on p. 22.

(9) Specific Features of the Deep-sea Macrobenthos in Mediterranean Seas

A 'mediterranean sea' communicates with the open ocean through one strait or several straits whose depth is less than the maximum depth of the Mediterranean Sea itself. The most important feature of a mediterranean sea is that its waters are homothermal down to depths below the average depth of the strait. The temperature of the homothermal deep layer generally corresponds to the surface temperature prevailing during the coldest month.

In general, the deep-sea benthos of Mediterranean seas has been insufficiently investigated. While most data presumably come from the European Mediterranean Sea, they are even so insufficient for presenting a comprehensive and detailed picture of the deep-sea macrobenthos.

(a) The Mediterranean Sea

Bathyal Mediterranean Sea assemblages are very similar to those described from neighbouring areas of the Atlantic Ocean, although they are generally impoverished both in number and abundance of species.

On hard substrates the deep-sea coral assemblage described above from the eastern Atlantic Ocean slope (p. 480) also exists in the Mediterranean Sea at comparable depths, except possibly in some areas of the eastern basin (e.g. isolated small basins). This assemblage is generally conspicuously impoverished compared to that occuring in the Atlantic Ocean, particularly as regards the associated fauna. Sometimes the colonies of the pilot species *Lophelia prolifera* and *Madrepora oculata* are dead; in other places with important input of fine sediments—possibly only since historic times, due to progressive forest destruction by man—ramified corals become increasingly buried into fine silt, compensating for such burying through vertical growth. There are colonies as high as 50 cm that are completely buried except for their top parts, which feature some scattered living polyps. The surface of the dead and buried colonies is covered with a blackish crust.

A less impoverished bathyal coral assemblage in the Mediterranean Sea seems to occur on the north African coast from the Straits of Gibraltar to Cape Bon, possibly due to the fact that the Atlantic surface water flows along this coast. The influence of this current on deep-sea corals remains to be investigated; the current may improve the fertility of surface layers owing to Atlantic water input. In fact, on the eastern shores of Corsica, where sediment input is negligible and the waters relatively oligotrophic, the deep-sea coral assemblage seems to be particularly poor. Insufficient primary production and a failing dispersion of resources at all trophic levels by the currents might be among the past and present causes of the (possibly progressive) decline of this Mediterranean Sea assemblage. Its generally low plankton productivity has been well documented.

One may also assume that the general degeneration of this assemblage may have been affected by sea-level fluctuations, hydrological characteristics and the deep-current regime which influenced the Mediterranean Sea at the end of the Tertiary and during the Quaternary. For example, during the Sicilian period the Mediterranean Sea level was about 100 m higher, the intermediate and deep-water temperatures lower, and the general circulation much more intensive than at present, and the current system in the Straits of Gibraltar was possibly the reverse of that prevailing today. Such a situation must have been eminently suitable for both the penetration of species from Atlantic Ocean slope assemblages and their subsequent thriving in the Mediterranean Sea. At that time, the deep-sea coral assemblages as well as the bathyal mud assemblage and even the abyssal assemblage may be presumed to have developed. On the other hand, during the Tyrrhenian period, the sea level was 30 m lower than at present, the temperature of intermediate and bottom waters higher, and only the current system in the Straits of Gibraltar was the same as today. Such conditions must have been adverse to all deep-sea assemblages previously established in the Mediterranean Sea.

Moreover, one or more closures of the Mediterranean Gibraltar, suggested by thick salt deposits below the abyssal mud in the western basin, may have exposed all deep-sea assemblages to excessive stress periods and induced local extermination.

On soft substrates, the bathyal mud assemblage in the Mediterranean Sea is similar to that on the north-eastern Atlantic Ocean slope; both were discussed on p. 483ff. However, the bathyal mud assemblage always exhibits a fairly marked impoverishment, both in specific richness and abundance. As regards specific richness, it is obvious that some Atlantic species cannot at present—and/or could not formerly—penetrate into the Mediterranean Sea owing to the insufficient depth of the Gibraltar sill (maximum depth 300 m) and the subsurface current of Mediterranean intermediate water (with high salinity and temperature) flowing towards the Atlantic Ocean. Even where some individuals are or were able to overcome these hindrances (insufficient depth, counter current, high temperature and salinity) and successfully crossed the straits of Gibraltar, they may not have been able to survive because of (i) the high and almost constant temperature (about 13 °C) prevailing in the whole Mediterranean Sea below approximately 200 m, or (ii) the food scarcity due to the relative oligotrophy of the Mediterranean Sea except in some limited areas.

Among the less oligotroph bathyal areas are those which benefit either from indirect—i.e. via plankton and benthos shelf assemblages—fertilizing river (Ebro, Rhône, Po, etc), or by local upwelling or doming processes. Moreover, as mentioned above with regard to the white-coral assemblage (p. 480), Atlantic surface water input renders the bathyal mud assemblage of the Algerian coast less impoverished than in most other areas of the Mediterranean Sea, compared with the Atlantic Ocean slope. The influence exerted by these various environmental peculiarities (often restricted to limited areas) makes the bathyal Mediterranean Sea mud assemblage much more heterogenous from place to place than its counterpart in the Atlantic Ocean, where large areas of the slope often exhibit rather homogeneous features at comparable depths.

Obviously, the consequences of both present and past environmental peculiarities for the composition and density of benthic deep-sea assemblages must also be taken into account for the abyssal zone. The Mediterranean Sea was formerly assumed to be devoid of a true abyssal fauna, but recent French investigations have corrected this view. On the basis of CHARDY and co-authors (1973a,b,c), LAUBIER (1972a), REYSS (1971b, 1972, 1973b,c), our present knowledge of the Mediterranean Sea abyssal macrofauna, restricted to animals restrained on a 0·5 mm screen and sampled with the epibenthic sled (complemented by samples obtained with Reineck corer and trawl), may be summarized as follows.

Although the number of abyssal pelecypod species in the Mediterranean Sea is estimated at about 50, the French expedition 'Polymede I' collected only numbers of nine live species, none of them exclusively abyssal. All were also recorded from the bathyal zone and may be considered as eurybathic bathyal species, because they also occur in the North Atlantic Ocean bathyal zone. The faunistic diversity of pelecypods is low, i.e. there are only 2–3 different species in each sample (instead of the 9–10 found in the North Atlantic Ocean abyssal zone). However, some populations may be rather dense: up to more 200 *Nuculoma tenuis* individuals were found in a single sample. None of the pelecypod species hitherto recorded in the Mediterranean deep-sea benthos seems to be endemic.

Polychaetes are much better represented and may be divided into three groups of species with different bathymetric and geographical distributions: Group 1 lives in the Atlantic Ocean and in the Mediterranean Sea, inhabiting either the bathyal or the abyssal zones (*Paraonis gracilis*, *Laonice cerrata*, *Cirrophorus branchiatus*, *Glycera rouxi*) or purely abyssal areas (*Fauveliopsis brevis*, *Tharyx marioni*, *Prionospio cirrifera*). Group 2 consists of endemic species that live in the bathyal as well as in the abyssal zone (*Arcidea annae*, *A. monicae*, *Paraonis lyra*) or only in the abyssal zone (e.g. *Macellicephala annae*, *M. laubieri*, *Aricidea aberrans*, *A. abyssalis*, *A. trilobata*, *Aedicira mediterranea*); Group 3 includes only one species of the genus *Macellicephaloides* which to date has been recorded only from the hadal zone.

The cumaceans were studied by REYSS (1972, 1973b,c). If we exclude the most eurybathic species which occur in shelf assemblages but often also down to 3000 m, we can recognize three groups among the forms never observed above 200 m: (i) a group of ten species restricted to the Aegean Sea not deeper than 1500 m and consisting mainly of *Macrokylindrus aegus*, *Diastylis charcoti*, *Campylaspis vitrea*, and *Leucon mediterraneus*; (ii) a group of 13 species inhabiting depths of 500–2500 m, largely composed of what appear to be endemic forms, whose most abundant members seem to be *Diastylis jonesi*, *Leptostylis bacescoi* and *Procampylaspis bacesoi*; (iii) a group of ca 10 species inhabiting both the eastern Atlantic and the Mediterranean Sea from the shelf edge down to 3000 m. These species must be considered as genuinely eurybathic forms, of which the most abundant and widespread are *Macrokylindrus longipes*, *Bathycuma brevirostris*, *Procampylaspis armata* and *Platysympus typicus* (REYSS, 1973c). The latter group dominates numerically in the Mediterranean Sea where the composition of the cumacean fauna does not change continuously with increasing depth as in the Atlantic Ocean. There is a superimposition, at depths below 2000 m, of a typical deep Mediterranean fauna upon an eurybathic fauna (from 200–500 to 3000 m and more). Most species of the latter also occur in the Atlantic Ocean.

Quantitatively, the macrofauna abundance varies considerably in different areas. For example, in the abyssal plain of the western basin (2000–3000 m) the average number of individuals per haul is 750 (min. 457; mas. 2131); polychaetes represent 60–70% of the total, cumaceans 10%, pelecypods and isopods less than 10%. For comparison, in the same depth range the abundance is 1000–2000 individuals in the Bay of Biscay and 5000–10000 on the Gat Head–Bermuda transect. In the latter, abundances as low as those reported from the western Mediterranean Sea may be observed only between 4000 and 5000 m. In the western basin abyssal plain the specific diversity is fairly low in the four main taxa mentioned above (polychaetes, cumaceans, pelecypods and isopods). In general, two or three species account for more than 70% of the total number of individuals. Compared to the Atlantic Ocean, the decrease in biomass on the deepest bottoms of the Mediterranean Sea is even more marked than the decrease in numbers of individuals, owing to the smaller average size of Mediterranean species.

In the Ionian Sea off Messina the abyssal (3500–4000 m) fauna is much poorer: 6–31 individuals per haul, whereas the Matapan region, which is closer to the shore, appears a little richer. Amazingly, in this latter region the abundance increases with increasing depth: 62 individuals per haul at 1664 m, 71 at 3174 m, and 139 at 4690 m. Numerous circular tracks of never-sampled *Echiurus abyssalis* (?), observed from the bathyscaphe in the Matapan deep (PICARD, 1968), suggest that the abyssal fauna might be somewhat richer than indicated by the above data.

It must now be emphasized that the bathyal and abyssal soft-bottom assemblages of the Mediterranean Sea tend to mix in some places. Such mixing arises partly from the rather high percentage of eurybathic bathyal species which can extend downwards to abyssal depths, since they are not hindered by decreasing temperatures with increasing depth. However, since true abyssal species (closely related to oceanic abyssal species) are known to exist on the deepest bottoms of the Mediterranean Sea, it is amazing to notice that some of these abyssal forms may extend upwards to the bathyal zone in some areas. Possibly, these formerly cold-loving forms may have become progressively adapted to the temperature of ca 13 °C near the abyssal bottom and are therefore no longer prevented from rising up by a thermal barrier.

An example of such a phenomenon was observed by CHARDY and co-authors (1973b) who with an epibenthic sled investigated the northern areas of the Aegean Sea, i.e. those located north of the 300 m deep sill stretching between the northern Sporades up to the Marmara Sea through the island of Lemnos. This sill separates the southern open sea from the northern Aegean basin. This basin comprises two deeps separated by a 500–600 m deep sill. The western deep measures 1500 m, the eastern 1300 m. The macrofauna (retained on a 0·5 mm screen) is of moderate abundance (average: 340 inds per haul). These values are similar to those (average: 350 inds per haul) obtained by REYSS (1971a, 1973a) at comparable depths in canyons off Banyuls (western part of French Mediterranean coast) but below the average abundance (750 inds per haul) recorded from the western basin between 2000 and 3000 m. The abundance is similar in both deeps of the Aegean Sea; maximum values were recorded between 400 and 800 m. Polychaetes largely predominate in almost all samples. Characteristic species occur only in the western basin below 2000 m and in the Ionian Sea down to 4000 m. Among the latter the presence of *Macelliphala annae* is particularly striking because the genus *Macellicephala* is exclusively abyssal in the oceans. Next in abundance are amphipods, followed by ostracods, cumaceans, isopods, tanaids and aplacophores. Echinoderms are scarce and almost only represented by small-sized individuals of the ophiuroid *Amphilepis norvegica* which is characteristic of the bathyal mud assemblage (p. 485). Almost all macrofauna representatives collected in the northern Aegean Sea deeps are significantly undersized (<10 mm) in comparison with those obtained from the Gulf of Lions slope and even from the western basin in general. In fact, dwarf forms are a general characteristic of the eastern basin, but the size decrease seems to be particularly marked in the Aegean Sea. Unfortunately, samplings could not be obtained on the southern side of the sill. This should be done as soon as possible in order to elucidate whether the characteristics of the northern Aegean Sea deep-sea macrofauna summarized above are of local significance, or whether whole Aegean Sea deep assemblages are impoverished in comparison with those of other Mediterranean regions.

To summarize our present knowledge on the deep-sea macrobenthos of the Mediterranean Sea, we conclude that despite the rather important percentage of eurybathic species a truly abyssal fauna is present. In the western basin whose depth generally does not exceed 3000 m, abyssal forms are always less abundant than eurybathic and bathyal species; in contrast, in the Matapan deep, abyssal species seem to dominate. The Mediterranean deep-sea macrobenthos is less diversified than in neighbouring areas of the Atlantic Ocean and also less abundant, owing to the less favourable trophic conditions prevailing at any depth in the Mediterranean Sea. Most animals collected during the two French cruises 'Polymede' I and II were small-sized (0·5–10 mm) except for such echinoderms as the asteroid *Plutonaster bifrons* and the holothurioid *Pseudostichopus occul-*

tatus. The elasipod holothurioids formerly found in the Mediterranean were not collected again during these cruises, belonging as they probably do to the bathyal zone.

Compared with the general rarity of the Mediterranean deep-sea macrobenthos—especially in the abyssal zone—it is amazing that the average abundance of deep-sea meiobenthos considered here as comprising animals smaller than 0·5 mm amounts to 54 individuals $10 \, cm^{-2}$ (min. 33; max. 78) in the western basin at depths of 2116–2855 m, i.e. it is of the same order of magnitude as that in the poorest areas of the Indian and Atlantic Oceans (DINET and co-authors, 1973; see also p. 563).

(b) Sea of Japan

The question of whether there is a true abyssal fauna in the Sea of Japan is similar to that discussed above for the Mediterranean Sea. To LEVENSTEIN and PASTERNAK (1976) and PASTERNAK and LEVENSTEIN (1978) data collected from depths of 99 down to 3850 m suggested that a specific abyssal fauna is present and differs from that of the slope.

Down to 2000 m very abundant populations of the eurybathic *Leptognathia armata* and *L. graulis* occur. Below 3000 m a new species of *Herpotanais* was found as well as several others (taxonomically still insufficiently investigated), together with ca 20 eurybathic species of the slope or of the shelf fauna. According to PASTERNAK and LEVENSTEIN (1978, p. 592), the number of endemic abyssal species in the Sea of Japan accounts for 21% of the total number of species found in the abyssal.

Unlike those of the Mediterranean Sea, the abyssal depths of the Sea of Japan do not seem to support genuinely abyssal species, i.e. species recorded elsewhere in abyssal oceanic depths or belonging to genera or families restricted to the abyssal zone. The Sea of Japan was formerly separated from the Pacific Ocean; hence, the macrofaunal species which presently inhabit the bathyal and abyssal zones may have emigrated from the shelf and progressively populated the depths. The long period available for this process seems to have allowed the emigrants to adapt to the new environmental conditions met on the deeper bottoms, thus initiating speciation. In conclusion, the deepest fauna of the Sea of Japan cannot be regarded as being genuinely abyssal, but should be called, rather, 'pseudo-abyssal'.

PASTERNAK and LEVENSTEIN (1978) noticed significant differences between the two basins separated by the Yamato Rise. They ascribed the differences between the western and eastern parts of the Sea of Japan to local hydrological peculiarities. Polychaetes predominate in the two basins, especially the carnivores *Harmothoea derjugini* and *H. imparmarinae* which, in places, comprise up to 90% of the total benthos biomass. In the north-western basin they are associated mainly with nudibranchial gastropods and hydroids (e.g. *Oplorhiza*). The pelecypod *Delectopecten randolphi* and *Ophiura leptoctenia* predominate in the south-eastern basin. On the peak and slopes of the Yamato Rise associations of sponges and brightly colored hydrocorals (*Allopora, Errynopora*) develop. The biomass averages 1 and $2·8 \, g \, m^{-2}$ in the north-western and south-eastern basins, respectively.

(c) Red Sea

The Red Sea is hydrographically characterized by an isothermal (21·7 °C) and isohaline (40·6°/oo) layer ranging from ca 200 m to the maximum depth of 2300 m, except for the hot-brine areas in which both temperature and salinity may be much higher. The Red Sea deep benthos has not yet been sufficiently investigated, despite the very recent and unpublished results of Dr. H. THIELE, who used a large variety of gears for investigating the deep-sea benthos in the central Red Sea, from ca 500 to 2000 m, i.e. in the bathyal zone. The sedentary macrobenthos (>0·5 mm) is very poor: maximum biomass was 0·5 g m^{-2}; polychaetes largely predominate, followed by foraminiferans. Fishes (five species) and decapod crustaceans (*Parapandalus* sp., *Periclimenes* sp., and crabs of the genus *Charybdis*) have also been collected; pelecypods and gastropods are represented by only two species each, and echinoderms seem to be extremely rare. Baited traps provided many more individuals of amphipods (31) and isopods (261) than for any other group, suggesting that these highly motile scavengers are best adapted to such an ultra-oligotrophic environment.

(d) Norwegian and Greenland Seas

To some extent the Norwegian and Greenland Seas, which are separated from the North Atlantic Ocean by the Wyville Thomson Ridge (average depth ca 500 m), may be considered as forming a mediterranean sea, because depths in the Norwegian and the Greenland Basins exceed 3500 m. According to DAHL and co-authors (1976) and LAUBIER (personal communication), who investigated benthos assemblages at 11 stations from 2500 to 3800 m, the deep-sea macrofauna in the Norwegian and Greenland Seas exhibits several striking characteristics:

(i) Except for the two westernmost stations investigated during the 'Norbi' expedition (13° W and 10° W) the faunistic composition is remarkably homogeneous over the 1300 m depth range sampled; hence, qualitatively, bathyal and abyssal zones cannot be distinguished.

(ii) The specific diversity is amazingly low. For example, not more than ten echinoderm species have been recorded, most of them holothurioids; ophiuroids, often common in oceanic depths, are extremely rare. Polychaetes comprise less than ten species. Crustaceans also exhibit a great decrease in species number: only two decapod species (*Bythocaris leucopis*, and *Hymenodora glacialis*) and few isopods (although the large-sized *Glyptonotus megalurus* is common). Amphipods are the best represented crustacean group with ca 30 species, 15 of them large-sized (for comparison, there are 100 in the Bay of Biscay at comparable depths). Other important crustacean groups widespread on deep-sea bottoms, such as pagurids, galatheids, lithodids and penaeid shrimps, are totally absent. Molluscs are represented by less than 20 species. Echiurids exist only in the Spitsbergen and Greenland basins. Sipunculids and ascidians seem to be absent. Fishes are limited to three species: *Rhadichthys regina*, *Lycodes frigidis* and *Paraliparis bathybius*; macrurids are totally absent.

(iii) The number of species in common with the north-eastern Atlantic Ocean is extremely low. On the other hand, the deep-sea fauna collected in the four deeper basins reveals narrow affinities at the generic level with the Arctic shelf fauna; as examples, we may mention species of *Halirages* (amphipod), *Bythocaris* (decapod), *Boreomysis* (mysid), and *Elpidia* (elasipod holothurioid). This suggests that the deep-sea benthos of the Norwegian and Greenland Seas is the youngest in the whole world ocean. The deeper basins were formed approximately 50–60 million years ago, while the littoral fauna at the same latitudes was a temperate one. The progressive cooling which ended about 3 million years ago probably killed all the ancient deep-sea fauna which was then gradually replaced by eurybathic cold-loving species from the shelf.

(iv) According to quantitative epibenthic sled data, approximate densities range from $1 \cdot 76$ to $40 \cdot 65$ individuals m^{-2}, whereas in beam-trawl samples, the abundance ranges from $0 \cdot 48$ to $3 \cdot 17$ individuals m^{-2} (mean $1 \cdot 23$ inds m^{-2}). Comparing their own data with the average value of $0 \cdot 024$ individual m^{-2} obtained in the Bay of Biscay from eight trawl samples between 2700 and 3200 m, DAHL and co-authors (1976) conclude that while the diversity is lower in the Norwegian and Greenland Seas, the abundance is much higher. Macrofaunal densities from core samples range from 40 to 6440 individuals m^{-2} (average value 1584 inds m^{-2}), i.e. they are much higher than those obtained by the two other gears. With all gears used, macrofaunal densities are much higher in the Greenland Basin than at corresponding depths in the Norwegian Basin: this might be due to a convection phenomenon. Some dominant species such as *Elpidia glacialis* may reach an approximate density of one individual m^{-2}; the large amphipod *Eurythenes gryllus* (individual weight about $2 \cdot 5$ g) may reach a density of one individual 50 m^{-2}.

(v) The trophic net seems to be very specific: below 3000 m, where decapod crustaceans and fishes are totally absent, carnivores and scavengers seem to be represented only by amphipods of the family Lysianassidae. Considering the average biomass of $0 \cdot 05$ g m^{-2} mentioned above for the single species *Eurythenes gryllus*, the whole biomass of the assemblage might amount to several g m^{-2}. However, such extrapolation must be considered with caution, because WOLFF (1977) noticed a striking abundance of lysianassid amphipods (together with scavenging fishes) in areas with an extremely low standing crop. He therefore assumed that ubiquitous scavengers, such as lysianassid amphipods (and also fishes in oceanic depths)

'rely on large food parcels descending from above, for example dead bodies of fish and marine mammals and food fragments from surface-feeding fish. Once on the bottom the food is quickly located and consumed.' (WOLFF, 1977, p. 785).

Finally, WOLFF suggests that the faeces dispersed over the sea floor by scavengers probably represent a significant energy source for deposit feeders, thus contributing to the nutritional support of life in the deep sea.

(vi) DAHL and co-authors (1976) recorded no mature *Eurythenes gryllus*. They ascribed this fact to a change in trophic behaviour of mature females. In addition WOLFF (1977, p. 785) observed that 'amphipods caught in baited traps in the North Pacific and the Phillipine Trench were all without broodpouch plates or eggs in spite of their large size'. He assumes that the attainment of reproductive maturity may be initiated by the intake of a large meal or may be postponed until such a meal is available.

It is interesting to compare the two 'mediterranean seas' whose deep-sea macrobenthos has been sufficiently investigated: the Mediterranean Sea and the Norwegian and

Table 8-3

Comparison between deep-sea macrobenthos of the Mediterranean Sea and Norwegian and Greenland Seas (Original)

Variable	Mediterranean Sea (MS)	Norwegian and Greenland Seas (NGS)	Remarks
Deep-water temperature	Homothermal (ca 13° C) layer below ca 200 m	400 m: 0 °C below 800 m: −0·5 °C	The higher the temperature the higher the faunistic diversity; hence MS might be favoured in this respect. Reduced growth with decreasing temperature might slightly favour MS production per unit of biomass
Plankton production in the euphotic layer	Very low except for some privileged areas	Rather high in spring and summer	This largely favours NGS because the organic 'rain' may be important during the productive seasons, inasmuch as the bacterial degradation of sinking organic material is reduced due to low temperature
Interzonal migrant zooplankton on a seasonal scale	Insignificant	Very common	In NGS, over-wintering (downward migration) of many zooplankters increases the food supply at about 600 to 2000 m and, in turn, organic rain from these layers
Nekton populations (especially fishes) in surface and subsurface waters	Poor	Rather rich	Medium or large-sized dead bodies sinking towards the bottom are much more abundant in NGS than in MS
Deep-sea fauna composition	Many groups represented in the bathyal zone sometimes with several or few tens of species; in the abyssal zone, specific richness decreases. Some endemic species present	Some groups totally absent. Specific richness very low	MS: Absence of some groups or species in comparison with the open ocean may be ascribed either to inability to inhabit the warm homothermal intermediate and deep waters or to insufficient or inadequate food supply. NGS: All existing taxa are poorly represented; several others are completely absent either because some of them (e.g. penaeids) do not exist in the Arctic shelf fauna from which the NGS deep sea benthos originated, or because some taxa were represented by species unable to adapt to deep-sea conditions, possibly for reasons related to food sources (lithodids, ascidians); fishes seem to be eliminated from depths over 3000 m due to competition with abundant lysianassid amphipods

Abundance	Irregular but generally low	High—even compared with some oceanic areas—and possibly rather uniformly distributed	—
Biomass	Irregular but generally low	Presumably high and rather uniformly distributed	At equal numerical abundance the smaller mean size of MS animals results in decreased biomass compared to NGS
Vertical zonation	In general, bathyal and abyssal zones comprise different assemblages except in some isolated basins	Bathyal and abyssal zones cannot be distinguished possibly because specific richness is too low	—
Specific diversity	Of mean value	Exceptionally low	MS: 3 to 4 times lower than in Atlantic Ocean at similar depths; partly due to paleo-oceanographic events which affected MS over a rather brief period (geologic time scale), resulting in alternate impoverishment and repopulation, each generally leaving some species which could become adapted. Higher species diversity in MS arises partly from its most diversified shelf fauna: the more diversified this fauna, the more numerous might be the species sufficiently eurybathic. NGS: exceptionally low because once the original deep-sea macrofauna was destroyed (probably totally) the depths could be repopulated only through adaptation by few sufficiently eurybathic species of the poorly-diversified Arctic shelf fauna.

Greenland Seas respectively. Nine criteria have been selected: one parameter related to the environment and eight aspects of the deep-sea macrofauna (Table 8-3).

The composition and specific diversity of deep-sea assemblages in a mediterranean sea mainly depends on the sea's history, the diversity and adaptability of the immigrating faunistic stocks and on the ability of the species to adapt to the new environment. Abundance and biomass mainly depend on the productivity of the mediterranean sea and the distribution of the production by currents.

Literature cited (Chapter 8)

ANCELLIN, J. (1957). Observations sur la faune et les fonds de pêche de quelques secteurs de la Manche et des mers nordiques. *Recl Trav. Inst. Pêches marit. Fr.*, **21**, 449–484.

ANDRIASHEV, A. P. (1967). Microflora and microfauna of Antarctic fast-ice. (Russ.) *Zool. Zh.*, **46**, 1585–1593.

ANSELL, A. D., SIVADAS, P., NARAYANAN, B., SANKARANARAYANAN, V. N. and TREVALLION, A. (1972). The ecology of two sandy beaches in south-west India. I: Seasonal changes in physical and chemical factors and in the macrofauna. *Mar. Biol.*, **17**, 38–42.

ARNAUD, P. M. (1965). Nature de l'étagement du benthos marin algal et animal dans l'Antarctique. *C.r. hebd. Séanc. Acad. Sci., Paris* (Sér. D), **261**, 265–266.

ARNAUD, P. M. (1970). Frequency and ecological significance of necrophagy among the benthic species of Antarctic coastal waters. *Antarctic Ecol.*, **1**, 259–267.

ARNAUD, P. M. (1971). Les moulières à *Mytilus* et *Aulacomya* des Iles Kerguelen. (Sud de l'Océan Indien). *C.r. hebd. Séanc. Acad. Sci., Paris* (Sér. D), **272**, 1423–1425.

ARNAUD, P. M. (1972). Observations faites au cours des plongées 3 et 4. In *Bathyscaphe "Archimède"*. Campagne 1966 à Madère. Campagne 1969 aux Açores. *Publ. CNEXO, Ser. Résult. Campagnes à la Mer*, **3**, 53–57.

ARNAUD, P. M. (1974). Contribution à la bionomie marine benthique des régions antarctiques et subantarctiques. *Téthys*, **6**, 465–656.

ARNTZ, W. E. (1971). Biomasse und Produktion des Makrobenthos in den tieferen Teilen der Kieler Bucht in Jahr 1968. *Kieler Meeresforsch.*, **17**, 36–72.

ARUDPRAGASAM, K. D. (1970). Zonation on two shores on the west coast of Ceylon. *J. mar. biol. Ass. India*, **12**, 1–14.

AUSTIN, A. P. (1960). Observations on *Furcellaria fastigiata* (L.) Lam. f. *aegagropila* Reinke in Danish waters together with a note on other unattached algal forms. *Hydrobiologia*, **14**, 255–277.

BACESCU, M. (1961). Cercetari fizico-chimice si biologice rominesti la marea neagra, effectuate in perioada 1954–1959. *Hidrobiologia*, **3**, 17–46.

BAKUS, G. J. (1966). Some relationships of fishes to benthic organisms on coral reefs. *Nature, Lond.*, **210**, 280–284.

BAKUS, G. J. (1967). The feeding habits of fishes and primary production at Eniwetok, Marshall Island. *Micronecica*, **3**, 135–149.

BAKUS, G. J. (1969). Energetics and feeding in shallow marine waters. In W. J. L. Felts and R. J. Harrison (Eds), *International Review of General and Experimental Zoology*, Vol. IV. Academic Press, New York. pp. 275–369.

BAL, D. V. and BHATT, Y. M. (1959). The intertidal regime in Bombay waters. *First Int. oceanogr. Congr.* (Abstr. papers), **1**, 327.

BALLARD, R. D. and GRASSLE, J. F. (1979). The incredible world of the deep sea. *Nat. geog. Mag.*, **156**, 680–703.

BARDACH, J. E. (1959). The summer standing crop of fish on a shallow Bermuda reef. *Limnol. Oceanogr.*, **4**, 77–85.

BARHAM, E. G., AYER, N. J. and BOYCE, R. E. (1967). Macrobenthos of the San Diego Trough: photographic census and observations from bathyscaphe Trieste. *Deep Sea Res.*, **14**, 773–784.

BARNARD, J. L. and HARTMAN, C. (1959). The sea bottom off Santa Barbara, California: biomass and community structure. *Pacif. Nat.*, **1**, 1–15.

BARNARD, J. L. and ZIESENHENNE, F. C. (1961). Ophiuroid communities of southern Californian coastal bottoms. *Pacif. Nat.*, **2**, 132–152.

BELAYEV, G. M. (1969). Ultraabyssal fauna. (Russ.) In L. A. Zenkevich (Ed.), *Pacific Ocean*, 2, Biology of Pacific Ocean; Deep-sea bottom fauna—Pleiston. Izdat. Nauka, Moskva. pp. 217–334.

BELAYEV, G. M. and USHAKOV, P. V. (1957). Some views on the quantitative repartition of the benthic fauna in antarctic waters. (Russ.) *Dokl. Akad. Nauk. SSSR*, **112**, 137–40.

BELAYEV, G. M. and VINOGRADOVA, N. G. (1961). Quantitative repartition of the benthic fauna in the northern half of the Indian Ocean. (Russ.) *Dokl. Akad. Nauk. SSSR*, **138**, 1191–1194.

BELAYEV, G. M., VINOGRADOVA, N. G. and FILATOVA, Z. A. (1960). Investigations on the benthic fauna in the trenches of the South Pacific Ocean. (Russ.) *Trudy Inst. Okeanol*, **41**, 106–122.

BELLAN-SANTINI, D. (1962). Etude floristique et faunistique de quelques peuplements infralittoraux sur substrat rocheux. *Recl Trav. Stn mar. Endoume*, **41**, 237–298.

BELLAN-SANTINI, D. (1966). Eléments de bionomie quantitative des peuplements de l'étage infralittoral sur substrat rocheux en Méditerranée nord-occidentale. *Second Int. oceanogr. Congr.* (Abstr. papers), **2**, 29–30.

BELLAN-SANTINI, D. (1970a). Salissures biologiques de substrats vierges artificiels immergès en eau pure, durant 26 mois, dans la région de Marseille (Méditerranée nord-occidentale). I. Etude qualitative. *Téthys*, **2**, 335–356.

BELLAN-SANTINI, D. (1970b). Salissures biologiques de substrats vierges artificiels immergés en eau pure, durant 26 mois, dans la région de Marseille (Méditerranée nord-occidentale). II. Résultats quantitatifs. *Téthys*, **2**, 357–364.

BENNET, I. and POPE, E. C. (1960). Intertidal zonation of the exposed rocky shores of Tasmania and its relationship with the rest of Australia. *Aust. J. mar. Freshwat. Res.*, **2**, 182–221.

BERGQUIST, P. L. (1960). Notes on the marine algal ecology of some exposed rocky shores of Northland, New Zealand. *Botanica Mar.*, **1**, 86–94.

BIRSHSTEIN, J. A. (1959). The ultrabyssal fauna of the Pacific Ocean. *First Int. oceanogr. Congr.* (Abstr. papers), **1**, 390.

BIRSHTEIN, J. A. and SOKOLOVA, M. N. (1960). The benthic fauna of the Bougainville Trench. (Russ.) *Trudy Inst. Okeanol.*, **41**, 128–131.

BOYD, D. W., KORNICKER, L. S. and REZAK, R. (1963). Coralline Algal Microatolls near Cozumel Island, Mexico. *Contrib. Geol.*, **2**, 105–108.

BRATTEGARD, R. (1967). Pogonophora and associated fauna in the deep basin of Søgenfjorden. *Sarsia*, **29**, 299–306.

BROEKHUYSEN, G. J., and TAYLOR, H. (1959). The ecology of South African estuaries. VIII, Kosi Bay estuary systems. *Ann. S. Afr. Mus.*, **44**, 279–296.

BRUNEL, P. (1971). Aperçu sur les peuplements d'Invertébrés marins des fonds meubles de la baie de Gaspé, *Naturaliste can.*, **97**, 679–710.

BRUUN, A. F. (1957). Deep sea and abyssal depths. In J. W. Hedgpeth (Ed.), *Treatise on Marine Ecology and Paleoecology*. The Geological Society of America. (*Mem. geol. Soc. Am.*, **67**, 641–672.)

BUCHANAN, J. B. (1958). The bottom fauna communities across the continental shelf off Accra, Ghana (Gold Coast). *Proc. zool. Sco. Lond.*, **103**, 1–56.

BUCHANAN, J. B. (1960). On *Julienella* and *Schizammina*, two genera of arenacean Foraminifera from the tropical Atlantic with a description of a new species. *J. Linn. Soc. Lond.*, **44**, 270–277.

BUCHANAN, J. B. (1963). The bottom fauna communities and their sediment relationships off the coast of Northumberland. *Oikos*, **14**, 154–175.

BULLIVANT, J. S. (1959). Photographs of the bottom fauna in the Ross Sea. *N.Z. Jl Sci.*, **2**, 485–497.

BULLIVANT, J. S. (1961). Photographs of Antarctic bottom fauna. *Polar. Res. G.B.*, **10**, 505–508.

BULLIVANT, J. S. (1967). Ecology of the Ross Sea benthos. *The fauna of the Ross Sea*, **5**, 49–75.

BULLIVANT, J. S. and DEARBORN, J. H. (1967). The fauna of the Ross Sea. 5: General account, station lists and benthic ecology. *N. Z. Dept. Sci. Industr. Res. Bull.*, **176**, 1–77.

BURROWS, E. M. (1958). Sublittoral algal population in Port Erin Bay, Isle of Man. *J. mar. biol. Ass. U.K.*, **37**, 687–703.

CABIOCH, J. (1969). Les fonds de Maërl de la baie de Morlaix et leur peuplement végétal. *Cah. Biol. mar.*, **10**, 139–161.

CABIOCH, L. (1961). Etude de la répartition des peuplements benthiques au large de Roscoff. *Cah. Biol. mar.*, **2**, 11–40.

CABIOCH, L. (1968). Contribution à la connaissance des peuplements benthiques de la Manche occidentale. *Cah. Biol. mar.*, **9**, 493–720.

CARPINE, C. (1958). Recherches sur les fonds à *Peyssonnelia polymorpha* (Zan.) (Schmitz) de la région de Marseille. *Bull. Inst. océanogr. Monaco*, **1125**, 50.

CARPINE, C. (1972). Mission 1969 du Bathyscaphe Archimède aux Açores. Observations faites au cours des plongées 10, 11 et 15. *Publ. CNEXO, Ser. Résult. Campagnes à la Mer*, **3**, 59–63.

CHARDY, P. LAUBIER, L., REYSS, D. and SIBUET, M. (1973a). Données préliminaires sur les résultats biologiques de la campagne Polymède. I. Dragages profonds. *Rapp. P.-v. Réun. Commn int. Explor. scient. Mer Méditerr.*, **21**, 621–625.

CHARDY, P., LAUBIER, L., REYSS, D and SIBUET, M. (1973b). Dragages profonds en mer Egée. Données préliminaires. *Rapp. P.-v. Réun. Commn int. Explor. scient. Mer. Méditerr.*, **22**, 107–108.

CHARDY, P., LAUBIER, L., REYSS, D. and SIBUET, M. (1973c). Dragages profonds en Mer Ionienne. Données préliminaires. *Rapp. P.-v. Réun. Commn int. Explor. scient. Mer Mediterr.*, **22**, 103–105.

CHAVE, K. E., SMITH, S. V. and ROY, K. J. (1972). Carbonate production by coral reefs. *Mar. Geol.*, **12**, 123–140.

CLAUSADE, M. (1970). Importance et variations du peuplement mobile des cavités, au sein des formations épiréciafales et modalités d'échantillonnage en vue de son évaluation. *Recl Trav. Stn mar. Endoume*, **10** (Suppl.), 107–109.

CLAUSADE, M., GRAVIER, N., PICARD, J., PICHON, M., ROMAN, M. L., THOMASSIN, B., VASSEUR, P., VIVIEN, M. L. and WEYDERT, P. (1971). Morphologie des récifs coralliens de la région de Tuléar (Madagascar). *Téthys*, **2** (Suppl.), 1–74.

CLEMENTS, F. E. and SHELFORD, V. E. (1952). *Bio-ecology*, Wiley, New York.

COLLIGNON, J. (1960). Observations faunistiques et écologiques sur les Mollusques testacés de la Baie de Pointe-Noire (Moyen Congo). *Bull. Inst. fr. Afr. noire* (Sér. A), **22**, 411–464.

COSTA, S. and PICARD, J. (1956). Recherches sur la zonation et les biocoenoses des grèves de galets et de graviers des côtes méditerranéennes. *Rapp. P.-v. Réun. Commn int. Explor. scient. Mer Méditerr.*, **14**, 449–451.

CROSSLAND, C. (1905). Oecology and deposits of the Cape Verde marine fauna. *Proc. zool. Soc. Lond.*, **1**, 170–186.

DAHL, E. (1953). Some aspects of the ecology and zonation of the fauna on sandy beaches. *Oikos*, **4**, 176–230.

DAHL, E., LAUBIER, L., SIBUET, M., and STOMBERG, J. O. (1976). Some quantitative results on benthic communities of the deep Norwegian Sea. *Astarte*, **9**, 61–79.

DAYTON, P. K., ROBILLIARD, G. A., and DE VRIES, A. L. (1969). Anchor ice formation in McMurdo Sound, Antarctica, and its biological effects. *Science, N.Y.*, **163**, 273–274.

DAYTON, P. K., GORDON, A., ROBILLIARD, G. A. and PAINE, R. T. (1970). Benthic faunal zonation as a result of anchor ice at McMurdo Sound, Antarctica. *Antarctic Ecol.*, **1**, 244–258.

DAYTON, P. K., ROBILLIARD, G. A., PAINE, R. T. and DAYTON, L. B. (1974). Biological accommodation in the benthic community at McMurdo Sound, Antarctica. *Ecol. Monogr.*, **44**, 105–128.

DEARBORN, J. H. (1967). Stanford University Invertebrate Studies in the Ross Sea 1958–61: General Account and Station List. In *The Fauna of the Ross Sea*, Vol. 5. New Zealand Dept. of Science & Industry Research, *Bulletin* **176**, pp. 31–47.

DELL, R. K. (1972). Antarctic benthos. *Adv. mar. Biol.*, **10**, 1–216.

DE QUATRE-FAGES, A. (1854). *Souvenirs d'un Naturaliste*, Charpentier, Paris.

DERIJARD, R. (1965). Contribution à l'étude des peuplements des sédiments sablovaseux intertidaux compactés ou fixés par la végétation dans la région de Tuléar. *Recl Trav. Stn mar. Endoume*, **3**, 1–94.

DERJUGIN, K. (1915). Fauna of the Kola Gulf and its environmental conditions. (Russ.) *Zap. Akad. Nauk.SSSR*, **8**, 929.

DERJUGIN, K. and SOMOVA, N. (1940). Contribution to quantitative estimate of the benthonic population Peter the Great Bay (Japan Sea). (Russ.) *Invest. Far-Eastern Seas*, **1**, 13–36.

DINET, A., LAUBIER, L., SOYER, J. and VITIELLO, P. (1973). Résultats biologiques de la campagne Polymède II. Le méiobenthos abyssal. *Rapp. P-v. Réun Commn int. Explor. scient. Mer Méditerr.*, **21**, 701–704.

DISALVO, L. H. (1971). Regenerative functions and microbial ecology of coral reefs: labelled bacteria in a coral reef microcosm. *J. exp. mar. Biol. Ecol.* **7**, 123–136.

EKMAN, S. (1953). *Zoogeography of the Sea*, Sidgwick and Jackson, London.

ELLIS, D. V. (1959). The benthos of sea bottom in Arctic North America. *Nature, Lond.*, **184**, 79–80.

ELLIS, D. V. (1961). Marine infaunal benthos in Arctic North America. *Tech. Pap. Arct. inst. N. Am.*, **5**, 1–53.

EMERY, K. O. (1960). *The Sea off Southern California*, Wiley, New York.

ERNST, J. (1966). Données quantitatives au sujet de la répartition verticale des Laminaires sur les côtes nord de la Bretagne. *C.r. hebd. Séanc. Acad. Sci., Paris* (Ser. D), **262**, 2715–2717.

EYRE, J. and STEPHENSON, T. A. (1938). The South African intertidal zone and its relation to ocean currents. V:A sub-tropical Indian Ocean shore. *Ann. Natal Mus.*, **9**, 1–47.

FELDMANN, J. (1937). Recherche sur la végétation marine de la Méditérannée. La côte des Albères. *Revue algol.*, **10**, 339.

FELDMANN, J. (1951). Ecology of marine algae. In G. M. Smith (Ed.), *Manual of Phycology: an introduction to the algae and their biology.* Chronica Botanica Company, Waltham, Mass. pp. 313–334.

FENCHEL, T. (1970). Studies on the decomposition of organic detritus derived from the turtle grass *Thalassia testudinum. Limnol. Oceanogr.*, **15**, 14–20.

FILATOVA, Z. A. (1959). Deep sea bottom fauna communities (complexes) of the northern Pacific Ocean. *First Int. oceanogr. Congr.* (Abstr. papers), **1**, 313.

FILATOVA, Z. A. (1969). Quantitative distribution of deep-sea benthic fauna. (Russ.) In L. A. Zenkevich (Ed.), *Pacific Ocean*, 2, Biology of the Pacific Ocean; Deep-sea bottom fauna— Pleuston. Izdat. Nauka, Moskva. pp. 202–216.

FILATOVA, Z. A. (1973). On the problem of bottom fauna biocoenoses in the bathyal zone of the Alaska Gulf. (Russ.) *Trudy Inst. Okeanol.*, **91**, 80–92.

FILATOVA, Z. A. and BARSANOVA, N. G. (1964). Bottom fauna communities in the western parts of the Bering Sea. (Russ.) *Trudy Inst. Okeanol.*, **69**, 6–97.

FISHELSON, L. (1971). Ecology and distribution of the benthic fauna in the shallow waters of the Red Sea. *Mar. Biol.*, **10**, 113–133.

FORD, E. (1923). Animal communities in the level sea-bottom in the adjacent Plymouth. *J. mar. biol. Ass. U.K.*, **13**, 531–539.

FREUDENTHAL, H. D. (1962). *Symbiodinium* gen. nov. and *Symbiodinium adriaticum* sp. nov., a zooxanthella: taxonomy, life cycle and morphology. *J. Protozool.*, **9**, 45–52.

GAIN, L. (1912). La flore algologique des régions antarctiques et subantarctiques. *Deux. Exp. Ant. Franiç. 1908–1910*, Paris. pp. 1–218.

GAULD, D. T. and BUCHANAN, J. B. (1959). The principal features of the rock shore fauna in Ghana. *Oikos*, **10**, 121–132.

GERLACH, S. A. (1958). Die Mangroveregion tropischer Küsten als Lebensraum. *Z. Morph. Ökol. Tiere*, **46**, 636–730.

GESSNER, F. (1971). The water economy of the sea grass *Thalassia testudinum. Mar. Biol.*, **19**, 258–260.

GESSNER, F. and SCHRAMM, W. (1971). Salinity: plants. In O. Kinne (Ed.), *Marine Ecology*, Vol. 1, Environmental Factors, Part 2. Wiley, London. pp. 705–820.

GIBB, D. C. (1957). The free-living of *Ascophyllum nodosum*. *J. Ecol.*, **45**, 49–83.

GILET, R., MOLINIER, R. and PICARD, J. (1954). Etudes bionomiques littorales sur les côtes de Corse. *Recl Trav. Stn mar. Endoume*, (Bull. 35), **13**, 25–53.

GLÉMAREC, M. (1964). Bionomie benthique de la partie orientale du Golfe du Morbihan. *Cah. Biol. mar.*, **5**, 33–96.

GLÉMAREC, M. (1965). La faune benthique dans la partie méridionale du Massif Armoricain. Etude préliminaire. *Cah. Biol. mar.*, **6**, 51–66.

GLÉMAREC, M. (1969). *Les Peuplements Benthiques du Plateau Continental Nord-Gascogne*. Thèse Doct., Université Paris.

GLÉMAREC, M. (1973). The benthic communities of the European North Atlantic continental shelf. *Oceanogr. mar. Biol. A. Rev.*, **11**, 262–289.

GLYNN, P. W. (1973a). Aspects of the ecology of coral reefs in the western Atlantic region. In O. A. Jones and R. Endean (Eds), *Biology and Geology of Coral Reefs II. Biology I*. Academic Press, New York. pp. 271–324.

GLYNN, P. W. (1973b). Ecology of a Caribbean coral reef. The *Porites* reef-flat biotope. II. Plankton community with evidence for depletion. *Mar. Biol.*, **22**, 1–21.

GOERING, J. J. and PARKER, P. L. (1972). Nitrogen fixation by epiphytes on seagrasses. *Limnol. Oceanogr.*, **17**, 320–323.

GOLIKOV, A. N. and SCARLATO, O. A. (1967). Ecology of bottom biocoenoses in the Possjet Bay (the Sea of Japan) and the peculiarities of their distribution in connection with physical and chemical conditions of the habitat. *Helgoländer wiss. Meeresunters.*, **15**, 193–201.

GOREAU, T. F. (1959). The ecology of Jamaican coral reefs. I. Species composition and zonation. *Ecology*, **40**, 67–90.

GOREAU, T. F. (1964). Mass expulsion of zooxanthellae from Jamaican Reef communities after hurricane Flora. *Science, N.Y.*, **145**, 383–386.

GRAVIER, N., HARMELIN, J. G., PICHON, M., THOMASSIN, B., VASSEUR, P., WEYDERT, P. (1970). Les récifs coralliens de Tuléar (Madagascar): morphologie et bionomie de la pente externe. *C.r. hebd. Séanc. Acad. Sci., Paris* (Ser. D) **270** 1130–1133.

GRUA, P. (1964a). Sur la structure des peuplements de *Macrocystis pyrifera* (L.). C. Ag. observés en plongée à Kerguelen et Crozet. *C.r. hebd. Séanc. Acad. Sci., Paris*, **259**, 1541–3.

GRUA, P. (1964b). Plongées aux îles Saint Paul et Nouvelle-Amsterdam. Plongées en eaxfroides. In R. Carrick, M. W. Holdgate and J. Prevost (Eds), *Biologie Antarctique*, Hermann, Paris. pp. 279–282.

GRUZOV, E. N., PROPP, M. V. and PUSHKIN, A. F. (1967). Biological communities in the littoral zone of the Davis Sea according to underwater observations. (Russ.) *Inf. Byull. Sov. antarkt. Eksped.*, **65**, 124–141.

GRUZOV, E. N. and PUSHKIN, A. F. (1970). Bottom communities of the upper sublittoral of Enderby Land and the South Shetland Islands. *Antarctic Ecol.*, **1**, 235–238.

GUILER, E. R. (1959a). Intertidal belt-forming species on the rocky coasts of northern Chile. *Pap. Proc. R. Soc. Tasm.*, **93**, 33–58.

GUILER, E. R. (1959b). The intertidal ecology of the Montemar area, Chile. *Pap. proc. R. Soc. Tasm.*, **93**, 165–183.

GUILLE, A. and SOYER, J. (1976). Prospections bionomiques du Plateau continental des îles Kerguelen, Golfe du morbihan et Golfe des Baleiniers. *Com. nat. franç. Rech. ant., Paris*, **39**, 49–82.

GURJANOVA, E. F., LIU, J. Y., SCARLATO, O. A., and USHAKOV, P. V. (1958). A short report on the intertidal zone of the Shantung Peninsula (Yellow Sea). (Russ.) *Bull. Inst. mar. Biol. Acad. sin.*, **1**, 1–42.

GURJANOVA, E. F. (1959). Investigations on the intertidal zone of the China Seas. (Russ.) *Ind.-vo. Ak. Nauk. SSSR*, (sér. Biol.), **5**, 741–759.

HABE, T. (1959). Zonal arrangement of intertidal benthic animals in the Tanabe Bay. *Rec. oceanogr. Wks Japan*, Spec. No. **2**, 43–49.

HAGMEIER, A. (1951). Die Nahrung der Meerestiere III–IV. In H. Lübbert, E. Ehrenbaum and A. Willer (Eds), *Handbuch der Seefisherei Nord-Europa*, Band 1. E. Schweizerbart, Stuttgart. pp. 87–242.

HARLIN, M. M. (1971). Translocation between marine hosts and their epiphytic algae. *Plant. Physiol.*, **47** (Suppl.), 41.

HARMELIN, J. G. (1964). Etude de l'endofaune des "mattes" d'herbiers de *Posidonia oceanica* Delile. *Recl. Trav. Stn mar. Endoume*, (Bull. 35), **51**, 43–106.

HARMELIN, J. G. and SCHLENZ, R. (1964). Contribution préliminaire à l'étude des peuplements du sédiment des herbiers de Phanérogames marines de la Méditerranée. *Recl. Trav. Stn mar. Endoume*, (Bull. 31), **47**, 149–151.

HARMELIN-VIVIEN, M. L. (1979). *Ichthyofaune des Récifs Coralliens de Tuléar (Madagascar): Ecologie et Relations Trophiques*. Thèse Doct., Universite Aix-Marseille.

HEDGPETH, J. W. (1953). An introduction to the zoogeography of the north-western Gulf of Mexico with reference to the invertebrate fauna. *Publs Inst. mar. Sci. Univ. Tex.*, **3**, 109–224.

HEDGPETH, J. W. (1954). Bottom communities of the Gulf of Mexico. *Fish. Bull. Fish Wildl. Serv. U.S.*, **55** (89), 203–214.

HEDGPETH, J. W. (1969). Preliminary observations of life between tidemarks at Palmer Station 64° 45′ S. 64° 05′ W. *Antarct. J. U.S.*, **4**, 106–107.

HIATT, R. W. (1957). Factors influencing the distribution of corals on the reefs of Arno Atoll, Marshall Islands. *Proc. Pacif. Sci. Congr.*, **3a**, 929–970.

HOBSON, E. S. (1972). Activity of Hawaiian reef fishes during the evening and morning transitions between daylight and darkness. *Fish. Bull.*, *U.S.*, **70**, 715–740.

HOLME, N. A. (1961). The bottom fauna of the English Channel. *J. mar. biol. Ass. U.K.*, **41**, 397–461.

HOLME, N. A. (1966). The bottom fauna of the English Channel. *J. mar. biol. Ass. U.K.*, **46**, 401–493.

HURLEY, D. E. (1964). Benthic ecology of Milford Sound. *N.Z. Oceanogr. Inst. mem.*, **17**, 79–89.

HUVÉ, H. (1960). Sur l'envahissment récent d'une portion de l'étang de Berre (étang de Vaine) par une Céramiacée du genre *Spermothamnion*. *Rapp. P.-v. Réun Comm. int. Explor. scient. Mer Méditerr.*, **15**, 141–145.

HUVÉ, P. (1957). Contribution préliminaire à l'étude des peuplements superficiels des côtes rocheuses de Méditerranée Orientale. *Recl Trav. Stn mar. Endoume*, (Bull. 12), **21**, 50–62.

JACQUOTTE, R. (1961). Affinités des peuplements des fonds de maërl de Méditerranée. *Rapp. P.-v. Réun Comm. int. Explor. scient. Mer Méditerr.*, **16** (2), 141–235.

JACQUOTTE, R. (1962). Etude des fonds de maërl en Méditerranée. *Recl Trav. Stn mar. Endoume* (Bull. 26), **41**, 141–235.

JOHANNES, R. E. (1967). Ecology of organic aggregates in the vicinity of a coral reef. *Limnol. Oceanogr.*, **12**, 189–195.

JOHANNES, R. E., ALBERTS, J., D'ELIA, C., KINZIE, R. A., POMEROY, L. R., SOTTILE, W., WIEBE, W., MARSH, G. A., Jr., HELFRICH, P., MARAGOX, J., MEYER, J., SMITH, S., CRABTREE, D., ROTH, A., MCLOSKEY, L. R., BETZER, S., MARSHALL, N., PILSON, M. E. Q., TELEK, G., CLUTTER, R. I., DUPAUL, W. D., WEBB, K. L. and WELLS, J. M., Jr. (1972). The metabolism of some coral reef communities: a team study of nutrient and energy flux at Eniwetok. *BioScience*, **22**, 541–543.

JOHANNES, R. E. and GERBER, R. (1974). Import and export of net plankton by an Eniwetok coral reef community. In *Proceedings of 2nd. International Symposium on Coral Reefs*, **1**, 97–104.

JOHNSON, M. W. (1949). Zooplankton as an index exchange between Bikini Lagoon and the open sea. *Am. Geophys. Union Trans.*, **30**, 238–244.

JOHNSON, M. W. (1954). Plankton of Northern Marshall Islands. *Prof. Pap. U.S. geol. Surv.*, **260–F**, 301–314.

KAIN, J. M. (1966). The role of light in the ecology of *Laminaria hyperborea*. In R. Bainbridge (Ed.), *Light as an Ecological Factor*. Blackwell, Oxford. pp. 319–334.

KENNY, R. and HAYSOM, N. (1962). Ecology of rocky shore organisms at Macquarie Islands. *Pacif. Sci.*, **16**, 245–263.

KIKUCHI, T. (1961). An ecological study on animal community of *Zostera* belt in Tomioka Bay, Amakusa, Kyushu. Community composition (1) Fish fauna. *Rec. oceanogr. Wks Japan, Spec. No.* **5**, 211–219.

KIKUCHI, T. (1962). An ecological study on animal community of *Zostera* belt in Tomiaka Bay,

Amakusa, Kyushu. Community composition (2) Decapod crustaceans. *Rec. oceanogr. Wks Japan, Spec. No.* **6**, 135–146.

KIKUCHI, T. (1966). An ecological study on animal communities of the *Zostera marina* belt in Tomioka Bay, Amakusa, Khyshu. *Publ. Amakusa Mar. Biol. Lab.*, **1**, 163–192.

KIKUCHI, T. (1974). Japanese contributions on consumer ecology in eel-grass (*Zostera marina L.*) beds, with special reference to trophic relationships and resources in inshore fisheries. *Aquaculture*, **4**, 145–160.

KIKUCHI, T. and ARAGA, C. (1968). The fauna and underwater view of the coast of Amakusa Islands. In *Rep. Underwater Survey of the Coastal Area Proposed for Marine Park*. Kumamoto Prefecture. pp. 1–18.

KIKUCHI, T. and PERES, J. M. (1977). Consumer ecology of sea grass beds. In C. P. McRoy and C. Helfferich (Eds), *Seagrass Ecosystems*. Marcel Dekker Inc., New York, Basel. pp. 147–193.

KINSMANN, D. J. J. (1964). Reef coral tolerance of high temperatures and salinity. *Nature, Lond.*, **202**, 1280.

KNEELAND, J. and McNULTY, J. (1962). Ecological effects of sewage pollution in Biscayne Bay, Florida. Sediments and the distribution of benthic and fouling macro-organisms. *Bull. mar. Sci. Gulf Caribb.*, **11**, 394–447.

KNOX, G. A. (1960). Littoral ecology and biogeography of the Southern Oceans. *Proc. R. Soc.*, **152**, 577–624.

KUZNETSOV, A. P. (1959b). Distribution of the bottom fauna in the North-Kuril waters. (Russ.) *Trudÿ Inst. Okeanol.*, **36**, 236–258.

KUZNETSOV, A. P. (1961). Data on the zoogeography in waters of the Pacific Ocean near Kamchatka and North-Kuril Islands. (Russ.) *Dokl. Akad. Nauk SSSR*, **2**, 32–52.

LABOREL, J. (1960). Contribution à l'étude directe des peuplements sciaphiles sur substrat rocheux en Méditerranée. *Recl Trav. Stn mar. Endoume*, (Bull. 20), **33**, 117–173.

LABOREL, J. (1966). Contribution à l'étude des Madréporaires des Bermudes: systématique et répartition. *Bull. Mus. natn. Hist. nat., Paris*, (Ser. 2), **38**, 281–300.

LABOREL, J. (1969a). Les peuplements de Madréporaires des côtes tropicales du Brésil. *Ann. Univ. Abidjan* (Ecologie), **2**, 1–260.

LABOREL, J. (1969b). Madréporaires des côtes tropicales du Brésil: systématique, biologie, répartition. *Annls Inst. océanogr., Paris*, Nlle Ser. **47**, 171–230.

LABOREL, J. and VACELET, J. (1958). Etude des peuplements d'une grotte sous-marine du Golfe de Marseille. *Bull. inst. océanogr. Monaco*, **1120**, 1–20.

LABOREL, J. and VACELET, J. (1959). Les grottes sous-marines en Méditerranée, *C.r. hebd. Séanc. Acad. Sci., Paris*, **248**, 2619–2621.

LABOREL, J., PERES, J.-M., PICARD, J. and VACELET, J. (1961). Etude direct des fonds des parages de Marseille de 30 à 300 m avec la soucoupe plongeante Cousteau. *Bull. Inst. océanogr. Monaco*, **1206**, 16.

LAFOND, E. C. (1967). Movements of benthonic organisms and bottom currents as measured from the bathyscaph Trieste. deep-Sea photography. *Oceanogr. Stud.*, **3**, 295–302.

LAGARDERE, J. P. (1966). Recherches sur la biologie et l'écologie de la macrofaune des substrats, meubles de la côte des Landes et de la côte Basque. *Bull. Cent. Étud. Rech. scient. Biarritz*, **6**, 143–209.

LAGARDERE, J. P. (1977). Recherches sur la distribution verticale et sur l'alimentation des Crustacés Décapodes bernthiques de la pente continentale du Golfe de Gaseogne. Analyse des groupements carcinologiques. *Bull. Cent. Étud. Rech. Scient. Biarritz*, **11**, 367–440.

LAUBIER, L. (1966). Le coralligène des Albères. Monographie biocénotique. *Annls Inst. océanogr., Paris*, **43**, 137–316.

LAUBIER, L. (1972a). Découverte du genre abyssal *Fauveliopsis* (Annélide Polychète) en Méditerranée occidentale. *C.r. hebd. Séanc. Acad. Sci., Paris*, **274**, 697–700.

LAUBIER, L. (1972b). Mission 1969 du Bathyscaphe Archimède aux Açores. Observations faites au cours des plongées 1 et 2. *Publ. CNEXO Ser. (Résuld.) Campagnes à la Mer*, **3**, 65–71.

LAWSON, G. W. (1966). The littoral ecology of West Africa. *Oceanogr. mar. Biol. A. Rev.*, **4**, 405–448.

LE DANOIS, E. (1948). *Les profondeurs de la Mer*, Payot, Paris.

LEDOYER, M. (1962). Etude de la faune vagile des herbiers superficiels de Zostèracées et de quelques biotopes d'algues littorales. *Recl Trav. Stn mar. Endoume*, (Bull. 25), **39**, 117–235.

LEDOYER, M. (1965). Notes sur la faune vagile des grottes sous-marines obscures. *Rapp. P.-v. Reun. Commn. int. Explor. scient. Mer Méditerr.*, **18**, 121–124.

LEMCHE, H. (1956). Lower coastal animals. In A. F. Bruun, S. Greve, H. Mielche and R. Sparck (Eds), *The "Galathea" Deep Sea Expedition*. Allen and Unwin, London. pp. 119–133.

LEVENSTEIN, R. Ya. and PASTERNAK, F. A. (1976). Quantitative distribution of the bottom fauna in Japan Sea. (Russ.) *Trudy Inst. Okeanol*, **99**, 197–210.

LEWIS, J. B. (1960). The coral reefs and coral communities of Barbados W.I. *Can. J. Zool.*, **38**, 1133–1145.

LEWIS, J. B. (1964). *The Ecology of Rocky Shores*, English University Press Ltd., London.

LEWIS, J. B. (1965). A preliminary description of some marine benthic communities from Barbados, West Indies. *Can. J. Zool.*, **43**, 1046–1074.

LEWIS, M. S. and TAYLOR, J. D. (1966). Marine sediments and bottom communities of the Seychelles. *Phil. Trans. R. Soc.*, (A). **259**, 279–290.

LIE, U. and KISKER, D. S. (1970). Species composition and structure of benthic infauna communities off the coast of Washington. *J. Fish. Res. Bd Can.*, **27**, 2273–2285.

LONGHURST, A.R. (1958). An ecological survey of the West African marine benthos. *Col. Office, Fish. Publs, Lond.*, **11**, 1–102.

LONGHURST, A. R. (1959). Benthos densities off tropical West Africa. *Cons. perm. int. Explor. Mer*, **25**, 21–28.

LOYA, Y. (1972). Community structure and species diversity of hermatypic corals at Eilat, Red Sea. *Mar. Biol.*, **13**, 100–123.

McGINITIE, G. E. (1939). Littoral marine communities. *Am. Midl. Nat. Monogr.*, **21**, 28–55.

McGINITIE, G. E. (1955). Distribution and ecology of the marine invertebrates of Point Barrow, Alaska. *Smithson. misc. Collns*, **128**, 1–201.

McGINITIE, G. E. (1959). Marine Mollusca of Point Barrow, Alaska. *Proc. U.S. natn. Mus.*, **109**, 59–208.

McINTYRE, A. D. (1971). Marine zoobenthos in the light of recent research. In *The Ocean World Proceedings*. Joint Oceanographic Association, Tokyo, pp. 139–141.

McKNIGHT, D. G. (1969a). An outline distribution of the New Zealand shelf fauna: Benthos survey, station list and distribution of the Echinoidea. *Bull. N.Z. Dep. scient. ind. Res.*, **195**, 1–91.

McKNIGHT, D. G. (1969b). Faunal benthic communities of the New Zealand continental shelf. *N.Z. Jl mar. Freshwat. Res.*, **3**, 409–444.

McLEAN, J. H. (1962). Sublittoral ecology of kelp beds of the open coast area near Carmel, California. *Biol. Bull. mar. biol. Lab., Woods Hole*, **122**, 95–114.

McNAE, W. and KALK, N. (1958). The general ecology of the shores of Inhaca Island. In W. McNae and N. Kalk (Eds), *A Natural History of Inhaca Island Moçambique, Johannesburg*. Withwatersrand University Press, Johannesburg. pp. 31–45.

McROY, C. P. and BARSDATE, R. J. (1970). Phosphate absorption in eel-grass. *Limnol. Oceanogr.*, **15**, 6–13.

McROY, P. and McMILLAN, C. (1977). Production ecology and physiology of sea grasses. In C. P. McRoy and C. Helfferich (Eds), *Seagrass Ecosystems*. Marcel Dekker Inc., New York and Basel. pp. 53–87.

MAGNE, F. (1957). Sur un biotope marin favorable aux Chrysophycées benthiques. *C.r. hebd. Séanc. Acad. Sci., Paris.*, **245**, 983–985.

MARSH, G. A., Jr. (1970). Primary productivity of reef-building calcarous red algae. *Ecology*, **51**, 255–263.

MARSH, G. A. (1973). The *Zostera* epifaunal community in the York River, Virginia. *Chesapeake Sci.*, **14**, 87–97.

MASSÉ, H. (1964). Contribution à l'étude des Céphalocordés de la côte occidentale de Madagascar. *Recl Trav. Stn mar. Endoume*, **35**, 269–273.

MASSÉ, H. (1971). *Contribution à l'Etude Quantitative et Dynamique de la Macrofaune des Peuplements des Sables Fins Infralittoraux des Côtes de Provence.* Thèse Doct. Université Aix-Marseille.

MENZIES, R. J., EWING, M., WORKEL, J. L., and CLARKE, A. H., Jr. (1959). Ecology of the recent Monoplacophora. *Oikos*, **10**, 168–182.

MIYADI, D. (1941). Marine benthic communities of the Beppu-wan. *Mem. Imp. Marine Observatory*, **7**, 483–485.

MOHR, J. L., WILIMORSKY, N. J. and DAWSON, E. Y. (1957). An Arctic Alaskan kelp bed. *Arctic J. Arct. Inst. N. Amer.*, **10**, 45–52.

MOKIEVSKY, O. B. (1953). On the littoral fauna of the Okhotsk Sea. (Russ.) *Trudy Inst. Okeanol.*, **7**, 167–197.

MOKIEVSKY, O. B. (1960a). Littoral fauna of the northwestern coast of the Japan Sea. (Russ.) *Trudy Inst. Okeanol.*, **34**, 242–328.

MOKIEVSKY, O. B. (1960b). Geographical zonation of marine littoral types. *Limnol. Oceanogr.*, **5**, 389–396.

MOKIEVSKY, O. B. (1962). The biogeocoenotic system of the marine littoral zone. (Russ.) *Oceanologiya*, **9**, 211–223.

MOLINIER, R. (1955). Deux nouvelles formations organogènes construites en Méditerranée occidentale. *C.r. Hebd. Seanc. Acad. Sci., Paris*, **240**, 2166–2168.

MOLINIER, R. and PICARD, J. (1952). Recherche sur les herbiers de phanérogames marines du littoral Méditerranéen français. *Annls Inst. océanogr. Paris*, **27**, 157–234.

MOLINIER, R. and PICARD, J. (1953a). Notes biologiques à propos d'un voyage d'étude sur les côtes de Sicile. *Annls Inst. océanogr., Monaio* (Ser. 2), **28**, 163–187.

MOLINIER, R. and PICARD, J. (1953b). Recherches analytiques sur les peuplements littoraux méditerranéens se développant sur substrat solide. *Recl Trav. Stn mar. Endoume*, **9** (Bull. 4), 1–18.

MORGANS, J. F. C. (1957). Notes on the analysis of shallow-water substrata. *J. Anim. Ecol.*, **25**, 366–387.

MORGANS, J. F. C. (1959). The North Kenya Banks. *Nature, Lond.*, **184**, 259–260.

MORGANS, J. F. C. (1962). The benthic ecology of False Bay. II. Soft and rocky bottoms observed by diving and sampled by dredging and the recognition of grounds. *Trans. R. Soc. S. Afr.*, **36**, 287–334.

MUSCATINE, L. and CERNICHIARI, E. (1969). Assimilation of photosynthetic products of zooxanthellae by a reef coral. *Biol. Bull. mar. biol. Lab., Woods Hole*, **37**, 505–523.

MUSCATINE, L., POOL, R. R. and CERNICHIARI, E. (1972). Some factors influencing selective release of soluble organic material by zooxanthellae from reef corals. *Mar. Biol.*, **13**, 298–308.

NAUMOV, D. V. (1959). Coral reefs of the Hainan Island. (Russ.) *Priroda*, **9**, 83–90.

NESIS, K. N. (1962a). Soviet investigations of benthos in Newfoundland. Labrador fishing areas. (Russ.) *Sov. Ryb. Issled. v severo zapad. Atlant. Ok.* (Vniro-Pinro) **1962**, 219–225.

NESIS, K. N. (1962b). Scleractinians and Antipatharians indicators of the hydrological regime. (Russ.) *Okeanologiya*, **2**, 705–714.

NESIS, K. N. (1965). Biocenoses and biomass of the benthos in the Newfoundland Region. (Russ.) *Trudy Inst. Okeanol*, **57**, 453–489.

NONATO, E. and PERES, J. M. (1961). Observations sur quelques peuplements intertidaux de substrats durs dans la région d'Ubatuba (Etat de São Paulo). *Cah. Biol. mar.*, **2**, 263–269.

NORTH, W. N. (1959). Studies on the influence of water clarity may have on the giant kelp *Macrocystis pyrifera* and its associated organisms. *First Int. oceanogr. Congr.* (Abstr. papers), **1**, 347–348.

NORTH, W. N. (1971). The biology of giant kelp (*Macrocystis*) in California. *Nova Hedwigia, Beiheft*, **32**, 1–611.

ODUM, H. T. (1957). Primary production of eleven Florida springs and a marine turtle-grass community. *Limnol. Oceanogr.*, **2**, 85–97.

ODUM, H. T. (1963). Productivity measurements in Texas turtle-grass and the effects of dredging an intracoastal channel. *Publ. Inst. Mar. Sci. Texas*, **9**, 45–58.

ODUM, E. P. and ODUM, H. T. (1955). Trophic structure and productivity of a windward coral reef community on Eniwetok Atoll. *Ecol. Monogr.*, **25**, 291–320.,

O'GOWER, A. K. and WACASEY, J. W. (1967). Animal communities associated with *Thalassia*, *Diplanthera*, and sand beds in Biscayne Bay. I. Analysis of communities in relation to water movements. *Bull. mar. Sci.*, **17**, 175–210.

OLIVIER, S. R., PATTERNOSTER, I. K. and BASTIDA, R. (1966a). Esudios biocenoticos en las costas de Chubut (Argentina) I. Zonacion biocenologica de Puerto Pardelas (Golfo Nuevo). *Bol. Inst. Biol. Mar.*, **10**, 1–74.

OLIVIER, S. R., ESCOFET, A., ORENSANZ, J. M., PEZZANI, S. E., TURRO, S. M. and TURRO, M. E. (1966b). Contribucion al conocimiento de las comunidades benticas de Mar del Plata. *An. Com. Invest. Cient.*, Bs. As., **7**, 185–206.

OLIVIER, S. R., BASTIDA, R. and TORTI, M. R. (1968). Resultados de las campañas oceanograficas Mar del Plata I–V. Contribucion al trazado de una carta bionomica del area de Mar del Plata. Las asociaciones del Sistema litoral entre 12 y 70 m de profundidad. *Bol. Inst. Biol. Mar.*, **16**, 1–85.

OLIVIER, S. R., CAPEZZANI, D. A. A., CARRETO, J. I., CHRISTIANSEN, H. E., MORENO, V. J., AIZPUN DE MURENO, J. E. and PENCHASZADEH, P. E. (1971). Estructura de la comunidad, dinamica de la poblacion y biologia de la Almeja amarilla (*Mesodesma mactroides* Desh. 1854) en Mar Azul (Argentina). *Contrib. Instituto Biol. Mar. Mar del Plata*, **122**, 1–90.

ORTH, R. J. (1971). The effect of turtle-grass, *Thalassia testudinum*, on the benthic infauna community structure in Bermuda. *Bermuda Biol. Stn, Spec. Publ.* **9**, 18–38.

ORTH, R. J. (1973). Benthic infauna of eel-grass, *Zostera marina*, beds. *Chesapeake Sci.*, **14**, 258–269.

PARENZAN, P. (1940). Biocenologia bentonica dei fondi marina à fango. *Bol. Idrobiol. Caccia e Pesca Afr. or.*, **1**, 117–42.

PARENZAN, P. (1969). Il fondo a *Cladophora prolifera* e possibilita di una sua valorizzazione economica. *Thalassia Salentina*, **3**, 35–46.

PASTERNAK, F. A. (1968). Bottom fauna studies at the greatest depths in the Romanche Trench made from the Research Vessel Akademik Kurchatov. (Russ.) *Okeanologiya*, **8**, 312–316.

PASTERNAK, F. A. and LEVENSTEIN, R. Ya. (1978). New data on the distribution of deep-water bottom fauna in the Sea of Japan. *Oceanology*, **18**, 590–593.

PAUL, A. Z. and MENZIES, R. J. (1974). Benthic ecology of the high Arctic deep sea. *Mar. Biol.*, **27**, 251–262.

PENHALE, P. A. and SMITH, N. O., Jr. (1977). Excretion of dissolved organic carbon by eelgrass (*Zostera marina*) and its epiphytes. *Limnol. Oceanogr.*, **22**, 400–407.

PENRITH, M. L. and KENSLEY, B. (1970a). The constitution of the fauna of rocky intertidal shores of south West Africa. II. Rocky Point. *Cambebasia, A.S.W. Afr.*, **1**, 244–258.

PERES, J. M. (1959). Deux plongées au large du Japon avec le bathyscaphe français FNRS 3. *Bull. Inst. oceanogr. Monaco.*, **1134**, 1–28.

PERES, J. M. (1966). Aperçu sur les résultats de deux plongées effectuées dans le ravin de Puerto-Rico par le bathyscaphe "Archimède". *Deep Sea Res.*, **12**, 883–891.

PERES, J. M. (1967a). Les Biocoenoses benthiques dans le système phytal. *Recl Trav. Stn mar. Endoume*, **58**, 3–113.

PERES, J. M. (1967b). The Mediterranean benthos. *Oceanogr. mar. Biol. A. Rev.*, **5**, 449–533.

PERES, J. M. (1972a). Observations effectuées dans les parages de Madère au cours de deux plongées du bathyscaphe "Archimède". In: *Bathyscaphe Archimède*; Campagne 1966 à Madère; Campagne 1969 aux Açores. *Publ. CNEXO, Ser. Résult. Campagnes à la Mer*, **3**, 27–40.

PERES, J. M. (1972b). Observations faites au cours des plongées 12, 13, 14. In *Bathyscaphe Archimède*; Campagne. 1966 à Madère; Campagne 1969 aux Açores. *Publ. CNEXO, Sér. Résult. Campagnes à la Mer*, **3**, 73–78.

PERES, J. M. and PICARD, J. (1952). Les corniches calcaires d'origine biologique en Méditerranée occidentale. *Recl Trav. Stn mar. Endoume*, **4**, 2–33.

PERES, J. M. and PICARD, J. (1964). Nouveau manuel de bionomie benthique de la Méditerranée. *Recl Trav. Stn mar. Endoume*, **31**, 1–137.

PERES, J. M. and PICARD, J. (1969). Réflexions sur la structure trophique des édifices récifauz. *Mar. Biol.*, **3**, 227–232.

PERES, J. M., PICARD, J. and RUIVO, M. (1957). Résultats de la campagne de recherches du bathyscaphe F.N.R.S. 3 sur les côtes du Portugal. *Bull. Inst. océanogr. Monaco* **1092**, 7–30.

PETERSEN, C. G. J. (1914). Valuation of the Sea. II: The animal communities of the sea-bottom and their importance for marine zoogeography. *Rep. Dan. biol. Stn*, **21**, 1–68.

PETERSEN, C. G. J. (1915). On the animal communities of the sea bottom in the Skagenak, the Christiania Fjord and the Danish waters. *Rep. Dan. biol. Stn*, **23**, 31–38.

PETERSEN, C. G. J. (1918). The sea bottom and its production of fish food. *Rep. Dan. biol. Stn*, **25**, 1–62.

PEYROT-CLAUSADE, M. (1977). *Faune Cavitaire Mobile des Plauers Coralliens de la Region de Tuléar (Madagascar)*, Thèse Doct. Université Aix-Marseille.

PICARD, J. (1965). Recherches qualitatives sur les biocoenoses marines des substrats meubles dragables de la région marseillaise. *Recl Trav. Stn mar. Endoume*, **52**, 3–160.

PICARD, J. (1967). Essai de classement des grands types de peuplements marins benthiques tropicaux d'après les observations effectuées dans les parages de Tuléar (sud-ouest de Madagascar). *Recl Trav. Stn mar. Endoume* (*Trav. Stn Mar. Tuléar*), **6** (Suppl.), 3–24.

PICARD, J. (1968). Observations biologiques effectuées à bord du bathyscaphe Archimède dans l'une des fosses situées au Sud du Cap Matapan. *Annls Inst. Océanogr., Paris*, **46**, 47–51.

PICHON, M. (1964). Contribution à l'étude de la répartition des Madréporaires sur le récif de Tuléar (Madagascar). *Recl Trav. Stn mar. Endoume* (*Trav. Stn Mar. Tuléar*), **2** (Suppl.), 78–203.

PICHON, M. (1967): Caractères généraux des peuplements benthiques des récifs et lagons de l'ile Maurice (Océan Indien). *Cah. ORSTOM*, (Océanogr.), **5**, 31–45.

PICHON, M. (1974). Dynamics of benthic communities in the coral reefs of Tulear, Madagascar. In *Proceedings of Second International Symposium on Coral Reefs*. The Great Barrier Reef Committee, Brisbane. pp. 55–68.

PICHON, M. (1978). Recherches sur les peuplements à dominance d'Anthozoaires dans les récifs coralliens de Tuléar, Madagascar. *Atoll Res. Bull.*, **222**, 7–447.

PICHON, Mir. (1967). Contribution à l'étude des peuplements de la zone intertidale sur sables fins et sables vaseux non fixés dans la région de Tuléar (Madagascar). *Recl Trav. Stn mar. Endoume* (*Trav. Stn Mar. Tuléar*), **7** (Suppl.), 57–100.

PLANTE, R. (1964). Contribution à l'étude des peuplements de hauts niveaux sur substrats solides non récifaux dans la région de Tuléar. *Recl Trav. Stn mar. Endoume* (*Trav. Stn Mar. Tuléar*), **2** (Suppl.), 209–312.

PLANTE, R. (1967). Etude quantitative du benthos dans la région de Nosy-Bé; note préliminaire. *Cah. ORSTOM*, (Océanogr.), **5**, 95–108.

POR, F. D. and LERNER-SEGGEV, R. (1966). Preliminary data about the benthic fauna of the Gulf of Elat (Aqaba), Red Sea. *Israel J. Zool.*, **15**, 38–50.

PROPP, M. V. (1964). Upper sublittoral of the Western Murman at the Ainov Islands (results of investigations with aqualung. (Russ.) *Trudy murmansk. morsk biol. Inst.*, **5**, 57–60.

PROPP, M. V. (1970). The study of bottom fauna at Haswell Islands by Scuba diving. In M. W. Holdgate (Ed.), *Antarctic Ecology*, Vol. 1. Academic Press, pp. 239–241.

QASIM, S. Z. and SANKARANARAYANAN, V. N. (1970). Production of particulate organic matter by the reef on Kavaratti Atoll. (Laccadives). *Limnol. Oceanogr.*, **15**, 574–578.

QASIM, S. Z., BHATTATHIRI, P. M. A. and REDDY, C. V. G. (1972). Primary production of an Atoll in the Laccadives. *Int. Revue ges. Hydrobiol.*, **57**, 207–225.

RANDALL, J. E. (1965). Grazing effect on sea grasses by herbivorous reef fishes in the West Indies. *Ecology*, **46**, 255–260.

RANDALL, J. E. (1967). Food habits of reef fishes of the West Indies. *Stud. trop. Oceanogr.*, **5**, 665–847.

RASMUSSEN, E. (1973). Systematics and ecology of the Isefjord marine fauna (Denmark). *Ophelia*, **11**, 1–507.

REIMER, A. A. (1971). Observations on the relationships between several species of tropical zoanthids (Zoanthidea, Coelenterata) and their zooxanthellae. *J. exp. mar. Biol. Ecol.*, **7**, 207–217.

RENAUD-MORNANT, J., SALVAT, B. and BOSSY, C. (1971). Macrobenthos and meiobenthos from the closed lagoon of a Polynesian atoll: Maturei Vavao (Tuamotus). *Biotropica*, **3**, 36–55.

REYS, S. and REYS, J. P. (1965). Répartition quantitative du Benthos de la région de Tuléar (Madagascar) zone intertidale. *Trav. Stn Mar. Tuléar*, **5**, 71–86.

REYSS, D. (1971a). Les canyons sous-marins de la mer catalane: le rech du Cap et le rech Lacaze-Duthiers. 3: Les peuplements de macrofaune benthique. *Vie Milieu*, **22**, 529–613.

REYSS, D. (1971b). Résultats scientifiques de la campagne Polymède. II. Polychètes Aphroditidae de profondeur en Méditerranée. *Vie Milieu*, **22** (2A), 243–258.

REYSS, D. (1972). Résultats scientifiques de la campagne du "Jean Charcot" en Méditerranée Occidentale, main, Júin, juillet 1970, Cumacés. *Crustaceana*, **3** (Suppl.), 362–377.

REYSS, D. (1973a). Les canyons sous-marin de la mer catalane. Le rech du Cap et le rech Lacaze-Duthiers. 4: Etude synécologique des peuplements de macrofaune benthique. *Vie Milieu*, **23** (1B), 101–142.

REYSS, D. (1973b). Résultats scientifiques de la campagne Polymède 2 en mer Ionienne et en mer Egée. Cumacés. *Crustaceana*, **27**, 216–223.

REYSS, D. (1973c). Distribution of Cumacea in the deep Mediterranean. *Deep Sea Res.*, **20**, 1119–1123.

RIEDL, R. (1964). Die Erscheinungen der Wasserbewegung und ihre Wirke auf Sedentarier im mediterranean Felslitoral. *Heligoländer wiss. Meeresunters*, **10**, 155–186.

RIEDL, R. (1971). Water movement. Animals. In O. Kinne (Ed.), *Marine Ecology*, Vol. II, Physiological Mechanisms, Part 2. Wiley, London. pp. 1123–1149.

ROBERTS, C. D. (1975). Investigations into a *Modiolus modiolus* (L.) (Mollusca: Bivalvia) community in Strangford Lough, N. Ireland. *Rep. Underwat. Ass.*, (NS), **1**, 27–49.

RODRIGUEZ, G. (1959). The marine communities of Margarita Island, Venezuela. *Bull. mar. Sci. Gulf Caribb.*, **9**, 237–80.

ROWE, G. T. (1971). Observations on bottom currents and epibenthic populations in Hatteras submarine canyon. *Deep Sea Res.*, **18**, 569–581.

SALDANHA, L. (1972). Observations faites au cours de la plongée 7. *In* Bathyscaphe "Archimède"; Campagne 1966 à Madère; Campagne 1969 aux Açores. *Publ. CNEXO, Ser. Résult. Campagnes à la Mer*, **3**, 89–91.

SALVAT, B. (1964). Prospections faunistiques en Nouvelle-Calédonie dans le cadre de la mission d'étude des récifs coralliens. *Cah. Pacif.*, **6**, 77–119.

SCARLATO, O. A., GOLIKOV, A. N., VASILENKO, S. V., TSVETKOVA, N. L., GRUZOV, E. N., and NESIS, K. N. (1967). Composition, Structure and distribution of Bottom Biocoenoses in the Coastal Waters of the Possjet Bay, Japan Sea. (Russ.) In B. E. Birobskii (Ed.), *Biotsenozy zaliva Possjet, Japonskogo Moriya*. Izdat. Nauka. pp. 5–61.

SEED, R. (1969). The ecology of *Mytilus edulis* L. (Lamellibranchiata) on exposed rocky shores. I. Breeding and Settlement. *Oceologia (Berl.)*, **3**, 277–316.

SEGAL, E. (1970). Light: animals. Invertebrates. In O. Kinne (Ed.), *Marine Ecology*, Vol. I, Environmental Factors, Part 1. Wiley, London. pp. 159–211.

SIBUET, M. (1977). Répartition et diversité des échinodermes (holothurides et astérides) en zone profonde dans le Golfe de Gascogne. *Deep Sea Res.*, **24**, 549–563.

SNELI, J. A. (1968). The Lithothamnion community in Nord-Möre, Norway with notes on the epifauna of *Desmarestia viridis* (Müller). *Sarsia*, **311**, 69–74.

SOKOLOVA, M. N. (1959). On the distribution of deep-water bottom animals in relation to their feeding habits and the character of sedimentation. *Deep Sea Res*, **6**, 1–4.

SOKOLOVA, M. N. (1976). Large scale division of the World Ocean by trophic structure of deep-sea macrobenthos. *Trudy Inst. Okeanol.*, **99**, 20–30.

SOROKIN, Yu. I. (1972). A study of filtrational and osmotic feeding in corals. *Zh. Obshch. Biol.*, **33** (2), 123–128.

SOROKIN, Yu. I. (1978). Decomposition of organic matter and nutrient regeneration. In O. Kinne (Ed.), *Marine Ecology*, Vol. IV, Dynamics. Wiley, Chichester. pp. 501–616.

SOURIE, R. (1954). Contribution à l'étude écologique des côtes rocheuses du Sénégal. *Mém. Inst. Fr. Afr. Noire.*, **38**, pp. 342.

SOUTHWARD, A. J. and SOUTHWARD, E. (1958). On the occurrence and behaviour of two little-known Barnacles. *Hexelasma hirsutum* and *Verruca recta* from the continental slope. *J. mar. biol. Ass.*, **37**, 633–647.

SPÄRCK, R. (1929). Preliminary survey of the results of quantitative bottom investigations in Iceland and Faroe waters. *Rapp. P. v. Reun. Cons. perm. int. Explor. Mer*, **47**, 1–28.

SPÄRCK, R. (1937). The benthonic animal communities of the coastal waters. *The Zoology of Iceland*, **1**, 1–45.

SQUIRES, D. F. (1959). Results of the Puritan American Museum of Natural History expedition to western Mexico. VII. Corals and coral reefs in the Gulf of California. *Bull. Am. Mus. nat. Hist.* **18**, 367–432.

STEPHENSON, T. A. and STEPHENSON, A. (1949). The universal features of zonation between tide-marks on rocky coasts. *J. Ecol.*, **37**, 289–305.

STEPHENSON, W. and SEARLES, R. B. (1960). Experimental studies on the ecology of intertidal environments at Heron Island. *Aust. J. mar. Freshwat. Res.* **11**, 241–267.

STETSON, T. R., SQUIRES, D. F. and PRATT, R. M. (1962). Coral banks occurring in deep water on the Blake Plateau. *Am. Mus. Novit.*, **2114**, 1–39.

STODDART, D. R. (1969). Ecology and morphology of recent coral reefs. *Biol. Rev.*, **44**, 433–498.

STRIPP, K. (1969). Associations in the Benthos in Heligoland Bight. *Veröff. Inst. Meeresforsch. Bremerhaven*, **12**, 95–142.

SWEDMARK, B. (1956). Etude de la microfaune des sables marins de la région de Marseille. *Arch. Zool. exp. gen.*, **93**, 70–95.

TAYLOR, J. D. (1968). Coral reef and associated invertebrate communities (mainly (molluscan) around Mahé, Seychelles. *Phil. Trans. R. Soc.* (B. Biol. Sci.), **793**, 129–206.

TENORE, K. R. (1977). Utilization of aged detritus derived from different sources by the polychaete *Capitella capitata*. *Mar. Biol.*, **44**, 51–55.

THAYER, G. W., ADAMS, S. M. and LACROIX, M. W. (1976). Structural and functional aspects of a recently established *Zostera marina* community. In: L. E. Cronin (Ed.), *Estuarine Research*, Vol. 1, Chemistry, biology and the estuarine system. Academic Press, New York. pp. 528–540.

THOMASSIN, B. (1969). Peuplements de deux biotopes de sables coralliens sur le Grand Récif de Tuléar, Sud-Ouest de Madagascar. *Recl Trav. Stn mar. Endoume*, **19** (Suppl.), 59–134.

THOMASSIN, B. (1971). Les faciès d'épiflore et d'épifaune des biotopes sédimentaires des formations coralliennes dans la région de Tuléar (S. W. de Madagascar). In D. R. Stoddart and C. M. Yonge (Eds), *Regional Variation in Indian Ocean Coral Reefs*. Symp zool. Soc. Lond., **28**, 371–396.

THOMASSIN, B. (1972). Les biotopes de sables coralliens récifaux de la région de Tuléar, Madagascar. In C. Mukundan and C. S. G. Pillai (Eds) *Proceedings of Symposium on Corals and Coral Reefs*. Marine Biological Association, Cochin, India. pp. 291–313.

THOMASSIN, B. (1973). Peuplements des sables fins sur les pentes internes des récifs coralliens de Tuléar (S. W. de Madagascar): Essai d'interprétation dynamique des peuplements de sables mobiles infralittoraux dans un complexe récifal soumis ou non aux influences terrigènes. *Téthys*, **5** (Suppl.), 157–220.

THOMASSIN, B. (1976). Feeding behaviour of the felt, sponge, coral-feeder sea stars, mainly *Culcita schmideliana*. *Helgoländer wiss. Meeresunters.*, **28**, 51–65.

THOMASSIN, B. (1978). *Peuplements des Sédiments Coralliens de la Région de Tuléar (S.W. de Madagascar) et leur Insertion dans le Contexte Côtier Indo–Pacifique*, Thèse Doct., Université Aix-Marseille.

THOMASSIN, B. and GALENON, P. (1977). Molluscan assemblages on the boulder tract of Tuléar coral reefs. In D. L. Taylor (Ed.), *Proceedings of Third International Coral Reef Symposium*, Vol. 1. Rosenstiel School of Marine and Atmospheric Science, Miami. pp. 247–252.

THORSON, G. (1933). Investigations on shallow water animal communities in the Franz Joseph Fjord (East Greenland) and adjacent waters. *Medd. om Grønland*, **100**, 1–68.

THORSON, G. (1934). Contributions to the animal ecology of the Scoresby Sound fjord complex (East Greenland). *Medd. om Grønland*, **100**, 1–67.

THORSON, G. (1957). Bottom communities (sublitoral or shallow shelf). In: J. W. Hedgpeth (Ed.), *Treatise on Marine Ecology and Paleoecology*. The Geological Society of America. (*Mem. geol. Soc. Am.*, **67**, 461–534.)

TOKIOKA, T. (1953). Invertebrate fauna of the intertidal zone of the Tokara Islands. *Publ. Seto mar. biol. Lab.*, **3**, 123–148.

TOMMASI, L. R. (1969). Nota sobre os fundos detriticos do circalitoral inferior da plataforma continental brasileira ao sul do Cabo Frio (R.J.). *Bol. Instituto Oceanogr.*, **18**, 1–91.

TRUE, M. A. (1970). Etude quantitative de quatre peuplements sciaphiles sur substrat rocheux dans la région marseillaise. *Bull. Inst. océnogr. Monaco*, **1401**, 1–48.

TRUE-SCHLENZ, R. (1965). Données sur les peuplements des sédiments à petites phanérogames marines (*Zostera nana* Roth et *Cymodocea nodosa* Ascherson) comparés à ceux des habitats voisins dépourvus de végétation (côtes de Provence). *Recl Trav. Stn mar. Endoume*, **55**, 95–125.

USHAKOV, P. V. (1955). *Atlas of the Invertebrates of the Far-East Seas*, (Russ.) Izdat. Akad. Nauk SSSR, Moskva.

USHAKOV, P. V. (1951). The littoral of the Okhotsk Sea. (Russ.) *Dokl. Ak. Nauk SSSR*, **76**, 127–130.

USHAKOV, P. V. (1963). Quelques particularités de la Bionomie benthique de l'Antarctique de l'Est. *Cah. Biol. mar.*, **4**, 81–90.

USHAKOV, P. V. (1965). Bionomical peculiarities of the littoral zone of the Guinean Republic (Western Africa). (Russ.) *Okeanology*, **5**, 501–517.

USHAKOV, P. V. (1970). Observations sur la répartition de la faune benthique du littoral guinéen. *Cah. Biol. mar.*, **11**, 435–457.

VACELET, J. (1964). Etude monographique de l'éponge calcaire Pharétronide de Méditerranée *Petrobiona massiliana*. Vacelet et Levi. les Pharétronides actuelles et fosiles. *Recl Trav. Stn mar. Endoume*, **50**, 3–131.

VASSEUR, P. (1964). Contribution à l'étude bionomique des peuplements sciaphiles infralittoraux de substrat dur dans les récifs de Tuléar (Madagascar), *Recl Trav. Stn mar. Endoume (Trav. Stn Mar. Tuléar)*, 2 (Suppl.), 5–75.

VATOVA, A. (1949). La fauna bentonica dell' Alto et Medio Adriatica. *Nova Thalassia*, **1**, 42, 110.

VATOVA, A. (1961). Sulla zoocenosi *Lentidium* delle acque pecioaline del Mediterraneo. *Acc. Naz. Atti. Acad. naz. Lincei (Cl. Sci. fis. mat. nat., Ser. 8)*, **31** (5), 314–315.

VINOGRADOVA, N. G. (1958). Vertical repartition of the abyssal fauna of the ocean. (Russ.) *Trudȳ Inst. Okeànol*, **27**, 87–122.

VINOGRADOVA, N. G. (1969). Vertical repartition of the deep sea benthic fauna. (Russ.) In L. A. Zenkevich (Ed.), *Pacific Ocean*, 2, Biology of the Pacific Ocean; Deep-sea bottom fauna—Pleuston. Izdat. Nauka, Moskva. pp. 128–153.

VIVIEN, M. L. (1973a). Contribution à la connaissance de l'éthologie alimentaire de l'ichtyofaune du platier interne des récifs coralliens de Tuléar (Madagascar). *Téthys*, **5** (Suppl.), 221–308.

VIVIEN, M. L. (1973b). Régimes et comportements alimentaires de quelques poissons des récifs coralliens de Tulear, Madagascar. *La Terre et la Vie*, **27**, 551–577.

VIVIEN, M. L. (1974a). Ecology of the fishes of the inner coral reef flat in Tulear, Madagascar. *J. mar. biol. Ass. India*, **15**, 20–45.

VIVIEN, M. L., (1974b). Ichtyofaune des herbiers de phanérogames marines du Grand Récif de Tulear (Madagascar). I. Les peuplements et leur distribution écologique. *Téthys*, **5**, 425–436.

VOSS, G. L. and VOSS, N. A. (1960). An ecological survey of the marine invertebrate of Bimini, Bahamas, with a consideration of their zoogeographical relationships. *Bull. mar. Sci. Gulf Caribb.*, **10** (*Contr. Mar. Lab. Univ. Miami*, **260**.)

VOZZHINSKAJA, V. B. (1964). Macrophytes of the marine littoral of Sakhalin Island. (Russ.) *Trudȳ Inst. Okeanol*, **69**, 330–440.

WADE, B. A. (1972). A description of a highly diverse soft-bottom community in Kingston Harbour, Jamaica. *Mar. Biol.*, **13**, 57–69.

WALTON-SMITH, F. G. (1948). *Atlantic Reef Corals*, University of Miami Press, Coral Gables, Florida.

WALTON-SMITH, F. G. (1954). Biology of the commercial sponges. *Fish. Bull. Fish Wildl. Serv. U.S.*, **55** (89), 233–266.

WANDERS, J. B. W. (1976). The role of benthic algae in the shallow reef of Curaçao, Netherlands Antilles. II. Primary productivity of the Sargassum beds on the north-east coast submarine plateau. *Aquatic Botany*, **2**, 327–335.

WEBB, J. E. and THEODOR, J. (1969). Irrigation of submerged marine sands through wave action. *Nature, Lond.*, **220**, 682–683.

WELCH, B. L. (1965). Succession in the Caribbean *Thalassia* community. *Ocean Sci. Ocean Engng*, **1**, 297.

WELLS, J. W. (1954). Recent corals of the Marshall Islands, Bikini and nearly atolls II. Oceanography (Biological). *U.S. Geol. Survey Prof.*, **260**, 385–486.

WELLS, J. W. (1957). Coral Reefs. In J. W. Hedgpeth (Ed.), *Treatise on Marine Ecology and Paleoecology*. The Geological Society of America. (Mem. geol. Soc. Am., **67**, 609–631.)

WHITE, M. G. and ROBINS, M. W. (1972). Biomass estimates from Borge Bay, Signy Island, South Orkney Islands. *Br. antarct. Surv.*, **31**, 45–50.

WOLFF, T. (1977). Diversity and faunal composition of the deep-sea benthos, *Nature, Lond.*, **267**, 780–785.

YONGE, C. M. (1963). The biology of coral reefs. *Adv. mar. Biol.*, **1**, 209–260.

ZAVODNIK, D. (1965). A brief survey of knowledge of Echinoderm fauna in the Northern Adriatic. *Acta Adriat*, **11**, 285–288.

ZENKEVICH, L. A. (1958). Deep-sea echiurids from the northwestern areas of the Pacific Ocean (2nd part). (Russ.) *Trudy Inst. Okeanol.*, **27**, 192–203.

ZENKEVICH, L. A. (1963). *Biology of the Seas of the URSS*, G. Allen & Unwin Ltd., London.

ZENKEVICH, L. A., BELAYEV, G. M., BIRSHTEIN, J. A. and FILATOVA, Z. A. (1959). Qualitative and quantitative features of the deep-sea bottom fauna in the Ocean. (Russ.) In: L. A. Zenkevich (Ed.), *Itogi Nauki, Dostizhenya Okeanologii*; 1, Uspekhi y izuchenii okeanicheskih glubin. Izdat. Akad. Nauk. SSSR, Moskva. pp. 106–147.

ZENKEVICH, L. A., BIRSHTEIN, J. A. and BELAYEV, G. M. (1955). Investigations on the bottom fauna of the Kuril–Kamchatka trench. (Russ.) *Trudy Inst. Okeanol.*, **12**, 345–381.

ZENKEVICH, L. A. and FILATOVA, Z. A. (1958). Short account on the qualitative composition and quantitative repartition of the benthic fauna in the Far-East soviet waters and northwestern region of the Pacific Ocean. (Russ.) *Trudy Inst. Okeanol.*, **27**, 154–160.

ZIBROWIUS, H. (1972). Observations faites au cours de la plongée 8. In Bathyscaphe "Archimède"; Campagne 1966, Madère; Campagne 1969, Açores. *Publ. CNEXO, Ser. Résult. Compagnes à la Mer*, **3**, 93–97.

ZIBROWIUS, H. (1976). *Les Scléractiniaires de la Méditerranée et de l'Atlantique nord-oriental*, Thèse Doct., Université Aix-Marseille (II).

Marine Ecology Vol. 5, Part 1
Edited by Otto Kinne

9. SPECIFIC BENTHIC ASSEMBLAGES

J. M. PERES

(1) Size Categories of Benthic Organisms

The generally adopted system classifies benthos organisms into three size categories: micro-, meio- and macrobenthos (p. 50). Such classification is not entirely satisfactory, however, because the size of bottom-living marine species comprises a continuous spectrum from the smallest microorganisms to the largest invertebrates and demersal fishes. Moreover, the ability of an organism to pass through a sieve of a given mesh depends not only on its size but also on its shape and ability to avoid getting caught.

Investigations on small-sized benthos are mainly concerned with species inhabiting soft bottoms. Hence, only these can be considered in depth in the following pages.

Microbenthos includes viruses, bacteria, yeasts, fungi, mostly the cyanophytes and other algae—both belonging to the microphytobenthos—and small protozoans. In soft substrates, bacteria are of paramount importance in all the three ecosystems distinguished below, not only because bacterial cells are always more abundant than representatives of other microbenthic groups, but also from a functional point of view. Many aspects of the ecology of marine baceria (and fungi) have been reviewed extensively in Volumes I, II and IV. It is therefore sufficient to recall briefly the most important features of bacterial life on and within marine soft substrates.

(i) Soft substrates contain autotrophic and heterotrophic bacteria. The former may be either photoautotrophic or chemoautotrophic. Chemoautotrophic bacteria (Volume II: SCHLEGEL, 1975) satisfy their energy requirements through oxidation of various inorganic substances and thus are usually aerobic. Heterotrophic bacteria may be aerobic or anaerobic; they satisfy their energy requirements through oxidation of organic matter, which they decompose into mineral substances or into smaller organic molecules; the latter are subsequently mineralized by chemoautotrophic bacteria. (For a general classification of heterotrophic organisms see Volume II: PANDIAN, 1975, p. 61).

(ii) In soft substrates bacterial biomass ranges from 0·1 to 10 g C m^{-2} (up to several hundred million cells g^{-1} wet mud). This means that bacterial biomass may be about equal to, or even larger than, the standing crop of the fauna. Maximum bacterial abundance generally occurs in the superficial film of the sediment, which contains the highest amount of decaying organic material.

(iii) The smaller the mean grain size of the sediment, the higher the bacterial abundance tends to be. This is so because muddy sediments contain more organic matter adsorbed to clay particles and because most bacteria live epiphytically on mineral or organic particles. However, the bacterial abundance also depends on the nature of the organic matter contained in the sediment. Bacteria can attain high abundancies and activities only if the organic matter is easily metabolizable. In sediments which contain refractory organic substances, e.g. in deep-sea bottoms, bacteria are less abundant and often inactive. Bacterial populations, especially in the upper layers of the sediment, may exhibit seasonal changes in the respective percentages of different functional groups, mainly correlated to changes in the nature of the decaying organic matter deposited on the bottom. Bacterial abundancies in the upper layers of soft bottoms depend, to a certain degree, on the organic-material input from the pelagial, provided the organics were neither fully mineralized nor transformed into substances resisting bacterial decomposition prior to their deposit. This means that, to some extent, bacterial abundance depends on both plankton production and the duration of descent, which increases with depth. This is documented in Table 9-1.

Table 9-1

Bacterial abundance and biomass per litre of wet sediment in different benthos areas of the Pacific Ocean (Based on data provided by SOROKIN, 1971, 1972, 1973a, 1973b, 1974, 1978, 1979; modified)

Type and location of bottom	Depth (m)	Bacterial number ($\times 10^6$)	Bacterial biomass (mg)
Coral sand, Fanning Atoll	25	910	226
Japan continental slope	2725	96	94
Tahiti continental slope	1100	105	51
Red clay, South Pacific Ocean	5330	7·1	3·4

(iv) With increasing sediment depth, the bacterial abundance tends to decrease as does the numerical ratio aerobes: anaerobes. In general, below 0·4 to 0·7 m, anaerobes predominate. The distribution of aerobes and anaerobes largely depends on the position of the redox-potential discontinuity layer. Below 1 m, the bacterial abundance is very low and does not exhibit significant changes. However, living bacteria have been recorded from cores as deep as 8 m.

(v) Bacteria participate in many ecological processes in the benthal including the micro- and meiobenthal. Together with heterotrophic bacterioplankton, they assist in the regeneration of nutrient salts and make these available for photoautotrophic organisms (Volume IV: SOROKIN, 1978). Soft-bottom bacteria further participate in many other processes, mainly of a geochemical nature, such as the precipitation of calcium carbonate and heavy-metal hydroxides (possibly also in the formation of polymetallic nodules) and formation of humic substances. They may even develop magnetic properties: some marine bacteria are attracted by the south-pointing pole of a magnet bar and repelled by the north-pointing pole. Microbenthic bacteria also control the oxygen content of the interstitial water and its pH. Their most important role in the ecosystem, apart from nutrient–salt recycling, is to provide food for the benthic micro-, meio- and macrofauna by transforming, with an efficiency of 30 to 40%, non-living organic matter (particulate or dissolved; in the latter case through aggregates formation) into living matter which may be assimilated by microphagous consumers.

The **meiobenthos** consists practically only of animals, namely the largest-sized protozoans (mainly ciliates and foraminiferans) and small-sized metazoans.

Functionally, micro- and meiobenthic organisms are so intimately linked with each other and within the assemblage that we must treat them both together.

(2) Sediment as Environment

Irrespective of the diversity of sediment characteristics due to differences in granulometry, exposition, depth, geographical location, etc., which result in locally different ecological conditions for the micro- and meiobenthos, all sediments are heterogeneous from the surface to the subsurface and deeper layers. Within the sediment, most environmental entities exhibit vertical gradients, sometimes interrupted by discontinuities.

The sediment's inhabitability by an organismic assemblage as well as the assemblage composition, depend on the porosity of the sediment, i.e. on internal water and oxygen circulations and size of interstitial spaces. Granulometric characteristics and patterns of bottom-water movement are of particular significance.

The main boundary within a sediment corresponds to the depth at which the oxygen content of the interstitial water is totally depleted owing to consumption by biological activities and biochemical processes. The layer from the surface down to this level (inversion of the redox potential) is known as the oxygenated layer, that below as the reduced layer. On exposed shores, the reduced layer usually begins below 5 to 10 cm in offshore sand (STEELE and co-authors, 1970). But at the low-water mark water circulation within the interstitial environment (sublittoral pumping due to wave-induced pressure changes) is higher. Here, the net rate of downward water flow is a result of tidal movement.

While the oxygenated layer may be several centimetres thick in sandy substrates, it is very thin on muddy bottoms. In sandy substrates micro- and meiobenthic organisms can live both on and in the sediment (interstitial fauna), whereas on muddy substrates they are restricted to the uppermost layer. This system is often called the psammic system and its living component the psammobios.

The reduced layer supports the 'sulphide system'. It is inhabited by a particular assemblage called 'thiobios' (p. 570).

Finally, we must consider a third system near the sea–land boundary. Here, sea-water circulation in sediment interstices originates from percolation processes at higher beach levels caused by breakers, but also by subsoil water from land. In other words, in this 'phreatic' system advective circulation partly combines with percolation. This 'narrow zone of brackish water found in a sandy beach where fresh subsoil water from the land mixes with salt water from the sea' (SWEDMARK, 1964, p. 2) is inhabited by phreatic fauna.

The three micro- and meiobenthic systems, psammic, sulphide and phreatic, are highly complex. The major physical and chemical factors involved have not yet been sufficiently investigated and this also applies to the ecological requirements and tolerances of their inhabitants. Hence, proper delimitation of true assemblages—as attempted in the descriptive part devoted to the macrobenthic system—is still extremely difficult. Consequently each of the micro- and meiobenthic systems will be treated in a more general way.

(3) The Psammic System

(a) General Aspects

Among the three systems the psammic system alone contains photoautotrophic bacteria or algae, the latter constituting the microphytobenthos, which is largely restricted to the most superficial layers of the sediment. Animals of the psammic system—most of them belonging to the meiobenthos—are able to inhabit deeper layers, providing the interstitial water of these layers is sufficiently oxygenated. Since meiobenthic animals also exist in the sulphide and phreatic systems, another term should be used for characterising those restricted to the psammic system. NICHOLLS (1935) introduced the term 'interstitial fauna' later adopted by SWEDMARK (1964), whereas REMANE (1940) proposed the term 'mesopsammon'. Although interstitial fauna and mesopsammon are not completely synonymous they are generally used as such as is the term 'meiofauna', all three referring to the whole meiobenthos inhabiting the psammic system.

(b) Microphytobenthos

In principle, the term 'microphytobenthos' should include all small-sized plant species living on hard or soft substrates. Species living on and within hard substrates are distinguished as 'epilithic' and 'endolithic' microphytobenthos respectively. In practice, 'microphytobenthos' without either adjective is usually employed to characterize only small-sized algae on and within soft substrates.

While on muddy substrates motile or immotile microphytobenthos is generally restricted to the thin film at the water–sea bottom interface; in sandy substrates diatoms, for example, live both on and under the interface. The more frequently the sand is disturbed, the thicker the sand layer is populated with living microphytobenthic species. Very abundant microphytobenthic populations may become macroscopically visible. Sometimes they form a fairly continuous sheet, a 'veil', covering the sea bottom.

The microphytobenthos comprises several large taxonomic units: cyanophytes, euglenophytes and pyrrophytes (Dinophyceae and Cryptophyceae)—all flagellated species, as well as Chrysophyceae, Bacillariophyceae (diatoms) and Xanthophyceae. Living microphytobenthos has been observed in much greater depth than phytoplankton (except for the olive-green cells and similar bodies, p. 40). For example, PLANTE-CUNY (1969) collected living diatoms down to 380 m off the Gulf of Marseilles (France). Spores and resting stages have been sampled from sediments as deep as 7000 m.

Most investigations on microphytobenthos have been carried out in nearshore areas with waters of low transparency and maximum light penetration in the wavelength range 550 to 625 nm ('green radiation'). According to these investigations, the requirements for radiant energy seem to be comparable in neritic phytoplankton and microphytobenthos. Many microphytobenthic species have accessory pigments (e.g. biliproteins in cyanophytes and Cryptophyceae and carotenoids in phytoflagellates and diatoms). All these accessory pigments absorb radiant energy at wavelengths different from those of chlorophyll a absorption. For example, diatom fucoxanthin has an absorption peak between 430 and 560 nm, thus approximately filling the gap between the two absorption peaks of chlorophyll *a*. Accessory pigments seem to be of paramount importance for microphytes occasionally buried in the sediment, as, for example, after gales.

In sandy bottoms photoautotrophic microphytobenthos may also live beneath the sediment surface. For example, in tidal flats of the western Wadden Sea (The Netherlands) CADEE and HEGEMAN (1974) recorded a large quantity of functional chlorophyll below the sediment surface down to 10 cm, only 25% of the total amount being contained in the upper 0- to 1-cm layer. Living algae have also been found down to 15 cm in subtidal sands, where photosynthesis is impossible (FENCHEL and STRAARUP, 1971). Photosynthetic organisms buried in dark sediment may subsist transiently by heterotrophy until waves mix and redistribute the sediment layers again and thus bring them back to the illuminated layer. The sediment acts as light filter. The thickness of the layer populated by microphytobenthos depends on both transparency of the interstitial water and the characteristics of the sediment itself, especially the granulometry, cohesiveness and mineralogical nature (mainly transparency) of the particles. Some, if not the majority of, microphytobenthic species should apparently be considered as sciaphilic.

In intertidal areas TAYLOR (1964) and TAYLOR and GEBELEIN (1964) recorded 10% of the surface illumination at a depth of 3 mm and 1·2 mm in coarse sand and fine sand respectively; GOMOIU (1967) also measured 10% of the sediment surface illumination at a depth of 7 mm in a very coarse sand of northern Black Sea coasts. The data of FENCHEL and STRAARUP (1971) are more accurate; they determined 0·1% of the surface illumination to persist at depths of 3 mm and 6 mm in sediments of mean grain size of 62 to 120 μm and 500 to 1000 μm, respectively. According to the same authors, wavelengths with maximum penetration into clean sand are infrared, next in rank being red, green and blue. Therefore, one may assume that soft-bottom microphytes exhibit multiple adaptations to light quantity and quality.

Photosynthetic bacteria prefer to utilize infrared rays (750–900 nm) and hence always occur below the algal level. Light absorption by pigments of the sandy-bottom microphytobenthos is below 10 to 20%, i.e. lower than in other plant communities (FENCHEL and STRAARUP, 1971). In sediments emerging at low tide the thickness of the photic layer is lower at low tide than at high tide, the refraction index of quartz being greater in air than in water.

In general, the vertical distribution of microphytobenthos depends on three factors: (i) sediment displacement by storms; (ii) grazing pressure from meiofaunal herbivores, also exerted on bacteria; (iii) active migration. Several authors have documented that some dinoflagellates and diatoms exhibit rhythmic activities as a function of the tidal cycle. At low tide they tend to descend below the interface and come again at or nearer the interface, but only when the flow reaches the beach level they inhabit (i.e. where the moistening is sufficient). Two stimuli seem to control these migrations: a light stimulus and a tide stimulus (see also p. 528).

Biological and Ecological Comments on Different Algal Groups

Diatoms

Diatoms (Bacillariophyceae) are by far the best-known component in the soft-bottom photoautotrophic microflora. They may also be the most important among the microflora (bacteria excluded) in terms of abundance, diversity, biomass and production. In the Black Sea coastal microphytobenthos, diatoms largely predominate; BODEANU (1971) estimated from 229,000 to 1.84×10^6 cells m^{-2} in abundance and from 35·2 to 96·5 mg m^{-2} in biomass, on sandy and muddy bottoms. On the north-western shelf of Madagascar, PLANTE-CUNY (1978) recorded 128 to 173×10^6 cells m^{-2}. Biomass and production are more fully discussed on p. 530ff for all microphytobenthos.

Most benthic diatoms belong to the Pennatae and most planktonic diatoms to the Centricae. However, planktonic diatoms often mix with benthic species, mainly on soft shelf bottoms usually in the form of resting stages during periods of lower primary and production in the pelagial. Inoculation of a well-illuminated nutritive medium with a small quantity of sediment rapidly resulted in the premature development of the phyto-plankton spring outburst of many planktonic diatoms (KASHKIN, 1964), e.g. *Skeletonema costatum*, *Thalassiosira gravida*, *T. nordenskjoldi*, *Chaetoceros socialis*, *C. septentrionalis*, together with such planktonic algae as *Phaeocystis poucheti* (Chrysophyte). According to BODEANU (1971) planktonic forms (resting stages or sedimented live cells) represent about 11% of the total abundance of diatoms on the soft shelf bottoms of the Black Sea.

Soft-bottom diatoms on other substrates (rocks, various benthic organisms) may be either sessile or free living, i.e. moving between sand grains. Sessile diatoms attach to sediment particles by an adhesive substance. Free-living species (apparently only Pen-natae) exhibit jerky, zig-zag movements owing to friction of cytoplasmic currents along their raphé. Light requirements and tolerances of diatoms have been well documented. As early as 1942, off Plymouth (UK) on a 70-m deep sandy bottom, MARE observed that a small-sized species of *Cocconeis* was thriving between March and May, and sug-gested that benthic diatoms have a 'compensation point' much lower than phytoplank-ters. According to TAYLOR (1964) and TAYLOR and GEBELEIN (1964) the optimum energy level for diatoms is 0·2 cal cm^{-2} min^{-1}; at 0·0125 cal cm^{-2} min^{-1}, diatoms perform at about 35% of their maximum photosynthetic capacity (see also CADE and HEGEMAN, 1974). In general, it may be assumed that the energy value corresponding to the com-pensation point would be 0·002 cal cm^{-2} min^{-1} for microphytobenthos and 0·15 cal cm^{-2} min^{-1} for phytoplankton, diatoms being predominant in both the systems.

As regards vertical distribution, diatoms follow the same general pattern discussed above, especially in intertidal areas; migration behaviour seems to be common. The nature of the stimulus (or stimuli) inducing vertical movement requires further research. On tidal flats of an estuarine area in Scotland, PERKINS (1960) observed diatoms of the genus *Pleurosigma* rising up to the sediment surface by day and moving to deeper layers at night. He believes that this migration is strictly photoperiodic without any linkage to the tidal cycle.

In the Gulf of Marseilles diatoms attain peaks in diversity and abundance during April and May (PLANTE-CUNY, 1969). She noticed that in shelf areas the specific richness of diatom assemblages is higher than that in the plankton. This might be ascribed to the higher diversity of 'microniches' in the benthal, as well as to the higher and more permanent supply of mineral nutrients, thanks to the mineralizing activities of heterotrophic bacterial populations which are particularly abundant in many soft shelf bottoms. PLANTE-CUNY (1978) also investigated diatom populations off Nosy-Bè (Madagascar) from 0 to 60 m depth and found numerous (about 40) species, mainly large-sized (up to 50–60 μm) Pennatae; the most common genera were *Caloneis*,

Nitzschia, Pleurosigma, Navicula and *Diploneis*. In general, diversity increased with depth; the average cell size was inversely related to grain size, i.e. larger in sand than in muddy substrates.

Phytoflagellates

In general phytoflagellates of the microphytobenthos have been insufficiently investigated. In intertidal sands near Roscoff (France) DRAGESCO (1965) found representatives of coccolithophorids (*Pontosphaera roscoffensis*) euglenoids (*Euglena viridis* and *Trachelomonas abrupta*) and, in particular, dinoflagellates: *Exuviella*, *Amphidinium* spp., *Thekadinium kofoidi*, *Polykrikos lebourae*, *Gymnodinium* spp. Dinoflagellate populations may be very dense, up to 1000 to 2000 individuals cm^{-2} (standing crop about 2 g m^{-2}); several often constitute 'permanent' unispecific patches. According to FENCHEL (1969), species of *Gymnodinium* seem to be photosynthetic and were found close to, or at, the sediment surface. They may be much more abundant than the populations investigated by DRAGESCO: about 6000 individuals cm^{-2} in masses of purple sulphur bacteria (*Beggiatoa*) in Nivå Bay (Denmark). In cleaner sands, species of *Amphidinium*, often colourless, seem to predominate. According to FENCHEL, euglenoids are particularly common in sulphureta (*Euglena*, *Eutreptia*). He also found green species and colourless ones in cleaner sands, but usually only in small numbers.

Cyanophytes

Cyanophytes are a common constituent of the microphytobenthos. According to FENCHEL (1969, p. 71), 'filamentous cyanophyceans were especially found in great numbers close to the surface of sediments with high content of H_2S . . .'. Common genera are *Oscillatoria* and *Lyngbya*. They also live among sulphur bacteria and in mats of *Vaucheria*. In intertropical areas, e.g. within or near mangrove swamps, filamentous cyanophytes may develop thick mats on reducing estuarine sediments. Unicellular cyanophytes also exist together with filamentous forms in such environments, but also live in clean sands, sometimes attached to sand grains.

In the inner lagoon of Moorea Atoll (French Polynesia) on fine sands at 0·2 to 0·8 m depth, SOURNIA (1976) discovered large bottom areas covered with green patches of the blue-green alga *Oscillatoria limosa*. During hours of maximum illumination, this alga disappears beneath the uppermost few millimetres of sediment. Since the local tidal range is extremely small, SOURNIA assumes that this migration involves negative phototaxis rather than an endogenous rhythm. *O. limosa* is not uniformly distributed over the bottom but occurs in green patches, 10 to 50 cm in diameter, separated by colourless areas. In the uppermost 0- to 3-cm layer, the chlorophyll *a* content of the green patches is 127 μg cm^{-2} and primary production 236 μg C cm^{-2} d^{-1}; colourless areas contain only 69 μg chl. *a* cm^{-2} and exhibit a primary production of 53 μg C cm^{-2} d^{-1}. Since the green patches correspond only to about one third of the total surface area of the bottom, the average global estimates for coloured and colourless areas are 900 mg chl. *a* m^{-2} and the gross and net primary production approximately 1·14 g C m^{-2} d^{-1} and 0·62 g C m^{-2} d^{-1} respectively. As is well known, several blue-green algae can fix nitrogen. POTTS and WHITTON (1977) observed nitrogen fixation by heterocystous and non-heterocystous species in intertidal sediments of the lagoon of Aldabra, Indian Ocean.

Biomass and Production

Biomass

Measuring the sediment content in chlorophyll *a* seems to be a better and possibly more accurate way of evaluating microphytobenthic biomass than counting the cells or determining their dry weight. Colorimetric measurements also provide data on phaeophytin *a* and phaeophorbid *a*.

Chlorophyll *a* values recorded from different biotopes vary considerably. Examples are: 227 to 530 mg chl. *a* m^{-2} in the 1·1 cm superficial layer on intertidal sands at Barnstable Harbour, Massachusetts (USA) (MARSHALL and co-authors, 1971); 3·8 to 11·3 µg chl. *a* g^{-1} dry sediment on tidal flats in the Wadden Sea (The Netherlands) (CADEE and HEGEMAN (1974)); 24 to 64 mg chl. *a* m^{-2} in the 1 cm superficial sand layer between 2·5 and 12 m water depths in the Gulf of Marseilles (France) (COLOCOLOFF, 1972). In an atoll lagoon of French Polynesia, SOURNIA (1976) demonstrated a decrease in chlorophyll *a* content with increasing water depth: 236 to 907 mg m^{-2} at 0·5 to 1 m and 56 to 140 mg m^{-2} from 10 to 17 m. Furthermore, the higher the percentage of cyanophytes in the microphytobenthos, the higher the chlorophyll *a* content. However, off Nosy-Bè (Madagascar) PLANTE-CUNY (1978) recorded on 3- to 60-m deep sandy and muddy bottoms maximum chlorophyll *a* values at 10 to 20 m. In general, sandy bottoms contain more chlorophyll *a* than phaeopigments, whereas the reverse is true for muddy bottoms. The average values PLANTE-CUNY gave for all bottoms she studied (3–60 m) are 38·78 mg ± 4·95 chl. *a* m^{-2}, and 49·93 ± 6·60 mg phaeopigments m^{-2} respectively. The thickness of the sediment layer in which chlorophyll *a* occurs decreases with increasing water depth, but phaeopigments exhibit the reverse pattern. Below 300 to 350 m, chlorophyll *a* is lacking in the sediment and phaeopigments alone can be found.

As the sediment becomes coarser the chlorophyll *a* layer gets thicker. Hence, the photosynthetic potential of coarser, and more or less stirred, sediments is often higher than that of finer sediments in more sheltered places, notwithstanding that the latter often exhibit a much higher plant biomass. PLANTE-CUNY (1978) further emphasizes the large variability per square metre due to microphytobenthic patchiness, especially in sediments disturbed by casual storms. Seasonal changes in biomass are also related to alternating dry and rainy seasons. In fact, it appears that chlorophyll *a* content, and consequently the biomass, depend mainly on two factors: stirring of the bottom and illumination. Where water turbidity is strongly increased, e.g. due to river floods or land drainage, illumination decrease controls the chlorophyll *a* content more than sediment stirring. In summary, the microphytobenthic biomass on the shelf ranges between a few tens and several hundreds mg chl. *a* m^{-2}.

Off Nosy-Bè, Madagascar, the biomass of the microphytobenthos is higher than that of phytoplankton of the upperlying waters between 0 and 15 m, but lower in waters deeper than 15 m (PLANTE-CUNY, 1978). Furthermore, the seasonal fluctuations tend to be reversed, because phytoplankton peaks result in increased light absorption by the water column and therefore decrease the amount of radiant energy which reaches the bottom.

Primary production

Primary production values from the literature vary considerably, and comparison between the data is difficult because they have been obtained sometimes via oxygen measurements and sometimes via ^{14}C measurements. For example, on subtidal sandy bottoms primary production ranges from 44–71 (2·5–12 m, Gulf of Marseilles: COL-OCOLOFF, 1972) to 300 g C $m^{-2} y^{-1}$ (lagoon of Majuro Atoll, Pacific Ocean; SOROKIN, 1973). On more muddy bottoms, primary production seems to be lower, e.g. ranging from 31 (intertidal Scotland shores; LEACH, 1970) to an average of 8 g C $m^{-2} yr^{-1}$ (Bay of Concarneau, South Britanny, 5–18 m; BOUCHER, 1972, 1975). Off Nosy-Bè, Madagascar, from depths of about 3 to 40 m, PLANTE-CUNY also observed that primary production on sandy bottoms is about twice that on muddy bottoms, the average global value for both being around 80 g C $m^{-2} yr^{-1}$. On mixed intertidal flats in temperate seas the primary production averages around 100 g C $m^{-2} yr^{-1}$ (Denmark: GRØNTVED, 1960, 1962, 1966; Wadden Sea, The Netherlands: CADEE and HEGEMAN, 1974).

Finally, PLANTE-CUNY's (1978) investigations reveal the following general points: (i) Sandy bottoms seem to be more productive than muddy bottoms, but dark assimilation is similar in both: the maximum value recorded on Nosy-Bè sandy bottoms was 150 g C $m^{-2} yr^{-1}$. (ii) Carbon assimilation always decreases with increasing depth, especially below 15 m (average global value for 0–30 m: 21 mg C $m^{-2} h^{-1}$). (iii) Dark assimilation exhibits only slight changes with depth: average global value is 5 mg C $m^{-2} h^{-1}$. (iv) Below 35 to 43 m the light and dark values of carbon assimilation are very similar. (v) The light–dark difference in carbon assimilation is most obvious in waters shallower than 20 m. (vi) At least at the stations investigated, primary production is clearly related to the chlorophyll *a* content of the sediment surface in shallow waters, but to illumination in deeper waters (Fig. 9-1).

(c) Microfauna and Meiofauna

General Aspects

Among the micro- and meiofauna at the bottom-water interface and in the interstitial environment, we must distinguish between permanent inhabitants, whose whole life-cycle takes place in this environment, and transient inhabitants, i.e. larval stages of macrobenthic species, often referred to as mixobenthos.

The following pages, restricted to permanent inhabitants, are largely based on the review of the interstitial fauna by SWEDMARK (1964). Practically all marine invertebrate taxa are represented in the mesopsammon. Some groups, which are preadapted for mesopsammal life due to their small size (copepods, tardigrades, mites) or their type of organization, comprise many species. Others, e.g. cnidarians, bryozoans, ascidians, consist of only a few species whose morphology deviates more or less from that of the group they belong to.

Among the interstitial protozoans, ciliates are by far the most important group. Except for a few species they belong to the microbenthos. Several hundred species have been described. The distribution of different ciliate taxa in relation to granulometric characteristics led FAURE-FREMIET (1950, 1951) to distinguish three groups of interstitial ciliates: (i) microporal species in sands with predominating grain sizes 0·4 to 0·12 mm;

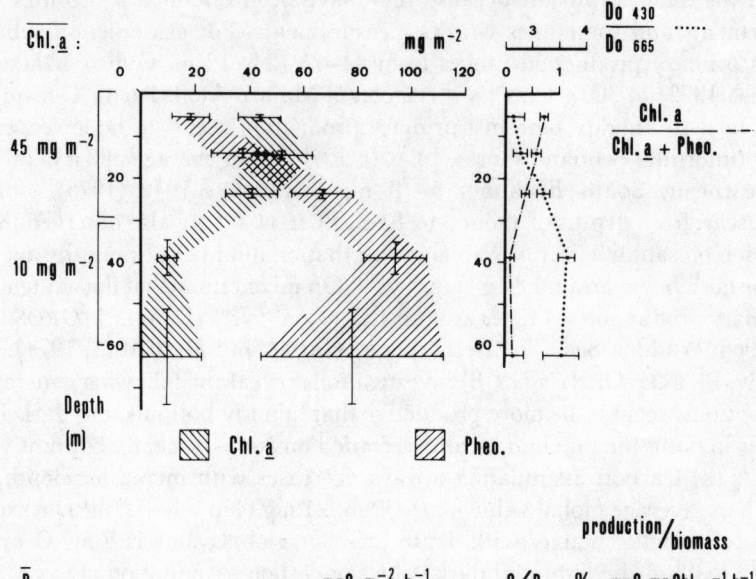

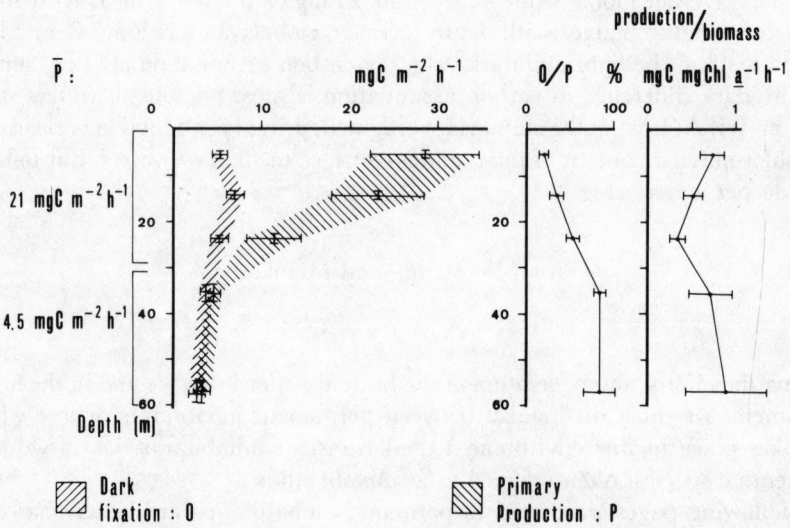

Fig. 9-1: Average distribution of parameters in relation to soft-bottom depth off Nosy-Bé (Madagascar). Upper: Functional chlorophyll *a* concentration and pigment ratio (143 sampling stations, 3–83 m deep). Lower: primary production, dark carbon fixation and production/biomass ratio (138 sampling stations, 3–60 m). (After PLANTE-CUNY, 1978; reproduced by permission of the author.)

(ii) mesoporal species in coarser sands with a dominant grain size of 1·8 to 0·4 mm; (iii) euryporal species which indifferently inhabit coarse and fine sands. In general, the species number in fine sands is about twice that in coarse sands. In White Sea psammophilous ciliate assemblages, true interstitial species predominate in fine-grained (<0·25 mm) sands, whereas eurytopic species (also present in algal ooze, water, kelp thalus) predominate in coarse-grainesd (>0·25 mm) sands (BURKOVSKIY, 1971).

Ciliates exhibit three different types of locomotion: ciliary gliding, swimming and creeping. In the interstitial environment, the first type is most characteristic for microporal species which usually have a threadlike or ribbon-shaped body. Sessile ciliates seem to be absent in the mesopsammon. Many species are cosmopolitans.

With regard to food habits, ciliates may be classified into four groups (FENCHEL, 1968a): (i) bacteriophagous species, some of them feeding only on bacteria of the sulphur cycle and thus mainly present in the thiobios (p. 571); (ii); herbivorous species feeding on diatoms, cyanophytes and phytoflagellates; (iii) carnivorous species, sometimes with venomous trichocysts, feeding on colourless flagellates, other ciliates and even small-sized metazoans, e.g. rotifers; (iv) histophagous species consuming living or dead tissues. Some ciliates are even able to ingest sand grains, digesting their epiphytic flora. In general, carnivorous and histophagous species are not abundant; each group accounts for less than 10% of the total ciliate population. Herbivorous species usually predominate and fully exploit algal resources. In some fine sands FENCHEL (1967, 1968b) observed four different size-classes of epiphytic diatoms, each being grazed upon by a species of the genus *Remanella* whose size turned out to be related to that of the plant cells. The size of the ciliate populations seems largely to depend on algal resources; hence, ciliate populations generally undergo a sharp decrease in winter. In the presence of food shortage, the ciliates can form cysts which may remain viable a long time but cannot tolerate desiccation.

In general, it seems that the most abundant ciliate populations occur in the midlittoral and upper infralittoral zones; supralittoral and 'subtidal' populations are poorer.

Interstitial foraminiferans have been as yet poorly investigated. Some sessile species have been described from coarse sediments. The feeding behaviour seems to be highly diverse: some species perform extracellular digestion, others are photoautotrophic thanks to endobiotic zooxanthellae. However, most species are phagotrophic, feeding on diatoms or other unicellular algae, ciliates, nematodes, nauplii, etc. In general, herbivorous species predominate.

At lower bathyal and abyssal depths a new group of rhizopod protozoans has recently been discovered: the giant-sized Xenophyophores. Their hemispheric colonies are extremely fragile, have anastomosing branches and measure several centimetres across. These foraminiferans are not interstitial; they live on the seabed—not beneath or in it. The density of their populations may sometimes attain a few colonies per square metre.

Hydrozoa are represented by few genera and species. *Halammohydra* and *Otohydra*, which may be regarded as actinulae, move by means of body cilia; *Otohydra* moves in the interstices 'with the help of its tentacles, which are sticky at the ends'; *Psammohydra* exhibits a caterpillar-like movement in the interstices by 'simultaneous contraction and adhesion to the substratum through sticky nematocysts on papillae around the mouth' (SWEDMARK, 1964, p. 16). The majority of hydrozoa seem to be predators.

In most reviews devoted to the interstitial fauna, only the small-sized mesopsammal scleractinian *Sphenotrochus* sp. is mentioned. Members of this free-living species, which frequently exhibit two opposite calices, have been found fairly frequently in coarse sediments with organogenous remnants in the northern areas of the Mediterranean Sea and on the northern coasts of Britanny at depths from 20 to 65 m. According to ZIBROWIUS (1976), this scleractinian cannot be considered as a real mesopsammal species, probably representing as it does the juvenile stage of the macrobenthic (10 mm) *Sphenotrochus andrewianus*.

Turbellaria are common in interstitial biotopes, particularly the Kalyptorhynchia which exhibit highly diversified adhesive organs and a very pronounced development of tactile cilia and sensory hairs, especially at their cephalic and caudal ends. Otoplanids, which inhabit the breaker zone, are quick-swimming predators, feeding on plankters washed up to upper beach levels.

The Gnathostomulida, generally considered as a separate phylum related to Turbellaria, are probably widespread in both psammic and phreatic systems. They have a flattened body ending in a tail with adhesive organs; the pharynx carries cuticular jaws.

Rotifers are rare in the interstitial marine environment; about ten species of the genera *Eucentrum* and *Proales* have been recorded. Some species have also been found in the phreatic system.

Gastrotrichs of the order Chaetonotoidea are less common in marine than in freshwater meiofauna. In contrast, species of the order Macrodasyoidea are exclusively marine. Locomotory mechanisms involve two different patterns: ciliary gliding with the help of ventral cilia and caterpillar-like body movements. Species of *Urodasys* have a tail appendix longer than their body which may act as an anchor and allow them to live semi-sessile, although even without detaching their tails they can exploit a considerable territory (SWEDMARK, 1964).

Nematodes are by far the most abundant (number of families, species and individuals) meiofaunal group, inhabiting almost all types of substrate. Most nematodes move with a writhing motion and are thigmotactic, i.e. requiring mechanical support by surrounding sand grains; they can also attach themselves with the help of caudal adhesive glands. Species of the families Desmoscolecidae and Epsilonematidae move like caterpillars. Pharynx organization may provide some information about the nutritional biology of the species (WIESER, 1953, 1960).

In the vicinity of Marseilles (France) the specific richness of nematodes has been investigated by VITIELLO (1972) in different biotopes from shallow-water muddy bottoms down to the upper bathyal zone (650 m). The species number turned out to be higher in biotopes with low sedimentation rates, e.g. muddy-detritic shelf areas (p. 472ff) and bathyal mud (p. 483ff) which supported 193 and 162 species respectively. The species number is lower in biotopes with high sedimentation rates, such as shallow-water muddy sands in sheltered areas (p. 433ff) and the terrigenous shelf mud which contained 64 and 52 species respectively.

Archiannelids are one of the most characteristic groups in the interstitial environment, although many species of this polyphyletic group are larger than the true meiofaunal species. The best represented families are Protodrilidae and Nerillidae, each of them with about 20 species. The families Dinophilidae and Saccocirridae (several species) are also well represented. In contrast to most other interstitial animals, members of the genus *Polygordius* go through a development with larval stages. Some archiannelids have a wide distribution: *Nerilla antennata* and *Protodrilus chaetifer*, for example, are found not only in Europe, but also on the Atlantic and Pacific coasts of North America (SWEDMARK, 1964).

Among the polychaetes, many members of Syllidae (genera *Syllis*, *Sphaerosyllis*, *Exogone*, etc.) inhabit the interstitial environment, especially fine, well-sorted sands. Specific adaptations to coarser and more mobile sands appear in *Eurysyllis brevipes*; with its broad body and marked capacity for adhesion it moves slowly and in close contact with the substratum (SWEDMARK, 1964). The family Pisionidae is typical of the sand meiofauna and probably cosmopolitan. In the European region, *Pisione remota* lives in organogenous sediments of the 'Amphioxus sand'. It has great adhesive ability and attaches itself with its parapodia (SWEDMARK). The Hesionidae (*Hesionides arenarius*) and the Eunicidae (*Ophryotrocha*) are among the families in which interstitial species are exceptional.

Two families of sedentary polychaetes are restricted to the sand meiofauna: Stygocapitellidae and Psammodrilidae. The latter includes only two species: *Psammodrilus balanoglossoides* and *Psammodriloides fauveli*. *P. balanoglossoides* lives in a thin, transparent mucous tube adhering to sand grains and pushes out its anterior part from the tube in different directions to exploit the surrounding territory. With its pharyngeal pump formed in the peristomium by transformation of longitudinal body muscles, this species can consume diatoms and other benthic microphytes. *Psammodriloides* remains free-living throughout its whole life-cycle and lacks the pharyngeal pump (SWEDMARK, 1964).

Meiobenthic oligochaetes have often been disregarded, but they may represent an important part of the total meiofauna. According to data from various authors (GIERE, 1975), oligochaetes represent 6 to 15% of the total meiofauna in Kiel Bight (FRG), and 3 to 13% in the western Baltic Sea. In surface samples from the wave-wash zone oligochaete abundance may amount to 20%. Calculated as weight data, the relation between oligochaetes and nematodes in some surface samples was about 40 : 60 (GIERE). Experiments with the interstitial enchytraeids *Marionina spicula* and *M. subterranea* indicate a seasonally varying generation time of about one to three months and an annual turnover rate of about three times the standing stock under field conditions. Applied to the richest station investigated in the wave-wash zone, GIERE suggests 'a maximum annual production for meiobenthic oligochaetes (under optimal conditions) of 130 to 150 g m^{-2} and an average value of about 40 to 70 g m^{-2} (25 to 50% of the maximum value)'.

While oligochaetes were formerly generally assumed to be detritus feeders, GIERE (1975) found that pennate diatoms and possibly bacteria and fungi as well constitute the main diet of such small interstitial oligochaetes as *Marionina subterranea* and many common naidids. Ciliates 'seem to represent only an additional, not a basic source of energy in the nutritional spectrum of oligochaetes' (GIERE, p. 173). Such trophic specialization suggests that oligochaete populations may sometimes be limited by scarcity of adequate foods.

In the interstitial environment the most important oligochaetes predators seem to be acarids and turbellarians. As far as the latter are concerned, alternating fluctuations between populations of oligochaetes and turbellarians have been demonstrated.

Nematodes might also be able to feed on oligochaetes, but this has not yet been confirmed definitely. However, in general predatory effects on oligochaetes seem moderate compared with their large numbers in littoral areas (GIERE, 1975). Macrobenthic predators (polychaetes, *Crangon*, nemertines, flatfishes, gobiids, etc.) seem to feed mainly on macrofaunal oligochaetes, but not on interstitial forms.

While most interstitial oligochaetes were recorded from shallow-water and estuarine areas, COOK (1970) found several species of the family Tubificidae on a transect between the United States coast and Bermuda both in the bathyal and the abyssal zones.

Tardigrades are less important in marine than in freshwater interstitial environments. About 20 species were recorded. The most common genus is *Batillipes* which bears clumps of pediculate adhesive discs at its extremities, whereas other tardigrades have claws. Most species seem rather stenohaline, disappearing where the water becomes brackish.

Psammic halacarids are common in shallow waters, but it is often difficult to decide whether they live on the sand or inhabit its interstices. To some extent, they are preadapted to interstitial life by their small size and their hard cuticle. Several species may be found in both the psammic and phreatic environments, e.g. *Halacarus anomalus*.

According to MONNIOT (1968), the major morphological adaptations to interstitial life in halacarids are the following. (i) Elongated, often spindle-shaped, body; disappearance of nodosities, spines and cuticular dorsal plates: the cuticule turns smooth or finely porous. (ii) The appendages become longer and finer. (iii) The body may be twisted, and can better penetrate the sand, due to the soft areas between the chitinous plates. (iv) Eyes frequently disappear. In general, halacarids appear to feed on unicellular algae or detritus.

Crustaceans are so numerous in the interstitial environment that we can present only a brief outline of major taxa here. Discrimination between psammic and phreatic species is sometimes difficult, as some species may exist in both these environments. The Mystacocarida (genus *Derocheilocaris*) (at present considered not as an order but as a very primitive subclass) exhibit regressive and specialized features; their unbranched thoracic appendages are reduced and intert, while the larger cephalic appendages— mainly antennae, mandibulae, and maxillulae—allow them to crawl on the sand grains, assisted by their thick setae. Telescoping of thoracic and abdominal segments plays an important role in locomotion, and the lateral or dorsoventral twisting of the abdomen facilitates directional changes (DELAMARE-DEBOUTTEVILLE, 1960). Mystacocarids are mentioned here because they have sometimes been observed in the mesopsammal, preferably inhabiting the phreatic environment.

Ostracods are common in the meiofauna, particularly in coarse sands. Many species seem to be cosmopolitans. Small in size, some ostracods exhibit morphological adaptations to interstitial life, for example, an elongated and wedge-shaped shell, e.g. the genera *Microcythere* and *Cytherura*.

Copepods are represented by several cyclopoid species of the subfamily Cyclopininae, but most of their meiofaunal representatives belong to the Harpacticoidea, which in psammic environments are usually second to nematodes in species number and abundance. Their cylindrical and elongated body (e.g. in *Cylindropsyllus*, *Paraleptastacus*) is the most striking adaptative peculiarity of mesopsammic harpacticoids.

Among the Isopoda, *Microcharon tessieri* found at depths of 45 to 65 m off Roscoff France) and *Microjaera anisopoda* which lives on both sides of the English Channel at about 20 m may be considered psammic species. The family Microcerberidae seems to be restricted to the phreatic environment. Microparasellids are generally assumed to be of marine origin, and the continental species might be regarded either as relicts from former seas, or as migrants from the sea through subsoil coastal water. A similar assumption has been made concerning the microcerberids whose biogeography suggests that they might be Tethys relicts.

Molluscs are common in the meiofauna. Solenogastrids are represented by the genera *Psammomenia* and *Lepidomenia*; *L. hystrix* regularly occurs in the 'Amphioxus' sand (SWEDMARK, 1964). However, gastropods are the predominant mollusc class in the mesopsammal, both in species number and abundance. Prosobranchial gastropods are represented by species of the family Caecidae. The most thoroughly investigated species (*Caecum glabrum*) is often found in the 'Amphioxus' sand. The adult shell, about 1 mm long, elongated tube-shaped and slightly bent, is well adapted for life in the interstitial environment (SWEDMARK, 1964). ARNAUD and POIZAT (1978), who studied the biology of several Caecidae species inhabiting the 'Amphioxus' sand in the Gulf of Marseilles (France), observed that some of them undertake vertical migrations within the sediment. Their ascent to the most superficial layer in spring is particularly striking. ARNAUD and POIZAT further noticed that massive kills of *C. subannulatum* and *C. auriculatum* may occur in summer, possibly owing to a decrease in interstitial-water oxygen content.

Opisthobranchs are much more diversified than the prosobranchial gastropods. They exhibit such aberrant features that some of them have been formerly described as turbellarians. All the interstitial opisthobranchs are small (1–3 mm), shell-less and gill-less; their bodies are relatively elongated and have 'a marked adhesive ability thanks to mucous secretion of epidermal glands' (SWEDMARK, 1964, p. 31). Most genera and species belong to the order Acochlidiacea with the genera *Microhedyle, Hedylopsis, Parhedyle, Ganitus* and *Unela*. All bear well-developed cephalic appendages and seem to be restricted to the psammic environment except for *Unela odhneri* which inhabits the phreatic environment. The order Philinoglossacea includes two psammic genera: *Philinoglossa* and *Pluscula* differing from the Acochlidiacea in several anatomical features. Both orders seem to be related to tectibranchial gastropods. They mainly inhabit coarse ('Amphioxus') sand, but some species have also been recorded from the wave-wash zone ('Otoplanid zone', p. 534).

Other interstitial forms may be related to nudibranchial gastropods of the order Aeolidiacea. Their typical features are easily recognized in the genus *Embletonia*, with its well-developed dorsal papillae; these are strongly reduced in *Pseudovermis*. It has been assumed that the latter genus feeds on interstitial hydroids.

The taxonomic status of the family Rhodopidae (genera *Rhodope* and *Rhodoplana*) is still under debate. Formerly related to opisthobranchial of the order Doridacea, these gastropods (whose tegument contains calcareous spicules) have come to be considered as Pulmonata. They are not strictly psammic, having also been recorded from algal biotopes in shallow waters. All the interstitial opisthobranchial gastropods mentioned above, together with the Rhodopidae, have a very wide geographical distribution.

Three other phyla that only contain one or two interstitial species must be mentioned: The insufficiently investigated *Gwynia* seems to be the single representative of brachiopods in the interstitial environment. The psammic bryozoan *Monobryozoon ambulans*, discovered by REMANE (1938) in sandy bottoms near Helgoland is much better known: it is a solitary, free-moving species, exhibiting about ten tube-shaped growths at the lower part of the cystid. While each growth can serve as stolon for reproduction by budding, the growths also serve as adhesive organs and are used in locomotion. According to SWEDMARK (1964), adhesion is achieved by terminal cells on the stolon and locomotion by contraction of stolonar muscles. Another species (*M. limicola*), featuring root-like stolons without terminal adhesive cells, has been described by FRANZEN (1960) from muddy bottoms of the Gullmar Fjord (Sweden). Echinoderms are only represented by the small-sized holothurioid *Leptosynapta minuta*.

Psammic ascidians deserve particular attention, because these protochordates, thoroughly investigated by MONNIOT (1965) and MONNIOT and MONNIOT (1975), exhibit striking adaptations to interstitial life. The body size is so reduced that the organs, although often very simplified, sometimes seem compressed. A small number of eggs is released and protected by incubation processes, allowing them to develop and sometimes even metamorphose in the maternal body; thus, once released, the larva can attach itself immediately (MONNIOT and MONNIOT, 1975). As adults all these species exhibit some remnants of larval features (e.g. a large-sized nervous ganglion) and hence must be considered neotenic. The interstitial ascidians thus far recorded (ca 12 species) can be divided into two groups. The first group comprises members with a hard and thick tunic adhering to the mantle; their body is rounded and their siphons, highly contractile, are very far apart, sometimes even opposite. These animals are free-living, although their rhizoids can attach to some sand grains. The second group includes species whose shape is not clearly defined. Their tunic, soft and translucent and attached only to the mantle in the vicinity of the siphons, does not carry rhizoids but can fix sand particles; the group members 'live attached between flat sand grains which they use to hide themselves, opening them like a hinge' (MONNIOT and MONNIOT, p. 115). These two groups correspond to species belonging to two different orders of Ascidiacea: Stolidobranchiata and Phlebobranchiata respectively.

MONNIOT (1965) emphasizes that interstitial ascidians are common only in bottoms which contain a high percentage of coarse particles. On the shelf they mainly inhabit the 'Amphioxus' sand (p. 450) quite devoid of silt or mud and with grains of irregular shape. In the deep sea they mainly live on bottoms where large-sized foraminiferans are common. They are generally associated with a rather diversified meiofauna: e.g. archiannelids, polychaetes, ostracods, copepods, and nematodes. Interstitial ascidians seem to become abundant only in bottoms with a high content of such heavy metals as schistore sand and gravels, and in areas of volcanism. MONNIOT noticed that the heavy-metals content is higher in the interstitial water than in the upperlying waters, and that the water-column copper content minimum falls in October, i.e. exactly during the period in which these interstitial ascidians exhibit maximum activity (fast growth, breeding). MONNIOT thus assumes that heavy metals such as copper and zinc may block some enzymatic mechanisms as soon as their content exceeds a well-defined threshold.

Interstitial ascidians are filter feeders like almost all the species in this class. They consume small algae, diatoms and others, and probably also bacteria.

Adaptations of Interstitial Animals

Morphological adaptations

The first and most obvious adaptation to interstitial life concerns body size. Except ciliates—where elongated forms may reach up to 4 mm in length (*Helicoprorodon*)—all interstitial animals which do not belong to 'preadapted groups' (i.e. those in which small size is the rule) are characterized by a decrease in the body size. In general, body size does not exceed 2 mm. Examples are the hydroid *Psammohydra nana* (1 mm), the archiannelid *Diurodrilus minimus* (0·35 mm), the prosobranchial gastropod *Caecum glabrum* (2 mm) and the synaptid holothurioid *Leptosynapta minuta* (1 mm) (SWEDMARK, 1964). Decrease in body size results in a correlative organizational regression in comparison with that of related forms with larger bodies. As an example, SWEDMARK mentions the reduction in the tentacle number from 36 to 7 in a series of *Halammohydra* species. Such 'regressive evolution towards smaller body dimensions is eventually bound to jeopardize the possibility of survival of an extremely simplified form' (SWEDMARK, p. 7). A body size of about 0·5 mm is all but the lowest limit for most invertebrate phyla. Regressive evolution may sometimes result in the disappearance of important organs characteristic for the adult stage; hence the resulting oversimplified interstitial form becomes similar to the juvenile stage. Such neoteny exists in the archiannelid *Psammodriloides*, which lacks pharyngeal sac and nephridia in adults of the less regressed genus *Psammodrilus*. Morphological peculiarities such as a ciliated epidermis (*Halammohydra* and *Otohydra*), or transverse ciliations, which are common in adult archiannelids, must also be regarded as retaining larval features. Even in groups like harpacticoid copepods, interstitial species may be smaller (0·3 mm) than their non-interstitial counterparts.

The body shape is frequently flattened and broad, but vermiform animals are obviously best adapted for life in the interstitial environment. A tendency towards vermiform body shapes is evident in such groups as turbellarians, nematodes, copepods, and ostracods, and becomes even much more pronounced in groups in which such a body shape is unusual, e.g. *Halammohydra vermiformis* among the coelenterates and *Pseudovermis* among the opisthobranchs (SWEDMARK, 1964). Vermiformity has also been recorded in halacarids, in particular among those that are truly interstitial; members of *Acarochelopodia* and *Actacarus*, for example, are spindle-shaped (MONNIOT, 1968).

Since many mesopsammic forms have a fragile body wall, they must develop other means for protecting themselves against the rugosity of sand grains. One such mechanism is based on a high ability for quick contraction, e.g. in some elongated ciliates (*Gelcia* and *Trachelocerca*) and may vermiform turbellarians, gastrotrichs and opisthobranchial gastropods. Another mechanism is reinforcement of the cuticle (nematodes, harpacticoids, ostracods) which sometimes differentiate spines or scales (some gastrotrichs). However, according to MONNIOT (1968), a reverse tendency prevails in interstitial halacarids.

Adhesive organs are a further characteristic feature of many interstitial species. In arthropods, adhesion to sand grains is often effected by the appendages which develop claws or hooks. In other groups adhesion is performed by epidermal glands. According to SWEDMARK (1964), the latter are either distributed over the body or concentrated in certain areas (some turbellarians) and often specialize in papillae or adhesive tubes connected to muscles (turbellarians, gastrotrichs). Static organs—obviously important in an environment which often tends to be disturbed—are common, e.g. in the hydroids *Halammohydra* and *Otohydra*, several archiannelids of the genus *Protodrilus*, nemerteans, opisthobranchial gastropods and the holothurioid *Leptosynapta minuta*. Eyes are invariably reduced and often almost absent. The body tends to become depigmented.

Biological adaptations

Locomotion. Semi-sessile and non-motile forms seem to be relatively rare in the mesopsammal. Among the latter there are some foraminiferans, the scleractinian *Sphenotrochus* and the brachiopod *Gwynia capsula*. The bryozoan *Monobryozoon ambulans* and some ascidians are examples of semi-sessile species. Semi-sessile and non-motile forms seem to exist only where interstitial-water circulation and hence food supply are satisfactory, but they are completely absent in the most wave exposed biotopes such as the surf zone.

Motile species predominate in the interstitial environment. In general, crustaceans move by crawling, whereas most nematodes and *Polygordius* species move by writhing, as does the elongated harpacticoid *Cylindropsyllus laevis*. Both crawling and writhing require mechanical support provided by sand grains. Locomotion by ciliary gliding is widespread, obviously in ciliates, but also in hydroids (*Halammohydra* and *Otohydra*), turbellarians, gastrotrichs, archiannelids, polychaetes and opisthobranchial gastropods. Caterpillar-like locomotion, some examples of which have been given above, is less widespread.

As regards **nutritional ethology**, SWEDMARK (1964) classified interstitial animals into four groups: (i) Predators, such as hydroids, many turbellarians (Otoplanidae, Kalyptorhynchia), some nematodes, some ciliates feeding on other ciliates, and even some foraminiferans which seem to be strictly carnivorous and consume ciliates, crustacean nauplii nematodes and on occasion other foraminiferans (WIESER, 1953). (ii) Diatom and epi-growth feeders that feed mainly on diatoms and flagellates but probably also on epiphytic bacteria. They were subdivided by REMANE (1933) into four types: browsers, e.g. ostracods, harpacticoids, archiannelids, molluscs; pump-suckers (some turbellarians, gastrotrichs and nematodes as well as the polychaete *Psammodrilus*); puncture-suckers (tardigrades); sand-lickers, i.e. some small-sized but possibly not really interstitial amphipods and cumaceans. (iii) Detritus feeders, e.g. some foraminiferans, gastrotrichs, nematodes and archiannelids. (iv) Suspension feeders, uncommon and always small semi-sessile and non-motile forms, comprise the bryozoan *Monobryozoon*, the brachiopod *Gwynia* and the ascidians. To these four groups may be added some foraminiferans, e.g. *Marginopora vertebralis* and *Amphisorus hemprichi*, which belong to group (ii) but must be considered to be partly photoautotrophic because their cytoplasm contains numerous zooxanthellae (*Gymnodinium obesum* and *G. rotundatum*) (ROSS, 1972); in the latter species PLANTE-CUNY (1978) found 200 to 250 μg of chl. a g^{-1} dry matter. Some foraminiferans exhibit extracellular digestion.

As regards **reproduction**, the most striking characteristic of the interstitial fauna is its low fecundity, presumably related to small body size. Many produce only one egg at a time and have relatively few spermatozoa, e.g. several gastrotrichs and archiannelids such as *Diurodrillus, Nerillidium* (SWEDMARK, 1964). Most mesopsammal species produce less than ten eggs at a time; the production of more than 100 eggs is highly exceptional.

Brood protection and a trend towards viviparity are important prerequisites for forms with low reproduction rates. Such adaptations may be found in interstitial invertebrate groups in which they are otherwise unknown (SWEDMARK, 1964). Examples are the hydroid *Otoplana vagans* which incubates one or two embryos in a pouch; the gastrotrich *Urodasys viviparus*; the brachiopod *Gwynia capsula* which incubates two to four embryos in its coelome; archiannelids of the family Nerillidae, whose embryos remain attached to the female until they have attained the four-body-segment stage capable of independent nutrition.

The eggs may either be released freely and individually into the ambient water (e.g. *Halammohydra*, gastrotrichs, *Diurodrilus, Trilobodrilus*) or quickly become sticky and attach themselves to sand grains, such adhesion favouring population recruitment within a limited area. In other species, such as many turbellarians, certain species of *Protodrilus* and opisthobranchial gastropods (*Philinoglossa, Pseudovermis, Microhedyle*. etc.) eggs are released in cocoons or glutinous envelopes, which immediately adhere to sand grains; within the cocoon development usually proceeds to advanced stage (SWEDMARK, 1964).

Pelagic larvae are unusual in interstitial animals and have been observed only in about 2% of the species (REMANE, 1952). Among these are members of the genera *Polygordius* and *Protodrilus*. After brief development within the cocoon, *Protodrilus adhaerens* transforms into a trochophore larva which is typical of the genus. Opisthobranchial gastropods of the genera *Philinoglossa, Pseudovermis* and *Microhedyle*, also begin to develop within the cocoon; the hatching veliger larvae is very similar to its planktonic counterparts but does not react to light stimuli. SWEDMARK therefore assumes that 'they remain in the interstitial environment in the area to which the population is restricted' (1964, p. 11). Benthic larval stages are also known from *Halammohydra* (planula) and *Psammodrilus* (trochophore). According to SWEDMARK, all 'benthic larvae contain more yolk and are more unwieldly and less active than the corresponding pelagic larvae'.

The reproductive period extends over most of the year. This may compensate to some extent for the limited production of gametes. However, this view requires qualification.

Firstly, in species whose uninterrupted reproduction is well documented, seasonal changes in reproductive rate may occur. For example, GERLACH and SCHRAGE (1971) found longer generation times in cultivated nematodes in winter. Very low temperatures induce a trend towards viviparity, the eggs being retained in the uterus during early development. GERLACH and SCHRAGE assume seasonal changes in the generation to be fairly widespread in meiobenthic animals.

Second, successional reproduction prevails in both sand-dwelling harpacticoids (LASKER and co-authors, 1970; HARRIS, 1972a,b) and in fine-sediment detritus living harpacticoids (HEIP, 1973). The presence of ovigerous harpacticoid females and nauplii throughout the year is no proof of year-round reproduction. Over a two-year period COULL and VERNBERG (1975) continuously investigated meiobenthic harpacticoids in a muddy and a sandy subtidal bottom of a South Carolina (USA) estuary and demonstrated that even if ovigerous females and nauplii were invariably present, they did not belong to the same species in the different seasons. They found a successional pattern of reproduction there, particularly evident in the two closely related *Halectinosoma* sp. and *Pseudobradya pulchella*. COULL and VERNBERG conclude that these species

'have effectively partitioned the available [food resources and] fractionated the niche temporally through alternating reproductive cycles . . . Thus, the generaliza- tion that year-round reproduction is the rule (or the forms with a low production of gametes) is not necessarily valid' (p. 292).

Year-round reproduction might be more widespread in intertropical meiobenthos, as, for herbivorous species, the food supply fluctuates less on an annual scale. In contrast, on temperate shores seasonal reproduction appears to be more common. However, in South Carolina (USA) COULL and VERNBERG (1975) observed two meiobenthic copepods (*Microarthridion littorales* and *Leptastacus macronys*) that breed throughout the year. They considered that this might be related to food-supply versatility, possibly in combination with eurythermy, i.e. their ability to exploit the food available over a large temperature range.

(d) Factors Affecting Meiofauna Abundance and Distribution

Granulometric Substrate Characteristics

The first general account of relationships between meiofauna composition and granulometric characteristics of the sediment was produced by WIESER (1959). He studied the distribution of meiofaunal metazoans in the sands of Puget Sound, Washing- ton State (USA). According to WIESER, a mean grain size of about 200 μm represents a critical threshold, with the exception of nematodes, which easily tolerate smaller average values. In general, species which glide in the interstices particles are more abun- dant in coarser sands, whereas 'burrowing' species, i.e. those able to displace the particles, predominate in finer sands. According to WIESER, approximate grain-size thresholds would be: *Protohydra* (burrowing) <200 μm; gastrotrichs (gliding) >100 μm; archiannelids <200 μm; ostracods in general >180 μm, juveniles >260 μm. Taking into account ciliates as well, FENCHEL (1969) lowered the upper threshold for meiofauna comprising only nematodes to 90–100 μm in well-sorted sands. According to him, fine sands (125–250 μm) would be mainly inhabited by ciliates (whose biomass often exceeds that of metazoans), and medium-sized and coarser sands by small-sized metazoans.

Among the sediment characteristics which directly control interstitial meiofauna abundance, the more important ones are porosity (CRISP and WILLIAMS, 1971) (aeration during low tide) and the size of interstices, which mainly controls the body size of the inhabitants. The abundance of epiphytic microflora available to epistrate feeders is also important. Granulometry seems to control nematode distribution, both directly as regards population abundance—and, to some extent, diversity—and indirectly by determining the type of food available: for example, the higher the sediment content in silt and clay, the more abundant the epistrate feeders. Moreover, nematodes with strong cuticular ornamentation are more abundant in coarser silt-free sediments (WARD,

1975). In the Mediterranean Sea grain-size sorting—depending on waves, tide and currents—mainly controls meiofaunal abundance on tidal beaches, whereas on atidal beaches competition and predation play the major role (HULINGS and GRAY, 1976).

Although postulated by many authors, a direct correlation between grain size and meiofauna distribution seems questionable. Rather, the size of interstices between the grain particles, which is poorly correlated with granulometric frequence (except for well-sorted sediments), significantly controls interstitial meiobenthos distributions. When the interstices have become increasingly clogged by silt and/or clay, meiofauna abundance decreases, while the percentages of nematodes increases. In coarser sands harpacticoids are more abundant and even predominate. Meiofauna diversity is higher in coarse sands but no species predominate.

Chemical Sediment Characteristics

Chemical characteristics of the sediment may influence meiobenthic distribution. RENAUD-DEBYSER (1963) observed that sand from coral-reef areas transferred to a beach of siliceous sand is colonized within a few days by a meiofauna which is quantitatively very similar to that inhabiting the beach, but qualitatively different. For example, harpacticoids are represented by only two species instead of 12 in the siliceous sand.

If not associated with chemical toxicants—e.g. surfactants, often present in domestic wastes—increasing contents in organic matter (up to about 4%) result in increasing meiofauna abundance; above 5 to 10% decreases. However, organic-matter content and abundance are not significantly related, which suggests that trophic relationships between meiofauna abundance and organic matter presumably depends as much on its quality as on its quantity, as well as on bacterial activities (see also p. 567).

Temperature

Temperature fluctuations are rapidly dampened with increasing sediment depth. Extensive temperature changes are largely restricted to intertidal biotopes. The ability of the different group or species to withstand such temperature changes is probably a major factor controlling the zonation of the fauna (HARRIS, 1972a). The effects of seasonal changes are considered on p. 546. For a comparison between intertidal and subtidal meiofauna. Observations in restricted areas with supranormal temperatures owing to thermal release from power plants indicate that these may induce decrease or disappearance of some harpacticoïd populations. Freezing usually does not result in catastrophic consequences except on shores where it constitutes a highly unusual phenomenon.

Salinity

Similar to temperature changes, salinity fluctuations are dampened with increasing sediment depth, even in intertidal areas. As with temperature, the ability of most meiofaunal species to perform vertical migrations reduce the detrimental effects of extreme salinities. Maximum salinity stress is encountered at the uppermost beach levels during rain periods. However, heavily (subsoil) freshwater input of continental origin can lead to a dangerous salinity decrease, at mid-beach levels as well (McINTYRE, 1969).

In estuarine areas, salinity stress can profoundly influence meiofauna composition, and its abundance tends to decrease (about 4 to 10 times; FENCHEL, 1969).

Other Parameters

Among the other parameters which may affect or even control vertical distribution of meiofauna populations the oxygen content of the interstitial water and the food supply should not be overlooked. In general, such animals as harpacticoids and ostracods tend to require well-oxygenated interstitial waters, while others, such as nematodes, are usually more tolerant to lower values of dissolved oxygen. Food supply has repeatedly been shown to control distributions and abundance. It receives special attention in the section devoted to the different meiofaunal assemblages and the trophic net.

(e) General Features of Meiofaunal Assemblages on Shelf and Upper Slope Soft Substrates

Intertidal Meiofauna

In intertidal areas wave exposure and tidal regime effectively control assemblage composition. Sediment sorting and fluctuations in temperature and salinity owing to periodic emergence combine in controlling the patterns of vertical and horizontal distributions of the meiofauna (see also p. 550).

Vertical distribution usually reveals maximum abundance values in the most superficial (a few cm) sand layer, which is the best oxygenated. Salinity decrease (by rain) and temperature increase (during air exposure) tend to force the meiofauna down. On tropical beaches where the warming of the sand is more marked, the abundance maximum may occur as deep as 20 to 25 cm. On tidal beaches, the meiofauna tends to migrate into deeper sediment layers along the lines which successively correspond to the tidally moving breaker zone. In muddy substrates, maximum abundance is restricted to the 0- to 1-cm layer for harpacticoids, whereas nematodes inhabit the deepest layers of the true psammic system.

Horizontal distribution on sandy beaches in comparatively sheltered areas reveals abundance maximum of the most common groups (nematodes, harpacticoids and ostracods) in the vicinity of the mean low-tide level. On wave-exposed beaches the abundance maximum usually occurs at the mean sea level. According to SWEDMARK (1955), the intertidal meiofauna of northern Britanny coasts—i.e. on beaches where the terrigenous elements always predominate in the sediment—exhibits two different ecological levels: (i) an upper level corresponding to the midlittoral zone (p. 18) with rather rare meiofaunal populations distributed in patches; (ii) a lower level seeming to correspond to the upper fringe of the infralittoral zone where the populations cover more extended areas and are more abundant and homogeneous.

Subtidal Shallow-Water Meiofauna

Except for emergence, the same factors control the distribution and abundance of the meiofauna in subtidal shallow bottoms. Sediment characteristics (granulometry and mobility) are of particular significance here.

In the coarse, organogenous sand, the vertical distribution depends on the thickness of the layer stirred by waves and bottom currents, i.e. of the well-oxygenated zone. A rich meiofauna may often be observed down to 20 to 25 cm beneath the sediment surface. In terrigenous fine sands and in muddy sands the meiofauna is abundant only in the uppermost sediment layer (about 0–2 cm).

Horizontal distributions on bottoms which mainly consist of terrigenous sands are similar to those described by SWEDMARK (1955) for the upper fringe of the infralittoral zone, i.e. abundancies are rather high, but diversities low. In coarser sands with a rather high percentage of organogenous remnants such as broken shells, bryozoans and calcareous algae, the meiofauna is much more diversified but of lesser abundance. On northern Britanny coasts, this latter assemblage begins at a depth of 15 to 20 m (SWEDMARK). On coasts with a small tidal range, the respective distribution of fine terrigenous and coarse organogenous assemblages depends primarily on both bottom-water movements and sediment composition.

Estuarine Areas

Meiofauna horizontal distributions in an estuarine area are exemplified here by referring to NYHOLM and JANSSON (1973), who investigated the Swedish west coast (Table 9-2).

Table 9-2

Meiobenthos (0·2–2 mm) distributions at polyhaline sampling stations of Kungsbackafjorden (Sweden), October 1964 to November 1965. N: Nematoda; F: Foraminifera; P: Podoplea (Harpacticoidea and Cyclopoidea) (After NYHOLM and JANSSON, 1973; modified; reproduced by permission of Zoon, Institute of Zoology, Uppsala)

Station	Depth (m)	Abundance (ind. m^{-2})	Wet weight (mg m^{-2})	Dominating groups			
				NP	NF	FP	Σ
60 A	3	4,530–38,730	24·5–271·9	7	2	0	9
60 C	4	16,330–87,470	128·9–691·6	2	6	1	9
63	16	16,100–147,800	102·9–637·2	8	1	0	9

(f) Seasonal Fluctuations and Quantitative Distribution of Interstitial Fauna
of Shelf and Upper-slope

Seasonal Fluctuations

In addition to the influence exerted by environmental factors (temperature, salinity, emergence), seasonal changes in meiofauna abundance and distribution may depend on biological factors.

On an exposed sand beach at Whitsand, Kent (UK), HARRIS (1972a,b) investigated the intertidal meiofauna down to a depth of 50 cm over a two-year period. Most species attained maximum abundancies in summer (Fig. 9-2), and vertical distributions

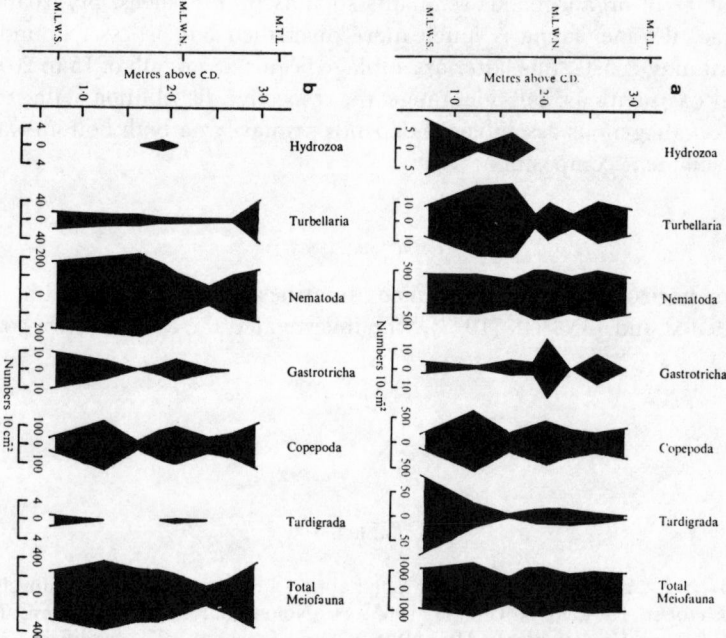

Fig. 9-2: Distribution of major meiofauna groups on an inter-
tidal transect at Whitsand Bay (East Cornwall) on 24
September 1968 (a) and 8 February 1970 (b). (After
HARRIS, 1972a; reproduced by permission of Marine Bio-
logical Association of the United Kingdom, Cambridge
University Press.)

revealed significant seasonal changes: for example, most copepods inhabited the upper
layers in summer but migrated deeper in winter (as soon as the temperature decreased

below 4 °C: PERKINS, 1958). Seasonal changes in abundance seem related to the direct influence of temperature on breeding, but also to seasonal fluctuations in the food supply, i.e. of diatoms and, possibly, also of bacterial abundancies. The breeding peak generally occurs when the population concerned has migrated to the highest habitable level. In general, nematodes exhibit less pronounced seasonal fluctuations. In other meiofaunal groups (e.g. copepods, ostracods) maximum abundance is often observed only when the nematode population declines. Among the crustaceans, copepods predominate in summer, ostracods over the rest of the year.

SKOOLMUN and GERLACH (1971) investigated seasonal fluctuations of nematodes on an intertidal sand flat in the Weser estuary, German Bight. Their 600-cm³ samples (8 cm deep) contained from 58 to 800 individuals, usually 90 to 290. The abundance was low in March and September to October, but high from April to August and from November to February. In summer, nematodes were most abundant in the top layer of the sediment while in December and January abundance was below 5 cm depth. The nematode fauna seemed to be somewhat impoverished. Since almost no juveniles were found, it may be assumed that the population studied received recruitment from elsewhere.

An intermediate situation between that in intertidal and subtidal areas is exemplified by investigations of HULINGS (1971) on Sindbad Beach (Lebanon). Here the meiofauna revealed two seasonal abundance peaks: in spring (March–May) and in late summer–autumn (August, September–November). In general, the second peak was less marked. Seasonality in the wave-wash zone tends to mark seasonal fluctuations of individual taxa. Turbellaria, Gastrotricha, Nematoda (except the family Epsilonematidae) and Oligochaeta exhibit one seasonal peak, but Epsilonematidae and Ostracoda exhibit two. The Harpacticoida do not appear to exhibit seasonality. Since increased bacterial abundance in this unpolluted beach coincided with part of the spring peak, HULINGS attributes it to an increase in the food supply. This assumption is supported by the fact that on Khalde Beach, which is polluted with sewage, the meiofauna is one or two orders of magnitude higher than on Sindbad beach, but does not exhibit seasonality. Nevertheless, the relationship between food supply and meiofauna abundance should not lead to an underestimation of the direct temperature influence, even though the relation temperature/meiofaunal abundance remains unresolved, as the increased abundance is associated with increasing water temperature (spring) and/or with the maximum temperature followed by a thermal decrease (autumn). 'This suggests a wide differential response to temperature as a triggering or controlling mechanism for reproduction for total meiofauna and certain taxa' (HULINGS, pp. 331–2).

In subtidal shelf areas, seasonal fluctuations in meiofaunal abundance tend to be less pronounced. However, even at 220 m deep in a Norwegian fjord with small temperature fluctuations, the breeding period of benthic calanoids seems to occur in winter when the temperature reaches its maximum value (McINTYRE, 1969).

In the German Bight, seasonal fluctuations in a sublittoral meiofauna assemblage, as regards the specific richness in nematodes (representing 93–99% of the total meiofauna), exhibits two maxima (August and April) and a minimum from September to February. The overall nematode population follows the same pattern. The average meiofaunal abundance in the cold months (November–May) and warm months (June–October) did not differ significantly (JUARIO, 1975).

Absence of seasonal meiofauna changes in sublittoral muddy bottoms has been reported and attributed to the higher content in organic matter. However, for coastal terrigenous mud (32·5 m depth, off Banyuls-sur-mer, Mediterranean Sea) DE BOVEE and SOYER, 1974) reported rather large abundance fluctuations in the total meiofauna from 2900 ind. 10 cm^{-2} during the colder months (January–March) to an average abundance of about 6000 ind. 10 cm^{-2} during the warmer season (June–December), with a peak of 8800 ind. 10 cm^{-2} at the end of June. Nematodes largely predominated and their seasonal variations followed the same pattern (Fig. 9-3). Some 97% of the total

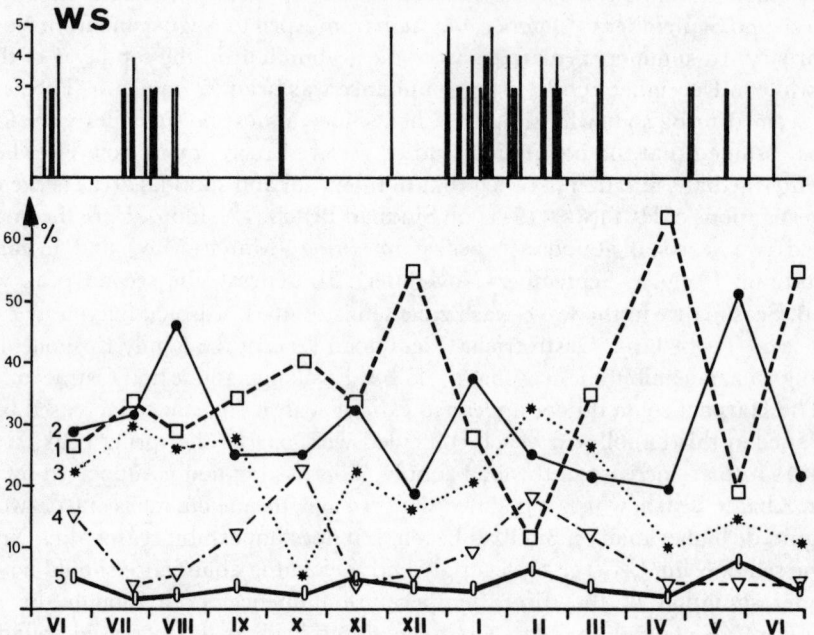

Fig. 9-3: Annual fluctuations of nematodes percentages in the 1- to 5-cm thick substrate layer in relation to sea heaving from eastwinds off Banyuls (Mediterranean Sea) from June 1971 to June 1972 (ws: wind strength in Beaufort scale 3–5). (After DE BOVEE and SOYER, 1974; reproduced by permission of the authors.)

nematode population occupied the layer from 0 to 5 cm, 86% of the harpacticoid population the uppermost centimetre. While similarly distributed, the kinorhynchs featured some individuals down to 8 cm. Analysing the data on the changes in vertical distribution—mainly with regard to nematodes—DE BOVEE and SOYER noticed that the changes do not seem to correspond to a seasonal rhythm, but might be related to periods of heavy storms, mainly east-winds (Fig. 9-4). Abundance decrease in the uppermost (0–1 cm) layer is combined with an increase in underlying layers which sometimes may become the most populated. It seems that such distributional restructuring originates

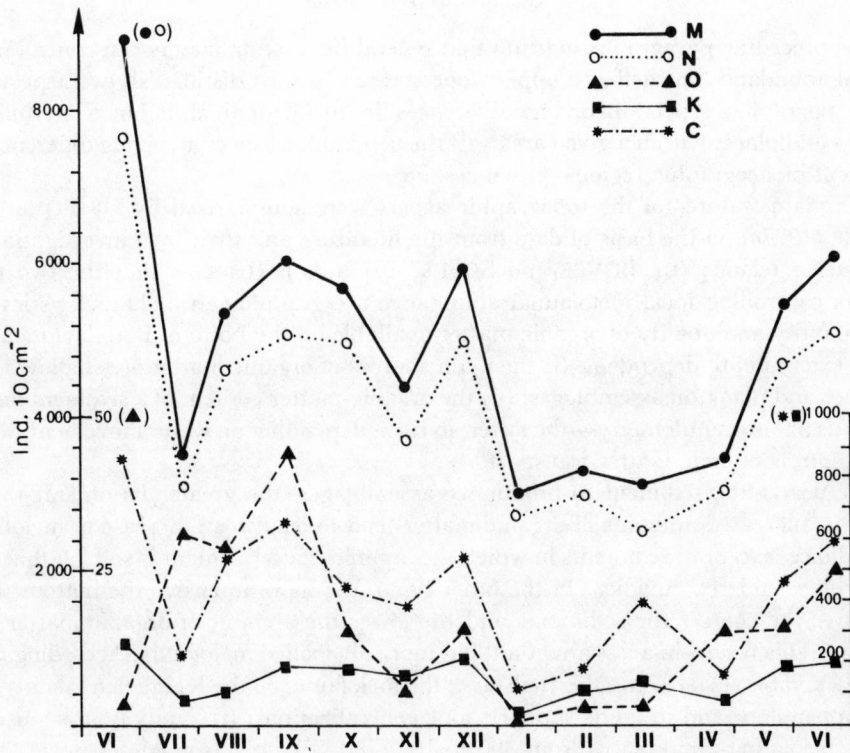

Fig. 9-4: Annual cycle of meiobenthos abundance (ind. 10 cm^{-2}) from June 1971 to
June 1972 off Banyuls (Mediterranean Sea). M: total meiobenthos; N: nematodes;
O: ostracods; K: kinorhynchs; C: copepods. (After DE BOVEE and SOYER, 1974;
reproduced by permission of the authors.)

from both stirring of superficial sediment layers (possibly down to 5 cm) and lowering of
the discontinuity layer of the redox potential.

The presence of seasonal differences in meiofaunal abundance from muddy bottoms
on the lower part of the Scottish shelf (101–146 m) led MCINTYRE (1964) to suggest
that the annual temperature range might be too small to induce a seasonal reproduction cycle. The discrepancies between the interpretations of MCINTYRE and
DE BOVEE and SOYER (1974) were considered by SOYER (1971) who analysed more
than 300 cores on soft bottoms from 35 to 550 m deep off Banyuls-sur-Mer, Mediterranean Sea. He noticed that the quantitative meiofauna distribution exhibits a sudden
decrease in abundance at around 55 m and attributed this to reduced seasonal temperature changes below 55 m. Not denying a direct thermal influence, the abrupt decrease in
microphytobenthos abundance reported by PLANTE-CUNY (1969) on lower shelf sediments of the Gulf of Marseilles may play a major role in meiofauna impoverishment.

Quantitative Distribution

The preceding paragraphs indicate that several interacting factors can control meio-faunal abundance on shelf and upper slope areas. We must distinguish two aspects: (i) the topographic aspect involving differences in meiofaunal abundance at different depths and places within a given area; (ii) the geographic aspect involving differences in different biogeographic regions, provinces, etc. .

The main features of the topographic aspect were summarized by DE BOVEE and SOYER (1977c) on the basis of data from the literature and their own investigations at Kerguelen Islands (DE BOVEE and SOYER, 1977a,b,c). It seems that the two main factors controlling local meiofaunal abundance are granulometric characteristics and the quantity and quality of organic matter available on the bottom. Both factors are to some extent depth dependent: (i) the main sources of organic matter are shallow-water benthos and plankton assemblages; (ii) the organic-matter content of a sediment largely depends on its granulometry—the latter, in turn, depending on water movement, which also controls organic-matter transport.

The nearer the sediment to productive assemblages, the greater its organic-matter content. Increased amounts of organic-matter tend to lead to an increase in meiofauna abundance, except in sediments in which the organic-matter content is so high that they must be considered euxinic. On the other hand, the more intensive the bottom water currents, the coarser the sediments and the lower the amount of organic matter they contain. This results in a less abundant but more diversified meiofauna. According to DE BOVEE and SOYER (1977c), the analysis of the meiofauna on the Kerguelen Islands shelf and upper slope and in fjords supports four generalizations: (i) Sandy bottoms usually carry meiofauna abundances from 300 to 1000 ind. 10 cm^{-2} (sometimes up to 2000); nematodes tend to represent about 80% of the total meiofauna. (ii) In muddy bottoms the abundance is always higher (500–3000 ind. 10 cm^{-2}, sometimes up to 8000); nematodes always represent more than 90% of the total meiofauna. (iii) The highly reduced euxinic mud with its very high sulphide content is severely impoverished and usually supports only from a few tens up to 100 (rarely 200 in less reduced sediments) ind. 10 cm^{-2}. (iv) From shallow-water shelf bottoms to the upper slope (200–250 m) the meiofauna abundance progressively decreases, partly due to the increasing speed of bottom currents and partly because the sediments are generally coarser (Fig. 9-5).

The geography of the interstitial meiofauna on shelf areas is at present insufficiently known. Most investigations have been carried out in the northern hemisphere, especially on temperate and cold-temperate shores of the North Atlantic Ocean.

In general, interstitial animals are assumed to be geographically widely distributed. However, this assumption, acceptable at the generic level, calls for taxonomic inquiry, in particular at the specific level.

Very few data are available from intertropical regions. MCINTYRE (1968) who inves-tigated both macro- and meiofauna at Porto Novo on the south-eastern coast of India, found below the mean sea level up to 3820 ind. 10 cm^{-2}. Nematodes, harpacticoids and turbellarians were the predominant groups. MCINTYRE attributed the observed decrease in meiofauna abundance in the upper levels to predation pressure by decapod crustaceans and gastropods. On 40-m deep poorly-sorted sands at Mururoa Atoll

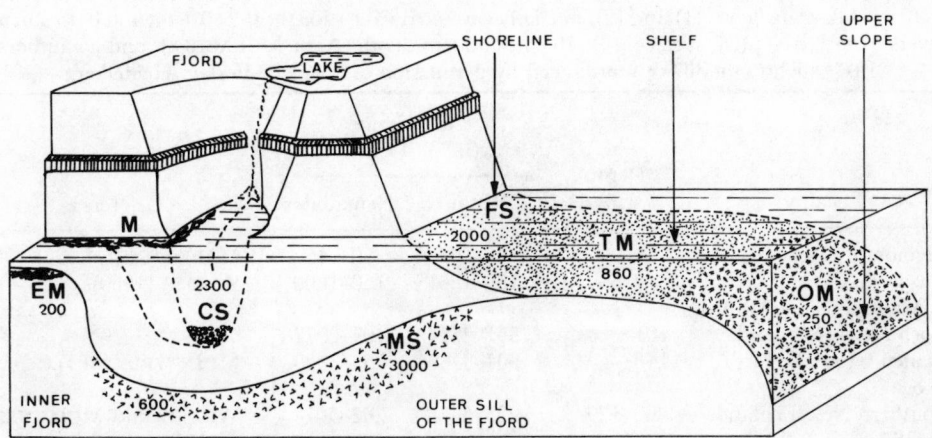

Fig. 9-5: Meiofaunal abundance (ind. 10 cm^{-2}) on the coasts of Kerguelen Islands. CS: coarse sand; EM: euxinic mud; FS: fine sand; M: *Macrocystis* beds; MS: mud mixed with sponge spicules; OM: organogenous mud; TM: terrigenous mud. (After DE BOVEE and SOYER, 1977c; reproduced by permission of CNFRA.)

(French Polynesia) SALVAT and RENAUD-MORNANT (1969) found a meiofauna featuring predominantly (foraminiferans excluded) nematodes, polychaetes and harpacticoids; in the surface layer the abundance was 1300 ind. 10 cm^{-2}. Table 9-3 suggests that the abundance does not differ significantly in temperate and intertropical regions. Differences in biogeographical origin, where they exist, seem less marked than those related to the ecological factors discussed in the preceding paragraph devoted to the topographical aspect.

(g) Deep-Sea Meiofauna

In the following pages, we shall consider qualitative and quantitative aspects together. Additional information is presented on the ratio macrofauna : meiofauna (M : m) in terms of biomass and/or individual numbers on p. 567ff.

Bathyal Zone

The most comprehensive study on bathyal meiofauna was conducted by TIETJEN (1971) in Atlantic Ocean areas off North Carolina (USA). From the lower shelf down to 500 m the sediments consists of medium-sized calcareous sands with relatively low organic carbon contents. The fauna mainly comprises nematodes predominant and large numbers of harpacticoid copepods, ostracods, benthic foraminiferans, small polychaetes, gastrotrichs*, nemerteans, as well as several other groups (turbellarians*,

Table 9-3

Meiofaunal abundances ($10\ cm^{-2}$) recorded by the authors listed, in general within the uppermost 0–10 cm substrate layer. (1) and (2): partially corrected values for the 0- to 10-cm and 0- to 35-cm layers, respectively; (3): surface; (4): 15 cm sediment depth (After MᴄLᴀᴄʜʟᴀɴ and co-authors, 1977; slightly modified; reproduced by permission of Springer-Verlag, Heidelberg)

Locality	Depth (m)	Abundance (ind. $10\ cm^{-2}$)		Reference
		Total	Nematodes	
Plymouth, England	45	87–202	50–116	Mᴀʀᴇ (1942)
Buzzards Bay, Massachusetts, USA	12–30	169–1861	150–1800	Wɪᴇsᴇʀ (1960)
Loch Nevis, Scotland	101	541–2224	405–2072	⎫
Fladen Ground, North Sea	146	904–3163	755–3020	⎬ MᴄINᴛʏʀᴇ (1964)
Southern New England, USA	69–179	235–537	202–507	Wɪɢʟᴇʏ and MᴄINᴛʏʀᴇ (1964)
Bermudas	2–13	123–1333	88–958	Cᴏᴜʟʟ (1970)
Off North Carolina, USA	50–100	352–849	157–593	Tɪᴇᴛᴊᴇɴ (1971)
Keralan coast, India	4	78–145	—	Dᴀᴍᴏᴅᴀʀᴀɴ (1972)
Kiel Bay	7·5–26	94–1191	—	Sᴄʜᴇɪʙᴇʟ and Nᴏᴏᴅᴛ (1975)
Western Baltic Sea	2–26	300–1995	178–1994	Sᴄʜᴇɪʙᴇʟ (1976)
Algoa Bay, South Africa	5–30 (1)	55–584	47–450	⎫ MᴄLᴀᴄʜʟᴀɴ and co-authors (1977)
	15–25 (2)	680–2090	570–1722	⎭
Chupinsky inlet, White Sea	intertidal,	319	239	⎫ Gᴀʟ'ᴛsᴏᴠᴀ (1971)
	subtidal	239	229	⎭
Mururoa Atoll, French Polynesia	40 (3)	300	130	⎫ Sᴀʟᴠᴀᴛ and Rᴇɴᴀᴜᴅ-Mᴏʀᴀɴᴛ (1969)
	(4)	360	190	⎭
Porto Novo, India	intertidal	3820	—	MᴄINᴛʏʀᴇ (1968)

tardigrades*, kinorhynchs*, halacarids*, hydrozoans*, gnathostomulids*, pelecypods*, cumaceans*). The meiobenthos is rather scattered; at a depth of 400 m the total number of individuals is about 150 to 100 ml^{-1} of sediment. Below 500 to 600 m, the sediment consists of clayey-silt with an organic-carbon content six to ten times higher than on the upper slope, whose inhabitants are very different. All the taxa marked with an asterisk (*) above disappear; the numbers of harpacticoids, ostracods, nemerteans and polychaetes become significantly reduced; the nematodes abundance slightly increases at 600 and 800 m but most of the species are different from those of the shallower, more sandy bottoms. At 2500 m, a depth which may have to be considered as the abyssal rise, the meiofauna is much more abundant (about 1200 ind. 100 ml^{-1}) than in the upper slope. Changes in the distribution of both total meiofauna and nematodes might be related to changes in sediment composition and accompanying qualitative and quantitative changes in food supply, although bottom-water temperature may also play some part.

Changes in the food available, correlated with changes in sediment quality, are probably of paramount importance. This may be exemplified by a study of nematode feeding types. From 50 to 500 m, deposit feeders are most abundant (about 50% of the total nematode fauna), but epigrowth feeders are also present in significant numbers (about 40%).

'A sufficient food supply was present for epigrowth feeders to a depth of 500 m in the form of (1) benthic micro-algae (to a depth of 100 m) and (2), planktonic foraminiferal remains (250–500 m) around which organic coatings may form. . . . Below 500 m, significant increases in organic matter, coupled with the disappearance of planktonic foraminiferal remains and the organic coatings around them, may have favoured the deposit feeders, and produced the significant changes seen below this depth' (TIETJEN, 1971, p. 954).

At depths of 600 to 2500 m, corresponding to a zone of deposition (as shown by photographs of the bottom), deposit feeders constituted about 80% of the nematode fauna, epigrowth feeders about 8%, and predators/omnivores about 12%. TIETJEN (1971) noticed that areas below 500 m are 10 °C colder than those above 500 m and reveal far fewer annual temperature changes. Thus, at least two ecological factors seem to be responsible for the maximum faunal change between 400 and 750 m: temperature and sediment composition (with accompanying change in available food).

Benthic foraminiferan distributions may provide indicator criteria for delimitating vertical meiofauna subzones on the continental slope. As is well known, arenaceous species largely predominate in abyssal assemblages; however, these may also be common in the bathyal zone of Arctic areas (p. 488). Calcareous benthic foraminiferans never live below 3500 to 4500 m. According to SAIDOVA (1963, 1970, 1971), who studied foraminiferan distributions in the whole Pacific Ocean, Buliminidae are abundant everywhere in the upper bathyal, but may also be found on the shelf in Arctic and subarctic regions, whereas Alabaminidae predominate everywhere in the lower bathyal.

In the same area as that investigated by TIETJEN (1971), COULL and co-authors (1977) observed total meiofauna numbers at 400 m (442 ind. 10 cm^{-2}) in the bathyal zone, about half those found at 800 m (892 ind. 10 cm^{-2}). The increase at 800 m is primarily attributable to increased numbers of nematodes and foraminiferans. Below 800 m, i.e. just beyond the transition from sandy to muddy sediments, 'most of the truly interstitial taxa (Gastrotricha, Turbellaria) disappear, and the two dominant taxa (nematodes and foraminiferans) account for 92·8% of the fauna' (COULL and co-authors, p. 238). Since the main flow of the Gulf Stream passes directly over the slope at depths of 400 to 500 m, we may assume that changes in sediment composition and food supply due to Gulf Stream current activities are primarily responsible for the reduced meiofauna abundance at 400 to 500 m. Other examples of the current's influence on sedimentation rate and food resources are discussed on p. 490. Further scattered bathyal meiofauna data—except for the Mediterranean Sea—are listed in Table 9-4.

Table 9-4

Soft-bottom bathyal meiofauna (ind. 10 cm^{-2}) and nematode biomass (mg 10 cm^{-2}, bracketed) in the 0- to 4-cm layer (After THIEL, 1975, slightly modified; reproduced by permission of the author)

| Depth (m) | Atlantic Ocean | | | | | | | Indian Ocean |
	Massachusetts slope (40–41° N)	Portugal slope (38° N)	Josephine Seamount (37° N)	West from Gibraltar (36° N)	North Carolina (34° N)	Great Meteor Seamount (30° N)	Mauritania (20–21° N)	Somalia slope (5–8° N)
193–250		—(1·8)	370(0·6)				—(1·0)	
250–750	—(0·3)	—(1·7)	330(0·4)	370(0·4)	—(1·6)	320(0·5)	—(2·9)	
750–1250		—(0·6)		298(0·4)	—(0·2)		—(2·2)	40(0·1)
1250–1750		—(1·1)	470(0·6)		—(0·2)	(0·1)	—(0·5)	380(0·6)
1750–2250			500(0·8)		—(0·2)			

Abyssal Zone

In the Atlantic Ocean off the north-western Spanish coast THIEL (1972c, 1973) investigated muddy bottoms of abyssal depths (5270–5340 m) concentrating both on the meiofauna (0- to 7-cm layer) and macrofauna (hauls with 6-m Agassiz trawl). Nematodes (excluding Desmoscolecidae) always turned out to be the most abundant group, constituting nearly 95% of the total meiofauna. They were followed by copepods (about 2%), Desmoscolecida, ostracods, nauplii and polychaetes. Total abundances in the 0- to 7-cm layer ranged from 402 to 733 ind. 25 cm^{-2} (160–290 10 cm^{-2}). The percentages of the total meiofauna in the 0- to 4-cm layer ranged from 92·3 to 71·8; practically only nematodes occur in the layers from 5 to 7 cm; the uppermost sediment centimetre contained 42·6% of the meiofauna, the deepest centimetre of the sampled layer only 3·7%. Average biomass was estimated at about 1 mg 10 cm^{-2}. THIEL noticed that the meiofauna decreases with increasing depth more slowly than the macrobenthos and estimated the biomass ratio macrobenthos : meiobenthos to be 1 : 1. Assuming similar metabolic adaptations to deep-sea conditions in macro and meiofauna, he estimated the total metabolism of meiofaunal invertebrates at more than 80%. Further assuming that most meiobenthic animals have a one-year life span, i.e. a P : B ratio of 1 (instead of possibly 3 in shelf meiofauna), the annual production of the abyssal meiobenthos would be 1 g m^{-2} and hence much higher than the macrobenthos production (about 0·2 g m^{-2} yr^{-1}).

Investigations along two transects in the Bay of Biscay by DINET and VIVIER (1977) over depths ranging from about 2000 to 4700 m, i.e. mainly covering the abyssal zone, add to the information provided by THIEL (1972c, 1973) which was restricted to a rather narrow depth range. Fig. 9-6 illustrates differences in total abundance related to depth

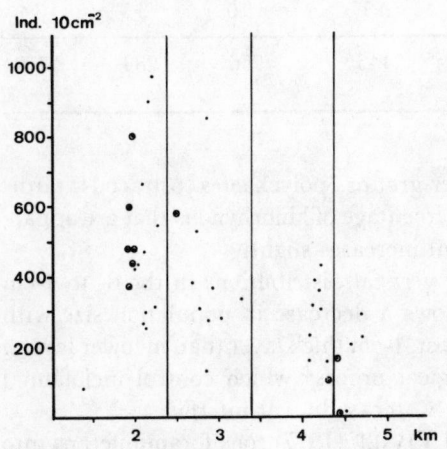

Fig. 9-6: Changes in meiofauna abundance with depth, Bay of Biscay, lower slope and abyssal plain. Consult text for depth (km) of stations. (After DINET and VIV-IER, 1977; reproduced by permission of the authors.)

and abundance values in each of the six stations sampled six times between October 1972 and October 1974 during different seasons. It appears that at the lesser depths of the abyssal zone the meiofauna is more abundant but its variability rather high. For example, on the northern transect individual numbers were 500 ± 263 and 407 ± 196 10 cm^{-2}, respectively at stations 1 (2000–2200 m depth) and 2 (2700–3000 m); they were only 305 ± 100 and 189 ± 120 cm^{-2}, respectively at stations 3 (4100–4400 m) and 4 (4700 m). However, the values were highly dispersed at station 4. Minimum in abundance and the maximum variability, 94 ± 82 10 cm^{-2}, occur at station 5 (4000 m) of the southern transect. The most homogenous granulometry values and carbon and nitrogen contents were recorded at station 3, together with a reduced organismic variability. DINET and VIVIER ascribe the variability at the other stations to their locations in front of a submarine canyon's mouth, holding local turbidity currents responsible for episodic meiofauna impoverishment.

As in Table 9-5, nematodes usually predominate. They are followed by nauplii and copepods. The percentage of nematodes increases with increasing depth due to the

Table 9-5

Percentages of meiofauna groups at six stations of different depth in the Bay of Biscay (After DINET and VIVIER, 1977; reproduced by permission of the authors)

Group	1 2000–2200 m	2 2700–3000 m	3 4100–4400 m	4 4700 m	5 4000 m	6 2000 m
Nematodes	86·0	88·0	91·2	92·5	85·9	90·3
Copepods	3·8	3·8	3·2	3·2	7·4	3·8
Polychaetes	1·2	1·2	1·0	0·3	1·1	1·2
Tardigrades	0·4	0·1	0	0	0	0·2
Kinorhynchs	0·2	0·2	0·2	0·7	0·4	0·3
Ostracods	0·3	0·2	0·2	0	1·1	0·2
Nauplii	7·4	6·1	4·1	3·0	3·5	3·4
Total number of individuals	5005	4071	2446	756	283	3779

decrease or even disappearance of several other groups (polychaetes, ostracods, tardigrades) at the deepest stations (3 and 4). The percentage of kinorhynchs that are apparently better adapted to the abyssal environment increases slightly.

DINET and VIVIER (1977) further analysed vertical distributions in the 0- to 4-cm layer along the northern transect. Fig. 9-7 shows a decrease in population size with increasing depth—more marked in the uppermost, 1-cm thick layer than in lower layers. This suggests that the most important biological process which control meiofaunal abundance are largely restricted to the bottom water–sediment interface.

Neither THIEL (1972c, 1973) nor DINET and VIVIER (1977) took foraminiferans into account. However, this group seems to play an important role in the meiofauna, not only in Arctic and subarctic shelf and deep-sea areas, but in other places also. According to COULL and co-authors (1977, p. 239) 'major differences in the meiofauna off North

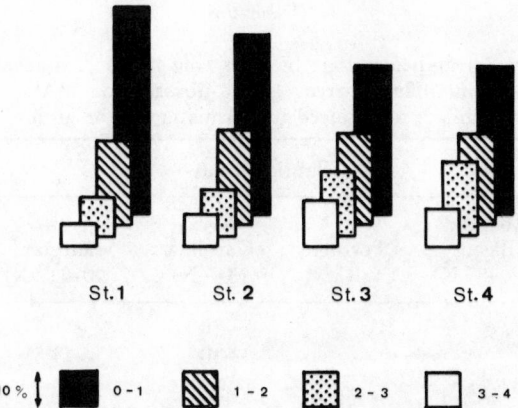

Fig. 9-7: Quantitative distribution for the 4 upper 1-cm thick layers expressed as percentage of total meiofauna at four stations of the Bay of Biscay. Consult text for depth of stations. (After DINET and VIVIER, 1977; reproduced by permission of the authors.)

Carolina occur in the greatly increased Foraminifera numbers'. Moreover, presence or absence of different foraminiferan taxa—and sometimes their relative percentage also —might be used for delimitating meiofaunal assemblages of the vertical zones. For example, SAIDOVA (1963) found abyssal Astrorhizida and Ammodiscida to be the most abundant in the Pacific Ocean. In the north-western Indian Ocean calcareous foraminiferans (Alabaminidae) live in deeper waters than in the Pacific Ocean and are distributed in the open parts of the study area from 2200 down to 4750 m (KHUSID, 1971; BURMISTROVA, 1971). They account for 90% of the total foraminiferal biomass $(0.6–1.1 \text{ g m}^{-2})$ down to 4500 m. A transition zone lies between 4400 and 4500 m where the calcareous species may still constitute more than 50% of the whole biomass. Below 4800 m all calcareous benthic foraminiferans disappear and only arenaceous forms (Astrorhizida, Ammodiscida) are found down to 6700 m. Owing to the large size of the latter, the biomass increases to values from 3.08 to 5.15 g m^{-2} (apparently inclusive of calcareous matter). All this information on the sharp change in foraminiferal coenoses at about 4500 m lends some support to the division of the abyssal zone into two subzones proposed by VINOGRADOVA (1958) for the macrobenthos. However, the increased solubility of lime with increasing depth also seems to be significant. Table 9-6 compares some data obtained from abyssal zones with those from the bathyal zone (Table 9-4).

Factors Controlling Meiofaunal Abundance in Deep-Sea Soft Bottoms

As pointed out above, meiofaunal abundance depends on sediment grain size, organic-matter content and bacteria serving as trophic link between dead organic mater-

Table 9-6

Meiofaunal (ind. 10 cm^{-2}) and nematode biomass (mg 10 cm^{-2}, bracketed) in the 0- to 4-cm sediment layer recorded in different areas of the abyssal zone (After THIEL, 1975; slightly modified; reproduced by permission of the author)

Depth (m)	Atlantic Ocean				Indian Ocean
	Iberian Basin (42–43° N)	Portugal (38° N)	North Carolina (34° N)	Mauretania (20–21° N)	Somalia deep (5–8° N)
2250–2750			—(0·1)		310(0·5)
2750–3250			—(0·1)	—(1·5)	200(0·3)
3250–3750					
3750–4250		—(0·5)			140(0·2)
4250–4750					110(0·2)
4750–5250	210(0·5)	—(0·3)			50(0·1)

ial and living animals. According to THIEL (1972b), owing to their large numbers in soft bottoms and their high division rates, bacteria appear to account for a large part of the metabolism of deep-sea sediments. These sediments may contain more food for invertebrates than the water above the bottom and thus offer an explanation for relatively high meiobenthic abundance in deep-sea assemblages. THIEL speculates that the rather rich meiobenthos explains the dominance of sediment feeders in abyssal sedentary macrofauna. His underwater photographs suggest that some macrofaunal suspension feeders, which are fixed on hard substrates scattered over the sediment surface, may bend down to the bottom in order to obtain food. Employing the Bathyscaphe I had already observed such behaviour in some octocorals, for instance, in *Umbellula*. Thus it seems that some deep-sea suspension feeders partly depend on resources from the sediment surface upon which they feed directly (see also p. 488).

With the restrictions specified above (p. 37) regarding the degradability of sinking detrital organic material reaching deep-sea bottoms, we may assume that the higher the sediment's content in organic matter, the more abundant the meiofauna. Consequently, we must consider the pathways of food transport to the deep sea: sinking of dead organisms, moults and faeces; vertical migration of organisms; formation of organic aggregates; and the availability of heterotrophic organisms other than bacteria. In some cases appreciable amounts of detrital organic material deposited on the sea bottom may originate from land (e.g. river input, salt marshes, etc.), providing the shelf is sufficiently narrow and the slope sufficiently steep; or from the shelf itself, providing it comprises highly productive assemblages.

SOKOLOVA (1970, 1976a,b), who compared meiofaunal abundance and fertility of the upperlying waters in different regions of the Pacific Ocean, noticed that in regions with an oligotrophic pelagos, the meiobenthic biomass is 10 to 30 times lower than beneath eutrophic pelagos regions corresponding to the equatorial current system, and 3 to 18 times lower than beneath transitional areas between oligotrophic and eutrophic regions

Table 9-7

Quantitative data on deep-sea meio- and macrobenthos in different regions of the Indian and Pacific Oceans (Based on information provided by SOKOLOVA, 1976a, b)

Trophic regions	Meiobenthos weight (g m^{-2})			Macrobenthos weight (g m^{-2}) sampled with Sigsbee trawl		
	Number of samples	Mean	Range	Number of samples	Mean	Range
Eutrophic regions						
Precontinental areas	47	2·617 ± 0·424	9·63–0·25	16	476·83 ± 109·49	1660–39·96
Oceanic areas	71	0·181 ± 0·030	1·20–0·02	11	28·94 ± 7·16	71·0–3·0
Equatorial areas	58	0·261 ± 0·073	4·08–0·02	5	39·88	140·78–2·45
Oligotrophic regions						
Northern hemisphere	40	0·034 ± 0·005	0·15–0	7	1·49 ± 0·39	6·09–0·08
Southern hemisphere	47	0·017 ± 0·002	0·07–0	10	0·39 ± 0·18	1·06–0·01

(Table 9-7). On an average, the biomass is 2·5 times higher in the northern than in the southern oligotrophic regions, and this exactly corresponds to the difference in the plankton fertility of the two regions. SOKOLOVA further found a direct relation—more marked in eutrophic than in oligotrophic regions—between meiobenthic biomass, production and activity of the heterotrophic microflora (Fig. 9-8).

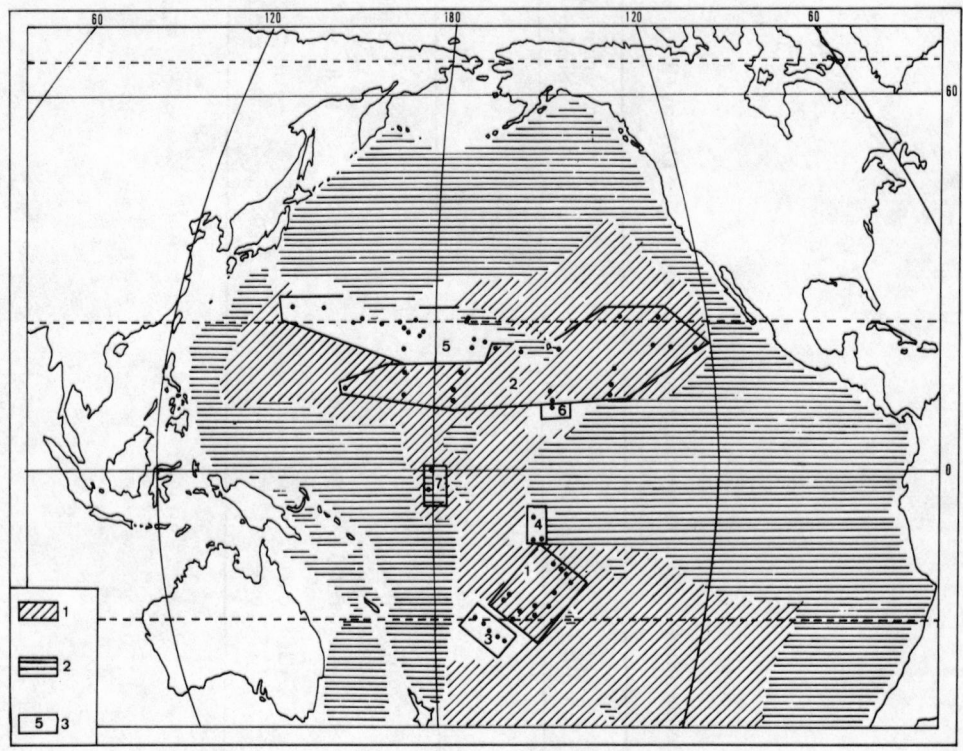

Fig. 9-8: Distribution of eutrophic (1), oligotrophic (2) and transitional (3) regions of meiofaunal abundance in the abyssal zone of the Pacific Ocean. Numbers indicate areas of more intensive sampling; they correspond to the following depths (m) and biomasses (g m^{-2}): 1 = 4700 to 5300, 0·007 ± 0·002; 2 = 4250 to 5700, 0·017 ± 0·0034; 3 = 5300 to 5700. 0·060 ± 0·013; 4 = 4700 to 5400, 0·024 ± 0·0004; 5 = 4500 to 6000, 0·81 ± 0·017; 6 = 4500 to 5000, 0·312 ± 0·126; 7 = 5000 to 5400, 0·47 ± 0·20. (Based on information provided by SOKOLOVA, 1970).

The influence of planktonic production on meiobenthic abundance is also exemplified by investigations of COULL and co-authors (1977) who found a very low population density (73·5 ind. 10 cm^{-2}, corresponding to a biomass of about 0·1 mg cm^{-2}) at 4000 m depth in the Sargasso Sea with its particularly low planktonic production. Micrometazoans were almost non-existent, except for nematodes, and the fauna was dominated by agglutinated foraminiferans.

In general, the meiofauna seems to be more abundant in bathyal depths at the rather high latitudes characterized by a very high planktonic production, restricted to a brief period (Table 9-8).

Table 9-8

Meiofauna abundance and nematode biomass on the slope off Svalbard (latitude 75° N) (After THIEL, 1972a, b; modified; reproduced by permission of the author)

Organisms	Depth (m)			
	250–750	750–1250	1250–1750	1750–2250
Meiofauna (ind. 10 cm^{-2})	550	1470	1170	610
Nematode biomass (mg 10 cm^{-2})	2·5	3·4	3·0	1·6

Increased meiofaunal abundance, originating from the increased summer fertility of polar and subpolar water masses is also witnessed by differences recorded on both sides of a submarine ridge or sill separating two deep basins, one of them receiving polar or subpolar waters. Such a situation prevails over the Iceland–Faroe Ridge (average depth: about 400 m), which separates the Norwegian Sea from the north-eastern Atlantic Ocean and was investigated by THIEL (1971). Above the sill, Atlantic surface water enters the Norwegian Sea while colder water—mainly Norwegian intermediate and deep water masses—transgresses the ridge and flows below the Atlantic surface water towards the Atlantic Ocean; on the south-western slope of the ridge, cold waters mix with central North Atlantic water and form the North Atlantic deep water. On the north-eastern side (Norwegian Sea), hydrological conditions below 400 m are invariable throughout the year (0 °C at about 400 m, ≤0·5 °C below 800 m; 34·9⁰/₀₀ S). Currents are very weak, hence sedimentation rate is high, as is the organic input to the bottom, owing to high planktonic spring production in the euphotic layer. Degradation of sinking organic matter is slowed down by low temperatures. Consequently, more food is available to benthic organisms. On the ridge—and, to some extent, on its south-western side—environmental conditions are just the reverse: currents are irregular and strong (up to 50 cm s^{-1}) and can overthrow the sediment. Hence the sediments are much coarser than on the Norwegian Sea slope and are mostly sandy. Suspended organic material can scarcely form deposits and most of it is transported farther and deeper thus enriching the north-eastern Atlantic Ocean abyssal zone, providing its mineralization rate is sufficiently slow (the temperature changes is unusually large there for bathyal depths: e.g. 7·5 °C from 600 to 1000 m). THIEL's 25 cm^{-2} samples were 7-cm thick. He evaluated meiofaunal abundance in 7 layers. In the two uppermost layers abundance is approximately inversely related to the average current's strength (Table 9-9).

The three lower layers (4–7 cm) contain less than 3% of the total meiofauna in the 0- to 7-cm samples. As usual, nematodes representing 83 to 97% of the meiofauna tend to predominate. Copepods and nauplii—which prefer coarser sediments—are next in abundance and account for 6·3% on the Atlantic slope of the ridge, 9·5% on the crest itself and only 4·1% on the Norwegian Sea slope. Minimum abundance occurred on the

Table 9-9

Vertical meiofauna distribution, expressed as percentage of total meiofauna; 0- to 4-cm sediment layers on both the sides of the Iceland–Faroe Ridge (After THIEL, 1971; modified; reproduced by permission of the author)

Locations	Layers (cm)			
	0–1	1–2	2–3	3–4
Atlantic slope	50	28·5	13·2	8·3
Ridge	31·2	24·4	28·1	16·3
Norwegian Sea slope	52·1	31·8	12·0	4·1

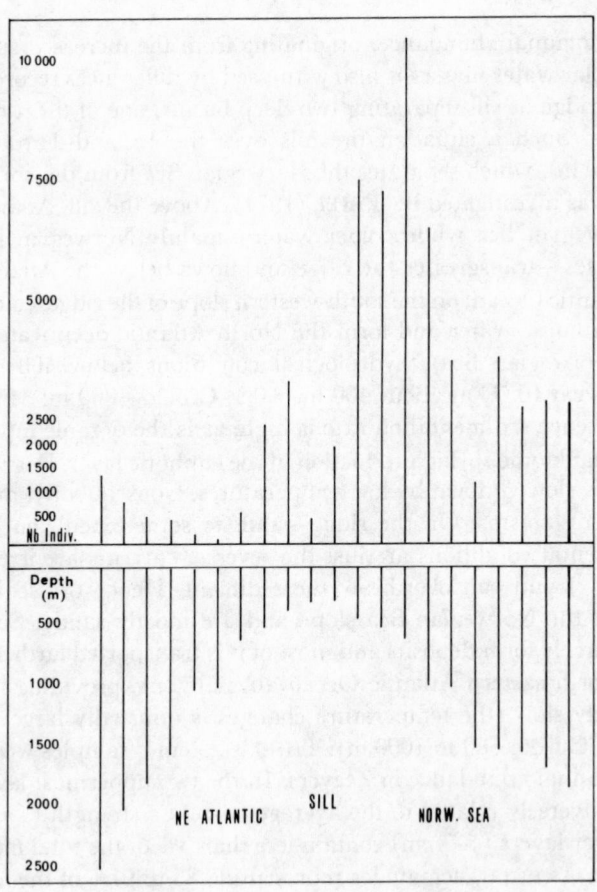

Fig. 9.9: Meiofauna (number of individuals and depth) on both sides of the Wyville–Thomson sill. (After THIEL, 1971; reproduced by permission of the author.)

Atlantic side of the crest at 1000 m. On the Norwegian Sea side, maximum abundance occurs at 600 m (Fig. 9-9). Individual numbers on the Atlantic slope are of about the same order as in other areas at the same, or shallower, depths (Table 9-10). The abundance of meiofauna in the Norwegian Sea is relatively high.

Table 9-10

Meiofauna abundance (total number of individuals under 10 cm sediment surface) and nematode biomass (mg 10^{-1} cm) on both the sides of the Iceland–Faroe Ridge (After THIEL, 1971, modified; reproduced by permission of the author)

Location	Meiofauna abundance	Nematode biomass
Atlantic slope (11 samples)		
Mean	388	0·042–2·738
Range	25–1264	
Summit of ridge (2 samples)	192 and 677	0·780 and 1·735
Norwegian Sea slope (14 samples)		
Mean	1623	1·611–7·136
Range	441–3594	

Another example of the influence exerted on meiofaunal abundance by hydrological features was provided by DINET (1973), who investigated the south-eastern Atlantic Ocean on both sides of the Walfish crest, whose main south-west–north-east axis separates the Angola Basin from the Cape Basin. The abundance values recorded are of the same order of magnitude (Table 9-11) as those recorded in the Wyville–Thomson crest

Table 9-11

Meiofauna and nematode abundance on both the aides of the Walfish Ridge (Data from DINET, 1973; reproduced by permission of the author)

Station	Depth (m)	Abundance (ind. 10 cm^{-2}) in the 0 to 5 cm layer	
		Total	Nematodes
Angola Basin 19°20, 1′ S–09°20, 2′ E	4660	316	303
Angola Basin 18°54, 0′ S–07°22, 4′ E	5170	313	294
Walfish Ridge 18°41, 2′ S–10°56, 0′ E	1440	367	343
Cape Basin 22°50, 6′ S–11°59, 2′ E	2800	1017	961
Cape Basin 22°01, 5′ S–09°16, 5′ E	4100	529	504

area. The higher abundance values recorded from abyssal bottoms (4100 m) in the Cape Basin, compared to those at comparable depths (4600 m) in the Angola Basin, seem ascribable to the higher trophic resources in the waters of Antarctic origin (Antarctic intermediate water and Antarctic bottom water) which fill up the intermediate and deep layers of the Cape Basin. However, the latter in particular can hardly extend northwards to the Angola Basin, as they are stopped by the Walfish crest.

Deep-Sea Meiobenthos in the Mediterranean Seas

Mediterranean Sea

In the upper levels of the bathyal zone off Banyuls (French Mediterranean coast) meiofauna abundance and biomass are very low (Table 9-12). The values obtained are

Table 9-12

Meiofauna abundance and biomass in the upper bathyal of the Mediterranean Sea (Based on SOYER, 1971)

Representation	Depth (m)			
	250	350	450	550
Abundance (ind. 10 cm^{-2})	55	45	49	38
Biomass (μg 10 cm^{-2})	51·7	39·8	51·5	50·2

comparable to those recorded by DINET and co-authors (1973), who studied the abyssal plain of the Western Basin (2100–2850 m): 33 to 78 ind. 10 cm^{-2}. Nematodes, harpacticoids and polychaetes occurred everywhere, kinorhynchs, tanaids and tardigrades only sporadically. Nematodes always predominate, representing 80 to 98% of the total meiofauna. Since the mesh of the smallest sieve employed by SOYER (1971) and DINET and co-authors was 88 μm and 50 μm respectively, we can conclude that meiobenthic abundance is very similar in the bathyal and abyssal zones and is unlikely to change greatly a few hundred metres below. The highest value (78 ind. 10 cm^{-2}) was recorded from a station off the Algerian coast, possibly owing to the Atlantic water current. According to THIEL (1972a), nematodes biomass in the Western Basin (based on data from DINET and co-authors, 1973) amounts to about 100 μg 10 cm^{-2} between 37 and 41° N at depths from 1750 down to 3250 m. At 36 to 37° N, exposed to the Atlantic water current, nematode biomass is much higher: 1·7, 0·9, 0·7 and 0·6 mg 10 cm^{-2} at depths of 750–1250, 1250–1750, 1750–2250 and 2250–2750 m respectively.

In the Eastern Basin, DINET (1976) discovered differences between the two North Aegean deeps, separated by an only 500 m deep sill (Table 9-13).

Table 9-13

Differences in meiofauna abundance (ind. 10 cm^{-2}) in North Aegean Deeps (Based on DINET, 1976)

Western deep					Eastern deep
1069 m	1209 m	1134 m	459 m	130 m	88 m
87	99	187	204	427	397

In the western deep, meiofauna density regularly decreases with increasing depth. The density ratio between deepest area/shelf area is about 1/5. More than 50% of the meiobenthos inhabits the 0- to 2-cm layer, sometimes even the uppermost centimetre. Presumably the higher abundance in the eastern deep is due to the outflow of Black Sea surface waters whose high planktonic production has been well documented. In the western deep, abundance decreases westwards with increasing distance to the sill, proportionately as the influence of Black Sea surface waters decreases.

In general, meiobenthic abundance in abyssal bottoms of the Mediterranean Sea appears to reach the same order of magnitude as in the open ocean only where surface waters attain above-average productivities. Elsewhere, abundance values tend to be always two to ten times lower than at comparable depths in open ocean.

Norwegian and Greenland Seas

The Norwegian and Greenland Seas may each be considered a 'mediterranean sea' since they are separated from the Atlantic Ocean by the Iceland–Faroe Ridge. According to DINET (unpublished), the abyssal meiofauna of these seas is largely dominated by nematodes (95·9%), but copepods, polychaetes and kinorhynchs are invariably found in the samples. The faunal composition appears homogeneous in the different basins. The richest area (720 ± 281 ind. 10 cm^{-2}) corresponds to abyssal depths (2700 m) of the north-eastern side of the Iceland–Faroe Ridge. The Spitzbergen and Greenland Basins are also rich: on an average (438 ± 244 ind. 10 cm^{-2} at depths of 3200–3700 m). The central part of the Norwegian Basin (3200 m) is very poor (<50 ind. 10 cm^{-2}), probably because deep waters are formed here. These deep waters originate from highly oxygenated dense surface waters, which enhance the decomposition of sinking organic material and thus result in a decrease of the non-refractory organic matter reaching the sediment. DINET compared meiobenthic abundance and the sediment's organic-matter content (organic carbon, total nitrogen, proteins); only the last parameter turned out to be correlated with meiofauna abundance; the other two and the C : N ratio do not exhibit any significant relation with meiobenthos density. Hence, DINET suggests that a large part of the organic material cannot be metabolized by microorganisms and thus cannot be utilized by the meiofauna.

Biomass Ratio Macrofauna: Meiofauna

On the shelf, biomass ratio macrofauna : meiofauna (M : m) usually ranges between 10 and 50, but it may sometimes be higher. For example, in sediments of Helgoland Bight, STRIPP (1969) found a range of 24 to 90. In general, the M : m ratio decreases with increasing depth (Table 9-14) to approximately 1 in the abyssal zone. The decreasing M : m with increasing water depth has been ascribed by THIEL (1975) to the parallel decrease in average body size of macrobenthic species (except for scavengers), apparently owing to the decrease in trophic resources. In other words, macrofaunal biomass decreases with depth faster than its abundance. THIEL's assumption may be supported by the smaller body size generally characterizing the Mediterranean Sea abyssal macrofauna compared with that of the eastern Atlantic Ocean: in the Mediterranean Sea, even in the bathyal zone, M : m is as low as 1·2 (GUILLE and SOYER, 1971). However, reductions in body size also involve reduction and simplification in organ systems:

Table 9-14

Macrofauna biomass (90% polychaetes and amphipods) compared with that of nematodes, and corresponding M : m ratios on both sides of the Wyville–Thomson crest (After THIEL, 1972b; modified; reproduced by permission of the author)

Depth (m)	Atlantic Ocean slope			Wyville–Thomson Ridge			Norwegian Sea slope		
	Biomass (g m^{-2})			Biomass (g m^{-2})			Biomass (g m^{-2})		
	Macrofauna (M)	Nematodes (m)	M : m	Macrofauna (M)	Nematodes (m)	M : m	Macrofauna (M)	Nematodes (m)	M : m
250–750	3·0	1·7	1·8	3·0	1·7	1·8	18·0	5·9	3
750–1250	1·4	0·1	14				17·3	3·6	5
1250–1750	1·8	0·6	3				12·4	1·9	6
1750–2250	2·3	0·5	4				3·7	1·8	2
2250–2750	0·8	1·1	1·5						

several macrofaunal taxa—for example some isopods, opisthobranchial gastropods and ascidians followed this evolutionary line and adjusted to meiofaunal life, whereas most others remained restricted to depths with more abundant food resources. An exception to the general rule that biomass M : m decreases with increasing depth is presented in Table 9-15.

Table 9-15

Macrofauna (data from KHRIPOUNOFF and co-authors, 1980) and meiofauna (data from DINET, 1980) biomasses and M : m biomass ratios in two different areas of the Atlantic Ocean (After (DINET, 1980; reproduced by permission of the author)

Group	Northern Bay of Biscay (2000 m depth)	Vema fault (5100 m depth)
Meiofauna (g m^{-2})	0·78	0·37
Macrofauna (g m^{-1})	0·223	0·053
M : m	3·50	6·98

(h) Trophic Net in the Meiofauna

The main features of the meiofaunal trophic network and its relationships with other compartments of the benthic ecosystem have been discussed by FENCHEL (1969) and MCINTYRE (1969, 1971). MCINTYRE (1971, p. 7) concludes:

'while meiofauna may be rich in areas where other zoobenthos is sparse, it is still rich even alongside dense populations of macrofauna, and the question arises as to whether there is any significant interaction between these two populations.'

He distinguishes between sandy and muddy grounds and between microphages and predators.

On sandy grounds the main food for microphages probably consists of bacterial epiphytes on sand grains or detrital organic particles (MCINTYRE and co-authors, 1970). Soluble organic materials made available via attached bacteria from water percolation through the beach sand are probably particularly important (MCINTYRE and MURISON, 1973). Among microphages, it is difficult to distinguish detritus feeders (which probably assimilate mainly, if not only, attached bacteria) from epistrate feeders or from true grazers feeding on free-living or sand-grains attached algae. Morphological peculiarities are of little help for such a distinction (WIESER, 1960): although many nematodes in sandy bottoms have strong buccal chitinous structures, most of them are epistrate feeders utilising bacteria and small algae—sometimes faeces of other species. Small-sized amphipods and cumaceans—whose appurtenance to the true meiofauna is somewhat questionable—are also epistrate feeders, whereas many species of harpacticoids, ostracods, opisthobranchial gastropods and some archiannelids appear to be primarily grazers. The availability of either food source (bacteria or algae) may cause seasonal fluctuations in the food consumed. According to MCINTYRE (1971), both copepods and gastrotrichs, some species of turbellarians and some nematodes feed on algae (possibly diatoms) in summer, but not in winter, when they appear to live on bacteria instead of algae.

Absent in muddy substrates, suspension feeders are so rare, even in sandy grounds, that they seem to play no significant role in the meiofaunal trophic net. However, this group is represented by several ascidians, the bryozoan *Monobryozoon* and the brachiopod *Gwynia*—all species which generally occur in coarse sands under strong bottom currents. There are some nematodes among the predators, such as members of the genus *Halichoanolaimus*, which feed on other nematodes. The hydroids *Halammohydra* and *Psammohydra* feed on ostracods but prefer copepods, nematodes and gastrotrichs—they sometimes also feed on diatoms, as do species of the related genus *Protohydra*. According to MCINTYRE (1969), *Protohydra* species may grow from 20,000 to 200,000 ind. m^{-2} within one month and thus control other meiofaunal populations. The nudibranch *Embletonia pallida* seems to feed mainly on *Psammohydra*.

Finally, some taxa have very diversified food habits, e.g. foraminiferans (p. 531) and halacarids. The latter suck on copepods and ostracods and may also utilize unicellular algae or detritus (MONNIOT, 1968). The feeding habits of turbellarians are more diverse: some feed on diatoms, small flagellates, protozoans and small crustaceans, others on nematodes and oligochaetes (mostly in brackish water), and a few are scavengers (MCINTYRE, 1969).

On muddy grounds where most macrofauna representatives are mainly 'detritus feeders', it seems that the meiofauna—which tends to be more restricted to the uppermost layers here—also feeds on detritus, in particular on nematodes (MCINTYRE, 1971). How can the striking ecological success of nematodes (both in abundance and specific richness) in the abyssal environment be explained?

According to KHRIPOUNOFF (1979), the calorific content of nematodes (5617 cal g^{-1}) is higher than that of copepods (3090 cal g^{-1}) and this almost irrespective of the depth and biogeographic region. This suggests that nematodes can convert the 'primary' energy—especially the sediment amorphous organic matter—with a high degree of efficiency (DINET, 1980).

As pointed out in Volume V, Part 2, meiofauna and macrofauna elements may compete nutritionally, especially if the food resources are insufficient, either owing to insufficient organic-matter input or to an explosive increase in meiofaunal abundance. In some cases macrofauna species may to some extent subsist at the expense of the meiofauna. Macrofauna predation on sandy bottom meiofauna was reviewed by MCINTYRE (1969).

The consumption of meiofauna by macrobenthos appears to be rather insignificant on sandy bottoms. However, in terms of the whole micro- and meiobenthos, i.e. including bacteria and algae, the microphagous macrobenthos *does* compete with the meiofauna for the food. The trophic role of protozoans, chiefly ciliates and foraminiferans is certainly significant (FENCHEL, 1969). Phytophagous ciliates may rapidly decrease the populations of small algae; since they also feed intensively on bacteria, they tend to keep these in the phase of exponential growth. It is very difficult to assess the significance of ciliates as a food source for secondary consumers because they leave practically no recognizable remnants in the gut of the predator. However, some observations suggest that ciliates are fed by some turbellarians, gastrotrichs and rotifers and, among the macrofauna, by *Arenicola* and some pelecypods (FENCHEL).

Nevertheless, keeping in mind the questionable trophic role of ciliates, with some reservations, we agree with MCINTYRE (1971) who suggested that in the sandy bottoms the meiofauna constitutes an almost closed ecological system. This means that on shallow sandy shelf bottoms the meiofauna tends to be relatively independent of the macrofauna. Ecologically its more important role would simply be a faster recirculation of nutrients. However, I think MCINTYRE's hypothesis pertains mainly to interstitial metazoans. Taking into account the micro- and meiobenthos in general, one cannot deny the relations between microphagous macrobenthos and microbenthos. For example, even on sandy substrates there are pelecypods, polychaetes and many other macroinvertebrates, such as Aspidochirota holothurioids and irregular echinoids, which feed on unicellular algae and also on bacteria and protozoans of the upper sediment centimetres.

Before attempting to evaluate the meiofaunal contribution to the food supply of macrobenthic carnivores it seems necessary to dispose of estimates regarding meiofaunal secondary production. Unfortunately, relevant data are rare, except for oligochaetes (p. 535). As far as nematodes are concerned, on an intertidal sand flat of the German Bight, SKOOLMUN and GERLACH (1971) inferred from size distribution that *Enoploides spiculohamatus* has two to three annual generations. According to GERLACH and SCHRAGE (1972), some species have only one annual generation and two of the species studied completed their life-cycles in only two years. In harpacticoid copepods variability seems to be similar: in the Gulf of Marseilles (France), ovigerous females of some species were observed throughout the year, whereas others exhibited two annual generations (NODOT, personal communication). On temperate shores, temperature fluctuations seem to be the most important factor controlling breeding, although photoperiod and food supply might play some part. As a provisional hypothesis, I suggest that the meiofauna annual production on temperate shores accounts for two to three times the standing crop—possibly for more in intertropical regions, but far less in colder environments, such as shallow-water sediments in high latitudes and the deep-sea bottoms.

The possible significance of the mixobenthos (p. 531) has been generally disregarded. While the mixobenthos is probably absent from deep-sea bottoms, at least beyond the abyssal rise, it has always exceeded 50% of the total biomass off Banyuls (France) at depths from 35 to 550 m (SOYER, 1971). In comparison, the real (i.e. permanent) meiofauna has always represented more than 80% of the total abundance. Energy-exchange patterns between plankton and interstitial environment remain to be investigated. Many larvae of benthic macroinvertebrates spend their early stages in the plankton and feed on pelagic resources. When settling, they should bring additional energy to the mesopsammal. In other words, the input of energy and matter from mixobenthos to permanent psammon seems to be greater than the output from meiofauna to macrofauna. However, such relations are questionable because the input is probably not available for most mesopsammal predators that are too small for feeding on macrobenthic post-larvae whose size is equal to or larger than their own. Nevertheless, the high density of empty post-larval mollusc shells, which can easily be observed when sorting sand samples, exemplifies a fair amount of organic-matter and energy input from the mixobenthos.

(4) The Sulphide System

(A) General Aspects

FENCHEL and RIEDL (1970) have discovered a peculiar ecosystem in reduced sediment layers covered by oxygen-containing substrate. Containing up to 700 mg of hydrogen sulphide, this black 'sulphide system' is inhabited by a highly specific organismic assemblage, the thiobios, which includes not only microorganisms, but also protophytes, protozoans and metazoans. The sulphide system appears to be distributed world-wide in soft bottoms, both sandy and muddy, from the highest levels inhabited by marine organisms down to at least the lower part of the continental slope. The boundary between the interstitial oxidized environment and the sulphide system is marked by the discontinuity layer of the redox potential. Above this layer the oxygen content is sufficient for oxidizing organic matter; below it, only anaerobiont organisms can exist. The finer the sediment and the more abundant the input of organic matter, the nearer to the sediment surface the discontinuity layer. Still insufficiently investigated, the thickness of the sulphide system depends on local conditions. In sheltered areas—where the oxidized layer is often only a few millimetres thick—metazoans may penetrate the thiobios down to 50 cm; deeper, only bacteria and protozoans can exist—bacteria possibly down to several metres.

(b) Geochemistry

The organic matter introduced to the sulphide system is decomposed in two steps. The first step involves fermentation by bacteria, yeasts and fungi, utilizing various organic compounds such as hydrogen acceptors, breaking down organic substances and releasing carbon dioxide. The second step involves bacteria which require small oxidized inorganic molecules such as NO_3^-, SO_4^{2-}, CO_2, H_2O, as hydrogen acceptors, reducing these to NH_3^+, CH_4, H_2 and in particular H_2S. These inorganics may spread upwards, but may also be utilized by other bacteria—mostly of the sulphur cycle—either chemoautotrophic (e.g. *Thiobacillus, Macromonas, Beggiatoa*) oxidizing H_2S into S or SO_4^{2-}, or photoautrophic utilizing H_2S for reducing CO_2. The latter process is more important when the *Eh* discontinuity layer occurs at less than 1 cm below the sediment surface. The reduced organic substances may be oxidized either by heterotrophic microorganisms in the presence of oxygen or by phototrophic anaerobionts (some algae, the bacteria of the Athiorhodacea group) which substitute these molecules for H_2S.

Consequently, below the bottom-water–sediment interface, three different layers can be distinguished: (i) the yellow oxidized layer containing Fe^{3+} substances with *Eh* values ranging between +400 mV at the surface and +200 mV; (ii) the grey layer which corresponds to the redox-potential discontinuity layer; (iii) the black reduced layer, with *Eh* values from −100 to −250 mV containing iron sulphides. The boundaries between these three layers may vary—even in the same place—as a function of season or owing to a circadian rhythm, relating to changes in oxygen production by photo-autotrophic organisms.

(c) Composition of the Thiobios

The composition of the thiobios is comparatively diversified, although several large taxonomic groups are completely absent, e.g. eukaryote plants and arthropods. Coelomate metazoans, too, are very rare. The thiobios includes two basically different components: transient hosts and endemic inhabitants.

Among the transient species—intruders from the upperlying oxidized zone—there are ciliates, nematodes macrodasyoid rotifers, etc. All these animals, which are possibly microaerophilic or facultative anaerobionts, 'dive' for relatively brief periods into the black layer.

The endemic assemblage is much more interesting. According to FENCHEL and RIEDL (1970) its main components are: (i) Cyanophytes, mainly of the genus *Oscillatoria*, which are highly diversified and abundant (up to 5000 colonies or filaments 100 cm^{-2}). (ii) Numerous species of Eubacteria and Spirochaeta. (iii) Protophytes represented by autotrophic but facultatively heterotrophic species, mainly chrysophytes and pyrrophytes; species of euglens, diatoms and dinoflagellates are of lesser importance. (iv) Yeasts and fungi, the latter belonging to the most primitive groups and insufficiently investigated. (v) Protozoans, mainly ciliates belonging to about a dozen different genera, generally predominate in numbers: 1000 to 30,000 ind. 100 cm^{-2}. Some foraminiferans were also recorded. (vi) Invertebrates: turbellaria are represented mainly by primitive Acoela. Gnathostomulids are among the predominant groups in the sulphide system; abundances as high as 100 ind. 100 cm^{-2} are not unusual. About 80 species have hitherto been described, but the total number is probably much higher; some species live immediately above the discontinuity layer but never farther away than a few centimetres. Nematodes, taxonomically still insufficiently investigated, are next in abundance: up to 1000 100 cm^{-2}. Some species of gastrotrichs and oligochaetes have also been recorded.

The distribution of the different components of the thiobios are controlled vertically by chemical gradients and horizontally by sediment characteristics and the position of the redox-potential discontinuity layer. In the upper beach levels, additional influence is exerted by sea-level fluctuations, especially on tidal shores.

(d) Biology and Adaptations of Thiobios Organisms

All anaerobic thiobios organisms seem to complete their life-cycles under identical conditions (OTT and SCHIEMER, 1973). Owing to the relatively high stability of chemophysical conditions in the anoxic environment, variations in the distribution related to seasonal or tidal cycles seem to be insignificant. Most endemic organisms of the sulphide system do not seem to be able to tolerate anaerobiosis but are adapted to permanent anaerobiotic conditions. In all probability, oxygen is toxic for them. The nematode *Paramonhystera* n.sp. is the first marine metazoan favourably affected by anoxic conditions compared with normal pO$_2$ conditions (WIESER and co-authors, 1974). The 'thiobiotic' gastrotrich *Thiodasys sterreri* survives anaerobic conditions and seems able to fix carbon dioxide (MAGUIRE and BOADEN, 1975).

While it is of considerable ecological significance to determine the role of anaerobic metabolism in benthic energy and nutrient flow and of general biological interest to understand the methods and evolutionary implications of physiological adaptations to anaerobic life, the metabolism of the obligate or facultative anaerobes has not yet been elucidated (MAGUIRE and BOADEN, 1975). These authors suggested that a reverse Krebs cycle might be involved and recalled that considerable integration exists between carbohydrate and aminoacid metabolism, stressing that a variety of amino-acids can serve as energy sources by deamination and transamination. The thiobios system is characterized by high levels of dissolved amino-acids and carbohydrates in the reduced interstitial water. These could be available to meiofauna organisms via active carrier-mediated absorption. The microvilli on the epidermis of *Gnathostomula genneri* may represent adaptations for absorbing dissolved organic substances.

Trophic relations within the thiobios have been insufficiently investigated. Ciliates feed on bacteria in the black layer but on cyanophytes, diatoms and other small unicellular algae in the grey layer. Invertebrates seem to feed on bacteria (mostly those of the sulphur cycle) and cyanophytes. Utilization of dead organic material is still a matter of debate.

Morphological adaptations of thiobios animals include body elongation, poor adhesion and sparse ciliation. Ciliates always exhibit a reduction of their cilia and sometimes have a rigid cell-surface structure. The whole tegument of Turbellaria and Gnathostomulida is monociliate. A striking peculiarity is the covering of the body with bacteria in some ciliates and with cyanophytes or bacteria in some nematodes of the family Stilbonematidae. Such covering possibly qualifies as symbiosis (OTT and SCHIEMER, 1973). Pigmented species often occur in the redox-potential discontinuity layer. The red colour of some thiobiotic turbellarians is due to haem compounds; their function may be respiratory or anti-oxidative. In the gnathostomulid genus *Haplognathia* red pigmentation has probably plesiomorphic and protective functions (BOADEN, 1977).

While thiobiotic taxa are widely distributed and inhabit a very stable environment, its metazoans exhibit a low species diversity in spite of its extreme age (BOADEN, 1977). This may be due to restrictions in food and respiratory resources leading to a general scarcity of available energy in the system. In addition, the anaerobic metabolism conveys obvious energetic disadvantages.

(e) Particular Features of the Sulphide System

The sulphide system, as interpreted by FENCHEL and RIEDL (1970), constitutes a somewhat isolated ecosystem with chemosynthetic production of organic carbon—through dioxide carbon reduction—at the beginning of a trophic chain leading to small invertebrates. Since it consists of archaic organisms, the sulphide system could shed some light on the structure of early ecosystems on earth prior to the development of the aerobic biosphere. It may represent a relict of the Earth's earliest metazoan biosphere (MAGUIRE and BOADEN, 1975).

Large coherent microbial communities were observed at about 50 to 280 m depth in H_2S containing sediments of shelf and upper slope off the coast of Chile GALLARDO, 1977). Most of the bacteria belong to the gliding genus *Thioploca* (Beggiatoaceae) which features large filaments. Mixed with other bacterial forms and with some blue-green algae of the *Oscillatoria*-type (probably *Chlorophlexis*), they seem typical of sulphide biota. The microbial biomass was 106 g 0·1 m⁻² at 60 m depth, whereas the benthic biomass in the same sample was only 11·5 g 0·1 m⁻² (GALLARDO). GALLARDO suggests that the large standing-crop of these filamentous bacteria plays an important role in the upwelling biome off south-western South America and might also be present in other areas with similar oceanographic conditions, for example in the Benguela current area, where slimy grass has been reported at about 90 to 130 m.

'The coincidence of the depth distribution of the microbial community off Chile and that of the principal shrimp (both penaeid and galatheid) and hake fishing grounds, strongly suggests a possible trophic relationships in view of the fact that most of the standing-crop therein seems to be made up of filamentous bacteria' (GALLARDO, 1977, pp. 331–2).

The existence of the thiobios has recently been questioned by REISE and AX (1979) who investigated an *Arenicola marina* tidal flat near the island of Sylt (North Sea), paying special attention to 'oxygen islands' embedded in the anaerobic sulphide system—'the brownish halo of oxic sand which surrounds permanent burrows of large infaunal polychaetes, nemertines, amphipods and others'. In these oxygen-carrying microhabitats REISE and AX found most of the thiobiotic meiofaunal species and supraspecific taxa listed by various authors. They stress that these species consistently attained higher numbers closer to the burrow than in the adjacent black sand (REISE and AX (p. 225)). Horizontal profiles through tail-shafts of lugworm burrows revealed a marked decrease in nematode abundance 10 mm away from the burrow wall, the same applying to several other species e.g. the gastrotrich *Megadasys sterreri* and species of gnathostomulids. The circumstances that increase the abundance of certain meiofauna elements along the burrow of infaunal animals differ from species to species. REISE and AX mention the following: (i) Burrows are oxic throughout and may even function as 'oxygen service stations' for meiofaunal species exploiting food resources in the black-sediment neighbourhood. (ii) Bacteria reducing sulphates and oxidizing sulphides (sulphur cycle) appear to be the dominant microorganisms in the burrow walls; they may constitute a major food source for many interstitial species. (iii) Predators among the burrow meiofauna can profit accidentally from plankton forced into the burrow by irrigation current. (iv) Worm burrows constitute a microenvironment free of hazards facing surface-dwellers. There is no desiccation threat, no extreme changes in temperature and salinity, no dislocation by currents, and no danger from epibenthic predators.

REISE and AX (1979, p. 230) conclude that 'no specific thiobiotic fauna of the sulphide system exists'. Only a few species managed to adapt to the extreme habitat of the oxygen deficient horizon and these are able to tolerate anaerobic intervals. In my opinion the lack of meiofauna in continuously oxygen-poor basins, e.g. Black Sea below 200 m and Baltic Sea below 100 m, supports this view. However, the conclusions drawn by REISE and AX have been challenged by BOADEN (1980) who insists that the case made by these authors is not proven. His arguments were, in turn, contested by REISE and AX (1980). It is impossible to side with either of these contrasting views without further data.

(5) The Coastal Subsoil Water System

(a) The Coastal Subsoil Environment

Apart from the true interstitial fauna reviewed above, there is a specific interstitial brackish-water fauna, closely related to the marine interstitial fauna, but restricted to coastal subsoil waters. Coastal subsoil waters occupy a narrow brackish zone in sandy beaches where fresh subsoil waters from the land mix with sea water (SWEDMARK, 1964). The fauna of coastal subsoil waters comprises, in addition to stenoecious forms, limnobiotic and terrestrial organisms (Collembola) as well as euryecious representatives of the neighbouring marine microbenthos. Physical features of the system in this 'mixing zone' depend on the conditions of freshwater and sea-water circulation within, tending to differ on atidal and tidal shores.

On atidal shores, fresh terrestrial subsoil waters slowly pass under the beach and progressively mix with the sea water. The brackish water thus formed percolates upwards through subtidal sands. Sea water moves in the opposite direction: (i) During gales, wave breaking at the higher beach levels as well as spray transport sea water and organic particles downwards through the upper beach sands. (ii) While the sea is smooth, sea water quickly percolates at the lower beach levels and flows back to the sea whereby part of the sea water passes into the mixing zone.

On tidal shores, the beach slope is often rather gentle down to the 'littoral furrow'—an elongated pool paralleling the shoreline. This furrow precedes a small rise with a summit approximately corresponding to the high water of mean spring tides. The beach slopes gradually down to a little below the level of a neap-tide low water, thereafter becoming very gentle. At high tide, sea water mainly percolates along the furrow and the beach part between the high-water neap tide level and the mean sea-level. Later, this sea water, mixed to a greater or lesser extent with subsoil fresh water, percolates partly along the concave beach belt and partly beyond, even in subtidal areas.

(b) Faunistic Composition

As the coastal subsoil system is a rather dark environment, there are no photo-autotrophic algae. Fungi have been reported but have been insufficiently investigated. The subsoil bios is often referred to as 'phreatic fauna' and was extensively studied by DELAMARE-DEBOUTTEVILLE (1960).

Protozoans, mainly ciliates, are numerous. These were sometimes insufficiently distinguished from those participating in the real interstitial fauna (p. 531). The foraminiferans appear to be next in importance among protozoans. Cnidaria, which are relatively common in the meiofauna, are usually only represented by *Protohydra leuckarti*. Polychaetes are rare, although several species of the family Nerillidae, which is well-represented in the meiofauna, may penetrate into the phreatic environment. Oligochaetes are often numerous, e.g. members of the genera *Michaelsenia* and *Aktedrilus*. Rotifers, represented by a few species of the bdelloid group, have been insufficiently investigated, as is the case with gastrotrichs. Among the latter several species of *Tur-*

banella have been recorded. Turbellaria, mainly members of the family Otoplanidae, are numerous.

Nematodes are once more the dominating taxon both in numbers of species and individuals. According to DELAMARE-DEBOUTTEVILLE (1960), the most common genera are *Syringolaimus, Dorylaimus, Ascolaimus, Theriscus, Eudemoscolex, Enoplus*. Crustaceans are the next in abundance and diversity: numerous copepods (mainly harpacticoids, together with some cyclopoids), and some ostracods. Small-sized peracarids are common, e.g. *Pseudoniphargus* and various species of the isopod families Microcerberidae and Microparasellidae. The most characteristic crustaceans are representatives of the apparently monogeneric (Derocheilocaris) subclass Mystacocarids, an archaic group formerly included within the Copepoda. Arthropods are further represented by the tardigrade *Stygarctus bradypus* and some Acari: the trombidiid *Nematalycus nematoides* and some halacarids. Opisthobranchial gastropods are less abundant than in the true mesopsammal (p. 537). French Mediterranean Sea beaches are inhabited by *Unela odhneri*.

(c) Adaptive Peculiarities of Phreatic Species

The biological and morphological adaptations of coastal subsoil-system inhabitants are similar to those of true interstitial species (p. 539). Gliding in the interstices predominates in the coarser sands (average grain size >200 μm), burrowing in finer sands. Gliding mechanisms are similar to those of mesopsammal species. Caterpillar-like moving prevails in some *Macrodasys* (*M. buddenbrocki*), in oligochaetes and in some nematodes. The microhedylid opisthobranch *Unela odhneri* moves by peristaltis. Mystacocarids move with their antennae, mandibules and maxillulae, assisted by maxillae; in addition, successive telescoping and sudden extension of the segmented body is applied whereby the latter is pushed against a sand particle by the furcal setae. Substrate contact is a necessity: isolated from the sand, mystacocarids quickly die from muscular fatigue.

The unpigmented body is often elongated and flattened. Decrease in size, already mentioned for the interstitial fauna, seems to be more marked in the phreatic fauna except in many gliding species, such as some ciliates (e.g. *Curtrophorella lanceolata*), the formaminiferan *Marenda nematoides*, the mite *Nematalycus*, and several crustaceans. All phreatic species seem to be blind. They have specialized sensory organs, which have still not been sufficiently investigated.

Phreatic species are highly eurythermal and euryhaline. For example, the mystacocarid *Derocheilocaris remanei* breeds at temperatures between 8 and 25 °C and salinities between 5 and 40⁰/∞ S (DELAMARE-DEBOUTTEVILLE, 1960). Nevertheless, salinities below 10⁰/∞ S appear to represent stress and at 20⁰/∞ S a temperature of 27 °C immobilizes *D. remanei* while 30 °C is lethal within 36 h (JANSSON, 1966). It seems that the upper thresholds found for *D. remanei* correspond to the temperature and salinity maxima tolerated by most other phreatic crustaceans. *Derocheilocaris typica* which inhabits beaches on eastern United States coasts is largely restricted to the upper half of the intertidal zone and mostly in exposed places (HALL, 1972). In summer, it tends to migrate to lower beach levels and deeper into the sediment. *D. typica* exhibits regular vertical migrations controlled by the tidal cycle, with a time lag of about three hours.

Most phreatic species seem mainly to consume organic detritus supplied by sea water percolating through the sands and portions of their bacterial coating. Food-supply never seems to be a limiting factor of population abundance (DELAMARE-DEBOUTTEVILLE, 1960).

As far as reproduction is concerned, the two trends mentioned for the interstitial fauna are even more marked in phreatic species: (i) Planktonic larval stages, already rare in the interstitial fauna, all but disappear in the phreatic fauna; the larvae tend to be more and more intimately linked to sand grains. The disappearance of larval stages results in a trend towards direct development with incubation (e.g. isopods of the family Microparasellidae) or without; the latter trend sometimes results in neoteny. (ii) Fecundity is very low; e.g. one to four eggs in *Parasellus* and never more than one in mystacocarids. No cyclic reproduction is known in the marine phreatic fauna, but it may be present in their freshwater counterparts. HALL (1972), for example, noted no significant variation in the abundance of *Derocheilocaris typica* on beaches of North American Atlantic coasts. However, there are exceptions to acyclic breeding: on Lebanon coasts HULINGS (1971) recorded a seasonality in the coastal subsoil system: harpacticoids exhibited a reproduction peak in June–July and mystacocarids and opisthobranchs two peaks in March–April and August.

(d) Distribution of the Phreatic Fauna

With increasing distance from the shoreline—providing the deep sand layers are sufficiently oxygenated—it becomes increasingly difficult to distinguish the phreatic fauna from the true interstitial fauna. However, on British coasts, SPOONER (1959) recorded phreatic crustaceans, associated with typical mesopsammal species, in a 40- to 50-m coarse sand and gravel substrate ('*Amphioxus* sand' p. 450). This suggests that phreatic species gradually rise up to the sediment upper layers with increasing water depth. Therefore the phreatic fauna can be investigated readily only on beaches and on shallow adjacent areas.

Quantitative data about the phreatic fauna are rare. On tidal beaches maximum abundances occur in the vicinity of the mean high-tide level at 20 to 60 cm substrate depth. However, phreatic species may also be encountered 1 m below the sediment surface.

Biographically many phreatic species seem to be widely distributed. Several species have been recorded from marine beaches and fresh contintental subsoil waters. This wide distribution together with the extreme age of many phreatic species suggest that the phreatic fauna has been isolated since very ancient times.

Literature Cited (Chapter 9)

ARNAUD, P. M. and POIZAT, C. (1978). Données écologiques sur des Caecidae (Gastéropodes Prosobranches) du Golfe de Marseille. *Malacologia*, (1979), **18**, 319–26.

BOADEN, P. J. S. (1977). Thiobotic facts and facies (aspects of the distribution and evolution of Anaerobic meiofauna). *Mikrofauna des Meeresbodens*, **61**, 41–63.

BOADEN, P. J. S. (1980). Meiofaunal thiobios and "the *Arenicola* negation": case not proven. *Mar. Biol.*, **58**, 25–9.

BODEANU, N. (1971). Données qualitatives et quantitatives sur le Microphytobenthos des fonds sablonneux et vaseux du littoral roumain de la Mer Noire. *Cercetor. marine (Institutul Roman de Cercetari Marine, Constantee)*, **1**, 27–58.

BOUCHER, D. (1972). Evaluation de la production primaire benthique en Baie de Concarneau. *C.r. hebd. Séanc. Acad. Sci., Paris (Ser. D)*, **275**, 1911–14.

BOUCHER, D. (1975). *Production Primaire Saisonnière du Microphytobenthos des Sables Envasés en Baie de Concarneau*, Thèse Doct., Universite Bretagne Occident Brest.

BOVEE, F., DE and SOYER, J. (1974). Cycle annuel quantitatif du meiobenthos des vases terrigènes côtières—distribution verticale. *Vie Milieu (Ser. B)*, **24**, 141–57.

BOVEE, F., DE and SOYER, J. (1977a). Le meiobenthos des Iles Kerguelen. Données quantitatives. I. Le Golfe du Morbihan. *Comité national francais des recherches antartiques (CNFRA)*, **42**, 237–47.

BOVEE, F., DE and SOYER, J. (1977b). Le Meiobenthos des Iles Kerguelen. Données quantitatives. II. Le Plateau continental. *Comité national francais des recherches antartiques (CNFRA)*, **42**, 249–58.

BOVEE, F., DE and SOYER, J. (1977c). Le Méiobenthos des Iles Kerguelen. Données quantitatives. III. Conclusions. *Comité national francais des recherches antartiques (CNFRA)*, **42**, 259–65.

BURKOVSKIY, I. V. (1971). The ecology of psammophilous ciliates in the White Sea. *Zool. Zh.*, **50**, 1285–302.

BURMISTROVA, I. I. (1971). Distribution of benthic foraminifera of North region of the Indian Ocean. In *Proceedings of the Symposium on the Indian Ocean and Adjacent Seas*. Marine Biological Association, Cochin, India. p. 64.

CADEE, G. C. and HEGEMAN, J. (1974). Primary production of the benthic microflora living on tidal flats in the Dutch Wadden Sea. *Neth. J. Sea Res.*, **8**, 260–91.

COLOCOLOFF, M. (1972). *Recherches sur la Production Primaire d'un Fond Sableux. Biomasse et Production*, Thèse Doct., Université Aix-Marseille (II).

COOK, D. G. (1970). Bathyal and abyssal Tubificidae (Annelida, oligochaeta) from the Gay Head Bermuda transect, with description of new genera and species. *Deep Sea Res.*, **17**, 973–82.

COULL, B. C. (1970). Shallow-water meiobenthos of the Bermuda platform. *Oecologia (Berl.)*, **4**, 325–57.

COULL, B. C. and VERNBERG, W. B. (1975). Reproductive periodicity of meiobenthic copepods: seasonal or continuous? *Mar. Biol.*, **32**, 289–93.

COULL, B. C., ELLISON, R. L., FLEEGER, J. W. and HIGGINS, R. P. (1977). Quantitative estimates of the meiofauna from the deep sea off North Carolina, USA. *Mar. Biol.*, **39**, 233–40.

CRISP, D. J. and WILLIAMS, R. (1971). Direct measurement of pore-size distribution on artificial and natural deposits and prediction of pore space accessible to interstitial organisms. *Mar. Biol.*, **10**, 214–226.

DALE, N. G. (1974). Bacteria in intertidal sediments: factors related to their distribution. *Limnol. Oceanogr.*, **19**, 509–17.

DAMODARAN, R. (1972). Meiobenthos of the mudbanks of Kerala coast. *Proc. Indian Acad. Sci. (Sect. B)*, **38**, 288–97.

DELAMARE-DEBOUTTEVILLE, C. (1960). *Biologie des eaux Souterraines Littorales et Continentales*, Herman, Paris.

DINET, A. (1973). Distribution quantitative du meiobenthos profond dans la région de la dorsale de Walvis (Sud-Ouest Africain). *Mar. Biol.*, **20**, 20–6.

DINET, A. (1976). Etude quantitative du meiobenthos dans le secteur nord de la mer Egée. *Acta adriat.*, **18**, 83–8.

DINET, A. (1980). *Répartition Quantitative et Écologie du Méiobenthos de la Plaine Abyssale Atlantique*, Thèse Doct., Universite Aix-Marseille (II).

DINET, A., LAUBIER, L., SOYER, J. and VITIELLO, P. (1973). Résultats biologiques de la campagne Polymède. II. Le meiobenthos abyssal. *Rapp. P.-v. Réun. Commn int. Explor. scient. Mer Méditerr.*, **21**, 701–4.

DINET, A. and VIVIER, M. H. (1977). Le meiobenthos abyssal du Golfe de Gascogne. I. Considérations sur les données quantitatives. *Cah. Biol. mar.*, **18**, 85–97.

DRAGESCO, J. (1965). Etude cytologique de quelques flagellés méso psammiques. *Cah. Biol. mar.*, **6**, 83–115.

FAURE-FREMIET, E. (1950). Ecologie des Ciliés psammophiles littoraux. *Bull. biol.*, **84**, 35–75.

FAURE-FREMIET, E. (1951). Ecologie des Protistes littoraux. *Année biol.*, **3**, 437–47.

FENCHEL, T. (1967). The ecology of marine microbenthos. I. The quantitative importance of ciliates as compared with metazoans in various types of sediments. *Ophelia*, **5**, 121–37.

FENCHEL, T. (1968a). The ecology of marine microbenthos. II. The food of marine benthic ciliates. *Ophelia*, **5**, 73–121.

FENCHEL, T. (1968b). The ecology of marine microbenthos. III. The reproductive potential of ciliates. *Ophelia*, **5**, 123–36.

FENCHEL, T. (1969). The ecology of marine microbenthos. IV. Structure and function of the benthic ecosystem, its chemical and physical factors and the microfauna communities with special reference to the ciliated protozoa. *Ophelia*, **6**, 1–182.

FENCHEL, T. and RIEDL, R. J. (1970). The sulfide system: a new biotic community underneath the oxidized layer of marine sand bottoms. *Mar. Biol.*, **7**, 255–68.

FENCHEL, T. and STRAARUP, B. J. (1971). Vertical distribution of photosynthetic pigments and the penetration of light in marine sediments. *Oikos*, **22**, 172–82.

FRANZEN, A. (1960). *Monobryozoon limicola* n. sp. a ctenostomatous bryozoan from the detritus layer on soft sediment. *Zool. Bidr. Upps.*, **3**, 135–47.

GALLARDO, V. A. (1977). Large benthic microbial communities in sulphide biota under Peru-Chili Subsurface Countercurrent. *Nature, Lond.*, **268**, 331–2.

GAL'TSOVA, V. A. (1971). Quantitative characteristic of meiobenthos in Chupinsky Inlet, White Sea (Russ.). *Zool. Zh.*, **50**, 641–7.

GERLACH, S. A. and SCHRAGE, M. (1971). Life cycles in marine meiobenthos. Experiments at various temperatures with *Monhystera disjuncta* and *Therisiotus pertenuis* (Nematoda). *Mar. Biol.*, **9**, 274–80.

GERLACH, S. A. and SCHRAGE, M. (1972). Life cycles at low temperatures in some free-living marine Nematodes. *Veröff. Inst. Meeresforsch. Bremerh.*, **14**, 5–10.

GIERE, O. (1975). Population structure, food relations and ecological role of marine oligochaetes with special reference to meiobenthic species. *Mar. Biol.*, **31**, 139–56.

GOMOIU, M. T. (1967). Some quantitative data on light penetration in sediments. *Helgoländer wiss. Meeresunters.*, **15**, 120–127.

GRØNTVED, J. (1960). On the productivity of microbenthos and phytoplankton in some danish fjords. *Medd. Kommn Danm. Fisk.-og Havunders.*, **3**, 55–92.

GRØNTVED, J. (1962). Preliminary report on the productivity of microbenthos and phytoplankton in the danish Wadden Sea. *Medd. Kommn Danm. Fisk.-og. Havunders.*, **3**, 347–78.

GRØNTVED, J. (1966). Productivity of the microbenthic vegetation in the danish Wadden Sea (Abstract). *Veröff. Inst. Meeresforsch. Bremerh.*, **2**, 275–6.

GUILLE, A. and SOYER, J. (1971). Contribution à l'étude comparée des biomasses du macrobenthos et du meiobenthos de substrat meuble au large de Banyuls-sur-Mer. *Vie Milieu*, **22**, 15–27.

HALL, J. R. (1972). Aspects of the biology of *Derocheilocaris typica* (Crustacea-Mystococarida). II. Distribution. *Mar. Biol.*, **12**, 42–52.

HARRIS, R. P. (1972a). The distribution and ecology of the interstitial meiofauna of a sandy beach at Whitsand Bay, East Cornwall. *J. mar. biol. Ass. U.K.*, **52**, 1–18.

HARRIS, R. P. (1972b). Seasonal changes in the meiofauna population of an intertidal sand beach. *J. mar. biol. Ass. U.K.*, **52**, 389–403.

HARRIS, R. P. (1972c). Seasonal changes in population density and vertical distribution of harpacticoid copepods on an intertidal sand beach. *J. mar. biol. Ass. U.K.*, **52**, 493–505.

HEIP, C. (1973). Partitioning of a brackish-water habitat by copepod species. *Hydrobiologia*, **41**, 189–98.

HULINGS, N. C. (1971). A comparative study of the sand beach meiofauna of Lebanon, Tunisia and Morocco. *Thalassia jugosl.*, **7**, 117–22.

HULINGS, N. C. and GRAY, J. S. (1976). Physical factors controlling abundance of meiofauna on tidal and atidal beaches. *Mar. Biol.*, **34**, 77–83.

JANSSON, A. M. (1966). Diatoms and microfauna—producers and consumers in the *Cladophora* belt. *Veröff. Inst. Meeresforsch. Bremerh.*, Sonderband **II**, 281–8.

JUARIO, J. U. (1975). Nematode species composition and seasonal fluctuation of a sublittoral meiofauna community in the German Bight. *Veröff. Inst. Meeresforsch. Bremerh.*, **15**, 283–337.

KASHKIN, N. I. (1964). On winter 'dormancy' of plankton algae on sublittoral bottoms. (Russ.) *Trudy Inst. Okeanol.*, **65**, 49–57.

KHRIPOUNOFF, A. (1979). *Relations Trophiques dans l'Écosystème Benthique Abyssal Atlantique*, Thèse Doct. Sci., Universite Pierre et Marie Curie, Paris.

KHRIPOUNOFF, A., DESBRUYERS, D. and CHARDY, P. (1980). Les peuplements benthiques de la faille Vema: données quantitatives et bilan d'énergie en milieu abyssal. *Oceanologica Acta*, **3**, 187–98.

KHUSID, T. A. (1971). The vertical distribution of living benthic foraminifera in North Western Indian Ocean. In *Proceedings of the Symposium on the Indian Ocean and Adjacent Seas*. Marine Biological Association, Cochin, India. p. 65.

LASKER, R., WELLS, J. B. L. and MCINTYRE, A. D. (1970). Growth, reproduction, respiration, and carbon utilization by the sand-dwelling harpacticoid copepod *Asellopsis intermedia*. *J. mar. biol. Ass. U.K.*, **50**, 147–60.

LEACH, J. H. (1970). Epibenthic algal production in an intertidal mudflat. *Limnol. Oceanogr.*, **15**, 514–21.

MCINTYRE, A. D. (1964). Meiobenthos of sub-littoral muds. *J. mar. biol. Ass. U.K.*, **44**, 665–74.

MCINTYRE, A. D. (1968). The meiofauna and macrofauna of some tropical beaches. *J. Zool. Lond.*, **156**, 377–92.

MCINTYRE, A. D. (1969). Ecology of marine meiobenthos. *Biol. Rev.*, **44**, 245–90.

MCINTYRE, A. D. (1971). Control factors on meiofauna populations. *Thalassia jugosl.*, **7**, 209–15.

MCINTYRE, A. D., MUNRO, A. L. S. and STEELE, J. H. (1970). Energy flow in a sand ecosystem. In J. Steele (Ed.), *Marine Food Chains*. Oliver and Boyd, Edinburgh. pp. 19–31.

MCINTYRE, A. D. and MURISON, D. J. (1973). The meiofauna of a flatfish nursery ground. *J. mar. biol. Ass. U.K.*, **53**, 93–118.

MCLACHLAN, A., WINTER, P. Ed. and BOTHA, L. (1977). Vertical and horizontal distribution of sublittoral meiofauna in Algoa bay, South Africa. *Mar. Biol.*, **40**, 355–64.

MAGUIRE, C. and BOADEN, P. J. S. (1975). Energy and evolution in the thiobios: an extrapolation from the marine gastrotrich *Thiodasys sterreri*. *Cah. Biol. mar.*, **16**, 635–46.

MARE, M. F. (1942). A study of a marine benthic community with special reference to the microorganisms. *J. mar. biol. Ass. U.K.*, **25**, 517–54.

MARSHALL, N., OVIATT, A. and SKAUEN, D. M. (1971). Productivity of the benthic microflora of shoal estuarine environments in southern New England. *Int. Revue ges. Hydrobiol.*, **56**, 947–56.

MONNIOT, F. (1965). Ascidies interstitielles des côtes d'Europe. *Mém. Mus. natn Hist. nat.*, Paris (*Ser. A*), **35**, 154.

MONNIOT, F. (1968). Les halacariens des sables. *Rapp. P.-v. Réun. Commn int. Explor. scient. Mer Méditerr.*, **19**, 185–6.

MONNIOT, C. and MONNIOT, F. (1975). Abyssal Tunicates: an ecological paradox. *Annls. Inst. océanogr.*, *Paris*, **51**, 99–129.

NICHOLLS, A. G. (1935). Copepods from the interstitial fauna of a sandy beach. *J. mar. biol. Ass. U.K.*, **20**, 379–406.

NYHOLM, K. G. and JANSSON, I. (1973). Seasonal fluctuations of the meiobenthos in an estuary on the Swedish West Coast. *Zoon*, **1**, 69–73.

OTT, J. and SCHIEMER, F. (1973). Respiration and anaerobiosis of free living nematodes from marine and limnic sediments. *Neth. J. Sea Res.*, **7**, 233–43.

PANDIAN, T. J. (1975). Mechanisms of heterotrophy. In O. Kinne (Ed.), *Marine Ecology*, Vol. II, Physiological Mechanisms, Part 1. Wiley, London. pp. 61–249.

PERKINS, E. J. (1958). Microbenthos of the shore at Whitstable (Kent). *Nature, Lond.*, **4611**, 181.

PERKINS, E. J. (1960). The diurnal rhythm of the littoral diatoms of the river Eden estuary, Fife. *J. Ecol.*, **48**, 725–8.

PLANTE-CUNY, M. R. (1969). Recherches sur la distribution qualitative et quantitative des diatomées benthiques de certains fonds meubles du Golfe de Marseille. *Recl Trav. Stn mar. d'Endoume*, **61**, 87–197.

PLANTE-CUNY, M. R. (1978). Pigments photosynthétiques et production primaire des fonds meubles néritiques d'une région tropicale (Nosy-Bé, Madagascar). *Trav. Doc. ORSTOM, Paris*, **96**, 1–359.

POTTS, M. and WHITTON, B. A. (1977). Nitrogen fixation by blue-green algal communities in the intertidal zone of the lagoon of Aldabra Atoll. *Oecologia*, **27**, 275–83.

REISE, K. and AX, P. (1979). A meiofaunal "Thiobios" limited to the anaerobic sulfide system of marine sands does not exist. *Mar. Biol.*, **54**, 225–37.

REISE, K. and AX, P. (1980). Statement on the thiobios-hypothesis. *Mar. Biol.*, **58**, 31–2.

REMANE, A. (1933). Verteilung und Organisation der benthonischen Mikrofauna der Kieler Bucht. *Wiss. Meeresunters., NF*, **21**, 161–221.

REMANE, A. (1938). Ergänzende Mitteilungen über *Monobryozoon ambulans* Remane. *Kieler Meeresforsch.*, **2**, 356–8.

REMANE, A. (1940). Einführung in die zoologische Ökologie der Nord- und Ostee. *Tierwelt d. Nord-u. Ostee, Leipzig*, pp. 1–238.

REMANE, A. (1952). Die Besiedlung des Sandbodens im Meere und die Bedeutung der Lebensformtypen für die Ökologie. *Verh. dt. zool. Ges.*, **1951**, 327–39.

RENAUD-DEBYSER, J. (1963). Recherches écologiques sur la faune interstitielle des sables. Bassin d'Arcachon, ile de Bimini, Bahamas. *Vie Milieu*, 15 (Suppl.), 1–143.

ROSS, L. A. (1972). Biology and ecology of *Marginopora vertebralis* (Foraminifera) Great Barrier Reef. *J. Protozool.*, **19**, 181–92.

SAIDOVA, Kh. M. (1963). On the quantitative zonation of bottom foraminiferans distribution in the Pacific Ocean. (Russ.) *Vop., Mikropaleont.*, **7**.

SAIDOVA, Kh. M. (1970). Benthic foraminiferans in the Kuril-Kamchatka trench. (Russ.) *Trudy Inst. Okeanol.*, **86**, 134–61.

SAIDOVA, Kh. M. (1971). On the extension of foraminiferans on the coasts of South America. (Russ.) *Okeanologiya*, **12**, 256–9.

SALVAT, B. and RENAUD-MORNANT, S. (1969). Etude écologique du macrobenthos et du meiobenthos d'un fond sableux du lagon de Mururoa. (Tuamotu—Polynésie). *Cah. Pacif.*, **13**, 159–79.

SCHEIBEL, W. (1976). Quantitative Untersuchungen am Meiobenthos eines Profils unterschiedlicher Sedimente in der westlichen Ostsee. *Helgoländer wiss. Meeresunters.*, **28**, 31–42.

SCHEIBEL, W. and NOODT, W. (1975). Population densities and characteristics of meiobenthos in different substrates in Kiel Bay. *Merentutkimuslait. Julk.*, **239**, 173–8.

SCHLEGEL, H. G. (1975). Mechanisms of chemo-autotrophy. In O. Kinne (Ed.); *Marine Ecology*, Vol. II, Physiological Mechanisms, Part 1. Wiley, London. pp. 9–60.

SKOOLMUN, P. and GERLACH, S. A. (1971). Jahreszeitliche Fluktuationen der Nematodenfauna im Gezeitenbereich des Weser Ästuars. (Deutsche Bucht). *Veröff. Inst. Meeresforsch. Bremerh.*, **13**, 119–38.

SOKOLOVA, M. N. (1970). Weight characteristics of meiobenthos from different parts of the deep-sea trophic regions of the Pacific Ocean. (Russ.) *Okeanologiya*, **10**, 348–56.

SOKOLOVA, M. N. (1976a). Trophic zonality of deep water macrobenthos as an element of the biological structure of the Ocean. *Oceanology*, **16**, 174–7.

SOKOLOVA, M. N. (1976b). Delimitation of large scale regions of the ocean with regard to the trophic structure of deep-sea macrobenthos. (Russ.) *Trudy Inst. Okeanol.*, **99**, 20–30.

SOROKIN, Yu. I. (1971). Role of microflora in the productivity of the biocoenosis of coral reefs. (Russ.) *Zh. Obshch. Biolog.*, **32**, 169–186.

SOROKIN, Yu. I. (1972). Bacteria as a food of the fauna of coral reef. (Russ.) *Okeanologiya*, **12**, 195–204.

SOROKIN, Yu. I. (1973a). Production characteristics of the microflora, periphyton and phyto-plankton of coral biocoenoses on Majuro Atoll (Marshall Islands). *Oceanology*, **13**, 404–8.

SOROKIN, Yu. I. (1973b). Trophic Role of Bacteria in the Ecosystem of the Coral Reef. *Nature, Lond.*, **242** (5397), 415–416.

SOROKIN, Yu. I. (1974). Bacteria as a component of the coral reef community. *Proceed. Second. Internat. Sympos. on Coral Reefs*. In A. M. Cameron and B. M. Campbell (Eds), Vol. 1, Brisbane, Australia. pp. 3–10.

SOROKIN, Yu. I. (1978). Decomposition of organic matter and nutrient regeneration. In O. Kinne (Ed.), *Marine Ecology*, Vol. IV, Dynamics. Wiley, Chichester. pp. 501–616.

SOROKIN, Yu. I. (1979). Microbial production in the coral-reef community. *Arch. Hydrobiol.*, **83**, 281–323.

SOURNIA, A. (1976). Ecologie et productivité d'une Cyanophycée en milieu corallien: *Oscillatoria limosa* Agardh. *Phycologia*, **15**, 363–6.

SOYER, J. (1971). Bionomie benthique du plateau continental de la côte catalane française. V. Densités et biomasses du meiobenthos. *Vie Milieu (Ser. B)*, **22**, 351–424.

SPOONER, G. M. (1959). New members of the british marine fauna. *Nature, Lond.*, **183**, 1695–96.

STEELE, J. H., MUNRO, A. L. S. and GIESE, G. S. (1970). Environmental factors controlling the epipsammic flora on beach and sublittoral sands. *J. mar. biol. Ass. U.K.*, **50**, 907–18.

STRIPP, K. (1969). Das Verhältnis von Makrofauna und Meiofauna in den Sedimenten der Helgoländer Bucht. *Veröff. Inst. Meeresforsch. Bremerh.*, **12**, 143–8.

SWEDMARK, B. (1955). *Recherches sur la Morphologie, le Développement et la Biologie de Psammodrilus balanoglossoides*, Thèse Doct., Faculté des Sciences, Paris.

SWEDMARK, B. (1964). The interstitial fauna of marine sand. *Biol. Rev.*, **39**, 1–42.

TAYLOR, W. R. (1964). Light and photosynthesis in intertidal benthic diatoms. *Helgoländer wiss. Meeresunters.*, **10**, 29–37.

TAYLOR, W. R. and GEBELEIN, C. D. (1964). Chromatography analyses of plant pigments in intertidal sediments. *Biol. Bull. mar. biol. Lab., Woods Hole*, **127**, 393.

THIEL, H. (1971). Häufigkeit und Verteilung der Meiofauna im Bereich des Island-Färöer-Rückens. *Ber. dt. wiss. Kommn Meeresforsch.*, **22**, 99–128.

THIEL, H. (1972a). *Quantitative Aspects of the Deep Sea Benthos*. Ocean Research Institute, Tokyo.

THIEL, H. (1972b). Die Bedeutung der Meiofauna in küstenfernen benthischen Lebensgemeinschaften verschiedener geographischer Regionen. *Verh. dt. zool. Ges.*, **65**, 37–42.

THIEL, H. (1972c). Meiofauna und Strukter der benthischen Lebensgemeinschaft des Iberischen Tiefseebeckens. *'Meteor' Forschungsergeb.*, D **3**, 36–51.

THIEL, H. (1973). Der Aufbau der Lebensgemeinschaft am Tiefseeboden. *Natur Mus., Frankf.*, **103**, 39–46.

THIEL, H. (1975). The size structure of the deep-sea benthos. *Int. Revue ges. Hydrobiol.*, **60**, 575–606.

TIETJEN, J. H. (1971). Ecology and distribution of meiobenthos off North Carolina. *Deep Sea Res.*, **18**, 941–57.

VINOGRADOVA, N. G. (1958). Vertical distribution of deep-sea bottom fauna. (Russ.). *Trudÿ Inst. Okeanol.*, **27**, 87–122.

VITIELLO, P. (1972). *Peuplements de Nématodes Marins des Fonds Envasés de Provence Occidentale*, These Doct., Universite Aix-Marseille (II).

WARD, A. R. (1975). Studies on the sublittoral free-living nematodes of Liverpool Bay. II. Influence of sediment composition on the distribution of marine nematodes. *Mar. Biol.*, **30**, 217–25.

WIESER, W. (1953). Die Beziehung zwischen Mundhöhlengestalt, Ernährungsweise und Vorkommen bei freilebenden marinen Nematoden. *Ark. Zool.*, **2**, 439–84.

WIESER, W. (1959). *Free-living Nematodes and Other Small Invertebrates of Puget Sound Beaches*, University of Washington Press, Seattle.

WIESER, W. (1960). Benthic studies in Buzzards Bay. The meiofauna. *Limnol. Oceanogr.*, **5**, 121–37.

WIESER, W., OTT, J., SCHIEMER, F. and GNAIGER, E. (1974). An ecophysiological study of some meiofauna species inhabiting a sandy beach at Bermuda. *Mar. Biol.*, **26**, 235–48.

WIGLEY, R. L. and McINTYRE, A. D. (1964). Some quantitative comparisons of offshore meiobenthos and macrobenthos south of Martha's Vineyard. *Limnol. Oceanogr.*, **9**, 485–93.

ZIBROWIUS, H. (1976). *Les Scléractiniaires de la Méditerranee et de l'Atlantique Nord Oriental*, Thèse Doct., Universite Aix-Marseille (II).

AUTHOR INDEX

Numbers in italics refer to those pages on which the author's work is stated in full.

TAXONOMIC INDEX

SUBJECT INDEX